Sustainable Environmental Engineering

Sustainable Environmental Engineering

Walter Z. Tang
Department of Civil and Environmental Engineering
College of Engineering and Computing
Florida International University
Miami, FL, USA

Mika Sillanpää
Department of Green Chemistry
School of Engineering Science
Lappeenranta University of Technology
Lappeenranta, Finland

WILEY

This edition first published 2019

Registered Office
John Wiley & Sons, Inc., 111 River Street, Hoboken, NJ 07030, USA

Editorial Office
111 River Street, Hoboken, NJ 07030, USA

For details of our global editorial offices, customer services, and more information about Wiley products visit us at www.wiley.com.

Wiley also publishes its books in a variety of electronic formats and by print-on-demand. Some content that appears in standard print versions of this book may not be available in other formats.

Library of Congress Cataloging-in-Publication Data

Names: Tang, Walter Z., author. | Sillanpää, Mika E. T., author.
Title: Sustainable environmental engineering / Walter Z. Tang, Mika Sillanpää.
Description: First edition. | Hoboken, NJ : John Wiley & Sons, 2019. | Includes index. |
Identifiers: LCCN 2017057648 (print) | LCCN 2018004508 (ebook) | ISBN 9781119085638 (pdf) | ISBN 9781119085584 (epub) | ISBN 9781119028376 (cloth)
Subjects: LCSH: Sustainable engineering. | Environmental engineering.
Classification: LCC TA170 (ebook) | LCC TA170 .T24 2019 (print) | DDC 628–dc23
LC record available at https://lccn.loc.gov/2017057648

Cover design: Wiley
Cover image: © MichaelUtech/Getty Images

Set in 10/12pt Warnock by SPi Global, Pondicherry, India

Printed in the United States of America

V10010526_052319

Contents

Preface

Sustainable Environmental Engineering (SEE) is to research, develop, design, build, operate, and maintain environmental engineering infrastructure systems (EEIS) that are economically feasible to reduce human health risk and minimize environmental damages so that man and the nature can coexist in harmony. EEIS include engineering systems of water distribution, sewer collection, water treatment plant (WTP) and wastewater treatment plant (WWTP), stormwater management gray infrastructure, and green infrastructure (GI). Twelve design principles (TDPs) have been developed to guide a designer to minimize footprint (FP) of human social and economic activities on the natural environments such as air, water, soil, and biosphere. FP on materials, energy, and water is typically quantified by life cycle assessment (LCA). In designing EEIS, a designer should apply the following TDPs: (i) integrated and interconnected alternatives, (ii) reliability on spatial scale, (iii) resiliency on temporal scales, (iv) economy of renewable materials, (v) efficiency of renewable energy, (vi) prevention strategies, (vii) recovery of materials and energy, (viii) early separation, (ix) effective treatment, (x) green retrofitting and remediation, (xi) optimization through modeling and simulation, and (xii) minimal life cycle costs and maximal benefits. The TDPs equip SEE designers with the knowledge and tools by quantifying uncertainty and sensitivity of the FP of EEIS alternatives. They provide a systematic and integrated approach to optimize the short-term capital investments and long-term operation cost while reducing the long-term detrimental impacts on human and environments. It is hoped that the TDPs will guide the next generation of sustainable environmental engineers to think critically in developing innovative and sustainable alternatives of EEIS.

In the past, EEIS were not designed according to the TDPs. For example, traditional processes such as activated sludge (AS) severely violate the TDPs. In an AS process, biodegradable organic pollutants are oxidized with excessive amount of sludge produced as solid waste. The sludge has to been either dewatered mechanically or anaerobically digested. However, aeration is energy-intensive process and consumes about one-half energy of a WWTP, and sludge disposal consists about one-third of WWTP operation cost. In fact, AS violates several TDPs because (i) organic pollutants in wastewater have their intrinsic energy content and should be utilized as energy resource and (ii) sludge incorporates free water into fixed cellular water, which is extremely energy intensive to be separated from the biosolids. Therefore, it should be avoided according to the fifth principle of prevention. One of retrofitting WWTP strategies is to divert biodegradable organic carbon to anaerobic digester to directly produce methane according to the seventh principle of recovery. Indeed, other tradition processes such as aluminum coagulation, lime neutralization, chlorination, and combined treatment of leachate with wastewater do not stand the test of the TDPs. After intensive research, development, and practice all over the world in the past decade, WWTP could have been retrofitted or designed as an energy-positive recovery center (WRRF) to recover water, energy, nutrients, and materials. For example, WWTP treating 10 million gallon wastewater per day for population equivalency of 250 000 at Strass im Zillertal, Innsbruck, in Austria became energy positive after a decade-long retrofitting effort in 2005. However, very few of 15 500 WWTPs in the United States and 4 800 WWTPs in China are currently energy neutral. The market sizes of retrofitting these WWTPs in the United States and China are estimated at 300 and 200 billion US dollars, respectively. In next three decades, paradigm will shift to design WWTPs as WRRFs at which water, energy, nutrients, and materials would be recovered to its maximal extent. It could be expected that a WRRF become energy production and material recovery center or even as a revenue generation asset rather than significant financial burden for a water utility. To successfully design WRRF, eight major design topics are covered to accelerate the paradigm shift toward new industrial trends in next decades. For example, Anammox and membrane biological reactor (MBR) are illustrated as the critical technology in designing energy-positive WWTP. Advanced oxidation processes (AOPs) are important to replace separation processes. UV disinfection is presented as critical technology to achieve the

US EPA second-stage disinfectant/disinfection by-product rules. Membrane filtration technologies such as microfiltration, nanofiltration, and reversed osmosis are key technologies in reclaiming water. To facilitate design of WRRF by using these technologies, Matlab codes for estimating capital and O&M costs of these technologies are presented for treatment, reclaiming, or recovery systems. As a result, life cycle cost and benefit of the design alternatives could be conducted.

Human has freely harnessed air, water, and food from the Earth and energy from the Sun in the past million years. In the modern society, however, the infrastructure to deliver water and electricity is expensive due to fit-for-all design philosophy in the past. Some of the flaws have not been challenged in the past hundred years since the first WWTP with AS was built. To meet these challenges and capture unique opportunities, these design flaws have to be corrected. The book starts with renewable resource and human FP in Chapter 1. Common environmental engineering (EE) issues such as air, water, and soil pollution under climate change are explained. Environmental quality indexes of air, water, and soil are defined and illustrated. Human FP and environmental laws are organized from the perspective of LCA. Chapter 2 shows how Crystal Ball could be applied to quantify the uncertainty and sensitivity during human health risk assessment using Monte Carlo simulation. Health risk of disinfection by-products (DBPs) in drinking water is used as an example. Chapter 3 defines and establishes the TDP SEE to guide the designers of EEIS. The topics covered include green chemistry, green engineering, regenerative design of WWTPs, life cycle cost and benefit analysis, and decision principles and metrics. New design challenges of SEE are explained under the context of the United Nations 17 Sustainable Development Goals (SDGs) with market size analysis of the United States and China. From Chapters 4 to 15, each design principle and the associated computer tools are explained. Chapter 4 lays the foundation of planet, people, and profit as the ultimate SEE design goals. Integrated management and best management practice (BMP) of EEIS are defined. The design hierarchy is ranked in the order of prevention, reduction, reuse, recycle, treatment, recovery, disposal, and remediation. Chapter 5 explores reliability of an EEIS on spatial scales such as residential house, community, and city. It shows how SEE design principles could be applied at different spatial scales and how laboratory and pilot data should be accurately extrapolated to full-scale treatment plants. Since UV disinfection is expected to replace chlorination in both WTP and WWTP, virus sensitivity index (VSI) and bacteria sensitivity index (BSI) are developed to estimate the UV fluence required by any given virus or bacteria from laboratory data through statistical analysis by SPSS. Chapter 6 focuses on resiliency on temporal scales. It demonstrates that population growth is the key element as unsteady flow and how to design an equalization tank to reduce the unsteady effect. Also, chemical kinetics and catalysts are discussed in detail to reduce reaction time. Chapter 7 emphasizes the economy of material using modern reactor design theory. Matlab codes are presented to compare the hydraulic characters of batch, continuous stirred tank reactor (CSTR), and plug-flow reactor (PFR). Chapter 8 demonstrates how to balance the embodied energy vs. the energy required for operation and maintenance. The energy auditing of WWTP for retrofitting purpose is illustrated. Six design hierarchies are presented sequentially from Chapters 9 to 14. To demonstrate the effectiveness of prevention, Chapter 9 explains why prevention is the number one priority in SEE design. For example, GI holds the key to prevent nonpoint source pollution, while rain harvest may be the most effective way to prevent contamination due to stormwater runoff and to save precious tap water from irrigation. Other GI such as green roof, bioswale, rain garden, septic tanks, and constructed wetlands are equally important in designing smart cities or communities. Chapter 10 provides unit cost and benefit analysis tools for water reclaiming systems because water is the major product of the highest market price to be recovered with huge demand in terms of irrigation, reuse, or recharge of underground aquifer. Chapter 11 critically examines precipitation as an unsustainable design because it requires large amount of chemicals, such as bases or acids. Even worse, it incorporates free water, which should have been the product, into sludge as the secondary waste. To avoid adding chemicals, a better design may be ion exchange. Ultra-, micro-, and nano-membrane filtrations are illustrated as an industrial trend to replace traditional tank processes such as coagulation, flocculation, and precipitation. Chapter 12 focuses on engineering design of AOP such as UV disinfection because it is expected that more and more UV disinfection will replace chlorination. Virus sensitivity index (VSI) and Bacteria sensitivity index (BSI) during UV disinfection are introduced. Membrane biological reactor (MBR) is introduced as major treatment technology to reduce FP of WWTPs. However, the energy FP of MBR is still relative large; as a result, it might not be suitable for financial restrained rural areas. Chapter 13 fully establishes the energy benchmarks in retrofitting WWTPs to energy-positive WRRF. For green remediation, a comprehensive software package to remediate contaminated air, water, and soil by the US EPA is presented. Chapter 14 examines the optimization of EEIS through modeling and simulation. By using Fenton treatment of leachate as an example, the optimal conditions are derived both theoretically modeled and validated experimentally. Chapter 15 offers the latest technologies in recovering N and P from wastewater. It reveals how different flow rates affect capital and operation cost of recovering N and P. It briefly introduces financing, accounting, and entrepreneurial concepts for SEE students. The information is critical for SEE designers to estimate costs of SEE alternatives according to regenerative, neutral, and negative design. Lastly, the potential of applying 3D printing, Internet of things, artificial intelligence, blockchain, big data, cloud computing, and data mining in SEE design is explored. It is to encourage SEE designers to think out of the box by defining the box and using multipliers to extend their design to exceed the physical or chemical limits.

The textbook materials are presented from simple to complicated, easy to difficult, and component to system. Fundamental sciences are presented in a simple and easy approach. For example, calculus is intentionally used as minimal as possible to inspire students with chemistry, biology, ecology, and physics to pursue their career in SEE. Critical information such as unit energy content, sustainable metrics, and benchmark rubrics are provided for students to make their own judgment in their engineering design. Starting from Chapter 4, each chapter illustrates one SEE design principle and presents step-by-step methods of its implementation. To make the TDPs practically and easily to be implemented in engineering design, connection between traditional three-phase submittals such as 30, 60, and 100% is established. Experiential learning is embedded in open-ended problems for students who could select their hometown. Students are encouraged to identify and prioritize EE issues in their hometown and communicate with its stakeholders to develop design alternatives through three steps. First, local air, water, and soil quality data have to be calculated to prioritize environmental issues by quantifying the corresponding quality indices. Second, these indices should be compared with governmental standards to prioritize the major environmental problems. Third, the local EE problems should be critically analyzed according to the TDPs by using tools presented in the book. Assessment rubrics are given to rank students' best sustainable design to serve their community-based and open-ended projects. By applying the TDPs together with innovative technologies, regenerative EE infrastructure can be designed by avoiding unsustainable design such as chlorination and leachate treatment with wastewater by AS. During the whole course project design, experiential teaching, critical thinking, entrepreneurial skills, and community involvement are intervened with traditional learning modules such as homework and assignments. As a result, students will be able to design sustainable EEIS adapted to the climate change in the next 20, 50, and 100 years.

Connecting dots of the human suffering as well as environmental damage due to unsustainable design in the past, we dream to empower new generations of SEE designers to help the less fortunate become healthier, richer, and better by designing and building sustainable communities across cities, states, countries, and the globe. As a result, we can hand a better world to our children and grandchildren. This book presents systematic design method of EEIS to achieve sustainable development by decoupling economic prosperity from environmental degradation. We hope that this book serves as a catalyst to transform EE education and practice for decades to come.

We would like to thank Mr. Jinze Li, who devoted his time and effort in developing part of the MatLab codes, figures and diagrams and Dr. Rena Chen for her contribution to part of Chapter 2. Students who took the courses of advanced treatment system, reactor design, open channel hydraulics and hydraulic engineering at Florida International University (FIU) are greatly appreciated. These students include Mrs. Angela Pierotti, Claudia Cardona, Natalie Duque, Annie Barahona, Katheline Tabuteau, and Mr. Carlos Fomoros. Also, original research projects have been carried out at FIU, Lappeenranta University of Technology (LUT), and Tallinn University of Technology (TUT) by our postdoctors include Dr. Shrawan Singh and Dr. Jean-Marie Fontmorin, Dr. Feiping Zhao, and Dr. Evgenia Iakovleva. WZT would like to thank the unwavering support from Dr. Kenneth Furton, Provost at FIU; Dr. John Volakis, Dean of the College of Engineering and Computing at FIU; Dr. Mark Weiss, Dr. Osama, Mohammed, and Dr. Anthony McGoron, Associate Deans of the College of Engineering and Computing at FIU; Dr. Atorod Azizinamini, the Chairman of Department of Civil and Environmental Engineering at FIU; and Professor Enn Loigu, Chairman of Department of Environmental Engineering at TUT, Tallinn, Estonia. We would also like to thank our editors, Mr. Michael Leventhal and Mrs. Grace Paulin and Shalisha Sukanya, for their excellent editing skills and timely professional service.

This book is dedicated to WZT's former advisor, Professor Pao-Chiu Hsu, of Tsinghua University, Beijing, China, on his 100th birthday anniversary in 2017 because this book is a continuation of his vision of water industry by educating the next generation of environmental engineers in China. Professor Hsu is the father of water treatment in modern China. After earned his Ph.D. in the United States in 1951, Dr. Hsu returned to Tsinghua University in 1955. Professor Hsu's third edition of textbook, *Water Treatment Engineering*, offered the best theoretical and engineering design of WTP in China. He catalyzed the transformation of water and wastewater engineering to water industry after he defined "water industry" as borrowing and returning clean water to nature scientifically in 1997. Since then, China integrated water and wastewater departments as water bureaus across the country. As a result, wastewater is treated and returned to the nature through about 4800 WWTPs in China.

Full-hearted appreciation goes to the FIU president, Dr. Mark Rosenberg, for his global vision and championship in protecting academic freedom. In 2005, WZT was going to add a hard copy of WZT's first book, *Physicochemical Treatment of Hazardous Wastes*, to his book collections authored by the FIU professors. WZT was not able to deliver a hard copy because Dr. Rosenberg was the Chancellor of Florida State University System in Tallahassee, Florida, from 2005 to 2008. Now, WZT will fulfill the promise by delivering the first hard copy of this book to his Ronald Reagan Presidential House at FIU. Under the president's leadership, FIU is ranked at 362 from 1000 best universities in the world by the U.S. News & World Report in 2017. Last, but not the least, this book is dedicated to our families. Their sacrifices and unconditional love made this period of life journey completed and fulfilled!

Walter Z. Tang
Department of Civil and Environmental Engineering
College of Engineering and Computing
Florida International University
Miami, FL, USA

Mika Sillanpää
School of Engineering Science
Lappeenranta University of Technology
Lappeenranta, Finland

1

Renewable Resources and Environmental Quality

Sun sustains every life on Earth.

OBJECTIVES
To present renewable energy such as solar, biomass, and wind To quantify hydrological, carbon, oxygen, nitrogen and phosphorus cycles To provide unit data on human demand and footprints To quantitatively describe the uncertainty of the carrying capacity of Earth by using SPSS To calculate C, N, P, water, energy, and ecological footprints To define nine planetary boundaries To explain peak oil and phosphorus in terms of mineral depletion To calculate air, water, and soil quality indexes

1.1 Renewable Resources and Energy

Free water and energy have sustained human life on Earth in the past million years because water can be harvested from the sky and energy can be produced from solar, biomass, and wind. Since the Industrial Revolution in 1781, however, nonrenewable fossil energy such as coal, oil, and natural gas has become the major power source for human economic activities. For example, environmental engineering infrastructure system (EEIS) such as centralized water treatment plant (WTP) and wastewater treatment plant (WWTP) are mostly powered by fossil fuels and have become symbols of modern life. Due to the economy of scale, centralized EEIS was designed to fit for all because unit cost of water and wastewater produced decreases with the increasing plant capacity. As EEIS ages, however, maintenance becomes more and more expensive. Under climate change and sea level rise, retrofitting existing WWTP may cost more than building new plants. For example, Miami-Dade Water and Sewer Department (MD WASD) plans spend $6 billion dollar to increase resiliency of three WWTPs with total average flow rate of 300 MGD in next 20 years which could be much more expensive than building new plants within the same budget.

Currently, the average electricity compositions in the United States contributed by coal, natural gas, nuclear, and renewables are 33, 33, 20, and 14, respectively. Among the other renewable energy sources, wind, biomass, solar, and geothermal energy consist of 4.7, 1.6, 1.0, and 0.4% of the total energy portfolio, respectively. Oil provides almost 100% of transportation energy. Solar energy is the only primary energy continuously arriving on Earth at a rate of 1361 W/m^2. Each day, about 174 peta-watts (10^{15} W) of sunlight hits the planet Earth. Assuming Earth to be a black body, its mean temperature without the greenhouse effect can be estimated from Example 1.1.

Example 1.1 Earth's radius $R = 6.37 \times 10^6$ m, solar constant at the average radius of Earth's orbit around the sun $S_o = 1363\ W/m^2$, the global fraction of incoming solar radiation that is reflected as the albedo constant, $\alpha_p = 0.3$.

Find: The mean global temperature without global warming gas.

Solution:

If the average global ***albedo*** is α_p,

$$\text{Absorbed energy flux per m}^2 = Q_{abs} = \frac{S_o \pi R^2 (1 - \alpha_p)}{4\pi R^2}$$

$$\text{or} \quad Q_{abs} = \frac{S_o}{4}(1 - \alpha_p)$$

Sustainable Environmental Engineering, First Edition. Walter Z. Tang and Mika Sillanpää.

At the equilibrium climate, the amount of solar radiation absorbed must be balanced by the emission of terrestrial radiation by the planet at the global mean:

$$Q_{abs} = Q_{out}$$

where Q_{out} is the longwave radiation emitted by Earth.

Assuming that Earth is a black body, then its equilibrium emission temperature can be estimated from the Stefan–Boltzmann law:

$$Q_{abs} = \sigma T_E^4$$

where σ is the infrared energy that Earth will emit per unit area, 5.670373×10^{-8} W/m^2 K^4 (m = meters, K = Kelvin), which is referred as the Stefan–Boltzmann constant.

At the equilibrium, the incoming solar energy equals the energy emitted by Earth:

$$\sigma T_E^4 = \frac{S_o}{4}\left(1 - \alpha_p\right)$$

Answer: $T_E = 255$ K

Comments: 255 K is the equilibrium emission temperature of Earth and would also be the equilibrium global mean surface temperature if Earth did not have an atmosphere. In other words, the average Earth temperature should be −19.5 °C (−3.1 °F). However, because of the greenhouse effect of the atmosphere air such as CO_2, the actual Earth average global temperature is about 14 °C (57 °F).

Mankind has been working on technologies of renewable energy since the onset of the Industrial Revolution. Table 1.1 lists the history of renewable energy and the associated technologies to deliver them.

Since all the EEIS needs energy to operate, many roadmaps of renewable energy have been proposed. Table 1.2 shows that the United States, European Union, and China are targeting 20% renewable energy by 2020. By 2050, 50% of the world energy will be renewable and 100% of the world energy will come from renewables by 2090.

Regardless which route of the roadmap, EEIS design should anticipate the change of energy composition and contribute positively to decouple economic development from deteriorating environments by renewable energy to substitute fossil fuels such as coal, oil, and gas. One way to achieve this is to produce energy from WWTPs. If all the WWTPs in the United States can be retrofitted to energy-positive water resource recovery facility (WRRF), about 3% of total electricity produced in the United States could be saved. SEE is to help

Table 1.1 History of commercial renewable energy.

Year	Goal
200 BC	Vertical waterwheels powered mills to crush grain, full cloth, tan leather, smelt and shape iron, saw wood, and carry out a variety of other early industrial processes
100 AC	Vertical carousel-type mills utilized the wind to grind corn and to raise water from streams to irrigate gardens. Their use soon spread to India, other parts of the Muslim world, and China, where farmers employed them to pump water, grind grain, and crush sugarcane
1590s	The Dutch built windmills for multiple uses to their fullest scale to ground the grain produced on the rich meadows and to saw wood
1830s	In 1860, thousands of distilleries churned out at least 90 million gallons of alcohol per year for lighting in the United States
1850s	The windmill became a popular water pumping tool for home and rail builders
1860	First solar power system developed in France to produce steam to drive machinery
1876	First demonstration of generating electricity directly from sunlight in a selenium solar cell
1882	First commercial-scale hydroelectric plant went into operation in Appleton, Wisconsin
1888	First windmill to generate electricity developed in Cleveland, Ohio
1892	World's first geothermal district heating system built in Boise, Idaho
1921	World's first geothermal power plant built in California
1927	First commercial wind turbines sold to generate electricity on remote farms
1935	Hoover Dam, the world's largest hydroelectric power plant, was built
1953	First silicon solar cell developed at Bell Laboratories
1960	General Electric (GE) developed hydrogen fuel cells to generate electricity for Apollo and Gemini Space missions
1960	First commercial-scale geothermal electric plants in the United States built in California
1970s	Solar cells began to lower in price and became cost-effective for use on land
1978	World's first solar-powered village; Tohono O'Odham Reservation, Arizona
1980	World's first wind farm built in New Hampshire
1981	Solar One: first large-scale solar thermal power plant began operation in Daggett, California
2008	First commercial cellulosic ethanol plant went into production in Wyoming
2014	World's largest concentrated solar power generation plant went online in Ivanpah, California

Source: http://alternativeenergy.procon.org/view.timeline.php?timelineID=000015 (accessed 25 September 2016).

Table 1.2 Renewable energy roadmap.

Year	Goal
2018	100% of US electricity comes from solar, wind, and other renewables (Gore's prediction) $255 billion spent per year (more than four times what is currently spent) on biofuels, wind power, solar photovoltaics, and hydrogen fuel cells, according to market research firm Clean Edge $150 billion invested by this date by the US government on climate-friendly energy development (Obama's plan)
2020	All new cars are hybrids, according to an anonymous survey of car industry executives by IBM Institute for Business Value 35 miles/gal is average for the US fleet 20% of the European Union's energy comes from renewables 15% of China's energy comes from renewables
2022	36 billion gallons of biofuels sold in the United States, up from 4.7 billion gallons in 2007
2030	50% increase in world energy demand from 2005 levels, according to the US Department of Energy (DOE) All new federal buildings are carbon neutral, as stated in the 2007 Energy Act 70% of Hawaii's energy comes from renewables, thanks in part to a ban on new coal plants One-fifth of US power comes from wind, the DOE predicts One-fourth of US workers wear a green collar, according to the American Solar Energy Society 20 million new jobs created by renewable industry according to a United Nations report
2050	50% of the world's energy comes from renewables, as claimed by the Energy [R]evolution Report
2090	100% of the world's energy comes from renewables, as claimed by the Energy [R]evolution Report

Source: http://alternativeenergy.procon.org/view.timeline.php?timelineID=000015 (accessed 25 September 2016).

designers to retrofit or design WRRFs as true energy production centers.

Solar radiation also powers the Earth's natural hydrological cycle to replenish water through precipitation, evaporation, transpiration, infiltration, and runoff. Since 97% of the water is saltwater in the oceans, only 2% is bound as ice and 1% is in lakes, rivers, and groundwater. To manage flood and drought effectively for sustainable development, integrated water resources management (IWRM) is critical for sustainable development as defined by the Global Water Partnership:

> IWRM is a process which promotes the coordinated development and management of water, land and related resources, to maximize the resultant economic and social welfare in an equitable manner without compromising the sustainability of vital ecosystems.

When water is abundant, the water quality index (WQI) is critical in assessing water pollution problems and developing water management strategies to sustain local economic and societal needs. When water is scarce, rain harvest and the reclamation of gray and black water may hold the key to sustainable economic development of the region. For example, in the Aries region, rain harvest and the reuse of gray and black water become critical, because climate change is causing more extreme weather and altering water availability in many regions. Therefore, adaptation is crucial to ensure that sufficient water of adequate quality is available for human and the environment. One way to adapt to the climate change is to design and construct integrated and interconnected green infrastructure (GI) to build sustainable community. The reason is that GI adapts to climate change better in the hydrological cycle such as precipitation, evaporation, transpiration, runoff, and storage during flood and drought. Indeed, GI is one of the critical EEISs to ensure sustainable development within the regional hydrological boundary as average global GDP continues to rise and the human population approaches 10 billion by 2100. SEE designers should sustainably manage the water cycle so that the water footprint (WFP) is less than the regeneration rate of the regional hydrological cycle.

On a global scale, water cycle includes annually 4.25×10^{20} g of water evaporating from the oceans; this is equivalent to a water layer of 117.8 cm thickness spread over the total ocean surface. A total of 0.397×10^{20} g is transferred from the oceans to the continents (11 cm). The continents receive precipitation of 1.081×10^{20} g annually, equivalent to a layer of 74.6 cm covering the total land area. Two-thirds of this amount evaporates and one-third returns through rivers or glaciers to the sea. The global ratio between evaporation and runoff is 0.357. Water budget is one of the critical aspects in the integrated water management of a given area. For a specific area, water budget can be simplified as follows:

$$Q_{\text{storage}} = Q_{\text{prep.}} - Q_{\text{EP}} - Q_{\text{IF}} - Q_{\text{runoff}} \tag{1.1}$$

For natural cover, only 10% is runoff, while evapotranspiration, shallow infiltration, and deep infiltration are 40, 25 and 25% respectively. If a city has 75–100% impervious cover, 40% is evapotranspiration, while the runoff could be as high as 55%. For example, May to October is the rainy season in Miami when precipitation is greater than evapotranspiration and the ponds and lakes rise. From November to April, precipitation is less than evapotranspiration and the reservoirs decrease. Example 1.2 illustrates the hydrological balance of water in Miami.

Example 1.2 For Miami, the following precipitation data is given (Table 1.3):

Miami-Dade County covers 2000 miles^2, which is 5.5×10^{10} ft^2. If evapotranspiration is 35 in/year and infiltration is 8 in/year, what would be the total precipitation and runoff, respectively?

Find: What percentage of pavement is considered impervious?

Solution:

The total precipitation in Miami by in/year is

$$Q_{prep.}(\text{in/year}) = 1.61 + 2.24 + 2.99 + 3.15 + 5.35 + 9.69$$
$$6.50 + 8.90 + 9.84 + 6.34 + 3.27 + 2.05$$
$$Q_{prep.}(\text{in/year}) = 61.93 \text{ in/year}$$

Annual volume precipitation rate in Miami in m^3/year is

$$\begin{aligned} Q_{prep.}(\text{m}^3/\text{year}) &= 61.93 \text{ in/year} \div 12 \text{ in/ft} \\ &= 5.16 \text{ ft/year} \times 5.5 \times 10^{10}\text{ft}^2 \\ &= 2.84 \times 10^{11}\text{ft}^3/\text{year} \div 35.31 \text{ ft}^3/\text{m}^3 \\ &= 8.04 \times 10^{9}\text{m}^3/\text{year} \end{aligned}$$

The total runoff in Miami by in/year is

$$Q_{runoff} = Q_{prep.} - Q_{EP} - Q_{IF} = 61.93 - 35 - 8 = 18.93 \text{ in/year}$$

The annual runoff in Miami in m^3/year is to multiply in/year by area:

$$\begin{aligned} Q_{runoff}(\text{m}^3/\text{year}) &= 18.93 \text{ in/year} \div 12 \text{ in/ft} \times 5.5 \times 10^{10} \text{ ft}^2 \\ &= 1.58 \text{ ft/year} \times 5.5 \times 10^{10} \text{ ft}^2 \\ &= 8.68 \times 10^{10} \text{ ft}^3/\text{year} \div 35.31 \text{ ft}^3/\text{m}^3 \\ &= 2.46 \times 10^{9} \text{ m}^3/\text{year} \end{aligned}$$

The runoff and precipitation ratio is

$$\text{Runoff ratio} = \frac{Q_{runoff}}{Q_{prep.}} = \frac{18.93}{61.93} = 30.56\%$$

Answer: Since the runoff to precipitation ratio is 30.56%, at least 75% area of Miami is impervious.

Comments: The conclusion on the percentage of impervious surface in Miami greatly depends upon the evapotranspiration data. There are large differences in many reported evapotranspiration data. Therefore, the better way to solve problem is to compile all the data and calculate the mean and standard deviation to quantify the uncertainty of the runoff ratio.

Powered by the sun, plants and phytoplankton fix atmospheric CO_2 into biomass through photosynthesis. CO_2 is then added to the atmosphere through cellular respiration of animals, and the burning of fossil fuels and volcanoes complete the carbon cycle. Autotrophs convert carbon dioxide to organic molecules that are used by heterotrophs. During the carbon cycle, carbon could also be stored through fossil fuels, soils, aquatic sediments, the oceans, plant and animal biomass, and the atmosphere (CO_2) (Example 1.3).

Example 1.3 Under one atmospheric pressure, the partial pressure of CO_2 (P_{CO_2}) is 0.0003 parts CO_2 in 1 part of air. According to Henry's law, the CO_2 content of rainwater is between 1 ppm at 0 °C and 0.5 ppm at 20 °C, if 1.08×10^{20} g of freshwater water falls annually as precipitation on the continents and a global mean precipitation temperature is 15 °C with 0.6 ppm CO_2 in the rainwater.

Find: What would be the annual flux of CO_2 with precipitation to the continent surface?

Solution:

Since the average global annual precipitation is 1.08×10^{20} g H_2O, the annual flux of CO_2 with precipitation to the continent surface would be

$$\begin{aligned} &0.6 \times 10^{-6}\text{CO}_2 \times 1.08 \times 10^{20} \text{ g H}_2\text{O} \\ &= 6.5 \times 10^{13} \text{ g CO}_2 \ (1.8 \times 10^{13} \text{ g C/year}) \end{aligned}$$

Answer: The annual flux of CO_2 with precipitation to the continent surface is 6.5×10^{13} g CO_2.

Table 1.3 Average monthly precipitation in Miami.

Month	1	2	3	4	5	6	7	8	9	10	11	12
Precipitation (inch)	1.61	2.24	2.99	3.15	5.35	9.69	6.50	8.90	9.84	6.34	3.27	2.05

Comments: Some CO_2 is instantaneously lost again into the atmosphere when the precipitation evaporates from rocks, soil, or vegetation. Industrial fossil-fuel burning increases the CO_2 content of air locally, and rain may contain larger CO_2 concentration compared with the global average. Therefore, actual measurements of CO_2 in rainwater may be much higher than expected from the equilibrium.

Phytoplankton contributes 50–85% of oxygen in the Earth's atmosphere through photosynthesis with oxygen as a by-product through the reduction of CO_2. Forest, grass, algae, and agriculture produce the rest. Phytoplankton has a short turnover time, which means that it has a small standing biomass compared with its production:

$$\text{Turnover time} = \frac{\text{standing crop biomass } (\text{mg/m}^2)}{\text{production } (\text{mg/m}^2/\text{day})} \quad (1.2)$$

The nitrogen cycle starts from the fixation of 80% non-bioavailable nitrogen gas (N_2) in the atmosphere. Through bacteria, nitrogen enters ecosystems. Organic matter is decomposed to ammonia, NH_4^+, in *ammonification* by bacteria. During *nitrification*, bacteria convert NH_4^+ to NO_3^-. Bacteria use NO_3^- for metabolism instead of O_2 and release N_2 back into the atmosphere in *denitrification*, which completes the nitrogen cycle.

Different from the nitrogen cycle, phosphorus can only become bioavailable to plants or animals through weathering from rock. Some phosphate is taken up by producers and incorporated into organic materials. Other phosphate leaches into groundwater or surface water and enters to the sea. Phosphorus is returned to soil or water through the decomposition of biomass or the excretion by animals.

1.2 Human Demand and Footprint

1.2.1 Human Demand

Human demand can be simply expressed as the air we breathe, the water we drink, and the food we eat. As a result, human generates waste such as feces and urine as follows:

$$\text{Food} + H_2O + O_2 = \text{feces} + \text{urine} + CO_2 + H_2O \quad (1.3)$$

On average, each person needs the following amounts of natural resources to survive. The quantity of air required is relatively constant. For example, the daily air inhalation rates (m^3/day) for adult, child, and infant are 22.8, 14.8, and 3.76, respectively. Only 2 kg water is required for drinking purposes, while 2–4 kg food is needed by an adult for a day. Water and food have large variations because food contains different amount of moisture. The US Environmental Protection Agency (EPA) Exposure Factors Handbook (US EPA, 1997) provides detailed data related to water and food variation in different categories. Table 1.4 shows that total dairy food and meats are 308.6 ± 5.3 and 172.2 ± 1.6 g/day, respectively. The average and standard deviation of grain intake is 200.0 ± 3.0 g/day.

Table 1.4 Average consumption of meat in the United States (US EPA, 1997).

Animal protein	Daily take rate (g/day)
Fresh cows' milk	253.5 ± 4.9
Eggs	26.9 ± 0.5
Beef and veal	87.6 ± 1.1
Pork	28.2 ± 0.6
Poultry	31.3 ± 0.8
Other	25.1 ± 0.4

In ecology, primary producers are organisms in an ecosystem that produce biomass from inorganic compounds (autotrophs). Gross primary production (GPP) is the amount of chemical energy as biomass that primary producers create in a given length of time. The amount of chemical energy in consumers' food that is converted to their own new biomass during a given time period is called the secondary production of an ecosystem. The efficiency of animals as energy transformers can be calculated using the following equation:

$$\eta_p = \frac{\text{net secondary production}}{\text{assimilation of primary production}} \quad (1.4)$$

where η_p is the production efficiency.

Net secondary production is the energy stored in biomass represented by growth, reproduction, and respiration. Thus, the fraction of food energy not used for respiration is the production efficiency. For example, birds and mammals generally have low production efficiencies of between 1 and 3% because they use so much energy to maintain a constant body temperature. Fishes and insects have production efficiencies of around 10 and 40%, respectively.

Trophic efficiency is the percentage of production transferred from one trophic level to the next. The energy lost through respiration, feces, and unconsumed organics at lower trophic levels causes low production efficiencies from 5 to 20%. Therefore, 80–95% of the energy available at one trophic level is not transferred to the next over a food chain. Usually, 10% of energy is transferred from primary producers to primary consumers, and 10% of that energy is transferred to secondary consumers. As a result, only 1% of net primary production (NPP) is available to the secondary consumers.

The dynamics of energy through ecosystems have important implications for healthy diet. For example, eating meat is an inefficient way of tapping photosynthetic products due to low trophic efficiency in the food chain. More importantly, vegetable protein is no less nutritious than animal protein. Indeed, worldwide agriculture could feed many more people if vegetarian diet were the dominant diet because primary consumers feeding on plants have much higher trophic efficiency. More importantly, diet containing more vegetable, fruits, and plant protein has much healthier benefit than red meats.

Although each person's demand for air, water, and food may not be huge, the total impacts of human on the environments such as air, water, and soil could be devastating. If EEIS were not designed sustainably, residential, industrial, and agricultural wastes may become a bottleneck for economic development of the future generations. In 2016, 7.2 billion people in the world may take three population trajectories to 8, 9.3, and 10.6 billion by 2050, depending upon low, medium, and high birth rates, respectively, according to the United Nations' (UN) projection as shown in Figure 1.1.

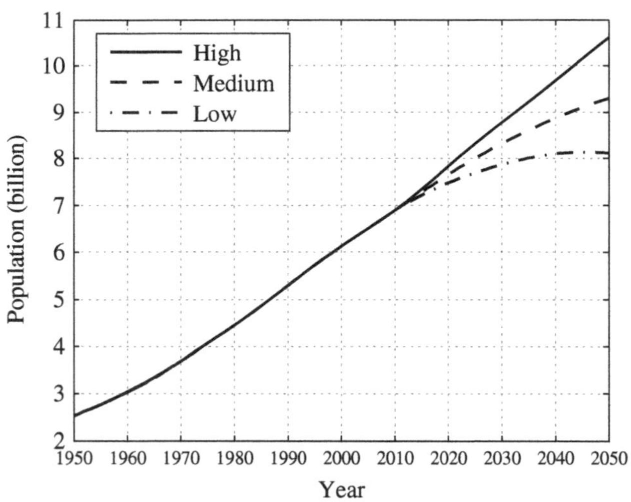

Figure 1.1 World population, 1950–2010, with projections to 2050.

In addition, life expectancy is predicted to increase from 69 to 75 between 2015 and 2050. Figure 1.2 shows that the life expectancy has also increased globally by almost 20 years. Twice as many people on the planet would live 40% longer, with each person consuming many times more animal protein than the average person did in the 1960s.

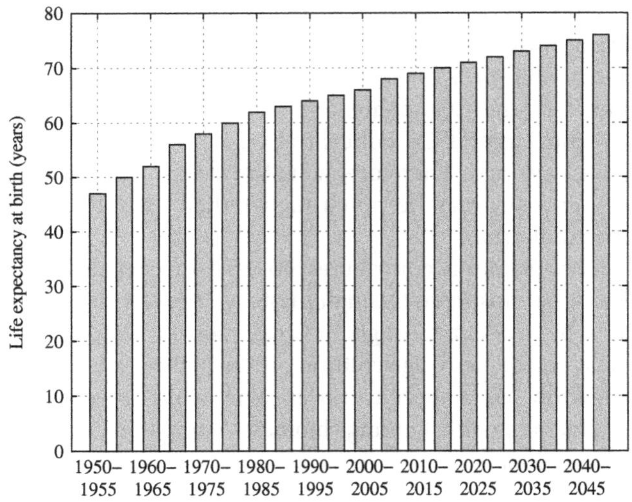

Figure 1.2 Life expectancy for the world, 1950–2005, with projection to 2050. (*Source:* esa.un.org/unpp.)

As living standards increase, animal protein consumption per person increases exponentially. From the data presented by the Earth Policy Institute, several major trends can be observed due to human protein consumption: (i) wild fish is declining due to depleting fish stocks from overfishing, (ii) beef production is also declining due to the stricter environmental regulations and change of diet, and (iii) pork and farmed fish are increasing due to insatiable demand from China. For example, the total number of pigs in China equals that of the rest of the world combined. Unfortunately, pork is rather fatty red meat, which directly correlates with the high death rate due to stroke and heart disease.

1.2.2 Human Footprints

Human footprint (FP) is measured in terms of the average land available for each person on planet Earth as ecological footprint (EF). Ewing et al. (2009) applied the EF to gauge how much land and water area a person requires to produce the resources it consumes and to absorb its carbon dioxide emissions based on current technology. Currently, the biocapacity of the Earth is 1.8 ha. In theory, the average EF of each person should not exceed 1.8 ha. Since 1 ha equals 20 000 m^2, each person on average has 40 000 m^2 land. However, the EF in the United States is 9 ha, while the global average EF is only 5 ha. Therefore, modern society such as the United States is ecologically overshooting in terms of EF. To allow nature to regenerate its stock sustainably for human consumption, about two additional planet Earths are needed if each person had EF of 9 ha. For this reason, sustainable cities have to be designed and built to reduce the ecological overshooting.

To estimate the loading factor, the unit value of the impact on the environment per capita is expressed as follows (Table 1.5).

Table 1.5 Waste production per capita per day.

Components	Standard	Unit
COD	180	g/capita/day
N	13	g/capita/day
P	2.1	g/capita/day
Flow rate	0.36	m^3/capita/day
COD	500	g/m^3/capita/day
N concentration	36	g N/m^3/capita/day
P concentration	5.8	g P/m^3/capita/day

1.2.2.1 Water Footprints

Water FP can be color coded as green for precipitation, blue for surface and groundwater, and gray, which refers to the volume required to assimilate pollutants to the background level. In the past, economic development is proportional to air, water, and land degradation. For example, before a forest land is developed, the percentages of precipitation by evaporation, seepage, and the runoff are 40, 50, and 10, respectively. After development, these percentages typically changed to 30, 15, and 55, respectively. The large runoff increases stormwater volume and flooding probability. In a typical year, one acre of land cleared for development may cause 10 tons of eroded sediment, while one acre of impervious cover leads to one million gallons of runoff. In addition, flooding could threaten lives and damage the habitat, ruining streams and rivers and decreasing biodiversity in water channels. Since runoff is also a major nonpoint source, pollution control is important during land development and construction to minimize its impact on the water quality. One of the effective strategies is to build green infrastructure to reduce runoff by absorbing the nutrients and increasing Earth's biomass and reduce water footprints of economic development.

1.2.2.2 Gray Water System

Gray water consists of all domestic water used except black water which is toilet wastewater. It does not contain urine, defecation, or pathogens. It is easier to treat than black water which has minute pathogens. The major treatment objectives for gray water are to (i) kill pathogens and protect public health; (ii) protect groundwater from contamination; (iii) enable safe reuse for urban agriculture, irrigation, aquifer recharge, and landscaping; and (iv) avoid damage to buildings and surrounding areas. Reusing gray water significantly reduces nutrients, energy, carbon, and water FP. The treated gray water can be used to irrigate crops and lawns (but cannot be used for drinking or cooking without advanced treatment), reduces potable water demand in urban agriculture, and increases river flows and aquifer recharge.

Gray water FP is defined as the freshwater volume required to assimilate gray water to its natural background water quality and is calculated according to the following equations:

$$\text{GWF} = \frac{L}{C_{\text{ambient}} - C_{\text{nature}}} \tag{1.5}$$

$$\text{GWF}_{\text{Nitrogen and phosphorus}} = \frac{\alpha \times Q \times \text{population}}{C_{\text{Nitrogen and phosphorus}} - 0} \tag{1.6}$$

where

$L = \alpha \times a_{\text{ppl}}$ is the loading rate (g/person/day)
α is the coefficient of location
a_{ppl} is the application rate of loading = rate Q (gallons or g/person/day) × population
C_{ambient} is the standard set for the ambient concentration in g/l (when trophic state of a water body cannot be determined, it suggests using the maximal allowable concentration (MAC))
C_{nature} is the background concentration in g/l and is usually zero

Since ambient water quality standards are used in the calculation, the water FP of gray water depends upon which ambient nutrient standards are used. Many different guidelines for the different reuse policies have been established by the World Health Organization (WHO, 2012) or the US EPA (Example 1.4).

Example 1.4 Please find the average income figure from the World Bank website for Opa-locka, Miami-Dade County, the United States, China, and Nigeria through the following websites:

http://waterfootprint.org/en/resources/interactive-tools/personal-water-footprint-calculator
http://data.worldbank.org/country

Find: Water footprint (WFP) of a person with average income in Opa-locka, Miami-Dade County, the United States, China, and Nigeria.

Solution:

WFP of a person with average income in Opa-locka:
Water footprint = 772 m^3/year = 27 263 ft^3/year.

Average income in Opa – Locka
= \$12 080.00 (from miamidade.gov)

WFP of a person with average income in Miami-Dade County:

$$\text{Water footprint} = 3081\ \text{m}^3/\text{year} = 108\ 805\ \text{ft}^3/\text{year}.$$

Average income in Miami – Dade
= \$49 900.00 (from miamidade.gov)

WFP of a person with average income in the United States:

$$\text{Water footprint} = 3341\ \text{m}^3/\text{year} = 117\ 986\ \text{ft}^3/\text{year}.$$

Average income in the United States = \$55 200.00

WFP of a person with average income in China:

$$\text{Water footprint} = 1404\ \text{m}^3/\text{year} = 49\ 582\ \text{ft}^3/\text{year}.$$

Average income in China = \$7400.00

WFP of a person with average income in Nigeria:

$$\text{Water footprint} = 727\ \text{m}^3/\text{year} = 25\ 674\ \text{ft}^3/\text{year}.$$

Average income in Nigeria = \$2970.00

The aforementioned WFP clearly shows that WFP depends upon average income. Figure 1.3 compares WFP by individual from Opa-locka, Miami-Dade County, the United States, China, and Nigeria (Example 1.5).

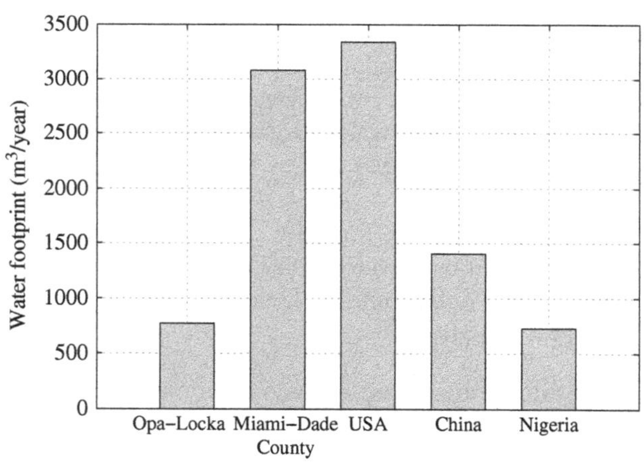

Figure 1.3 Comparison of water footprint.

Example 1.5 Please find the average N and P discharge per person per day in the United States, Japan, China, and Zimbabwe. Please calculate the gray water footprint for the aforementioned, respectively, using the following technical guideline book by Franke et al. (2013).

Find: The average N and P discharge per person per day and calculate the gray water footprint in the United States, Japan, China, and Zimbabwe. Table 1.6 lists the maximal allowable concentration of N and P for different eutrophication states.

Table 1.6 Maximal allowable concentration in μg/l.

Nutrients	Chemical formula	Maximum permitted concentration μg/l	Referenced guideline
Ammonium NH_3 (unionized as NH_3)	NH_3	19 unionized NH_3-N[a]	CCME
Nitrate as NO_3	NO_3	13 000	CCME
Nitrite as N	NO_2	60	CCME
Total phosphorus	Ultra-oligotrophic	4	CCME
	Oligotrophic	10	
	Meso-trophic	20	
	Meso-eutrophic	35	
	Eutrophic	100	

Note:
[a] The unionized ammonia guideline value is expressed as μg ammonia/l. This is equivalent to 16 μg ammonia-N/l (= 19 × 14.0067/17.35052).

Solution:

When trophic state of a water body cannot be determined, the tier 1 of gray water footprint guidelines suggests using the value of 20 μg/l (MAC) and 16 μg/l for nitrogen for meso-trophic water bodies to calculate the GWF.

For a natural/background concentration of nearly zero, Equation (1.5) can be used to calculate the GWF to be done:

$$\text{GWF} = \frac{L}{C_{\text{ambient}} - C_{\text{nature}}}$$

Average N and P discharge per person and per day and the grey water footprint in United States:

$$\text{N} : 18\ \text{g/person/day}$$

$$\text{P} : 4\ \text{g/person/day}$$

$$\text{GWF}_{\text{Nitrogen}} = \frac{18\ \text{g/person/day}}{16 \times 10^{-6}\ \text{g/l}} = 1\,125\,000\ \text{l/day/person}$$

$$\text{GWF}_{\text{Phosphorus}} = \frac{4\ \text{g/person/day}}{20 \times 10^{-6}\ \text{g/l}} = 200\,000\ \text{l/day/person}$$

Average N and P discharge per person and per day and the gray water footprint in Japan:

$$N: 8.5\ \text{g/person/day}$$

$$P: 1\ \text{g/person/day}$$

$$GWF_{\text{Nitrogen}} = \frac{8.5\ \text{g/person/day}}{16 \times 10^{-6}\ \text{g/l}} = 531\,250\ \text{l/day/person}$$

$$GWF_{\text{Phosphorus}} = \frac{1\ \text{g/person/day}}{20 \times 10^{-6}\ \text{g/l}} = 50\,000\ \text{l/day/person}$$

Average N and P discharge per person and per day and the gray water footprint in China:

$$N: 8\ \text{g/person/day}$$

$$P: 1.1\ \text{g/person/day}$$

$$GWF_{\text{Nitrogen}} = \frac{8\text{g/person/day}}{16 \times 10^{-6}\ \text{g/l}} = 500\,000\ \text{l/day/person}$$

$$GWF_{\text{Phosphorus}} = \frac{1.1\ \text{g/person/day}}{20 \times 10^{-6}\ \text{g/l}} = 55\,000\ \text{l/day/person}$$

Average N and P discharge per person and per day and the gray water footprint in Zimbabwe:

$$N: 7.8\ \text{g/person/day}$$

$$P: 1.2\ \text{g/person/day}$$

$$GWF_{\text{Nitrogen}} = \frac{7.8\text{g/person/day}}{16 \times 10^{-6}\ \text{g/l}} = 487\,500\ \text{l/day/person}$$

$$GWF_{\text{Phosphorus}} = \frac{1.2\text{g/person/day}}{20 \times 10^{-6}\ \text{g/l}} = 60\,000\ \text{l/day/person}$$

Comments: The water FP based upon N and P is different with FP due to N being much larger than P. WFP in the United States is significantly larger than other countries, which may be caused by the high animal protein diet.

Since the natural concentration of nitrogen and phosphorus would be the same for the same water body, GWF inversely proportional to the MAC. This can be clearly seen by the fact that if the nutrient status of a lake is different, the GWF will change significantly. For example, to maintain the ultra-oligotrophic status of 4 μg/l, the GWF would be 25 times greater than extra-eutrophic condition (Example 1.6).

Example 1.6 Phosphorus (P) ambient water quality standard can be defined as extra-trophic, meso-trophic, eutrophic, meso-eutrophic, and extra-eutrophic states as shown in Table 1.7.

Find: The gray water footprint if a person is discharging 1 g P/capita/day, while the background P concentration is 0.

Solution:

1) The problem can be solved using following equation:

$$GWF = \frac{L}{C_{\text{ambient}} - C_{\text{nature}}} \tag{1.5}$$

2) By substituting all the known variables for L, C_{ambient}, and C_{natural} into Equation 1.5, the GWF is listed as Table 1.7.

For example,

$$GWF = \frac{1\text{gP/capita/day}}{4 \times 10^{-6}(\text{g/l}) - 0} = 250\text{m}^3\text{/person/day}$$

Table 1.7 Footprint of phosphorus in m^3/person/day.

Ambient water quality standard	C_{ambient} in μg/l	$GWF = \frac{L}{C_{\text{ambient}} - C_{\text{nature}}}$ in m^3/person/day
Extra-trophic	4	250
Meso-trophic	10	100
Eutrophic	20	50
Meso-eutrophic	50	20
Extra-eutrophic	100	10

Comments: If ambient water quality standards are 4 and 100 μg/l, the GWF of P discharged by each person in a day would be 250 and 10 m^3/person/day, respectively. Therefore, the freshwater volume required to assimilate the P loading to ambient standards at 4 μg/l would be 25 times more than that at 100 μg/l. This is very significant because ambient standards were established by the receiving ecosystem. In the Everglades National Park (ENP), the native plants are so sensitive to P concentration, and the discharge standards of N and P for rehydration of the ENP were 0.27 and 0.005 mg/l, respectively. For recharging groundwater aquifer, however, the discharge standard concentration of N and P could be one to two orders of magnitude higher than that required by the ENP due to different receiving ecosystems because the aquifer soil could further purify the discharged water.

1.3 Challenges and Opportunities

In the developing world, about 1 billion people do not have access to modern sanitary facilities, 1.3 billion people lack electricity, and 2.7 billion people still rely on local biomass for food preparation, which results in deforestation, soil erosion, and human health deterioration (IEA, 2011). One way to reduce the health hazard of incomplete combustion is to increase application of high efficient stover with wood pellets. To improve sanitary facilities, composite toilets may serve as a transition for better environments. In the developed countries such as the United States, there is a trend to decentralize wastewater treatment infrastructure so that the treated wastewater could be close to irrigation sites for resue. In addition, aging water infrastructures such as WTP and WWTPs were assessed as grade D due to lack of retrofitting or improvement capital. For example, aging infrastructure caused lead poisoning in Wisconsin. There were many cases of sewage overflow during the wet season all over the major US cities. In Miami-Dade, thousands of gallons of sewer were discharged to the Biscayne Canal because a 64-inch sewer pipe broke at 840 NW 155 Lane on 30 June 2017. Since the leaked sewer could enter the Biscayne Canal and flowed into the interacoastal communities, a precautionary advisory to avoid recreational water activities including swimming was issued by the Miami-Dade County. As the WWTPs and sewer pipe system approach their design life, sewer leaks due to pipe breaks and the reduced performance of WWTPs suggest that design paradigm shift from fit-for-all to fit-for-purpose is the main challenge for the next-generation SEE designers. Indeed, some of the traditional design philosophy should be revisited as new knowledge and technologies are created and commercialized. For example, chlorine is used as a post-disinfectant in more than 70% of WWTPs in the United States, while almost 90% of WTPs in the United States use chlorine. The major health concern of chlorination is the disinfection by-products (DBPs) which are heavily regulated by the US EPA.

Table 1.8 History industrial productions of nitrogen fertilizers.

Year	Author	Contribution	References
1840	Justus von Liebig	Global biospheric cycles and of nitrogen in crop production	von Liebig (1840)
1860	Jean-Baptiste Boussingault	Nutritional value of fertilizers	Smil (2001)
1877	Théophile Schloesing	Bacterial origins of nitrification	Smil (2001)
1885	Ulysse Gayon	Denitrifying bacteria that can reduce nitrates and, via NO and N_2O, return N_2 to the atmosphere	Smil (2001)
1886	Hermann Hellriegel	Leguminous nodules as the sites of biofixation	Smil (2001)
1909	Fritz Haber	Synthesizing ammonia using iron-based catalyst	Smil (2001) and Stoltzenberg and Haber (2004)
1920	Carl Bosch	Mass production of ammonia inexpensive nitrogenous fertilizers	Smil (2001)

1.3.1 Excessive Nitrogen Runoff

Due to the exponentially increasing human population, the Haber–Bosch (HB) process is used in the industrial production of nitrogen fertilizers for agriculture. Without the HB process, nearly four times more land would be needed to produce the food needed for 7.2 billion people. Table 1.8 shows the major milestones of this process. Due to excessive application of fertilizer and nutrient runoff, however, the planetary boundary (PB) of nitrogen loading to the ecosystem is considered as being trespassed by human economic activities.

1.3.2 Phosphorus Depletion

Phosphorus is an essential component in adenosine triphosphate (ATP), an energy-bearing compound that drives biochemical processes in plants and animals. It is also a critical element in DNA, RNA, and phospholipids that are essential for the function of cellular membranes. Phosphorus, present as phosphate minerals in the soil, is a nonrenewable resource. Phosphate rock, or calcium phosphate, $Ca_3(PO_4)_2$, is the only important commercial source of phosphorus. The United States is the largest producer of phosphate rock in the world (USGS, 1997). In 1996, 13 300 000 metric tons of phosphate rock was mined in the United States, which counted to about one-third of the world's total. About 86% of phosphate rock comes from North Carolina and Florida. Ninety-one percentage of all the phosphate rock mined in the United States was used to make fertilizer. To meet the insatiable agricultural demand for phosphorus, the mining of phosphorus is depleting the reserve. With 7000 million tons of P reserve, it would be completely used up by 2061 and 2070 with annual consumption growth rates of 3 and 2.5%, respectively. If consumption growth rates were 2 and 1% annually (Wetzel, 1983; Steen, 1998), it would be depleted by 80 and 60% by the end of 2071, respectively. Detergents contain sodium tripolyphosphate (STPP) ($Na_5P_3O_{10}$). Over two million tons of phosphorus was used annually in detergents in the United States in 1983 alone (Wetzel, 1983). About 50% of the wastewater phosphorus nationwide came from detergents (Hammond, 1971). If excessive phosphorus is discharged to the natural water body, one pound of phosphorus could grow 700 lb of algae (Beeton, 1971) and leads to significant eutrophication in surface waters.

1.3.3 Carbon Pollution

Carbon dioxide concentration has been closely correlated with the global temperature. If current trends continue and no international collective efforts were implemented, even the best scenario on carbon reduction could barely stop the Earth from warming by less than 2 °C, which is considered as a tipping point by the end of this century although it has only increased by 0.83 °C currently. If international agreement developed in the Paris Conference in 2015 could be implemented, the deterioration trajectory might be reversed. China burns almost as much coal as the rest of the world combined and is the world's leading greenhouse gas emitter because its annual GDP had grown more than 10% in the past three decades. The major environmental issue rooted in China is its energy structure by using coal as the major energy source as shown in Figure 1.4.

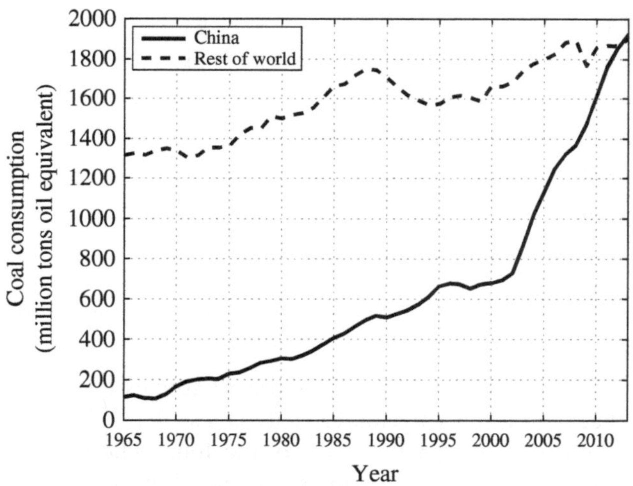

Figure 1.4 Coal consumption in China and the rest of the world in 1965–2013. (*Source:* Data from Statistical Review of World Energy June 2014, London.)

1.3.4 Peak Oil

Peak oil is the time when the rate of oil production worldwide is forecasted to decline. The International Energy Agency (IEA) released the results of a unique comprehensive analysis of 800 oilfields, including all 54 supergiant reserves (>5 Gbbl) in production (World Energy Outlook (WEO) of November 2008). As oil production peaks, liquid fuel prices and price volatility could increase dramatically, and, without timely mitigation, the economic, social, and political costs would be unprecedented (World Bank, 2012). For these reasons, leading countries such as Germany is planning to ban all the fossil fuels by 2030.

1.3.5 Climate Change

Extreme weather could be expected in the next decades due to combining factors such as greenhouse gas emission as well as solar minimum. Figure 1.5 shows that the observed global average temperature varied from −9 to +3 °C in the past 400 000 years (Petit et al., 1999). The mean temperature is less than 0 °C during 85% of the time with global average temperature greater than 0 degree, while only 15% of the interglacial represents warm holiday in this period of time. The major warning sign for climate is that the Earth should have been cooling much faster than it does now. Therefore, global warming due to CO_2 would be a major concern if CO_2 emission is out of control. Regardless of global warming or cooling, extreme weather will bring more intensive drought, flooding, and hurricanes. For these reasons, EEIS should be designed to be resilient and adaptive to both warm and cold climates in the next 100 years.

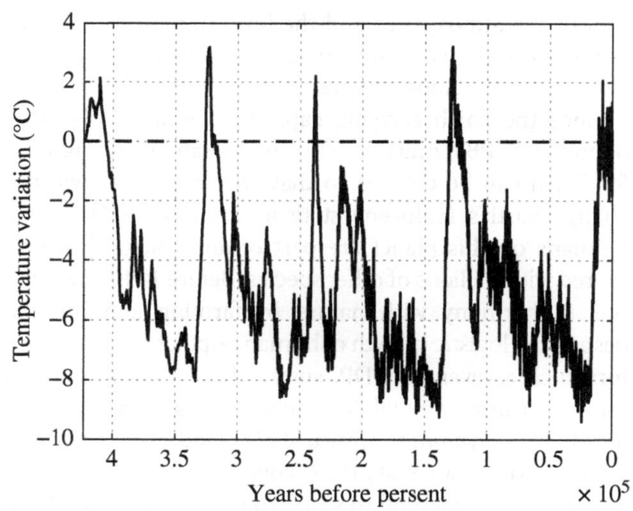

Figure 1.5 Historical isotopic temperature record from the Vostok Ice Core. (*Source:* Data from Petit et al. (1999).)

1.4 Carrying Capacity

Carrying capacity is the maximum number of organisms that can be supported in a given habitat. This pale blue dot planet Earth is home to 7.2 billion people in 2018. It is expected that 10 billion people will live on this planet by 2100. The question that begs an accurate answer is, how many people that the planet Earth can support without irreversible damage to its ecological system due to great pressure that human exerted on the Earth? What are the safe operating boundaries for humans to thrive without triggering the tipping point, at which Earth shifts from its Holocene to Anthropocene? To answer this question, the human EF were compared with Earth's carrying capacity by Ree and Wackernagel (2013). Another way to answer this question is through the primary biomass production of the Earth divided by the primary biomass to support a person. For example, Miller (1971) estimated that 1000 tons of grass as primary biomass would be needed to support a man for 1 year if he was fed on 300 trout, which must consume 90 000 frogs, which must consume 27 million grasshoppers. Net Primary Production (NPPP) is the amount of plant material produced on Earth and the net amount of solar energy converted to plant organic matter through photosynthesis. Carbon can be measured by the photosynthesis process (i.e. CO_2 exchange between atmosphere and biosphere). Twenty-four to thirty-two percentage of NPP is consumed by 7.2 billion people globally. If all the global NPP had been used, it could support 22–29 billion people. However, the global ecosystem would collapse as there would not be enough grass and other vegetation at the bottom of the food chain to regenerate the needed resource for humans to survive. In addition, it would also change the composition of the atmosphere and the level of biodiversity. For this reason, the carrying capacity was estimated around from 22 to 29 billion people.

Since the Earth carrying capacity is limited, planetary boundaries (PBs) must be set so that a safe operating zone for Earth can be defined so that human activity will not change Earth's Holocene state in the next 10 000 years. Humans on this planet are part of an ecological system shared with millions of other species. Before 1840, mankind was in harmony with nature by farming and hunting despite the logistic growth of human population. Many factors such as average GDP, sustainable technologies, and green consumption will determine Earth's carrying capacity. *World Dynamics* (Boyd, 1972) and *The Limits to Growth* (Meadows et al., 1972) concluded that Earth's economic system will tend to stop growing and collapse from reduced availability of resources, overpopulation, and pollution at some point. Technological innovation, population control, and resource availability could delay the collapse with the right policy of stabilizing material consumption.

Under the current technology and consumption pattern, about 10 billion people could be sustained at $10 000 GDP, respectively. If incomes in middle- and low-income countries were to catch up with incomes of the high-income countries (roughly $41 000 per capita), there would be a roughly 3.4-fold increase in global income further from $87 trillion to $290 trillion, which could increase if the high-income countries grow with the world population. If Brazil, Russia, India, China, and Southern Africa (BRICS) countries catch up with the developed countries and Africa catches up BRICS, there would be no new lands, fossil-fuel reserves, and groundwater. Even worse, Earth's ecosystems could not absorb the corresponding carbon dioxide from fossil fuels, nitrogen runoff from fertilizers, and toxic pollutants dissipated in the oceans and rivers. For these reasons, to ensure the world stays within PBs, sustainability has become a prerequisite for human development at all levels, from the local community to nations and the world. The world annual economy would double every two decades at the current growth rate of 4%. As a result, economic activities would break the PBs unless technological innovation could be much faster than economic development on a global scale. More importantly, all the EEIS could be designed, built, and operated sustainably. Table 1.9 lists the dramatically different estimates for the carrying capacity of Earth by 65 research projects.

Table 1.9 Different estimated carrying capacity of Earth.

Carrying capacity ≤ billions	Number of estimates
2	6
4	7
8	20
16	14
32	32
64	7
128	2
256	1
512	1
1024	1

Figure 1.6 shows that there are 32 studies estimating that the Earth's carrying capacity is less than 32 billion. The carrying capacities of two and four billion people appear not to be good estimates because the current population is 7.2 billion. Earth apparently supports us fairly well. At the other extreme, there is only one estimate that the carrying capacity is less than 256, 512, and 1024 billion, respectively. The different estimating methods with different assumptions are the major factors contributing to the widely different carrying capacity estimates. Food, water, and energy could be the limited constraining factors. The interaction between these factors could further complicates the estimates. For example, the natural reserves of fossil fuel could limit the fertilizer production, the pumping of irrigation water, and the use of farming machines. Therefore, the human population that could be supported by planet Earth is limited.

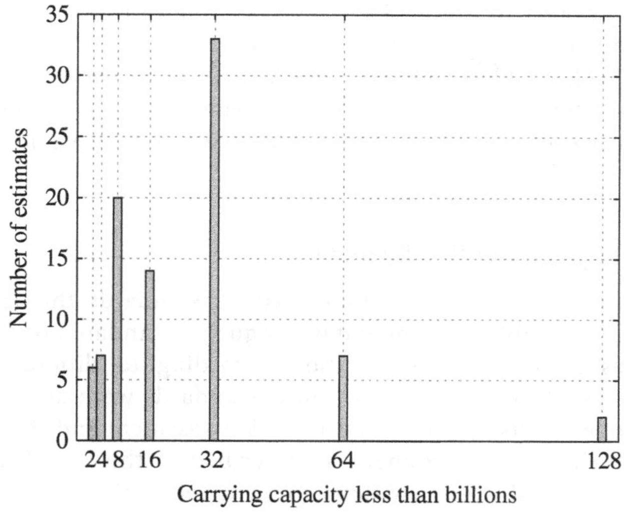

Figure 1.6 Carrying capacity estimates. (*Source:* Data from the UNEP (2012).)

The major difference arises from the constraint factors as well as the speed of technological advance and adoption. Innovation, leadership, entrepreneurship, finance, and policy will answer the challenges that climate change imposes upon us. At the same time, abundant opportunities are also presented to the next generation of EEIS designers.

Global environmental constraints have been quantified as carrying capacity and human FP to define safe operating space within specific PBs to avoid human impact cross tipping points (Rockström et al., 2009). To achieve this, innovation is the key in economic growth and human development. According to the sustainable development trajectory from 2015 to 2030, the world should cooperate within the PBs by adopting sustainable technologies, stabilizing the world's population, and protecting threatened species and ecosystems. The new global rules of the UN 2030 ensure an orderly and cooperative process, improving outcomes to achieve the sustainable development goals (SDGs) for 193 countries of the world. To achieve these sustainable targets, SEE designers would play a critical role in building new cities or villages. Smart cities, regenerative design, zero water, energy, and waste communities are all the new challenges and tasks for EEIS designers to decouple the economic development from environmental degradation.

For humans to thrive and coexist with nature, PB is defined as a safe operating space for humanity with respect to the functioning of Earth's system. Nine PBs are grouped in four categories: (i) the global biogeochemical cycles of nitrogen, phosphorus, carbon, and water; (ii) the major physical circulation systems of the planet such as the climate, stratosphere, and ocean system; (iii) biophysical features of Earth that contribute to the underlying resilience of its self-regulatory capacity such as marine and terrestrial biodiversity and land systems; and (iv) two critical features associated with anthropogenic global change such as aerosol loading and chemical pollution. The aims of the nine PBs are to quantify a safe global level of depleting nonrenewable fossil resources such as energy (coal, oil, gas) and fossil; use the living biosphere such as ecosystems, biodiversity, and land use; and absorb and dissipate human waste flows including carbon, nitrogen, phosphorus, and toxic chemicals. Currently, climate change, change rates of the global nitrogen cycle, and rate of biodiversity loss have apparently been transgressed by mankind. According to human FP, Rockström et al. (2009) determined seven quantitative PBs: "1) climate change (CO_2 concentration in the atmosphere <350 ppm and/or a maximum change of +1 W m^{-2} in radiative forcing); 2) ocean acidification (mean surface seawater saturation state with respect to aragonite ≥80% of pre-industrial levels); 3) stratospheric ozone (<5% reduction in O_3 concentration from pre-industrial level of 290 Dobson Units); 4) biogeochemical nitrogen (N) cycle (limit industrial and agricultural fixation of N_2 to 35 Tg N yr^{-1}) and phosphorus (P) cycle (annual P inflow to oceans not to exceed 10 times the natural background weathering of P); 5) global freshwater use (<4000 km^3 yr^{-1} of consumptive use of runoff resources); 6) land system change (<15% of the ice-free land surface under cropland); and 7) the rate at which biological diversity is lost (annual rate of <10 extinctions per million species)."

1.5 Air, Water, and Soil Quality Index

With the global picture in mind, an SEE designer should act locally by quantifying air, water, and solid indexes first. Once these indexes are quantified, environmental issues can be prioritized according to the relative scale of these indexes compared with local environmental standards.

1.5.1 Air Quality Standards

The US EPA established the following primary National Ambient Air Quality Standards (NAAQS) in Table 1.10 to protect human health.

1.5.2 Air Quality Index

The US EPA developed an air pollutant standard index (PSI) for introducing consistency in providing information on air quality throughout the United States. The system is based on a scale of 0–500. The computed index below 100 indicates that the air quality is within acceptable range. A value over 100 implies potential health

Table 1.10 National ambient air quality standards.

Air pollutants	Primary standards	Averaging times
Carbon monoxide	9 ppm (or 10 mg/m^3)	8 h
	35 ppm (or 40 mg/m^3)	1 h
Lead	1.5 µg/m^3	Quarterly average
Nitrogen dioxide	0.053 ppm (or 100 µg/m^3)	Annual (arithmetic mean)
Particulate matter (PM_{10})	50 µg/m^3	Annual (arithmetic mean)
	150 µg/m^3	24 h
Particulate matter ($PM_{2.5}$)	15.0 µg/m^3	Annual (arithmetic mean)
	65 µg/m^3	24 h
Ozone	0.08 ppm	8 h
	0.12 ppm	1 h
Sulfur oxides	0.03 ppm	Annual (arithmetic mean)
	0.14 ppm	24 h

problems. The alerts are issued at 200, 300, and 400 levels. Five pollutants (carbon monoxide, sulfur dioxide, total suspended particulate, ozone, and nitrogen dioxide) are included in the index. The index is based on the highest index value of any of the five pollutants. Table 1.11 presents the concentration and the corresponding index value.

Table 1.12 lists the likely health effects at different PSI levels as well as health advisory for different population with different vulnerabilities.

Similar to the US EPA approach, the China Department of Environmental Protection (DEP) developed Table 1.13 by using the concentration of the particulate matter with diameter less than 2.5 µm ($PM_{2.5}$) to count particles' contribution to the air quality index (AQI).

1.5.3 Water Quality Index

In addition to drinking water quality standards, the US EPA established ambient water quality standards such as N and P concentration according to different classes of water bodies. These standards were developed by using human health risk assessment and ecological risk assessment as described in the US EPA website (https://www.epa.gov/wqs-tech). Due to different environmental capacities of different states, the EPA authorized each state to establish its state standards of N and P concentration. In Florida, for example, Florida Department of Environmental Protection (FDEP) classifies water bodies in Florida Statute 62-302.400 according to its intended use, according to Table 1.14.

Since there are standards of many water quality parameters, a unified single index such as WQI could be used to assess ambient water quality under general conditions with the *Q*-values listed in Table 1.15 to convert water quality data to *Q*-values for each water body. Once converted, a weighting factor is used to quantify the WQI, which could be used to show the relative degree of water pollution of a water body (Example 1.7).

Table 1.11 US pollutant standard index.

Index value	Air quality level	TSP (24 h) 10^{-6} g/m^3	SO_2 (24 h) 10^{-6} g/m^3	CO (8 h) 10^{-3} g/m^3	O_3 (24 h) 10^{-6} g/m^3	NO_2 (24 h) 10^{-6} g/m^3
400–500	Significant harm	1000	2620	57.5	1200	3750
300–400	Emergency	875	2100	46	1000	3000
200–300	Alert	625	1600	34	800	2260
100–200	NAAQS	375	800	17	400[c]	1130
50–100	50% of NAAQS	260	365	10	235	[a]
0–50		75[b]	80[b]	5	80	[a]
0		0	0	0	0	[a]

Source: US EPA (2014).
Note:
[a] No index values reported at concentration levels below those specified by "alert level" criteria.
[b] Annual primary NAAQS.
[c] 400 × 10^{-6} gm/m^3 was used instead of the O_3 alert level of 200 × 10^{-6} g/m^3.

Table 1.12 Associated health effects with US pollutant standard index.

Index value	Health effect descriptor	General health effects	Cautionary statements
400–500	Hazardous	Premature death of ill and elderly. Healthy people will experience adverse symptoms that affect their normal activity	All persons should remain indoors, keeping windows and doors closed All persons should minimize physical exertion and avoid automobile traffic
300–400	Hazardous	Premature onset of certain diseases in addition to significant aggravation of symptoms and decreased exercise tolerance in healthy persons	Elderly and persons with existing diseases should stay indoors and avoid physical exertion General population should avoid outdoor activity
200–300	Very unhealthy	Significant aggravation of symptoms and decreased exercise tolerance in persons with heart or lung disease with widespread symptoms in the healthy population	Elderly and persons with existing heart or lung disease should stay indoors and reduce physical activity
100–200	Unhealthy	Mild aggravation of symptoms with susceptible persons, with irritation symptoms in the healthy population	Persons with existing heart or respiratory ailments should reduce physical exertion and outdoor activity
50–100	Moderate		
0–50	Good		

Source: US EPA (2014).

Table 1.13 Air quality sub-index levels and their corresponding air pollutant concentrations (Ministry of Environmental Protection of The People's Republic of China, 2012).

	Air pollutant concentration ($\mu g/m^3$)									
Air quality sub-index	SO_2 24 h	SO_2 1 h[a]	NO_2 24 h	NO_2 1 h	PM_{10} 24 h	CO 24 h	CO 1 h	O_3 1 h	O_3 8 h	$PM_{2.5}$ 24 h
50	50	150	40	100	50	2 000	5 000	160	100	35
100	150	500	80	200	150	4 000	10 000	200	160	75
150	475	650	180	700	250	14 000	35 000	300	215	115
200	800	800	280	1200	350	24 000	60 000	400	265	150
300	1600	[b]	565	2340	420	36 000	90 000	800	800	250
400	2100	[b]	750	3090	500	48 000	120 000	1000	[c]	350
500	2620	[b]	940	3840	600	60 000	150 000	1200	[c]	500

Note:

[a] The 1-h average concentrations of SO_2, NO_2, and CO are just used for real-time reports; the daily concentrations are acquired by 24-h average.

[b] The 1-h average concentration of SO_2 will not be included in the calculation of air quality sub-index if it is greater than 800 $\mu g/m^3$, and air quality sub-index of SO_2 is reported as 24-h average.

[c] The 8-h average concentration of O_2 will not be included in the calculation of air quality sub-index if it is greater than 800 $\mu g/m^3$, and air quality sub-index of O_2 is reported as 1-h average.

Table 1.14 Classification of surface waters.

Water body class	Description
I	Potable water supplies
II	Shellfish propagation or harvesting
III	Fish consumption; recreation, propagation, and maintenance of a healthy, well-balanced population of fish and wildlife
	Fish consumption; recreation or limited recreation; and/or propagation and maintenance of a limited population of fish and wildlife
IV	Agricultural water supplies
V	Navigation, utility, and industrial use

Table 1.15 *Q*-value tables for calculating water quality index.

DO (% saturation)	*Q*-value	Nitrate-N (mg/l NO_3-N)	*Q*-value	Total phosphate (mg/l P)	*Q*-value	pH (units)	*Q*-value
0	0	0	98	0	99	<2	0
10	8	0.25	97	0.05	98	2	2
20	13	0.5	96	0.1	97	3	4
30	20	0.75	95	0.2	95	4	8
40	30	1	94	0.3	90	5	24
50	43	1.5	92	0.4	78	6	55
60	56	2	90	0.5	60	7	90
70	77	3	85	0.75	50	7.2	92
80	88	4	70	1	39	7.5	93 (max)
85	92	5	65	1.5	30	7.7	90
90	95	10	51	2	26	8	82
95	97.5	15	43	3	21	8.5	67
100	99	20	37	4	16	9	47
105	98	30	24	5	12	10	19
110	95	40	17	6	10	11	7
120	90	50	7	7	8	12	2
130	85	60	5	8	7	>12	0
140	78	70	4	9	6		
>140	50	80	3	10	5		
		90	2	>10	2		
		100	1				
		>100	1				

Example 1.7 Lake A with a temperature of 21 °C has the following water quality data with the weighting factors given in Table 1.16.

Table 1.16 Measured data of lake A.

Parameter	Tested value	*Q*-value	Weighting factor	Calculation
Dissolved oxygen	80		0.3	
Nitrate	1.93		0.2	
Phosphate	0.4		0.2	
pH	6.5		0.3	
Totals				

Find: What is the water quality index of the lake?

Solution:

Dissolved oxygen: % saturation = 80

In the DO column, find the value equal to 80.

Write down the *Q*-value to the right of that number (*Q*-value = 88).

Nitrate: 1.93 mg/l

In the nitrate-N column, find the values closest to 1.93. This number falls between 1.5 and 2 in this column.

Since there is no value matching exactly to 1.93, estimate the *Q*-value between 92 and 90 (*Q*-value = 90.28).

Phosphate: 0.4 mg/l

In the total phosphate column, find the value equal to 0.4.

The *Q*-value corresponding to this concentration is 78.

pH: 6.5

In the pH column, find the values closest to 6.50. This number falls between 6 and 7.

Since there is no value matching exactly to 6.50, estimate the *Q*-value between 55 and 90 (*Q*-value = 72.5) (Table 1.17).

Answer: The water quality index of this lake is 81.8.

Table 1.17 Water quality index of lake A.

Parameter	Tested value	Q-value	Weighting factor	Calculation
Dissolved oxygen	80	88	0.3	88 × 0.3 = 26.4
Nitrate	1.93	90.28	0.2	90.28 × 0.2 = 18.056
Phosphate	0.4	78	0.2	78 × 0.2 = 15.6
pH	6.5	72.5	0.3	72.5 × 0.3 = 21.75
Totals				26.4 + 18.056 + 15.6 + 21.75 = 81.8

Comments: The WQI depends upon weighting factors, which could be adjusted according to the ecosystem of different intended use of the water resource. For surface water bodies such as a lake, it depends upon the eutrophication stage of the lake and whether it serves as a local drinking water resource or merely for recreational purposes. Many local and state environmental protection agencies provide local air, water, and soil quality data that can be used to quantify the WQI.

1.5.4 Soil Quality Index

The US EPA developed soil quality guidelines/standards for over 200 chemicals using environmental risk assessment. For example, the soil quality guidelines developed by Michigan Department of Environmental Quality (2004) are available at its website (http://www.deq.state.mi.us/documents/deq-rrd-part201-rules-Rule746Table.pdf). These soil quality standards can be used for calculating the soil quality index (SQI). Three factors are (i) scope (% of contaminants that do not meet their respective guidelines), (ii) frequency (% of individual tests of contaminants that do not meet their respective guidelines), and (iii) amplitude (the amount by which the contaminants do not meet their respective guidelines). The SQI can be used to compare different contaminated sites with similar types of contamination as well as to see if remediation of a particular site meets the required contaminate guidelines. A brief description of these factors and formulas for calculating the SQI is given as follows.

1.5.4.1 F_1 (Scope)

The factor F_1 (as calculated in Equation 1.7) represents the percentage of contaminants that do not meet their respective guidelines (failed contaminants) relative to the total number of contaminants that were measured (and selected for inclusion in the SQI calculation) at the site:

$$F_1 = \frac{\text{Number of failed contaminants}}{\text{Total number of contaminants}} \times 100 \quad (1.7)$$

1.5.4.2 F_2 (Frequency)

The factor F_2 (Equation 1.8) represents the percentage of individual tests that do not meet their respective guidelines (failed tests):

$$F_2 = \frac{\text{Number of failed tests}}{\text{Total number of tests}} \times 100 \quad (1.8)$$

1.5.4.3 F_3 (Amplitude)

The factor F_3 (Equation 1.12) represents the amount by which failed test values do not meet their respective guidelines (excursion from the guideline value). The relationship between F_3 and the amount by which the concentrations of contaminants depart from their guidelines could be calculated with the following three steps:

Step 1: Calculate the excursion of all tests in the dataset.

When the concentration of a contaminant is greater than (or less than, when the guideline is a minimum) the soil quality guideline, it is called an excursion. The magnitude of excursion of each test is calculated as follows.

When the test value must not exceed the guideline,

$$\text{Excursion}_i = \frac{\text{Failed test value}_i}{\text{Guideline}_i} - 1 \quad (1.9)$$

For cases in which the test value must not fall below the guideline,

$$\text{Excursion}_i = \frac{\text{Guideline}_i}{\text{Failed test value}_i} - 1 \quad (1.10)$$

Step 2: Calculate the average sum.

This refers to the average sum of excursion (ASE) by which individual tests are out of compliance and is calculated by adding together the excursion of all individual tests from their guidelines and dividing by the total number of tests that do not meet their guidelines as follows:

$$\text{ASE} = \frac{\sum_{i=1}^{n} \text{Excursion}_i}{\text{no. of failed tests}} \quad (1.11)$$

In the SQI (CCME, 2001), the "normalized sum of excursions" (NSE) was calculated by dividing the sum of excursions by the total number of tests (tests that meet as well as those that do not meet the guidelines). By using "NSE" instead of "ASE," factor F_3 becomes smaller and increases the value of the SQI. The impact of using "NSE" is that both F_3 and the final value of the SQI increase with an increase in the total number of tests. Therefore, for the purposes of the SQI, F_3 is modified to use "ASE," where the sum of excursions is divided by only the total number of tests that are not in compliance (Equation 1.11). The main reasons for this change are as follows:

As the quality of a contaminated site is primarily judged by the amount of excursions of various contaminants from their guidelines, the value of F_3 should appropriately reflect that in comparison with F_1 and F_2.

The effect of contaminants gets diluted by dividing the excursions by the total number of tests.

Any one of the contaminants that is not in compliance can cause severe limitations for the ecosystem or public health. Therefore, the impact of any contaminant should not be minimized in rating the contaminated site.

Step 3: F_3 is then calculated by an asymptotic function that scales the ASE to yield a range between 0 and 100 as follows:

$$F_3 = \frac{\text{ASE}}{0.01\,\text{ASE} + 0.01} \qquad (1.12)$$

1.5.4.4 Soil Quality Index (SQI)

Once the factors are quantified, the SQI can be calculated by adding together all the factors as if they were vectors as shown in Equation (1.13). This approach treats the index as a three-dimensional space defined by each factor along one axis. With this model, the index changes in direct proportion to changes in all three factors:

$$\text{SQI} = 100 - \sqrt{F_1^2 + \frac{F_2^2 + F_3^2}{1.732}} \qquad (1.13)$$

The divisor (1.732) normalizes the resultant values to a range between 0 and 100, where 0 represents a very high level of contamination or public concern and 100 represents a negligible amount of contamination or public concern. The value of the divisor is calculated as follows:

$$\frac{\sqrt{100^2 + 100^2 + 100^2}}{100} = 1.732 \qquad (1.14)$$

If a site is not tested more than once over time or space, the factor F_2 (frequency) will not be applicable and the divisor for the Equation (1.13) for calculating the SQI will be 1.414. Based on the final value of the index, contaminated sites can be divided into five different classes. Each class of contaminated site needs to be interpreted based on the level of concern for public and ecosystem health and the need for remediation. Sites with a high SQI score (e.g. 90–100) are of high quality, have a very low level of concern, and have low priorities for remediation. At the other end of the scale, a low SQI score (0–30) would indicate a very high level of concern due to contamination and therefore a great need for remediation (Table 1.18).

Table 1.18 Site classes or level of concern and soil ranking categories of the SQI.

Site classes or level of concern	Soil ranking categories of the SQI
Very low	90–100
Low	70–90
Medium	50–70
High	30–50
Very high	0–30

When reporting SQI scores for contaminated sites, users should also provide a list of major contaminants of public concern and a statement describing the need for soil remediation or soil management options where possible (Example 1.8).

Example 1.8 The US EPA provided a case study in Saskatchewan, which was contaminated with petroleum hydrocarbons and later remediated using a tier 1 approach. The soil samples were taken from four different locations and analyzed for benzene; ethylbenzene; toluene; xylenes; petroleum hydrocarbons (PHC) such as F1 (C6–C10), F2 (C10–C16), F3 (C16–C34), and F4 (C34–C50); and lead. The analytical data and applicable soil quality guidelines are presented in Table 1.19.
Find: The SQI for this site.

Solution:

$$F_1 = \frac{5}{9} \times 100 = 55.6$$

$$F_2 = \frac{6}{36} \times 100 = 16.7$$

F_3 :

$$\text{Excursion}_\text{i} = \frac{40}{5} - 1 = 7$$

$$\text{Excursion}_\text{ii} = \frac{82}{50} - 1 = 0.64$$

$$\text{Excursion}_\text{iii} = \frac{220}{50} - 1 = 3.4$$

$$\text{Excursion}_\text{iv} = \frac{2840}{1000} - 1 = 1.84$$

Table 1.19 Soil sample analysis of a commercial site contaminated with petroleum hydrocarbons.

Parameter	West wall (mg/kg)	Bottom (mg/kg)	North wall (mg/kg)	South wall (mg/kg)	Saskatchewan subsoil guidelines[a]
Depth (m)	2.4	5.3	2.4	3.0	—
Benzene	0.8	1.1	0.5	*40*	5.0
Ethylbenzene	27	0.8	4.7	*82*	50
Toluene	*100*	2.0	0.8	14	30
Xylenes	*180*	4.5	7.8	*220*	50
PHC F1	820	96	140	*2840*	1 000
PHC F2	130	8.9	21	180	3 000
PHC F3	5[b]	5[b]	5[b]	5[b]	5 000
PHC F4	5[b]	5[b]	5[b]	5[b]	10 000
Lead	10	8	8	16	1 000[c]

Note:
[a]Saskatchewan's interim criteria of BTEX and PHC fractions of soils for the year 2003.
[b]Not detected at level stated.
[c]Saskatchewan's "Risk Based Corrective Actions for petroleum contaminated sites," November 1995.

$$\text{Excursion}_{v} = \frac{100}{30} - 1 = 2.3$$

$$\text{Excursion}_{vi} = \frac{180}{50} - 1 = 2.6$$

$$\text{ASE} = \frac{17.78}{6} = 2.96$$

$$F_3 = \frac{2.96}{(2.96 \times 0.01 + 0.01)}$$

$$F_3 = \frac{2.96}{0.0396} = 74.8$$

$$\text{SoQI} = 100 - \frac{\sqrt{55.6^2 + 16.7^2 + 74.8^2}}{1.732}$$

$$\text{SoQI} = 100 - 56 = 45$$

Answer: SQI = 45

Comment: The level of concern is high because the soil is contaminated with benzene, ethylbenzene, toluene, xylenes, and PHC fraction F1 (C6–C10). It is recommended that the soil be remediated.

1.6 Air, Water, and Soil Pollution

Since the severe smog environmental disasters in London and Los Angeles in the 1950s and 1960s, respectively, China is now suffering the same fate as industrialized countries after 38 years of rapid economic growth since 1980.

1.6.1 Air Pollution

Figure 1.7 shows that the primary and secondary NAAQS in the United States are 12 and 15 μg/m^3 (annual average), respectively. The corresponding standards in China are 15 and 35 μg/m^3 (annual average), respectively. Yet, none of the top 10 provinces or cities in China meet the secondary Chinese standards of 35 μg/m^3, in fact, the minimum annual average of $PM_{2.5}$ is 55 μg/m^3, which is 57% higher than the Chinese secondary standards.

For 74 major cities in China with population greater than one million, the number of days when air quality parameters were above the Chinese secondary standards was also alarmingly high as shown in Figure 1.8. It suggests that about 44% of the days could meet the Chinese secondary (NAQQS) of 35 μg/m^3, while for the rest of 56% days, the air was mildly to severely polluted.

1.6.2 Water Pollution

Water pollution in China is even worse than its air pollution, especially in urban areas. In 2015 China has almost 110 cities with over one million people and will grow to more than 220 cities by 2025. China consumes four times more water for one US dollar gross domestic production (GDP) created than water required in the U.S. Two-thirds of China's 660 cities suffer from water shortages with the situation termed "severe" in 110 cities. About 700 million people drink water that is contaminated with animal and human waste. Water pollution poisons 190 million Chinese and causes an estimated 60 000 premature deaths every year. In 2005, the Chinese water supply system leaked an estimated 10 billion m^3, more than 20% of the total produced drinking water. China classifies water resource into (i) drinkable (Cat I–III), (ii) industry use only (Cat IV), and (iii) agricultural use only (Cat V). However, only a fraction of water resources satisfy Cat I–III.

China accounts for approximately 19.5% of the world population but has only 7% of the globe's freshwater resources. In northern China, 90% of the aquifers located in under Chinese cities are polluted. Over 75% of river water flowing through urban areas is considered unsuitable for drinking or fishing. Thirty percent of river water in China is regarded as unfit for agricultural or industrial use (Figures 1.9 and 1.10).

Figure 1.11 shows that only 12 and 28% of the groundwater in the country are excellent and good, respectively.

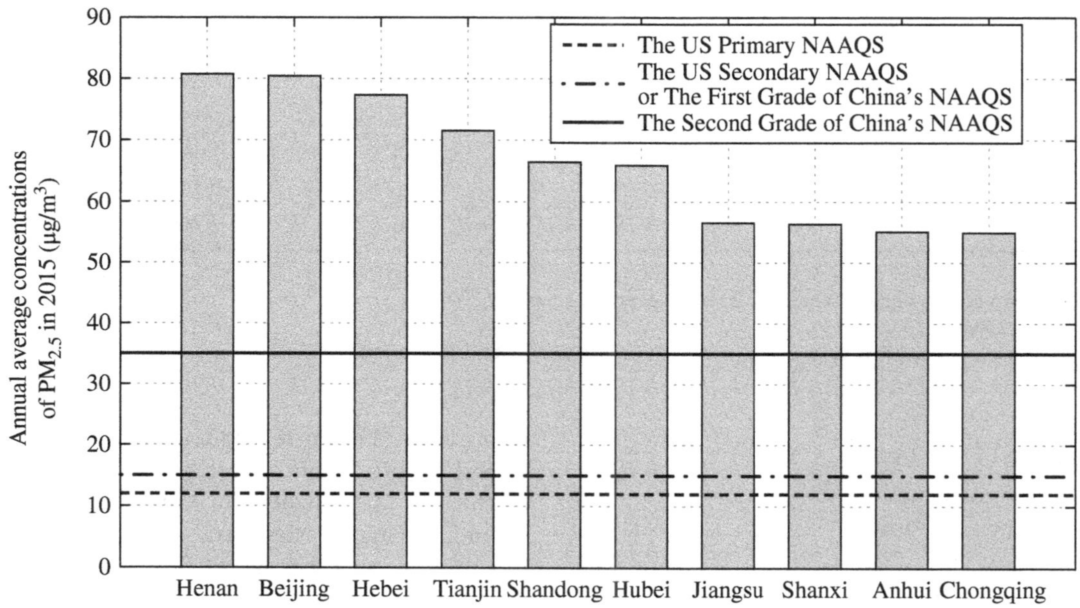

Figure 1.7 China's 10 most polluted provinces in 2015. Note: (1) US primary and secondary standards were from https://www.epa.gov/criteria-air-pollutants/naaqs-table. (2) The primary and secondary Chinese National Ambient Air Quality Standard (NAAQS) is 15 μg/m^3 (Annual average). Available from http://transportpolicy.net/index.php?title=China:_Air_Quality_Standards. (3) Chinese Ministry of Environmental Protection (MEP). *Ambient Air Quality Standards (GB 3095-2012).* (2012) (accessed 5 August 2015). Available at http://kjs.mep.gov.cn/hjbhbz/bzwb/dqhjbh/dqhjzlbz/201203/W020120410330232398521.pdf.

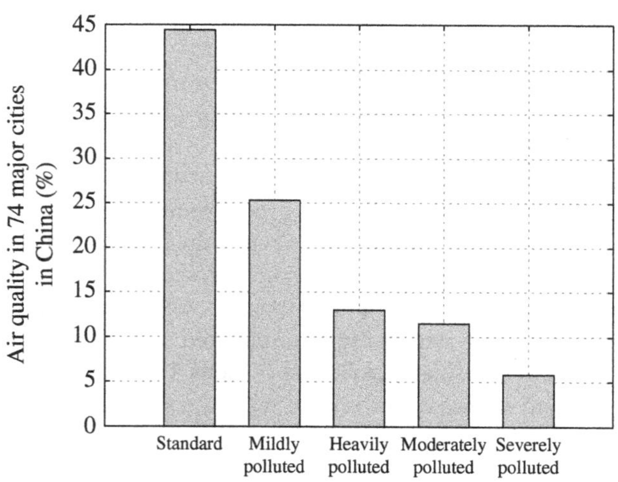

Figure 1.8 Air quality in 74 major cities in China in the first quarter of 2013. (*Source:* Department of Environmental Protection of the People's Republic of China, "Air quality in 74 major cities in China in the first quarter of 2013.")

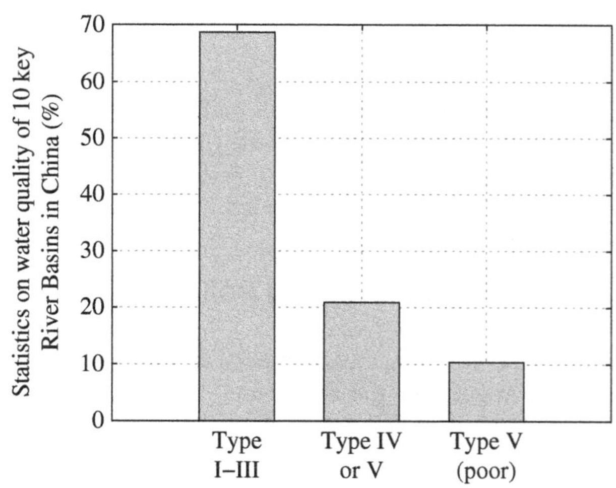

Figure 1.9 Statistics on water quality of 10 key river basins in China. (*Source:* "Communique on Land and Resources of China 2012," Bulletin of China's Environmental Conditions 2012, Ministry of Environmental Protection.)

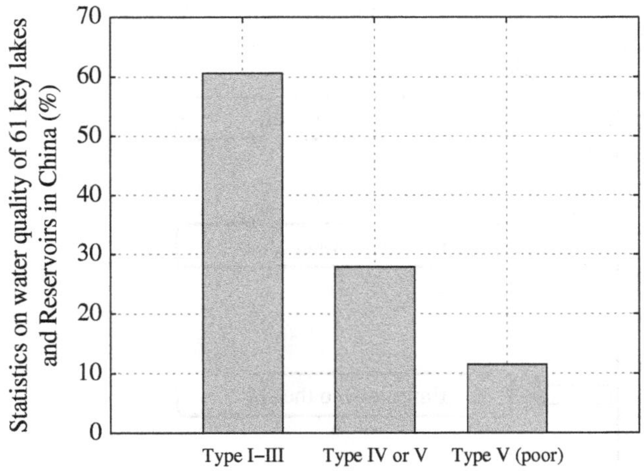

Figure 1.10 Statistics on water quality of 61 key lakes and reservoirs in China. (*Source:* "Communique on Land and Resources of China 2012," Bulletin of China's Environmental Conditions 2012, Ministry of Environmental Protection.)

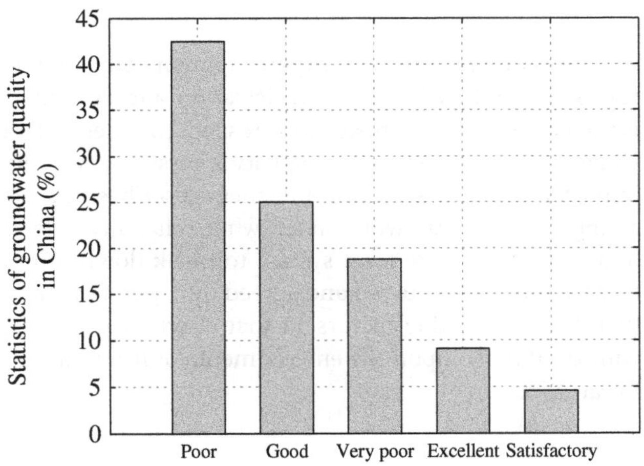

Figure 1.11 Statistics of groundwater quality in China. (*Source:* "Communique on Land and Resources of China 2012," Bulletin of China's Environmental Conditions 2012, Ministry of Environmental Protection.)

1.7 Life Cycle Assessment

One of the ways to quantify footprint of an EEIS is to conduct life cycle assessment (LCA). LCA is an evaluation method used to assess environmental impact associated with all the stages of a product's life cycle. LCA could be conducted from raw materials to finished goods, or cradle to grave, i.e. from raw materials to disposal of finished goods. LCAs are becoming a standardized protocol referred to by the International Organization for Standardization (ISO) (2006) with codes 14040–14048. Direct application of the ISO refers to product development and improvement, strategic planning, public policy making, marketing, and other processes through technique, economic, and social tools. The system should be isolated from the surroundings using system boundaries to define a specific area in the ISO process. LCA is the process of identifying the future consequences of a current or proposed action. The standardized categories of impact assessment include (i) global warming, (ii) smog formation, (iii) ozone depletion, (iv) acid rain, (v) human inhalation, (vi) ingestion toxicity, (vii) human carcinogenic inhalation, (viii) carcinogenic ingestion toxicity, and (ix) fish toxicity. Data quality is judged based upon its precision, completeness, representativeness, consistency, and reproducibility and should satisfy the assumptions and uncertainties made under the ISO. ISO has different orders to categorize the immediate impact by considering input and output methods usually used in LCA as shown in Figure 1.12.

A generic procedure of LCA according to the ISO 14040/14044 has the following five steps (ISO, 2006):

Step 1: Goal and scope definition

Goals of LCA should state the intended application, use, and audience as well as the reason for the study. The scope should define system boundaries to determine data collection. In the LCA of WWTPs, for example, following data should be collected: (i) concentration of each pollutant, (ii) flow rate, (iii) treatment and unit processing, and (iv) discharge standards and ambient water quality. The definition of goal and scope relates mainly to the identification of study type according to its intended application and audience as follows:

The objectives of the LCA assessment

Definition of a functional process

1) System boundaries defining the life cycle phases to be included
2) Procedure for coproduction or reuse and recycling in terms of mass, energy, or environmental significance
3) Choice of environmental impact categories, indicators, and characterization models
4) Data requirements, quality, uncertainty, type, and sources
5) Assumptions and limitations

Step 2: Inventory analysis and product model

The method of data collection and the quantification of input and output flows (e.g. materials or energy) depend on the assessed life cycle stages. Data collection, calculation, validation, and relation of the information to life cycle stages or objectives of assessment must be conducted for life cycle inventory analysis.

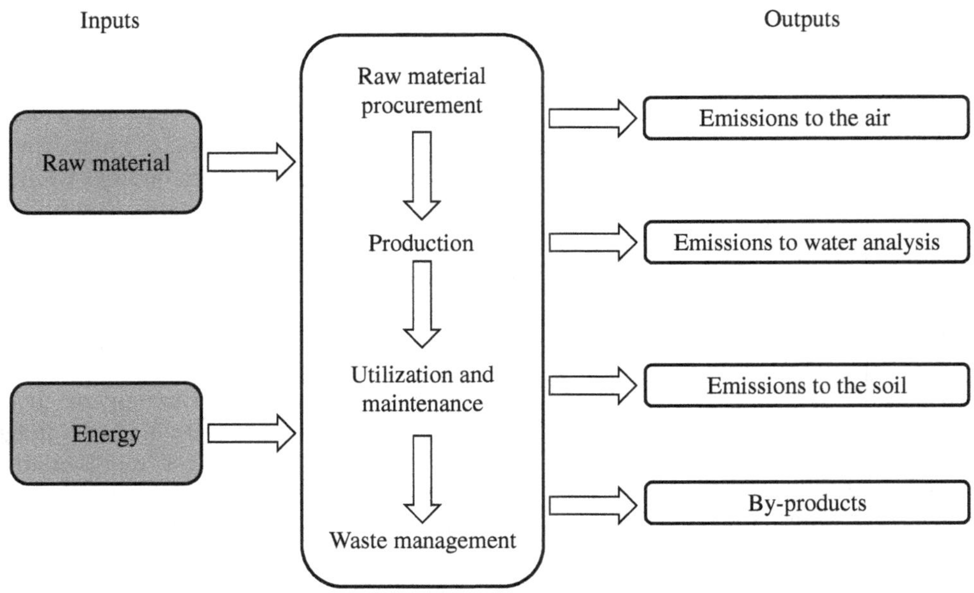

Figure 1.12 Input and output in life cycle assessment process.

Step 3: Impact assessment

Environmentally, technically, and scientifically valid models and indicators should be selected for impact assessment. Typical impact categories are water consumption, land-use change, human toxicity, ecotoxicity or carbon sequestration, etc.

Step 4: Interpretation and documentation

Hot spots contributing to environmental impacts should be identified with quantified uncertainty and sensitivity. Guidance on the normalization, grouping, or weighting should be established.

Step 5: Review

LCA helps satisfy government regulations, helps decrease the environmental impact of a given product or process, and identifies ways to improve sustainability. Therefore, an external critical review is necessary for all comparative LCA studies. LCA modeling has following five elements:

1) Data collection (quality control)
2) Model development (empirical and physical)
3) Calibration (establishment of standards)
4) Verification (confirmation that assumptions are correctly made)
5) Validation (accuracy of the calibrated equation)

One of the major uncertainties in LCA is to quantify the weights of different environmental impacts. Weighting factors are measured in a certain amount of units and can be assessed through expert opinion. Uncertainty can be quantified by standard deviation and sensitivity is the change rate of impact with respect to independent variables. Uncertainty and sensitivity analysis are two very important aspects in LCA. Robustness describes how big a range a system can work under, while reliability reflects how long it will take for a system to break down. All of these should be discussed and agreed by experts in quantifying the weighting factors so that alternatives can be compared based upon the environmental impact indexes for an EEIS.

1.7.1 LCA Tools

The complicated procedure in the LCA process as well as a vast amount of data makes manual calculation laborious, even impossible. Fortunately, many software packages have been developed for different assessment processes as shown in Table 1.20.

1.8 Environmental Laws

To achieve sustainable development without breaking the PBs, many environmental laws have been developed according to the concept of LCA. Many laws have different goals from different angles of human, social, and

Table 1.20 LCA and LCI software tools.

Tool	Vendor	URL
BEES 3.0	NIST Building and Fire Research Laboratory	http://www.bfrl.nist.gov/oae/software/bees.html
Boustead Model 5.0	Boustead Consulting	http://www.boustead-consulting.co.uk/products.htm
CMLCA 4.2	Centre of Environmental Science	http://www.leidenuniv.nl/cml/ssp/software/cmlca/index.html
DuboCalc	Netherlands Ministry of Transport, Public Works and Water Management	http://www.rws.nl/rws/bwd/home/www/cgi-bin/index.cgi?site=1&doc=1785
Ecoinvent 1.2	Swiss Centre for Life Cycle Inventories	http://www.ecoinvent.ch
Eco-Quantum	IVAM	http://www.ivam.uva.nl/uk/producten/product7.htm
EDIP PC-Tool	Danish LCA Center	http://www.lca-center.dk
eiolca.net	Carnegie Mellon University	http://www.eiolca.net
Environmental Impact Indicator	ATHENA™ Sustainable Materials Institute	http://www.athenaSMI.ca
EPS 2000 Design System	Assess Ecostrategy Scandinavia AB	http://www.assess.se/
GaBi 4	PE Europe GmbH and IKP University of Stuttgart	http://www.gabi-software.com/software.html
GEMIS	Öko-Institut	http://www.oeko.de/service/gemis/en/index.htm
GREET 1.7	DOE's Office of Transportation	http://www.transportation.anl.gov/software/GREET/index.html
IDEMAT 2005	Delft University of Technology	http://www.io.tudelft.nl/research/dfs/idemat/index.htm
KCL-ECO 4.0	KCL	http://www1.kcl.fi/eco/softw.html
LCAIT 4.1	CIT Ekologik	http://www.lcait.com/01_1.html
LCAPIX v1.1	KM Limited	http://www.kmlmtd.com/pas/index.html
MIET 3.0	Centre of Environmental Science	http://www.leidenuniv.nl/cml/ssp/software/miet/index.html
REGIS	Sinum AG	http://www.sinum.com/htdocs/e_software_regis.shtml
SimaPro 6.0	PRé Consultants	http://www.pre.nl/simapro.html
SPINE@CPM	Chalmers	http://www.globalspine.com
SPOLD	The Society for Promotion of Life-Cycle Assessment	http://lca-net.com/spold/
TEAM™ 4.0	Ecobalance	http://www.ecobalance.com/uk_lcatool.php
Umberto	ifu Hamburg GmbH	http://www.ifu.com/en/products/umberto
US LCI Data	National Renewable Energy Lab	http://www.nrel.gov/lci

economic development. From the LCA perspective, environmental laws governing petroleum-based fuel are categorized in Table 1.21.

In terms of LCA, environmental laws are categorized as in Table 1.22.

As regulations are becoming more and more stringent, water utilities must react to the new requirements in a relatively short time frame. Retrofits to reduce emissions or comply with new design criteria are costly. If SEE designers look beyond compliance to achieving the greatest energy efficiency, WWTP should be designed as energy positive using anaerobic ammonia oxidation (Anammox) and anaerobic membrane biological reactor (AnMBR). As a result, WWTP could produce energy and yet comply with environmental laws. Indeed, regenerative design saves money in materials and in energy, reduces the financial burden on consumers, and even increases cash flow of a WWTP.

Table 1.21 Environmental laws on onshore petroleum.

Purposes of laws	Environmental laws
Where to extract	National Park System Mining Regulation Act
	Federal Land Policy and Management Act
	National Forest Management Act
	Federal Onshore Oil and Gas Leasing Reform Act
	Clean Water Act
	Wild and Scenic Rivers Act
	National Wildlife Refuge System Act
How to extract	Clean Water Act
	Federal Onshore Oil and Gas Leasing Reform Act
	Resource Conservation and Recovery Act (RCRA)
	Endangered Species Act
	Safe Drinking Water Act
	National Environmental Policy Act
Transportation of petroleum	Federal Onshore Oil and Gas Leasing Reform Act
	Federal Land Planning and Management Act
	Department of Transportation Regulations
	CERCLA
	RCRA
Transportation and storage of petroleum	Mineral Management Service Regulations
	International Convention for the Prevention of Pollution from Ships
	National Environmental Policy Act
	1990 Oil Pollution Act
	Deepwater Port Act
	Shore Protection Act
Locate the refinery	Deepwater Port Act
	National Environmental Policy Act
	Endangered Species Act
	Wild and Scenic Rivers Act
	Migratory Bird Treaty Act
	Historic Preservation Act
	Clean Air Act
Refining process emissions	Clean Air Act
	Clean Water Act
	Occupational Safety and Health Act
Refining process	Toxic Substances Control Act

Table 1.21 (Continued)

Purposes of laws	Environmental laws
	Emergency Planning and Community Right-to-Know Act
	CERCLA
	Petroleum Marketing Practices Act
	Occupational Safety and Health Act
Use of gasoline	Clean Air Act
	Power Plant and Industrial Fuel Use Act
	Environmental taxes
	Transportation controls

Table 1.22 Environmental laws on onshore petroleum.

LCA	Purposes	Environmental laws
Resource extraction	How and where resource can be extracted	Clean Water Act
Production	How to refine the cruel oil	EPCRA
Emission	What are pollutants	Clean Air Act
Use	How products can be used	TCA, FIFRA
Disposal	Disposal and damage to the environments	CERCLA or RCRA

1.9 Exercise

1.9.1 Questions

What is an ecological footprint of a person, and what is the carrying capacity of the Earth?
How much larger is your ecological footprint in comparison with the rate the Earth can regenerate?
What does an "overshoot" of an ecological footprint mean?
What is the measurement unit of the ecological footprint?
What assumptions are made in order to calculate an ecological footprint?
How would be footprint on water change if people's diet reduce from red meat to vegetable protein?
What are the risks we face as humanity's demand for ecological resources of the Earth exceeds nature's supply?
Which carrying capacity do you think is the most probable estimate?
How technology advances would impact Earth's future carrying capacity?
What role of SEE designers can play to contribute to the UN SDGs?

1.9.2 Assignment

1) Go to http://www.myfootprint.org and read the intro. Most of it will cover what you just read in the aforementioned website. Enter the quiz and work through the questions.
2) Do the quiz and compare your average ecological footprint with the average footprint of 24 acres in the United States.
3) Go to http://www.rprogress.org/ecological_footprint/footprint_FAQs.htm and answer these questions.
4) Study SPSS tutorial manual.
5) On any specific day that you are working on this assignment, please select 20 cities in the United States, Europe, and China by real-time air quality index. Please go to http://aqicn.org/map/northamerica/ for major cities in America, http://aqicn.org/map/europe/ for major cities in the European Union, and http://aqicn.org/map/china/ for major cities in China. Please conduct the following analysis using SPSS and answer following questions:
   a) What are the mean and standard deviations of the AQI for the top 20 most polluted cities in these three regions?
   b) What does the mean and standard deviation tell you about the real-time AQI in these three regions?
   c) Please rank the three regions according to the mean of their AQIs on that day.
   d) According to the standard deviation, please quantify your rank in terms of confidence intervals using Monte Carlo simulations.
   e) Do you expect any seasonal change of this rank? Why?
6) From the real-time AQI that you get for that day, please estimate the total amount of pollutants that an adult will breathe in a lifetime of SO_2, NO_2, CO, and $PM_{2.5}$ by assuming that an adult lives 70 years and half of his/her life is spent outdoors. Please answer the following questions:
   a) Will this exposure data be representative data for your exposure assessment?
   b) How can you improve the exposure assessment?

1.9.3 Problems

1) For Example 1.1, if T_a and T_s are temperatures for the atmosphere and top of atmosphere and the atmosphere and surface are in equilibrium, the following equilibrium would be true:
   i) Surface:

$$\sigma T_s^4 = \sigma T_a^4 + \frac{S_o}{4}(1-\alpha_p)$$

   ii) Atmosphere:

$$2\sigma T_a^4 = \sigma T_s^4$$

   iii) Top of atmosphere:

$$\frac{S_o}{4}(1-\alpha_p) = \sigma T_a^4$$

   What are the temperatures for the atmosphere and top of atmosphere, T_a and T_s, respectively?
2) If a watershed of 2000 m^2 in Miami-Dade County has precipitation of 65 in/year, ET is 55 in/year, runoff from the canals is 1.023×10^6 acre-ft/year, groundwater withdrawal is 331 million gallons per day (MGD), wastewater discharge back to aquifer in the form of septic systems is 1 MGD, wastewater outfall to the ocean is 216.5 MGD, and wastewater injected into the Florida aquifer (deep well injection) is 97 MGD, please answer the following:
   a) What would be the difference between the inputs and outputs?
   b) Is the water budget in a steady state?
   c) What parameters are missing from this water budget?
3) A lake has water quality as follows: dissolved oxygen, 60; nitrate, 6.10 (mg/l); phosphate, 0.8 (mg/l); and pH, 5.5. Please answer the following:
   a) What is the WQI of this water if the weighting factors of the corresponding water quality parameter are 0.3, 0.2, 0.2, and 0.3, respectively?
   b) What is the WQI of this water if the weighting factors of the corresponding water quality parameter are 0.5, 0.2, 0.2, and 0.1, respectively?
   c) Why are they different? How would one determine the weighting factors of each water quality parameter?
4) A residential house in Miami receives precipitation as in Example 1.2, which has 1 acre of land with a 4356 ft^2 roof. If a rain harvest system is to be installed by collecting the rainwater from the roof, is the rainwater sufficient for a family of four people if water consumption is 200 gal/person/day?
5) Please determine whether the hydrologic budget is in deficit, in surplus, or in a steady state as change in storage by developing a hydrologic budget equation, if precipitation is 72 in/year, evapotranspiration is 59 in/year, surface water runoff is 9.6 in/year, groundwater inflow is 8 in/year, and groundwater pumping is 3.4 in/year.

1.9.4 Projects

1.9.4.1 Xiongan Project

China proposed to build an eco-friendly megacity, Xiongan New District, as a new special economic zone of China. The zone will cover Xiongxian, Rongcheng, and Anxin counties in Hebei province. Current population is 100 000. The city will be developed through ecological protection, clustered urban layout, and coordinated development. Within the new city, Baiyangdian is one of the largest freshwater wetlands in north China, which has more than 140 lakes and covers 366 km^2. The city will cover 100, 200, and 2000 km^2 to accommodate 1, 1.5, and 2.5 million people in the next 5, 10, and 50 years, respectively. As much as $348 billion of investment will be attracted to the new city over the next decade. To support the development of Xiongan, China Development Bank would loan $18.9 billion in infrastructure. As a smart city and a tech hub, it is planned to power the city with 100% renewable energy.

Please answer the following:

1) What are the air, water, and soil quality in the city?
2) What are the major remediation projects indexes of air, water, and soil that have to be carried out before the ambitious development?
3) According to the ecological footprint, what would be the average ecological footprint in the next 5, 10, and 20 years?
4) If the ecological footprint is greater than the biocapacity, what are the major strategies that could be employed to prevent ecological overshoot in the next 5, 10, and 20 years?
5) If all the WWTPs of the city were to be designed and built as energy positive WRRFs, how much energy have to be produced annually in WRRFs in the next 5, 10, and 20 years?

1.9.4.2 Community Project

1) Please collect raw data from the hometown where you originally came from:
   a) Land areas
   b) Current population
   c) GDP in the last twenty years
   d) Predict future population and GDP in the next 5, 10, 20, 50, and 100 years
   e) According to the predicted GDP, what are the ecological footprints of people in your hometown?
   f) What would be the ecological capacity of your hometown in the next 5, 10, 20, 50, and 100 years?
   g) Is the ecological FP of your hometown greater, equal, or less than the natural biological carrying capacity in the next 5, 10, 20, 50, and 100 years?
2) Please collect following relevant raw data from the hometown where you originally came from in the last 10 years:
   a) Industrial production
   b) Agricultural production
   c) Residential houses
   d) Unit emission factors from the US EPA manual
   e) Inventory of air, water, and soil pollution
3) Please calculate the following indexes of your city:
   a) Water quality index
   b) Air quality index
   c) Soil quality index
4) According to the calculated indexes, please rank which pollution is the most urgent environmental issue in your hometown and identify the major pollution sources in your hometown.

References

Beeton, A.M. (1971). Eutrophication of the St. Lawrence Great Lakes. In: *Man's Impact on Environment*, (ed. T.R. Detwyler), 233–245. New York: McGraw-Hill Book Co.

Boyd, R. (1972). World dynamics: a note. *Science*, 177(4048), 516–519.

Canadian Council of Ministers of the Environment (CCME) (2001). Canadian Water Quality Guidelines for the Protection of Aquatic Life: CCME Water Quality Index 1.0. Technical Report, Canadian Council of Ministers of the environment winnipeg, MB, Canada. http://www.ccme.ca/sourcetotap/wqi.html (accessed 15 December 2017).

Ewing, B., Goldfinger, S., Oursler, A., et al. (2009). *The Ecological Footprint Atlas 2009*. Oakland: Global Footprint Network.

Franke, N.A., Boyacioglu, H. and Hoekstra, A.Y. (2013) Gray Water Footprint Accounting: Tier 1 Supporting Guidelines. *Value of Water Research Report Series No. 65*. Delft: UNESCO-IHE.

Hammond, A.L. (1971). Phosphate replacements: problems with the washday miracle. *Science*, 172: 361–363.

International Energy Agency (IEA) (2011). World Energy Outlook 2011. Paris: International Energy Agency, OECD.

International Organization of Standardization, 14044 (ISO) (2006). Environmental management – Life cycle assessment – Requirements and guidelines.

von Liebig, J. (1840). *Chemistry in Its Applications to Agriculture and Physiology*, 85. Taylor & Walton: London.

Meadows, D.H., Meadows, D.L., Randers, J. et al. (1972), *The Limits to Growth: A Report for the Club of Rome's Project on the Predicament of Mankind*, Universe Books.

Michigan Department of Environmental Quality (2004). http://www.deq.state.mi.us/documents/deq-rrd-part201-rules-Rule746Table.pdf (accessed 20 February 2018).

Miller, G.T. Jr. (1971). *Energetics, Kinetics, and Life: An Ecological Approach*. Belmont, CA: Wadsworth Publishing Company.

Petit, J.R., Jouzel, J., Raynaud, D., et al. (1999). Climate and atmospheric history of the past 420,000 years from the Vostok ice core, Antarctica. *Nature* 399: 429–436.

Rees, W.E. and Wackernagel, M. (2013). The shoe fits, but the footprint is larger than earth. *PLoS Biology*. 11 (11): e1001701.

Rockström, J., Steffen, W., and Noone, K. et al. (2009). A safe operating space for humanity. *Nature*, 461: 472–475.

Smil, V. (2001). *Enriching Earth: Fritz Haber, Carl Bosch, and the Transformation of World Food Production*. Cambridge, MA, The MIT Press.

Steen, I. 1998. Phosphorus availability in the 21st century: Management of a non-renewable resource. *Phosphorus & Potassium* 217, 25–31.

Stoltzenberg, D. and Haber, F. (2004). *Chemist, Nobel laureate, Jew*. Philadelphia, Chemical Heritage Press.

United Nations Environment Program (UNEP) (2012). Global Environmental Outlook, GEO 5. Summary for Policy Makers. United Nations Environment Program, UNEP.

United States Geological Survey (USGS) (1997). *Phosphate Rock, Mineral Commodity Summaries*. (124–125), Washington, DC: US Department of the Interior, US Geological Survey, US Government Printing Office.

US Environmental Protection Agency (1997). Exposure Factors Handbook. Final Report, EPA/600/P-95/002F a-c. Washington, DC: US EPA.

US Environmental Protection Agency (2014). Air Quality Index, A Guide to Air Quality and Your Health, Office of Air Quality Planning and Standards, Research Triangle Park, NC. *EPA-456/F-14-002*.

Wetzel, R.G. (1983). *Limnology*, 2e. Philadelphia, PA: Saunders College Publishing.

World Health Organization (2012). China, WHO Statistical Profile. http://www.wpro.who.int/country_support/countries/chn/en/ (accessed 5 January 2018).

2

Health Risk Assessment

If the Earth were sick, nobody would be healthy.

OBJECTIVES

To explore environmental health problems and environmental standards
To identify the leadership role of SEE designer to achieve sustainable development goals of the United Nations
To explain the US EPA Stage 2 Disinfectants and Disinfection Byproducts Rules
To conduct human health risk assessment of carcinogenic and noncarcinogenic chemicals
To present method to calculate benchmark dose (BMD) of any pollutants by the US EPA BMDS 2.6.0.1 software
To describe qualitative and quantitative methods of characterizing the uncertainty in exposure assessments
To demonstrate regression analysis in developing QSAR models by using SPSS or SAS
To quantify uncertainty and sensitivity in health risk assessment using Monte Carlo simulation with Crystal Ball

2.1 Environmental Health

Human footprints on the nature significantly deteriorate the environmental quality of air, water, and soil, pose a great threat to human health and reduce human life expectancy. Taking air pollution as an example, the primary and secondary ambient air standards of particulate matter of diameter 2.5 μm ($PM_{2.5}$) in the United States are 12 and 15 μg/m^3, respectively (US EPA, 2016). The corresponding standards of $PM_{2.5}$ in China are set at 15 and 30 μg/m^3, respectively (China MEP, 2016). However, the average concentration of $PM_{2.5}$ for 10 provinces in China was greater than 55 (μg/m^3) in 2015. According to the data from the Administration of Energy of China, 64% of primary energy in China is from coal. China with 1.4 billion population burns about 47% of the world's total coal consumption. The annual average $PM_{2.5}$ for the top 10 most polluted provinces in China is at least twice as much as the secondary standard of 35 μg/m^3 as shown in Figure 2.1.

In China, mortality rates due to environmental damage are significantly greater than natural mortality rates. Lung, liver, and stomach cancers are the top three mortality causes in China. Epidemic studies indicate that lung cancer is associated with chemical smog due to coal burning, liver cancer with drinking water and strong liquor, and stomach cancers with contaminated food due to soil pollution. About 90% of the 161 cities whose air quality was monitored in 2014 were below Chinese official standards according to the National Bureau of Statistics of China (NBSC). More than half of China's surface water is so polluted that it cannot be used as drinking water resources. As a result, about 4000 people died every day in China due to coal burning as the likely principal cause. Due to its severity, air pollution is the leading cause of lung cancer and other respiratory infections. Figure 2.2 shows the percentage of leading causes of death in the world. In China, the most common causes of death are (i) lung cancer with 1.59 million deaths, (ii) liver cancer with 745 000 deaths, (iii) stomach cancer with 723 000 deaths, (iv) colorectal cancer with 694 000 deaths, (v) breast cancer with 521 000 deaths, and (vi) esophageal cancer with 400 000 deaths.

Worldwide, environmental pollution is also the leading cause of cancer and contributes to major premature mortality. WHO (2014) estimated that there were 3.1, 1.6, and 1.5 million people who died from lower respiratory infections, trachea bronchus lung cancers, and diarrhea diseases in the world as shown in Figures 2.2 and 2.3. There were approximately 14 million new cases, and

Sustainable Environmental Engineering, First Edition. Walter Z. Tang and Mika Sillanpää.

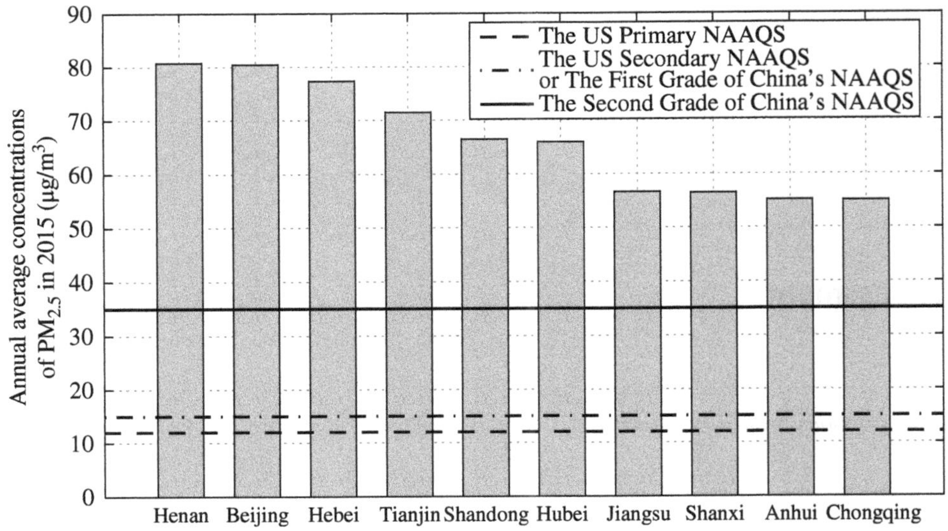

Figure 2.1 China's 10 most polluted provinces in 2015. (*Source:* From the US EPA and China MEP (2016).)

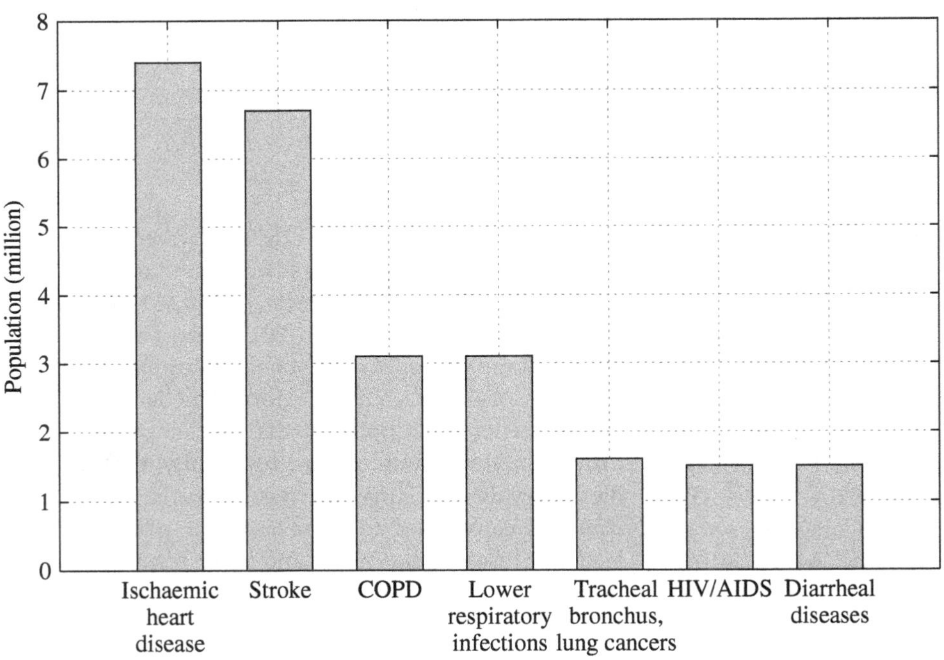

Figure 2.2 Seven leading causes of death in the world in 2012. (*Source:* Data from WHO (2014).)

the number of new cases is expected to rise by about 70% over the next two decades (WHO, 2014). In 2012, ischemic heart disease, stroke, and chronic obstructive pulmonary disease caused 7.4, 6.7, and 3.1 million deaths, respectively (NBSC, 2012).

Figure 2.4 shows that the leading causes of mortality due to malignant tumor is 168 and 159 deaths per 100 000 population in urban and rural China, respectively, which is about 28 and 25% of the leading causes of death in urban and rural China.

Further evidence of the environmental health issue can be seen from Figures 2.5 and 2.6. Zhao et al. (2010) showed that lung, liver, stomach, esophageal, and colorectal cancer were 32, 19, 18, 9, and 8% in urban China, while these data were 24, 23, 23, 15, and 5% in rural China, respectively, as shown in Figure 2.5.

To monitor the progress of protecting human health, the United Nation uses an improved sanitation facility as a metric. It is defined as piped sewerage, septic tanks, and pit latrines with slabs or composting toilets not shared with other households. From 1990 to 2015, the coverage of improved sanitation at these facilities rose from 54% to around 68% globally as shown in Figure 2.7. This number, however, missed the Millennium Development Goal (MDG) target by 9%. In 2015, about 946 million people were still practicing open defecation worldwide (WHO, 2016).

To achieve the Sustainable Development Goals (SDG) target, cities all over the world need to reduce open defecation, promote handwashing, and improve management and treatment of faucal wastes from both collected sewer and on-site facilities. For drinking water, the UN uses pipe water on premise and public standpipes, boreholes, protected wells, springs, and rainwater as indicators for improved drinking water. In 2015, there were 6.6 billion people using improved drinking water, while 0.7 billion still used an unimproved drinking water source or surface water (WHO, 2016), as shown in Figure 2.8.

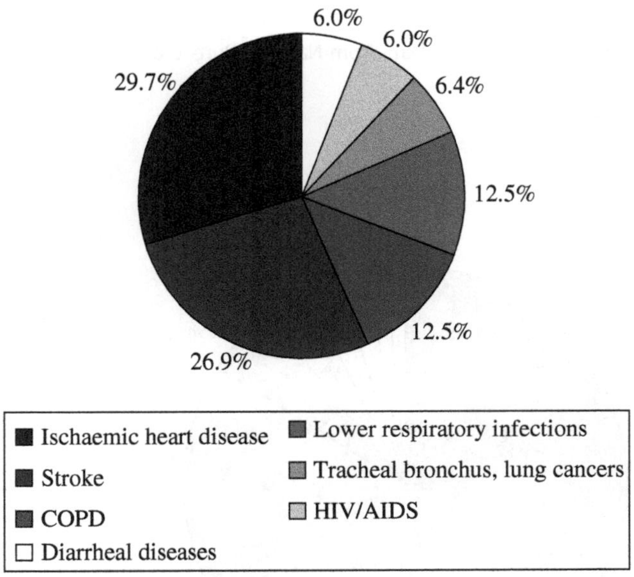

Figure 2.3 Seven leading causes of death (percent) in the world. (*Source:* Data from WHO (2014).)

About one-quarter of improved sources are contaminated with feces and approximately 1.8 billion people drink water containing such contamination (WHO, 2016). Therefore, only 68% in urban areas and 20% in rural areas have truly safe drinking water. Unsafe drinking water causes liver cancer due to disinfection by-products (DBPs) and other contaminants such as antibiotics in drinking water.

2.2 Environmental Standards

The US EPA establishes environmental standards using health risk assessment (HRA) according to carcinogenic or noncarcinogenic chemicals. For chemicals that are known or expected to cause adverse health effects, the EPA established an enforceable maximum contaminant level (MCL) or nonenforceable maximum contaminant level goal (MCLG). The Safe Drinking Water Act (SDWA) proclaims standards and health advisories (HAs) for DBPs. HRA quantifies factors such as adsorption during ingestion, pharmacokinetics, mutagenicity, reproductive and developmental effects, and carcinogenicity (Pontius, 1990). The MCLs are federally enforceable limits for contaminants in drinking water established as the national primary drinking water regulations (NPDWRs). The secondary MCLs are established under

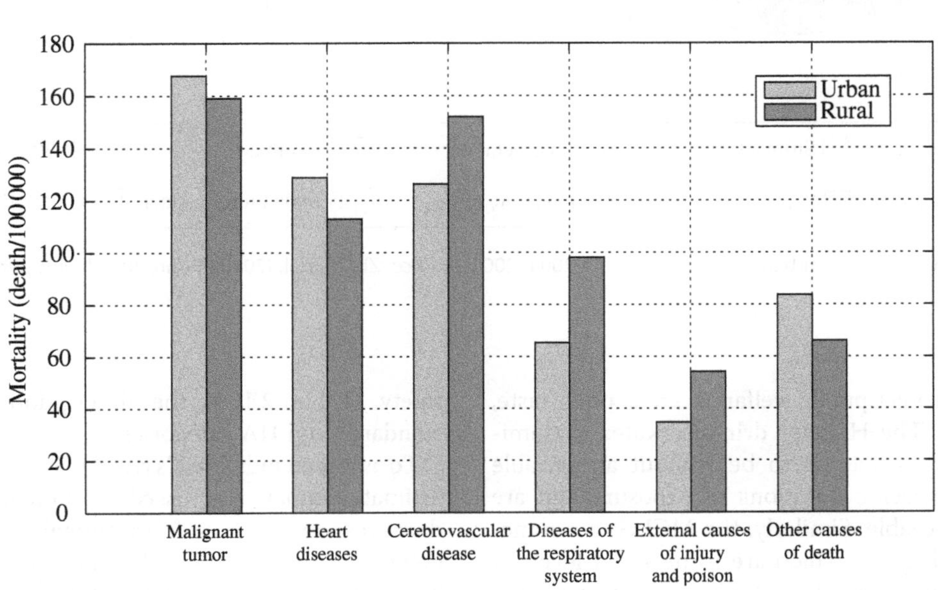

Figure 2.4 Leading causes of death in urban and rural China in 2009. (*Source:* Data from National Bureau of Statistics of China (2012).)

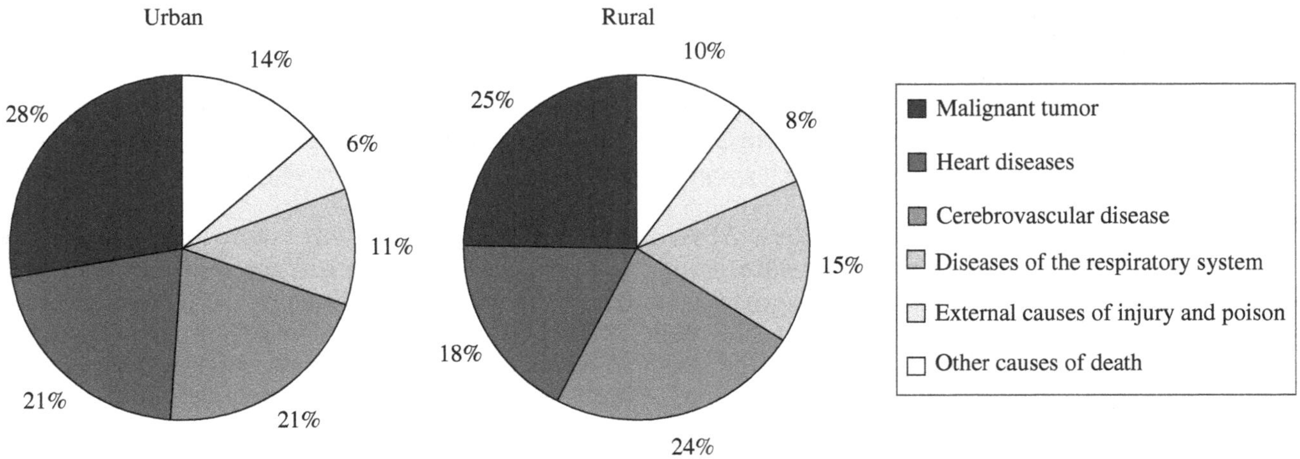

Figure 2.5 Leading causes of death (percent) in urban and rural China in 2009. (*Source:* Data from National Bureau of Statistics of China (2012).)

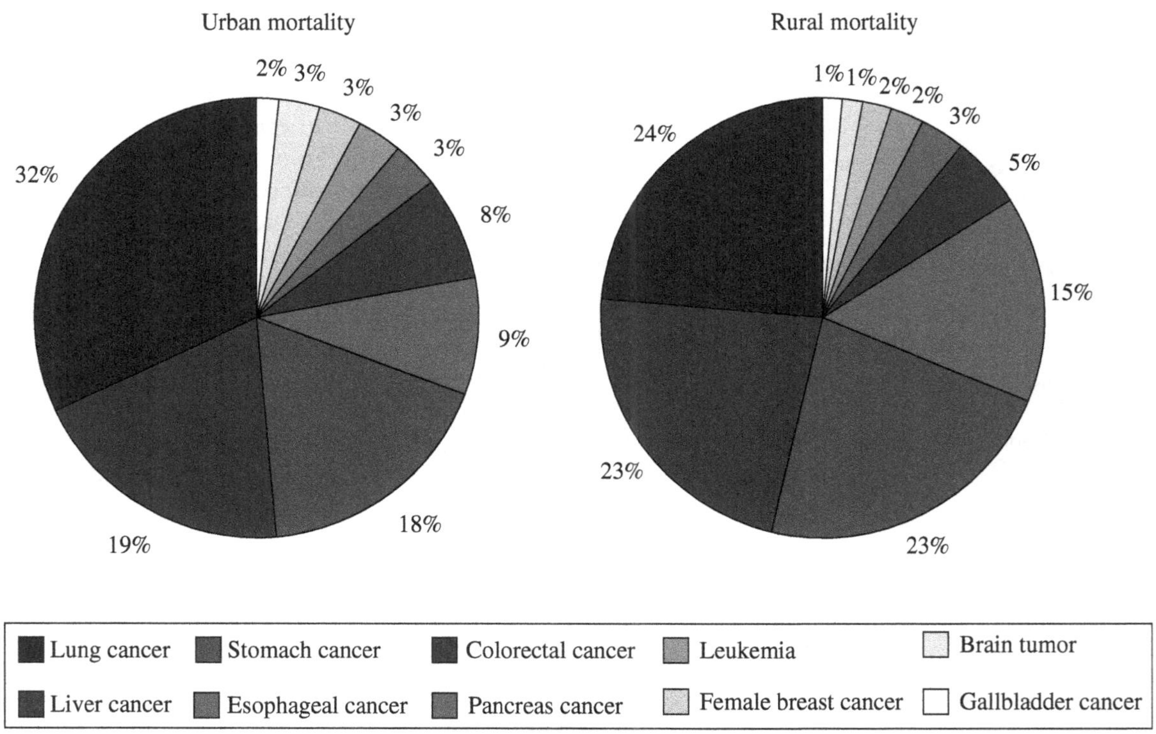

Figure 2.6 Top 10 cancers in urban and rural China in 2004–2005. (*Source:* Zhao et al. (2010). Reproduced with permission of Oxford University Press.)

the SDWA to protect public welfare such as odor, taste, and appearance. The HAs for drinking water contaminants are levels considered to be without appreciable health risk for specific durations of exposure and are not legally enforceable. Similarly, the MCLGs are nonenforceable health goals, which are to be set at levels at which no known or anticipated adverse effects on the health of persons occur with an adequate margin of safety. Table 2.1 is the description of the SDWA standards and HA categories.

To assess an individual's risk, bioassays are converted to estimate human risk based on human exposure. The dose–response curves from animal tests were used to determine the equivalent human dose–response curve. The MCLG is set using a three-category approach dependent upon the evidence of carcinogenicity. If the

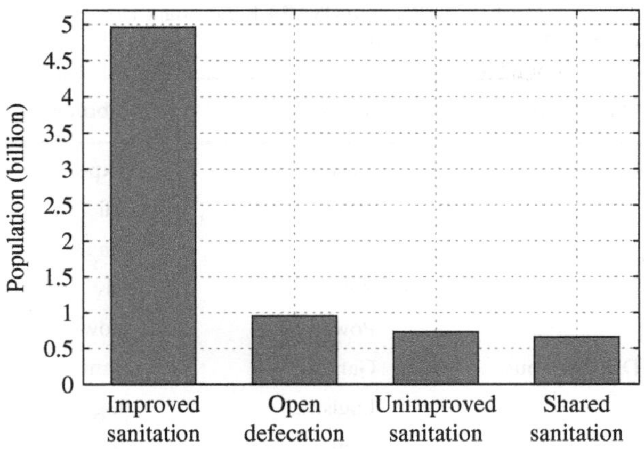

Figure 2.7 Global sanitation facilities in 2015. (*Source:* Data from WHO (2016).)

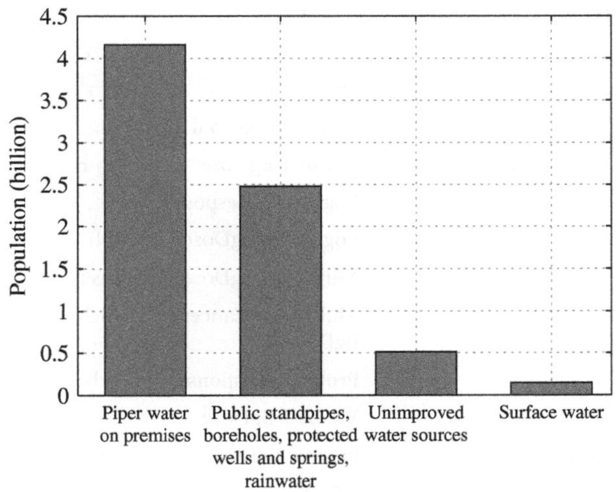

Figure 2.8 Global drinking water sources. (*Source:* Data from WHO (2016).)

evidence of carcinogenicity is strong, the MCLG is set to zero. If the evidence of carcinogenicity is limited, one of two methods is utilized to calculate the MCLG depending on the toxicity data available. For every chemical compound that has an MCLG, an MCL or treatment technique must be determined. Considering the cost of remediation, the MCL is based on the best available technology (BAT) and set close to the MCLG. If BAT cannot achieve zero, MCL cannot equal to zero. When the MCL is within a risk of 10^{-4} to 10^{-6}, the MCLG is set to zero.

In the past, reference dose (RfD) expressed in units of milligrams per kilogram of body weight per day (mg/kg/day) was used. The RfD is based on lifetime exposure level at which there is no significant risk to humans. It is calculated by dividing the no observed adverse effect level (NOAEL) or the lowest observed adverse effect level (LOAEL) by an uncertainty factor. The uncertainty factor accounts for differences between human and animal and differences within the human population and varies from 10 to 1000 depending upon the toxicity data available. RfD is determined using the following equation:

$$\text{RfD} = \frac{\text{NOAEL (or LOAEL or BMDL)}}{\text{UFs}} \tag{2.1}$$

Since an extremely high uncertainty factor is used, major limitations of RfDs are the following: (i) it is limited to one of the doses in the study and is dependent on study design, (ii) it does not account for variability in the estimate of the dose–response, (iii) it does not account for the slope of the dose–response curve, and (iv) it cannot be applied when there is no NOAEL, except through the application of an uncertainty factor (Crump, 1984; Kimmel and Gaylor, 1988).

Table 2.1 SDWA standards and health advisories.

Standards	
MCL (maximum contaminant level)	The enforceable concentration that is provided to public water system users
MCLG (maximum contaminant level goal)	The nonenforceable concentration that protects humans from adverse effects
Health advisories	
RfD (reference dose)	An estimation of daily human exposure without appreciable risk to adverse effects over a lifetime
DWEL (drinking water equivalent level)	The estimation of a lifetime exposure that protects humans from adverse noncancerous health effects, assuming the sole exposure source is from drinking water
One-day exposure	The drinking water concentration that is not expected to cause adverse noncarcinogenic effects if exposure continues for five consecutive days
10-Day exposure	The drinking water concentration that is not expected to cause adverse noncarcinogenic effects if exposure continues for 14 consecutive days
Long-term exposure	The drinking water concentration that is not expected to cause noncarcinogenic effects if exposure continues for 10% of the person's lifetime
Lifetime HA	The drinking water concentration that is not expected to cause adverse noncarcinogenic effects if exposure continues for a lifetime

Example 2.1 According to the US EPA *Exposure Factors Handbook* 2011 Edition (US EPA 2011), a chronic oral toxicity study of chloroform on dogs was used to derive the RfD of 0.01 mg/kg/day. This RfD is based on an LOAEL for hepatotoxicity and application of an uncertainty factor of 1000.
Find: The MCLG of chloroform.

Solution:

For a 70 kg adult consuming 2 l of tap water per day, applying a relative source contribution of 80% chloroform from drinking water.

$$\text{MCLG based on RfD for hepatotoxicity} = 0.01\ \text{mg/kg/day} \times 70\ \text{kg} \times 0.8/2\ \text{l/day} = 0.3\ \text{mg/l}$$

Answer: MCLG of chloroform in drinking water is 0.3 mg/l based on hepatotoxicity (U.S. EPA, 1994).
Comment: The MCLG of 0.3 mg/l based on hepatotoxicity (U.S. EPA, 1994) is lower than the values determined from the LED_{10}/ED_{10} approach for kidney tumorigenesis (0.6 or 1 mg/l) and is consistent with chloroform's putative mode of action involving the oxidative generation of reactive and toxic metabolites (phosgene and hydrochloric acid).

To overcome the high uncertainty factor of the RfD, the EPA developed benchmark dose (BMD) methods by fitting mathematical models to dose–response data and to select a BMD associated with a predetermined benchmark response (BMR), such as a 10% increase in the incidence of a particular lesion or a 10% decrease in body weight gain. Results from all models include a reiteration of the model formula and model run options chosen by the user, goodness-of-fit information, the BMD, and the estimate of the lower-bound confidence limit on the BMD (BMDL). The benchmark dose software (BMDS 2.6) by the US EPA (2016) has the nested models, parameter standard error reporting, and parameter initialization for continuous models. BMDS 2.6 contains thirty different models appropriate for the analysis of dichotomous (quantal) data, continuous data, nested developmental toxicology data, multiple tumor analysis, and concentration–time data. Typical models used in the software are shown in Table 2.2.

Table 2.2 Models used in the US EPA Benchmark Dose Software 2.6 (US EPA, 2015).

Model type	Model	Abbreviation
Continuous	Exponential	exp
	Hill	hil
	Linear	lin
	Polynomial	ply
	Power	pow
Dichotomous	Gamma	gam
	Logistic	log
	LogLogistic	lnl
	LogProbit	lnp
	Multistage	mst
	Multistage cancer	msc
	Probit	pro
	Weibull	wei
	Quantal linear	qln
	Dichotomous hill	dhl
Dichotomous alternative	Gamma-BgDose	gmb
	Logistic-BgResponse	
	LogProbit-BgDose	lpb
	Mutistage-BgDose	msb
	Multistage-Cancer-BgDose	mcb
	Probit-BgResponse	prb
	Weibull-BgDose	web
Nested	Nested logistic	nln
	NCTR	nct
	Rai and van Ryzin	rvr
Repeated response measures	ToxicoDiffusion	txd
Concentration × time	ten Berge	ten
Multitumor	MS_Combo	multi

The SDWA proclaims standards and HAs for DBPs (US EPA, 1997). However, setting the MCLs and MCLGs involves a lot of uncertainty, because discrepancies may exist for lifetime and longer-term exposure HAs due to conservative policies. For example, the uncertainty factor may vary from 5 to 5000 when the lifetime health advisory concentration is estimated (U.S. EPA, 1996). Human HRA and environments risk assessment (ERA) are used to develop both ambient and discharge standards by the US EPA. Risk assessment is a systematic approach to characterize the nature and magnitude of the risks associated with environmental or health hazards.

Since chlorination is the major disinfection process in the United States, regulation of DBP concentration in drinking water is one of the major challenges faced by the US EPA. Major human and financial resources have been devoted to identify, monitor, assess, and regulate the human health effect of DBPs. As a result, HRA guidelines of DBPs were developed to protect the public from both biological and chemical risks. For example, the US EPA developed the Stage 2 DBPR aiming to reduce peak DBP concentrations in the distribution system. When a water treatment plant (WTP) assesses its disinfection strategy, both the disinfectant effectiveness against the target pathogen and the DBPs formed as a result of the disinfectant must be considered in the decision-making process. Since no toxicity test could be performed on humans, animal test data are used with quantified uncertainty and variation involved. During HRA of DBP, information for accurate evaluation of DBP risk may not be complete, and the uncertainty factor in the assessment may be quite large (Cothern et al., 1986). For example, for total trihalomethanes (TTHM) and five haloacetic acids (HAA5), the drinking water standards were set at 80 and 60 μg/l as locational running annual average (LRAA) by the EPA (2006), respectively. If there was no specific toxicity information available, the MCLG was set up based upon a quantitative structure–activity relationship (QSAR) study. For example, the MCLG of 1,1-dichloroethylene and *cis*-1,2-dichloroethylene were developed using QSAR approach.

In addition to pathogenic bacteria and viruses, *Cryptosporidium* is one of the new public health concerns for the EPA. The EPA established the Long Term 2 Enhanced Surface Water Treatment Rule (LT2ESWTR) based on *Cryptosporidium* concentrations in source water and current treatment practices. Four bins were recommended corresponding to additional treatment requirements for filtered WTPs as shown in Table 2.3. WTPs with average *Cryptosporidium* concentrations less than 0.075 *oocysts* per liter (*oocysts*/l) are placed in bin 1 where no additional treatment is required. For concentrations of 0.075 *oocysts*/l or more, treatment beyond the existing processes is required. The additional treatment required for each bin, specified in terms of log removal, depends on the type of treatment that the WTP already uses. In setting the biological standards, the term "log" means the order of magnitude reduction in concentration; e.g. 2-log removal equals a 99% reduction, 3-log removal equals a 99.9% reduction, and 4-log removal equals a 99.99% reduction. Giardia and virus need 3-log and 4-log removal and/or inactivation, respectively.

To reduce biological and chemical risks simultaneously, UV disinfection appeared to be the best option for drinking water and treated wastewater effluent. Unlike chemical disinfectants, UV leaves no residual that can be monitored to determine UV dose and inactivation credit. To earn disinfection credits, however, a relationship between the required UV dose and these parameters must be established and then monitored at a WTP to ensure sufficient disinfection of microbial pathogens. The UV dose depends on the UV intensity (measured by UV sensors), the flow rate, and the UV transmittance (UVT). The US EPA (2006) recommended UV dose requirements (mJ/cm^2) in Table 2.4.

Table 2.3 Bin requirements for filtered PWSs (US EPA, 2006).

		And if the following filtration treatment is operating in full compliance with existing regulations, then the *additional* treatment requirements are:			
***Cryptosporidium* concentration (oocysts/l)**	**Bin classification**	**Conventional filtration treatment (includes softening)**	**Direct filtration**	**Slow sand or diatomaceous earth filtration**	**Alternative filtration technologies**
<0.075	1	No additional treatment	No additional treatment	No additional treatment	No additional treatment
≥0.075 and <1.0	2	1-log treatment	1.5-log treatment	1-log treatment	As determined by the state
≥1.0 and <3.0	3	2-log treatment	2.5-log treatment	2-log treatment	As determined by the state
≥3.0	4	2.5-log treatment	3-log treatment	2.5-log treatment	As determined by the state

Table 2.4 UV dose requirements: millijoules per centimeter squared (mJ/cm^2) (the US EPA, 2006).

	Log inactivation							
Target pathogens	**0.5**	**1.0**	**1.5**	**2.0**	**2.5**	**3.0**	**3.5**	**4.0**
Cryptosporidium	1.6	2.5	3.9	5.8	8.5	12	15	22
Giardia	1.5	2.1	3.0	5.2	7.7	11	15	22
Virus	39	58	79	100	121	143	163	186

2.3 Health Risk Assessment

Human health risk assessment (HRA) quantifies factors such as adsorption during ingestion, pharmacokinetics, mutagenicity, reproductive and developmental effects, and carcinogenicity (Pontius, 1990). It is to quantify the likelihood that adverse human health effects may occur or are occurring as a result of exposure to one or more stressors (US EPA, 1992). To reduce environmental health mortality, HRA is used to detect issues, identify hazards, characterize hazards, assess exposure, and manage and communicate risk. By using HRA, the EPA developed an Integrated Risk Information System (IRIS) that contains the human health effect databases of chemicals that may result from exposure to various chemicals in the environment. The IRIS 10^{-6} risk level is the contaminant concentration (in μg/l) in drinking water that would yield no greater than an additional risk of one in a million (10^{-6}) after a lifetime of drinking that water. The acute 10-day values apply specifically to acute toxic effects on children but are expected to be protective for adults. For noncarcinogenic chemicals, this value is typically the same as the MCLG. The chronic (lifetime) values for cancer are set at a level that should yield no greater than an additional 10^{-6} risk over a lifetime exposure. According to the IRIS, the EPA cancer risk is classified in six different categories: (i) H is carcinogenic to humans, (ii) L is likely to be carcinogenic to humans, (iii) L/N is likely to be carcinogenic above a specified dose, (iv) S is suggestive evidence of carcinogenic potential, (v) I is inadequate information to assess carcinogenic potential, and (vi) N is not likely to be carcinogenic to humans.

2.3.1 Hazard Identification

Hazard identification (HI) is to determine whether exposure to a stressor can cause an increase in the incidence of adverse health effect in humans, which reflects the capacity of an agent to cause adverse health effects in humans and other animals (US EPA, 1995). Qualitative description in HI is complemented with QSAR, genetic toxicity, pharmacokinetic, and the weight of evidence. For example, toxicokinetic data that deal with how the body absorbs, distributes, metabolizes, and eliminates specific chemicals are usually used in HI. In the HI of chlorination, chemical disinfectants such as chlorine and chloramines react with naturally occurring organic material (NOM) to produce DBPs that are potential carcinogens and development defects in laboratory animals. DBPs have been linked to potential health risks such as liver, kidney, or central nervous system problems and the increased risk of cancer. To quantify the toxicity of DBPs, QSAR analysis is one important method of estimating the carcinogenicity of DBP using animal toxicity tests. There are six major classes of DBPs, namely, halogenated alkane, halogenated alkene, halogenated aromatic, halogenated aldehyde, halogenated ketone, and halogenated carboxylic acid. A class of DBPs usually has the similar toxic mode or mechanisms. Table 2.5 lists these DBP classes and their definition.

In addition to the main molecular structures such as alkane, alkene, aromatic compounds, substituents such as chlorine and bromine within each DBP class contribute significantly to the toxicity of the corresponding DBPs. Table 2.6 compares the relative order of potency of haloacetonitriles in terms of different toxic mechanisms (Pontius, 1990). For example, haloalkanes are procarcinogens that are activated via reactions in which CYP450 acts as a catalyst in dehalogenation. Brominated compounds are expected to be more alkylating activity reactive than chlorinated compounds. Chlorine or bromine substitution at the α-carbon or terminal carbon in this group is expected to be potential alkylating agents. The active halogen at both ends of the aliphatic chain is

Table 2.5 Classification of DBPs.

DBP class	Definition
Trihalomethanes	Chemical compounds in which three of the four hydrogen atoms of methane (CH_4) are replaced by halogen atoms
Haloacetic acids	Carboxylic acids in which a halogen atom takes the place of a hydrogen atom in acetic acid
Haloacetonitriles	Small organic compounds containing nitrogen, chlorine, and/or bromine. Little data is available on the toxicity of the haloacetonitriles; however, animal studies suggest that dichloroacetonitrile is mutagenic and therefore potentially carcinogenic
Haloketone	A functional group consisting of a ketone group with an α-halogen substituent. The general structure is RR′C(X)C(=O)R where R is an alkyl or aryl residue and X any one of the halogens
Other halogenateds compounds	Chloropicrin, chloral hydrate, and cyanogen chloride
Aldehydes	Organic compounds that have an acyl group, R–C=O with a hydrogen bond to the carbonyl or acyl carbon (double-bonded carbon)
Inorganic compound	Chlorine, chloramines, chlorine dioxide, and bromate

Table 2.6 Comparison of chemical, biochemical, and biologic properties of haloacetonitriles.

Chemical/biochemical/biologic tests in assessing carcinogenic potential	Relative order of potency of haloacetonitriles tested
Alkylating activity 4-(*p*-nitrobenzyl)(pyridine reaction)	$Br_2 \gg BrCl > Cl \gg Cl_2 \gg Cl_3$
Inhibition of glutathione *S*-transferase	$Cl_3 > Br_2 > Cl_2 > Br > Cl$
Escherichia coli SOS chromotest	$BrCl > Br_2 > Cl_2 \gg Cl$ (inactive)
Ames or Ames fluctuation tests	$Cl_2 \approx BrCl > Cl_3 \approx Br_2 > Cl \approx Br$ (inactive)
	$Cl_3 > Cl > Cl_2 > Br_2 \approx Br$ (inactive)
DNA single-strand breaks in HeLa cell (comet assay)	$Cl_3 > BrCl > Br_2 > Cl_2 > Cl$
	$Br_2 > Cl_3 \approx Cl_2 \approx BrCl > Cl$
Sister chromatid exchanges in Chinese hamster ovary cells	$Br_2 > BrCl > Cl_3 \geq Cl_2 > Cl$
Newt micronucleus assay	$Br_2 \geq Cl_3 > Br > Cl_2 \gg Cl$
In vivo mouse micronucleus assay	$Br_2 \approx BrCl \approx Cl_3 \approx Cl_2 \approx Cl$ (all inactive)
Lung adenoma assay in strain A mice	$Cl \approx Cl_3 \approx BrCl > Br_2 \approx Cl_2$ (inactive)
Skin tumor initiation in SENCAR mice	$Br_2 \approx Cl \approx BrCl > Cl_3$ (inconsistent) $> Cl_2$ (inactive)

also expected to be cross-linking agents. The stability of several chlorinated ketones in aqueous solutions follows this order: 1,3-dichloro > pentachloro ≫ hexachloro. Among the mutagenic chloropropanols (mostly direct acting), the relative mutagenic potency follows the order 1,3- > 1,1,3,3- > penta- > 1,1,3- > 1,1,1- > 1,1-, with the potency of 1,3- being about 100–1000 times higher than that of 1,2-dichloroacetic acid (DCA) and trichloroacetic acid (TCA), which are known as mouse liver carcinogens and are by-products found in drinking water. In animal tests, DCA produces developmental, reproductive, neural, and hepatic effects. Aldehydes as an electrophilic compound may form DNA–protein cross-links and lead carcinogenesis.

2.3.2 Dose–Response Curves

A dose–response relationship describes how severity of adverse health effects (the responses) is related to the exposure amount to a chemical compound. The measured response usually increases as the dose increases. To establish drinking water standards, dose–response curves of animal toxicity tests are used to establish toxicity slopes or BMD and to document the dose–response relationship over the range of observed doses. With animal toxicity data, the health risk is extrapolated beyond the lower range of available observed data until the dose level begins to be adverse to human health. The shape of the dose–response relationship curve depends on the chemical, the kind of response, incidence of disease, and death. However, there are major gaps during the extrapolation from animal to human and from high dose to low dose to predict human health risk. Great uncertainty could be introduced during these extrapolations through either nonlinear or linear dose–response models for different modes of action. For example, animal tumor data is based on the development of the dose–response parameter as effect dose (ED) at 10% (ED_{10}) or lethal dose at 10% (LD_{10}). The actual dose heavily depends upon the exposure routes and animals. The EPA lists the general LD_{50} of different chemical compounds with great variations in Table 2.7. For dioxin (TCDD), the LD_{50} is as low as 0.001 mg/kg. For inhalation toxicity test, the LD_{50} for rats and mice are 293 and 137 mg/kg, respectively. Table 2.7 shows LD_{50} for different chemicals varied thousands of times.

2.3.2.1 Nonlinear Dose–Response Assessment

In nonlinear dose–response assessment, the threshold of toxicity is where the effects (or their precursors) begin to occur. The no observed adverse effect level (NOAEL) is the highest exposure level at which no statistically or biologically significant increases are seen in the frequency or severity of adverse effect between the exposed population and its control population. Different mathematical models were used to establish the bench mark dose (BMD) or benchmark dose lower confidence limit (BMDL) in the range from 1 to 10% depending on toxicity tests. The BMDL is a statistical lower confidence limit on the dose that produces the selected response. The lowest observed adverse effect level (LOAEL), NOAEL, or BMDL is used as the point of departure for extrapolation to lower doses. The EPA (2012) developed a guideline on using dose–response modeling to obtain BMD, i.e. dose levels corresponding to specific response levels, near the low end of the observable effect as shown in Figure 2.9. The fraction of animals affected in each group is indicated by the points with the error bars of 95% confidence intervals. In developing the BMD, the

Table 2.7 LD_{50} of typical chemicals based upon animal toxicity test (US EPA, 2006).

Chemical	LD_{50} (mg/kg)	Chemical	LD_{50} (with route and animal)
Ethyl alcohol	10 000	Caffeine	620 mg/kg – oral mouse
Sodium chloride	4 000		192 mg/kg – oral rat
			105 mg/kg – i.v. rat
			68 mg/kg – i.v. mouse
Ferrous sulfate	1 500	Chlorine (LC_{50})	293 ppm/1 h – rat
			137 ppm/1 h – mouse
Morphine sulfate	900	THC (from marijuana)	175 mg/kg – i.v. mouse
Strychnine sulfate	150		155 mg/kg – i.v. rabbit
			100 mg/kg – i.v. dog
Nicotine	1	Mercury(I) chloride	210 mg/kg – oral rat
			8 mg/kg – i.v. mouse
Black widow	0.55	Mercury(II) chloride	37 mg/kg – oral rat
			10 mg/kg – oral mouse
Curare	0.50	Arsenic acid (V oxidation state)	48 mg/kg – oral rat
Rattlesnake	0.24	Arsenic trioxide (III oxidation state)	20 mg/kg – oral rat
Dioxin (TCDD)	0.001	Dimethylarsenic acid (methylated arsenic form used as a cotton defoliant)	700 mg/kg – oral rat

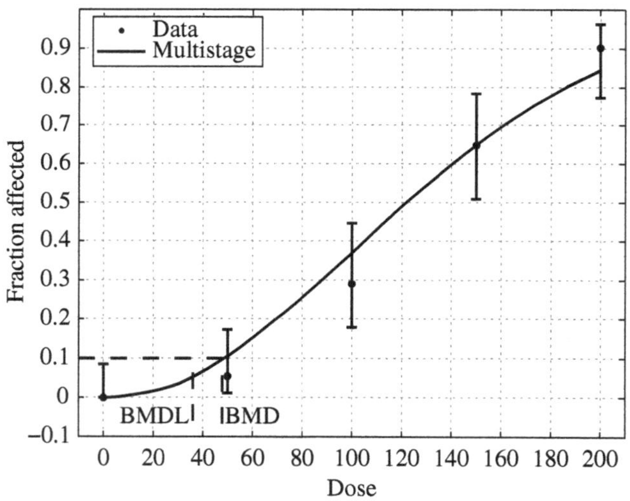

Figure 2.9 Example of a model fit to dichotomous data, with BMD and BMDL indicated.

EPA requires the following statistical information of the process: (i) rationale, (ii) estimation procedure, (iii) estimates of model parameters, (iv) goodness of fit such as log-likelihood and Akaike Information Criterion (AIC), and (v) standardized residuals. The US EPA (2012) provided excellent examples to illustrate some important aspects of computing benchmark doses (BMDs) and BMDLs from simple datasets and endpoints using EPA's BMDS package.

Example 2.2

Given: Table 2.8 lists dose–response data with the tumor endpoint of hepatocellular adenomas or carcinomas in a cancer bioassay using B6C3F1 mice exposed by gavage.

Find: The BMDL and BMD of chlorodibromomethane with 95% confidence interval.

Table 2.8 Dose–response data.

Administered dose (mg/kg/day)	Human equivalent dose (mg/kg/day)	Tumor incidence
0	0	5/50
50	2.83	10/49
100	5.67	19/50

NTP (National Toxicology Program) (1988).

Solution:

Although US EPA's cancer guidelines (US EPA, 2005) emphasize that the choice of benchmark response (BMR) should be independent of the extrapolation method, 10% extra risk is a typical BMR for standard cancer bioassay data when using linear extrapolation from the point of departure (POD). Therefore, a BMR of 10% extra risk was selected, as it was near the low end of the observable range. The one-sided BMDL was calculated at the 95% confidence level. EPA's cancer guidelines also recommend reporting an upper bound on the BMD, or a benchmark dose unit (BMDU), to measure uncertainty. Accordingly, the 95% one-sided BMDU was also estimated. Together the two limits provide a 90% two-sided confidence interval.

Model fitting (using the BMDS 2.6.0.1 (BMD software). Available from https://www.epa.gov/bmds/download-benchmark-dose-software-bmds#installing).

First, a 2nd-degree (i.e. $n - 1$) multistage model was fitted to the data. The model form is

$$P(\text{dose}) = \text{background} + (1 - \text{background}) \times \left[1 - \exp\left(-\text{beta1} \times \text{dose} - \text{beta2} \times \text{dose}^2\right)\right]$$

This model fits all three observations exactly (Table 2.9); hence, the x^2 goodness-of-fit p-value is undefined and the scaled residuals are all zero. The Akaike information criterion (AIC) was 158.7. The BMD and lower bounds (estimated by likelihood profile) estimates were

BMD (ED10) = 2.91 mg/kg/day

BMDL (LED10; 95% one-sided confidence limit) = 1.25 mg/kg/day

Table 2.9 Parameter estimates with standard errors for 2nd-degree multistage model.

Parameter	Maximum likelihood estimates (MLEs)	Standard error
Background	0.12	0.132665
Beta1	0.00930036	0.141898
Beta2	0.00925286	0.0246904

The Matlab code to plot the fit to 2nd-degree multistage model is

```
clear
clc
x=[0 2.83 5.67];
y=[0.12 0.204081633 0.38];
xBMD=[-0.6 2.91 2.91];
yBMD=[0.208045678 0.208045678 0];
xBMDL=[-0.6 1.25 1.25];
yBMDL=[0.208045678 0.208045678 0];
hold on
x4=[0:0.01:7];
z4=0.12+(1-0.12)*(1-exp(-0.00930036*(x4)
  -0.00925286*(x4).^2));
z3=0.12+(1-0.12)*(1-exp(-0.00930036*(x)
  -0.00925286*(x).^2));
plot(x,y,'ko','MarkerSize',3,'MarkerFace
  Color',[0 0 0],'MarkerEdgeColor',[0 0 0],
  'linewidth',1);
hold on
plot(x4,z4,'k-','LineWidth',1);
hold on
plot(xBMD,yBMD,'k--','LineWidth',1);
hold on
plot(xBMDL,yBMDL,'k--','LineWidth',1);
hold on
hold on
text(0.5,0.12,'BMDL','fontsize',11,
  'FontWeight','bold','FontName',
  'Times New Roman')
hold on
text(3,0.12,'BMD','fontsize',11,
  'FontWeight','bold','FontName',
  'Times New Roman')
hold on
R2=1-sum((y-z3).^2)/sum((y-mean(y)).^2)
set(gca,'xlim',[-0.6 6],'FontWeight',
  'bold','FontName', 'Times New Roman');
set(gca,'ylim',[0.1 0.4],'FontWeight',
  'bold','FontName', 'Times New Roman');
xlabel('Human Equivalent Dose (mg/kg/day)',
  'FontSize',14,'FontWeight',
  'bold','FontName', 'Times New Roman')
ylabel({'Tumor Incidence'},'FontSize',14,
  'FontWeight','bold','FontName',
  'Times New Roman')
legend('Data Means','Multistage',
  'Location','NorthWest');
box on
grid on
```

Next, a 1st-degree multistage model was fitted to the data to see if a more parsimonious model could also provide an adequate fit (Figure 2.10). The model form is

$$P(\text{dose}) = \text{background} + (1 - \text{background}) \times [1 - \exp(-\text{beta1} \times \text{dose})]$$

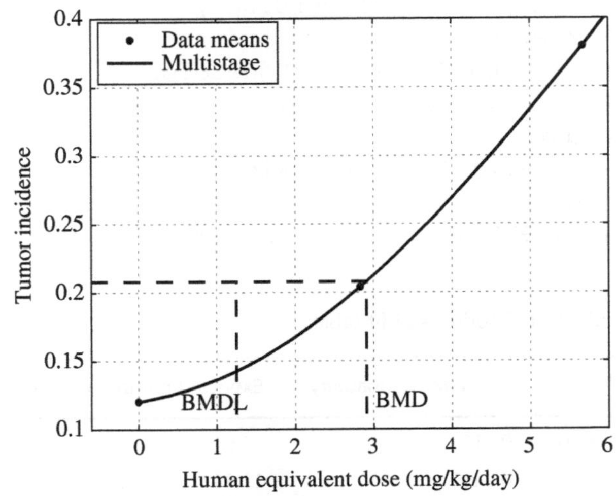

Figure 2.10 Fitted 2nd-degree multistage model and data means.

The 1st-degree multistage model also fitted the data adequately (see Tables 2.10 and 2.11), with a x^2 goodness-of-fit p-value of 0.4494 and scaled residuals, shown in Table 2.11, are not unusually large. The AIC was 157.3. The BMD and BMDL were

$$\text{BMD (ED10)} = 1.88 \text{ mg/kg/day}$$

$$\text{BMDL (LED10; 95\% one-sided confidence limit)} = 1.20 \text{ mg/kg/day}$$

Table 2.10 Parameter estimates with standard errors for 1st-degree multistage model.

Parameter	Maximum likelihood estimates (MLEs)	Standard error
Background	0.111488	0.111488
Beta1	0.120556	0.120556

The Matlab code to fitted 2nd-degree multistage model is

```
clear
clc
x=[0 2.83 5.67];
y=[0.12 0.204081633 0.38];
xBMD=[-0.6 1.88 1.88];
yBMD=[0.200245794 0.200245794 0];
xBMDL=[-0.6 1.2 1.2];
yBMDL=[0.200245794 0.200245794 0];
hold on
x4=[0:0.01:7];
z4=0.111488+(1-0.111488)*(1-exp
  (-0.0559807*(x4)));
z3=0.111488+(1-0.111488)*(1-exp
  (-0.0559807*(x)));
plot(x,y,'ko','MarkerSize',3,
'MarkerFace
Color',[0 0 0],'MarkerEdgeColor',
[0 0 0],'linewidth',1);
hold on
plot(x4,z4,'k-','LineWidth',1);
hold on
plot(xBMD,yBMD,'k--','LineWidth',1);
hold on
plot(xBMDL,yBMDL,'k--','LineWidth',1);
hold on
hold on
text(0.4,0.12,'BMDL','fontsize',11,
 'FontWeight','bold','FontName',
 'Times New Roman')
hold on
text(2,0.12,'BMD','fontsize',11,
'FontWeight','bold','FontName',
'Times New Roman')
hold on
R2 = 1 - sum((y - z3).^2)/sum((y - mean(y)).^2)
set(gca,'xlim',[-0.6 6],'FontWeight',
 'bold','FontName', 'Times New Roman');
set(gca,'ylim',[0.1 0.4],'FontWeight',
 'bold','FontName', 'Times New Roman');
xlabel('Human Equivalent Dose (mg/kg/
  day)','FontSize',14,'FontWeight',
 'bold','FontName', 'Times New Roman')
ylabel({'Tumor Incidence'},
 'FontSize',14,'FontWeight',
 'bold','FontName', 'Times New Roman')
legend('Data Means','Multistage',
 'Location','NorthWest');
box on
grid on
```

Answer: AIC can be used to compare models from different families using a similar fitting method (for example, least squares or a binomial maximum likelihood) (Figure 2.11). AIC is lower for the 1st-degree model suggesting that this is the preferred model. If L presents the log-likelihood at the maximum likelihood estimates (MLEs) for p estimated parameters, AIC is $-2L + 2p$ (Akaike, 1973; Linhart and Zucchini, 1986; Stone, 1998).

2.3.2.2 Linear Dose–Response Assessment

For carcinogens, if "mode of action" information is insufficient, then linear extrapolation is typically used as the default approach for dose–response assessment. A straight line is drawn from the point of departure for

Table 2.11 Goodness-of-fit table.

Dose	Estimated probability	Expected number responding	Observed number responding	Group size	Scaled residual
0.0000	0.1115	5.574	6	50	0.086
2.8300	0.2417	11.842	10	49	–0.205
5.6700	0.3531	17.657	19	50	0.118

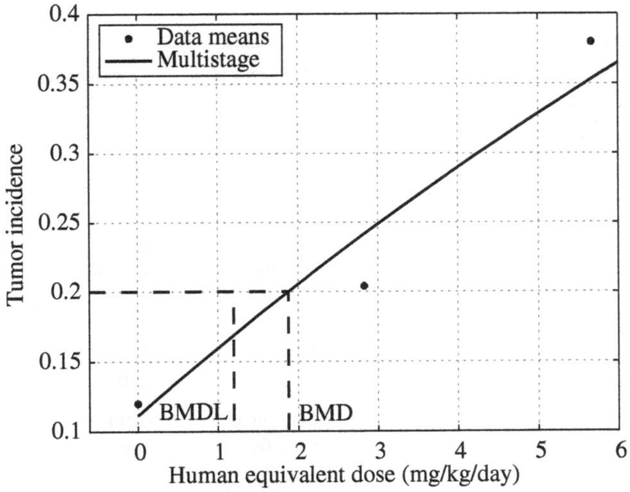

Figure 2.11 Fitted 1st-degree multistage model and data means.

the observed data (typically the BMDL) to the origin (where there are zero dose and zero response). The slope of this straight line is referred to as the slope factor (SF) or cancer SF that is used to estimate risk at exposure levels. When linear dose–response is used to assess cancer risk, excess lifetime cancer risk is calculated as follows:

$$\text{Cancer risk} = \text{exposure} \times \text{slope factor} \tag{2.2}$$

Total cancer risk is calculated by adding the individual cancer risks for each pollutant in each pathway of concern (i.e. inhalation, ingestion, and dermal absorption) by using reasonable maximum exposure (RME) and then adding together the risk for all pathways.

2.3.3 Exposure Assessment

Exposure assessment estimates the magnitude, frequency, and duration of human exposure to an agent in the environment or estimates future exposures for an agent that has not yet been released. Exposed concentration can be measured at the point of contact (the outer boundary of the body), estimated by separately evaluating the exposure concentration and the time through different scenarios, or reconstructed through internal indicators (biomarkers, body burden, and excretion levels) after the exposure. Table 2.12 lists the variables used in the US EPA Guidelines for Exposure Assessment (1992) for HRA.

The internal dose via the dermal route, μg/kg bw/day, can be calculated as follows:

$$D_{\text{int, derm}} = \frac{A_{\text{prod}} \times n \times (C/100) \times (B_{\text{derm}}/100) \times \text{RF} \times \text{CF}}{\text{BW}} \tag{2.3}$$

where

$D_{\text{int, derm}}$ = internal dose via the dermal route, g/kg bw/day

A_{prod} = amount of cosmetic/personal care product applied to skin, mg/event

n = frequency of product application, event/day

C = concentration of DBP in product, % (w/w)

B_{derm} = bioavailability via the dermal route, %

RF = retention factor

CF = conversion factor, 1000 g/mg

BW = adult body weight, 70 kg

2.3.3.1 Cancer Screening Calculation for Dermal Contaminants in Water

For a given cancer risk level at 10^{-6}, the following equations can be used to estimate dermal absorbed dose (DAD) in mg/kg/day:

For cancer risk

$$\text{DAD} = \frac{\text{Dermal cancer risk}}{\text{SF}_{\text{ABS}}} = \frac{(\text{Dermal cancer risk}) \times (\text{ABS}_{\text{GI}})}{\text{SF}_{\text{o}}} \tag{2.4}$$

For hazard quotient

$$\begin{aligned}\text{DAD} &= \text{Dermal hazard quotient} \times \text{RfD}_{\text{ABS}} \\ &= \text{Dermal hazard quotient} \times \text{RfD}_{\text{o}} \times \text{ABS}_{\text{GI}}\end{aligned} \tag{2.5}$$

Evaluate AD_{event}:

$$\text{AD}_{\text{event}} = \frac{\text{DAD} \times \text{BW} \times \text{AT}}{\text{EV} \times \text{ED} \times \text{EF} \times \text{SA}} \tag{2.6}$$

Evaluate permissible water concentration, C_{w}:

For organics

$$\text{If } t_{\text{event}} \le t^*, \text{ then: } C_{\text{w}} = \frac{\text{AD}_{\text{event}}}{2 \times \text{FA} \times K_{\text{p}} \sqrt{\dfrac{6\tau_{\text{event}} \times t_{\text{event}}}{\pi}}} \tag{2.7}$$

$$\text{If } t_{\text{event}} > t^*, \text{ then: } C_{\text{w}} = \frac{\text{AD}_{\text{event}}}{\text{FA} \times K_{\text{p}} \left[\dfrac{t_{\text{event}}}{1+B} + 2\tau_{\text{event}}\left(\dfrac{1+3B+3B^2}{(1+B)^2}\right)\right]} \tag{2.8}$$

For inorganics

$$C_{\text{w}} = \frac{\text{AD}_{\text{event}}}{K_{\text{p}} \times t_{\text{event}}} \tag{2.9}$$

Table 2.12 Specific parameters in health risk assessment.

Parameter	Definition	Default – child	Default – adult
TRL	Target risk level (unitless)	10^{-6}	10^{-6}
BW	Body weight (kg)	15	70
AT	Averaging time (year)	70	70
SFABS	Absorbed cancer slope factor $(mg/kg/day)^{-1}$	Chemical specific	Chemical specific
ED	Exposure duration (year)	6	30
EV	Event frequency (events/day)	1	1
EF	Exposure frequency (days/year)	350	350
FA	Fraction absorbed (unitless)	Chemical specific	Chemical specific
$t_{event\text{-}RME}$	Event duration (h)	1 (bathing)	0.58 (showering)
SA	Surface area (cm^2)	6 600	18 000
K_p	Permeability coefficient (cm/h)	Chemical specific	Chemical specific
ABSGI	Absorption fraction (unitless)	Chemical specific	Chemical specific
τ_{event}	Lag time per event (h)	Chemical specific	Chemical specific
SF_o	Oral cancer slope factor (mg/kg/day)	Chemical specific	Chemical specific
t^*	Time to reach steady state (h)	Chemical specific	Chemical specific
DAD	Dermal absorbed dose (mg/kg/day)	Site specific	Site specific
AD_{event}	Absorbed dose per event $(mg/cm^2/event)$	Site specific	Site specific
B	Dimensionless ratio of the permeability coefficient of a compound through the stratum corneum relative to its permeability coefficient across the viable epidermis (ve) (dimensionless)	Chemical specific	Chemical specific

Example 2.3 illustrates the steps used to calculate the cleanup level from dermal exposure to compounds in water given an acceptable risk of 10^{-6}. The default scenarios used in the calculations are (i) the adult 30-year exposure and (ii) an age-adjusted 30-year exposure incorporating a child bathing for 1 h/event (RME value), once a day, 350 days/year for 6 years and an adult showering at 35 min/event (RME value), once a day, 350 days/year for 24 years. The general equations could be applied to any compound, and the example gives the calculation for one compound in water with a cancer risk of 10^{-6}.

Example 2.3

Given: $SF_o = 5.2 \times 10^{-2} (mg/kg/day)^{-1}$, $K_p = 0.033$ cm/h, $ABS_{GI} = 1$, $t^* = 2.18$ h, $\tau_{event} = 0.91$ h, $t_{event} = 0.58$ h, FA = 1, DAD = 1.9×10^{-5} mg/kg/day.

Find: The concentration in water C_{water}.

The default scenarios used in the calculations are

1) The adult 30-year exposure.
2) An age-adjusted 30-year exposure incorporating a child bathing for 1 h/event (RME value), once a day, 350 days/year for 6 years and an adult showering at 35 min/event (RME value), once a day, 350 days/year for 24 years.

Solution:

Adult:

$$AD_{event} = DAD \times AT \left[\frac{BW_{adult}}{EV_a \times ED_a \times EF_a \times SA_a} \right]$$

$$= (1.9 \times 10^{-5}\ mg/kg/day) \times (25\,550\ days) \left[\frac{70\ kg}{1\ event/day \times 30\ years \times 350\ days/year \times 18\,000\ cm^2} \right]$$

$$= 1.8 \times 10^{-7}\ mg/cm^2/event$$

$$C_w = \frac{AD_{event}}{2 \times FA \times K_p \sqrt{\frac{6\tau_{event} \times t_{event}}{\pi}}}$$

$$= \frac{1.8 \times 10^{-7}\ \text{mg/cm}^2 - \text{event}}{2(1)(0.033\ \text{cm/h})\sqrt{\frac{6 \times 0.91\ \text{h/event} \times 0.58\ \text{h/event}}{\pi}}}$$

$$= 2.7 \times 10^{-6}\ \text{mg/cm}^3 = 2.7\ \mu\text{g/l} = 2.7\ \text{ppb}$$

Age adjusted:

$$AD_{event} = \frac{DAD \times AT}{\frac{EV_c \times ED_c \times EF_c \times SA_c}{BW_{child}} + \frac{EV_a \times ED_a \times EF_a \times SA_a}{BW_{adult}}}$$

Note the age-adjusted t_{event} for 6 years as child and 24 years as adult:

$$t_{event} = \frac{(6\ \text{years} \times 1\ \text{h/event}) + (24\ \text{years} \times 0.58\ \text{h/event})}{30\ \text{years}}$$

$$= 0.66\ \text{h/event}$$

$$AD_{event} = \frac{(1.9 \times 10^{-5}\ \text{mg/kg/day}) \times (25\ 550\ \text{day})}{\frac{(1\ \text{event/day}) \times (6\ \text{years}) \times (350\ \text{days/year}) \times (6660\ \text{cm}^2)}{15\ \text{kg}} + \frac{(1\ \text{event/day}) \times (6\ \text{years}) \times (350\ \text{days/year}) \times (18\ 000\ \text{cm}^2)}{70\ \text{kg}}}$$

$$= 1.6 \times 10^{-7}\ \text{mg/cm}^2/\text{event}$$

$$C_w = \frac{1.6 \times 10^{-7}\ \text{mg/cm}^2 - \text{event}}{2(1)(0.033\ \text{cm/h})\sqrt{\frac{6 \times 0.91\ \text{h/event} \times 0.66\ \text{h/event}}{\pi}}}$$

$$= 2.3 \times 10^{-6}\ \text{mg/cm}^3 = 2.3\ \mu\text{g/l} = 2.3\ \text{ppb}$$

2.3.3.2 Noncancer Screening Calculation for Contaminants in Residential Soil

The following equations are provided by the EPA in the exposure assessment process. The scenario to be evaluated is residential soil. Equations (2.10), (2.11), and (2.12) are used for calculating the soil concentration, C_{soil}:

Child or adult:

$$C_{soil} = \frac{THQ \times RfD \times BW \times AT \times 365\ \text{days/year} \times 10^6\ \text{mg/kg}}{ED \times EV \times EF \times SA \times AF \times ABS_d} \quad (2.10)$$

Age adjusted:

$$C_{soil} = \frac{THQ \times RfD \times AT \times 365\ \text{days/year} \times 10^6\ \text{mg/kg}}{EV \times EF \times SFS_{adj} \times ABS_d} \quad (2.11)$$

The age-adjusted, body-part-weighted dermal factor is as presented in equation

$$SFS_{adj} = \frac{(SA_{1-6}) \times (AF_{1-6}) \times (ED_{1-6})}{(BW_{1-6})} + \frac{(SA_{7-31}) \times (AF_{7-31}) \times (ED_{7-31})}{(BW_{7-31})} \quad (2.12)$$

For toxicity assessment, cancer SF can be derived based on absorbed dose:

$$SF_{ABS} = \frac{SF_o}{ABS_{GI}} \quad (2.13)$$

while RfD can be expressed as follows based on absorbed dose:

$$RfD_{ABS} = RfD_o \times ABS_{GI} \quad (2.14)$$

Example 2.4 Exposure risk assessment of noncarcinogen cadmium.

Given: Cadmium has both an oral reference dose (RfD) and ABS_d to allow for a quantitative evaluation. $ABS_d = 0.001$, cadmium in soil, and a level of concern equal to a hazard index of 1. All the other parameters are listed in Table 2.13.

Find: The soil concentration C_{soil} using child, adult, and age-adjusted scenarios to estimate exposure risk to noncarcinogen cadmium.

Solution:

$$SFS_{adj} = \frac{(SA_{1-6}) \times (AF_{1-6}) \times (ED_{1-6})}{(BW_{1-6})} + \frac{(SA_{7-31}) \times (AF_{7-31}) \times (ED_{7-31})}{(BW_{7-31})}$$

$$= \frac{(2800\ \text{cm}^2) \times \left(\frac{0.2\ \text{mg/cm}^2}{\text{event}}\right) \times (6\ \text{years})}{(15\ \text{kg})} + \frac{(5700\ \text{cm}^2) \times (0.07\ \text{mg/cm}^2 - \text{event}) \times (24\ \text{years})}{(70\ \text{kg})}$$

$$= 360\ \text{mg} - \text{years/kg/event}$$

To determine the dermal RfD, it suggests that the gastrointestinal (GI) adjustment for cadmium is either 5% for water or, more applicable for this example, 2.5% from

Table 2.13 Specific parameters in health risk assessment.

Parameter	Definition	Default – child	Default – adult	Default – age adjusted
THQ	Target hazard quotient (unitless)	1	1	1
BW	Body weight (kg)	15	70	—
AT	Averaging time (year)	6	30	30
RfD	Reference dose (mg/kg/day)	Chemical specific	Chemical specific	Chemical specific
ED	Exposure duration (year)	6	30	—
EV	Event frequency (events/day)	1	1	1
EF	Exposure frequency (days/year)	350	350	350
SA	Surface area (cm^2)	2800	5700	—
AF	Adherence factor (mg/cm^2/event)	0.2	0.07	—
ABS	Absorption fraction (unitless)	Chemical specific	Chemical specific	Chemical specific
SFS_{adj}	Age-adjusted dermal factor	—	—	360

food. Therefore, the dermal RfD is 3×10^{-5} (mg/kg/day) using Equation (2.14), the oral RfD of 1×10^{-3} from food, and a GI absorption of 2.5%. The oral RfD is calculated as follows:

$$RfD_{ABS} = RfD_o \times ABS_{GI}$$
$$= (1 \times 10^{-3}\ \text{mg/kg/day}) \times (0.025)$$
$$= 2.5 \times 10^{-5}\ \text{mg/kg/day}$$

Child:

$$C_{soil} = \frac{(1) \times (0.000025\ \text{mg/kg/day}) \times (15\ \text{kg}) \times (6\ \text{years}) \times (365\ \text{days/year}) \times (10^6\ \text{mg/kg})}{(6\ \text{years}) \times (1\ \text{event/day}) \times (350\ \text{days/year}) \times (2800\ \text{cm}^2) \times (0.2\ \text{mg/cm}^2/\text{event}) \times (0.001)}$$
$$= 700\ \text{mg/kg} = 700\ \text{ppm}$$

Adult:

$$C_{soil} = \frac{(1) \times (0.000025\ \text{mg/kg/day}) \times (70\ \text{kg}) \times (30\ \text{years}) \times (365\ \text{days/year}) \times (10^6\ \text{mg/kg})}{(30\ \text{years}) \times (1\ \text{event/day}) \times (350\ \text{days/year}) \times (5700\ \text{cm}^2) \times (0.07\ \text{mg/cm}^2/\text{event}) \times (0.001)}$$
$$= 4600\ \text{mg/kg} = 4600\ \text{ppm}$$

Age adjusted

$$C_{soil} = \frac{(1) \times (0.000025\ \text{mg/kg/day}) \times (30\ \text{years}) \times (365\ \text{days/year}) \times (10^6\ \text{mg/kg})}{(1\ \text{event/day}) \times (350\ \text{days/year}) \times (360\ \text{mg-year/kg/event}) \times (0.001)}$$
$$= 2200\ \text{mg/kg} = 2200\ \text{ppm}$$

2.3.4 DBP Health Advisory Concentration

Many benchmark concentrations could be correlated with E_{LUMO}, while E_{LUMO} significantly correlates with the number of chlorine for a given class DBP with a specific carbon number (Tang and Wang, 2010). Example 2.5 illustrates such correlation.

Example 2.5
Given: The number of chlorines, E_{LUMO}, and longer-term exposure for a 70 kg adult of chlorinated 1-carbon halogenated aliphatic compounds are shown in Table 2.14 (Jiang et al., 2000; Yang et al., 2000).

Table 2.14 Number of chlorines, E_{LUMO}, and longer-term exposure for 70 kg adult of chlorinated 1-carbon halogenated aliphatic compounds.

	Number of chlorines	E_{LUMO}	Longer-term exposure for 70 kg adult (mg/l)
Carbon tetrachloride	4	0.09465	0.3
Chloroform	3	0.12650	0.4
Dichloromethane	2	0.1662	
Chloromethane	1	0.21662	1

Find:

1) Regression between E_{LUMO} and the number of chlorines for 1-carbon halogenated aliphatic compounds by using Matlab.
2) The longer-term exposure concentration for 70 kg adult versus E_{LUMO} for 1-carbon halogenated aliphatic compounds by using Matlab.
3) How the number of chlorine in 1-carbon halogenated aliphatic compounds affects the longer-term exposure concentration for a 70 kg adult.

Solution:

Regression between E_{LUMO} and the number of chlorines for 1-carbon halogenated aliphatic compounds is conducted using the following Matlab code:

```
clear
clc
NumberofChlorines=[4 3 2 1];
ELUMO=[0.09465 0.12650 0.1662 0.21662];
hold on
n=1;
p=polyfit(NumberofChlorines, ELUMO, n)
x4=[0:0.01:10];
z3=polyval(p, NumberofChlorines);
z4=polyval(p, x4);
plot(NumberofChlorines,ELUMO,'ko',
  'MarkerSize',3,'MarkerFaceColor',
  [0 0 0],
  'MarkerEdgeColor',
  [0 0 0],'linewidth',1);
hold on
plot(x4,z4,'k-','LineWidth',1);
hold on
R2 = 1 - sum((ELUMO - z3).^2)/sum
  ((ELUMO - mean(ELUMO)).^2)
set(gca,'xlim',[0.5 4.5],'FontWeight',
  'bold','FontName', 'Times New Roman');
xlabel('Numberof Chlorines','FontSize',14,
  'FontWeight','bold','FontWeight',
  'bold','FontName', 'Times New Roman')
ylabel({'E_L_U_M_O'},'FontSize',14,
  'FontWeight','bold')
legend('Original Data','Trendline',
  'Location','NorthEast','FontWeight',
  'bold','FontName', 'Times New Roman');
box on
grid on
```

Figure 2.12 shows that E_{LUMO} has close correlation and decreased with the number of chlorine atoms according to the following equation:

$$E_{LUMO} = -0.0406 N_{Cl} + 0.2524 \quad r^2 = 0.9896$$

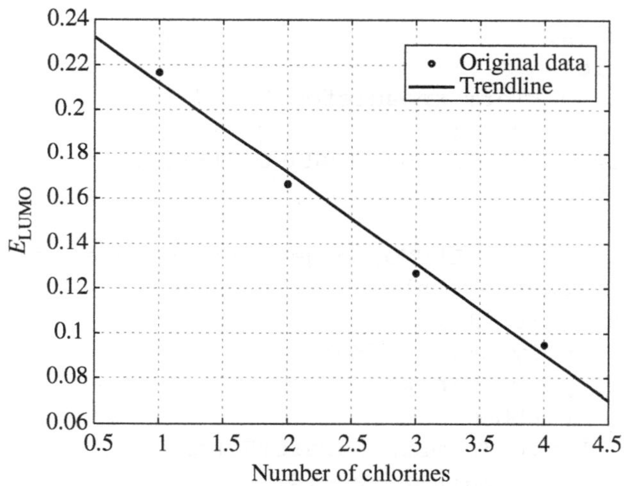

Figure 2.12 E_{LUMO} versus number of chlorines for 1-carbon halogenated aliphatic compounds.

For chlorinated 1-carbon halogenated aliphatic compounds, E_{LUMO} is shown in Figure 2.13.

Regression between longer-term exposure dose for a 70 kg adult and E_{LUMO} for 1-carbon halogenated aliphatic compounds:

Longer-term exposure dose for a 70 kg adult is closely related with E_{LUMO}:

$$D_{LTED,\,70} = 5.94\, E_{LUMO} - 0.3001 \quad r^2 = 0.9852$$

where $D_{LTED,\,70}$ is the longer-term exposure dose for a 70 kg adult and E_{LUMO} is the energy of the lowest unoccupied molecular orbital.

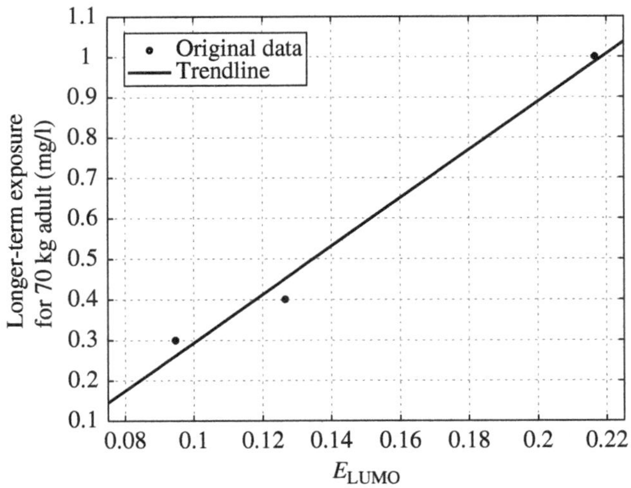

Figure 2.13 Longer-term exposure for 70 kg adult versus E_{LUMO} for 1-carbon halogenated aliphatic compounds.

```
clear
clc
LongerTermExposurefor70kgAdult=
  [0.3 0.4 1];
ELUMO=[0.09465 0.12650 0.21662];
hold on
n=1;
p=polyfit(ELUMO,LongerTermExposurefor
  70kgAdult,n)
x4=[0:0.01:10];
z3=polyval(p,ELUMO);
z4=polyval(p,x4);
plot(ELUMO,
  LongerTermExposurefor70kgAdult,
  'ko','MarkerSize',3,
  'MarkerFaceColor',[0 0 0],
  'MarkerEdgeColor',[0 0 0],'
  linewidth',1);
hold on
plot(x4,z4,'k-','LineWidth',1);
hold on
R2 = 1 - sum((LongerTermExposurefor
  70kgAdult - z3).^2)/sum((LongerTerm
  Exposurefor70kgAdult - mean(LongerTerm
  Exposurefor70kgAdult)).^2)
set(gca,'xlim',[0.075 0.225],
  'FontWeight','bold','FontName',
  'Times New Roman');
xlabel('E_L_U_M_O','FontSize',14,
  'FontWeight','bold','FontWeight',
  'bold','FontName', 'Times New Roman')
ylabel({'Longer Term Exposure';'for 70 kg
  Adult (mg/l)'},'FontSize',14,
  'FontWeight','bold')
legend('Original Data','Trendline',
  'Location','NorthWest','FontWeight',
  'bold','FontName', 'Times New Roman');
box on
grid on
```

For chlorinated 1-carbon halogenated aliphatic compounds, the longer-term exposure for a 70 kg adult increased with E_{LUMO} as shown in Figure 2.13

The slope, k_1, is 5.94 between the longer-term exposure dose for a 70 kg adult and E_{LUMO} for 1-carbon halogenated aliphatic compounds. The slope k_2 for E_{LUMO} versus the number of chlorines for 1-carbon halogenated aliphatic compounds is –0.0406.

$$\frac{\Delta \text{Longer-term exposure for 70 kg adult}}{\Delta E_{LUMO}} = k_1 = 5.94$$

$$\frac{\Delta E_{LUMO}}{\Delta \text{Number of chlorines}} = k_2 = -0.0406$$

$$\frac{\Delta \text{Longer-term exposure for 70 kg adult}}{\Delta \text{Number of chlorines}} = k_1 k_2 = -0.2412$$

Comments: For chlorinated 1-carbon halogenated aliphatic compounds, the corresponding longer-term exposure concentration for a 70 kg adult decreased with the number of chlorine atoms.

2.3.5 Risk Characterizations

Risk characterization conveys the risk assessor's judgment as to the nature and presence or absence of risks. How the risk was assessed should present where assumptions and uncertainties exist and where policy choices will need to be made. EPA recommends that the characterization fully and explicitly disclose the risk assessment methods, default assumptions, logic, rationale, extrapolations, uncertainties, and overall strength of each step in the assessment. After risk assessment, risk management and communication are key to protecting public health. Each component of the risk assessment (e.g. hazard assessment, dose–response assessment, exposure assessment) has an individual risk characterization for the key findings, assumptions, limitations, and uncertainties. Risk characterization also applies to both human HRA and ecological risk assessments.

2.4 QSAR Analysis in HRA

The US EPA permits and uses quantitative structure and activity relationship (QSAR) principles to classify and prioritize DBPs because more than 600 DBPs have been identified and cataloged by the US EPA but only small fraction of them have been studied on toxicity quantification. QSAR can be used to predict the toxicity of a specific DBP or molecular property of DBP such as log *P*, which reflects hydrophobicity of a chemical compound. Experts use the principles of mechanism-based structure and activity relationship (SAR) relative to known carcinogens, and mechanisms include structural analogy to known carcinogens, toxicokinetic and toxicodynamic factors, potency indicators for a structural analog, short-term test data, and metabolic activation. Therefore, QSAR analysis has its unique role in setting drinking water standards.

QSAR analysis is based upon a simple fact that similar chemical structures are expected to exhibit similar chemical behavior. With tens of thousands of chemical structures to assess and many different empirical tests, QSAR is often used in strategic screening in product development through HI and risk assessment. Given sufficient knowledge on structurally or functionally related compounds, QSAR may be used for screening well-defined biological, toxicological, or pharmacological endpoints of interest and associated kinetic characteristics such as absorption, distribution, metabolism, and excretion. In QSAR analysis, chemical structure is quantitatively correlated with a well-defined process, such as biological activity, chemical reactivity, or toxicity of a chemical compound. The US EPA uses QSAR analysis to prioritize DBPs using three general types of predictive models:

1) Mathematical models such as SARs and QSARs use descriptors and mathematical relationships to derive predictions. Examples of SAR/QSAR models are ECOSAR and select modules contained in EPISuite™ like WSK_{OWWIN}.
2) Fragment-based models such as the BioWin module within EPISuite evaluate the features of molecular fragments present on the molecule to make predictions.
3) Expert systems, like OncoLogic™, use rule-based decision trees to mimic an expert's judgment. Other expert systems utilize artificial neural networks and molecular models.

There are hundreds of chemicals that have been identified as DBPs, a large proportion of which fall into the general category of halogenated DBPs. Due to evidence of carcinogenicity and developmental effects, halogenated DBPs are of particular interest. Two major molecular descriptors can be used in predicting the toxicity of DBP. One is log *P* and the other is the number of carbon and halogen atoms. For QSAR analysis, halogenated DBPs are classified into eight classes as illustrated in Figure 2.14.

1. Halogenated alkane:
2. Halogenated alkene:
3. Halogenated aromatic:
4. Halogenated aldehyde:
5. Halogenated ketone:
6. Halogenated carboxylic acid:
7. Heterocycle:
8. Others:

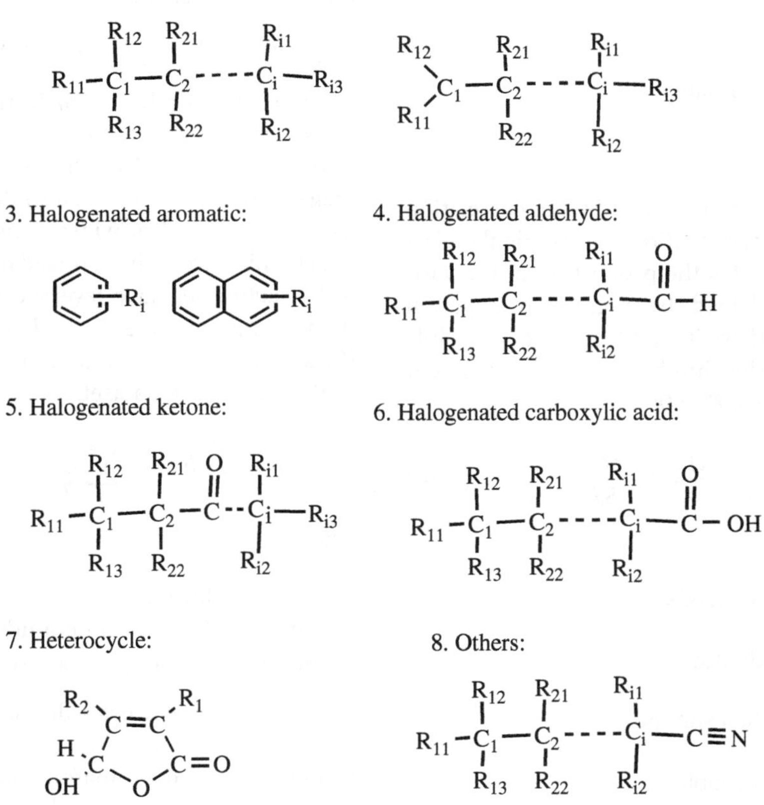

Figure 2.14 Classification of DBPs.

According to the molecular structure of DBPs, Pierotti (1999) developed a DBP database containing more than one thousand DBP compounds for the eight classes of DBPs. The database includes more than 20 different physical and chemical characteristics for each DBP compound. Major variables are chemical name, address, subclass, CAS number, SMILES, disinfection process, molecular formula and structure, molecular weight, log P, cLog P, log *P* reference, melting point (°C), boiling point (°C), vapor pressure (mm of Hg), solubility (mol/l), E_{HOMO}, E_{LUMO}, surface area, MCLG (mg/l), MCL (mg/l), 10-day exposure for a 10 kg child (mg/l), long-term exposure for 10 kg child (mg/l), long-term exposure for a 70 kg adult (mg/l), mg/l at 10^{-4} cancer risk, cancer group, SDWA reference, and toxicological effect. The database is used in developing QSAR models in the following section to illustrate statistical methods such as regression, outlier detection, quantification of uncertainty and sensitivity, and validation of QSAR models.

2.4.1 Multiple Linear Regression (MLR)

Multiple linear regression (MLR) is a common algorithm in deriving QSAR models. It relates the dependent variable y to a number of independent (predictor) variables, x_j, by using a linear equation as follows:

$$\hat{y} = b_0 + b_1x_1 + b_2x_2 + \cdots + b_jx_j \quad (2.15)$$

where

$\hat{y}$ = calculated dependent variable
x_j = predictor variable
b_j = regression coefficient

To assess goodness of fit quantified by the correlation coefficient of multiple determination, (R^2) is calculated by Equation (2.17). R^2 estimates the proportion of the variation of y that is explained by regression equation (Massart et al., 1997). If there is a perfect fit, $R^2 = 1$, while if there is no linear relationship between the dependent and independent variables, $R^2 = 0$:

$$R^2 = \frac{SS_{reg}}{SS_T} = \frac{(SS_T - SS_{res})}{SS_T} = 1 - \frac{SS_{res}}{SS_T} \quad (2.16)$$

where

$SS_T = \sum_i (y_i - \bar{y})^2$ = total sum of squares

$SS_{res} = \sum_i (y_i - \hat{y})^2$ = residual sum of squares

$SS_{reg} = \sum_i (\hat{y} - \bar{y})^2$ = residual sum of squares

y_i = observed dependent variable

$\bar{y}$ = mean value of the dependent variable

$\hat{y}$ = calculated dependent variable

In predicting the toxicity of a chemical compound, two important molecular descriptors are hydrophobicity such as log *P* and electronic properties such as the energy of the lowest unoccupied molecular orbitals (E_{LUMO}). Log *P* reflects the hydrophobicity of molecules, which often correlates well with the bioactivity of chemicals (Leo and Hansch, 1999). The logarithm of the partition coefficient (log *P*), also referred to as log K_{ow}, describes the distribution of a compound between organic (usually *n*-octanol) and water phases by the following equation:

$$\log P = \log\left(\frac{C_{org}}{C_{water}}\right) \quad (2.17)$$

where

C_{org} = concentration of the nonionized solute in the organic phase

C_{water} = concentration of the nonionized solute in the water phase

If log *P* is greater than zero (0), the compound has a greater solubility in the organic phase; if log *P* is less than zero (0), the compound has a greater solubility in the aqueous phase. Chen et al. (2016) reported that log *P* can be predicted accurately by the E_{LUMO}, the number of carbon (N_C), and the number of chlorine (N_{Cl}). A general MLR model is to predict log *P* is

$$\log P = b_0 + b_1E_{LUMO} + b_2N_{Cl} + b_3N_C \quad (2.18)$$

To develop a robust QSAR model, outliers have to be detected to improve the correlation coefficiency of the QSAR model. The Hotelling test and the associated leverage statistics can be used to detect outliers. The leverage of a chemical provides a measure of the distance of the chemical from the centroid of its active set. Chemicals in the active set have leverage values between 0 and 1. A warning leverage (h^*), defined in Equation (2.19) (Eriksson et al., 2003), is a critical value to cut off the outliers from the dataset:

$$h^* = 3 \times h = 3 \times \sum_i \frac{h_i}{n} = 3 \times \frac{p}{n} \quad (2.19)$$

where

h^* = warning leverage
n = number of active compounds
p = number of descriptor plus one

The William plot has a double-ordinate Cartesian plot of cross-validation residuals (first ordinate), standard residuals (second ordinate), and leverage (Hat diagonal: abscissa) values (h), which defined the domain of applicability of the model as a squared area within ±2 band for residuals and a leverage threshold, h^*.

Example 2.6

Given: Table 2.15 lists the halogenated alkane compounds selected for QSAR model development. The table consists of the list number, the chemical name, and molecular formula. The $\log P$ values were obtained from Hansch's QSAR database (Hansch et al., 1995). Molecular descriptor, E_{LUMO}, was calculated using the latest edition of Spartan software. N_{Cl} and N_C are presented in Table 2.15 with reference to their molecular structures.

Find: QSAR model to predict $\log P$ of DBPs of halogenated alkanes using the LMR method.

Solution:

Outlier detection by leverage approach

Statistical software SAS can be used to analyze the DBP datasets. Statistical parameters such as R^2, adjusted R^2, mean square error (MSE), and F-value are calculated by SAS. Analyzing the model AD in the William plot using the data in Table 2.16, four DBP compounds are identified as outliers as labeled in Figure 2.15. Table 2.15 shows that the compounds are 1-bromo-3-chloropropane (4), chlorotribromomethane (20), 1,1,1,2-tetrachloroethane (44), and hexachloroethane (46). These compounds have a high leverage that are greater than h^* value, 0.26087. Compared with other DBP compounds in the dataset, these molecules may have some structural anomalies that are not well modeled by the selected descriptors for the outliers or that are too particular for the influential chemicals. For example, 1-bromo-3-chloropropane (4) and chlorotribromomethane (20) consist of bromide atom in their molecular structure, whereas 1,1,1,2-tetrachloroethane (44) and hexachloroethane (46) have a high number of chlorine in the molecular structure.

Multivariable linear regression

After four outliers were removed, QSAR model was developed with correlation coefficient R^2 of 0.891 as follows:

$$\log P = 2.631 - 11.967E_{LUMO} + 0.101N_{Cl} + 0.534N_C \tag{2.20}$$

($n = 42$, $R^2 = 0.891$, RMSE = 0.310, PRESS RMSE = 0.385, $F = 103.936$)

Statistical results for halogenated alkanes after removal are summarized in Table 2.17. It shows that the best model requires three variables including E_{LUMO}, N_{Cl}, and N_C after the removal of the outliers. All the statistical indicators are improved for the QASR model of predicting $\log P$ of halogenated alkanes.

Figure 2.16 is a box plot that shows the contribution of each variable such as E_{LUMO}, N_{Cl}, and N_C to the standardized coefficients after the outliers were removed. It suggests that N_C contributes to $\log P$ most positively, while E_{LUMO} contributes to $\log P$ negatively to a lesser extent. The number of Cl and N_{Cl} contributes to $\log P$ positively but less than both N_C and E_{LUMO}.

Figure 2.17 shows that all the predicted $\log P$ values are within the boundary of 95% confidence of the observed $\log P$ values.

2.4.2 Validation of QSAR Models

QSAR models should be balanced between the two extremes of overfitted versus underfitted through model validation. Model validation consists of internal model performance (goodness of fit and robustness) and external model performance (predictivity). Cross-validation, bootstrapping, response randomization test, and training/test set splitting are recommended by the OECD (2007). In cross-validation, a number of modified datasets are deleted. In each case, one or a small group of compounds from the data are processed in such a way that each object is removed one at a time. From the original dataset, a reduced dataset (training set or active set) is used to develop a partial model, while the remaining data (validation set) are used to evaluate the model predictivity (Efron, 1983; Osten, 1988).

The leave-one-out (LOO) method is the simplest cross-validation procedure. Each compound is removed one at a time. For given n compounds, n-reduced models are developed. Each of these models is developed with the remaining $n - 1$ compounds and used to predict the response of the deleted compound. The predictive power of the model is calculated as the sum of squared differences between the observed and estimated responses. This LOO method is accurate to quantify each compound's impact on the developed QSAR model. However, the predictive power is often too optimistic, particularly with a large dataset compound because the perturbation of one compound is often insignificant.

The leave-many-out (LMO) cross-validation method is to remove more than one compound each time. The dataset is divided into a number of blocks (referred to as cancelation groups). At each time, all the compounds belonging to a block are left out from the derivation of the model. Compared with the LOO method, the LMO method gives a more realistic estimation of predictive power because it introduces a larger perturbation in the dataset. There is no standard rule for splitting the data for a block, so it is normally defined by model users. The random clustering method in which one or more compounds are randomly selected as a subset of compounds in the cancelation block and left out for the deviation of the QSAR model can also be used.

Table 2.15 Halogenated alkane compounds used in QSAR analysis.

List no.	DBP chemical name	Molecular formula	log *P*	E_{LUMO}	N_{Cl}	N_C
1	Bromochloromethane	CH_2BrCl	1.39	0.14	1	1
2	1-Chlorobutane	C_4H_9Cl	2.52	0.22	1	4
3	1-Chlorohexane	$C_6H_{13}Cl$	3.58	0.22	1	6
4	1-Bromo-3-chloropropane	C_3H_6BrCl	1.85	0.16	1	3
5	2-Bromo-2-methylpropane	C_4H_9Br	2.53	0.17	1	4
6	2-Chlorobutane	C_4H_9Cl	2.52	0.21	1	4
7	1-Bromopentane	$C_5H_{11}Br$	3.19	0.19	1	5
8	2-Chloro-2-methylbutane	$C_5H_{11}Cl$	2.92	0.20	1	5
9	1-Chloroheptane	$C_7H_{15}Cl$	4.11	0.22	1	7
10	Chloromethane	CH_3Cl	0.94	0.22	1	1
11	1-Chloropropane	C_3H_7Cl	1.99	0.21	1	3
12	2-Chloropropane	C_3H_7Cl	1.99	0.21	1	3
13	1-Chloropentane	$C_5H_{11}Cl$	3.05	0.22	1	5
14	1-Chlorooctane	$C_8H_{17}Cl$	4.64	0.22	1	8
15	1-Chlorodecane	$C_{10}H_{21}C$	5.7	0.22	1	10
16	2-Chlorohexane	$C_6H_{13}Cl$	3.581	0.21	1	6
17	3-Chlorohexane	C_6H1_3Cl	3.581	0.22	1	6
18	Chloroethane	C_2H_5Cl	1.47	0.22	1	2
19	Bromochloroiodomethane	CHBrClI	2.62	0.08	1	1
20	Chlorotribromomethane	CBr_3Cl	3.295	0.07	1	1
21	1,2-Dichloroethane	$C_2H_4Cl_2$	1.46	0.19	2	2
22	1,2-Dichloropropane	$C_3H_6Cl_2$	1.99	0.19	2	3
23	Bromochloromethane	CH_2BrCl	1.39	0.14	1	1
24	1,5-Dichloropentane	$C_5H_{10}Cl_2$	2.77	0.20	2	5
25	Dichloromethane	CH_2Cl_2	1.25	0.17	2	1
26	*trans*-1,2-Dichlorocyclohexane	$C_6H_{10}Cl_2$	3.3	0.18	2	6
27	1,1-Dichloroethane	$C_2H_4Cl_2$	1.78	0.17	2	2
28	1,2-Dichlorobutane	$C_4H_8Cl_2$	2.52	0.19	2	4
29	2,2-Dichloropropane	$C_3H_6Cl_2$	2.31	0.17	2	3
30	Dichlorodifluoromethane	CCl_2F_2	2	0.10	2	1
31	Dichloroiodomethane	$CHCl_2I$	2.482	0.08	2	1
32	2,3-Dichlorobutane	$C_4H_8Cl_2$	2.52	0.18	2	4
33	1,2-Dichloro-2-methylbutane	$C_5H_{10}Cl_2$	2.91	0.18	2	5
34	1,1,1-Trichloroethane	$C_2H_3Cl_3$	2.48	0.13	3	2
35	1,1,2-Trichloroethane	$C_2H_3Cl_3$	2.05	0.16	3	2
36	1,2,3-Trichloropropane	$C_3H_5Cl_3$	1.98	0.16	3	3
37	Fluorotrichloromethane	CCl_3F	2.44	0.09	3	1
38	1,1,2-Trichloropropane	$C_3H_5Cl_3$	2.58	0.16	3	3
39	1,1,3-Trichloropropane	$C_3H_5Cl_3$	1.98	0.16	3	3
40	1,2,2-Trichloropropane	$C_3H_5Cl_3$	2.58	0.15	3	3
41	1,1,1-Trichloropropane	$C_3H_5Cl_3$	3.01	0.13	3	3
42	1,1,2,2-Tetrachloroethane	$C_2H_2Cl_4$	2.64	0.15	4	2
43	Carbon tetrachloride	CCl_4	2.88	0.09	4	1
44	1,1,1,2-Tetrachloroethane	$C_2H_2Cl_4$	3.03	0.26	4	2
45	Pentachloroethane	C_2HCl_5	3.63	0.12	5	2
46	Hexachloroethane	C_2Cl_6	4.61	0.11	6	2

Table 2.16 Leverage values of halogenated alkanes.

List no.	Standardized residuals	Leverage h	Leverage threshold h^*	Cutoff value (−)	Cutoff value (+)
1	−0.21192	0.09885	0.26087	−1.93218	1.93218
2	0.32864	0.17935	0.26087	−1.93218	1.93218
3	1.27523	0.18739	0.26087	−1.93218	1.93218
4	2.51973	0.21456	0.26087	−1.93218	1.93218
5	0.39921	0.1208	0.26087	−1.93218	1.93218
6	−0.96939	0.04923	0.26087	−1.93218	1.93218
7	0.26173	0.06586	0.26087	−1.93218	1.93218
8	0.26173	0.06586	0.26087	−1.93218	1.93218
9	0.51915	0.05641	0.26087	−1.93218	1.93218
10	−0.39389	0.04334	0.26087	−1.93218	1.93218
11	0.3323	0.04671	0.26087	−1.93218	1.93218
12	0.32604	0.04487	0.26087	−1.93218	1.93218
13	−0.05965	0.04323	0.26087	−1.93218	1.93218
14	0.58972	0.05059	0.26087	−1.93218	1.93218
15	0.66029	0.06233	0.26087	−1.93218	1.93218
16	0.47556	0.06114	0.26087	−1.93218	1.93218
17	0.66241	0.06233	0.26087	−1.93218	1.93218
18	0.73086	0.09166	0.26087	−1.93218	1.93218
19	0.80143	0.13855	0.26087	−1.93218	1.93218
20	0.94257	0.28508	0.26087	−1.93218	1.93218
21	−0.53812	0.06966	0.26087	−1.93218	1.93218
22	−0.25567	0.0907	0.26087	−1.93218	1.93218
23	0.39272	0.12862	0.26087	−1.93218	1.93218
24	−0.77241	0.05482	0.26087	−1.93218	1.93218
25	−0.46755	0.03743	0.26087	−1.93218	1.93218
26	−0.70184	0.03166	0.26087	−1.93218	1.93218
27	−0.39698	0.02278	0.26087	−1.93218	1.93218
28	−1.03817	0.03157	0.26087	−1.93218	1.93218
29	−0.63127	0.02607	0.26087	−1.93218	1.93218
30	−0.81812	0.02411	0.26087	−1.93218	1.93218
31	−0.96760	0.0393	0.26087	−1.93218	1.93218
32	−1.04442	0.04036	0.26087	−1.93218	1.93218
33	−1.27073	0.07417	0.26087	−1.93218	1.93218
34	−0.09936	0.09095	0.26087	−1.93218	1.93218
35	−0.32045	0.04486	0.26087	−1.93218	1.93218
36	−0.67173	0.04238	0.26087	−1.93218	1.93218
37	−1.87347	0.03703	0.26087	−1.93218	1.93218
38	−0.60116	0.03703	0.26087	−1.93218	1.93218
39	−1.87347	0.03703	0.26087	−1.93218	1.93218
40	−0.78801	0.03857	0.26087	−1.93218	1.93218
41	−0.24988	0.05228	0.26087	−1.93218	1.93218
42	0.24379	0.11295	0.26087	−1.93218	1.93218
43	−0.19734	0.08404	0.26087	−1.93218	1.93218
44	2.68499	0.34074	0.26087	−1.93218	1.93218
45	0.75155	0.16316	0.26087	−1.93218	1.93218
46	2.05293	0.27957	0.26087	−1.93218	1.93218

Example 2.7

Given: The developed QSAR model of Equation (2.20) is as follows:

$$\log P = 2.631 - 11.967E_{\text{LUMO}} + 0.101N_{\text{Cl}} + 0.534N_{\text{C}} \quad (2.20)$$

($n = 42$, $R^2 = 0.891$, RMSE = 0.310, PRESS RMSE = 0.385, $F = 103.936$)

Find: Please validate the QSAR model of Equation 2.20 using the leave-one-out (LOO) and leave-many-out (LMO) cross-validation methods, respectively.

Solution:

1) The whole dataset is divided so that it has approximately the same intervals of log *P* values for the active and validation sets, with approximately 20% of the initial compounds set aside for external validation. The active set and validation set in each DBP class are split based on the available compound data after the outliers are removed and randomly grouped into validation sets by computer. The validation sample set for HA includes 37 active sets (n_c) and 5 validation sets (n_v). The QSAR model will be first validated by the LOO method, followed by the LMO method. The results obtained by the LOO and LMO methods are then compared.

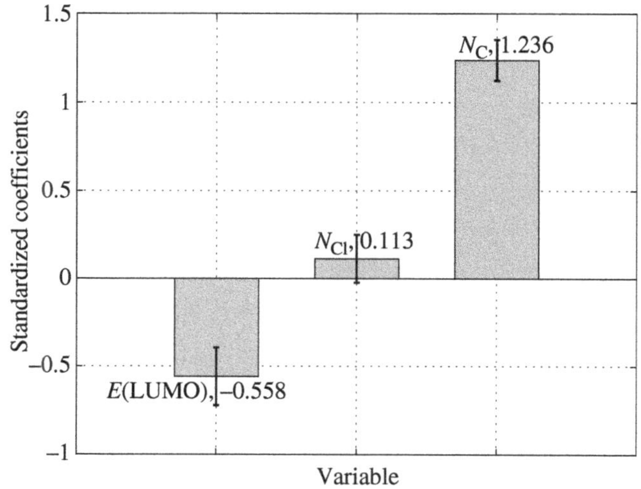

Figure 2.16 Box plot of each variable contribution to log *P* in halogenated alkane after removal of outliers. (*Source:* Reproduced with permission of Springer.)

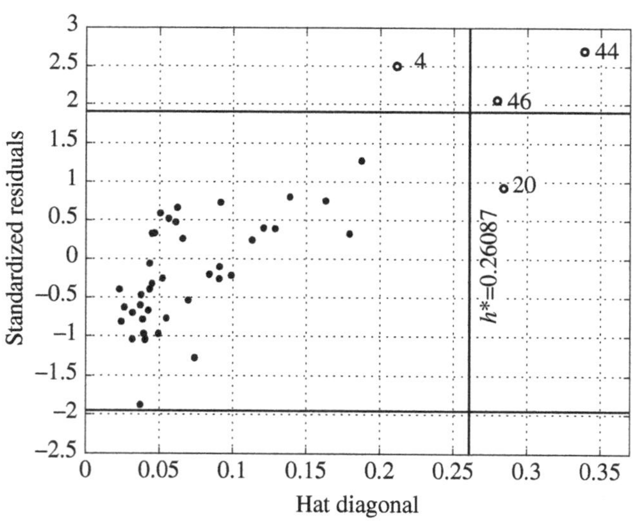

Figure 2.15 Williams plot for detecting outliers in halogenated alkane. (*Source:* Reproduced with permission of Springer.)

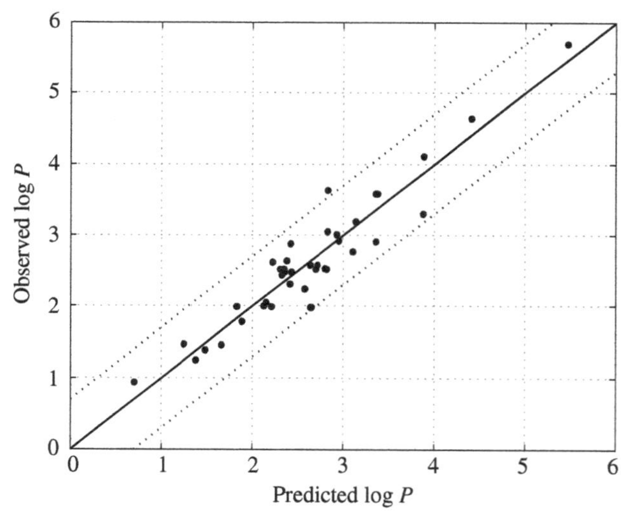

Figure 2.17 Measured log *P* versus predicted log *P* for halogenated alkane after removal of outliers. (*Source:* Reproduced with permission of Springer.)

Table 2.17 Statistics of QSAR models for log *P* of halogenated alkane after removal of outliers.

No. of variables	Variables	MSE	R^2	Adjusted R^2	Mallows' C_p	Akaike's AIC	Schwarz's SBC
1	N_C	0.280	0.667	0.642	78.506	−51.508	−48.033
2	E_{LUMO}/N_C	0.101	0.883	0.874	4.838	−93.538	−88.324
3	$E_{\text{LUMO}}/N_{\text{Cl}}/N_C$	0.096	0.891	0.883	4.000	−94.563	−87.612

2) LOO Cross-validation
Each DBP compound in the halogenated alkane class is singled out in turn as a tested or validated DBP, and the MLR regression is repeated for each of the remaining DBP compounds. The results of the LOO method are illustrated in Figure 2.18. The numbers 1, 2, 3, ..., 45, 46 represent the sequence numbers corresponding to the studied compounds as listed in Table 2.15. From the radar graph of Figure 2.18, the values of R^2 for the LOO validation test of the QSAR model are between 0.85 and 0.92, which shows the model's robustness. The DBP compound 15, 1-chlorodecane, is closer to the radar center because its R^2 value is relatively lower. Therefore, the removal of 1-chlorodecane would significantly decrease the R^2, so it is an important data for this QSAR model. The average of the calculated square of the cross-validation coefficient is 0.891, indicating both the robustness and high predictive ability of the model.

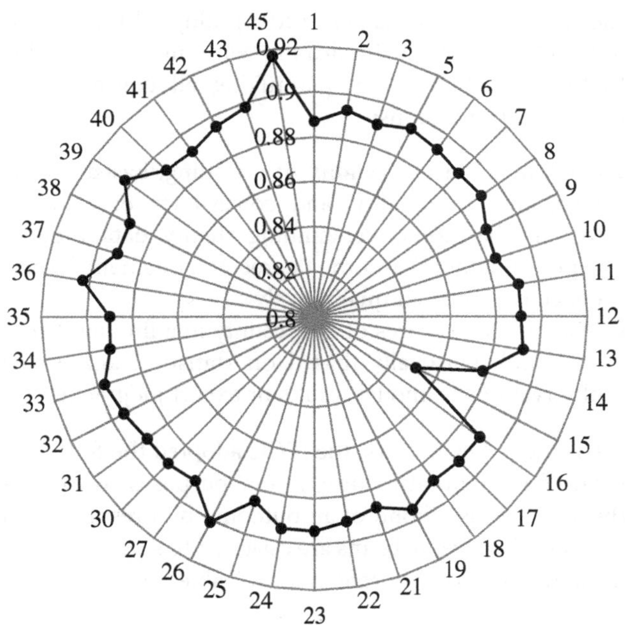

Figure 2.18 Radar graph of R^2 for halogenated alkane by LOO CV. (*Source:* Reproduced with permission of Springer.)

3) LMO Cross-validation
The dataset of 42 sampling halogenated alkane chemicals are randomly split by computer, resulting in 37 compounds in the active set and 5 compounds in the validation set. Similar to the QSAR model development, an equation of MLR for predicting $\log P$ is shown as follows:

$$\log P = 2.284 - 10.820E_{\text{LUMO}} + 0.133N_{\text{Cl}} + 0.530N_{\text{C}} \tag{2.21}$$

($n = 42$, $R^2 = 0.898$, MSE = 0.101, PRESS RMSE = 0.370)

Figure 2.19 presents the validation results by a plot of the observed $\log P$ values versus predicted $\log P$ values for the active and validation set. All five validation datasets fall within the boundary of 95% confidence interval, thus indicating that the model is very robust and has high predictive power of $\log P$ for halogenated alkanes.

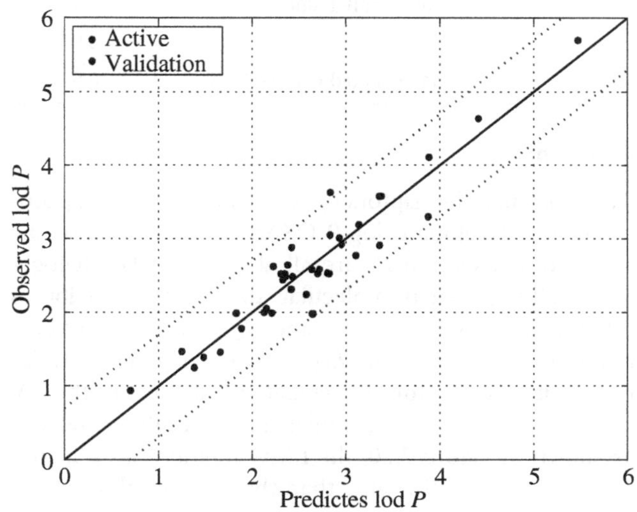

Figure 2.19 Predicated $\log P$ versus measured $\log P$ for halogenated alkane validation. (*Source:* Reproduced with permission of Springer.)

Answer: QSAR models of $\log P$ in relation to E_{LUMO}, N_{Cl}, and N_{C} for halogenated alkanes as one class of DBPs were developed and validated.

Comments: Statistic validation methods can provide information about how robust the developed QSAR models are. However, the ultimate validation method should be validation by external data that were not used during the developing the QSAR models.

Example 2.8
Given: The leave-one-out (LOO) and leave-many-out (LMO) cross-validation (CV) methods were used to assess the goodness of fit, robustness, and predictive ability of QSAR models for the DBP classes. Six classes of DBPS include halogenated alkane, halogenated alkene, halogenated aromatic, halogenated aldehyde, halogenated ketone, and halogenated carboxylic acid. The key statistical results from the two methods are compared in Table 2.18.

Find: The robustness of QSAR models for six DBP classes.

Table 2.18 Statistical parameters of LOO and LMO CV for QSAR models.

Group no.	DBP class	R^2 LOO	R^2 LMO	R^2 Diff (%)	MSE LOO	MSE LMO	MSE Diff (%)
1	Halogenated alkane	0.891	0.903	1.35	0.096	0.095	−1.04
2	Halogenated alkene	0.948	0.951	0.32	0.100	0.119	19.00
3	Halogenated aromatic	0.819	0.830	1.28	0.144	0.142	−1.39
4	Halogenated aldehyde	0.952	0.951	−0.11	0.091	0.099	8.79
5	Halogenated ketone	0.976	0.980	0.41	0.044	0.041	−5.91
6	Halogenated carboxylic acid	0.871	0.947	8.70	0.090	0.046	−48.67

Solution:

The molecular descriptors E_{LUMO}, N_{Cl}, and N_C are appropriate for developing $\log P$ QSAR models because the values of correlation R^2 are all above 80%. For halogenated alkanes, each molecular descriptor contributes differently to $\log P$. Both N_{Cl} and N_C have a positive correlation with $\log P$, while E_{LUMO} has a negative correlation with $\log P$ for halogenated alkane. Both CV methods have produced similar model results. However, as shown in Figure 2.20, the LMO method has a higher R^2 and a lower MSE value than the LOO method.

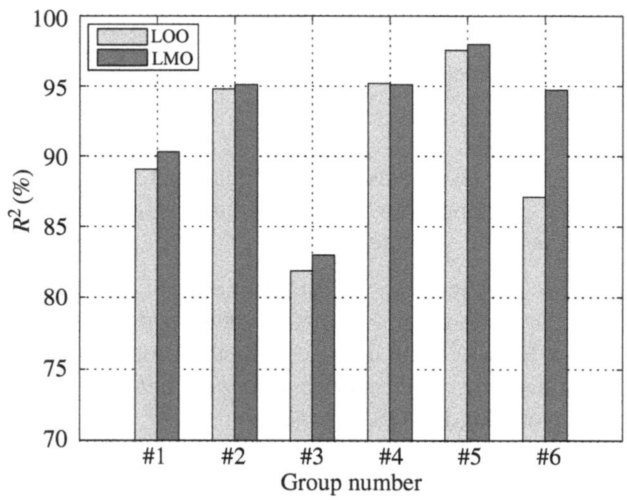

Figure 2.20 Comparison of R^2 obtained by LOO and LMO CV. (*Source:* Reproduced with permission of Springer.)

Comments: The LMO CV method shows higher percentage of R^2 than LOO CV method. For halogenated carboxylic acid, the MSE difference between LOO and LMO CV method is as high as −48.67%.

2.5 Quantification of Uncertainty

All the models have intrinsic uncertainty that has to be quantified to communicate to the public about risks. Constraints, uncertainties, and assumptions having an impact on the risk assessment should be explicitly considered at each step in the risk assessment. Therefore, the quantitative description of uncertainty in HRA, carrying capacity, and climate change are critical for policy makers and the general public. For example, the IPCC Fifth Assessment (2015) adopted the following technical terms in dealing with uncertainty due to the uncertainties in quantifying all consequences of different emissions:

- Confidence in the validity of a finding, based on the type, amount, quality, and consistency of evidence (e.g. mechanistic understanding, theory, data, models, expert judgment) and the degree of agreement. Confidence is expressed qualitatively.
- Quantified measures of uncertainty in a finding expressed probabilistically (based on statistical analysis of observations or model results or expert judgment).

For the policy makers, the IPCC adopted the following terms for quantitative description purposes to express the assessed likelihood of an outcome or a result. Seven different descriptive terms are assigned to seven different likelihood probability ranges as in Table 2.19.

Table 2.19 Technical terms used for describing the probability of an outcome (IPCC, 2013).

Term	Likelihood probability of an outcome (%)
Virtually certain	99–100
Very likely	90–100
Likely	66–100
About as likely as not	33–66
Unlikely	0–33
Very unlikely	0–10
Exceptionally unlikely	0–1

To quantify different uncertainty, the US EPA identifies scenario, parameter, and model uncertainty. In each category of uncertainty, the sources of uncertainty are identified with a specific example in Table 2.20.

Table 2.20 Type of uncertainty with sources and examples (US EPA).

Type of uncertainty	Sources	Examples
Scenario uncertainty	Descriptive errors	Incorrect or insufficient information
	Aggregation errors	Spatial or temporal approximations
	Judgment errors	Selection of an incorrect model
	Incomplete analysis	Overlooking an important pathway
Parameter uncertainty	Measurement errors	Imprecise or biased measurements
	Sampling errors	Small or unrepresentative samples
	Variability	In time, space, or activities
	Surrogate data	Structurally related chemicals
Model uncertainty	Relationship errors	Incorrect inference on the basis for correlations
	Modeling errors	Excluding relevant variables

Furthermore, the US EPA lists different quantification methods of uncertainty with examples in Table 2.21.

2.5.1 Quantification of QSAR Model's Uncertainty

For a QSAR model in which r is function of j variables $X_1, X_2, \ldots, X_j$:

$$r = r(X_1, X_2, \ldots, X_j) \tag{2.22}$$

The uncertainty due to errors in variable can be expressed as follows (Coleman and Steele, 1989):

$$U_r^2 = \left(\frac{\partial r}{\partial X_1}\right)^2 U_{X_1}^2 + \left(\frac{\partial r}{\partial X_2}\right)^2 U_{X_2}^2 + \cdots + \left(\frac{\partial r}{\partial X_j}\right)^2 U_{X_j}^2 \tag{2.23}$$

$$U_r = \sqrt{\left(\frac{\partial r}{\partial X_1}\right)^2 U_{X_1}^2 + \left(\frac{\partial r}{\partial X_2}\right)^2 U_{X_2}^2 + \cdots + \left(\frac{\partial r}{\partial X_j}\right)^2 U_{X_j}^2} \tag{2.24}$$

When the above uncertainty equation is applied to log P as function r, while E_{LUMO}, N_{Cl}, and N_{C} are the three variables, the uncertainty equation of $\log P$ can be expressed in Equations 2.25 and 2.26:

$$U_{\mathrm{Log\,P}}^2 = \left(\frac{\partial(\log P)}{\partial b_0}\right)^2 U_{b_0}^2 + \left(\frac{\partial(\log P)}{\partial E_{\mathrm{LUMO}}}\right)^2 U_{E_{\mathrm{LUMO}}}^2 + \left(\frac{\partial(\log P)}{\partial N_{\mathrm{Cl}}}\right)^2 U_{N_{\mathrm{Cl}}}^2 + \left(\frac{\partial(\log P)}{\partial N_{\mathrm{C}}}\right)^2 U_{N_{\mathrm{C}}}^2 \tag{2.25}$$

Table 2.21 Approaches to quantitative analysis of uncertainty.

Approach	Description	Examples
Sensitivity analysis	Changing one input variable at a time while leaving others constant, to examine effect on output	Fix each input at lower (then upper) boundary while holding others at nominal values (e.g. medians)
Analytical uncertainty propagation	Examining how uncertainty in individual parameters affects the overall uncertainty of the exposure assessment	Analytically or numerically obtain a partial derivative of the exposure equation with respect to each input parameter
Probabilistic uncertainty analysis	Varying each of the input variables over various values of their respective probability distributions	Assign probability density function to each parameter; randomly sample values from each distribution and insert them in the exposure equation (Monte Carlo)
Classical statistical methods	Estimating the population exposure distribution directly, based on measured values from a representative sample	Compute confidence interval estimates for various percentiles of the exposure distribution

$$U_{\log P} = \sqrt{\left(\frac{\partial(\log P)}{\partial b_0}\right)^2 U_{b_0}^2 + \left(\frac{\partial(\log P)}{\partial E_{\text{LUMO}}}\right)^2 U_{E_{\text{LUMO}}}^2 + \left(\frac{\partial(\log P)}{\partial N_{\text{Cl}}}\right)^2 U_{N_{\text{Cl}}}^2 + \left(\frac{\partial(\log P)}{\partial N_{\text{C}}}\right)^2 U_{N_{\text{C}}}^2} \quad (2.26)$$

When the data reduction expression is very complex and the task of computing the partial derivatives in the above equations is extremely laborious, Monte Carlo simulation (MCS) that is the most efficient quantification of uncertainty according to distribution of independent variables can be used (Cox et al., 2001). MCS is used as an example to quantify the uncertainty of the predicted log P using E_{LUMO}, N_{Cl}, and N_{C} as three independent variables.

2.5.2 Monte Carlo Simulation

Monte Carlo Simulation (MCS) has been extensively used for quantifying uncertainty of linear equations (Tang et al., 2009). Estimating propagation of error distributions by MCS is based on theoretical principles and supports a fully consistent and transferable estimation of measurement uncertainty. In general case, a linear model equation is to measure output Z indirectly obtained from the input variables $X_1, X_2, \ldots, X_n$ by a functional relationship F:

$$Z = F(X_1, X_2, \ldots, X_n) \quad (2.27)$$

The knowledge about the values that may be reasonably attributed to quantities X_i, considered as continuous random variables, is expressed by their probability distribution function (PDF), $P_{x_i}(x_i)$, within the corresponding domain. An expectation or best estimate for the value of $X_i\, E(X_i)$ and the uncertainty associated with this value, $\mu(X_i)$, assimilated to the standard deviation $\sigma(X_i)$, are obtained from the PDF:

$$E(X_i) = \mu_i = \int_{-\infty}^{\infty} x_i P_{x_i}(x_i)\,\mathrm{d}x_i \quad (2.28)$$

$$\mu^2(X_i) = \sigma^2(X_i) = E\left[(X_i - \mu_i)^2\right] = \int_{-\infty}^{\infty} (x_i - \mu_i)^2 P_{x_i}(x_i)\,\mathrm{d}x_i \quad (2.29)$$

If the equation is linearized by means of a Taylor expansion about the point, the estimated uncertainty of the measured output $\mu(Z)$ is from the input $\mu = (\mu_1, \mu_2, \ldots, \mu_n)$ while the second- and higher-order terms are neglected, the estimated uncertainty of $\mu(Z)$ from the input variables is calculated as follows:

$$Z \approx F(\mu) + \sum_{i=1}^{n} \left(\frac{\partial F}{\partial x_i}\right)(x_i - \mu_i). \quad (2.30)$$

Taking $\mu(Z)$ as the values $F(\mu) = F(\mu_1, \mu_2, \ldots, \mu_n)$ leads to the well-known law of propagation of uncertainty:

$$\mu^2(Z) = \sum_{i=1}^{n} \left(\frac{\partial F}{\partial x_i}\right)_u^2 E\left[(x_i - \mu_i)^2\right] + \sum_{i=1}^{n-1} \sum_{j=i+1}^{n} \left(\frac{\partial F}{\partial x_i}\right)_u \times \left(\frac{\partial F}{\partial x_j}\right)_u E\left[(x_i - \mu_i)(x_j - \mu_j)\right]$$

$$\mu^2(Z) = \sum_{i=1}^{n} \left(\frac{\partial F}{\partial x_i}\right)_u^2 \mu^2(X_i) + \sum_{i=1}^{n-1} \sum_{j=i+1}^{n} \left(\frac{\partial F}{\partial x_i}\right)_u \times \left(\frac{\partial F}{\partial x_j}\right)_u \times \operatorname{cov}(X_i, X_j) \quad (2.31)$$

In the Guide to the Expression of Uncertainty in Measurement (GUM, 1993), which is the internationally accepted master document for the evaluation of uncertainty, the combined standard uncertainty $\mu(Z)$ is evaluated from the standard uncertainty of the input variables, $\mu(X_i)$, and the covariances between correlated ones, $\operatorname{cov}(X_i, X_j)$, if all the input variables are independent, $(X_i, X_j) = 0$. The essence of the MCS is to simulate sampling to a target population with a given expectation μ_z and variance $\sigma^2(Z)$. A random sample of size M is obtained from the simulation of M independent and identically distributed random variables $Z_1, Z_2, \ldots, Z_M$. According to the central limit theorem (Martinez, 2002), the distribution of the sum is approximately normal, with expectation $M_{\mu z}$ and variance $M_{\sigma 2}$:

$$\xi_N = Z_1 + Z_2 + \cdots + Z_M \quad (2.32)$$

According to the rule of "two sigmas," the probability

$$P\left(M\mu_Z + 2\sigma\sqrt{M} < \xi_N < M\mu_Z + 2\sigma\sqrt{M}\right) \approx 0.954 \quad (2.33)$$

By dividing Equation (2.33) by M on both sides, the expression becomes

$$P\left(\mu_Z + \frac{2\sigma}{\sqrt{M}} < \frac{\xi_N}{M} < \mu_Z + \frac{2\sigma}{\sqrt{M}}\right) \approx 0.954/M \quad (2.34)$$

By substituting $\bar{Z} = \frac{\xi_N}{M} = \frac{\sum_{j=i}^{M} Z_j}{M}$ into Equation (2.34), the expression becomes

$$P\left(|\bar{Z} - \mu_Z| < \frac{2\sigma}{\sqrt{M}}\right) \approx 0.954/M \quad (2.35)$$

This relationship is the foundation of the MCS because it establishes the rule to evaluate the error of set $\mu_Z = \bar{Z}$ as $2\sigma/\sqrt{M}$. The sample variance may be best estimated using σ^2:

$$\sigma^2(Z) = \frac{\sum_{j=1}^{M} (Z_j - \bar{Z})^2}{M - 1} \quad (2.36)$$

After defining the coverage probability, P, the confidence interval for the result is evaluated as $\left[Z_{(1+P)_{M/2}} - Z_{(1-P)_{M/2}}\right]$, where extremes correspond to the 2.5 and 97.5% percentiles of the sorted Z values. When the skewness value of the Z forecast discrete distribution is near zero, the confidence interval becomes symmetric and expanded uncertainty $U(Z)$ is estimated by Equation (2.38):

$$U(Z) = \frac{\left[Z_{(1+P)_{M/2}} - Z_{(1-P)_{M/2}}\right]}{2} \tag{2.37}$$

MCS is a powerful tool in quantifying uncertainty in addition to experimental, analytical, or other numerical methods. MCS protocol can replace point estimates with random variables drawn from probability density functions. To perform MCS, a computer program is used to generate the pseudorandom numbers to simulate the values of the variables within a given PDF. Several commercial software programs such as Crystal Ball, LabVIEW, @RISK, and Analytica are the most popular for this purpose. There are three major steps if Crystal Ball by Oracle (2016) is used to carry out the MCS. First, a probability distribution is incorporated into a spreadsheet cell, and each time the spreadsheet is recalculated, a new value of the random variable is selected from the distribution and used for calculations. Second, the entire simulation is run at least 10 000 times to satisfy the required high number of trials. Each time new values of the random variables are selected, a new estimate of the final target is generated. Third, the results of simulations are summarized in a user-friendly interface such as a table and a figure.

A step-by-step procedure for uncertainty assessment of regression QSAR model for halogenated DBPs is illustrated in Figure 2.21. The first step is the compilation of the DBP data. Three variables such as E_{LUMO}, the number of chorine (N_{Cl}), and the number of carbon (N_C) were used. The characterization of the probability distributions is carried out by statistic software SAS 9.4 to obtain the data average and standard deviation. All these parameters are fed into the MCS that gives the results in a probability distribution around a mean value that is used to carry out a detailed sensitivity analysis. A sensitivity analysis is applied to identify which parameters had the most impact on the predicted log P. If small modifications of one parameter characterized by a probability distribution strongly influenced the final result, it may be concluded that the sensitivity of the variable is very high. Sensitivity is crucial in determining what variables are the most important in predicting a

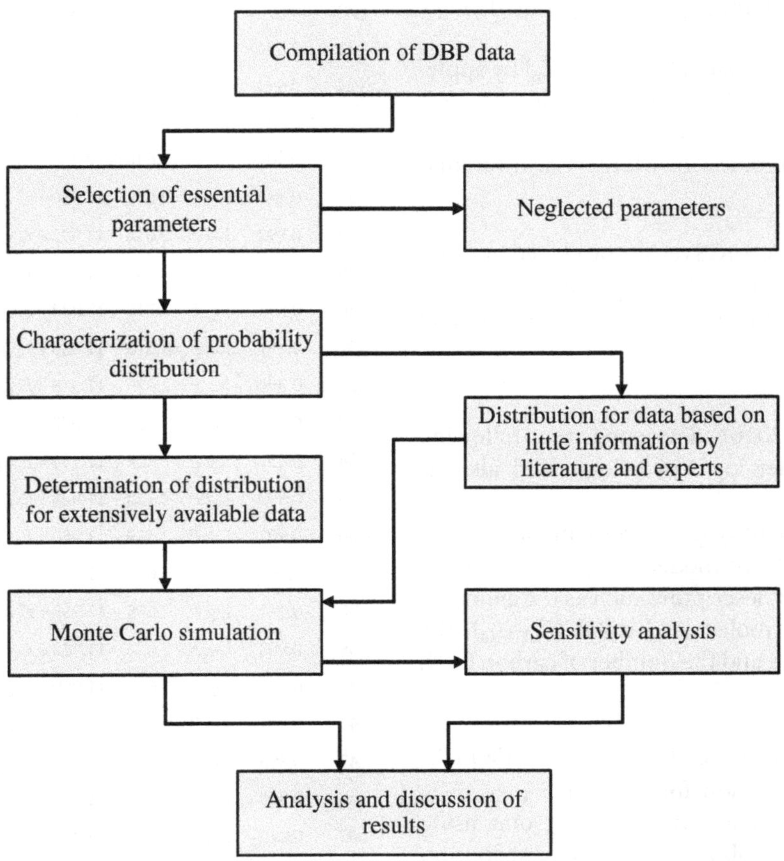

Figure 2.21 Procedure for Monte Carlo simulation in uncertainty assessment. (*Source:* Adapted from the US EPA (1999).)

dependent outcome. This can be analyzed by displaying the sensitivity as a percentage of the contribution from each parameter to the variance of the final result. Crystal Ball presents contribution in terms of percentage for each independent variable with the sum of percentage contribution to 100%. A general procedure in the uncertainty quantification of the predicted log P of a class DBP through MCS is outlined as follows:

1) Selection of significant sources of uncertainty. Molecular descriptors E_{LUMO}, N_{Cl}, N_C, the intercept b_0, and the coefficiencies b_1, b_2, and b_3 are the significant sources contributing to the model's uncertainty.
2) Identification of the probability density function corresponding to the uncertainty sources selected. All uncertainty sources, e.g., E_{LUMO}, N_{Cl}, N_C, b_0, b_1, b_2, and b_3 are analyzed by Crystal Ball to obtain its best fit probability distribution functions. The results of the probability functions for each variable are then used in the best fit MLR equation to predict log P.
3) Selection of the number M of Monte Carlo Simulation trials. For example, MCS was carried out 10 000 times in order to have a sufficiently high number of trials.
4) Simulation of M samples $\{X_{i1}, X_{i2}, \ldots, X_{in}\}$ for each X_i (E_{LUMO}, N_{Cl}, N_C, b_0, b_1, b_2, and b_3) uncertainty source, which was considered a random variable with a probability density function $p(E_{LUMO})$, $p(N_{Cl})$, $p(N_C)$, $p(b_0)$, $p(b_1)$, $p(b_2)$, $p(b_3)$.
5) Computation of the M results $\{Z_1, Z_2, \ldots, Z_n\}$ by applying Equation (2.27) or (2.37) to M samples for each variable X_i (E_{LUMO}, N_{Cl}, N_C, b_0, b_1, b_2, and b_3).
6) Analysis and interpretation of Monte Carlo Simulation results.

Example 2.9 presents the MCS results produced by Crystal Ball software. The methodology can be used for the quantification of uncertainties of any regression curve in general.

Example 2.9

Given: Chen et al. (2016) developed the following QSAR models to predict log P of halogenated alkanes (Table 2.22).

Find: Using MCS to quantify the uncertainty and sensitivity of the following QSAR models for predicting log P of DBPs using molecular descriptors such as the energy of the lowest unoccupied molecular orbital (E_{LUMO}), the number of chlorine (N_{Cl}), and the number of carbon (N_C).

Solution:

Table 2.23 lists MLR equations developed by LOO CV. A probability density function for each of E_{LUMO}, N_{Cl}, N_C, b_0, b_1, b_2, and b_3 is analyzed one by one using Crystal Ball software. The above probability distribution

Table 2.22 List of MLR equations for halogenated alkanes.

No.	R^2	MLR equation
1	0.887	$\log P = 2.679 - 12.068 * E_{LUMO} + 0.092 * N_{Cl} + 0.531 * N_C$
2	0.893	$\log P = 2.641 - 12.143 * E_{LUMO} + 0.104 * N_{Cl} + 0.536 * N_C$
3	0.890	$\log P = 2.634 - 11.995 * E_{LUMO} + 0.103 * N_{Cl} + 0.531 * N_C$
5	0.894	$\log P = 2.696 - 12.207 * E_{LUMO} + 0.089 * N_{Cl} + 0.535 * N_C$
6	0.892	$\log P = 2.635 - 12.084 * E_{LUMO} + 0.103 * N_{Cl} + 0.535 * N_C$
7	0.890	$\log P = 2.626 - 11.951 * E_{LUMO} + 0.101 * N_{Cl} + 0.533 * N_C$
8	0.891	$\log P = 2.632 - 11.967 * E_{LUMO} + 0.100 * N_{Cl} + 0.534 * N_C$
9	0.885	$\log P = 2.630 - 11.918 * E_{LUMO} + 0.103 * N_{Cl} + 0.528 * N_C$
10	0.884	$\log P = 2.664 - 12.535 * E_{LUMO} + 0.106 * N_{Cl} + 0.547 * N_C$
11	0.891	$\log P = 2.637 - 12.137 * E_{LUMO} + 0.104 * N_{Cl} + 0.537 * N_C$
12	0.891	$\log P = 2.636 - 12.130 * E_{LUMO} + 0.104 * N_{Cl} + 0.537 * N_C$
13	0.893	$\log P = 2.638 - 12.071 * E_{LUMO} + 0.103 * N_{Cl} + 0.534 * N_C$
14	0.878	$\log P = 2.625 - 11.833 * E_{LUMO} + 0.103 * N_{Cl} + 0.526 * N_C$
15	0.850	$\log P = 2.614 - 11.606 * E_{LUMO} + 0.102 * N_{Cl} + 0.518 * N_C$
16	0.890	$\log P = 2.631 - 11.979 * E_{LUMO} + 0.103 * N_{Cl} + 0.531 * N_C$
17	0.890	$\log P = 2.632 - 11.986 * E_{LUMO} + 0.103 * N_{Cl} + 0.531 * N_C$
18	0.889	$\log P = 2.652 - 12.345 * E_{LUMO} + 0.105 * N_{Cl} + 0.542 * N_C$
19	0.895	$\log P = 2.363 - 10.840 * E_{LUMO} + 0.136 * N_{Cl} + 0.532 * N_C$
21	0.888	$\log P = 2.612 - 11.780 * E_{LUMO} + 0.102 * N_{Cl} + 0.530 * N_C$
22	0.891	$\log P = 2.616 - 11.836 * E_{LUMO} + 0.103 * N_{Cl} + 0.533 * N_C$
23	0.894	$\log P = 2.587 - 11.713 * E_{LUMO} + 0.106 * N_{Cl} + 0.532 * N_C$
24	0.894	$\log P = 2.592 - 11.809 * E_{LUMO} + 0.107 * N_{Cl} + 0.536 * N_C$
25	0.885	$\log P = 2.632 - 11.887 * E_{LUMO} + 0.100 * N_{Cl} + 0.531 * N_C$
26	0.901	$\log P = 2.640 - 12.288 * E_{LUMO} + 0.106 * N_{Cl} + 0.548 * N_C$
27	0.889	$\log P = 2.632 - 11.930 * E_{LUMO} + 0.100 * N_{Cl} + 0.532 * N_C$
28	0.892	$\log P = 2.618 - 11.895 * E_{LUMO} + 0.103 * N_{Cl} + 0.534 * N_C$
29	0.891	$\log P = 2.634 - 11.963 * E_{LUMO} + 0.101 * N_{Cl} + 0.533 * N_C$
30	0.891	$\log P = 2.718 - 12.333 * E_{LUMO} + 0.092 * N_{Cl} + 0.533 * N_C$
31	0.891	$\log P = 2.613 - 11.885 * E_{LUMO} + 0.102 * N_{Cl} + 0.533 * N_C$
32	0.894	$\log P = 2.626 - 11.938 * E_{LUMO} + 0.103 * N_{Cl} + 0.535 * N_C$
33	0.897	$\log P = 2.633 - 12.077 * E_{LUMO} + 0.104 * N_{Cl} + 0.540 * N_C$
34	0.891	$\log P = 2.628 - 11.954 * E_{LUMO} + 0.100 * N_{Cl} + 0.534 * N_C$
35	0.890	$\log P = 2.623 - 11.928 * E_{LUMO} + 0.103 * N_{Cl} + 0.533 * N_C$
36	0.903	$\log P = 2.569 - 11.768 * E_{LUMO} + 0.123 * N_{Cl} + 0.534 * N_C$
37	0.891	$\log P = 2.613 - 11.877 * E_{LUMO} + 0.101 * N_{Cl} + 0.534 * N_C$
38	0.891	$\log P = 2.628 - 11.958 * E_{LUMO} + 0.102 * N_{Cl} + 0.534 * N_C$
39	0.903	$\log P = 2.572 - 11.788 * E_{LUMO} + 0.123 * N_{Cl} + 0.534 * N_C$
40	0.892	$\log P = 2.624 - 11.960 * E_{LUMO} + 0.104 * N_{Cl} + 0.534 * N_C$
41	0.891	$\log P = 2.627 - 11.926 * E_{LUMO} + 0.099 * N_{Cl} + 0.533 * N_C$
42	0.895	$\log P = 2.703 - 12.223 * E_{LUMO} + 0.078 * N_{Cl} + 0.535 * N_C$
43	0.898	$\log P = 2.599 - 11.640 * E_{LUMO} + 0.083 * N_{Cl} + 0.533 * N_C$
45	0.917	$\log P = 2.888 - 12.354 * E_{LUMO} + 0.080 * N_{Cl} + 0.527 * N_C$

Table 2.23 List of variables in QSAR model of halogenated alkanes.

Variable	Distribution type	Defining characteristics
b_0	Student's T	Middle point 2.635, scale 0.020, degree of freedom 1.001
b_1	Logistic	Mean −11.970, scale 0.131
b_2	Logistic	Mean 0.102, scale 0.007
b_3	Logistic	Mean 0.533, scale 0.002
E_{LUMO}	Beta	Minimum 0.05, maximum 0.22, alpha 1.682, beta 0.6348
N_{Cl}	Discrete uniform	Minimum 1, maximum 5
N_C	Discrete uniform	Minimum 1, maximum 10

for each variable is defined to forecast log *P* in the Crystal Ball. A trial value of $M = 10\,000$ is selected, and the results from Crystal Ball are shown in Figures 2.22 and 2.23. Figure 2.22 shows that log *P* has a beta bell curve with a mean value of 2.61 and a standard deviation of 1.24. The forecasted distribution of the predicted log *P* has the skewness of 0.0422. The model's uncertainty could also be calculated by Equation (2.37). The best probability distribution model for predicting or forecasting log *P* ranked by Anderson–Darling method is beta, while the normal distribution is ranked as the second best fit.

The sensitivity chart (Figure 2.23) shows that N_C contribute to positive 91.7%, N_{Cl} has a negative 8.2% influence on the log *P* predicted, and E_{LUMO} has almost no influence, which suggests that log *P* and E_{LUMO} are two independent variables.

The above matrix is a correlation map that allows a viewer to visually identify patterns in correlations. The closer the points conform to the regression line, the stronger the correlations exist among the variables. Figure 2.24 shows that log *P* has the highest correlation coefficient of 0.9006 with N_C, while it has the lowest correlation coefficient of 0.1003 with N_{Cl}. The very low correlation coefficients among E_{LUMO}, N_{Cl}, and N_C suggest that the three molecular descriptors are independent.

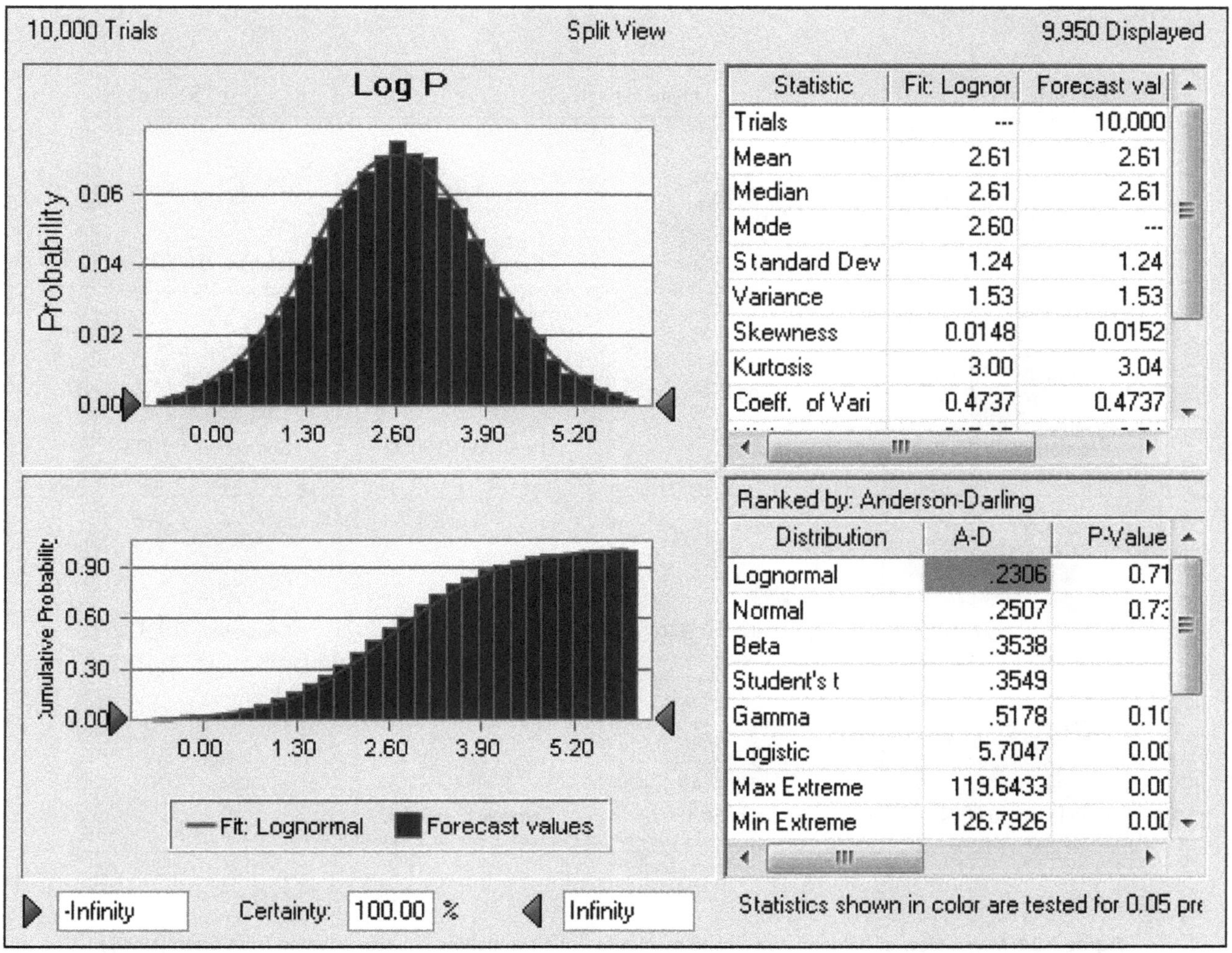

Figure 2.22 Log *P* output of halogenated alkane by Monte Carlo simulation. (*Source:* Adapted from Chen (1999).)

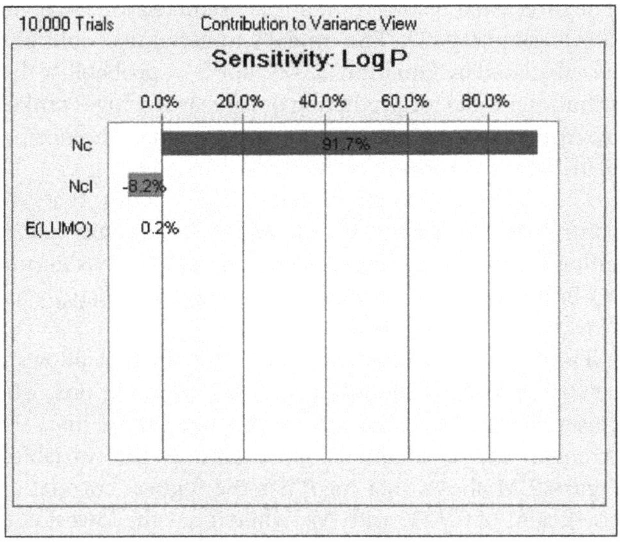

Figure 2.23 Contribution to variance view of halogenated alkane by Monte Carlo simulation. (*Source:* Adapted from Chen (1999).)

2.5.3 Comparison of Uncertainties of Different QSAR Models

The uncertainty of log *P* could also be estimated by point estimate method (PEM). In most cases, uncertainty quantified by MCS should be more accurate than that calculated by PEM because MCS counts the distributions and propagation of the uncertainty. The uncertainty results obtained from the PEM method and the MCS for QSAR models are compared in Example 2.10.

Example 2.10

Given: Chen (1999) quantified uncertainties in terms of standard deviation of DBP QSAR models obtained though point estimate method (PEM) and MCS method, respectively. The results are listed in Table 2.24.

Find:

1) Which uncertainty estimated by point estimate (PE) versus MCS is more reliable?
2) What are the maximal and minimal errors when the uncertainties are compared?

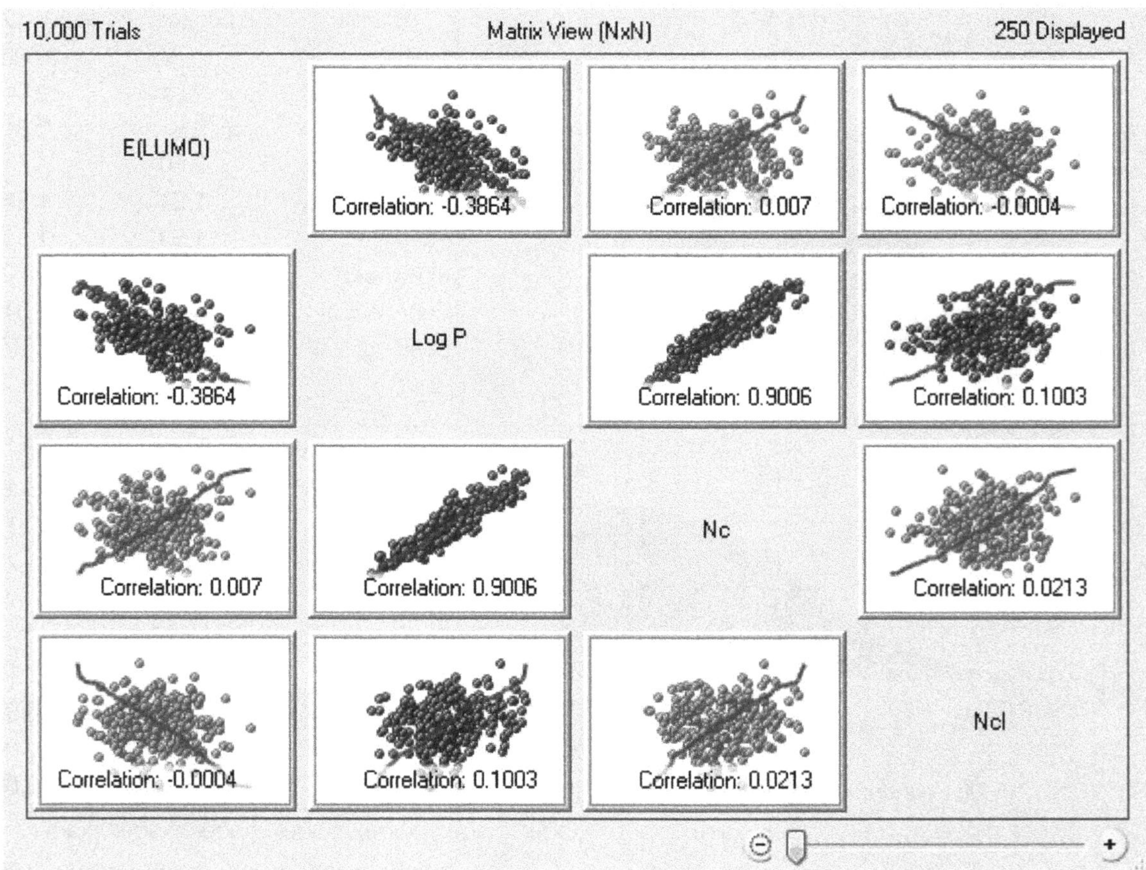

Figure 2.24 Matrix view of halogenated alkane by Monte Carlo simulation. (*Source:* Adapted from Chen (1999).)

Table 2.24 Uncertainties obtained by point estimate and MCS.

No.	DBP class	Point estimate method	Monte Carlo method
1	Halogenated alkane	1.81	2.48
2	Halogenated alkene	2.46	1.82
3	Halogenated aromatic	1.74	2.06
4	Halogenated aldehyde	2.15	2.98
5	Halogenated ketone	2.32	2.68
6	Halogenated carboxylic acid	2.14	2.68

Solution:

1) The uncertainty results from two different methods are different as shown in Figure 2.25. The uncertainties computed by the MCS for all the DBP classes, except halogenated alkene, are relatively higher than those computed by the PEM. Uncertainty results from MCS are more reliable than the results from PEM because MCS simulates the full spectra of the variables, while PEM only takes consideration of the first order of the model errors.
2)
The maximal error is the uncertainty of the QSAR model for $\log P$ of halogenated aldehyde. The PEM underpredicts the $\log P$ by −27.85%.
The minimal error is the uncertainty of the QSAR model for $\log P$ of halogenated ketone. The PEM underpredicts the $\log P$ by −13.43%.

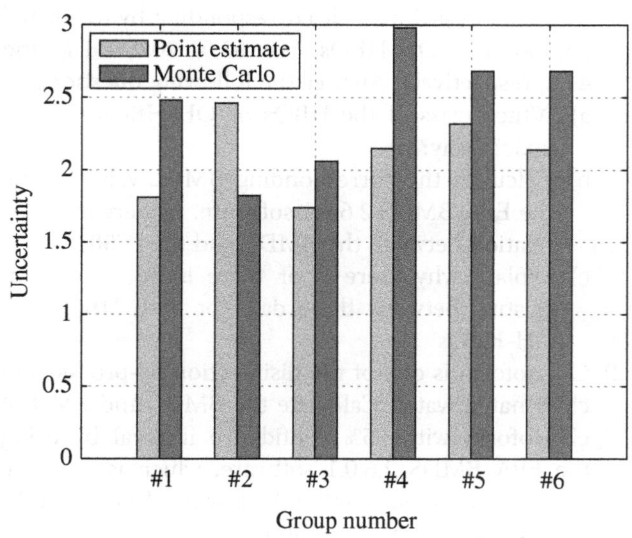

Figure 2.25 Uncertainty obtained by point estimate and Monte Carlo methods. (*Source:* Adapted from Chen (1999).)

Comments: In general, MCS would be more accurate than PEM if the probability density function models assumed are correct.

2.5.4 Sensitivity Analysis by Monte Carlo Simulation

The influence of different variables on the outcome of model's prediction can be visually presented through sensitivity analysis in Crystal Ball. Table 2.25 summarizes the sensitivity results for each DBP class. Figure 2.26 demonstrates that N_{Cl} is the most influential molecular descriptor in predicting $\log P$ of all DBP classes, except halogenated alkane.

Table 2.25 shows the major difference between sensitivities of single bond compounds, such as halogenated alkanes. $\log P$ of chlorinated alkane changes significantly with the length of the carbon chain, N_C. However, for all

Table 2.25 Sensitivity of various descriptors to $\log P$ by DPB classes.

No.	Compound class	E_{LUMO} (%)	N_{Cl} (%)	N_C (%)
1	Halogenated alkane	−16.0	0.8	83.2
2	Halogenated alkene	−44.0	53.9	2.1
3	Halogenated aromatic	0.6	90.6	8.7
4	Halogenated aldehyde	−7.3	64.9	27.8
5	Halogenated ketone	0.0	93.0	7.0
6	Halogenated carboxylic acid	36.5	55.6	7.9

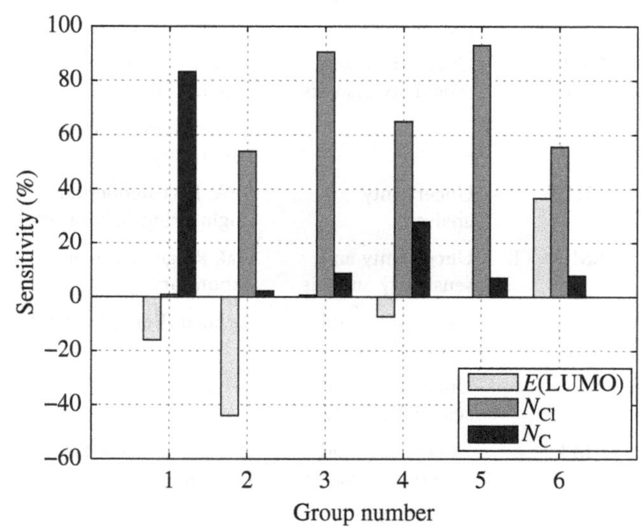

Figure 2.26 Sensitivity of various descriptors to $\log P$ by DPB classes. (*Source:* Adapted from Chen (1999).)

the other classes of DBPs containing unsaturated π bond, log P will change mostly with the number of chlorine, N_{Cl}. Log P reflects the amount of hydrogen bonding between the chemical compound and the hydrogen in the water molecule in the absence of oxygen in the chlorinated alkanes. It appears that when chlorine attaches to carbon, it only has influence on the electron cloud of the attached carbon and such electronic interaction is tightly bonded; therefore, the number of chlorine on the chlorinated alkane carbon chain does not significantly influence the strength of hydrogen bonds. However, with DBP compounds containing unsaturated π bonds, the chlorine atom would attract electron cloud to itself since most compounds containing unsaturated bonds may have a resonance structure through which the influence of chlorine will be transmitted. As a result, the strength of the hydrogen bond between the chemical compound and water molecule would be significantly influenced by the number of chlorine, N_{Cl}. Therefore, log P is more sensitive to the number of chlorine.

2.5.5 Computer Software for Quantitative Risk Assessment

There are many computer software available to conduct risk assessment. Table 2.26 lists other computer software for quantitative risk assessment.

Table 2.26 Software for quantitative risk assessment.

Software	Type of analysis	Creator
@RISK	Uncertainty and risk analysis	Palisade Corporation
Analytica	Uncertainty and risk analysis	Lumina Decision Systems, Inc.
GENII/ SUNS	Uncertainty and sensitivity analysis	Sandia National Laboratories
		Pacific Northwest Laboratory
MOUSE	Uncertainty analysis	EPA, Risk Reduction Engineering Laboratory
ORMONTE	Uncertainty and sensitivity analysis	Oak Ridge National Laboratory
Risk Calc	Uncertainty and risk analysis	Applied Biomathematics
SimLab	Uncertainty and sensitivity analysis	Simlab
TAM3	Uncertainty and sensitivity analysis	Oak Ridge National Laboratory
Uncertainty Analysis	Uncertainty analysis	Integrated Sciences Group
Crystal Ball®	Uncertainty and sensitivity analysis	Oracle

2.6 Exercise

2.6.1 Questions

1) What is the meaning of typical MD such as log P, E_{LUMO}, and E_{HOMO}?
2) What is quantitative structure–activity relationship (QSAR)?
3) How can QSAR be used in HRA?
4) What are the typical molecular descriptors?
5) What EPA QSAR tools can be used for HRA?
6) What are the four steps in risk assessment?
7) What are the four different ways to identify the hazards of a chemical compound?
8) How do you calculate RfD?
9) How do you calculate BMD and BMDL?
10) What are the differences between assessment procedures of carcinogenic and noncarcinogenic chemical compounds?
11) How is the carcinogenic risk of a chemical compound calculated?

2.6.2 Calculation

1) Wang et al. (2014) reported that halobenzoquinones (HBQs) as a class of DBPs are of likely to be carcinogenic and found that the IC_{50} of HBQs are always less than their counter-hydroxylated products such as halo-hydroxyl-benzoquinones (OH-HBQs). For toxicity to the Chinese hamster ovary CCL-61 (ATCC, Manassas, VA) cell (CHO-K1), the IC_{50} in μM for 24-hour incubation of DCBQ, DCMBQ, TriCBQ, and DBBQ are 27.3, 11.4, 45.5, and 19.8, respectively. The IC_{50} in μM of their corresponding hydroxylated compounds, OH-HBQs, are 61.0, 20.4, 64.4, and 42.8, respectively. Answer the following questions:
   a) Which class of the HBQs or OH-HBQs is more toxic? Why?
   b) Calculate the corresponding BMDL values using the EPA BMDS 2.6.0.1 software. Is there any correlation between the BMDL and the IC50?
   c) Explain why there is or there is no correlation existing between these data for both HBQs and OH-HBQs.
2) Chloroform is one of the disinfection by-products of chlorinated water. Calculate the BMDL and BMD of chloroform with 95% confidence interval by using the EPA BMDS 2.6.0.1 software, which is available at https://www.epa.gov/bmds/download-benchmark-dose-software-bmds#installing
3) The current FDA guideline for methyl mercury in fish is 1 ppm (or 1 mg of methyl mercury/kilogram of fish). Answer the following:

a) Calculate the dose of mercury (mg/kg body weight of the person) if a person were to eat a tuna fish sandwich (with 3 oz of tuna fish meat) with methyl mercury levels of 0.4 ppm.
b) How does the amount of methyl mercury in a 4-oz sandwich compare with the maximum amount that a pregnant woman should eat on a daily basis, according to the EPA official guidelines? List your citation.
c) How many 4-oz tuna fish sandwiches would a person need to eat in order to have a 50% probability of dying from acute methyl mercury poisoning? Assume that humans and rats (or mice) have equal sensitivity to methyl mercury, that oral exposure in rats (or mice) is comparable with oral exposure in humans, and that all mercury in the fish is in the form of methyl mercury.

4) Study EPA example of human health risk assessment of isodecyl acrylate. Answer the following:
   a) What uncertainty could be introduced in the HRA?
   b) How would you outline an MCS procedure to quantify the uncertainty?

2.6.3 Assignment

1) Study the Crystal Ball user manual.
2) Study the SPSS user manual.
3) Study EPA guidelines on Monte Carlo simulation.
4) Go to the website http://apps.who.int/gho/data/node.main.1?lang=en and download the following Excel database on World Health Statistics:
   - Mortality and global health estimates
   - Cause-specific mortality and morbidity
   - Selected infectious diseases
   - Health service coverage
   - Risk factors
   - Health systems
   - Health equity monitor
   - Demographic and socioeconomic statistics

Under the heading of "Mortality and global health estimates," you will be able to download the following database:

Life expectancy for different countries. Use SPSS to conduct the following:

1) Statistical analysis of mean and standard deviation of life expectancy for developed and developing countries. What does this mean and standard deviation suggest for the two groups of countries when their mean life expectancy is compared?
2) Income and environmental indicators are the two major indicators. Conduct correlation analysis between life expectancy and income and air quality indicator, or both. Answer the following:
   a) Is there any correlation between life expectancy and income and air quality indicator, or both?
   b) Using Crystal Ball to analyze sensitivity and state, which one is more important predictor for life expectancy?

2.6.4 Projects

2.6.4.1 Xiongan Project

Collect water quality and air quality data of Xiongan and conduct health risk assessment on the major primary pollutants of the following. Toxicity data could be found in the US EPA *Exposure Factors Handbook* 2011 Edition:

1) Air pollutants: O_3, SO_2, NO_2, $P.M._{2.5}$, and $P.M._{10}$
2) Water pollutants: trihalomethanes (TTHM) and five haloacetic acids (HAA5)
3) Soil pollutants: (a) toxic metals (Hg, Cr, and As), (b) pesticides (parathion, dichlorvos, and aldrin), and (c) herbicides (glyphosate, atrazine, and flazasulfuron)

Answer the following questions:

1) What are the $\log P$ and E_{LUMO} and E_{HOMO} for water pollutants?
2) How do these molecular descriptors contribute to their toxicity?
3) What are the health risks of these pollutants?
4) Estimate lung cancer and liver cancer number of Xiongan due to air and water pollutants in the next 5, 10, 20, and 50 years.

2.6.4.2 Community Project

1) Do the same as above as your hometown and estimate lung cancer and liver cancer number of your city due to air, water, and soil pollutants in the next 5, 10, 20, and 50 years.
2) Rank the health risk of air, water, and soil pollutants and identify major contributor of major pollutants.
3) Make recommendations on priority of major EEIS to be built to the local environmental protection agencies to reduce these risks less than 10^{-6} in the next 5, 10, 20, and 50 years.

References

Akaike, H. (1973). Information theory and an extension of the maximum likelihood principle. *Proceedings of the Second International Symposium on Information Theory* (ed. B.N. Petrov and F. Csaki), 267–281. Budapest: Akademiai Kiado.

Chen, R.Z. (1999). Development, Validation, and Uncertainty Analysis of Quantitative Structure and Activity Relationship Models for log P of Disinfection By-products. PhD.dissertation, Florida International University, Miami, FL.

Chen, R.Z., Tang, W.Z., and Sillanpää, M. (2016). Prediction of log P of halogenated alkanes by their E_{LUMO} and number of chlorine and carbon. *Environmental Processes* 1: 73–91.

Chinese Ministry of Environmental Protection (MEP). (2016). Ambient Air Quality Standards. http://english.mep.gov.cn/Resources/standards/Air_Environment/quality_standard1/201605/t20160511_337502.shtml (accessed 29 November 2016).

Coleman, H.W. and Steele, Jr., W.G. (1989). Planning an experiment: General uncertainty analysis. *Experimentation and Uncertainty Analysis for Engineers*. New York: Wiley Blackwell.

Cothern, C.R., Coniglio, W.A., and Marcus, W.L. (1986). Estimating risk to human health. *Environmental Science and Technology* 20 (2): 111–116.

Cox, M.G., Dainton, M.P., and Harris, P.M. (2001). Software support for metrology best practice guide. Uncertainty and statistical modeling. *Technical Report*. Teddington: National Physical Laboratory.

Crump, K. (1984). A new method for determining allowable daily intakes. *Fundamental and Applied Toxicology* 4: 854–871.

Efron, B. (1983). Estimating the error rate of a prediction rule: improvement on cross – validation. *Journal of American Statistical Association* 78 (382): 316–331

EPA (U.S. Environmental Protection Agency) (1995). Guidelines for Neurotoxicity Risk Assessment. *EPA/630/R-95/001F*. Risk Assessment Forum, U.S. (EPS), Washington, DC [online]. www.epa.gov/ncea/raf/pdfs/neurotox.pdf (accessed 30 March 2005).

Eriksson, L., Jaworska, J., Worth, A.P., et al. (2003). Methods for reliability and uncertainty assessment and for applicability evaluations of classification and regression-based QSARs, *Environmental Health Perspectives* 111: 1361–1375.

GUM (1993). *Guide to the Expression of Uncertainty in Measurements*. Geneva: ISO.

Hansch, C., Leo, A., and Hoekman, D. (1995). *Exploring QSAR—Fundamentals and Applications in Chemistry and Biology*. Washington, DC: American Chemical Society.

IPCC (2013). Summary for Policymakers. In: *Climate Change 2013: The Physical Science Basis. Contribution of Working Group I to the Fifth Assessment Report of the Intergovernmental Panel on Climate Change* (ed. T.F. Stocker, D. Qin, G.-K. Plattner, et al.). Cambridge/New York: Cambridge University Press.

IPCC (2015). Fifth assessment. *AR5*.

Jiang Z., Yang, H., Guan, Y. et al. (2000). Chlorine effect on molecular descriptors of disinfection by-products. *Environmental Science* 21(5): 51–54.

Kimmel, C. and Gaylor, D. (1988). Issues in qualitative and quantitative risk analysis for developmental toxicology. *Risk Analysis* 8: 15–21.

Leo, A. and Hansch, C. (1999). Role of hydrophobic effects in mechanistic QSAR. *Perspectives in Drug Discovery* 17: 1–25.

Linhart, H. and Zucchini, W. (1986). *Model Selection*. New York: Wiley.

Martinez, W.L. (2002). *Computational Statistics Handbook with MATLAB*. Boca Ratón: Chapman & Hall/CRCnetBase.

Massart, D.L., Vandeginste, B.G.M., Buydens, L.M.C., et al. (1997). *Handbook of Chemometrics and Qualimetrics: Part A*. Amsterdam: Elsevier Science.

National Bureau of Statistics of China (2012). *China Statistical Yearbook*. Beijing: China Statistics Press. http://www.stats.gov.cn/tjsj/ndsj/2012/indexeh.htm (accessed January 2018).

NTP (National Toxicology Program) (1988). Toxicology and carcinogenesis studies of chlorodibromomethane. *CAS No. 124-48-1* in F344/N rats and B6C3F1 mice (gavage studies). TR-282. https://ntp.niehs.nih.gov/results/pubs/longterm/reports/longterm/index.html (accessed 21 March 2018).

OECD (2007). Guidance document on the validation of (Quantitative) Structure–Activity Relationships [(Q)SAR] models. Paris: Organization for Economic Co-Operation and Development.

Osten, D. (1988). Selection of optimal regression models via cross-validation. *Journal of Chemometrics* 2: 39–48

Pierotti, A.J. (1999). Chlorine effect on molecular descriptors of disinfection by-products. Master of Science Degree thesis. Florida: Florida International University.

Pontius, F.W. (1990). Toxicology and drinking water regulations. *Journal of American Water Works Association* 90: 17–19.

Stone, M. (1998). Akaike's criteria. In: *Encyclopedia of Biostatistics* (eds. P. Armitage and T. Colton). New York: Wiley.

Tang, W.Z. and Wang, F. (2010). Chlorine effect on quantum molecular descriptors of disinfection by-products-chlorinated alkanes. *Chemosphere* 78: 914–921.

Tang, W.Z., Wang, F., Miralles-Wilhelm, F., and Damisse, E. (2009). Uncertainty analysis of rating equations of submerged orifice flow at gated spillway. Conference on Reliability and Quality in Design, the International Society of Science and Applied Technologies (ISSAT) and the IEEE Reliability Society, San Francisco, CA (6–8 August 2009), 165–169.

The US EPA (1997). Guiding principles for Monte Carlo analysis. *EPA/630/R-97/001*.

The US EPA (2016). NAAQS table. https://www.epa.gov/criteria-air-pollutants/naaqs-table (accessed 29 November 2016).

U.S. Environmental Protection Agency (EPA) (1994). Drinking water maximum contaminant level goals and national primary drinking water regulations for lead and copper. *Federal Register* 59(125): 33860–33864.
US EPA (1992). Guidelines for environment exposure assessment. *Federal Register* 57: 22888–22936.
U.S. EPA (1996). Safe drinking water act standards and health advisories. Office of Water 4305. *EPA-822-B-96-002.*
US EPA (1997). National primary drinking water regulations: disinfectants and disinfection byproducts notice of data availability; proposed rule. *EPA-815-Z-98-005.*
US EPA (2005). Guidelines for carcinogen risk assessment. *Federal Register 70, 66, 177650-18717.* http://www.epa.gov/raf/pubalpha.htm (accessed 15 December 2017).
US EPA (2006). Ultraviolet disinfection guidance manual for the final long term 2 enhanced surface water treatment rule. *EPA 815-R-06-007,* November 2006.
US EPA (2011). Exposure factors handbook 2011 edition (Final). *EPA/600/R-09/052F.* Washington, DC: US Environmental Protection Agency.
US EPA (2012). Benchmark dose technical guidance, risk assessment forum. *20460, EPA/100/R-12/001.* Washington, DC: US Environmental Protection Agency, June 2012.
US EPA (2015). User's Manual for Benchmark Dose Software 2.6. Washington, DC: US Environmental Protection Agency.
US EPA (2016). Benchmark dose tools. https://www.epa.gov/bmds (accessed 20 November 2016).
Wang, W., Qian, Y.C., Li, J.H., et al. (2014). Analytical and toxicity characterization of halo-hydroxylbenzoquinonesas stable halobenzoquinone disinfection byproducts in treated water. *Analytical Chemistry* 86: 4982–4988.
World Health Organization (WHO) (2014). The world cancer report. Geneva: WHO.
World Health Organization (WHO) (2016). World health statistics 2016: monitoring health for the SDGs, sustainable development goals. Geneva: WHO.
Yang, H., Jiang, Z., Guan, Y., et al (2000). Correlations between SDWA standards and health advisories concentrations of DBPs and molecular descriptors. *Water and Wastewater Engineering* 26 (1): 22–25.
Zhao, P., Dai, M., Chen, W., and Li, N. (2010). Cancer trends in China. *Japanese Journal of Clinical Oncology* 40 (4): 281–85.

3

Twelve Design Principles of Sustainable Environmental Engineering

A compass is more important than a map.

OBJECTIVES
To define the twelve design principles (TDPs) and 36 design rules of sustainable environmental engineering (SEE)
To identify major design challenges of environmental engineering infrastructure system (EEIS)
To compare different sustainable development goals (SDGs) of the developed and developing countries and explore strategies to achieve those SDGs of the United Nations
To design alternatives by integrating EEIS of water, wastewater, stormwater, solid waste, and energy to comply environmental regulations
To apply system approach in terms of prevention, reduction, reuse, recycle, recovery, treatment, disposal, and remediation
To evaluate design alternatives according to material and energy footprints, life cycle assessment, and benefit and cost ratio by using LCA tools
To lay out SEE education roadmap through horizontal and vertical integration in environmental engineering (EE) curriculum

3.1 Sustainability

In the United States, the National Environmental Policy Act (NEPA) in 1969 requires that "the Federal Government use all practicable means and measures and to create and maintain conditions under which man and nature can exist in productive harmony, and fulfill the social, economic, and other requirements of present and future generations." The concern for future generation is the core foundation of sustainable development, which was introduced at the Rio Summit in Brazil in 1984 to the world stage as shown in Table 3.1.

Environment is one of the three pillars of sustainability among economic development and social equality. Environmental engineers should have been the leaders in designing environmental engineering infrastructure system (EEIS) such as water treatment plant (WTP), wastewater treatment plant (WWTP), stormwater, solid wastes, and air pollution control systems in a sustainable manner. The American Society of Civil Engineers' (ASCE) established the following criteria for true sustainability (Doughty and Hammond, 2003):

1) Finite materials such as fossil fuels should not be extracted at a faster rate than they can be redeposited in the Earth's crust.
2) Artificial materials such as plastics should not be produced at a faster rate than they can be broken down by natural processes.
3) The biodiversity of ecosystems should be maintained, while renewable resources should only be consumed at a slower rate than they can be naturally replenished.
4) Basic human needs must be met in an equitable and efficient manner.

Due to the long degradation half life of some wastes such as plastics in nature, ideal sustainable conditions might never be met. Therefore, to achieve a sustainable society realistically, the water, carbon, energy, and nutrient footprints (FPs) of each person should be less than our Earth's carrying capacity. EEIS must be designed to reduce the human FP so that the ecological FP is less than the biocapacity of the Earth as follows:

1) Sustainable design: Ecological FP should be less than biocapacity.

Sustainable Environmental Engineering, First Edition. Walter Z. Tang and Mika Sillanpää.

Table 3.1 History of sustainability.

Year	Milestone	Target
1969	National Environmental Policy Act (NEPA)	Requires the federal government to use all practicable means and measures and to create and maintain conditions under which man and nature can exist in productive harmony and fulfill the social, economic, and other requirements of present and future generations
1984	Rio Summit in Brazil	Sustainable development concepts were introduced
1992	Rio Summit in Brazil	Define societal, economic, and environmental as three pillars
2000	Paris Summit in France	Millennium development goals to reduce unsafe drinking water and no sanitation to half were established
2012	UN Conference on Sustainable Development in Rio+20	17 Sustainable development goals were established
2015	Paris Summit in France	17 Sustainable development goals were approved by the UN

2) Neutral design: Ecological FP should be equal to the Earth biocapacity.
3) Unsustainable design: Ecological FP is greater than the Earth biocapacity.

If natural environments are considered as biosphere and human society as technosphere, the following two loops of energy and materials should be in balance to achieve sustainable equilibrium as shown in Figure 3.1.

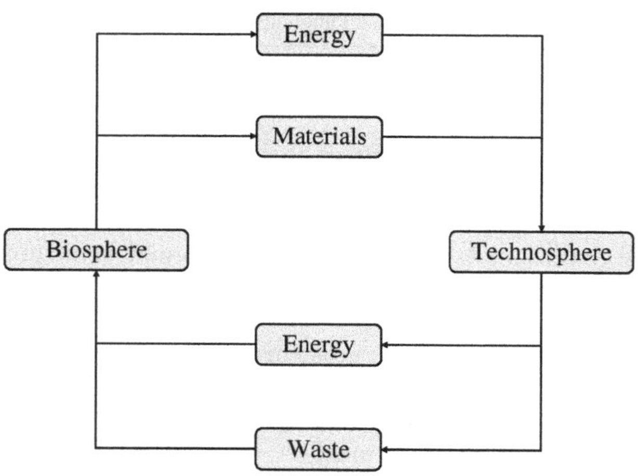

Figure 3.1 Achieving material sustainability in natural cycles.

To achieve material sustainability within natural cycles, the minimal goal for a sustainable environmental engineering (SEE) designer should achieve zero water, waste, and energy (3Z). Zero waste means no discharge of air, water, and solid waste into the biosphere. In assessing a zero discharge EEIS, life cycle assessment (LCA) could be used to compare environmental impacts on human health and the environment using the US Environmental Protection Agency (EPA) software such as TRACI 2. Since the wastes should be considered as renewable materials and energy, all the materials and energy could be recovered because there are no wastes in a functioning ecosystem. For example, regenerative design of EEIS is the best alternative because energy-positive and material recovery will ensure that EEIS is sustainable with a safe margin for humans to coexist with nature. The major assessment criteria are how much materials and energy could be recovered and what are the benefits and cost ratio in dollar terms. The least desirable design alternative is the negative design due to financial restraint. In the past decades, activated sludge (AS) process resulted in almost all energy negative WWTPs. SEE designers are facing great challenges to develop retrofitting strategies to achieve regenerative design with innovative technologies. If a negative design alternative could not be avoided, minimal FP on energy, water, and materials could be the design objective. Therefore, EEIS should then be assessed according to their FTs. Table 3.2 lists technical terms that are used to measure sustainability.

3.1.1 The United Nations Sustainable Development Goals

After the United Nations (UN) Millennium Development Goals ended in 2015, the UN established 17 sustainable development goals (SDGs) for the next 15 years. SEE design principles provide a systematic approach to design EEIS for sustainable sanitation and wastewater management to reach the UN goals by 2030. SDGs have different meanings for developed versus developing countries due to their economic development stages. SEE designers should play a critical role in achieving the 17 SDGs. In the international development agenda, poverty eradication, ending hunger, and healthy living are the first three major priorities. For developed countries, however, combating climate change, renewable energy, and sustainable consumption and production are three top priorities. To switch to more sustainable economy and growth

Table 3.2 Definition of sustainable environmental engineering.

Technical terms	
Sustainable environmental engineering	To design EEI so that the rates of renewable resource harvest, pollution creation, and nonrenewable resource depletion could be continued indefinitely
Emission factor	A representative value that attempts to relate the quantity of a pollutant released to the atmosphere with an activity associated with the release of that pollutant
Emission intensity	The average emission rate of a given pollutant from a given source relative to the intensity of a specific activity
Energy intensity	A measure of the energy efficiency of a nation's economy. It is calculated as units of energy per unit of GDP
Green living	A lifestyle that tries in as many ways as it can to bring into balance the conservation and preservation of the Earth's natural resources, habitats, and biodiversity with human culture and communities
Green gross domestic product	An index of economic growth with the environmental consequences of that growth factored into a country's conventional GDP. Green GDP monetizes the loss of biodiversity and accounts for costs caused by climate change

pathway, the goal of greater equality and to better protect oceans and terrestrial ecosystems are secondary aims in the developed countries. These 17 SDGs have been ranked as different priorities for developed versus developing countries as shown in Table 3.3 with typical projects for SEE designer.

3.2 Challenges and Opportunities

3.2.1 Challenges

Human social and economic activities have enormous negative impacts on the global environment due to waste and fossil fuels that are proportional to a country's economic scale measured by gross domestic product (GDP). Annual GDP is the total market value of all final goods and services produced in a country in a given year. Due to different stages of population and economic growth, current GDP ranking may very likely change as predicted in Table 3.4.

In calculating GDP, if market exchange rates are used, it is referred to as a nominal method. On the other hand, purchasing power parity (PPP) is used to compare economies and the incomes of people by adjusting for differences in prices in different countries. The International Monetary Fund (IMF) outlook in October 2016 projected the GDPs of 190 countries/economies at current prices (US dollars) for 2016 and 2020. Gross world product in 2016 is projected at $75.21 trillion, and its GDP in terms of PPP is forecasted at $119.1 trillion. The global economy is 1.58 times greater in PPP terms than in nominal terms. The United States and China are the largest economies in either nominal or PPP terms, respectively. In US dollar terms, the United States shares 24.7% of global wealth. However, in PPP China ranks the first and shares 17.9%. In nominal data, 15 economies have a GDP above $1 trillion, 61 have above $100 billion, and 176 have above $1 billion. The top 5 economies account for approximately 54.28%, top 10 accounts for approximately 67.44 %, while the 153 lowest ranked constitutes only 10% of total. In PPP data, 24 economies, which are 9 more than nominal, have GDP above $1 trillion. Eighty-one economies have GDP greater than $100 billion and 181 have greater than $1 billion. The top 5 economies add up to over 48% of world's economy, the top 10 to over 61%, and the top 20 to over 75%. Table 3.5 shows that major growth of GDP would come from developing countries such as Brazil, Russia, India, and China (BRIC) with a total GDP of $66.694 trillion in 2030.

Table 3.5 can be used to divide the top GDP countries into two groups using the PPP in terms of international dollar (ID). The developed countries include the United States, Japan, Germany, and the United Kingdom (AJGE) with a total GDP of $33.79 trillion. The emerging economies include Brazil, India, Indonesia, Russia, China, and Mexico (BRIICM) with a total GDP of 75.79 ID. According to these GDP predictions, major opportunities for SEE designers to make differences exist in BRIICM because there are many EEISs to be designed and built. In the developed countries such as AJGE, retrofitting WWTPs might be the biggest challenging market to make EEIS sustainable.

In the past, energy and materials used to produce GDP were mostly nonrenewable. About 86% of energy and 96% of organic chemicals came from depleting fossil fuels such as coal, oil, and gas. The United Nations Environmental Protection (UNEP) estimated that it may take about 50–80 years for mankind to reach a sustainable level by slowing climate change, preventing pollution, and renewing natural resources from 2015. Over $70 billion of biodiversity is lost due to ecological degradation and water

Table 3.3 Ranking of SDGs for the developed countries.

SGDs number	Ranking	Sustainable development goals	Examples of a SEE designer
1	1.8	End poverty in all its forms everywhere	Green jobs integrated with air, water, and land management such as recycle, reuse, and recovery
2	2.3	End hunger, achieve food security and improved nutrition, and promote sustainable agriculture	Locally produced gray water reuse and decentralized WWTP design
3	1.5	Ensure healthy lives and promote well-being for all at all ages	UV disinfection and oxidation
4	2.5	Ensure inclusive and equitable quality education and promote lifelong learning opportunities for all	Grassroots movement of GEE design by involving high school seniors and STEM students
5	2.5	Achieve gender equality and empower all women and girls	Green economy and GEE design involving girls and green jobs for woman
6	2.5	Ensure availability and sustainable management of water and sanitation for all	Sustainable regenerative water and wastewater design
7	6.4	Ensure access to affordable, reliable, sustainable, and modern energy for all	Renewable energy application in WTPs, WTWPs, GI, and residential houses
8	2.7	Promote sustained, inclusive, and sustainable economic growth; full and productive employment; and decent work for all	Green jobs, renewable energy
9	2.1	Industry, innovation, and infrastructure	Sustainable WWT design for industries
10	3.6	Reduce inequality within and among countries	Water and carbon footprints in trade
11	2.6	Make cities and human settlements inclusive, safe, resilient, and sustainable	Green infrastructure design to protect air, water, and land
12	6.3	Ensure sustainable consumption and production patterns	Reduce meat in diet and encourage more recycle
13	7.1	Take urgent action to combat climate change	Environmental engineering infrastructure to reduce carbon and energy footprints
14	4.4	Conserve and sustainably use the oceans, seas, and marine resources for sustainable development	Use renewable energy in EE infrastructure design to reduce coal combustion
15	2.7	Protect, restore, and promote sustainable use of terrestrial ecosystems; sustainably manage forests; combat desertification; and halt and reverse land degradation and halt biodiversity loss	Biomass production from WWTPs and GIs
16	2.7	Promote peaceful and inclusive societies for sustainable development, provide access to justice for all, and build effective, accountable, and inclusive institutions at all levels	Leadership in SEE design at all local and global scales and take active role in consultation to decision makers for sustainable development
17	NA	Partnership for the goals	Form alliance to reduce inequality of water, carbon, material, and energy footprints during trades

Source: From Osborn et al. (2015).

Table 3.4 Top 10 countries of highest GDP in US dollar.

GDP rank in 2030	Country	Predicted GDP in trillion USD in 2030	GDP rank in 2020	GDP in trillion USD in 2020	GDP rank in 2016	GDP in trillion USD in 2016
1	United States	23.475	1	21.927	1	18.561
2	China	20.350	2	16.458	2	11.391
3	Japan	5.375	3	5.506	3	4.730
4	India	5.036	5	3.297	7	2.251
5	Germany	4.133	4	4.008	4	3.495
6	United Kingdom	3.458	6	2.928	5	2.650
7	Brazil	2.505	8	2.214	7	1.770
8	Mexico	1.688	15	1.325	15	1.063
9	Indonesia	1.686	16	1.274	16	0.941
10	Russia	1.600	12	1.598	12	1.268

Table 3.5 Top 10 countries of highest GDP in international dollar in 2030.

GDP rank in 2030	Country	Predicted GDP in trillion USD in 2030	GDP rank in 2020	GDP in trillion USD in 2020	GDP rank in 2016	GDP in trillion USD in 2016
1	China	38.008	1	29.348	2	17.927
2	United States	23.475	1	21.927	1	18.561
3	India	19.511	3	12.842	3	8.720
4	Japan	5.606	4	5.483	4	4.932
5	Indonesia	5.424	7	4.119	8	3.028
6	Russia	4.736	6	4.309	6	3.745
7	Germany	4.707	5	4.583	5	3.979
8	Brazil	4.439	8	3.631	7	3.135
9	Mexico	3.661	11	2.800	11	2.307
10	United Kingdom	3.638	9	3.244	9	2.788

scarcity annually (UNEP, 2016). In terms of gigatons CO_2 equivalent ($GtCO_2e$), current global GHG 42 $GtCO_2e$ per annum, which is five times higher than the Earth can absorb (WWF), is due to human activities. Figure 3.2 shows the top 10 percentages of CO_2 emissions by country.

3.2.2 Opportunities

Two of the seventeen UN SDGs are universal access to clean drinking water and sanitation. The UN-Water (2015) defines population using a basic sanitation facility at household level and not shared with other households. In the past, most EEIS were designed to meet discharge standards as required by environmental regulations. The ASCE (2016) rated American EEIS such as WTP and WWTPs as grade "D^+" because WTPs and WWTPs had suffered significant deterioration after more than five decades service. The combined sewer system exhibited significant deficiencies and caused many overflow incidents of raw sewage into nearby water bodies. In 2009, the EPA reported to Congress that the states had assessed 16% of America's stream miles and found that 36% of those miles were unfit for use by fish and wildlife, 28% were unfit for human recreation, 18% were unfit for use as a public water supply, and 10% were unfit for agricultural use. Over the next two decades, \$298 billion of capital investment is needed for the nation's water, wastewater, and stormwater systems. Three-quarters of this requirement is pipes for sanitary sewer overflows, combined sewer overflows (CSOs), and other pipe-related issues. About one-quarter of the \$298 billion is required for the treatment plants. Specifically, \$188 billion is needed for wastewater treatment, pipe repairs, and new pipes; \$64 billion to correct CSOs; and \$42 billion for stormwater management. Therefore, great opportunities exist for SEE designers to design SEEI

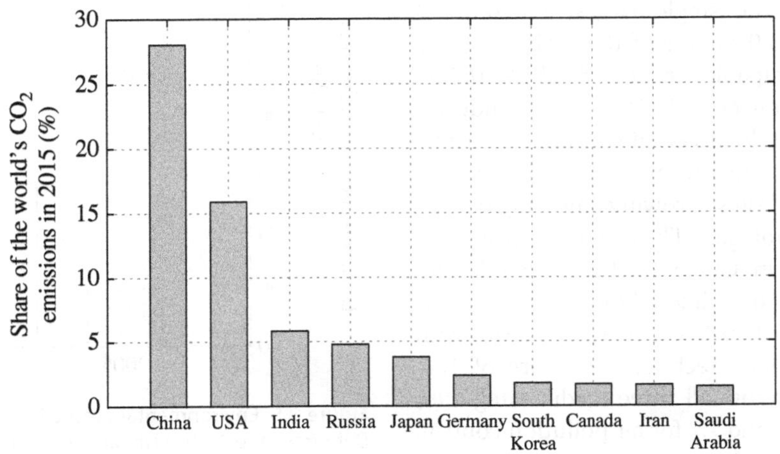

Figure 3.2 Top 10 CO_2 emitters worldwide. (*Source:* Data from the United Nations, 2016.)

in the coming decades. For example, about 3% of the total electricity generated in the United States is currently consumed by EEI such as WWTP and WTPs. Almost all 15 000 traditional WWTPs need to be retrofitted from energy negative to energy positive with an estimate of $600 billion capital investment required by the ASCE in the coming two decades. To show the energy saving potential in retrofitting WWTPs, Figure 3.3 shows that most energy is consumed by aeration, while pumping and anaerobic digestion consumes 13%.

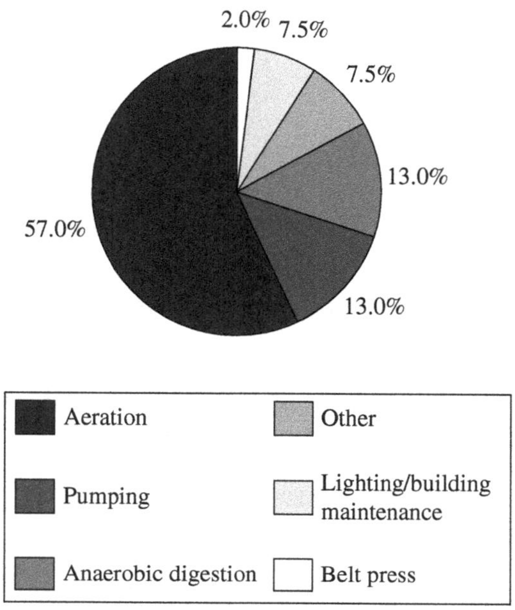

Figure 3.3 Energy breakdown at water resource recovery facilities (WRRFs).

During operation, sludge and aeration require 25 and 23% cost, respectively. Staff account for 19% of the operating cost (Figure 3.4).

According to the World Trade Organization (WTO, 2017), the global market for environmental technologies, goods, and services reached USD 1.05 trillion in 2015. The United States is the single largest market and accounted for USD 303.0 billion of the global market. US environmental companies exported $51.2 billion worth of goods and services in 2015. The US industry for environmental technologies employed approximately 1.6 million people and had revenues of $320.4 billion. China is the largest and fastest-growing emerging market for environmental technologies. The overall environmental technology market in China (including goods and services) is valued at $60.7 billion (2016). China ranks first overall on the 2016 Top Markets Study (TMS) with a composite environmental technologies score of 100. China also ranks first across all three media categories, with scores of 47.4, 44.9, and 7.7 for air pollution control, water, and waste and recycling markets, respectively.

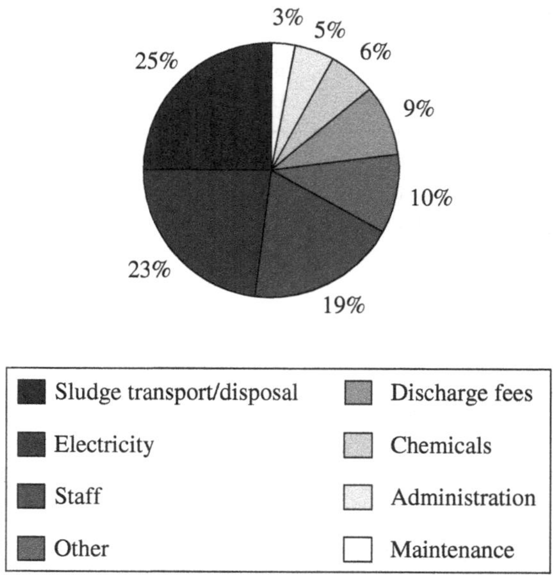

Figure 3.4 Water resource recovery facilities' (WRRFs) operating cost breakdown.

In 2016, China signed the Paris Agreement on a climate change action plan during the G20 Summit in Hangzhou, China (G20 Hangzhou Summit, 2016). More and more investment will be allocated to air pollution control for the reduction of CO_2 emissions, especially coal-burning or clean coal technology. The Chinese government is committed to green economy with green finance. China established its Asian Infrastructure Investment Bank (AIIB) with green metrics as key criteria for its investments in European and Asian countries. As new regulations are established, billions of dollars have been allocated to water and wastewater as shown in Figure 3.5.

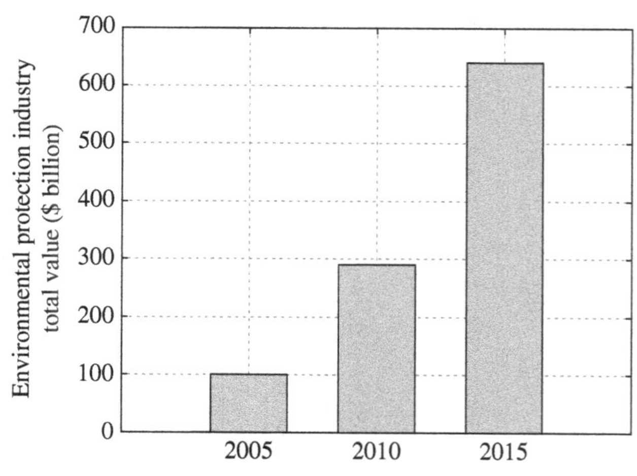

Figure 3.5 Environmental protection industry total value. (*Source:* Data from the 10th, 11th, and 12th Five-Year Plan for the Energy Conservation and Environmental Protection Industry.)

The major markets are water, wastewater, and sludge treatment and disposal, and an estimated \$1.4 trillion would be required in China by 2020. The biggest growth potential lies in soil remediation, solid waste, wastewater, and sludge treatment and disposal. Goldman Sachs reported that the 12th Five-Year Plan only set aside \$4.8 billion to address soil pollution, while \$277 billion was allocated by the State Council to reduce air pollution in 2013–2017 (Figures 3.6 and 3.7).

In China alone, the capital expenditure for water and wastewater is projected to grow exponentially (Figures 3.8 and 3.9).

Figure 3.10 shows that equipments such as pipes, automatic control system, pumps, valves, and aeration devices are projected to grow exponentially to \$5.6, 3, 2.5, 2.4, and 1.4 billion, respectively, by 2018.

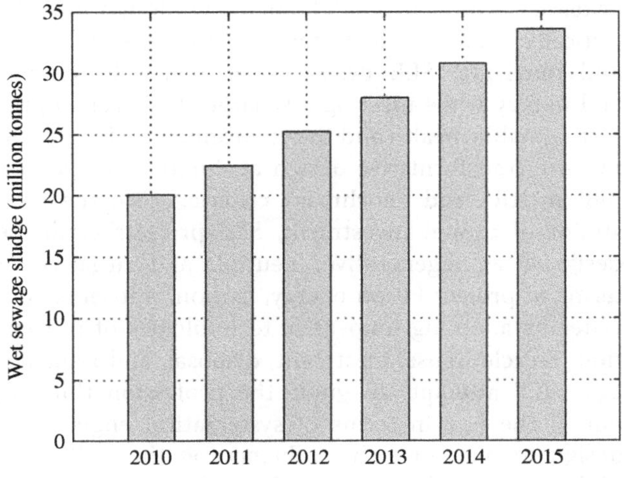

Figure 3.6 Sludge generation by wastewater treatment plants in China (water content 80%). (*Source:* Survey data from h2o-china.com and quarterly figures from Ministry of Housing and Urban–Rural Development: "2013–2017 In-depth Survey and Investment Strategy Planning Analysis on Sludge Treatment and Disposal in China.")

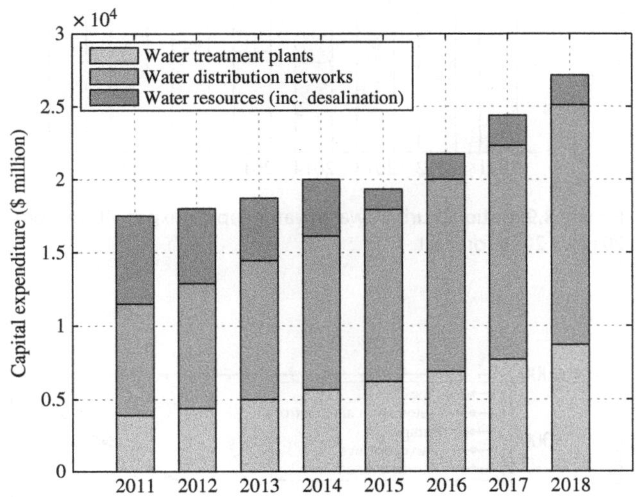

Figure 3.8 National urban water supply capital expenditure from 2011 to 2018 forecast.

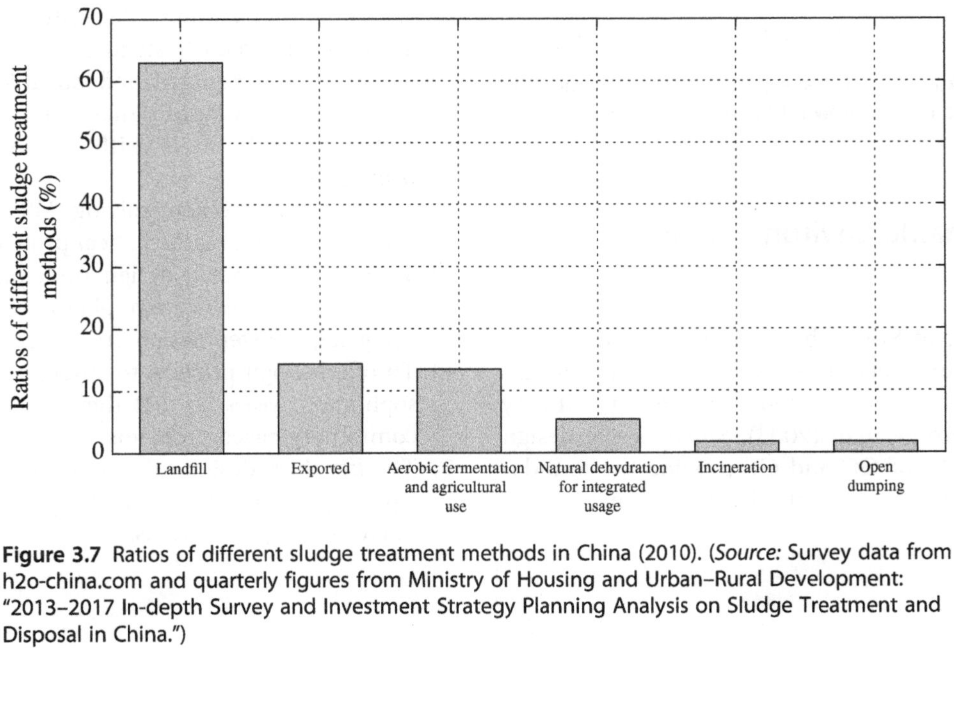

Figure 3.7 Ratios of different sludge treatment methods in China (2010). (*Source:* Survey data from h2o-china.com and quarterly figures from Ministry of Housing and Urban–Rural Development: "2013–2017 In-depth Survey and Investment Strategy Planning Analysis on Sludge Treatment and Disposal in China.")

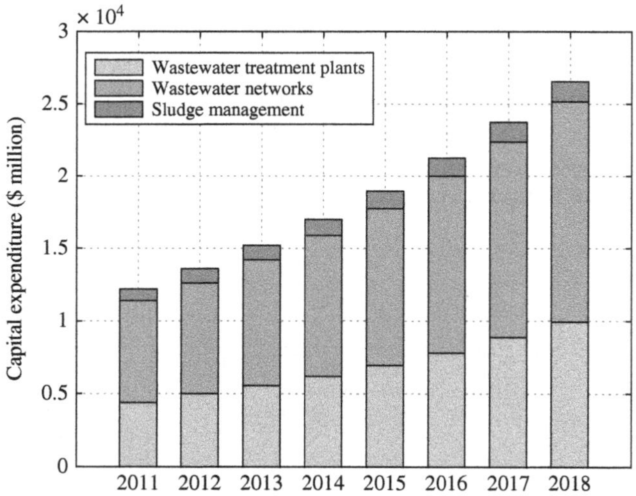

Figure 3.9 National urban wastewater capital expenditure from 2011 to 2018 forecast.

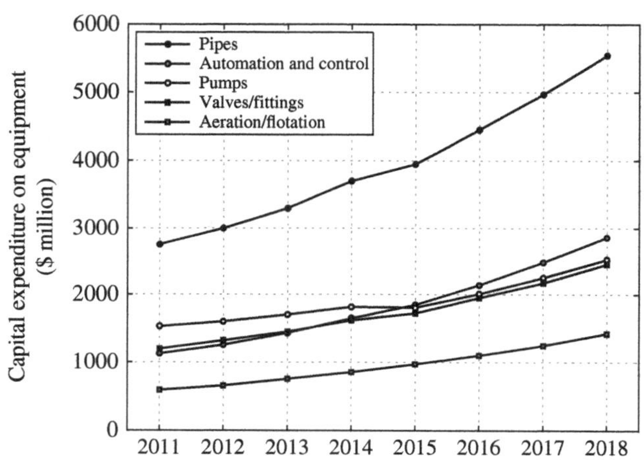

Figure 3.10 Breakdown of capital expenditure on equipment by type and expected market growth from 2011 to 2018.

3.3 Sustainable Environmental Engineering

The framework of SEE is built upon the transformative concepts of green chemistry (GC) by Anastas and Warner (1998) and green engineering (GE) by Anastas and Zimmerman (2003). Since twelve design principles (TDPs) of GC and GE provide the foundation of SEE, the history of GC is summarized in Table 3.6.

Twelve SEE principles have been developed by applying GC and GE principles to EEI design. Therefore, SEE is to establish a systematic design method in conceiving, researching, building, and operating sustainable EEI. SEE design principles are formulated to design EEI in harmony with nature using FP as sustainability design criteria to protect human health and the environment at minimal cost. Design theory and tools of critical technologies are defined to achieve design standards or benchmarks. According to these design criteria, traditional WWTP and WTP design must be critically examined in terms of FP of energy, water, and nutrients. SEE design saves materials, energy, and money while meeting discharge standards in protecting public health and the environment. SEE aims to nurture transformative design at different levels with unit metrics and benchmark criteria. With the constraint of capital investment, SEE projects could be designed as regenerative, neutral, and negative in terms of project FP on energy, carbon, nutrients, and water by adopting innovative technologies of prevention, recycle, reuse, treatment, disposal, and remediation. SEE attempts to guide the profession thinking out of the box in terms of systematical engineering design to maximize the influence on the well-being of human health and the planet Earth (Tables 3.7 and 3.8). Therefore, the TDPs are developed to achieve the following:

1) To critically examine past unsustainable engineering design practice and provide sustainable design solutions if possible.
2) To establish metrics or benchmarks for SEE design to lay a solid foundation for legislators or governmental agencies to establish standards or criteria for SEE design at local, regional, and global levels.
3) To align with traditional engineering 30, 60, and 100% submittals with different SEE design principles.
4) To emphasize problem-solving skills such as modern design tools and mathematical proficiency.
5) To offer sustainable design principles and tools to achieve ABET education objectives and outcomes using step-by-step design procedure.
6) To offer design projects to students as early as their sophomore years so that they can stay longer with community-based problems.
7) To offer three design rules to make each design principle applicable, effective, and appropriate with quantified FP using LCA tools.

Table 3.6 History of pollution prevention and green chemistry.

Year	Background on pollution prevention and green chemistry movement
1962	Rachel Carson – writer, biologist, and environmental conservation icon – published the first of three installments of *Silent Spring*, a literature that is historically tied to the launch of the environmental movement. The publication helped spread public awareness of the hazards of environmental pollution and pesticides to the environment
1969	President Richard Nixon establishes Citizens' Advisory Committee on Environmental Quality and a Cabinet-level Environmental Quality Council (www.presidency.ucsb.edu). Later that year, Nixon expanded his environmental efforts by appointing the White House committee to determine whether an environmental agency should be developed
1970	The Environmental Protection Agency (EPA) is born
1976	The Toxic Substances Control Act (TSCA) of 1976 provides EPA with authority to require reporting, recordkeeping and testing requirements, and restrictions relating to chemical substances and/or mixtures. Certain substances are generally excluded from TSCA, including, among others, food, drugs, cosmetics, and pesticides
1980s/ 1988	Shift from end-of-pipeline control to pollution prevention is recognized, leading to the Office of Pollution Prevention and Toxics in 1988
1990	The Pollution Prevention Act under the George H.W. Bush Administration is passed. Recognizing the need to shift from traditional approach of controlling, treatment, and abatement, Congress passed the Pollution Prevention Act in 1990. The Act established a "national policy to prevent or reduce pollution at its source whenever feasible"
1993	The EPA implements the Green Chemistry Program, which serves as a precedent for the design and processing of chemicals that lessen the negative environmental impact
1995/ 1996	In 1995, President Bill Clinton established the Presidential Green Chemical Challenge Awards, which served to encourage those involved with the manufacture and processes of chemicals to incorporate environmentally sustainable design and processes in their practices. The following year, the first recipient receives the award, the only award issued by the president that honors work in chemistry (Source: http://portal.acs.org/)
1997	The Green Chemistry Institute is launched. Its vision is "...to advance the broader chemistry enterprise and its practitioners for the benefit of Earth and its people" (Source: http://portal.acs.org/)
1998	"Twelve Principles of Green Chemistry" is published by Paul Anastas (of the EPA) and John Warner
2000s	In the past decade, advances in green chemistry policy have been realized including the California Green Chemistry Initiative. Governor Arnold Schwarzenegger signed the bill in 2008, which serves to develop policy options for green chemistry (Source: www.dtsc.ca.gov). One year later, President Obama nominated Paul Anastas (of Yale) as head of Research and Development at the EPA
2007	The US House of Representatives passed legislation seeking to improve federal coordination, dissemination, and investment in green chemistry research and development. The Green Chemistry Research and Development Act of 2007 (HR 2850) aims to provide safer, more sustainable technological options to replace traditional products and processes

Table 3.7 Definition of related concepts to green chemistry for environmental management.

Concept	Definition
Eco-efficiency	The delivery of competitively priced goods and services that satisfy human needs and bring quality of life, while progressively reducing ecological impacts and resource intensity throughout the life cycle, to a level at least in line with the Earth's estimated carrying capacity (WBCSD, 1993)
Industrial ecology	An integrated systems-perspective examination of industry and environment, which conceptualizes the industrial system as a producer of both products and wastes and examines the relationship between producers, consumers, other entities, and the natural world (Sagar and Frosch, 1997)
Cleaner production	The continuous use of an integrated and preventive environmental strategy, applied to processes, products, and services to increase the eco-efficiency and reduce risks to the population and the environment (Rigola, 1998)
Ecodesign	Designing products to minimize their direct and indirect environmental impacts at every possible opportunity (Lewis et al., 2001)
Green engineering	The design, commercialization, and use of processes and products that are feasible and economical while minimizing pollution at the source and risk to human health and the environment (Kirchhoff, 2003)
Life cycle thinking	Addressing environmental issues and opportunities from a system or holistic perspective. This way of thinking involves evaluating a product or service with the goal of reducing potential

Table 3.8 Green engineering and green chemistry principles.

Principle	Green engineering	Green chemistry
1	Inherent rather than circumstantial	Prevention
2	Prevention instead of treatment	Atom economy
3	Design for separation	Less hazardous chemical use
4	Maximize efficiency	Design for safer chemicals
5	Output pulled versus input pushed	Safer solvents and auxiliaries
6	Conserve complexity	Design for energy efficiency
7	Durability rather than immortality	Use renewable feedstock
8	Meet need, minimize excess	Reduce use of derivatives
9	Minimize material diversity	Catalytic reagents rather than stoichiometric reagents
10	Integrate material and energy flows	Design for degradation
11	Design for commercial afterlife	Use real-time analysis for pollution prevention
12	Renewable rather than depleting	Use safer chemistry to prevent accidents

8) To select the best alternative by ranking rubrics in terms of regenerative, neutral, and negative design as reflected by material, water, and energy FP.

The TDPs of SEE have six dimensions and six design strategies as shown in Figure 3.11.

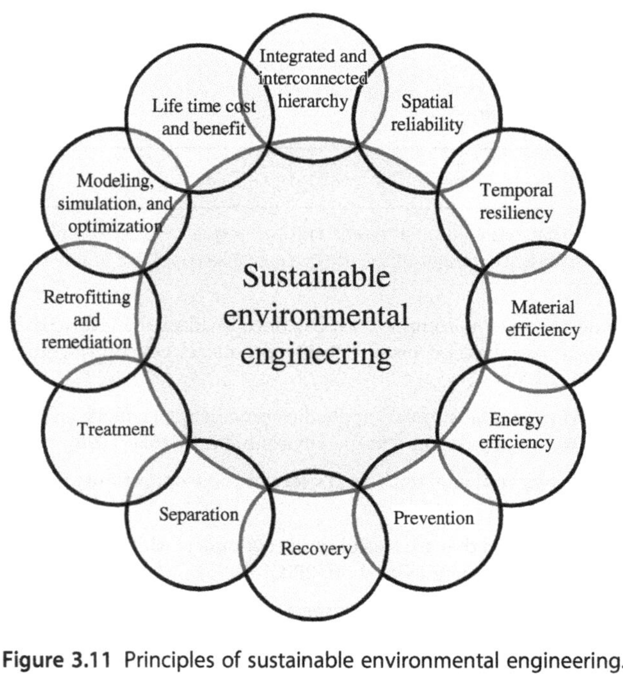

Figure 3.11 Principles of sustainable environmental engineering.

3.3.1 SEE Metrics

Global climate change, energy production, food production, and resource depletion need paradigm shift of EEIS design philosophy. For example, green infrastructure (GI) could be designed generatively to decouple economic development from environmental pollution and to prevent pollution at source according to ecological principles. If all the communities were designed as sustainable communities, environmental stress would be significantly reduced. The best metrics to gauge sustainability, therefore, are material, water, and energy FP. SEE engineers should design EEIS that meet technical and cost objectives based on the criteria of protecting human health and the biosphere. For example, wastewater management is fundamentally linked with a natural system's capacity to achieve sustainability. The US EPA developed the following performance metrics in Table 3.9.

Table 3.9 The US EPA unit impact metrics.

Material efficiency	kg inputs/kg product(s)
Carbon efficiency	Product C × 100/reactant C
Effective mass yield	kg product/kg nonbenign reactants
Quantity of virgin material consumed	kg virgin/kg product + coproduct
Water consumption	l/kg product + coproduct
Atom economy	kg desired product/kg reactants
Embedded energy	Energy used to produce and transport raw materials to chemical transformer gate
Waste produced	E-factor = kg waste/kg (product + coproducts)
Quantity release to environment posttreatment	kg waste − kg destroyed
Hazard characteristics	E-factor × (sum of waste hazard quotients)
EcoScale	100 − (sum of 6 parameter penalty points)
Carbon footprint	GHG emitted

These metrics can be applied in defining SEE design as shown in Table 3.10.

SEE aims to empower designers with tools, models, techniques, and procedures that offer them the ability to solve complex problems involved in embracing the huge market in the coming decades. Stakeholders such as industry, consumers, workers, citizens, and government at all levels must be involved in the early stage of projects. Economic, legal, institutional,

Table 3.10 Sustainable environmental engineering design principles, technology, and metrics.

Design principles	Problems	Technology	Metrics
Integrated design and hierarchy	Solid waste Air pollution Water pollution	IWRM ISWM IAPM Systems approach Synergy	Regenerative, neutral, and negative design photosynthesis energy, biomass Leadership in stakeholders involvement, interdisciplinary, and grassroots movement
System defined by spatial scale	Air, water, solid, and hazardous waste treatment in north versus south	GIS in planning	Design by PE
Resilient at temporal scale	Climate warming and cooling	Underground WWTPs	Ecological footprint of WWTP
Renewable material efficiency	Tank process Mass transfer limit in AS Disinfection and oxidation	Pipe reactor UV disinfection AnMBR	Trade-off capital and operating cost Head loss per reactor or unit process Increase mass transfer efficiency
Renewable energy efficiency	Plant head loss Pressured versus gravity	Pump efficiency Pipe reactor AnMBR VUV	Head loss per reactor or unit process Increase efficiency and save energy E-factor
Prevention	Solid waste Activated sludge Leachate treatment	Green engineering, green infrastructure, and decentralized WWT	Human health Percentage recycled
Recovery	N and P energy	Struvite Biogas	Benchmarks
Separation	Food wastes Leachate	Food waste to bioenergy Leachate treatment	Separated treated leachate Separation for recycle Separation for reuse
Treatment	Cl_2	Nontoxic materials, UV, catalysts, AOP Renewable energy, solar, biogas photosynthesis	Regenerative, neutral, and negative design Reduce risk of exposure, release, explosions, and fires Increase human health Catalysts minimize or eliminate chemicals added
Green retrofitting and remediation	Air, water, soil, sludge	EPA tool	Footprint
Optimization Simulation	Process efficiency	Reactor and system design SCADA	Production/m^3
Balance between capital and operation costs	Tank processes in WTP and WWTP Product design Concrete WTP and WWTP tank design Commercialization	Modular design Pipe reactor WTP and WWTP pipe modular design AutoCAD SolidWorks 3D printing	Degree of modular component used Low FP products Regenerative design

technology-based, and societal-based barriers must be identified to achieve the desired transformational goals: problem formulations, knowledge of options or solutions, and the ability to evaluate alternative courses of action. SEE designers will be able to develop alternatives that are most resilient under the uncertainty caused by climate change and sea level changing. Alternative solutions other than AS and chlorination should be designed. Adverse effects of technologies such as chlorination and aluminum coagulation should be avoided by promoting vacuum ultraviolet (VUV). To quantify environmental footprints of an SEEI design, LCAs are usually applied and include extraction/synthesis, design, process, product, and disposal/reuse with aid of following assessment tools in Table 3.11.

Table 3.11 Assessment tool summary.

Tool	Objective	Life cycle stage
Cradle to cradle	Product certification	Product
GreenScreen	Chemical hazard assessment	Design
GreenWERCS	Chemical hazard assessment	Design/product
QCAT	Simplified hazard assessment	Design
Paris III	Solvent assessment	Design
TEST	Toxicity evaluation	Design
WAR	Waste generation impacts	Design/process

3.4 SEE Design Principles

EEIS includes WTPs, WWTPs, water distribution systems, sewer collection systems, stormwater control systems, solid waste management systems such as landfills, and air pollution management systems such as incinerators or energy recovery facilities. To achieve the maximum objectives of environment, economy, and equality (3E), EEIS should be designed according to the following 12 principles:

1) Integrated and interconnected alternatives should be developed to protect human and ecological health for sustainable economic development involving stakeholders.
2) Design alternatives should be reliable spatially within the system boundary.
3) The EEIS should be resilient temporally with quantified uncertainty and sensitivity.
4) Renewable materials should be maximally utilized with minimal wastes.
5) Renewable energy should be efficient according to energy benchmarks.
6) Prevention, reduction, and minimization strategies should be implemented as early as possible.
7) Recovery should be prioritized according to scarcity of materials and the benefit/cost ratio.
8) Separation should consume minimal materials and energy with early implementation.
9) Oxidation is the best way to avoid adding chemicals or use minimal nontoxic chemicals and renewable energy.
10) Retrofitting and remediation should recover material, energy, or land with minimal FPs.
11) Component, process, and system should be optimized through modeling and simulation under both global warming and cooling scenarios within minimal FP.
12) Standardized and modularized components or systems should be designed with real-time monitoring and controlling during operation to optimize capital and operation costs.

In short, the ultimate goal for EE designers is to reduce chemical, biological, and radioactive risks to protect human health and to minimize FT on ecological systems to protect the air we breathe, the water we drink, and the food we eat.

3.4.1 Principle 1: Integrated and Interconnected System Hierarchy

SEE aims to protect people, the planet, and profit (3P). Therefore, designers should play a leadership role to involve all the major stakeholders. Sustainable EEIS alternatives should be developed by actively engage communities and stakeholders. Implementing and feedback loops of SEE alternatives should be intertwined in achieving environmental, economic, and employment (3E) objectives. The hierarchy to achieve 3E objectives is prevention, reduction, reuse, recycling, recovery, treatment, disposal, and remediation. To rank the effectiveness of these strategies, the relevant data on human and industrial generation of wastes and current EE system is needed to develop waste management engineering alternatives. The life cycle of materials should be taken into consideration to minimize the FT of EEIS. Design alternatives should be assessed and compared in terms of reduction of chemical, biological, and radioactive risks and minimal FP on air, water, soil, and energy. The state-of-the-art design tools should be used to keep the capital cost of EEIS at minimal budget.

Principle 1 Integrated and interconnected alternatives should be developed to protect human and ecological health for sustainable economic development by involving stakeholders:

1) Environmental, economic, and employment goals need to be integrated and prioritized.
2) Air, water, and solid waste management systems should aim at regenerative or zero discharge through integrated and interconnected design.
3) Blue, gray, and black water management may be decentralized depending upon the system boundary.

Traditional EE design usually has negative FP in the past. For a WWTP using AS process, it is very difficult to bring the operation to energy neutral due to energy intensive aeration process and the huge amount of sludge produced. Typically, performance metrics including the FP of C, N, P, and energy within the design life expectancy are negative. Regenerative design integrates external energy into the system and becomes more sustainable and resilient. Regenerative design requires EEIS to produce more than the waste exergy through utilization of solar energy such as photosynthesis or photovoltaic. In addition, regenerative design should be developed together with the green economy by creating green jobs such as bioenergy, solar energy, and recovery materials. Regenerative design usually combines different innovative technologies by using tools (Table 3.12). For example, tools for integrated water resources management (IWRM) could be used to design regenerative through integrated and interconnected EEIS:

1) EEIS should satisfy governmental regulations to protect human and environment health.
2) Air, water, and solid waste engineering systems should be integrated to achieve zero discharge.
3) EEIS should be interconnected to achieve synergistic functions with minimal material, water, nutrient, and energy FP.
4) EEIS should be designed to fit for purpose and could be centralized or decentralized systems depending upon system boundary.
5) Different water quality such as blue, gray, and black should be considered together for sustainable management of water.
6) GI should be interconnected.

3.4.2 Principle 2: Reliability on Spatial Scale

A well-defined EEIS is critical to identify 3E objectives on the spatial scale. EEIS could be defined at residential, community, city, regional, and global scales. Stakeholders within the system should be identified and engaged in the planning and implementing phases. The key for an EEIS to be reliable is to fit the purpose and be adaptive to local situations in terms of renewable materials and energy.

Table 3.12 SEE design general metrics.

Design	Problems	Regenerative	Neutral	Negative
Air	SMOG from energy production	Photosynthesis rate is greater than CO_2 releasing rates	Photosynthesis rate equals the CO_2 releasing rates	Photosynthesis rate is slower than CO_2 releasing rates
Water	Retrofitting WWTP AS	Energy produced is greater than the power needed by the plant	Energy neutral	WWTP needs to buy more than 80% energy
	Recovery nutrients and water	Recovery value is greater than the cost	Recovery value equal the cost	Recovery value is less than the cost
	Separate treatment of leachate	Benefits is higher than combined treatment with wastewater	Benefits equals the combined treatment with wastewater	Benefits is less than the combined treatment with wastewater
	Chlorination	The carbon FP of alternative is less than the carbon FP of chlorination	The carbon FP of alternative equals the carbon FP of chlorination	The carbon FP of alternative is greater than the carbon FP of chlorination
	Coagulation	The energy FP of alternative is less than the energy carbon FP of coagulation	Energy FP of alternative equals the energy carbon FP of coagulation	Energy FP of alternative is greater than energy carbon FP of coagulation
Land	Sludge	Energy recovered is greater than the energy input	Energy recovered equals the energy input	Energy recovered is less than the energy input
	Food waste separation	Energy recovered is greater than the energy input	Energy recovered equals the energy input	Energy recovered is less than the energy input
	Leachate	Energy recovered is greater than the energy input	Energy recovered equals the energy input	Energy recovered is less than the energy input
Ecological	Biodiversity	Biospecies is more than the baseline	Biospecies equals the baseline	Biospecies is less than the baseline
	Eutrophication	N and P FP is less than the natural capacity	N and P FP equals the natural capacity	N and P FP is greater than the natural capacity

Principle 2 Design alternatives should be reliable to achieving given health and environmental risks with quantified uncertainty and sensitivity within the system:

1) EEI design alternatives should be clearly defined on residential, community, city, regional, and global spatial scales.
2) Molecules, particles, single and multiphase system, reactor, process, and plants should be analyzed in designing modern EEIS.
3) Design alternatives should have spatial efficiency.

EEIS such as WTPs, WWTPs, and GIs should be sustainably designed with consideration of local situation on its spatial scale. Due to geographic differences, renewable energy resources are distributed differently, for example, solar energy in the south and wind energy in north. Treatment will heavily depend upon scale in terms of population equivalent (PE). In general, decentralized WWTP is more suitable for rural areas than metropolitan cities. For residential houses, on-site wastewater treatment processes such septic tanks and constructed wetlands may hold the key to achieve regenerative design. If the land of a residential house is large enough, zero water, energy, and wastes should be achievable. For a community greater than 1000 PE, energy-neutral design may be more realistic. For large cities, if energy neutrality cannot be achieved, designers should target maximum recovery and reuse to minimize FP. New urban communities and cities have the opportunity to design and build source-separating systems optimized for cost-effective resource recovery using rain harvest and food waste separation from municipal solid waste. In addition, following issues should be considered on spatial scales:

1) Economy of scale should be assessed at the individual, household, community, city, state, and country levels to avoid fit-for-all design.
2) Regenerative or zero discharge design should have priority, while negative design should have minimal FP.
3) GI should be implemented as much as possible at residential house, community, city, state, and country levels with different weights of blue, gray, and black water management strategies.
4) Laboratory-, pilot-, and full-scale testing should be validated with known uncertainty to increase the reliability of EEIS. Laboratory-scale data could be directly used to design the full-scale system if kinetics and mechanisms were correctly modeled and simulated to eliminate a prototype.

3.4.3 Principle 3: System Resiliency on a Temporal Scale

On temporal scales, EEIS should be resilient at 20, 50, and 100 years depending upon design life expectancy. Furthermore, the uncertainty and sensitivity of the EEIS performance should be quantified under different temporal scales.

Principle 3 Design alternatives should be resilient with the quantified uncertainty and sensitivity at different life expectancies:

1) Both maximum warming and cooling scenarios should be considered for climate change within design life expectancy.
2) Steady versus unsteady system should be clearly identified to avoid overcapacity due to using a safety factor in designing EEIS.
3) Design alternatives should be resilient on specific temporal scales such as 10-, 20-, and 50-year life expectancies.

EEIS such as WTPs, WWTPs, and GIs should be resilient to more serious drought and flooding due to climate change. Minimum temperature in the north and maximum temperature in the south should be considered in designing EEIS. At minimum temperature, biological activity will decrease by half for every ten-degree drop. The global warming will have dire consequence for human society due to sea level rising (SLR) and saltwater intrusion (Table 3.13). Therefore, on a different temporal scale, the following issues should be considered:

1) Short-term versus long-term operation conditions should be considered.
2) Operating cost and benefit should be quantified to determine the net profitability of EEIS within the project life expectancy.
3) LCA should also consider short-term and long-term energy mix.
4) Reaction kinetics is the determining factor in reactor design.
5) Blue, gray, and black water management should be resilient to climate change in 20, 50, and 100 years.

3.4.3.1 Principle 4: Efficiency of Renewable Material

A paradigm shift from fossil fuel as feedstock to renewable material calls for the efficiency of renewable materials in designing EEIS. Renewable material alternatives should

Table 3.13 Spatial and temporal scales.

Design	Problems	Spatial	Temporal	Scale
Air	Coal power plant	Locations of different renewable energy resources	Winter seasons as worst design scenarios	Household Community City State Country
Water	WWTPs	Central versus decentralized	Winter as lowest biomass production season	Same as the above
	WTPs	Separate water quality supply	Algae bloom seasons as worst time	Same as the above
Land	Landfill	Residential versus public spaces for landfill sites	Mobile integrated leachate treatment system	Same as the above
	Leachate	Equalization pond with water plants for biomass production	BOD–COD ratios for young, medium, and mature landfill leachates	Same as the above
Ecological	River delta hypoxic zone	Nutrient recovery	Summer as the worst season due to algae bloom	Same as the above
	Wildlife extinction	Location of endangered species	Winter as most vulnerable seasons for wildlife	Same as the above

be regenerative or neutral. Negative design in terms of FP on air, water, and soil should be avoided.

Principle 4 Mass efficient alternatives should be designed to achieve the minimal footprints on air, water, and soil:

1) Mass balance, kinetics, and stoichiometry are three pillars in designing efficient reactors.
2) Mass efficiency requires maximum renewable mass incorporated into the final products such as water and nutrients.
3) Component, unit, process, and system should be designed with maximum material efficiency.

Regenerative design is the key for sustainable development by decoupling economic development from environmental deterioration. To reach this sustainable goal, regenerative design is the most practical way to provide a technical approach so that EEIS can be designed to satisfy the above conditions. Regenerative design requires the following elements:

1) Reactor, process, and system should be designed to achieve the maximum material efficiency.
2) Raw materials should be renewable with minimal FP on the environment.
3) Plants and communities should be designed to achieve regenerative or zero water, energy, and waste.

Renewable materials should be efficient and utilized to their maximal physical limits or best practical technical benchmarks. Ideally, every effort should be made to avoid adding chemicals, for example, using UV disinfection to avoid adding chlorine to drinking water. Green chemistry principles suggest that atom economy, less hazardous chemical use, and safer chemicals are key ways of achieving mass efficiency. For water and wastewater design, the following are typical enabling technologies:

1) UV disinfection should be used to avoid adding chlorine to drinking water or treated wastewater.
2) Ion exchange could be used to replace precipitation or lime softening.
3) Synthesized magnetic adsorbents could be used to replace other adsorbents for easy separation.

Principle 5 Energy-efficient alternatives should be regenerative or neutral and avoid negative design in terms of energy:

1) Renewable energy such as biomass, solar, or wind energy should have priority in designing EEI projects.
2) Energy efficiency demands each unit or process to have maximum energy efficiency of the renewable energy.
3) Pumps should work at maximum efficiency to reduce operating costs to achieve maximal efficiency of the system.

Thermodynamic limits or technical benchmarks should be used to identify opportunities in the most potential gap in material and energy efficiency to maximize the benefit of design alternatives. Maximum heat recovery will be dictated by thermodynamics such as exergy, primary biomass by ecological principles, and nutrient recovery by natural biocapacity. Technical benchmarks should be used to identify feasible recovery criteria from waste using the best available technologies. Designers should calculate thermodynamic potential or exergy as well as other limits for each process. Embedded entropy is an investment when recycle, reuse, or recovery is assessed. In general, total energy to be recovered could not exceed the exergy of content organic pollutants from wastewater. After the theoretical limits are known, technical benchmarks should be used to identify the maximum potential saving gap for materials or energy by employing state-of-the-art technologies. The following factors should be implemented in the design process:

1) Theoretical limits should be calculated by thermodynamic properties.
2) The exergy of wastewater or wastes should be calculated to obtain realistic limits.
3) Potential gaps between limits and achievable metrics should be quantified to identify maximum SEE design opportunities using technical benchmarks.
4) Thermodynamic laws will put a limit on which level the design can achieve. On the other hand, the choice of regenerative, neutral, and negative engineering design may very likely depend upon benefit and cost ratio within the project budget. Renewable energy sources such as biomass should be selected to have minimal FP on the environment.
5) WWTPs should be designed to be energy positive as water resource recovery facilities (WRRFs) to generate a net profit on the investment.
6) Plants and communities should be designed to be zero energy within thermodynamic limits.

Renewable energy feedstock should be efficient and utilized to the thermodynamic maximum or best practical technical benchmarks. When regenerative design cannot be achieved due to a thermodynamic limit, an external energy source such as solar or wind power should be sought. For WRRF design, the following are typical successful approaches:

1) Codigestion of fat, oil, and greases (FOG) from kitchen waste.
2) Codigestion with glycerin waste generated from biodiesel production.
3) Algae growth in oxidation ponds of wastewater treatment.
4) Solar panels for additional electricity.
5) Anaerobic membrane biological reactor (AnMBR).
6) Anaerobic ammonia oxidation processes (Anammox).
7) Microsieving to separate and divert organic matter to anaerobic process to produce biomethane (Table 3.14).

Table 3.15 provides examples of applying Principle 5 for sustainable design solutions at each design scale.

3.4.4 Principle 6: Prevention

Prevention is the most effective way to ensure the sustainability of EEI design. For example, one gallon of gasoline is a useful fuel, while it would cost thousands of dollars to clean up if it leaked into groundwater as evidenced by many underground tank cleanup programs in the United States. On every design scale, an opportunity exists to prevent waste rather than treat it after it is generated.

Table 3.14 System and component efficiency.

Design	Problems	System	Components	Efficiency
Air	Open burning	Compact wood pellets	Combustion efficient	Incinerator efficient
Water	Tank process	Pipe reactor	MBR separating hydraulic retention from cell residence time	Hydraulic efficiency
Land	Landfills	Bioreactor landfills	Bioreactor landfill methane production	Utilization efficient of methane from bioreactor landfill
Ecological	Food wastes leachate problems	Codigestion with sludge	Methane production from anaerobic digester	Electricity and heat generation efficiency from methane production from anaerobic digester
	Urban heat	GI	Green roof and green garden	Primary biomass production efficiency of GIs

Table 3.15 SEE design metrics.

Design	Problems	Mass limits	Energy limits
Air	Incomplete combustion	Mass conservations	Efficiency of any conversion process is less than 100%
Water	AS in WWTP	Create excessive sludge	Energy intensive and not productive
Land	Pollutants in runoff	GI to reduce soil erosion and pollutant runoff	Regenerative design for biomass production
Ecological	Energy efficiency in food chain	Biomass for habit	Higher energy efficiency process

Principle 6 Prevention, reduction, and minimization strategies should be implemented as early as possible:

1) Source reduction, recycling, and minimization on- or off-site should be maximized.
2) Chemicals such as chlorine, lime, and aluminum should be avoided as much as possible.
3) Waste such as sludge should be prevented or minimized at all costs.

Waste requires unnecessary expenditure of capital, energy, and resources and has no benefit. The following technologies are important in prevention:

1) The TDPs of green chemistry should be applied to prevent the generation of chemical waste.
2) Anammox should be designed to prevent sludge production in the AS process whenever nutrients such as nitrogen are to be removed.
3) Enhanced coagulation using biodegradable coagulants should be used to prevent lime and aluminum sludge.

Prevention opportunities exist as presented in Table 3.16 in air, water, and soil media.

3.4.5 Principle 7: Recovery

Recovery serves dual purposes. It protects human and ecosystem health by removing the "contaminants" of wastewater while recovering them as useful resources such as water, energy, or materials. In addition, the recovered water can be used for agricultural irrigation, energy to power WRRFs, and nutrients as fertilizer in agriculture. Furthermore, resalable products could bring cities sizeable economic, social, and environmental benefits.

Principle 7 All waste should be recovered in terms of benefit and cost ratio within an allowable budget:

1) Water, energy, and nutrients should be recovered in SEEI design.
2) Material and energy recovery should reduce operation costs.
3) Life cycle benefit and cost ratio should be the major factor determining to what material is to be recovered and to what extent.

Table 3.16 Prevention.

Design	Problems	Materials	Energy
Air	Coal power plants	Reduction of carbon dioxide to fuel	Renewable energy
Water	Chlorination	Vacuum ultraviolet disinfection (VUV) to prevent disinfection by-products during chlorination	Energy efficiency, UV
Land	Food wastes	Recovery of oil from production of biodiesel	Codigestion of food wastes with sludge
	Soil degradation	Constructed wetlands	GI for biomass production
Ecological	Leachate	Emerging pollutants such as antibiotics and pharmaceutical chemicals	Codigestion with sludge to produce methane

Extensive energy and material are the major costs for EEIS and result in waste, inefficiency, and FT on the environments. Recovery is the most important design strategy to achieve regenerative design. For WWTPs, the most valuable products are reclaimed water, energy, and nutrients. Table 3.17 presents examples designed for recovering nutrients.

Table 3.17 Examples of nutrient recovery across design scales.

Design scale	Current practice	Recovery
Molecular	Excess reagent such as methanol for nitrification	Methanol is not needed as carbon source
Process	Nitrification and denitrification, excessive aeration is needed	Struvite in Struvite precipitation
Product	No recovery	Struvite as marketable fertilizer
System	Nitrification and denitrification tanks	Modularized struvite reactors

Specific rules for recovery are as follows:

1) Recovery is to reuse and recycle mass and energy from components, reactors, and processes of EEIS.
2) Energy and materials such as water, and nutrients in wastewater should be considered recoverable.
3) Solid waste refining should be considered as an alternative in solid waste management systems.

The key technologies for effective recovery are:

1) Anaerobic digestion of sludge and energetic food wastes.
2) Recovery of N and P, for example, through precipitation of struvite.
3) Ion exchange to recover inorganic elements such as precious metals or rare earth.
4) Membrane technology such as ultra-, micro-, and nanofiltrations and reverse osmosis.
5) Microbiological fuel cell.
6) Molten salt technology for energy, heat, and hydrogen recovery.

All materials should be considered as recoverable resources. The recovery extent should be determined by the constraints of technology and costs as shown in Table 3.18.

3.5 Principle 8: Separation

Separation should be implemented at the initial design to recover and reuse materials for different processes in different media as shown in Table 3.19. Separation may serve different purposes and the more separation categories, the better. In Japan, as many as eight different categories of solid waste bins are installed in major public places. Separation may also serve treatment. For water reuse, water can be separated into blue, gray, and black water so that they can serve different purposes. Water of different qualities should be provided to residential houses because it wastes a lot of resource in irrigation by using water of drinking quality.

Principle 8 Separation should be implemented as early as possible at every scale to minimize the consumption of both materials and energy:

1) Waste should be separated according to its nature such as recyclability or biodegradability.
2) Separation technologies such ultra-, micro-, and nanofiltrations should be selected according to the size of pollutants.
3) Phase transfer technologies such as activated carbon should be the secondary choice only if destructive technologies such as advanced oxidation process are not feasible to recover inorganic elements such as precious metals or rare earth.

Table 3.18 Recovery.

Design	Problems	Mass	Energy
Air	Landfill CH_4	Bioreactor landfills	CH_4
	Carbon dioxide from coal power plants	Carbon sequestration	Methanol
Water	Nutrients	N and P	Struvite
	Waste water	Water reuse	Reclaimed water
Land	Soil erosion	Regenerative GI	Biomass
Ecological	Algae boom	Constructed wetlands	Blue algae for biodiesel

Table 3.19 Separation.

Design	Problems	For recycle	For reuse	For treatment
Air	Hazardous wastes incineration	Hazardous wastes such as acid and base recycle	CO_2 sequestration	Catalyst
Water	Leachate combined with WW	Blue, gray, and black water	Dripping irrigation	BMR, MF, NF, RO
Land	Food wastes	Fat, oil, and grease (FOG) codigestion with sludge	Kitchen oil for biodiesel	Compositing
Ecological	Invasive plants and aquatic species	Isolate and destroy the invasive plants and aquatic species	Aquatic plants as animal feeding stock	Green algae for biodiesel production

For example, enhanced coagulation should increase the removal efficiency of COD so that the organic matter can be maximally diverted to anaerobic process to produce methane. As a result, less aeration is required and less sludge will be generated. Food waste should be separated from solid waste so that they can be codigested with sludge. Following are typical separation cases:

1) Separate urine from feces.
2) Enhanced primary sedimentation or separation for diversion of COD to anaerobic digestion.
3) Ion exchange for ammonia removal.
4) Food waste should be separated from solid waste and codigested with sludge.
5) Leachate should be treated separately from wastewater.

3.5.1 Principle 9: Treatment

When pollutants are neither recoverable nor separable, destructive technologies such as advanced oxidation process (AOP) are better than phase transfer technologies. In treatment, material and energy inputs should be renewable and from readily available sources throughout all life cycle stages.

Principle 9 Treatment should avoid adding chemicals or use minimal nontoxic renewable chemicals and energy to reduce footprints:

1) No toxic chemicals such as chlorine or aluminum should be added to drinking water.
2) Destructive technologies such as AOPs should be the first choice over phase transfer technologies such as activated carbon.
3) Waste such as sludge produced per unit of product such as water should be minimal.

In most cases, particle size determines the physical treatment process, while the nature of pollutants determines the adoption of either biological or physicochemical treatment technologies. For WWTP, regenerative design is the key to ensure a plant able to operate sustainably. Renewable and nontoxic materials should be used whenever possible. Catalysts should be used to avoid chemicals. Immobilized catalysts should be used to eliminate separation processes. Design for unnecessary capacity should be considered a design flaw because fit for all may not be a sustainable solution (Table 3.20).

Table 3.20 Treatment.

Design	Problems	Nature of pollutants	Concentration
Air	VOC	Halogenated versus nonhalogenated	
Water	Organic	Nonbiodegradable, physical/chemical Biodegradable	Less than 0.1%, AOPs 0.1–5%, wet oxidation Greater than 5%, thermal treatment
Land	Contaminated land	Biodegradable, *in situ* bioremediation Nonbiodegradable, *in situ* physicochemical Toxic metals, EDTA extraction	Less than 1 ppm
Ecological	Biodiversity	Eutrophication	Greater than 10 ppm, recovery

The following oxidation technologies play a critical role in regenerative design:

1) Anaerobic ammonia oxidation (Anammox) to reduce the aeration intensity of AS.
2) AOPs to replace activated carbon adsorption.
3) UV disinfection and oxidation to replace chlorination.

3.5.2 Principle 10: Retrofitting and Remediation

Due to the unsustainable design of EEIS in the past, 95% of WWTPs were energy negative and consumed about 3% of total electricity generated in the United States. Retrofitting these WWTPs is a major challenge for SEE designers.

> **Principle 10** Retrofitting and remediation should recover material, energy, or land with minimal footprints:
>
> 1) WWTPs should be systematically retrofitted to become energy-positive water resource recovery facilities (WRRFs).
> 2) WTP should gradually adopt ultraviolet disinfection and degradation to replace chlorination.
> 3) Remediation of air, water, and soil should have minimal footprints on air, water, soil, and energy.

In the next two decades, one of the most challenging engineering issues is to effectively retrofit the old energy-negative WWTPs to be energy-positive WRRFs all over the world. The US EPA and the Water Environment Research Foundation have produced many design manuals and presented many successful examples. To retrofit WWTPs, the first step is energy audit to identify the greatest opportunity to improve energy efficiency in a WWTP. Second is to replace old devices with new technology to reduce energy consumption. Third is to reconfigure processes so that energy consumption can be reduced by adopting following technologies:

1) Traditional AS should be coupled with Anammox to reduce sludge.
2) MBR could be added in the aeration chamber to increase treatment efficiency.
3) AnMBR could be added in anaerobic digester to increase methane production.

Green remediation principles developed by the US EPA should be implemented to minimize the remediation FP of resources and energy. The Excel-based software developed by the US EPA will assist the designer to significantly reduce the FP of major retrofitting projects. Green remediation must consider contaminated media, the nature of pollutants, and concentration (Table 3.21):

1) Renewable energy resources could be used to regenerate biomass.
2) Green plants and algae could be used in water remediation.
3) Solar energy could be used for sludge dewatering.
4) Renewable energy should be used in remediation to reduce FP.

3.5.3 Principle 11: Optimization through Modeling and Simulation

The most economical way to optimize EEIS design is through mathematical modeling, which could be then used to simulate. Opportunities of increasing the performance of the system and decreasing costs could be identified through computer modeling and simulation. For example, after the detail mechanism and kinetics of Anammox were established, full-scale Anammox reactor were

Table 3.21 Green remediation.

Design	Problems	Nature of pollutants	Concentration
Air	SMOG	Reduce volatile organic pollutants Reduce coal combustion	VOC, nitroxide, and $PM_{2.5}$ should be below ambient standards
Water	Organic	Nonbiodegradable, physical/chemical Biodegradable, biological	Less than 0.1%, AOPs 0.1–5%, wet oxidation Greater than 5%, thermal treatment
Land	Contaminated land	Biodegradable, *in situ* bioremediation Nonbiodegradable, *in situ* physicochemical Toxic metals, EDTA extraction	
Ecological	Biodiversity	Eutrophication	Less than 1 ppm Greater than 10 ppm

successfully designed and built without pilot prototype (van der Star et al., 2007).

Principle 11 The component, process, and system of EEIS should be modeled, simulated, and optimized under both global warming and cooling scenarios:

1) Models should be calibrated, verified, and validated for component, reactor, process, and system before their application in simulation and optimization.
2) Simulation could be conducted with computer tools such as Simulink so that time series performance data can be obtained for different alternatives.
3) Optimization should be carried out using optimization software such as Monte Carlo simulations to quantify uncertainty and sensitivity of the outcome.

Since pilot- and prototype-scale experiments are expensive, computer modeling and simulation could be used to directly scale up a process without design, construction and operation of the corresponding pilot plant. Real-time monitoring and control are critical to ensure that the performance of the system will not deteriorate during operation. Important technologies are:

1) Supervisory control and data acquisition (SCADA).
2) AutoCAD and SolidWorks for sustainable materials.
3) 3D printing for pilot-scale testing.
4) Simulink for computer simulation by Matlab.

Optimization of designed systems should be conducted by using computer software to reduce cost (Table 3.22).

3.5.4 Principle 12: Balance Between Capital and Operating Costs

Ideally, future EEIS should be designed and constructed as standardized modules so that the cost and performance of the components, reactors, processes, and systems could be sustainable. Also, standardized modules may be reused when the EEIS is no longer needed. For example, leachate treatment systems after 30-year service may not be needed; the system could be used at the other landfill sites. Modular design makes products, processes, and systems reusable or recyclable after design life expectancy. Therefore, designers should aim at durability, not immortality.

Principle 12 Standardized and modularized components or systems should be designed with real-time monitoring and control during operation to reduce both capital and operating costs:

1) Components and units should be standardized and modulated for mass production.
2) Final design alternatives should be determined by balancing capital and operating costs according to years of return for an EEIS.
3) Constructability should meet LEED standards in terms of material, water, and energy consumption.

Regenerative design can ensure beneficial use of recycled components, while standardized modules are the best route to reducing the FP of modern EEIS. Standardized modules should have minimal capital and operation cost as well as minimal environment FP during their life cycle. To achieve sustainability, LCA of components,

Table 3.22 Model and simulation.

Design	Problems	Model	Simulation
Air	SMOG	How SMOG is formed, transported, and dispersed What are the corresponding health effects under different concentrations of SMOG	Dynamic SMOG models
Water	Groundwater contamination	How contaminants transport through saturated versus unsaturated zones	Contaminant transportation models
	Surface water pollution	How contaminants transport and disperse in lakes or rivers	Water pollutant transportation models
Land	Soil erosion	How contaminants transport and disperse kinetics in soil	Transportation and accumulation of metals and organic pollutants in soil
Ecological	Eutrophication	Total maximal daily load	Models of nonpoint source runoff of N and P

reactors, processes, and systems should be quantified during all the design phases and operation during its life time by using inherently safe and benign material and energy (Table 3.23).

Innovative approaches are required to create engineering solutions beyond current or dominant technologies. To achieve sustainability requires SEE designer to improve, innovate, and invent new technologies with passion. When TDPs are systematically applied in the early stage of the design, significant cost saving could be recognized by new EEIS design. Staying with problems longer is an excellent way to make a significant contribution. Creative thinking, leadership, and entrepreneurship should be implemented in design phases as shown in following design strategies of innovative technologies:

1) Innovation through new science and technologies is the key for designing sustainable EEIS.
2) Rapid prototyping such as 3D printing and computer simulation is crucial to test the reliability of the product or processes.
3) Market penetration through entrepreneurial spirit would accelerate the adoption of innovative technologies.
4) Incorporating SEE into EEIS design will transform the industry through innovation and entrepreneurship in all the design phases.
5) Innovative technologies should be used as soon as they have been proven on a commercial scale within tested reliability.
6) Entrepreneurial spirit should be reflected in identifying environmental challenges and providing the best solutions for the markets in air, water, and soil quality management (Table 3.24).

3.6 Implementation of the SEE Design Principles

3.6.1 Procedure to Implement SEE Design Principles

The driving force for sustainability is the educated SEE human capital to design engineering systems in a sustainable manner. The TDP could be integrated in engineering design phases as follows:

Step 1: Define objectives on 3E in addition to sponsored goals.
Step 2: Collect related data on air, water, and land background information related to a project.
Step 3: Assess the alternatives in terms of FP on material and energy using LCA and FP tools.
Step 4: Develop alternatives by targeting regenerative design or neutral design, and avoid deficient design under cost constraints.

SEE design can be integrated into three phases of traditional engineering design. Therefore, in phase I for 30% submittal, the following steps are recommended:

1) Collect relevant data for air, water, and soil quality index.
2) Set up a design goal to meet regulatory requirements.
3) Get all stakeholder input and continuously involve them in the design process.
4) Collect sustainable metrics of the unit process in terms of FP on materials and energy.
5) Develop design alternatives that can achieve productive, neutral, or negative energy efficiency.

Table 3.23 Standardized and modular system for cost reduction.

Design	Problems	Modular	Standards
Air	SMOG due to coal power plants	Solar photovoltaic modules	Standards
Water	Coagulation and sedimentation tank	Static mixer UF, MF, NF, and RO	Standards
Land	Solid waste management	Incineration for energy recovery	Standards
Ecological	Eutrophication	Modular nutrient recovery products	Standards

Table 3.24 Innovation and entrepreneur.

Design	Problems	Innovation example	Entrepreneur
Air	SMOG	Facial mask	Marketing facial mask
Water	Drinking water contamination	Drinking water pen	Market penetration of DWP
Land	Mining site reclamation	Regenerative public park	Sports activity on the park
Ecological	Sludge	Sludge application to land for vegetation	Green economy

In the second phase of 60% submittal, the following steps are recommended:

1) Select two design alternatives with stakeholder input according to the TDP of SEE.
2) Quantify material and energy FP.
3) Quantify unit efficiency and compare with design metrics or benchmarks.
4) Estimate project cost.
5) Detail the design of the system.

In the third phase of 100% submittal, the following steps are recommended:

1) Optimize the system.
2) Improve the system by identifying potential for saving energy and cost without reducing efficiency.
3) Develop detailed engineering drawings for constructability.

3.6.2 Integration of SEE into Undergraduate Education

SEE may transform EE education through horizontal and vertical integrations in the 4-year curriculum with examples, problems, and case studies. Horizontal refers to an interdisciplinary approach, while vertical refers to the integration of SEE throughout the 4-year curriculum of EE. The TDP could be taught as modules for a multidisciplinary and freshman-level introduction to engineering or upper-level WTP and WWTP design. Using SEE principles at the start of the design process could shift the philosophy of EE design towards a sustainable future. Horizontal integration refers to SEE concepts that should be learned together with a specific course in the SEE curriculum according to Table 3.25.

Sustainability has many facets and requires an extremely broad list of possible inquiries related to policy analysis and design. In addition, the final alternative is subject to many constraints such as the economic, environmental, sustainability, manufacturability, ethical, health and safety, social, and political limits. To achieve true SEE design, teaching strategies may require collaboration between many disciplines. For example, chemistry, biology, electrical engineering, and information technology all have their place in the integrated regenerative design of future WRRF. Inter- and multidisciplinary pedagogy should be promoted. Multidisciplinary teaching brings together several disciplinary perspectives. For example, regenerative design needs SEE and ecological principles to produce biomass such as algae to produce biodiesel from wastewater. Many facets of sustainability require the breaking of economic, legal, institutional, chronological, and societal barriers to achieve the desired transformational goals. To these ends, SEE should be vertically integrated into the EE curriculum as in Table 3.26.

To accelerate the adoption of SEE, education and training are effective ways to penetrate the SEEI market. SEE could follow the success of GC and GE all over the world. Course projects could be based upon a real EEIS design project at residential, community, or city scales. These projects could be conducted to improve information flow on TDPs and examples. Effective economic and sustainability tools and criteria could be developed to aid the design process. Benchmarks that lead to regenerative products, processes, and systems of EEIS could be developed in the integrated air, water, and land management network. One of effective way is to promote a grassroot movement towards SEE education and practice. Grassroot movement requires students to collect their relevant EEIS data to develop sustainable design alternatives by following the TDPs. Grassroot movement from all the SEE designers will help implement the SEE design principles throughout the EE industry during all design phases. For example, if 15 000 WWTPs in the United States are retrofitted into energy-positive WRRFs, market drivers will clearly also demand more financially sound design alternatives, so retrofitting the current WWTPs and designing new WRRFs will provide the best sustainable design solutions. At the same time, regulatory frameworks should be created to reduce uncertainty for the industry in adopting SEE TDPs. Institutions such as the UN should adapt the SEE principles to achieve the SDGs. To achieve 17 SDGs, SEE designers may lead engineering design communities across the spectrum by forming alliance with GC and GE. In addition, SEE designers should also team up with mechanical and electrical engineering to design sustainable components needed for automatic and standardized regenerative WWTPs. When three major education and working forces of GC, GE, and SEE come together, innovative technologies could be quickly developed through interdisciplinary research and true sustainable EEIS could be designed for decades to come.

Vertical integration with other curricula is crucial to promote multidisciplinary and transdisciplinary thinking. Government and academic institutions need to come together to create a new generation of trained experts in SEE, building sustainable communities and cities, regenerative WRRFs, and stand-alone leachate treatment systems. It also needs to create educational, information, and networking tools for innovators to speed up the development of SEE solutions. Students can use the LCA tools of SEE to design new processes and retrofit existing WWTPs. According to the TDP of SEE, the challenges of critical SEE technologies should be solved through research and development and marketed. As a result, innovative technologies such as Anammox, AnMBR, and UV disinfection and oxidation could penetrate markets all over the world.

Table 3.25 Conceptual mapping of SEE with undergraduate EE curriculum.

SEE topic	Chapter in SEE	Corresponding course in EE curriculum
Renewable and nonrenewable natural capitals	Natural capital and carrying capacity	Earth Science, first year
Water, carbon, nitrogen, phosphorus, energy, and ecological footprints Risks and fossil fuel depletions	Human demand and footprint	Earth Science, first year
Sustainable development goals Regenerative, neutral, and negative EE design	Introduction to SEE	Environmental Engineering, 4th semester
Integrated management of air, water, and land	Integrated design and hierarchy	Environmental Engineering, 4th semester
Decentralized WWTP Separate water quality supplies GIs such as separate gray and blue water reuse	Spatial and temporal scales	Environmental Engineering, 4th semester
Mass and energy conversion laws Thermodynamic limits	Constraints by fundamental laws and cost	Thermodynamics, 5th semester
Inefficiency in energy conversion	Mass and energy efficiency	Chemistry Environmental Engineers, 5th semester
Renewable energy Intrinsic toxicity	Prevention	Environmental Engineering, 4th semester
Food waste separation from solid wastes Leachate separation from WWTP Enhanced coagulation for anaerobic digestion	Separation	Solid and Hazardous Waste Management, 6th semester
UV disinfection and oxidation Leachate treatment systems Ultra-, micro-, and nanofiltration and reverse osmosis	Treatment	Water Supply Engineering, 5th semester Sewerage and Wastewater Treatment, 6th semester
Water, carbon, nitrogen, phosphorus, and energy recovery in WWTP and combustion heat or pressure recovery calculations	Recovery	Solid and Hazardous Waste Management, 6th semester Sewerage and Wastewater Treatment, 6th semester
Renewable energy Green infrastructure after remediation	Green remediation	Solid and Hazardous Waste Management, 6th semester Sewerage and Wastewater Treatment, 6th semester
WTP and WWTP design Design GI systems Green remediation projects Specific optimization and simulation tools such as Simulink	Optimization and simulation	Water Supply Engineering, 5th semester Sewerage and Wastewater Treatment, 6th semester
LEED certification of green building in WWTPs and WTPs	Constructability Reliability	Environmental Engineering Senior Design, 8th semester
SEE product design and development Case studies of drinking water pen and book, pipe reactors Marketing of SEE products	Innovation and entrepreneurship	Environmental Engineering Senior Design, 8th semester

All the TDPs can be applied to design EEIS of air, water, solid waste, and soil management. The most important is to develop the best EEIS solution to achieve the best performance and minimal FP on the environments at the minimal costs. The major environmental engineering challenges for the SEE designers include residential water and solid wastes in urban and rural areas, industrial wastes, and agricultural wastes. One of the effective

Table 3.26 Vertical integration of twelve design principles in environmental engineering curriculum.

Introduction to civil and environmental engineering	Green environmental engineering topic
Freshman engineering	Food waste separation
	Introduction to environmental regulations
	Introduction to life cycle assessment
Sophomore engineering	Life cycle assessment of WWTP
	Environmental regulations
Material and energy balances	Emissions terminology/calculations
	Renewable material and energy balances
Mass transfer/equilibrium stage separations	Mass separation
	Risk assessment
Material science	Estimation of properties
	Life cycle assessment
Heat transfer	Introduction to exergy
Chemical thermodynamics	Estimation of chemical properties
Separation processes	Pollution prevention strategies
	Separation process integration
Mass and energy efficiency of reactor	Pollution prevention strategies
	Green chemistry
Unit operations	Tank processes versus pipe reactor
Process/plant design	Heat integration and mass integration
	Flowsheet analysis
	Life cycle assessment
Process dynamics and control	Pollution prevention modeling and control
Design for pollution prevention	Heat and mass integration
	Process analysis
Senior engineering design project	Community environmental engineering project

way for the SEE designers is to establish benchmark data so that governmental agencies can enforce the TDPs in their bidding documents. With these benchmark data, governments could accelerate the adoption of SEE technologies with effective incentives policies to reward SEE design and construction. State and federal policies could pass legislation to design and advocate innovations that increase the supply and demand for SEE solutions. Funding, incentives, and/or prizes should be awarded to SEE research and commercialization. Through state and federal funding, national SEE research or demonstration centers should be set up to transfer technology and technical know-how. To ensure the success of the governmental programs, the benefits of SEE to communities and cities should be quantified and publicized to society. SEE business, economic and health benefits, and opportunities and funding should be disseminated. Opportunities and strategies should be identified to jointly develop design criteria and SEE solutions. SEE progress at city, county, state, country, and global scales should be tracked. Why certain SEE products, technologies, and designs have succeeded or failed in the marketplace should be analyzed. Revenues, green job, economic benefits, and trends using SEE TDPs should be documented. Environmental or public works agencies should monitor the effectiveness and usefulness of sustainable EEIS.

3.7 Exercise

3.7.1 Questions

1) What are the six design hierarchies and six dimensions in the TDPs of the SEE?
2) What are the major barriers and drivers in designing sustainable EEIS?
3) What partnerships will have to be built, policies put in place, educational needs met, and investments made to implement SEE at undergraduate and graduate education?
4) What role should the SEE designers take to achieve the 17 UN SDGs?
5) What should the scale of innovation in SEE be?
6) How can the importance of SEE in education and research be evaluated?
7) What are smart policies that support SEE markets, research, and innovation?
8) What is the most important design rule that you would add to each SEE design principle?

3.7.2 Calculation

The paradigm shift in designing sustainable WRRF rooted in the view that wastewater is a resource and water, energy, and nutrients could be recovered. Table 3.27 lists the typical wastewater characters in Miami-Dade County.

1) If each person in the county is producing 60 $gBOD_5$ (biochemical oxygen demand consumed over 5 days) in wastewater, which is equivalent to 100 gCOD/

Table 3.27 Typical wastewater characters in Miami-Dade County.

Wastewater characters	Concentration
BOD_5, mg/l	223
COD, mg/l	333
TN as N, mg/l	50.5
NH_4–N, mg/l	26.4
Organic N, mg/l	23
NO_2–N, mg/l	0.02
NO_3–N, mg/l	1.12
TP as P, mg/l	36.68
TSS, mg/l	134
Alkalinity as $CaCO_3$, mg/l	205
pH	7.8
TDS, mg/l	379
Sulfate, mg/l	33
Fecal coliform, # col/100 ml	55 385
Conductivity, μΩ/cm	719
Temperature, °C	
Max	31
Min	22
TOC, mg/l	110.47

person/day, and if the maximal energy content is 14.7 kJ/gCOD, please answer the following:

a) What would be the theoretical maximal energy content in the wastewater if the Miami-Dade County population is 2.6 million?
b) In reality, Miami-Dade Water and Sewer Department operates three WWTPs with average flow rate of 300 MGD, what are the maximal energy contents in the flow daily?
c) How much differences are there between the answers of (a) and (b) and why?
d) On unit volume of cubic meter wastewater, what would be the theoretical unit energy content for biodegradable versus nonbiodegradable portion of organic pollutants in the wastewater of Miami-Dade County in kWh/m^3?
e) How are your two answers compared with the maximal unit energy contents of 0.70 and 1.23 kWh/m^3 estimated by McCarty et al. (2011)? Why are there differences?
f) If 30% of the maximal energy content were to be recovered as electricity, what would be the unit energy recovered in kWh/m^3/day?

2) If EEISs of air, water, solid, and soil quality management are to be designed according to the TDPs, what are the major quantitative indicators such as benchmark of unit energy consumption and chemical efficiency that are needed for each EEIS? In the United States, 15 000 WWTPs need to be retrofitted to energy positive. If benchmarks of unit energy consumption and chemical efficiency were to be established, how would you help your state to develop achievable benchmark data so that retrofitting WWTPs would be practical within realistic budget during a specific time period?
3) To achieve a high atom economy is important for the sustainability of a process. For example, COD can be oxidized to CO_2. If COD is to be converted to methane (CH_4), please answer the following:
   a) How many moles of oxygen is needed to oxidize 1 g COD?
   b) How many grams of COD are there in 1 g CH_4?
   c) Based upon Hess's law, what would be the overall enthalpy of the following reaction?

$$CH_4 + 2O_2 \rightarrow CO_2 + 2H_2O$$

   d) What would be the theoretical heat content per gram of COD?

3.7.3 Projects

3.7.3.1 Xiongan Project

Please conduct statistical analysis of different technologies such as microsieving to divert organic matter to anaerobic digester, coanaerobic digestion of FOG and glycerin wastes, Anammox, and algae to biodiesel in terms of unit energy consumption per unit cubic meter wastewater treated or COD removed. Please establish technical benchmarks of these metrics at top 10% best performance for air, water, and solid waste management so that Xiongan EEIS could be sustainably designed to accommodate future growth in the next 5, 10, 20, and 50 years.

3.7.3.2 Community Projects

1) Collect data of past 10 years on emission of the community to air, water, and soil.
2) Conduct regression analysis of the emission with time using SPSS.
3) Predict future discharge to air, water, and soil.
4) Design three alternatives for the community to achieve zero air, water, and waste in 10 years by using TDPs.

3.7.3.3 Proposal Development

During this course, you need to form a group to develop a group project proposal to address a sustainability challenge at residential, community, and city scales. It is designed to help you to gain experience of working in an interdisciplinary team and synthesizing concepts and ideas learned during the course. Please follow the steps outlined in the procedure to apply SEE TDPs during the course project.

The US EPA People, Prosperity and the Planet (P3) is to research, develop, and design solutions to real-world challenges involving sustainability. The P3 competition highlights the use of scientific principles in creating innovative projects focused on sustainability. Please develop a proposal according to the TDPs to improve quality of life, economic prosperity, and protection of P3. On the EPA website, http://www.epa.gov/ncer/P3, sustainability challenge project proposals are classified into the following categories. Please refer to the requirements from the website, and select one project to develop a group proposal:

1) Water (e.g. water quality, quantity, conservation, availability, and access).
2) Materials and chemicals (e.g. materials conservation; renewable, bio-based feedstocks; inherently benign materials and chemicals through GE and GC; biotechnology; recovery and reuse of materials through product, process, or system design).
3) Energy (e.g. reduction in air emissions through innovative strategies for energy production and energy distribution; energy conservation; inherently benign energy through GC and GE; biotechnology).
4) Built environment (e.g. environmental benefits through innovative green buildings, transportation and mobility strategies, and smart growth as it results in reduced vehicle miles traveled or reduces stormwater runoff).
5) Agriculture (e.g. irrigation practices, reduction, or elimination of pesticides).
6) Information technology (e.g. delivery of and access to environmental performance, technical, educational, or public health information-related environmental decision making).

The environmental components in the US EPA sustainable matrices are as follows:

1) Ecosystem services: Protect, sustain, and restore the health of critical natural habitats and ecosystems. Example could be innovative nutrient management techniques (GI).
2) Green engineering and chemistry: Design chemical products and processes to eliminate toxic hazards, reuse or recycle chemicals, and reduce total life cycle costs. Example could be LCA in molecular design.
3) Air quality: Attain and maintain air quality standards, and reduce the risk from toxic air pollutants. Example could be to investigate potential greenhouse gas emission reduction strategies.
4) Water quality: Reduce exposure to contaminants in water systems and infrastructure (including protecting source waters), and optimize aging systems and next-generation treatment technologies and approaches. Example could be purpose-driven water reuse and innovative treatment technologies.
5) Stressors: Reduce effects by stressors (e.g. pollutants, greenhouse gas emissions, genetically modified organisms) to the ecosystem and vulnerable populations. Example could be to determine the fate of modified nanoparticles in aqueous media.
6) Resource integrity: Reduce adverse effects by minimizing waste generation to prevent accidental release and future cleanup. Example could be to develop innovative technologies and processes to prevent environmental impact.

Once you determined the worst pollution in terms of air, water, and soil in Xiongan and your community, you need to develop EEIS proposal to be listed in the corresponding governmental agency's 5-year plan. Please develop proposal for air, water, and solid waste management EEIS for Xiongan and your city, respectively, according to the US EPA's requirements.

References

Anastas, P.T. and Warner, J.C. (1998). *Green Chemistry: Theory and Practice*. Oxford: Oxford University Press.

Anastas, P. and Zimmerman, J. B. (2003). Design through the 12 principles of green engineering. *Environmental Science & Technology* 37: 94A–101A.

ASCE (American Society of Civil Engineers) (2016). Failure to Act: Closing the Infrastructure Investment Gap for America's Economic Future, ASCE, Reston, Virginia.

Doughty, M.R.C. and Hammond, G.P. (2003). Cities and sustainability. *Towards an Environment Research Agenda*. A. Winnett and A. Warhurst, 81–105. Basingstoke: Palgrave Macmillan.

G20 Hangzhou Summit (2016). http://www.g20chn.org/English/ (accessed 15 December 2017).

Kirchhoff, M. (2003). Promoting green engineering through green chemistry. *Environmental Science and Technology* 37: 5349–5353.

Lewis, H., Gertsakis, J., Grant, T. et al. (2001). *Design + Environment. A Global Guide to Designing Greener Goods*. Sheffield: Greenleaf Publishing.

McCarty, P.L., Bae, J., and Kim, J. (2011). Domestic wastewater treatment as a net energy producer: can this be achieved?. *Environmental Science and Technology* 45: 7100–7106.

Osborn, D., Cutter, A., and Ullah, F. (2015). Universal Sustainable Development Goals: Understanding the Transformational Challenges for Developed Countries. Report of a Study by Stakeholder Forum (Stakeholder Forum, 2015).

Rigola, M. (1998). Producció + neta, Barcelona.

Sagar, A.D. and Frosch, R.A. (1997). A perspective of Industrial Ecology and its application to a metals-industry ecosystem. *Journal of Cleaner Production* 5: 39–45.

van der Star, W.R., Abma, W.R., Blommers, D. et al. (2007). Startup of reactors for anoxic ammonium oxidation: experiences from the first full-scale Anammox reactor in Rotterdam. *Water Research* 41 (18): 4149–4163.

UNEP (2016). UNEP Releases GEO-6 Regional Assessments. http://sdg.iisd.org/news/unep-releases-geo-6-regional-assessments/ (accessed 26 February 2018).

UN-Water (2015). Metadata on Suggested Indicators for Global Monitoring of the Sustainable Development Goal 6 on Water and Sanitation. http://www.unwater.org/fileadmin/user_upload/unwater_new/docs/Goal%206_Metadata%20Compilation%20for%20Suggested%20Indicators_UN-Water_v2016-04-01_2.pdf (accessed 26 February 2018).

World Business Council for Sustainable Development (WBCSD) (1993). Eco-efficiency. World Business Council for Sustainable Development, Congress in Antwerp, Belgium.

World Trade Organization (2017). World Trade Statistics. http://stat.wto.org/Home/WSDBHome.aspx?Language=E (accessed 26 February 2018).

4

Integrated and Interconnected Systems

All the things are eternally connected with everything.

OBJECTIVES
To define integrated and interconnected EEIS and to achieve high air, water, and soil quality indices within a specific time frame To illustrate WRRF as key enabling EEIS for sustainable food–energy–water (FEW) nexus of local economy To optimize environmental, economy, and equality (3E) goals of integrated EEIS by identifying and involving stakeholders To calculate waste generation factors for element management in integrated solid waste management To apply integrated system approach using blue, green, gray, brown, and black water management To design zero energy and zero waste for integrated solid waste management To understand integrated air quality management with renewable energy instead of fossil fuels such as coal To compare and select the best integrated EEIS using LCA, FT, and life cycle cost and benefit analysis

4.1 Principle 1

Population growth, increasing living standards, and urbanization under climate change require integrated and interconnected sustainable EEIS. EEISs should be integrated and interconnected for sustainable development (NSF et al., 2015). For example, integrated plan is at the core of food–energy–water (FEW) nexus. WRRFs could be integrated with the power infrastructure for the benefit of producing energy. More importantly, the integration would make it possible for WRRFs to operate independently of the electrical grid to increase the resiliency of the EEIS. WRRFs could also be integrated with other water management infrastructures to provide reliable water supply together with reclamation and desalination. In most cities, WRRFs locate close to sanitary landfills that also produce biogas. Fat–oil–grease (FOG) from restaurants could be codigested with sludge to produce more methane. Other organic wastes in the solid waste might be diverted to anaerobic digester to produce energy and save space in landfills. Ultimately, integrated plan and design would efficiently utilize all forms of wastes to generate reclaimed water, energy, or even nutrients such as N and P.

As human population continues the path to nine billion by 2050, more than half of the population would be in the middle class. Future cities have to be designed eco-friendly to accommodate the population growth and rising living standards. Ideally, a sustainable and resilient city should be in balance with the nature in terms of water, energy, and critical elements such as C, H, O, N, and P. Therefore, urban EEIS should be integrated and interconnected with the ecological system. With the paradigm shifts from wastes to resource, a new concept is sustainable element management. WRRFs could be designed in such a way that all the essential elements such as C, H, O, N, and P should be returned to the nature in its best original form. The integration and interconnection would bring synergy for EEIS and increase reliability on a spatial and resiliency on a temporal scale that will be explored in more detail in the next two chapters.

To sustainably manage essential elements such as C, H, O, N, and P, designers should develop alternatives to achieve environmental, economy, and equality (3E) goals using integrated approaches by engaging and interacting all the stakeholders within implementing feedback loops

Sustainable Environmental Engineering, First Edition. Walter Z. Tang and Mika Sillanpää.

to achieve 3E objectives. To achieve specific air, water, and soil quality indices and to decouple economic growth from environmental pressure, stakeholders within communities should be actively participating in the development of SEE solutions. The interconnections between engineering constraints and the economic and social domain to achieve sustainability are illustrated by the Venn diagram in Figure 4.1.

The ultimate goal of SEE is to design EEIS to meet demand both from the human and environment for generations to come. The major challenge faced by SEE designers is to optimize the conflicting interests of different stakeholders in achieving 3E goals. For example, using potable water for irrigation is not sustainable and wastes material and energy. Therefore, WWTP cannot be fit for all because transportation of sewage is expensive and reclaimed water is not close to its demands. For these reasons, fit for all would be intrinsically unsustainable. Therefore, SEE designers should target zero emission of air, water, and soil discharged to the environment after raw materials were used for human activities. One of the most important strategies is to manage waste by color codes such as blue, green, grey, brown, and black to achieve zero water design, while all the EEIS should be integrated and interconnected.

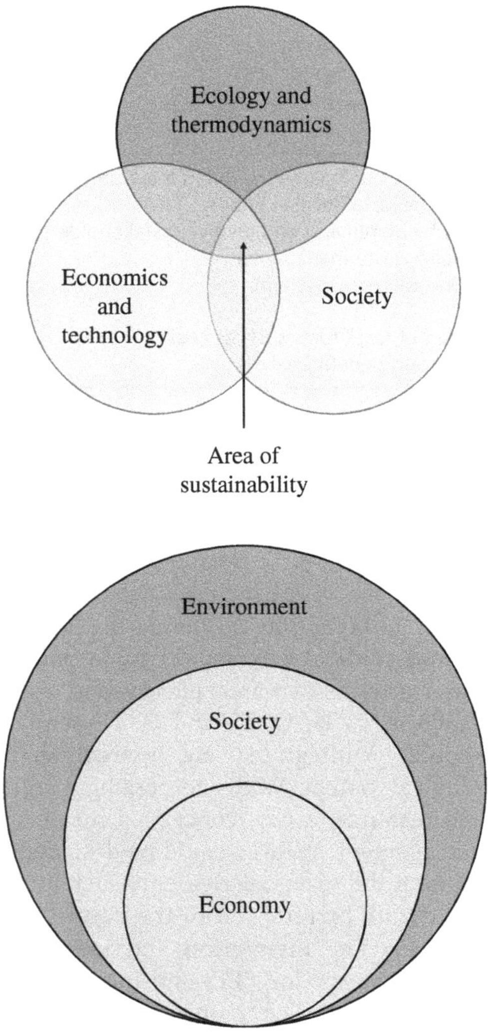

Figure 4.1 Sustainability "Venn" diagram for SEE.

Principle 1 Integrated and interconnected alternatives should be developed to protect human and ecological health for sustainable economic development by involving stakeholders.

1) Environmental, economic, and employment goals need to be integrated and prioritized.
2) Air, water, and solid waste management systems should aim at regenerative or zero discharge through integrated and interconnected design.
3) Blue, grey, and black water management may be decentralized depending upon the system boundary.

Integrated management plan is to decouple economic growth from environmental pressures. Many nations, e.g. the Netherlands, Sweden, Norway, Finland, and Denmark, showed that GDP increase could be decoupled from environmental deterioration through integrated design approach. Figure 4.2 clearly shows that the increasing GDP in the Netherlands has been decoupled from environmental degradation since 1990 (Bilthoven, 2007). These environmental effects include climate, eutrophication, acidification, and waste management.

The dual objectives of protecting human health and the biosphere require designers to find the best SEE planning strategies by thinking of the system as whole. Therefore, an integrated air, water, and land management plan should develop alternatives that minimize material and energy footprint (FP) using life cycle assessment (LCA) tools. Stakeholders should be closely consulted during development of the integrated and interconnected design alternatives. The US EPA recommended the following strategies to engage stakeholders:

1) A network of community opinion leaders should cover the existing boards or other governing bodies to provide stakeholder perspectives. Community

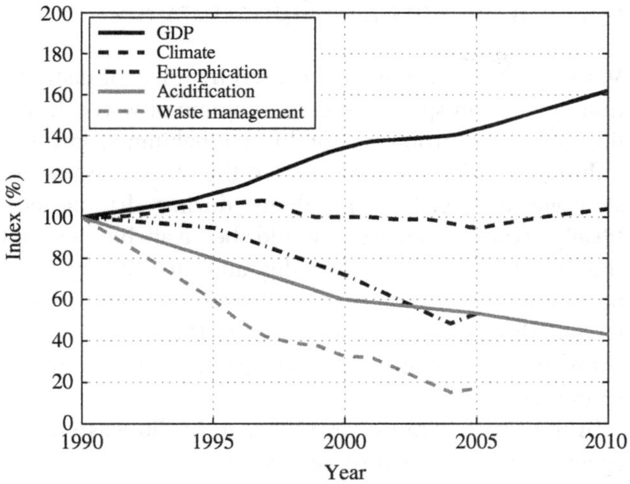

Figure 4.2 Comparing the Netherlands' economic growth and reduction of environmental impacts, Netherlands Environmental Assessment Agency, Environmental Balance 2007, Netherlands Environmental Assessment Agency. (*Source:* Adapted from Bilthoven (2007).)

priorities or utility sustainability objectives should be identified to prioritize SEE objectives, projects, values, and ideas for the short and long term.

2) Stakeholder steering committees should be engaged on key planning decisions, such as setting long-range goals and establishing project selection criteria and/or weighting schemes.
3) Traditional and social media should be used to keep the community informed about SEE design process and strategies.
4) Public meetings on milestones in SEE design should be held for the stakeholders and the public.

Integrated system approach (ISA) is a process whereby engineers analyze and optimize the whole technical system, which is composed of components, attributes, and relationships to achieve a specified goal. Components, attributes, and relationships, in an engineering sense, are defined as follows:

1) *Components* are the operating parts of a system consisting of input, process, and output. Each system component may assume a variety of values to describe a system state as set by some control action and one or more restrictions.
2) *Attributes* are the properties or discernible manifestations of the components of a system. These attributes characterize the system.
3) *Relationships* are the links between components and attributes.

ISA involves an interdisciplinary or team approach throughout the system design and development process to ensure that all design objectives are addressed in an effective and efficient manner. Different design disciplines should be involved to use different methods, techniques, and tools to implement the systems engineering approach. ISA helps ensure that engineers examine the many choices that are available to meet the specific needs of society. To reduce energy and material FP, designers should:

1) Explore all options and question all assumptions.
2) Analyze the whole problem breadthwise before exploring parts of the solution in detail.
3) Understand the primary system level before exploring the subsystem.
4) Determine the functionality of the whole system before developing a physical prototype.

Effective ISA design optimization in the early stages of design projects provides significant economic, social, and environmental benefits. Decisions made early in the design process have an enormous impact on life cycle system costs, both economic and environmental. Front-end design (FED) can lead to better-considered decisions, lower life cycle costs, and fewer late changes through concentration of design activity and decisions in the earliest phases, where changes cost the least. FED creates easier, more rapid integration and testing by avoiding many of the problems normally encountered in these phases to develop sustainable solutions. By reducing risk early in the design process, the overall result is a saving in both time and cost, with a higher quality system design. Systems engineering has a positive effect on cost compliance and the quality of a project. In addition to the direct costs associated with the project, the cost of making design changes escalates as system development progresses. Figure 4.3 shows that the cost of making design changes is lowest during the initial design phase, is 10-fold higher during the preproduction phase, and is more than 80-fold higher during the production phase. Approximately 60% of life cycle costs are determined in the concept phase, and a further 20% are determined in the design phase.

In the FED phase, EEIS should be integrated and interconnected according to design hierarchy – prevention, reduction, reuse, recycle, treatment, recovery, disposal, retrofit, and remediation – as shown in Figure 4.4 with input of stakeholders.

Figure 4.4 shows that prevention is the first choice among all the SEE design alternatives. If waste could not be prevented, reduction and minimization could be the second consideration. Once materials enter the

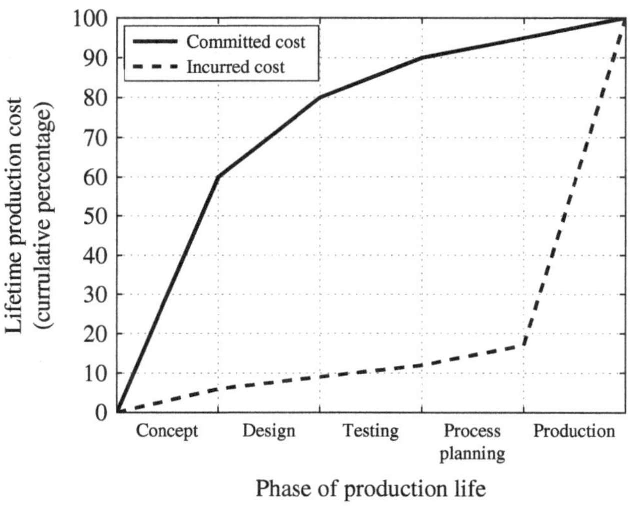

Figure 4.3 Comparison of the incurred costs and committed costs for each phase of system development. (*Source:* Adapted from Anderson (2008).)

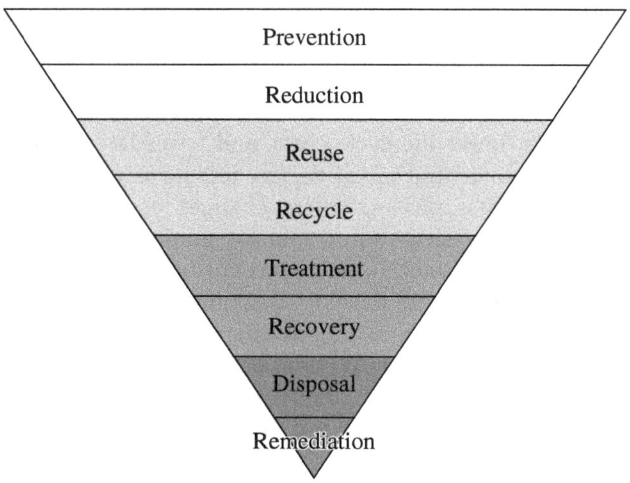

Figure 4.4 Design hierarchy for air, water, and land management.

economic process, recycling within the economy should be assessed to promote circular economy. To reduce the operating cost of EEIS, regenerative design should maximize the 3E objectives so that the EEIS produces more than the exergy of wastes. One of the effective ways to achieve this is photosynthesis. Ideally, green economy should target at creating green jobs such as bioenergy, solar energy, and recovery materials to achieve 3E goals. To assess the effectiveness of prevention strategies, design alternatives should be assessed and compared in terms of the reduction of chemical, biological, and radioactive risks and minimal FP on air, water, soil, and energy within the proposed budget.

4.2 Challenges and Opportunities

Water problems are mainly caused by unequal distribution of water in space. Excessive volume of water during floods creates problems for local inhabitants, farmers, industries, and municipalities. On the other hand, water shortage also creates problems during dry periods. Ideally, water resources should be available without significant changes in the quality and quantity throughout the year. The quality of water resources can be protected through strict wastewater (WW) treatment, elimination of soil erosion, point or non-point pollution sources.

In the United States, air, water, and solid waste management are governed by separated departments within a county. Miami-Dade County is a typical example. The Water and Sewer Department (WASD) manages water and WW, while the Department of Solid Waste is in charge of solid waste collection and disposal at landfills. Since landfill leachate is discharged to the southern Dade WWTP, reclaiming treated WW became too expensive to build the water reclaiming plant at full scale.

In China, about 1.5% of all deaths, or 64 000 persons/year, can be attributed to water pollution related to diseases. Intelligence quotient (IQ) loss in children due to ingestion of water and food contaminated with lead, mercury, and other heavy metals is the highest cost of water pollution damage. Each year, about seven million children are affected, losing on average 6.5 points on the IQ scale. In the next decades, US companies can benefit from cooperation between the United States and China in many areas. For example, the prevention and control of air emissions from electricity generation, industrial sources, vehicles, off-road machinery, ports, and vessels are the priority. Other major markets include management of hazardous waste and solid waste, persistent organic pollutants, and other toxic substances; environmental science, technology, and standards; emergency response; environmental threats to human health and to ecosystems; ecosystems restoration and recovery; environmental policy and management; environmental education and public awareness; and environmental law development, implementation, compliance, and enforcement.

4.2.1 Market Size of Solid Waste Management in China

Rapid urbanization in China is creating huge markets for solid waste management. More importantly, the government is opening this market to private investors through public–private partnership (PPP). The major indicators to measure effectiveness of solid waste management are percentage of detoxification rate, sanitary landfill, and incineration. Table 4.1 shows that the detoxification rate of municipal solid waste (MSW) increased from 63.5 to 90% at city level. At county level, the rate increased from 27.4 to 79.4%. If every city and county is to achieve adequate detoxification treatment capacity, an additional $19.17 billion is needed in the next 5 years. Table 4.2

Table 4.1 Main indicators of domestic waste treatment facilities in the 13th FYP.

Serial number	Main indicators		2010	2015	12th FYP	Goal
1	Detoxification rate (%)	Cities and counties	63.5	90.2		Achieved
		Cities	77.9	94.1	90	Achieved
		Counties	27.4	79.04	70	Achieved
2	Number of detoxification treatment facilities	Cities and counties	1076	2077		
		Cities	628	890		
		Counties	448	1187		
	Sanitary landfill	Cities and counties	919	1748		
		Cities	498	640		
		Counties	421	1108		
	Waste incineration plant	Cities and counties	119	257		
		Cities	104	220		
		Counties	15	37		
	Others	Cities and counties	38	72		
		Cities	26	30		
		Counties	12	42		
3	Detoxification treatment capacity (10 000 tons/day)	Cities and counties	45.7	75.8	87.1	Partially achieved
		Cities	38.8	57.7	65.3	Partially achieved
		Counties	6.9	18.1	21.8	Partially achieved
	Sanitary landfill	Cities and counties	35.2	50.2		
		Cities	28.9	34.4		
		Counties	6.2	15.8		
	Waste incineration plant	Cities and counties	8.9	23.5		
		Cities	8.4	21.9		
		Counties	0.46	1.6		
	Others	Cities and counties	1.5	2.1		
		Cities	1.3	1.4		
		Counties	0.25	0.7		
4	Ratio of incineration	Ratio of incineration		31	35	Partially achieved
		Ratio of incineration in east area		48	48	achieved
5	Cities and counties with detoxification treatment capacity	There are 43 cities and 367 counties without detoxification treatment capacity		There are 43 cities and 367 counties without detoxification treatment capacity	Every city and county has detoxification treatment capacity	Partially achieved
6	Investment (billion $)			18.485	37.657	−19.171 (needed)

Table 4.2 Treatment capacity of domestic waste treatment facilities in the 13th FYP (10 000 tons/day).

Serial number	Region	Treatment capacity in 2015	13th FYP Established treatment capacity	13th FYP New treatment capacity	13th FYP Closured treatment capacity	Treatment capacity in 2020
1	Beijing	2.38	0.93	0.99	0.50	3.80
2	Tianjin	1.04	0	1.24	0.53	1.75
3	Hebei	3.98	0	0.97	0.60	4.35
4	Shanxi	2.35	0	0.90	0.08	3.17
5	Inner Mongolia	1.90	0.17	0.74	0	2.81
6	Liaoning	2.34	0.11	1.85	1.01	3.29
7	Dalian	0.40	0	0.70	0.15	0.95
8	Jilin	1.58	0	0.74	0	2.32
9	Heilongjiang	1.58	0.40	1.20	0.69	2.49
10	Shanghai	2.05	0.50	1.10	1.02	2.63
11	Jiangsu	6.28	0.79	1.46	0.49	8.04
12	Zhejiang	5.12	0.74	1.10	0.33	6.63
13	Ningbo	0.80	0.23	0.32	0.06	1.29
14	Anhui	2.47	0.58	2.34	0.76	4.63
15	Fujian	2.32	0.44	0.45	0.07	3.14
16	Xiamen	0.38	0	0.36	0	0.74
17	Jiangxi	1.43	0.47	1.11	0.59	2.42
18	Shandong	4.30	0.20	0.85	0.80	4.55
19	Qingdao	0.54	0.18	0.18	0.08	0.82
20	Henan	3.73	0.23	1.53	0.40	5.09
21	Hubei	2.67	0.10	1.75	0.47	4.05
22	Hunan	3.63	0.03	1.60	0.49	4.77
23	Guangdong	6.11	3.10	4.16	2.83	10.54
24	Shenzhen	1.57	0	1.03	0.43	2.17
25	Guangxi	1.45	0.34	0.73	0.33	2.19
26	Hainan	0.55	0.05	0.38	0.27	0.71
27	Chongqing	1.19	0.69	1.11	0.26	2.73
28	Sichuan	3.20	0.60	1.90	0.55	5.15
29	Guizhou	1.36	0.98	0.58	0.72	2.20
30	Yunnan	1.63	0.18	0.84	0.51	2.14
31	Tibet	0	0.03	0.12	0	0.15
32	Shaanxi	2.70	0.10	0.60	0.26	3.14
33	Gansu	0.85	0.31	0.95	0.30	1.81
34	Qinghai	0.33	0	0.68	0.12	0.89
35	Ningxia	0.46	0.11	0.25	0.11	0.71
36	Xinjiang	1.16	0.26	0.96	0.50	1.88
37	Xinjiang Production and Construction Corps	0	0.05	0.19	0	0.24
38	Heilongjiang Land Reclamation	0	0	0.11	0	0.11
	Total	75.83	12.90	38.07	16.31	110.49

indicates that for major cities, treatment capacity is planned to increase 60% from 23.8 to 38 ktons for Beijing from 2015 to 2020. In terms of total treatment capacity, Guangdong will have the highest of 105.4 ktons/day, while Tibet will only have 1.5 ktons/day by 2020.

Figure 4.5 shows that the top five current treatment capacities are Jiangsu, Guangdong, Zhejiang, and Hebei in 2015.

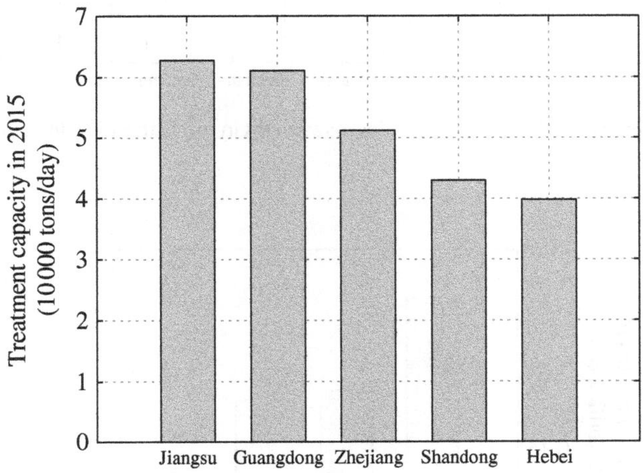

Figure 4.5 Treatment capacity of domestic waste treatment facilities in 2015 (10 000 tons/day).

By 2020, Guangdong will have 105.4 kton/day followed by Jiangsu, Zhejiang, Sichuan, and Henan at 80, 66.3, 51.5, and 47.7 ktons/day, respectively (Figures 4.6, 4.7, 4.8, 4.9, 4.10, 4.11, 4.12, 4.13, 4.14, 4.15, and 4.16).

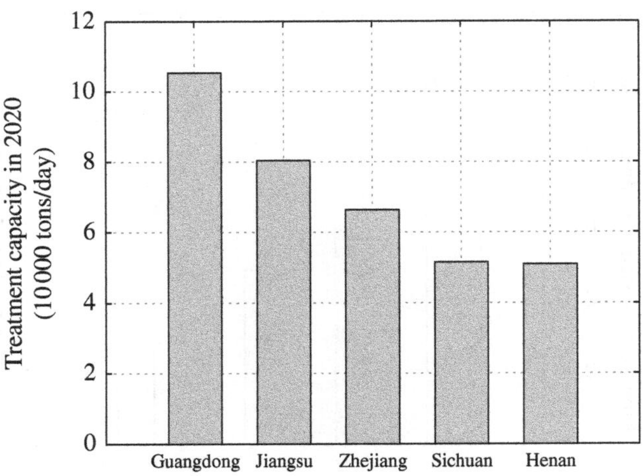

Figure 4.6 Treatment capacity of domestic waste treatment facilities in 2020 (10 000 tons/day).

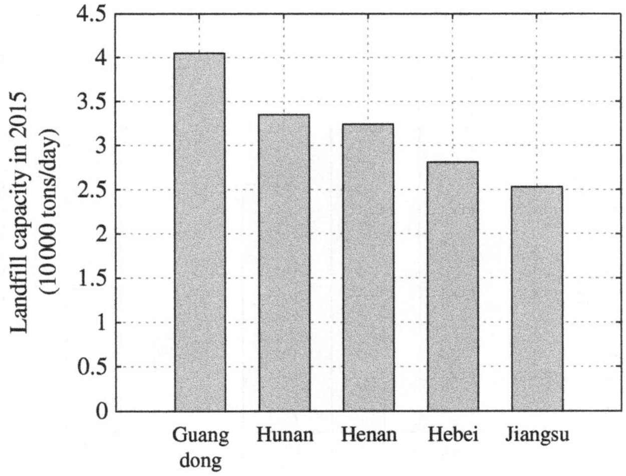

Figure 4.7 Landfill capacity of domestic solid waste treatment facilities in 2015 (10 000 tons/day).

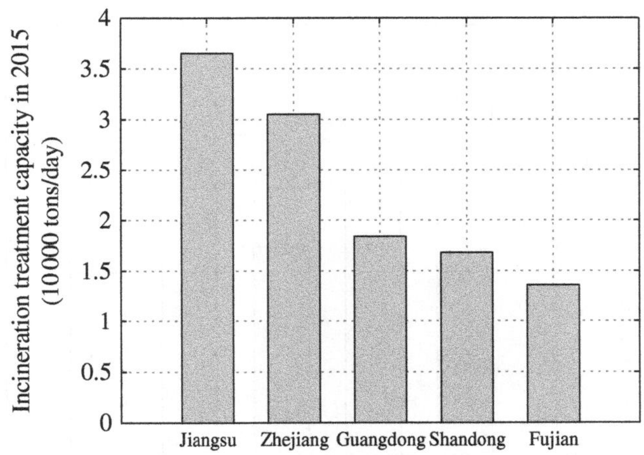

Figure 4.8 Incineration capacity of domestic waste treatment facilities in 2015 (10 000 tons/day).

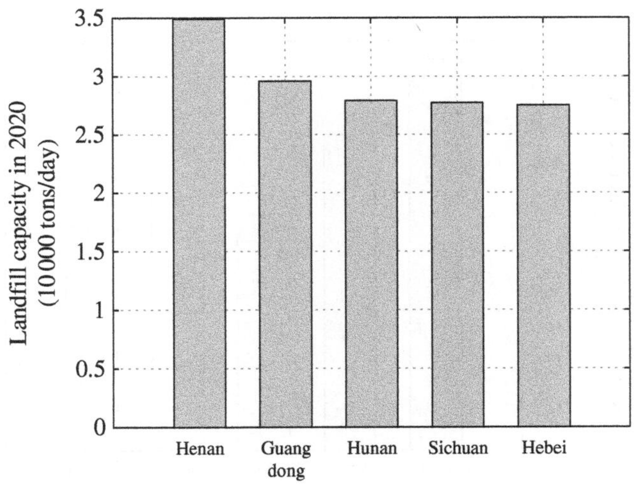

Figure 4.9 Landfill of domestic waste treatment facilities in 2020 (10 000 tons/day).

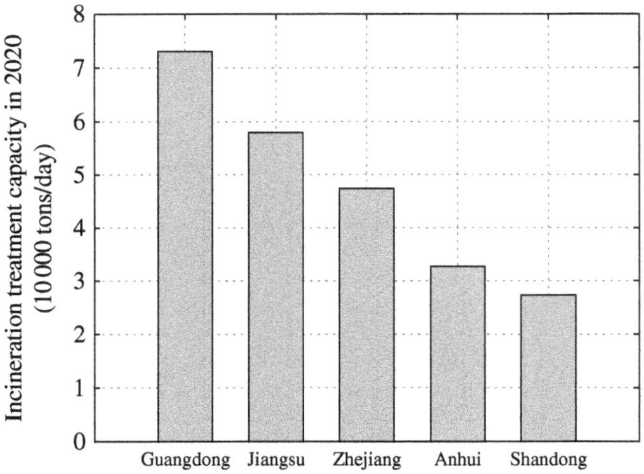

Figure 4.10 Incineration capacity of domestic waste treatment facilities in 2020 (10 000 tons/day).

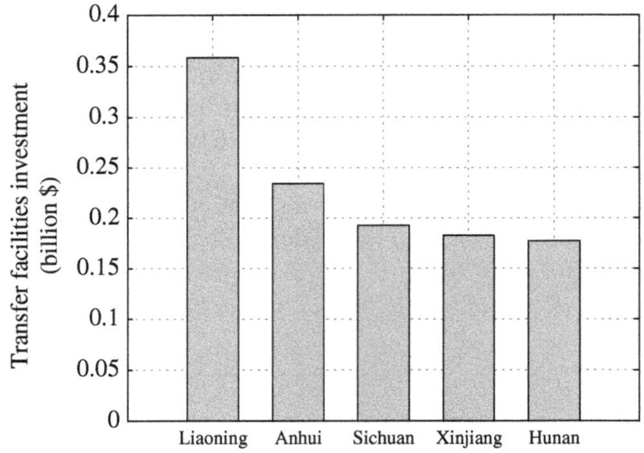

Figure 4.13 Transfer facilities investment in the 13th FYP (billion $).

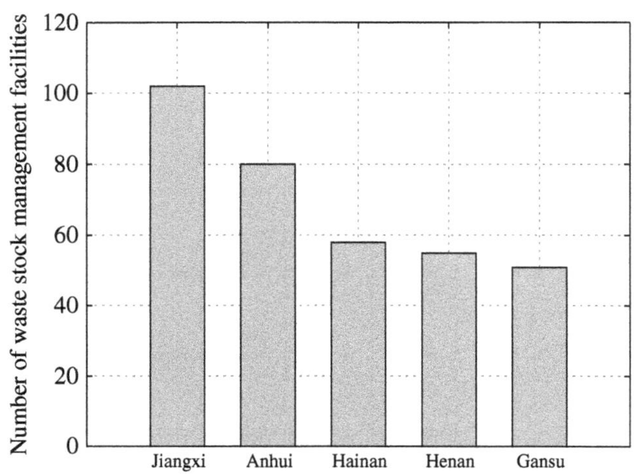

Figure 4.11 Number of landfills in the 13th FYP.

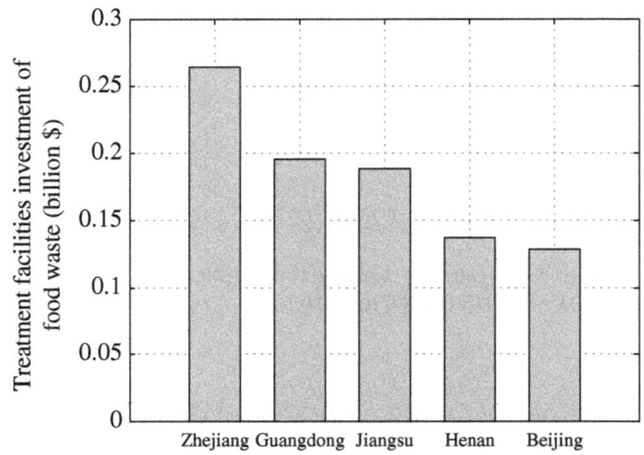

Figure 4.14 Treatment facilities investment of food waste in the 13th FYP (billion $).

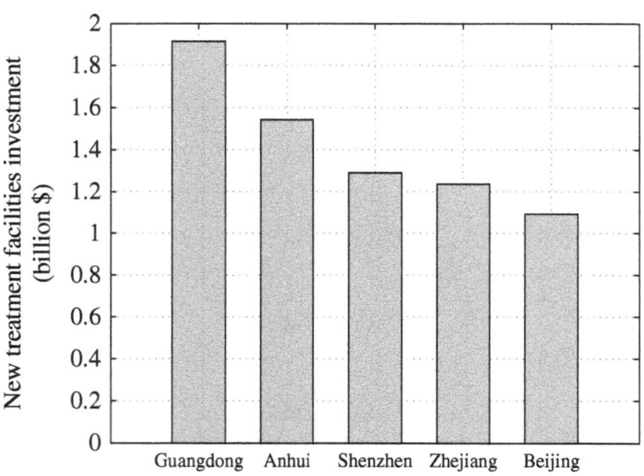

Figure 4.12 New treatment facilities investment in the 13th FYP (billion $).

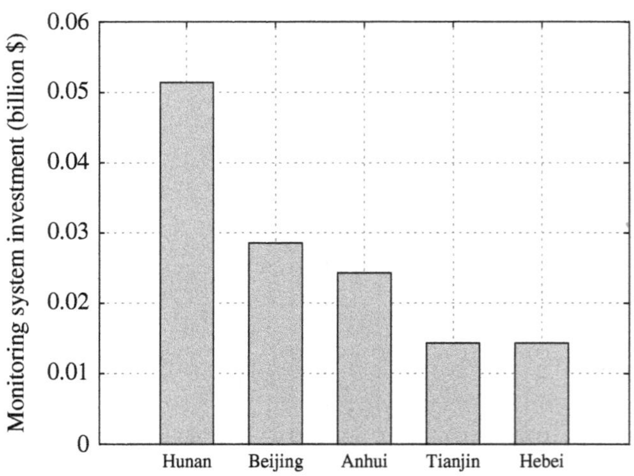

Figure 4.15 Monitoring system investment in the 13th FYP (billion $).

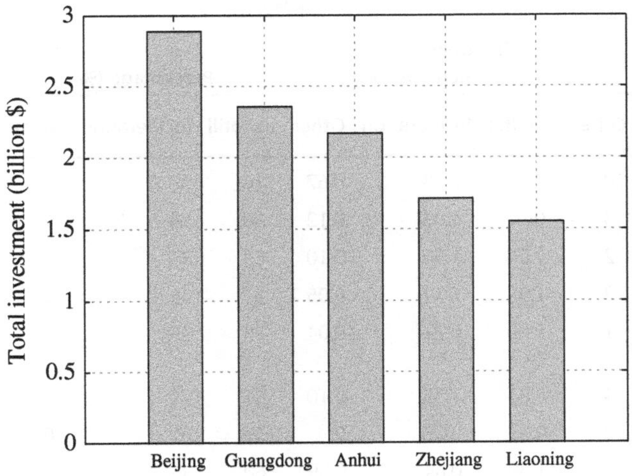

Figure 4.16 Total investment of domestic solid waste treatment facilities in the 13th FYP (billion $).

4.3 Integrated Solid Waste Management

4.3.1 Integrated Solid Waste Management Market in China

There are huge market and investment opportunities in China in integrated solid waste management (ISWM). Currently, most municipal solid waste (MSW) were open dumped in the city surrounding. Satellite images show that many cities are surrounded by domestic wastes that catastrophically cause deterioration in air, water, and soil quality and ultimately pose threat to human health. These opportunities can be seen from the 13th Five-Year Plan (FYP) of China. Since there is limited land available for waste, incineration of solid wastes is the major trend. Taking Beijing as an example, the percentage of incineration of municipal solid waste (MSW) is 44% in 2015. The 13th FYP targets at 67% incineration which is 27% increase. There are many benefits for incineration of MSW: (i) it harvests energy for the MSW, (ii) it occupies much less land, (iii) it does not produce leachate that needs continuous treatment, and (iv) it does not produce odor, fliers and birds. As a result, property value of properties close to the facilities will not significantly depreciate as those close to landfills. For other major cities and provinces, the same increasing trend of incineration is planned (Table 4.3).

The market opportunities can be exemplified in Table 4.4, which lists the new collection and transport management and treatment facilities of food waste in the 13th FYP of China.

Table 4.5 shows the monetary values of these MSW markets. A total of $35 billion investment is needed for the next 5 years to achieve the planned targets of MSW management. The major trend in China's EEIS markets is PPP mechanism that opens great opportunities for EEIS designers to reduce the environmental impacts of MSW in China.

4.3.2 Strategy of ISWM

The hierarchy of ISWM is reduction, reuse, recycling, recovery, landfill or incineration, and controlled dump by the best available technologies. For example, waste reduction includes longer-lasting and reusable products and decreased consumption. Waste collection should use alternative nonfossil fuels. Recycling/materials recovery uses materials recovery facilities (MRFs) to process source-separated materials or mixed waste. MRFs use a combination of manual and mechanical sorting options. Materials should be separated as a recyclable source to reduce contamination. Organic material should be composted after digestion to produce a useful soil conditioner and avoid landfill disposal. Finished compost applied to soils is also an important method to reduce GHG emissions by reducing nitrogen requirements and associated GHG emissions. Waste can be incinerated to generate energy, while last and least option is landfill. Landfill sites should capture the methane that they generate or use it as a renewable energy resource. The US EPA (2012) recommended that an ISWM plan should include the following sections (Table 4.6).

Figure 4.17 shows the US EPA proposed solid waste management hierarchy.

In short, ISWM aims to prevent, reduce, recycle, recover, and dispose of wastes through:

1) Ecodesign
2) Eco-innovation
3) Resource efficiency
4) Cleaner production
5) Sustainable production and consumption
6) Product as service
7) Life cycle approach
8) Cradle to cradle
9) Green economy

There are three perspectives in ISWM, namely, source, stakeholders, and life cycle. Figure 4.18 shows that reduction, reuse, and recycle (3Rs) are the most important strategies in minimizing residential, industrial, and service solid wastes. Separating hazardous waste from MSW is critical in SEE design. From a stakeholder perspective, the "three Rs" are important for residential, industrial, and service stakeholders. Governments should establish the "three Rs" rules which are guided by the SEE design principles. The innovative solutions including energy

Table 4.3 Landfill and incineration of domestic solid waste in the 13th FYP.

		2015						2020					
		Treatment capacity (10 000 tons/day)			Percentage (%)			Treatment capacity (10 000 tons/day)			Percentage (%)		
Serial number	Region	Landfill	Incineration	Other	Landfill	Incineration	Other	Landfill	Incineration	Other	Landfill	Incineration	Other
1	Beijing	0.86	1.04	0.48	36	44	20	0.60	2.53	0.67	16	67	17
2	Tianjin	0.53	0.48	0.03	51	46	3	0	1.63	0.12	0	93	7
3	Hebei	2.81	1.07	0.10	71	27	2	2.75	1.50	0.10	63	34	3
4	Shanxi	1.76	0.54	0.05	75	23	2	1.98	1.14	0.05	62	36	2
5	Inner Mongolia	1.74	0.14	0.02	92	7	1	2.06	0.71	0.04	73	25	2
6	Liaoning	2.24	0	0.10	96	0	4	1.83	1.36	0.10	56	41	3
7	Dalian	0.22	0.18	0	55	45	0	0.37	0.58	0	39	61	0
8	Jilin	1.11	0.38	0.09	70	24	6	1.39	0.84	0.09	60	36	4
9	Heilongjiang	1.20	0.18	0.20	76	11	13	0.96	1.32	0.21	39	53	8
10	Shanghai	1.12	0.83	0.10	55	40	5	0.50	1.83	0.30	19	70	11
11	Jiangsu	2.53	3.65	0.10	40	58	2	2.24	5.80	0	28	72	0
12	Zhejiang	2.02	3.05	0.05	39	60	1	1.74	4.74	0.15	26	72	2
13	Ningbo	0.15	0.65	0	19	81	0	0.14	1.15	0	11	89	0
14	Anhui	1.66	0.78	0.03	67	32	1	1.18	3.27	0.18	25	71	4
15	Fujian	0.94	1.36	0.02	40	59	1	0.89	2.25	0	28	72	0
16	Xiamen	0.22	0.16	0	58	42	0	0.23	0.40	0.11	31	54	15
17	Jiangxi	1.41	0.02	0	99	1	0	1.41	0.98	0.03	58	41	1
18	Shandong	2.42	1.68	0.20	56	39	5	1.62	2.73	0.20	36	60	4
19	Qingdao	0.30	0.24	0	56	44	0	0.22	0.60	0	27	73	0
20	Henan	3.24	0.49	0	87	13	0	3.49	1.60	0	69	31	0
21	Hubei	1.42	1.15	0.10	53	43	4	1.82	2.03	0.20	45	50	5
22	Hunan	3.35	0.16	0.12	92	5	3	2.79	1.72	0.26	58	36	5
23	Guangdong	4.05	1.84	0.22	66	30	4	2.96	7.30	0.28	28	69	3
24	Shenzhen	0.84	0.73	0	54	46	0	0.43	1.74	0	20	80	0
25	Guangxi	1.30	0.15	0	90	10	0	1.57	0.62	0	72	28	0
26	Hainan	0.31	0.24	0	56	44	0	0.14	0.57	0	20	80	0
27	Chongqing	0.83	0.36	0	70	30	0	0.92	1.78	0.03	34	65	1
28	Sichuan	2.22	0.95	0.03	69	30	1	2.77	2.37	0.01	54	46	0
29	Guizhou	1.08	0.26	0.02	79	19	2	1.01	0.78	0.41	46	35	19
30	Yunnan	1.00	0.60	0.03	61	37	2	1.14	0.98	0.02	53	46	1
31	Tibet	0	0	0	—	—	—	0.15	0	0	100	0	0
32	Shaanxi	2.49	0.15	0.05	93	6	2	2.35	0.75	0.04	75	24	1
33	Gansu	0.85	0	0	100	0	0	1.18	0.60	0.03	65	33	2
34	Qinghai	0.33	0	0	100	0	0	0.67	0.22	0	75	25	0
35	Ningxia	0.45	0.01	0	98	2	0	0.62	0.09	0	87	13	0
36	Xinjiang	1.15	0	0.01	99	0	1	1.34	0.53	0.01	71	28	1
37	Xinjiang Production and Construction Corps	0	0	0	—	—	—	0.19	0.05	0	79	21	0
38	Heilongjiang Land Reclamation	0	0	0	—	—	—	0.06	0.05	0	55	45	0
	Total	50.15	23.52	2.16	66	31	3	47.71	59.14	3.64	43	54	3

Table 4.4 New collection and transport, management, and treatment facilities of food waste in the 13th FYP.

Serial number	Region	Collection and transport capacity (10 000 tons/day)	Solid waste management	Treatment capacity of food waste (10 000 tons/day)
1	Beijing	1.39	7	0.15
2	Tianjin	0.95	3	0.08
3	Hebei	1.84	22	0.16
4	Shanxi	0.75	7	0.06
5	Inner Mongolia	1.02	19	0.05
6	Liaoning	4.62	17	0.07
7	Dalian	0.85	6	0.02
8	Jilin	0.55	21	0.05
9	Heilongjiang	1.11	33	0.08
10	Shanghai	0.86	2	0.13
11	Jiangsu	1.87	17	0.22
12	Zhejiang	0.87	10	0.31
13	Ningbo	0.57	3	0.15
14	Anhui	3.13	80	0.10
15	Fujian	1.04	0	0.08
16	Xiamen	0.40	0	0.02
17	Jiangxi	1.62	102	0.08
18	Shandong	1.40	10	0.08
19	Qingdao	0.20	0	0.01
20	Henan	1.73	55	0.16
21	Hubei	2.30	9	0.09
22	Hunan	2.14	41	0.11
23	Guangdong	0	48	0.34
24	Shenzhen	1.28	3	0.07
25	Guangxi	1.21	22	0.05
26	Hainan	0.21	58	0.02
27	Chongqing	1.34	8	0.15
28	Sichuan	2.70	39	0.11
29	Guizhou	1.16	10	0.06
30	Yunnan	0.94	36	0.08
31	Tibet	0.17	0	0.01
32	Shaanxi	0.85	20	0.05
33	Gansu	0.70	51	0.06
34	Qinghai	0.48	11	0.01
35	Ningxia	0.17	22	0.04
36	Xinjiang	1.33	11	0.12
37	Xinjiang Production and Construction Corps	0.29	0	0.01
38	Heilongjiang Land Reclamation	0.18	0	0
	Total	44.22	803	3.44

Table 4.5 Domestic solid waste treatment facilities investment in the 13th FYP (billion $).

Serial number	Region	New treatment facilities	Transfer facilities	Treatment facilities of food waste	Landfills	Classification facilities	Monitoring system	Total
1	Beijing	1.093	0.1	0.129	1.464	0.071	0.029	2.886
2	Tianjin	0.736	0.068	0.057	0.014	0.071	0.014	0.961
3	Hebei	0.514	0.131	0.094	0.057	0.029	0.014	0.840
4	Shanxi	0.511	0.052	0.051	—	0.028	0.014	0.658
5	Inner Mongolia	0.351	0.073	0.043	0.050	0.029	0.014	0.560
6	Liaoning	1.041	0.358	0.056	0.053	0.029	0.014	1.551
7	Dalian	0.284	0.103	0.011	0.179	0.071	0.014	0.663
8	Jilin	0.389	0.039	0.043	0.072	0.028	0.014	0.585
9	Heilongjiang	0.786	0.079	0.064	0.034	0.029	0.014	1.006
10	Shanghai	0.536	0.061	0.074	0.006	0.071	0.014	0.763
11	Jiangsu	0.754	0.167	0.189	0.050	0.029	0.014	1.202
12	Zhejiang	1.237	0.125	0.264	0.041	0.029	0.014	1.711
13	Ningbo	0.321	0.140	0.126	0.014	0.054	0.014	0.670
14	Anhui	1.543	0.234	0.084	0.247	0.036	0.024	2.169
15	Fujian	0.270	0.074	0.069	—	0.029	0.014	0.456
16	Xiamen	0.311	0.040	0.017	—	0.028	0.014	0.411
17	Jiangxi	0.652	0.115	0.051	0.038	0.028	0.014	0.901
18	Shandong	0.728	0.142	0.064	0.128	0.028	0.014	1.107
19	Qingdao	0.205	0.014	0.008	0.000	0.028	0.014	0.271
20	Henan	0.901	0.122	0.137	0.117	0.028	0.014	1.321
21	Hubei	0.918	0.164	0.075	0.050	0.028	0.014	1.251
22	Hunan	1.007	0.177	0.092	0.100	0.042	0.051	1.471
23	Guangdong	1.914	—	0.195	0.202	0.028	0.014	2.355
24	Shenzhen	1.290	0.091	0.047	0.030	0.028	0.014	1.501
25	Guangxi	0.388	0.085	0.042	0.050	0.028	0.014	0.610
26	Hainan	0.261	0.014	0.017	—	0.028	0.014	0.335
27	Chongqing	0.704	0.095	0.120	0.038	0.071	0.014	1.044
28	Sichuan	1.074	0.192	0.080	0.065	0.028	0.014	1.455
29	Guizhou	0.471	0.082	0.040	0.028	0.028	0.014	0.665
30	Yunnan	0.370	0.067	0.045	0.072	0.028	0.014	0.598
31	Tibet	0.134	0.074	0.008	0.007	0.028	0.014	0.267
32	Shaanxi	0.364	0.061	0.034	0.017	0.028	0.014	0.520
33	Gansu	0.552	0.050	0.042	0.107	0.028	0.014	0.795
34	Qinghai	0.775	0.054	0.004	0.028	0.028	0.014	0.905
35	Ningxia	0.130	0.012	0.030	0.048	0.028	0.014	0.264
36	Xinjiang	0.584	0.182	0.105	0.034	0.054	0.014	0.975
37	Xinjiang Production and Construction Corps	0.098	0.02	0.005	—	0.014	0.014	0.152
38	Heilongjiang Land Reclamation	0.068	0.012	—	—	0.014	0.014	0.110
	Total	24.275	3.682	2.621	3.448	1.344	0.604	35.977

Table 4.6 The US EPA procedure to develop a strategic plan (The US EPA, 2012).

Step	Actions
1	List municipal policies, aims, objectives, and initiatives
2	Identify the character and scale of the city, natural conditions, climate, development, and distribution of population
3	Collect the MSW waste generation data such as MSW composition, moisture content, and density (dry weight), present and predicted of both recent years and projections over the lifetime of the plan (usually 15–25 years)
4	Identify all proposed alternatives for waste collection, transportation, treatment, and disposal of the defined types and quantities of solid waste
5	Assess alternatives of the ISWM by assessing all technical, environmental, social, and financial issues
6	Specify the amount, scale, and distribution of collection, transportation, treatment, and disposal systems of different alternatives
7	Identify institutional reforms and regulatory arrangements needed to support the plan
8	Analyze both investment and recurrent costs associated with the proposed facilities and services over the lifetime of the plan
9	Develop and operate the final alternative including estimated subsidy transfers and user fees
10	Develop what facilities are required, who will provide them and the related services, and how such facilities and services will be paid for
11	Implement the plan covering a period of at least 5–10 years, with an immediate action plan detailing actions set out for the first 2–3 years
12	Outline public consultations carried out during preparation of the plan and proposed in future
13	Outline the detailed program to be used to site key waste management facilities, e.g. landfills, compost plants, and transfer stations
14	Assess GHG emissions and the role of MSW in the city's overall urban environmental plans

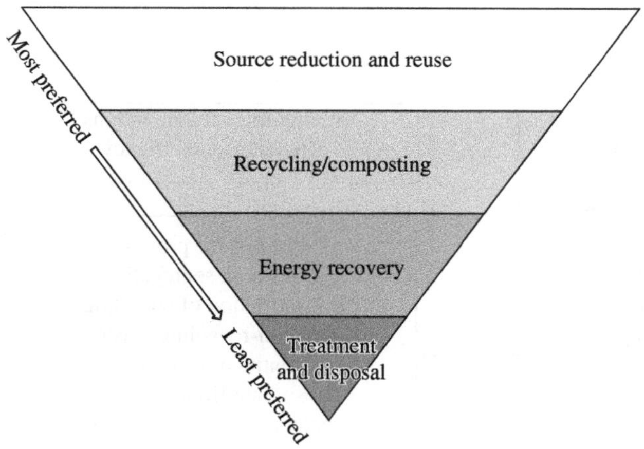

Figure 4.17 EPA solid waste management hierarchy.

recovery, green jobs, and recycling materials were illustrated by the UNEP (2013) in Figure 4.19.

From an LCA perspective, when materials are extracted from the nature, the reduction of solid waste should be emphasized. In the consumption phase, reuse and recycle are the key. After consumption, proper treatment and disposal should be designed so that human health and the environment are protected. For developed countries, sustainable consumption holds the key to reduce overshooting in terms of ecological FP. For developing countries, the recycling rate should be significantly increased in SEE design (Figure 4.20).

MSW per capita generation directly correlates with the country's GDP. Hoornweg and Bhada-Tata (2012) divided countries into four groups according to the national gross production as low, lower middle, upper middle, and high income at GDP calculated using the World Bank Atlas method of \$1 045 or less in 2014; middle-income economies are those with a gross national income (GNI) per capita of more than \$1 045 but less than \$4 125, upper-middle income is from \$4 125 to \$12 736, and high-income economies are those with a GNI per capita of \$12 736 or more. The MSW collection rates in countries with GNIs high, upper middle, low middle, and low change from 98, 85, 69 to 42%, respectively (World Bank, 2012). The MSW generation data for different income countries are estimated in 2010 and 2025 as shown in Tables 4.7 and 4.8.

The composition of MSW also depends upon income levels. The MSW can be classified as organic, paper, plastic, glass, and metal. Specifically, organic includes food scraps, garden (leaves, grass, brush) waste, wood, and process residues. Paper includes paper scraps, cardboard, newspapers, magazines, bags, boxes, wrapping paper, telephone books, shredded paper, and paper beverage cups. Plastic includes packaging, containers, bags, lids, and cups. Glass includes bottles, broken glassware, light bulbs, and colored glass. Metal includes cans, foil, tins,

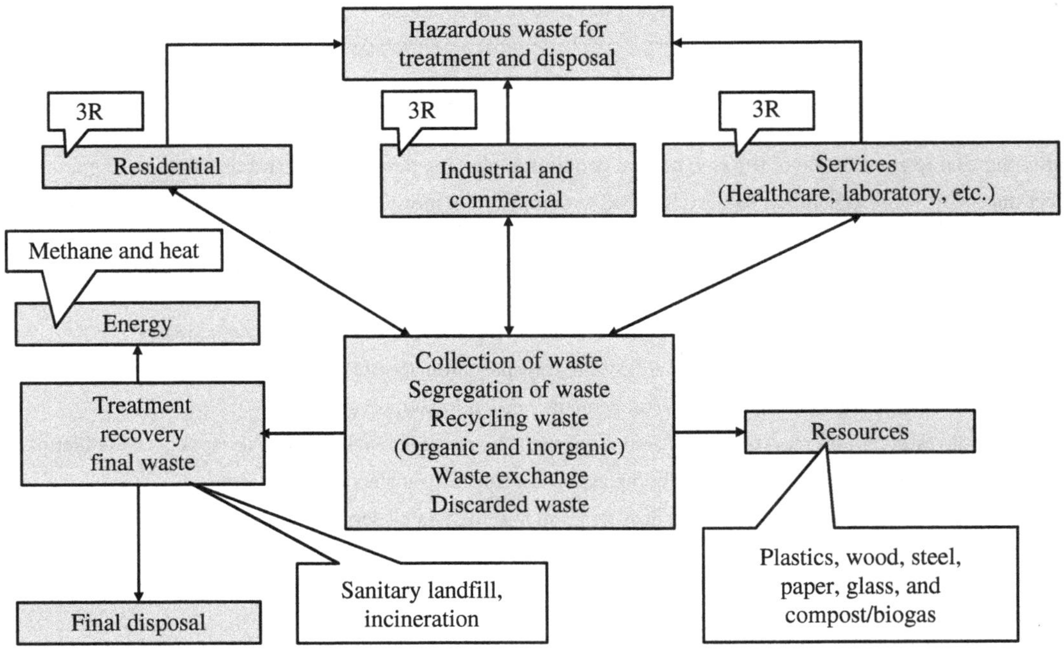

Figure 4.18 Generation source perspective of integrated solid waste management. (*Source:* Adapted from UNEP (2013).)

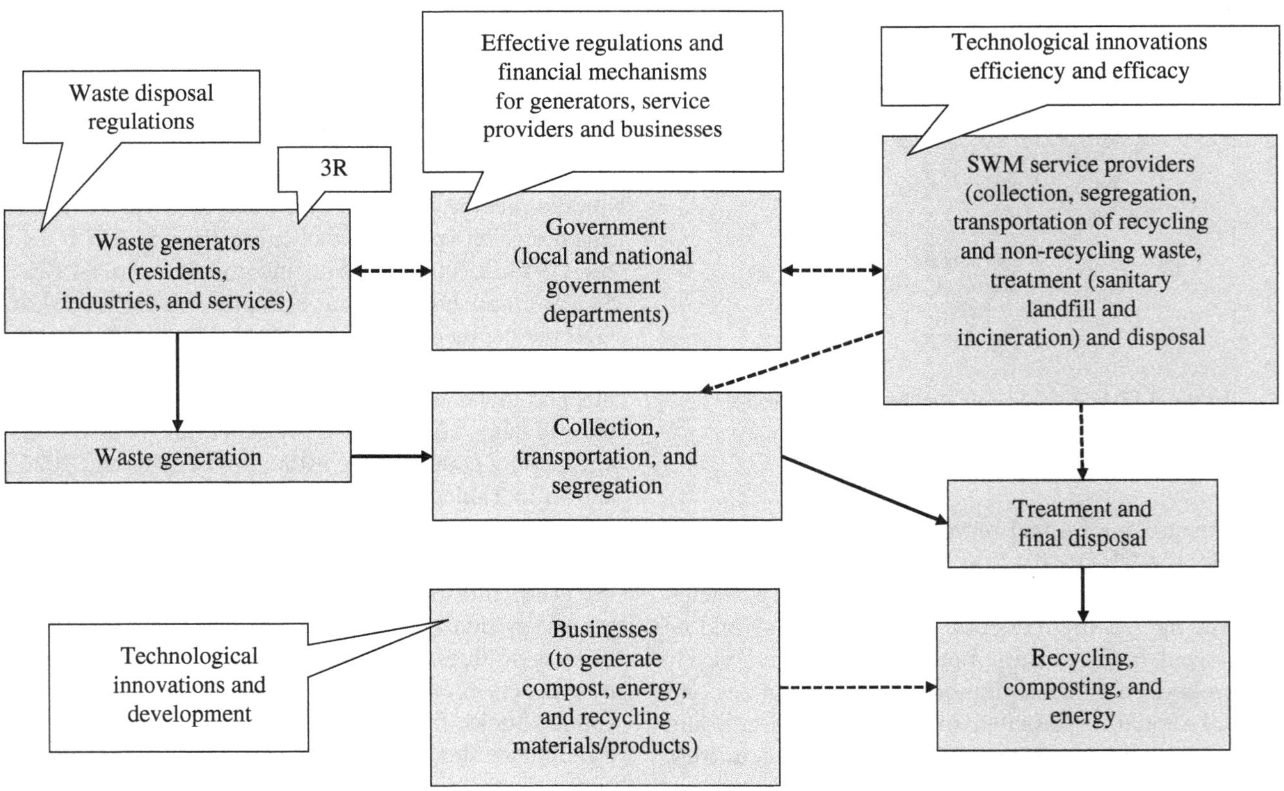

Figure 4.19 Stakeholders/management perspective of integrated solid waste management. (*Source:* Adapted from UNEP (2013).)

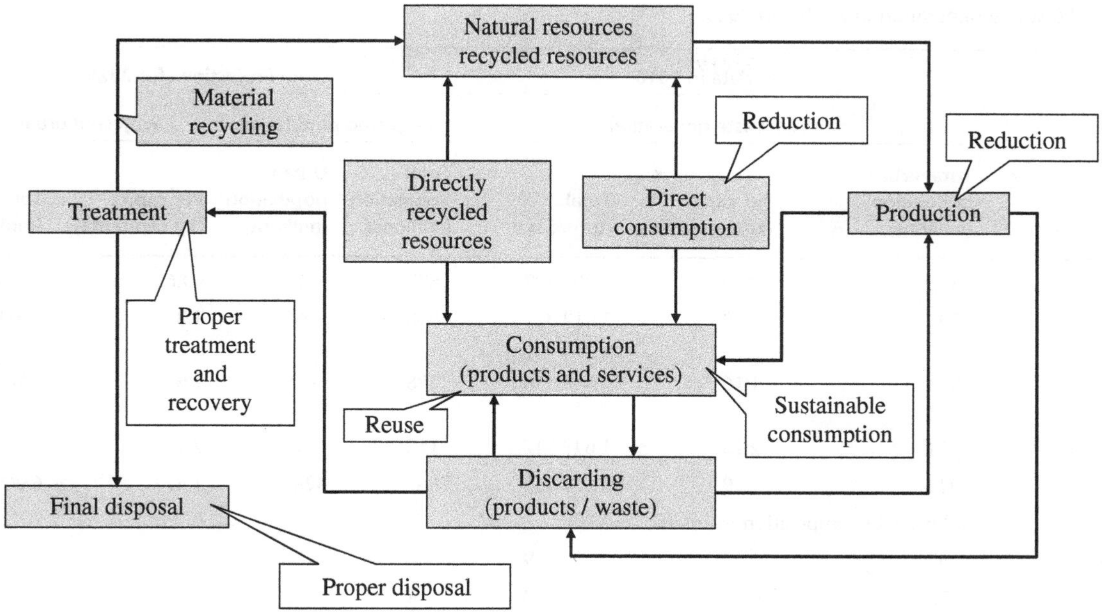

Figure 4.20 Life cycle perspective of integrated solid waste management. (*Source:* Adapted from UNEP (2013).)

Table 4.7 Estimated solid waste management costs in 2010 and 2025.

Country income group	MSW management cost in 2010 (billion \$)	MSW management cost estimated in 2025 (billion \$)
Low income	1.5	7.7
Lower middle income	20.1	84.1
Upper middle income	24.5	63.5
High income	159.3	220.2
Total global cost (\$)	205.4	375

Source: World Bank (2012).

nonhazardous aerosol cans, appliances (white goods), railings, and bicycles. Other includes textiles, leather, rubber, multilaminates, e-waste, appliances, ash, and other inert materials. Table 4.9 lists the distribution of these classes according to countries' income groups. The per capita weight and total solid waste generated will continue to increase with urbanization. In 2002, there were 2.9 billion urban residents who generated about 0.64 kg of MSW per person per day (0.68 billion tons/year). In 2010, Hoornweg and Bhada-Tata (2012) estimated that about three billion residents generate 1.2 kg/person/day (1.3 billion tons/year). By 2025 this will likely increase to 4.3 billion urban residents generating about 1.42 kg/capita/day of MSW (2.2 billion tons/year). The increasing trend of MSW requires ISWM.

4.3.3 LCA on Footprint of Solid Waste Recycle

Example 4.1

Given: Miami-Dade solid waste composition and amounts recycled are listed in Tables 4.10 and 4.11. Please use the life cycle assessment calculator provided by http://www.sustainability.vic.gov.au/publications-and-research/research/life-cycle-assessment/life-cycle-assessment-calculator.

Find: Please estimate the footprint reductions on CO_2, energy, and water if the total amount of solid waste in Miami-Dade is 4 000 000 tons/year and the county is going to recycle 20, 40, 60, and 75% of total solid waste, respectively.

Please answer following questions:

1) What are footprint CO_2 reduction, footprint energy reduction, and footprint water use reduction?
2) Plot footprint reduction versus percentage recycled.

Solution:

1) Percentage recycled (2006–2011)

$$\text{Percentage recycled} = \frac{\text{Total amount of waste products recycled}}{\text{Total amount of waste products managed}} \quad (4.1)$$

For 2006,

$$\text{Percentage recycled (2006)} = \frac{918696 \text{ tons}}{4878337 \text{ tons}} = 18.83\%$$

Table 4.8 Solid waste generation in 2010 and 2025.

	MSW data in 2010			Projections for 2025			
		Urban waste generation		Projected population		Projected urban waste	
Region	**Total urban population (millions)**	**Per capita (kg/capita/day)**	**Total (tons/day)**	**Total population (millions)**	**Urban population (millions)**	**Per capita (kg/capita/day)**	**Total (tons/day)**
Low income	343	0.60	204 802	1637	676	0.86	584 272
Lower middle income	1293	0.78	1 012 321	4010	2080	1.3	2 618 804
Upper middle income	572	1.16	665 586	888	619	1.6	987 039
High income	774	2.13	1 649 547	1112	912	2.1	1 879 590
Total	2982	1.19	3 532 256	7647	4287	1.4	6 069 705
	Solid waste composition estimated in 2025						
Low income	62	6	9	3	3		17
Lower middle income	55	10	13	4	3		15
Upper middle income	50	15	12	4	4		15

Source: World Bank (2012).

Table 4.9 Solid waste composition in 2010 and 2025.

	Solid waste composition in 2010					
Region	**Organic**	**Paper**	**Plastic**	**Glass**	**Metal**	**Other**
Low income	64	5	8	3	3	17
Lower middle income	59	9	12	3	2	15
Upper middle income	54	14	11	5	3	13
High income	28	31	11	7	6	17
Region	Solid waste composition estimated in 2025					
Low income	62	6	9	3	3	17
Lower middle income	55	10	13	4	3	15
Upper middle income	50	15	12	4	4	15
High income	28	30	11	7	6	18

Source: World Bank (2012).

Equation (4.1) was used to calculate the percentage recycled (2006–2011). The results are shown in Table 4.12.

Recycled tons/total recycled tons

$$\text{Recycled tons/total recycled tons} = \frac{\text{Amount of a kind of waste products recycled}}{\text{Total amount of waste products recycled}} \tag{4.2}$$

For aluminum cans in 2006,

$$\text{Recycled tons/total recycled tons} = \frac{1355\ \text{tons}}{918696\ \text{tons}} = 0.15\%$$

Equation (4.2) can be used to calculate the recycled tons/total recycled tons (2006–2011). The results are shown in Table 4.13.

2) Average value of recycled tons/total recycled tons

$$\frac{\text{Average value of recycled tons}}{\text{Total recycled tons}} = \frac{\text{The sum of recycled tons/total recycled tons}}{6} \tag{4.3}$$

Table 4.10 Amount of waste products managed through recycling, incineration, and landfilling (tons).

	2006	2007	2008	2009	2010	2011
Recycling	918 696	823 229	801 935	668 776	836 844	805 079
Incineration	669 410	636 387	750 893	940 979	629 497	967 462
Landfilling	3 290 231	2 930 608	2 216 855	2 237 275	1 963 818	1 796 489
Total	4 878 337	4 390 224	3 769 683	3 847 030	3 430 159	3 569 030

Table 4.11 Waste products recycled (tons) by material type.

Materials type	2006	2007	2008	2009	2010	2011
Aluminum cans	1 355	2 024	1 634	2 461	1 191	1 072
Construction and demolition debris	80 598	37 828	48 268	127 777	36 656	15 713
Corrugated paper	139 920	153 762	115 491	112 094	186 459	190 922
Ferrous metals	230 728	199 666	185 433	144 111	179 362	204 021
Food waste	0	0	0	0	0	0
Glass	22 428	23 346	10 614	15 456	20 605	7 146
Miscellaneous	30 304	30 619	30 805	30 744	31 043	31 293
Newspapers	70 822	56 259	33 902	42 502	56 027	37 616
Nonferrous metals	38 768	35 356	32 249	20 212	22 804	64 389
Office paper	3 386	12 810	25 387	18 431	22 376	12 702
Other paper	24 358	26 189	13 502	14 308	71 347	16 116
Other plastics	1 057	1 238	2 123	2 414	2 493	4 553
Plastic bottles	7 967	7 967	2 712	4 885	5 680	3 329
Process fuels	67 469	93 048	125 485	82 044	49 273	54 959
Steel cans	2 315	2 085	49 635	11 815	1 328	474
Textiles	0	0	0	236	1 371	84
Tires	2 132	14 528	17 862	0	18 697	22 769
White goods	43 948	38 032	35 263	204	34 164	38 861
Yard trash	151 141	88 472	71 570	39 082	95 968	99 060
Total	918 696	823 229	801 935	668 776	836 844	805 079

Table 4.12 Percentage recycled (%).

	2006	2007	2008	2009	2010	2011
Recycling	918 696	823 229	801 935	668 776	836 844	805 079
Incineration	669 410	636 387	750 893	940 979	629 497	967 462
Landfilling	3 290 231	2 930 608	2 216 855	2 237 275	1 963 818	1 796 489
Total	4 878 337	4 390 224	3 769 683	3 847 030	3 430 159	3 569 030
Percentage recycled	18.83	18.75	21.27	17.38	24.40	22.56

Table 4.13 Recycled tons/total recycled tons by material type (%).

Materials type	2006	2007	2008	2009	2010	2011
Aluminum cans	0.15	0.25	0.20	0.37	0.14	0.13
Construction and demolition debris	8.77	4.60	6.02	19.11	4.38	1.95
Corrugated paper	15.23	18.68	14.40	16.76	22.28	23.71
Ferrous metals	25.11	24.25	23.12	21.55	21.43	25.34
Food waste	0.00	0.00	0.00	0.00	0.00	0.00
Glass	2.44	2.84	1.32	2.31	2.46	0.89
Miscellaneous	3.30	3.72	3.84	4.60	3.71	3.89
Newspapers	7.71	6.83	4.23	6.36	6.70	4.67
Nonferrous metals	4.22	4.29	4.02	3.02	2.73	8.00
Office paper	0.37	1.56	3.17	2.76	2.67	1.58
Other paper	2.65	3.18	1.68	2.14	8.53	2.00
Other plastics	0.12	0.15	0.26	0.36	0.30	0.57
Plastic bottles	0.87	0.97	0.34	0.73	0.68	0.41
Process fuels	7.34	11.30	15.65	12.27	5.89	6.83
Steel cans	0.25	0.25	6.19	1.77	0.16	0.06
Textiles	0.00	0.00	0.00	0.04	0.16	0.01
Tires	0.23	1.76	2.23	0.00	2.23	2.83
White goods	4.78	4.62	4.40	0.03	4.08	4.83
Yard trash	16.45	10.75	8.92	5.84	11.47	12.30
Total	100.00	100.00	100.00	100.00	100.00	100.00

For aluminum cans,

$$\frac{\text{Average value of recycled tons}}{\text{Total recycled tons}} = \frac{0.15\% + 0.25\% + 0.20\% + 0.37\% + 0.14\% + 0.13\%}{6} = 0.21\%$$

Equation (4.3) can be used to calculate the average value of recycled tons/total recycled tons (2006–2011). The results are shown in the following table.

3) Amount of waste products recycled should be

$$\text{Amount of waste products recycled} = \frac{\text{Average value of recycled tons}}{\text{Total recycled tons}} \times \text{percentage recycled} \times \text{total amount of waste products managed} \quad (4.4)$$

For aluminum cans (20% percentage recycled),

$$\text{Amount of waste products recycled} = 0.21\% \times 20\% \times 4000000 = 1654.09\text{tons}$$

Equation (4.4) can be used to calculate the amount of waste products recycled for different percentage recycled. The results are shown in Table 4.13.

4) Footprint CO_2 reduction, footprint energy reduction, and water footprint reduction for aluminum cans (20% recycled) can be obtained by entering 1654.09 ton of aluminum in the calculator.

   The same method can be used to calculate footprint CO_2 reduction, footprint energy reduction, and footprint water use reduction. The results are shown in Tables 4.14, 4.15, 4.16, 4.17, 4.18, 4.19, and 4.20.

   Plot footprint reduction versus percentage recycled for aluminum cans (Figures 4.21, 4.22, and 4.23).

Table 4.14 Average value of recycled tons /total recycled tons (%) (2006–2011).

Materials type	2006	2007	2008	2009	2010	2011	Ave
Aluminum cans	0.15	0.25	0.20	0.37	0.14	0.13	0.21
Construction and demolition debris	8.77	4.60	6.02	19.11	4.38	1.95	7.47
Corrugated paper	15.23	18.68	14.40	16.76	22.28	23.71	18.51
Ferrous metals	25.11	24.25	23.12	21.55	21.43	25.34	23.47
Food waste	0.00	0.00	0.00	0.00	0.00	0.00	0.00
Glass	2.44	2.84	1.32	2.31	2.46	0.89	2.04
Miscellaneous	3.30	3.72	3.84	4.60	3.71	3.89	3.84
Newspapers	7.71	6.83	4.23	6.36	6.70	4.67	6.08
Nonferrous metals	4.22	4.29	4.02	3.02	2.73	8.00	4.38
Office paper	0.37	1.56	3.17	2.76	2.67	1.58	2.02
Other paper	2.65	3.18	1.68	2.14	8.53	2.00	3.36
Other plastics	0.12	0.15	0.26	0.36	0.30	0.57	0.29
Plastic bottles	0.87	0.97	0.34	0.73	0.68	0.41	0.67
Process fuels	7.34	11.30	15.65	12.27	5.89	6.83	9.88
Steel cans	0.25	0.25	6.19	1.77	0.16	0.06	1.45
Textiles	0.00	0.00	0.00	0.04	0.16	0.01	0.03
Tires	0.23	1.76	2.23	0.00	2.23	2.83	1.55
White goods	4.78	4.62	4.40	0.03	4.08	4.83	3.79
Yard trash	16.45	10.75	8.92	5.84	11.47	12.30	10.96
Total	100.00	100.00	100.00	100.00	100.00	100.00	100.00

Table 4.15 Amount of waste products recycled (tons) for different percentage recycled (%).

Materials type	20	40	60	75
Aluminum cans	1 654.09	3 308.19	4 962.28	6 202.85
Steel cans	11 571.85	23 143.71	34 715.56	43 394.45
Corrugated paper	148 088.95	296 177.90	444 266.84	555 333.55
Office paper	16 130.49	32 260.99	48 391.48	60 489.35
Other paper	26 910.99	53 821.98	80 732.97	100 916.21
Yard trash	87 652.46	175 304.91	262 957.37	328 696.71
Glass	16 348.89	32 697.79	49 046.68	61 308.35
Other plastics	2 339.43	4 678.85	7 018.28	8 772.85
Plastic bottles	5 327.79	10 655.58	15 983.37	19 979.22
Construction and demolition debris	59 766.94	119 533.88	179 300.82	224 126.02
Ferrous metals	187 753.71	375 507.42	563 261.13	704 076.41
Food waste	0.00	0.00	0.00	0.00
Miscellaneous	30 737.11	61 474.23	92 211.34	115 264.18
Newspapers	48 657.34	97 314.67	145 972.01	182 465.01
Nonferrous metals	35 041.57	70 083.13	105 124.70	131 405.87
Process fuels	79 035.81	158 071.62	237 107.43	296 384.29
Textiles	279.40	558.81	838.21	1 047.76
Tires	12 382.11	24 764.23	37 146.34	46 432.93
White goods	30 321.06	60 642.13	90 963.19	113 703.99
Total	800 000.00	1 600 000.00	2 400 000.00	3 000 000.00

Table 4.16 Footprint CO_2 reduction (tons) for aluminum cans (20% percentage recycled).

Materials type	Total recovery (tons)	Total CO_2 eq (kg)	Total CO_2 eq (tons)
Metals			
Aluminum cans	1654.09	28 119 530	28 120
Packaging steel (steel cans)	0	0	0
Paper/cardboard			
Cardboard/paper packaging (including liquid paperboard)	0	0	0
Printing and writing paper	0	0	0
Other (mixed paper)	0	0	0
Organics			
Garden organics	0	0	0
Glass			
Glass containers	0	0	0
Plastics			
PET	0	0	0
HDPE (average) (milk bottles)	0	0	0
Total	1654.09	28 119 530	28 120

Table 4.17 Footprint CO_2 reduction (million tons) for different percentage recycled (%).

Materials type	20	40	60	75
Metals				
Aluminum cans	0.0281	0.0562	0.0844	0.1054
Packaging steel (steel cans)	0.0197	0.0393	0.0590	0.0738
Paper/cardboard				
Cardboard/paper packaging (including liquid paperboard)	0.0252	0.0504	0.0755	0.0944
Printing and writing paper	0.0210	0.0419	0.0629	0.0786
Other (mixed paper)	0.0121	0.0242	0.0363	0.0454
Organics				
Garden organics	0.0202	0.0403	0.0605	0.0756
Glass				
Glass containers	0.0087	0.0173	0.0260	0.0325
Plastics				
PET	0.0028	0.0056	0.0084	0.0105
HDPE (average) (milk bottles)	0.0044	0.0087	0.0131	0.0164
Total	0.1420	0.2841	0.4261	0.5327

Table 4.18 Footprint energy reduction (terajoule [TJ]) for different percentage recycled (%).

Materials type	20	40	60	75
Metals				
Aluminum cans	347.36	694.72	1042.08	1302.60
Packaging steel (steel cans)	173.58	347.16	520.73	650.92
Paper/cardboard				
Cardboard/paper packaging (including liquid paperboard)	69.60	139.20	208.81	261.01
Printing and writing paper	−10.97	−21.94	−32.91	−41.13
Other (mixed paper)	9.69	19.38	29.06	36.33
Organics	0.00	0.00	0.00	0.00
Garden organics	−27.17	−54.35	−81.52	−101.90
Glass				
Glass containers	73.57	147.14	220.71	275.89
Plastics				
PET	128.67	257.34	386.01	482.51
HDPE (average) (milk bottles)	269.43	538.87	808.30	1010.38
Total	1033.76	2067.52	3101.28	3876.60

Table 4.19 Footprint water use reduction (megaliter [ML]) for different percentage recycled (%).

Materials type	20	40	60	75
Metals				
Aluminum cans	48	96	144	180
Packaging steel (steel cans)	486	972	1 458	1 823
Paper/cardboard				
Cardboard/paper packaging (including liquid paperboard)	1629	3258	4 887	6 109
Printing and writing paper	177	355	532	665
Other (mixed paper)	296	592	888	1 110
Organics				
Garden organics	500	999	1 499	1 874
Glass				
Glass containers	15	31	46	58
Plastics				
PET	161	323	484	605
HDPE (average) (milk bottles)	120	241	361	451
Total	3433	6866	10 299	12 874

Table 4.20 Variables.

Variable	Explanation
S	Tonnage associated with a sector during a particular time period
Q	Quantity of waste encountered in the study
TPE	Tons per employee
j	Designation of a particular material
i	Designation of a particular sample
r	Ratio of material weight to total sample weight for an individual sample
m	Weight of a material in an individual sample
w	Total weight of an individual sample
V	The variance associated with the estimate for a material's percent in a group of samples
n	Number of samples in the group
p	A weighting factor given to a segment of the waste stream, where the sum of all the values of *p* is 1
G	Designation of a size subgroup within a segment of the waste stream – usually used for generator samples

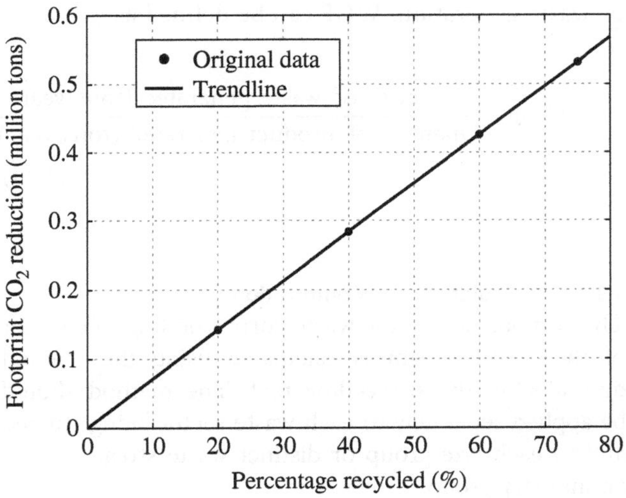

Figure 4.21 Total footprint CO_2 deduction (million tons).

4.3.4 ISWM Data Analysis

Data analysis plays a critical role in ISWM because EEIS design needs design parameters such as waste generation factors (WGF).

4.3.4.1 Calculations for Measuring Quantity

1) Quantifying a waste sector on vehicle surveys
   If the annual tonnage of all waste disposed of at the facility is known, the vehicle survey data can be used

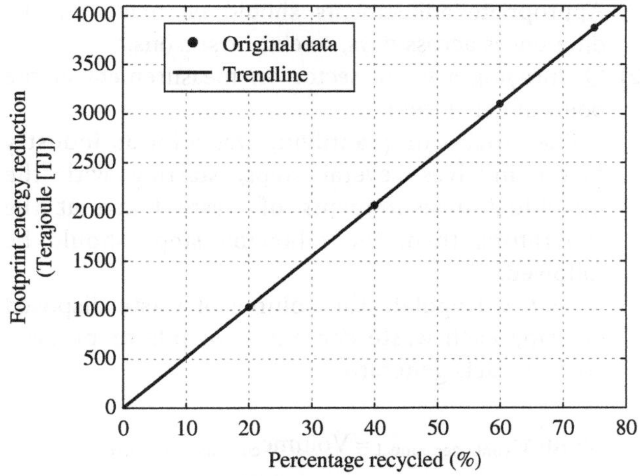

Figure 4.22 Total footprint energy reduction (terajoule [TJ]).

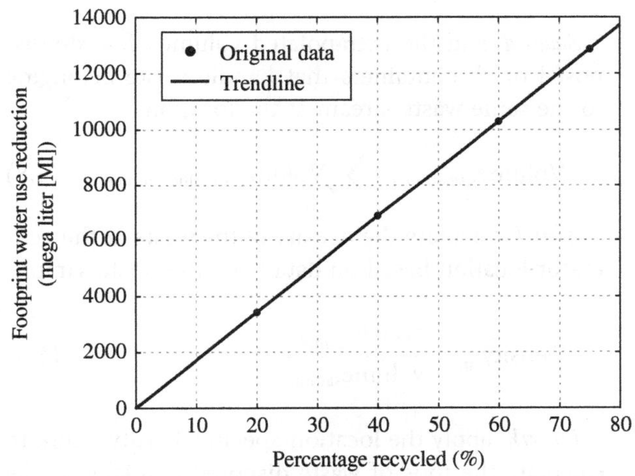

Figure 4.23 Total water footprint reduction (megaliter [Ml]).

to determine the portion of annual disposal corresponding to the waste sectors being studied. For a given waste sector, *S*, the sector tonnage can be calculated from the tonnage, of an individual vehicle:

$$\text{Sector (tons)} = \frac{\sum Q_{S,\text{survey period}}}{\sum Q_{\text{all,survey period}}} \times \sum Q_{\text{all,annual}} \tag{4.5}$$

If the annual tonnage of all waste disposed at the facility is not known, the corresponding tons that were counted during the vehicle survey can be used to extrapolated:

$$\text{Sector (tons)} = \sum Q_{S,\text{survey period}} \times \frac{\text{Operating in year}}{\text{Days in survey period}} \tag{4.6}$$

Appropriate adjustments should be made for the differences across days, weeks, or seasons.

2) Quantifying a waste sector on measurement at the point of generation

The process of quantifying waste for an industry sector involves several steps, starting with the individual measurements of waste taken at the generators, then, the following steps should be followed:

First, extrapolate the volume of waste disposed of using each waste container (or pile or process, etc.) at each generator:

$$\text{Volume}_{\text{container,annual}} = \text{Volume}_{\text{container,measured}} \times \frac{\text{generation time}_{\text{annual}}}{\text{generation time}_{\text{measured}}} \tag{4.7}$$

Second, add the extrapolated volume of waste disposed of all containers that handle waste belonging to the same waste stream at the location:

$$\text{Volume}_{\text{site,annual}} = \sum \text{Volume}_{\text{container,annual}} \tag{4.8}$$

Third, calculate the density of the waste at the generator location based on data from the waste sample:

$$\text{Density}_{\text{site}} = \frac{\text{weight}_{\text{sample}}}{\text{volume}_{\text{sample}}} \tag{4.9}$$

Fourth, apply the location-specific density figure to calculate the tons of waste disposed annually by the generator:

$$\text{Tons}_{\text{site,annual}} = \text{volume}_{\text{sample}} \times \text{density}_{\text{site}} \tag{4.10}$$

Fifth, calculate a "scale-up factor" for waste generation by the industry and size group. For many commercial sectors, the appropriate scale-up factor is proportional to the number of employees. For most agricultural sectors, it is according to number of crop acres or number of animals. The following example shown involves calculating tons per employee (TPE) for a given industry. It draws upon data reflecting the tone disposed:

$$\text{TPE}_{\text{annual,size group}} = \frac{\sum_{\text{visited sites}} \text{Tons}_{\text{site,annual,size group}}}{\sum_{\text{visited sites}} \text{Employees}_{\text{site,annual,size group}}} \tag{4.11}$$

Sixth, calculate the tons disposed of from the entire industry. The following example draws upon data reflecting the total number of employees in the larger population (e.g. countywide, statewide, etc.) of industry members in the appropriate size group:

$$Q_{\text{site,annual}} = \text{TPE}_{\text{annual,size group}} \times \text{industry-wide employment in size group} \tag{4.12}$$

Seventh, add the results for the size groups to calculate total tons disposed of by the industry:

$$Q_{\text{industry}} = \sum Q_{\text{size group}} \tag{4.13}$$

3) Quantifying based on waste generation factor (WGF)

In calculating and projecting waste quantities, especially from the service and industrial sector, WGF should be determined. Then it will be easy to extrapolate the waste generation rates for the industries and services. WGF depends on size of operation, waste management practices, and process technology. Therefore, the information for this sector should have the size of production and process technology with respect to nonhazardous and hazardous waste generation. WGF can be defined as

$$\text{WGF} = \frac{\text{Quantity of waste generated (tons/year)}}{\text{Quantity of product generated (tons/year)}} \tag{4.14}$$

4.3.4.2 Calculations for Composition

The composition of the waste corresponding to a sector of the waste stream is calculated using the method described in the succeeding text. The method should be applied separately to each waste sector being studied and to each size group or distinct waste stream within an industry group.

1) Calculating the mean estimate

For a given material, j, in all of the relevant samples, i, calculate the ratio, r, of the material weight, m, to the total sample weight, w:

$$r_j = \frac{\sum_i m_{i,j}}{\sum_i w_{i,j}} \tag{4.15}$$

The calculation should be repeated for each material.

2) Calculating the error range

For each mean estimate, r_j, calculated, the confidence interval (error range) surrounding the mean estimate is calculated as follows. First, calculate the variance, V_{r_j}, of the mean estimate:

$$V_{r_j} = \frac{1}{n} \times \frac{1}{w^2} \times \frac{\sum_i \left(m_{i,j} - r_j w_i\right)^2}{n-1} \tag{4.16}$$

where n is the number of samples and mean sample weight $w = \Sigma_i w_i / n$.

Confidence level is $\pm\left(t \times \sqrt{V_{r_j}}\right)$, where t depends on the number of samples, n, and the desired confidence level. The value of t can be estimated from t-static.

3) Volume-to-weight conversion factors and net weight of waste
Combining the composition estimates for two or more segments of the waste stream requires the use of the weighted averages method. The result for each segment of the waste stream is weighted according to the relative size of that segment in the larger waste stream being studied.
4) Calculating the weighting factors when combining waste sectors
A specific weighting factor should be calculated for each sector or segment of the waste stream being studied. The weighting factor, P_G, for each segment or size group, G, within the waste stream is calculated as follows:

$$P_G = \frac{t_{G,\text{annual}}}{t_{\text{all sectors annual}}} \tag{4.17}$$

A weighting factor should be calculated for every waste sector, so the sum of all the values of P_G should equal one.
5) Calculating the mean estimate for combining waste sectors
The mean estimate for a given material, j, in a combination of segments (1, 2, 3...) of the waste stream can be found as follows:

$$r_{j,\text{combined}} = \left(p_1 \times r_{j,1}\right) + \left(p_2 \times r_{j,2}\right) + \left(p_3 \times r_{j,3}\right) + \cdots \tag{4.18}$$

6) Calculating variance and confidence level for combined sectors
After a mean estimate for combined waste sectors is calculated as shown earlier, the variance surrounding the estimate can be calculated as follows:

$$V_{r_{j,\text{combined}}} = \left(p_1^2 \times V_{r_{j,1}}\right) + \left(p_2^2 \times V_{r_{j,2}}\right) + \left(p_3^2 \times V_{r_{j,3}}\right) + \cdots \tag{4.19}$$

Confidence level is $\pm\left(t \times \sqrt{V_{r_{j,\text{combined}}}}\right)$.

4.3.5 Determining Waste Composition

4.3.5.1 Moisture Content

Moisture content is a very important factor that influences decisions on converting organic waste into compost and biogas, using solid waste as a fuel and designing landfills or incineration plants. Currently there are various types of moisture meters available to check moisture content. However, the traditional test could also be done on certain types of materials. The moisture content is measured by heating the sample at 105°C in an oven until the weight loss stabilizes. The weight of the sample before and after gives the moisture content. The different fractions of the waste stream should have their moisture content measured separately.

4.3.5.2 Calorific Value

The energy value of the waste components depends on its calorific value, which is influenced by the moisture content and hydrogen content of the waste. The formula for determining the calorific value of waste components is

$$\text{CV}_{\text{raw}} = \left((1-\text{MC}) \times \left(\text{CV}_{\text{upper}} - (2241^* \times 9) \times \text{H}\right)\right) - 2241 \times \text{MC} \tag{4.20}$$

where

CV = calorific value ("raw" is real "as delivered" value, and "upper" is value for dried material) in kJ/kg
MC = % moisture content (by weight)
H = % hydrogen content (from literature)
*Vaporization enthalpy of water (2441 kJ/kg at 25°C)

To determine the calorific value of a waste stream, the following steps must be carried out:

1) Sample to be sorted and analyzed into fractions.
2) CV_{upper} is applied from literature.
3) CV_{upper} is analyzed for unknown fractions.
4) % hydrogen is applied from literature.
5) Moisture of fractions is determined.
6) The values for CV_{raw} are calculated.

Default higher calorific values and hydrogen contents for solid waste are shown in Table 4.21. Moisture content varies by location (climatic variation) and by season and leads to a directly proportional change in the CV_{raw}.

4.3.5.3 Chemical Composition

For organic and inorganic waste, information on chemical composition is quite important to design recycling, composting, energy recovery, and disposal. The

Table 4.21 Moisture content, hydrogen content, and calorific values for MSW.

Material in MSW	Moisture (%)	Hydrogen content (%)	CV_{upper} (kJ/kg)	CV_{raw} (kJ/kg)
Eligible components				
Kitchen organics – vegetable	80.9	6.2	19 800	1 540
Kitchen organics – meat	52.9	9.4	11 900	3 340
Municipal garden organics	46.5	6	16 800	7 140
Paper composite	12	7.5	21 450	17 130
Mixed paper	29.7	5.8	15 150	9 030
Liquid paper board	4.5	7.5	21 450	12 520
Newspaper	7.2	6.1	17 330	14 660
Magazines	5	5.1	13 500	11 640
Cardboard	6.7	5.9	18 670	16 050
Disposable nappies	55	6.4	22 900	41 902
Wood (timber)	19	6	20 630	15 070
Noneligible components				
Textiles[4]	26.8	6.4	16 780	10600
Liquid paper board	4.5	7.5	21 450	6 360
Disposable nappies	55	6.4	22 900	4 140
Compounds (radios)	10	5.1	12 000	9 570
Mixed plastics	10	10	39 000	32 880
Plastic composite	<1	10	37 100	34 900
Plastic film	<1	10	40 000	37 800
Polystyrene (PS)	<1	8.4	40 000	38 150
Polyethylene (PE)	<1	14.2	45 000	41 880
Polyvinyl chloride (PVC)	<1	5.6	25 000	23 770
Polyethylene terephthalate (PET)	<1	6	25 000	23 680
Polypropylene (PP)	<1	14	44 000	40 920
Rubber	18.7	8.7	23 100	16 770

Table 4.22 Quantity of residential waste for city X.

Daily samples			Aggregated data		
Day	Samples	Tonnage per sample	Day	Tonnage per 3 samples	Total (daily)
1	1	2	1	7	46.7
1	2	2			
1	3	3			
2	4	4	2	7	46.7
2	5	2			
2	6	1			
3	7	2	3	7	46.7
3	8	2			
3	9	3			
4	10	2	4	6	40.0
4	11	3			
4	12	1			
5	13	2	5	7	46.7
5	14	2			
5	15	3			
6	16	3	6	7	46.7
6	17	2			
6	18	2			
7	19	1	7	6	40.0
7	20	2			
7	21	3			
8	22	2	8	6	40.0
8	23	3			
8	24	1			
9	25	2	9	8	53.3
9	26	4			
9	27	2			
				61	406.7
				Average (daily)	45.2

important elements are carbon, hydrogen, oxygen, nitrogen, sulfur, and ash. That data may be utilized to calculate the composition of the collected waste.

Example 4.2

Given: For a small city "X" in the United States, the data collection and analysis has to be done to quantify and characterize residential solid waste (Table 4.22). Based on historical data or pilot surveys, the major waste components are identified as food, paper, cardboard, plastics, textiles, rubber, leather, garden, wood, glass, metal, dirt, and ash. To collect the data for quantification and characterization of current solid waste, the number of samples should be determined.

Find: solid waste composition of this city

Solution:

The confidence level required depends on the utility of the data. If recycling units were to be determined, 70% confidence level could be used.

Analytical methodology of the samples depends whether all the wastes were being collected and transported to the facility. Assume that samples would be taken directly from the vehicles at the disposal facility and at different times of the day and there were no seasonal variations in residential solid waste. The dry weight can be calculated by subtracting moisture content from the wet weight.

For example, the weight of raw food waste is 4.1 tons and moisture content is 70%. The dry weight would be $4.1-(4.1\times 0.7)=1.23$ tons.

Based on Tables 4.23 and 4.24, the composition of various types of solid waste can be calculated. For example, food waste: dry weight = 4.08 − (4.08 × 0.7) = 1.2 tons; $C = 1.2 \times 0.48 = 0.6$ tons. Table 4.25 shows the composition for all the waste components.

Note: C, H, O, N, and S are carbon, hydrogen, oxygen, nitrogen, and sulfur, respectively.

Table 4.24 Typical data for dry waste in city X.

	Percent by weight (dry basis)					
Component	**Carbon**	**Hydrogen**	**Oxygen**	**Nitrogen**	**Sulfur**	**Ash**
Food waste	48.0	6.4	37.6	2.6	0.4	5.0
Paper	43.5	6.0	44.0	0.3	0.2	6.0
Cardboard	44.0	5.9	44.6	0.3	0.2	5.0
Plastics	60.0	7.2	22.8			10.0
Textiles	55.0	6.6	31.2	4.6	0.2	2.5
Rubber	78.0	10.0		2.0		10.0
Leather	60.0	8.0	11.6	10.0	0.4	10.0
Yard waste	47.8	6.0	38.0	3.4	0.3	4.5
Wood	49.5	6.0	42.7	0.2	0.1	1.5
Glass	0.5	0.1	0.4	0.1		98.9
Metals	4.5	0.6	4.3	0.1		90.5

4.3.5.4 Calorific Values

$$CV_{raw} = \left((1-MC)\times\left(CV_{upper}-(2241^{*}\times 9)\times H\right)\right) - 2241\times MC \tag{4.21}$$

where

1) CV = calorific value ("raw" is real "as delivered" value, and "upper" is value for dried material) in kJ/kg
2) MC = % moisture content (by weight)
3) H = % hydrogen content (from literature)
4) *Vaporization enthalpy of water (2441 kJ/kg at 25°C).

Table 4.23 Raw weight and moisture content of collected MSW in city X.

Component	Raw weight (tons)	Moisture content, MC (%)	Dry weight (tons)
Food waste	4.1	70.0	1.2
Paper	15.4	5.9	14.5
Cardboard	2.7	5.0	2.6
Plastics	3.2	1.4	3.2
Textiles	0.9	10.0	0.8
Rubber	0.2	0.0	0.2
Leather	0.2	20.0	0.2
Yard waste	8.4	64.9	2.9
Wood	0.9	20.0	0.7
Inorganic	9.1	3.0	8.8
Total	45.2		35.1

Table 4.25 Waste composition for city X.

			Composition					
Component	**Wet weight tons**	**Dry weight tons**	**C**	**H**	**O**	**N**	**S**	**Ash**
Food waste	4.1	1.2	0.6	0.1	0.5	0.0	0.0	0.1
Paper	15.4	14.5	6.3	0.9	6.4	0.0	0.0	0.9
Cardboard	2.7	2.6	1.1	0.2	1.2	0.0	0.0	0.1
Plastic	3.2	3.1	1.9	0.2	0.7	0.0	0.0	0.3
Textiles	0.9	0.8	0.4	0.1	0.3	0.0	0.0	0.0
Rubber	0.2	0.2	0.2	0.0	0.0	0.0	0.0	0.0
Leather	0.2	0.2	0.1	0.0	0.0	0.0	0.0	0.0
Yard waste	8.4	2.9	1.4	0.2	1.1	0.1	0.0	0.1
Wood	0.9	0.7	0.4	0.0	0.3	0.0	0.0	0.0
Total	36.1	26.4	12.4	1.6	10.4	0.2	0.1	1.6

4.3.5.5 Data Presentation

Similar tables may be produced for other sectors. Thereafter, similar waste components from different sectors could be grouped for different purposes, including the overall available amount of certain components for recycling or for final disposal (Tables 4.26, 4.27, 4.28, 4.29, and 4.30).

Table 4.26 Calorific values for waste components in city X.

Component	CV_{raw} kJ/kg	CV_{upper} kJ/kg	Hydrogen (H) (%)	Moisture content (MC) (%)
Food	3 809	19 800	6.	70
Paper	14 905	17 330	6	6
Cardboard	16383	18 670	5.9	5
Plastics	37 020	39 000	7.2	1
Textiles	13 553	16 780	6.6	10
Rubber	41 803	44 000	10	0
Yard waste	3 832	16 800	6	65
Wood	14 961	20 630	6	20

CV is calorific value; "raw" is real "as delivered" value of the collected waste, while "upper" is value for dried material from literature.

Table 4.27 Solid waste generation in city X (167tpd nonhazardous and 33tpd hazardous).

Sector estimated	Ratio (%age)	Estimated tonnage (per day)	
		Nonhazardous	Hazardous
Residential	22.6	45.2	0.0
Commercial	17.4	34.8	0.0
Construction and demolition	9.0	16.0	1.0
Healthcare	0.5	1.0	1.0
Industrial	50.0	70.0	30.0
Sludge	0.5	0.0	1.0

4.3.6 Zero Waste

In 2007, the City Council of San José (City) developed an ISWM Zero Waste Strategic Plan for 2040. To provide resource conservation, waste reduction, pollution prevention, and a sustainable economy, the ISWM Zero Waste Strategic Plan addresses the following key components:

1) Strengthening recycling programs.
2) Identifying infrastructure requirements for reusing, recycling, and composting.
3) Establishing effective waste prevention programs, incentives, and fee structures.
4) Identifying economic development opportunities from expanding solid waste processing facilities and industries using recycled materials as feedstock.

Example 4.3

Given: The City aims 75% waste diversion by 2013 and zero waste by 2022. Generation, disposal, and diversion tonnages through 2040 were estimated and characterized by generator sector. Food waste composting, reducing packaging, extended producer responsibility, and commercial solid waste system design are the key considered. Energy generated from selected waste materials is also critical in the plan (Table 4.31).

Find:

1) To determine the volume of material generated, disposed of, and recycled from the residents and businesses of the City, with waste projections through 2040.

Table 4.28 Residential waste in city X.

Component	Wet weight kg	Dry weight kg	Composition							
			MC	CV	C	H	O	N	S	Ash
Food waste	4.1	1.2	70	3 809	0.6	0.1	0.5	0.0	0.0	0.1
Paper	15.4	14.5	6	14 905	6.3	0.9	6.4	0.0	0.0	0.9
Cardboard	2.7	2.6	5	16 383	1.1	0.2	1.2	0.0	0.0	0.1
Plastic	3.2	3.1	1	37 020	1.9	0.2	0.7	0.0	0.0	0.3
Textiles	0.9	0.8	10	13 553	0.4	0.1	0.3	0.0	0.0	0.0
Rubber	0.2	0.2	0	41 803	0.2	0.0	0.0	0.0	0.0	0.0
Leather	0.2	0.2	20		0.1	0.0	0.0	0.0	0.0	0.0
Yard waste	8.4	2.9	65	3 832	1.4	0.2	1.1	0.1	0.0	0.1
Wood	0.9	0.7	20	14 961	0.4	0.0	0.3	0.0	0.0	0.0
Glass	4.0	4.0								
Metals	5.1	5.1								

MC, CV, C, H, O, N, and S are moisture content, calorific value, carbon, hydrogen, oxygen, nitrogen, and sulfur, respectively.

Table 4.29 Aggregated data on waste components.

Components	Residential	Commercial	Construction	Industrial	Healthcare	Total
Food waste	4.1	5.2	0.0	2.0	0.3	11.6
Paper	15.4	5.0	0.2	7.0	0.1	27.7
Cardboard	2.7	3.0	1.0	9.0	0.0	15.7
Plastic	3.2	3.3	1.0	10.0	0.1	17.6
Textiles	0.9	1.2	0.0	5.0	0.2	7.3
Rubber	0.2	1.0	1.0	5.0	0.1	7.3
Leather	0.2	1.0	1.0	5.0	0.1	7.3
Yard waste	8.4	4.0	1.0	5.0	0.1	18.5
Wood	0.9	2.0	4.0	5.0	0.0	11.9
Glass	4.0	4.0	2.0	7.0	0.0	17.0
Metals	5.1	5.1	3.0	10.0	0.0	23.2
Hazardous	0.0	0.0	1.0	30.0	1.0	32.0
Total	45.2	34.8	15.2	100.0	2.0	197.2

Table 4.30 Time series data and projections (1991–2010) for residential waste in city X.

	Residential				Residential		
Year	Organic	Inorganic	Total	Year	Organic	Inorganic	Total
1991	20.0	15.0	35.0	2001	14.0	28.0	42.0
1992	19.0	16.0	35.0	2002	14.0	29.0	43.0
1993	18.0	17.0	35.0	2003	13.8	30.0	43.8
1994	17.0	20.0	37.0	2004	13.6	31.0	44.6
1995	17.0	22.0	39.0	2005	13.4	31.8	45.2
1996	16.0	23.0	39.0	2006	13.4	33.0	46.4
1997	15.0	24.0	39.0	2007	12.9	34.2	47.1
1998	15.0	25.0	40.0	2008	12.5	35.4	47.9
1999	14.5	26.5	41.0	2009	12.0	36.6	48.6
2000	14.5	27.5	42.0	2010	11.6	37.8	49.4

Organic waste covers food waste and yard waste that can be converted into compost.

Table 4.31 Goal to bring greenhouse gas (GHG) emissions below 1990 levels.

Goal to bring GHG emissions below 1990 levels	
By year	Percentage below 1990 levels
2012	25
2015	30
2020	35
2030	50
2045	80

2) To identify significant material not currently being diverted, by generator sector and material type, with waste projections through 2040.
3) To provide waste characterization data by generator sector.
4) To provide recommendations regarding portions of the waste stream that could be targeted with new or improved programs.
5) To provide recommendations on increasing waste diversion through the CDDD program.

Solution:

To develop the ISWM strategic plan, waste materials generated in the City that are not currently being diverted should be documented. Diversion programs for different waste streams to target need the following data: city demographics and its projections, waste stream generation and trends, disposal tonnage, diversion tonnage, and material flow. Wastes are characterized according to the following sectors: single-family dwellings (SFD), multifamily dwellings (MFD), commercial sector, city facilities, construction and demolition (C&D) debris, nonfranchised hauler waste (self-haul) material, and alternative daily cover (ADC). The percentages of each waste sector's disposal are (i) recyclable, (ii) compostable, (iii) potentially recyclable, and (iv) nonrecyclable. The single-family residential section lists the material types that are in each group (Table 4.32).

Solid waste composition is given for different wastes in Figures 4.24 and 4.25.

Divertability of disposed mixed C&D is analyzed and presented in Figure 4.26.

The previous ratios are used to carry out the ADC analysis in Table 4.33.

The overall solid waste production data and the ADC data are used to determine the annual facility capacity, and the projected capacity need (tons) is estimated in Table 4.34.

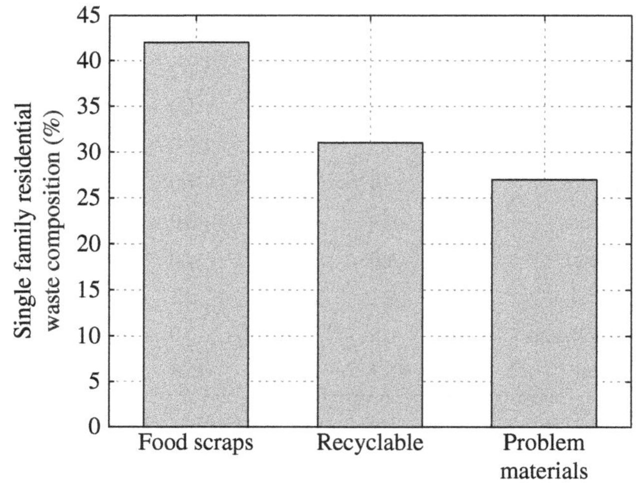

Figure 4.24 Single-family residential waste composition.

Table 4.32 Data sources.

Waste stream	Data source for tonnage	Data source for waste composition
Single family	City data for fiscal year 2006/2007, including disposal, recycling, and yard trimmings collection	Cascadia Consulting Group's City of San José Waste Characterization Study, May 2008
Multifamily	City data for fiscal year 2006/2007, including disposal and composting collection	CIWMB 2004 Statewide Waste Characterization Study for Multifamily Waste
Commercial	City data for fiscal year 2006/2007 and HF&H Consultants' Commercial Redesign White Paper of November 2007, including disposal, recycling, and yard trimmings collection	Cascadia Consulting Group's City of San José Waste Characterization Study, May 2008
City facilities	City data for fiscal year 2006/2007 for both disposal and recycling	CIWMB 2006 Statewide Waste Characterization Study for Commercial Waste
Construction and demolition debris	City data for fiscal year 2006/2007 and HF&H Consultants' Commercial Redesign White Paper for both disposal and recycling	CIWMB 2006 Statewide Waste Characterization Study for Construction and Demolition Debris Waste
Nonfranchised hauler waste	Calculated based on subtracting all other known waste quantities from the total waste quantity delivered to landfills in 2006 for disposal quantity only (no data available for nonfranchised diversion)	CIWMB 2004 Statewide Waste Characterization Study for Self-haul Waste
Alternative daily cover and other "beneficial use" at disposal facilities	CIWMB website totals for ADC, as well as more detailed data on AIC, cover material, and inerts from Santa Clara County	All ADC was reported by waste type, so there was no need to use additional studies to estimate waste composition
Overall city-wide waste stream	Tonnage data was a calculated sum of all of the waste streams. CIWMB official disposal data for calendar year (CY) 2006 was used for total disposal, and CIWMB official generation total for CY 2006 was used for generation	Composition data for the overall waste stream was calculated as a weighted average of all of the individual sector waste streams combined

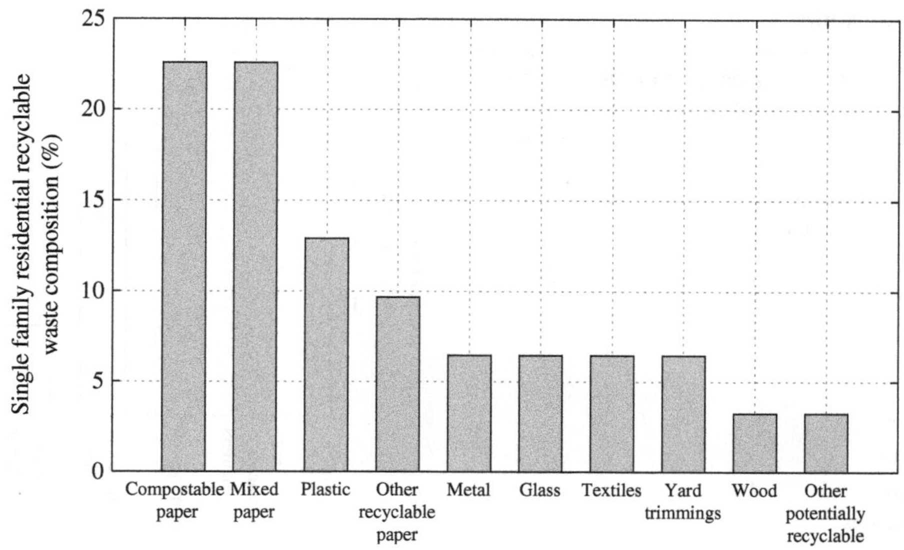

Figure 4.25 Single-family residential recyclable waste composition.

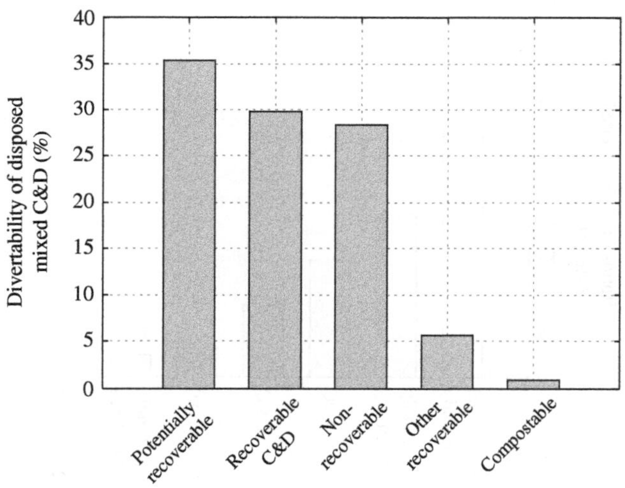

Figure 4.26 Divertability of disposed mixed construction and demolition (C&D).

Table 4.33 Alternative daily cover analysis.

San Jose ADC type	Tonnage (2006)
C&D debris	105 059
Green waste	41 818
Sludge	1 257
Mixed waste	16 875
Other	77
Total ADC	165 086

Table 4.34 Annual facility capacity and projected capacity need (tons).

	Facility capacity (2007)		Capacity need	
Facility type	**Permitted**	**Unused**	**2007/2008**	**2040**
Landfills or other disposal	4 274 400	2 423 840	153 356	227 588
Recycling and transfer	1 475 448	580 115	198 274	294 247
Composting	1 054 560	422 068	196 793	292 051
Construction and demolition	4 859 400	3 417 514	176 807	262 390

According to the previous solid waste composition, the City of San Jose attempts major effort to bring waste to zero (Table 4.35).

The strategies ensure the city to achieve the following diverted ratios for different waste streams (Figures 4.27, 4.28, and 4.29).

ISWM decision should be based upon life cycle cost and benefit analysis. Table 4.36 lists the facility host fees that could be used for estimate benefit of any policy in terms of monetary benefit.

Comment: One of the major strategies missing is the interconnected design. For example, to achieve the true sustainable waste management, the solid wastes should be incinerated to generate heat or electricity to achieve zero energy (Table 4.37).

Table 4.35 Zero waste initiatives.

Zero waste initiative	2008–2013 to 75% diversion	2013–2022 to zero waste
1. Commercial redesign	√	
2. Food scraps composting and recycling (including anaerobic digestion)	√	√
3. CDDD redesign and enhancements	√	
4. Residential system enhancements	√	√
5. Landfill regulations	√	√
6. Generator mandates		√
7. Extended producer responsibility programs	√	√
8. Mixed waste recycling	√ (Commercial)	√ (Single-family residential)
9. Future development of energy conversion technologies		√

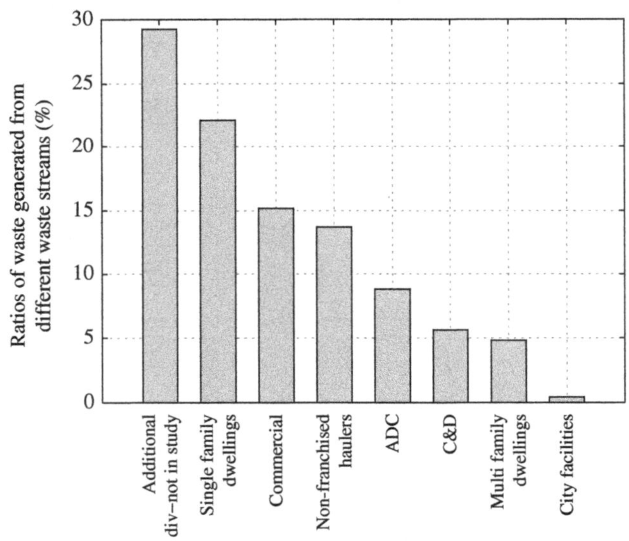

Figure 4.27 Ratios of waste generated from different waste streams.

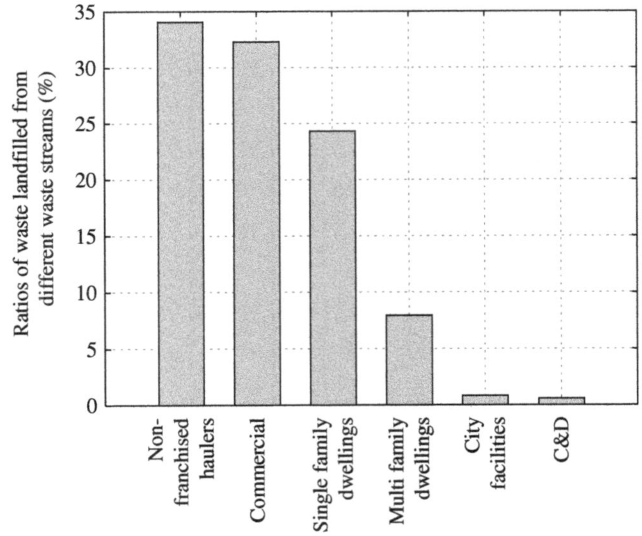

Figure 4.28 Ratios of waste landfilled from different waste streams.

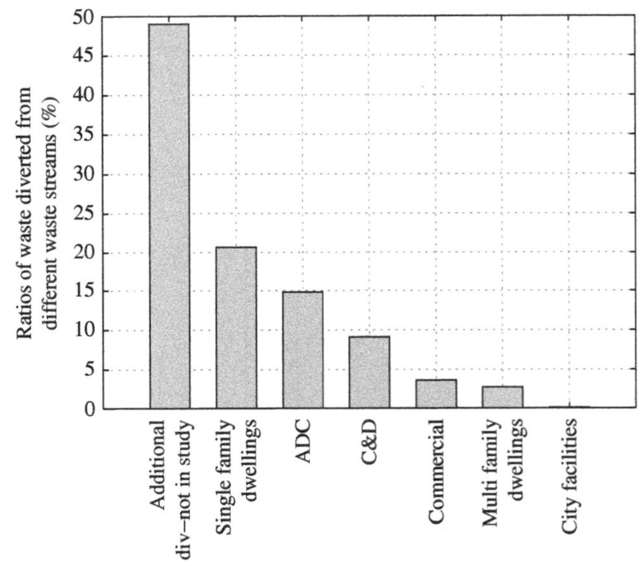

Figure 4.29 Ratios of waste diverted from different waste streams.

4.3.7 Integrated Water Resource Management (IWRM)

"IWRM is a process that promotes the coordinated development and management of water, land and related resources in order to maximize the resultant economic and social welfare in an equitable manner without compromising the sustainability of vital ecosystems" (GWP, 2017). IWRM could enhance water resources management toward economic efficiency, equity, and environmental sustainability. IWRM should be based on a participatory approach, involving users, planners, and policy makers at all levels. Water has an economic value in all its competing uses and should be recognized as an economic good.

To establish IWRM objectives, the EPA lays out the following procedure:

1) Goal setting: Establish sustainability goals that reflect utility and community priorities.
2) Objectives and strategies: Establish explicit, measurable objectives for each sustainability goal and identify strategies for meeting the objectives.
3) Alternatives analysis: Based on sustainability goals and objectives, set explicit and consistent evaluation criteria to analyze a range of infrastructure alternatives.

Table 4.36 Facility host fees.

Facility	Permitted (tons/day)	"Host" fee	Based on	Equivalent fee/assessed per ton
Transfer stations and recycling facilities				
Downey Area Recycling and Transfer (DART)	1 500	$1.32 per ton	Disposal tons and recycling sales	$1.32
Rainbow Transfer Station and MRF	2 800	2% of gross receipts	Public dumping charges ("paid disposal")	$0.64
Waste Management Transfer Station	2 000	$400 000 per year	Flat fee	$0.64 or more, depending on actual throughput
South Gate Transfer Station	1 000	$1.26 per ton	All inbound tons	$1.26
Interior Removal Specialist – MRF	3 000	$1.25 per ton	All inbound tons	$1.25
Hanson Aggregate C&D Recycling	1 000	2% of gate revenue	All inbound tons	$0.14
Landfills				
Olinda Alpha Sanitary Landfill	8 000	1.50 per ton[a]	All out-of-city tons	1.50 per ton[a]
Colton Sanitary Landfill	3 100	1.00 per ton[b]	Disposal tons	$1.00
Scholl Canyon Landfill	3 400	25% of gate fee	All out-of-city tons	$10.02
Frank R. Bowerman Sanitary Landfill	8 500	$1.50 per ton[a]	All out-of-city tons	$1.50 per ton[a]
Puente Hills Landfill	13 200	$1.00 per ton	Disposal tons	$1.00
		$1.00 per ton	Disposal tons	$1.00
		10% of gate fee	Disposal tons	$2.94
San Timoteo Sanitary Landfill	1 000	$1.00 per ton[b]	Disposal tons	$1.00
Mid-Valley Landfill	7 500	$4.98 per ton	Disposal tons	$2.49 to each city
Victorville Sanitary Landfill	1 600	$.50 per ton[b]	Disposal tons	$0.50

[a] The host fee is a fixed amount until 2014. Then the fee converts to $1.50 per ton.
[b] Landfill mitigation fund – form mitigation of hosting landfill, road repair, and illegal dumping.

Table 4.37 Summary projections of landfilled and diverted tons.

Waste stream	Actual 06/07	% of total generated	Projected			
			2010	2020	2030	2040
Single-family dwelling						
Landfilled	183 132	9.8	190 427	218 007	242 973	271 776
Diverted	230 721	12.3	239 911	274 658	306 112	342 400
Total	413 853	22.1	430 338	492 665	549 084	614 175
Diversion rate	55.7%		55.7%	55.7%	55.7%	55.7%
Multifamily dwelling						
Landfilled	59 938	3.2	62 326	71 352	79 523	88 951
Diverted	30 188	1.6	31 390	35 937	40 052	44 800
Total	90 126	4.8	93 716	107 289	119 576	133 751
Diversion rate	33.5%		33.5%	33.5%	33.5%	33.5%

(Continued)

Table 4.37 (Continued)

			Projected			
Waste stream	**Actual 06/07**	**% of total generated**	**2010**	**2020**	**2030**	**2040**
Commercial – franchised collection						
Landfilled	243 407	13.0	253 102	289 760	322 943	361 226
Diverted	40 117	2.1	41 715	47 757	53 226	59 535
Total	283 524	15.2	294 817	337 516	376 169	420 761
Diversion rate	14.1%		14.1%	14.1%	14.1%	14.1%
City facilities						
Landfilled	6 236	0.3	6 484	7 423	8 273	9 254
Diverted	1 591	0.1	1 654	1 894	2 111	2 361
Total	7 827	0.4	8 138	9 317	10 384	11 615
Diversion rate	20.3%		20.3%	20.3%	20.3%	20.3%
C&D – franchised collection						
Landfilled	4 278	0.2	4 448	5 092	5 676	6 348
Diverted	101 010	5.4	105 033	120 245	134 016	149 903
Total	105 287	5.6	105 033	125 338	139 691	156 251
Diversion rate	95.9%		95.9%	95.9%	95.9%	95.9%
Nonfranchised haulers						
Landfilled	256 759	13.7	266 986	305 655	340 658	381 041
Diverted	—	0.0	—	—	—	—
Total	256 759	13.7	266 986	305 655	340 658	381 041
Diversion rate	0.0%		0.0%	0.0%	0.0%	0.0%
ADC						
Landfilled	—	0.0	—	—	—	—
Diverted	165 086	8.	171 662	196 524	219 030	244 995
Total	165 086	8.8	171 662	196 524	219 030	244 995
Diversion rate	100.0%		100.0%	100.0%	100.0%	100.0%
Totals						
Total landfilled	753 749	40.3	783 773	897 289	1 000 046	1 118 596
Total diverted	568 713	30.4	591 366	677 015	754 546	843 994
Additional diversion not tracked by this study	546 741	29.2	568 520	650 860	725 396	811 388
CIWMB total generation	1 869 203	100.0	1 943 659	2 225 164	2 479 988	2 773 978
Diversion rate	59.7%		59.7%	59.7%	59.7%	59.7%
Growth rate versus 06/07			3.98%	19.04%	32.68%	48.40%

4) Financial strategy: Implement a financial strategy including adequate revenues so that new infrastructure and operational investments – as well as the overall system – are sufficiently funded, operated, maintained, and replaced over time on a full life cycle cost basis, with appropriate considerations for disadvantaged households.

The potential sustainable goals and the corresponding examples are listed in Tables 4.38, 4.39, 4.40, 4.41, 4.42, and 4.43.

For small utilities, the EPA's strategic planning describes seven steps to develop a "strategic road map" as follows.

Table 4.38 Integrated planning for WTPs and WWTPs.

Sustainable goals	Examples
Improve compliance	Collaborate with neighboring utilities to increase or maintain technical, managerial, or financial capacity
Reduce energy cost	Invest in more energy efficient equipment or explore operational changes that can enhance energy optimization (such as pumping at night when the rate is lower)
Reduce overall infrastructure costs	Coordinate infrastructure projects such as rain gardens
Extend the projected adequacy of current water supplies	Implement water metering, fix distribution system leaks, or make use of reclaimed water Implement a mix of nontraditional infrastructure alternatives such as green infrastructure solutions with integrated stormwater and combined sewer overflow control
Preserve critical ecological areas	Adopt management programs for septic systems to reduce nutrient loadings to water bodies or green treatment such as GI
Improve the economic vitality	Target water infrastructure projects to encourage redevelopment
Enhance community livability	Incorporate green space or recreational opportunities into projects
Reduce long-term system operational costs	Use wetlands to reduce the input of energy and chemicals for treatment or reuse water treatment solids
Improve operational resilience	Understand operational, financial, and potential climate vulnerabilities and incorporate them into alternatives analysis to reduce risk
Reduce vulnerability to water supply disruption or contamination	Conduct real-time water quality monitoring, install isolation shutoff values, or provide connections to alternative water supplies

Source: The US EPA (2012).

Table 4.39 The US EPA procedure for identifying sustainable goals.

Step	Actions
1	Identify sustainability priorities and potential opportunities for the utility
2	Identify community sustainability priorities
3	• Engage the community about its sustainability priorities • Participating in direct discussions with other planning institutions and community bodies early in their planning processes • Getting involved with existing community-wide planning efforts • Aligning utility planning with existing community plans • Convening a stakeholder process for water planning
4	Identify and document sustainability goals

Source: The US EPA (2012).

1) Identify criteria.
2) Establish a scale (e.g. −3 to +3) for each criterion.
3) Assign a weight factor to each criterion.
4) Score each criterion.
5) Multiply each score by the criterion's weighting factor.
6) Sum weighted scores across all criteria.

Table 4.40 The US EPA procedure to develop a strategic plan.

Step	Actions
1	Develop a strategic roadmap
2	Define the service area
3	Assess the system's technical, managerial, and financial capabilities
4	Identify alternatives
5	Assess alternatives
6	Implement an action plan
7	Evaluate alternatives

Source: The US EPA (2012).

7) Identify the alternative with the highest calculated score.

4.3.8 Water Resource Recovery Facilities (WRRF)

Recovery of water, energy, and nutrients is the key of WRRF. It enables WRRF to be independent of the grid and adds revenue through recovering water, energy,

Table 4.41 Step-by-step guide for conducting baseline analysis.

Step	Actions
1	Describe the water storage and delivery system, including the size of the physical system, the number of people and connections, services, land use, demographics, and any unique characteristics that affect supply or demand
2	Inventory the water supply system, including sources of water supply, the status of water rights, and any limits on system capacity
3	Estimate present water demand (e.g. with information from current billing records)
4	Estimate future water demand based on population growth projections and other relevant information
5	List and rank water problems, including high per capita use, significant losses, constraints on system capacity, and/or insufficient water rights
6	List and analyze potential solutions, including water conservation through infrastructure investments (e.g. repairing leaks, replacing old lines and tanks, etc.) and/or demand reduction

Table 4.42 Sustainable criteria.

Number	Sustainable criteria
1	Ecological and economic impacts, such as the extent to which projects damage (or create) important habitat, or create green space and recreation opportunities
2	Preference for treatment or operational functions that rely on natural systems for lower life cycle operating costs through reduced energy and chemical inputs
3	Reduced reliance on the energy grid through greater energy efficiency or self-generation of energy
4	The extent to which projects focus on sustainability of infrastructure in a utility's existing service area
5	Cost-effectiveness based on an assessment of full life cycle costs

Table 4.43 Procedure in assessing alternatives.

Step	Action
1	Identify alternatives
2	Develop specific sustainability criteria
3	Assess the benefits of each alternative
4	Assess the full costs of each alternative using life cycle assessment
5	Compare and select alternatives
6	Document the alternatives analysis

and nutrients. Reclaimed water can be integrated into power plants for cooling, irrigation in food production, and recharging ecosystems. For example, biogas could be used to heat sludge or generate electricity. Digested sludge could be then used as fertilizer (Figures 4.30, 4.31, 4.32, 4.33, and 4.34) (NSF et al., 2015)

The US Water Research Foundation developed major tools in integrated water management as follows (Table 4.44).

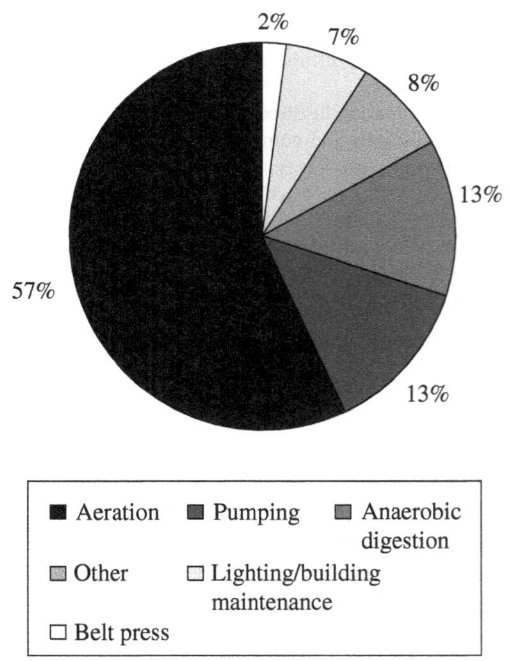

Figure 4.30 Energy breakdown at water resource recovery facilities (WRRFs).

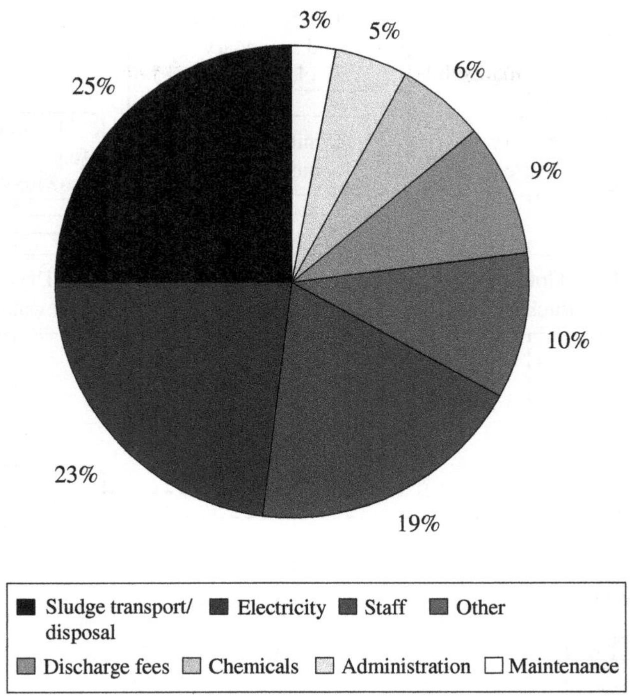

Figure 4.31 Water resource recovery facilities' (WRRFs) operating cost breakdown.

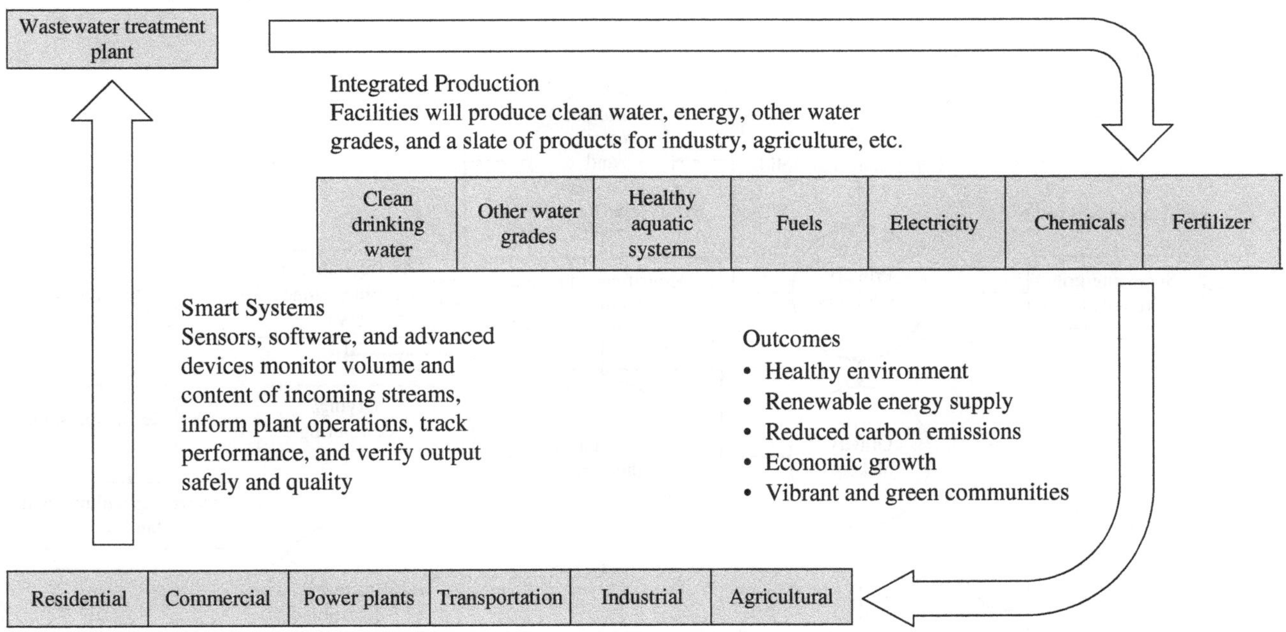

Figure 4.32 Water resource recovery facility of the future.

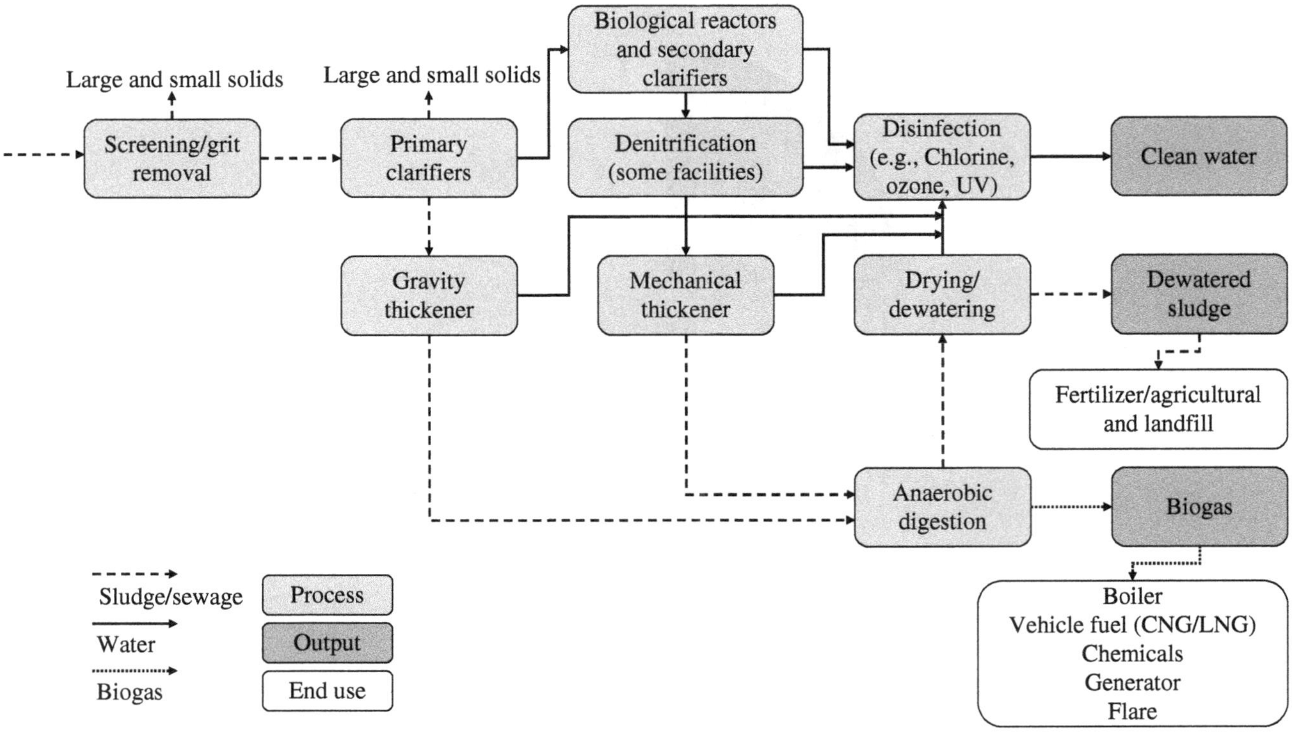

Figure 4.33 A simplified view of water resource recovery facility operations.

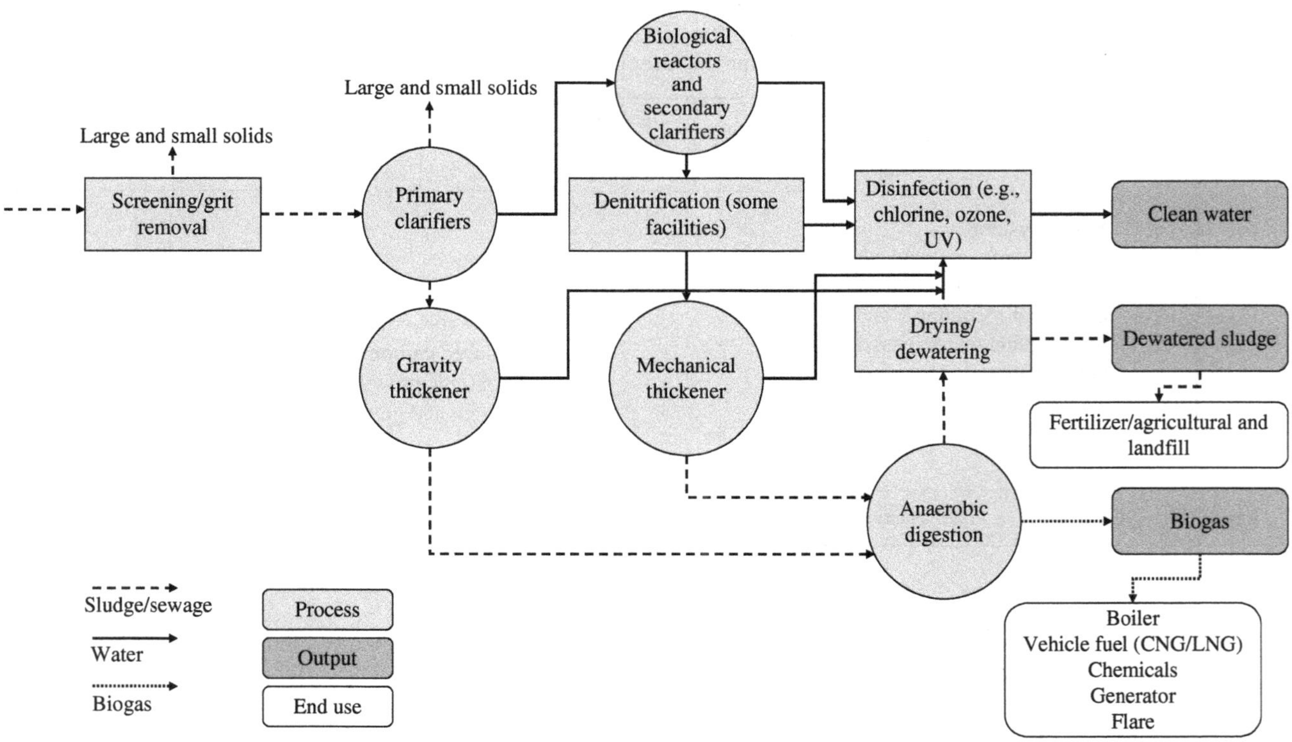

Figure 4.34 A simplified view of water resource recovery facility operations.

Table 4.44 Integrated planning tools (The US Water Research Foundation).

Design tools	Functions	Sponsor agencies
Integrated Water Management with Urban Planning and Design	How a full spectrum of water services can be integrated with land-use development and provides recommendations to encourage integration	WE&RF, 2017
Economic Pathways and Partners for Water Reclamation and Stormwater Harvesting	How effective economic conditions can be developed to promote the beneficial use and develops the pathways to utilize reclaimed water and stormwater reuse as a commodity for communities	WE&RF, 2017
Integrating Land Use and Water Resources Planning to Support Water Supply Diversification	User-friendly resources that can help advance the integration of water resource and land-use planning	WRF, 2018
Framework for Evaluating Alternative Water Supplies: Balancing Cost with Reliability, Resilience, and Sustainability	Integration of reliable, resilient, and sustainable water supply into a framework to support water supply planning	WRF, 2018
Using Graywater and Stormwater to Enhance Local Water Supplies: An Assessment of Risks, Costs, and Benefits	Technical, economic, regulatory, and social issues associated with gray water and stormwater for nonpotable urban uses, irrigation, and groundwater recharge	WE&RF, WRF, NAS, 2016
Water and Electric Utility Integrated Planning	Tools, knowledge, and recommendations for water and electric utility collaboration in strategic and project-level integrated planning	WRF, AWWA, & NYSERDA, 2017
Opportunities and Barriers for Distributed Energy Resource Development at Water and Wastewater Utilities	Opportunities for distributed energy resource (DER) development at water and wastewater utilities; (2) assess the legislative, regulatory, and policy environment impacting DER developments; and (3) undertake case studies of implemented DER utility projects	WE&RF, WRF, 2019
Toolbox for Completing an Alternatives Analysis as Part of an Integrated Planning Approach to Water Quality Compliance	A user's guide to help municipalities and utilities determine whether to pursue US EPA's Integrated Planning Framework (IPF) and how to use current tools and information to prepare a successful integrated water plan	WE&RF, 2017
Pathways to One Water: A Guide for Institutional Innovation	Guidance for organizations to move toward "One Water" management based on a literature review of major challenges and how organizations take action	WE&RF, WRF, 2017

AWWA, American Water Works Association; NAS, National Academy of Sciences; NYSERDA, New York State Energy Research and Development Authority; WE&RF, Water Environment and Reuse Foundation; WRF, Water Research Foundation.

4.4 Integrated Air Quality Management (IAQM)

There are two types of air quality standards. Primary standards are established to protect human health, while secondary standards are set to reduce environmental damage. The secondary standards are mostly the same as the primary standards. In addition, there are no secondary standards for carbon monoxide. Black carbon, a component of particulate matter, and ozone both have adverse impacts on human health, leading to premature deaths worldwide. Ozone is also the major air pollutant responsible for reducing crop yields and thus affects food security. Black carbon darkens snow and ice surfaces and increases their absorption of sunlight, which, along with atmospheric heating, exacerbates the melting of snow and ice around the world, including in the Arctic, the Himalayas, and other glaciated and snow-covered regions. This affects the water cycle and increases risks of flooding and drought. They disturb tropical rainfall and regional circulation patterns such as the Asian monsoon, affecting the livelihoods of millions of people.

A major aim of IAQM is to reduce black carbon from the combustion of coal, oil, and gas. Emissions reduction measures targeting black carbon, and ozone precursors could immediately begin to protect the climate, public health, water and food security, and ecosystems.

Measures include the recovery of methane from coal, oil, and gas extraction and transport, methane capture in waste management, the use of clean-burning stoves for residential cooking, diesel particulate filters for vehicles, and the banning of field burning of agricultural waste. If these measures were implemented by 2030, they could halve the potential increase in global temperature projected for 2050. Global warming would be reduced by 0.5°C, and the rate of regional temperature increase would also be reduced. As a result, the chances of keeping the Earth's global mean temperature increase within the UN's 2°C target would increase significantly. For these reasons, renewable energy must be utilized in SEE design whenever it is feasible.

4.5 Exercise

4.5.1 Questions

1) On stakeholder's input and support
   a) Why is stakeholders' involvement important?
   b) What are the likely stakeholders for EEIS of a water utility?
   c) How to identify and get the stakeholders involved in both planning and implementation loops?
   d) How to get the support from the stakeholders for a timely and effective implementation?

2) On goal setting
   a) What was the internal process a local utility undertook to identify its sustainability opportunities? What opportunities did the utility identify?
   b) What community plans or information sources did the local utility consult to identify community sustainability priorities?
   c) If applicable, how did a local utility consult with other community members or community planning organizations about utility and community sustainability priorities and the relationship between them?
   d) If applicable, how did a local utility consult with neighboring utilities about potential partnership opportunities to share information or services?
   e) What sustainability goals did a local utility set and why?
   f) How were the local utility's sustainability goals documented and communicated internally and externally?
   g) How will the community and others consulted be kept informed of subsequent decisions and developments?

3) On objectives
   a) How much water is available from the water supply source(s)?
   b) What are the legal and regulatory implications for water withdrawals while maintaining ecological flows?
   c) What are the water supply needs and demands of the community, including energy and industry, and projected growth?
   d) How much storage capacity is built in to the water supply?
   e) Does the utility have backup or alternative sources and interconnections with other water systems in case of extreme weather events, such as droughts and floods?
   f) Does the utility have a conservation plan in case of a water shortage?
   g) Is the water supply susceptible to saltwater intrusion from over withdrawals of groundwater or climate change?
   h) Do the community's land use plan and zoning include provisions for determining adequate water supply production and protection of drinking water sources and environmentally sensitive areas?
   i) Does the water utility have a source water protection plan?
   J) Does the water supply have natural filters and barriers (e.g. riparian buffers, land conservation, and wellhead protection) in place to prevent pollution, or are there opportunities to implement them?

4) On strategies
   a) How was each of a local utility's sustainability goals reflected in specific, measurable objectives?
   b) In what ways were the utility's sustainability objectives articulated consistent with the *SMART* principles?
   c) For each sustainability objective, what kind of baseline analysis did you conduct to assess your current status?
   d) What types of tools and resources did you use for the baseline analysis? Are there monitoring programs already in place to generate data for baseline analysis and to monitor progress toward objectives?
   e) For each sustainability objective, what traditional and nontraditional strategies did the utility identify?
   f) How and where were the sustainability objectives described and codified in a planning document?

g) What is your plan for measuring and tracking the accomplishment of sustainability objectives over time?

5) On assessing alternatives
   a) Did you describe, analyze, and rank all alternatives?
   b) How to assess different alternatives using the TDPs of SEE?
   c) What are the other methods for analyzing alternatives and the criteria for ranking them?
   d) Were all sustainability objectives reflected in the specific ranking criteria or in the alternatives analyzed? How?
   e) How were alternatives ranked according to the criteria? In what ways did the ranking process reflect specific consideration of integrated and interconnected natural systems?
   f) Were alternatives all assessed on a full life cycle cost basis?
   g) Was the alternatives analysis transparent, and were the approach, rationale, and results communicated to community members?
   h) To what extent was the community involved in, or kept up-to-date on, the alternatives considered and selected?

6) On finance WTPs and WWTPs
   a) Were a full range of capital financing options considered, and were their interest, acquisition, and implementation costs fully identified and thoroughly compared?
   b) Does the capital financing strategy keep capital acquisition and interest costs as low as possible and keep the repayment schedule (principal and interest) consistent with utility revenue capacity (cash flow)?
   c) What was considered in determining whether to use cash versus debt financing?
   d) Are rates, fees, and charges sustainable, and do they generate sufficient revenue to fully cover long-term full life cycle costs of the selected project alternatives?
   e) Are costs allocated fairly/appropriately (e.g. reliability costs to current customers, cost recovery for industrial wastewater permitting and treatment, growth costs to new development, rates for disadvantaged households)?
   f) Does the rate structure create appropriate customer incentives consistent with the utility's objectives (e.g. conservation pricing)?
   g) Does the financial strategy maintain or improve the bond rating, debt coverage ratio, or capital financing reserves where relevant?

4.5.2 Calculation

1) In the past decades, the average solid waste collection rate is about 1.8 ton/person/year in Miami-Dade County. If the solid waste contains 15% food waste, what could be the total amount of food waste diverted to anaerobic digester of Miami-Dade County of 2.6 million residents?
2) Miami-Dade County is required to achieve 60% of reuse of wastewater treated, which is about 120 MDG. Currently, however, there are only 7% treated wastewater that have been reused. The current water rate in Miami-Dade is $3.34/ft^3$ for water and $6.58/ft^3$ for sewer if a household water uses rate ranging from 6 to 9 ft^3. Please calculate the following:
   i) What monetary benefits could bring to the county if a WRRF of 120 MGD is designed and built annually?
   ii) What would be the annual saving for an average single family residential house that consumes 108 ft^3/year, if all the benefit of the WRRF is transferred to the consumers in the Miami-Dade County?
3) In three months from July to September 2017, Miami-Dade County recycled 1652 tons of cardboard/paper, 4204 tons of newspaper, 1502 tons of mixed paper, 201 tons of plastics, 150 tons of tin cans, and 150 tons of aluminum. Please answer the following:
   i) How many metric tons of CO_2 emission are avoided?
   ii) If 4805 tons of residue were sent to the Miami-Dade waste-to-energy (WTE) facility and generated 12.49 million kWh electricity, what is the energy conversion efficiency of this WTE facility?

4.5.3 Projects

4.5.3.1 Community Projects

1) Please contact director of air, water, and solid waste management in your city and collect following inventory data in the last 10 years:
   a) Air emission data in terms of $P.M._{2.5}$, CO_2, SO_x, NO_x, O_3, and volatile organic compounds (VOC). Also, please go to the website: http://aqicn.org/map/world/. Please report the following:
      i) Air quality index.
      ii) From the UNEP website below to do the following:

      http://uneplive.unep.org/country/data/CN#data_tab, From the Excel file of cn_ambient_air_quality_by_key_cities_2013, please answer the following:

i) Historical average AQI of the city.
ii) What is projected average AQI using crystal ball?
iii) What are the major air pollutants contribute to the high AQI in your city?
iv) What are the major emitters contributing to the high AQI?
v) Please develop outline of two alternatives of integrated and interconnected air quality management plan for your city for the next 5, 10, and 20 years.

b) Please do the same as above on water and solid wastes.

4.5.3.2 Xiongan Projects

Xiongan is located in Hebei province. Please use the related data to do the same as above.

References

Anderson, D.M. (2008). *Design for Manufacturability & Concurrent Engineering, How to Design for Low Cost, Design in High Quality, Design for Lean Manufacture, and Design Quickly for Fast Production*. CIM Press.

Bilthoven (2007). Netherlands Environmental Assessment Agency (2007). Environmental Balance 2007, Netherlands Environmental Assessment Agency (MNP), Bilthoven, the Netherlands.

Global Water Partner, http://www.gwp.org/en/GWP-CEE/about/why/what-is-iwrm/ (accessed December 2017)

Hoornweg, N. and Bhada-Tata, P. (2012). *What a waste, a global review on solid waste management*. Washington, DC: Urban Development & Local Government Unit, World Bank.

NSF (National Science Foundation), US DOE (US Department of Energy), and US EPA (US Environmental Protection Agency) (2015). Energy-Positive Water Resource Recovery Workshop Report. Workshop date 28–29 April 2015, Arlington, VA.

The US EPA (2012). Planning for Sustainability: A Handbook for Water and Wastewater Utilities. Washington, DC: US EPA.

UNEP (2013). cn_ambient_air_quality_by_key_cities_2013, http://uneplive.unep.org/country/data/CN#data_tab (accessed 20 January 2018).

World Bank (2012). What a waste: a global review of solid waste management. Urban development series knowledge papers no. 15. Washington, DC: World Bank.

5

Reliable Systems on a Spatial Scale

Simplicity is prerequisite for reliability.

OBJECTIVES
To define reliability design strategies on residential house, community, city, country, regional, and global scales
To design fit for purpose by selecting centralized versus decentralized WWTPs on different spatial scales
To compare integrated system approach with conventional design approach
To explore major technical issues in scale-up from collimated beam test (CBT) laboratory data to pilot and full scale in ultraviolet water disinfection
To predict log inactivation (log *I*) of pathogenic microorganisms during UV disinfection by using virus or bacteria sensitivity indexes
To quantify the uncertainty and sensitivity of extrapolation from laboratory data to full-scale UV disinfection system by using Matlab
To define electrical energy per order (EE/O) in UV disinfection

5.1 Principle 2

A system is a region where EEIS are intended to serve. On a spatial scale, a system can be classified as residential, community, city, country, or global level. Reliability is defined as the duration of an EEIS to function normally and provide the required service to protect human health and the environments under normal conditions. Reliability usually is measured by the time period during which the system will not fail. Therefore, it can be quantified as the time from starting service to failure of the system. EEIS should be designed in such a way that their reliability is quantified with known uncertainty and sensitivity on different spatial scales. To reduce probability of failure is to design EEIS as simple as possible because propagation of failure increases with increasing components.

Principle EEIS alternatives on different scales should be reliable with quantified uncertainty and sensitivity.

1) A system could be defined within residential, community, city, country, regional, or global boundaries.
2) Different design strategies should be sought for systems on different scales with front-end design using integrated system approach (ISA).
3) Laboratory, pilot, and full-scale testing should be validated with known uncertainty to increase the reliability of EEIS.

Reliability on spatial scales must be quantified in designing EEIS such as WTPs, WWTPs, and GIs. Local geography, resource, economy, and culture could also be weighted to increase reliability of EEIS. For a household, if the land is large enough, on-site WWT processes such as septic tanks and constructed wetlands could be the most economical way to achieve regenerative design. For communities of population greater than 1000 PE, energy-neutral design may be more realistic. For large cities, if energy-neutral design cannot be achieved, designers should target maximum reuse of materials and energy to reduce the FP on water, energy, and nutrients. In general, decentralized WWT is more suitable for rural areas than metropolitan cities. New urban communities and cities have the opportunity to design and build source-separating systems optimized for cost-effective resource recovery due to established technologies and design strategies. EEIS should be designed to support sustainable industries, agriculture, and social

Sustainable Environmental Engineering, First Edition. Walter Z. Tang and Mika Sillanpää.

institutions. Engineers should design EEIS adapted to local geographical conditions, because local resources and geographical settings significantly influence the availability of water, energy, and nutrients. For example, in hilly settings, WTP and WRRF should be sited on a slope from 1 to 3% when gravity flow is used to save energy needed for pumping. For WRRF, reliability is affected by the variability of system performance under normal operation and impact of mechanical failure. For WTP, two components determining the system reliability are the probability of being online and the number of parallel treatment units (AWWA, 2012). To achieve reliability at different spatial scales, following design strategies are recommended:

1) To achieve reliability, design philosophy should switch from fit for all to fit for purpose.
2) Economy of scale should be quantified in terms of unit cost on individual, household, community, city, state, and country levels.
3) Zero discharge and energy-independent residential house should be designed when sufficient land or space is available.
4) Green infrastructure (GI) should be implemented as much as possible on household, community, city, state, and country levels to prevent nonpoint pollution.
5) Laboratory, pilot, and full-scale testing should be validated with quantified uncertainty to increase the reliability of EEIS. Laboratory-scale data could be used to design the full-scale system if kinetics and mechanisms are fully understood and validated through computer simulation.
6) Separated management of blue, green, gray, brown, and black water could increase reliability on a specific spatial scale.
7) Decentralized WRRF could increase reliability of water utility for small to medium size cities due to its approximate to irrigation and other water reuse demands.

In designing an EEIS, basic research on molecular reaction, applied research on materials and components, and the development of devices and products should consider the life cycle impact of the EEIS on the environments. Figure 5.1 shows that basic and applied research, product development, and manufacturing are corresponding to molecules, materials, components, and products. Life cycle assessment should be used to quantify and minimize the footprints on the environments at each stage. Since all the stages interact with environments, increasing reliability should not increase environmental impacts.

5.1.1 Central Versus Decentralized WWTP

Sedlak (2014) considered that water and wastewater engineering have went through four stages: (i) transportation of water for drinking and irrigation in the Roman Empire 2000 years ago until 1800, (ii) water treatment and the collection of wastewater and discharge into nature in England until 1900, (iii) centralized wastewater treatment plants from 1900 to today, and (iv) decentralized wastewater treatment that may offer ecological and economic benefits from now on as the fourth stage. Indeed, aging EEIS such as WWTPs causes great problems due to increased population, infiltration, and inflow in sewer system. For example, from January 2007 to May 2013, Miami-Dade County reported 211 sewer overflows totaling in excess of 51 million gallons, including at least 29 million gallons of raw sewage. In June 2013, the Environmental Protection Agency (EPA) and the state of Florida imposed a fine of $978 000 on the Miami-Dade County

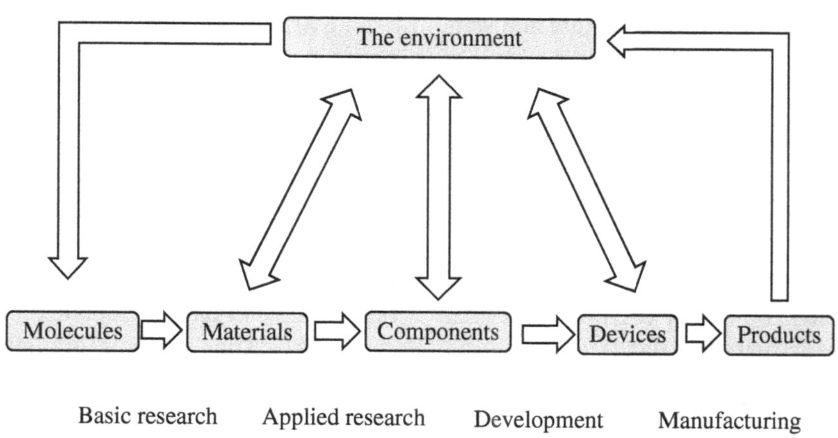

Figure 5.1 Schematic of sustainable environmental engineering system development.

for wastewater violations. The county has to sell a $13.6 billion bond to finance the required repairs of three WWTPs, ending ocean outfalls, upgrading pump stations, and constructing deep wells for the injection of treated wastewater. While the county has made progress toward improving operations since the penalty was issued, many proposed projects are not necessarily sustainable in terms of increasing system reliability. For example, $4 billion is proposed to design and construct two deep wells to dispose treated effluent after high level chlorination at the North and Central WWTPs. Many stakeholders and county commissioners are expressing their concerns for wasting tax payers' money. Indeed, to increase reliability of the system, innovative technologies have to be assessed to reclaim the effluent for reuse and avoid the costly deep well injection.

In developing countries, decentralized wastewater treatment appears to be even more important. Rudimentary sanitation facilities such as stand-alone septic tanks, composting or pit toilets, may be one of the best solutions for low-income areas. However, the potential of contaminating the ground or surface water makes it necessary to employ decentralized WWTP, because conventional on-site treatment has proved to be of low community acceptance while conventional sewage is too costly. Oxidation ponds, constructed wetlands, sand filters, and biological disks could be effectively applied in small communities as decentralized treatment technologies. A decentralized wastewater management system is defined as a wastewater treatment system that collects wastewater from the source, conveys to a treatment system, and disperses the water near the point of origin. It includes treatment facilities in rural, satellite, cluster, on-site, and small WWTPs. Small WWTPs refer plants with flow rates ranging from 1 000 to 50 000 GPD. There are also privately owned WWTPs such as mobile home parks, camp sites, recreational facilities, apartments, schools, and privately owned public facilities such as condominiums, shopping centers, and parks. Major advantages of these systems include the protection of public health and the environment, cost effectiveness, recharging of local aquifers, and the provision of water close to wastewater generation. Therefore, they are appropriate for low-density communities with varying site conditions and are especially suitable for ecologically sensitive areas. Small WWTP systems can be optimized to achieve low cost and high performance with several criteria. They should be easy to be constructed using low-precision construction techniques, minimize moving parts, operate without electricity, provide feedback on the performance of every step of the treatment process, operate without complicated calculations, and use chemical dosages that can be set directly by a few persons.

5.1.2 Best Practice for Small WWTPs

US EPA recommends the following procedure as the best practice to increase the reliability of small WWTPs:

1) Preparing an asset inventory and system maps.
2) Developing a condition assessment and rating system.
3) Assessing remaining useful life by consulting projected-useful-life tables or decay curves.
4) Determining asset values and replacement costs.
5) Analyzing current and anticipated customer demand and satisfaction with the system.
6) Understanding current and anticipated regulatory requirements.
7) Writing system's performance targets.
8) Using level of service standards to track system performance over time.
9) Conducting a failure analysis (root cause analysis, failure mode analysis).
10) Determining the probability of failure and listing assets by failure type.
11) Analyzing failure risk and consequences.
12) Reviewing and updating system's vulnerability assessment.

Financing assets for rehabilitation, repair, and replacement through borrowing or other financial assistance should be analyzed in asset management plan (long term). The asset management program should assess performance, evaluate progress, adopt new best practices, and take action based on review results to ensure that assets are managed to ensure reliable EEISs.

5.2 Integrated System Approach

5.2.1 The EPA Tools

There are many tools available to facilitate the decision-making process of different treatment options. Table 5.1 lists analysis tools that the US EPA provided to design an integrated system by selecting.

Operations and maintenance (O&M), personnel, and the capital budget account for an estimated 85% of a typical water system's expenses. Asset management enables a system to determine the lowest cost options for providing the highest level of reliable service over time. The work by O&M crew should be optimized on what are they doing, where they are doing it, and why the tasks have to be performed as listed by the US EPA in Table 5.2.

5.2.2 Integrated Engineering Design Example

Example 5.1

Given: The Australia Natural Edge Project (Stasinopoulos et al., 2008; TNEP, 2017) presents an

Table 5.1 Analysis tools.

Tools	Description
Check Up Program for Small Systems (CUPSS)	This EPA-developed software provides a simple comprehensive approach to asset management for small water and wastewater utilities in an easy-to-use, no-cost package
Financing Alternatives Comparison Tool (FACT)	A financial analysis tool developed by EPA to identify the most cost-effective method to fund a wastewater or drinking water management project
The Wastewater Information System Tool (TWIST)	This computer-based tool allows state and local health departments to inventory and manage small wastewater treatment systems by tracking information on residences and facilities served, permits, site evaluations, types of systems, inspections, and complaints in their jurisdictions
Water and Wastewater Treatment Technologies Appropriate for Reuse (WAWTTAR)	This software identifies appropriate water and wastewater treatment options. It estimates local performance and costs associated with each system based on available material and manpower resources

Table 5.2 Five core questions for SWWTPs.

Step	Action
1	What is the current state of my system's assets?
2	What is my required "sustainable" level of service?
3	Which assets are critical to sustained performance?
4	What are my minimum life cycle costs?
5	What is my best long-term funding strategy?

IES to design on-site WWTP for a single-family house located in Adelaide, South Australia. AN on-site secondary wastewater treatment and reuse systems for a four-bedroom house of five residents is to be designed by using conventional and ISA, respectively. Wastewater is from the toilet, bathroom, kitchen, and laundry appliances.

Find:

1) *Conventional design solution*: Conventional on-site system design
2) *Integrated Engineering Design (IED) solution*: Improved system using the *ISA Design*
3) *Performance comparison*: Comparison of the economic and environmental costs and benefits

Solution:

1) *Conventional design solution*

The conventional solution incorporates

a) Standard water-consuming household appliances.
b) An on-site wastewater treatment includes a septic tank and a slow sand filter as in Figure 5.2.
c) A water reuse system includes a drip irrigation.

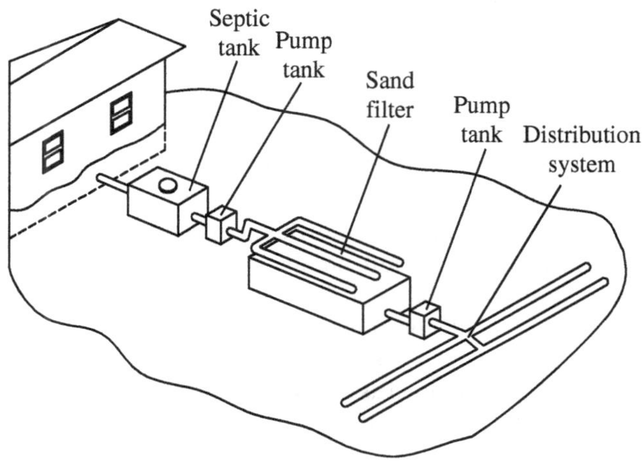

Figure 5.2 Components of a conventional on-site wastewater treatment and reuse system.

Standard Water-consuming Household Appliances

Several appliances contribute to household water consumption and wastewater generation. The standard water-consuming appliances' capital cost is

$$\text{Capital cost} = \$300 + \$50 + (\$80)(10) + \$700 + \$800 = \$2650$$

The installation costs of standard appliances and water-efficient appliances (Table 5.3).

The standard appliances' water consumption in a household of five people is

$$\text{Water consumption} = (150\ \text{l/person/day})$$
$$(5\ \text{people}) \times (91\ \text{days/quarter})$$
$$\text{Water consumption} = 68\,250\ \text{l/quarter}$$

This consumption is below the quarterly water consumption threshold set by the state water utility and therefore not subject to excess water consumption rates.

Table 5.3 Daily water consumption for standard domestic appliances.

Waste source	Allowance (l/person/day)
Toilet	50
Bath and shower	50
Handbasin tap	10
Kitchen	10
Tap	7
Dishwasher	3
Laundry	30
Tap	5
Washing machine	25
Total	150

The regular rate in Australia is assumed to be \$0.47/kl. The total water cost is

$$\text{Water supply} = (\$0.47/\text{kl})\,(68250\text{l/quarter}) \times (0.001\text{l/kl})\,(4\ \text{quarters/year})$$

$$\text{Water supply} = \$129/\text{year}$$

Standard water-consuming appliances can last for at least 20 years if operated correctly.

On-site Water Treatment System: Septic System for Primary Treatment

Septic systems are used for primary wastewater treatment and are available as both above- and below-ground systems. Figure 5.3 shows a typical septic tank. Wastewater enters the tank by pumping or action of gravity. The wastewater then settles and separates into three main zones. Substances such as grease, fats, and oils that float to the top are called scum. Substances that settle in the bottom zone are called sludge. Between the scum and sludge is relatively clear water that still contains bacteria and dissolved chemicals. The solids are anaerobically decomposed by bacteria in the septic tank.

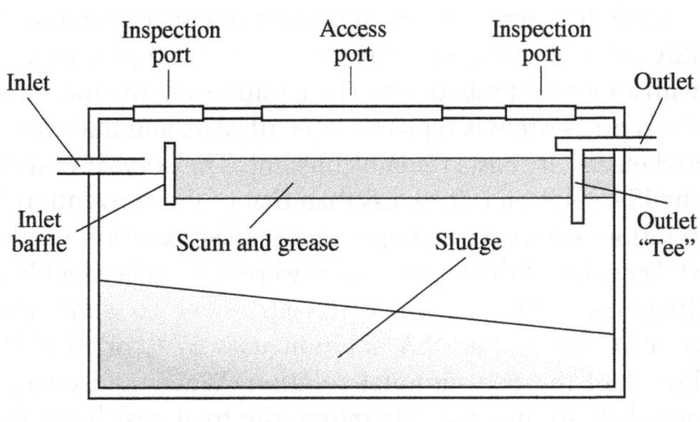

Figure 5.3 Cross section of a single-compartment septic tank.

2) *The ISA solution*

The ISA solution is exemplified by the Australia TNEP (2017):

a) Water-efficient household appliances.
b) An on-site wastewater treatment system.
c) A water reuse system consists of a drip irrigation pipe network.

Water-efficient Appliances

There are water-efficient models of every common household water-consuming appliances. The water-efficient appliances' capital cost is

$$\text{Capital cost} = \$300 + \$80 + (\$100)(10) + \$1000 + \$1000 = \$3380$$

The installation costs of standard appliances (in a conventional solution) and water-efficient appliances are comparable, so these costs are not considered and compared (Table 5.4).

3) *Performance comparisons*

A performance comparison reveals that while the capital costs of conventional and ISA solutions are about the same, the long-term cost of the ISA solution is

Table 5.4 Daily water consumption for water-efficient domestic appliances.

Waste source	Allowance (l/person/day)
Toilet	33
Bath and shower	19
Handbasin tap	1
Kitchen	3
Tap	1
Dishwasher	2
Laundry	11
Tap	1
Washing machine	10
Total	64

substantially lower. Table 5.5 compares the capital costs, the water consumption, and the running cost of standard and water-efficient appliances. The results indicate that, in this example, the water-efficient appliances consume 57% less water than standard appliances.

Table 5.5 Comparing the costs and water consumption of standard and water-efficient appliances.

Water-consuming appliances	Capital costs (not installed)	Water consumption	Running costs (water only)
Standard appliances	$2650	273 kl/year	$129/year
Toilet	$300		
Shower head	$50		
Taps	$800		
Dishwasher	$700		
Washing machine	$800		
Water-efficient appliances	$3380	116 kl/year	$55/year
Toilet	$300		
Shower head	$80		
Taps	$1000		
Dishwasher	$1000		
Washing machine	$1000		

Table 5.6 compares the capital and operating costs of the conventional and ISA water treatment and reuse systems.

Figures 5.4 and 5.5 summarize the component costs of the conventional and ISA solutions.

Table 5.6 Comparing the capital and running costs of the water treatment and reuse systems.

Water treatment and reuse system	Capital costs (installed)	Running costs
Conventional solution	$10 836	$640/year
Septic system	$6 500	$240/year
Slow sand filter system	$1 500	$400/year
Subsurface drip irrigation system	$2 836	
WDS solution	$10 701	$367/year
Biolytix system	$8 829	$367/year
Subsurface drip irrigation system	$1 872	

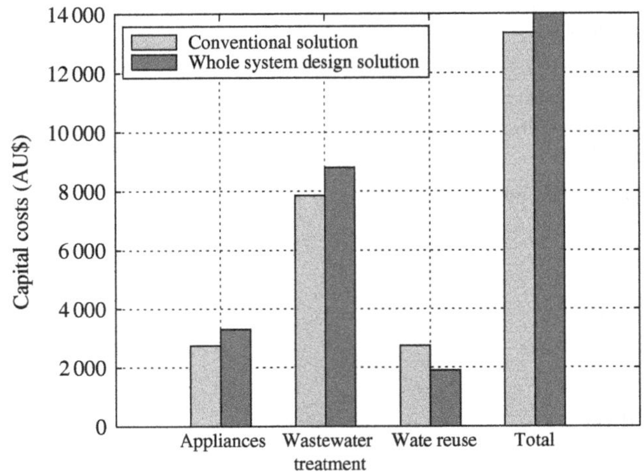

Figure 5.4 Comparison of the capital costs of components.

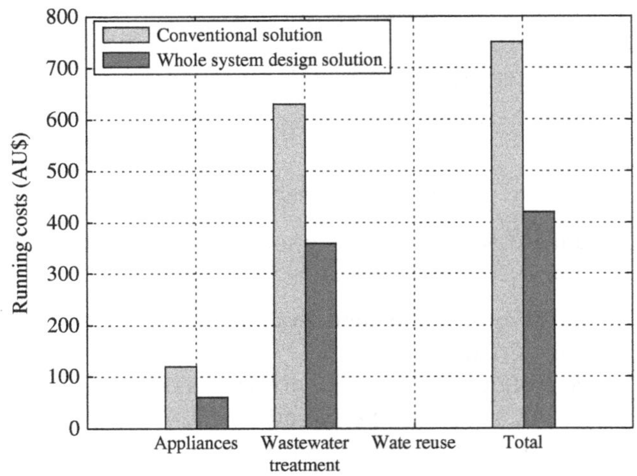

Figure 5.5 Comparison of the operation costs of components.

Table 5.7 compares the total cost of conventional and ISA solutions. The comparison is over a 20-year period with an interest rate of 6%. The comparison assumes that the septic system is replaced after 10 years and that water and electricity costs remain constant. The 20-year cost of the ISA solution is 29% less than that of the conventional solution. Figure 5.6 suggests that the ISA solution would still cost less if the septic system were not replaced within 10 years. Table 5.7 also compares water consumption over 20 years. The ISA solution uses 57%, or 3140 kl, less than the conventional solution. Water and energy are likely to increase. Therefore, the total cost becomes more favorable for the ISA solution.

Table 5.7 Comparing the total cost of conventional and IED systems over 20 years.

Solution	Capital costs	O/M costs	20-year cost	20-year water consumption
Conventional	$13 486 + $3 630 in 10 years	$769/year	$25 741	5460 kl
Whole system design	$14 081	$422/year	$18 311	2320 kl

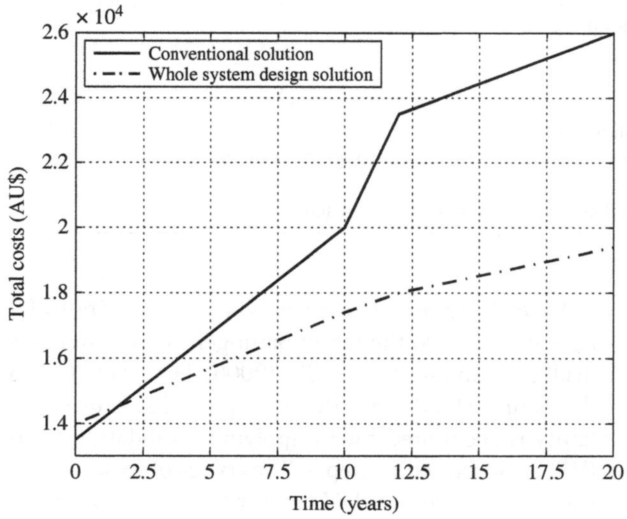

Figure 5.6 Comparison of the total cost of the conventional and integrated engineering design systems over 20 years.

In addition, the ISA solution has several other benefits over the conventional solution:

a) Water-efficient appliances that use hot water can also save on energy costs since less hot water is heated.
b) Water-efficient appliances may have a longer useful life than standard water-consuming appliances due to less wear on components.
c) The subsurface drip irrigation system in the ISA solution requires less time for residents to inspect because it is smaller.
d) The subsurface drip irrigation system in the ISA solution is less likely to be damaged because of its peak flow of storm with pollutants and better site design.

5.3 Scale-up of Laboratory or Pilot Design to Full-scale Plant

Innovative technologies in EEIS design start from laboratory research. Therefore, laboratory data are the major source for scaling up to pilot and full scale. For example, to ensure the reliability of a UV disinfection system to achieve a specific inactivation level of log inactivation (log I), the US EPA requires utilities to validate their UV processes so that the credit of a specific log I can be granted.

5.3.1 Minimum Requirements for Validation Testing

Reliability must be proven according to the US EPA protocol of UV disinfection from laboratory data to the full-scale reactor. In addition, uncertainty associated with the scale-up must be quantified so that the system can perform as expected to earn the US EPA disinfection credit under different flow rate or water quality such as the transmittance of water. Although UV sensors can measure the intensity of UV light, they cannot measure the dose delivered to the microorganisms as they pass through the reactor at different trajectories. Therefore, to receive treatment credit for inactivating *Cryptosporidium*, *Giardia*, or viruses using UV light, the LT2ESWTR requires water systems to use UV reactors that have undergone validation testing. Table 5.8 summarizes the regulatory requirements for all LT2ESWTR requirements related to UV disinfection, including minimum dose, validation, monitoring, and reporting requirements.

5.3.1.1 Collimated Beam Test

Collimated beam test typically has the following steps (US EPA, 2006):

1) The microbe of interest is cultured under controlled and reproducible laboratory conditions.
2) This culture is then prepared for testing, again under controlled and reproducible conditions: (i) there is a known concentration and/or distribution of the cultured pathogen, and (ii) the medium in which the culture is to be presented is of a known UVT.
3) The culture is then exposed to a known UV intensity, of known wavelength, for a fixed period of time, delivering a known UV dose to the known area of the presentation plate; hence dose and intensity are measured

Table 5.8 Summary of LT2ESWTR validation requirements.

Requirement	Conditions
Validated operating conditions must include	• Flow rate • UV intensity as measured by a UV sensor • UV lamp status
Validation testing must include	• Full-scale testing of a reactor that conforms uniformly to the UV reactors used by the water system • Inactivation of a test microorganism whose dose–response characteristics have been quantified with a low-pressure mercury-vapor lamp
Validation testing must account for	• UV absorbance of the water • Lamp fouling and aging • Measurement uncertainty of online sensors • UV dose distributions arising from the velocity profiles through the reactor • Failure of UV lamps or other critical components • Inlet and outlet piping or channel configurations of the UV reactor

per cm^2 (area) rather than cm^3 (volume), which is referred as a collimated beam apparatus.

4) The exposed content is recultured to see how many of the microbes survived.
5) This procedure is replicated many times at systematically increasing doses to build a dose–response curve. This curve enables the log survival (and by inference, log reduction) for the microbe of interest to be determined at any given UV dose.
6) After the various dose–response curves have been constructed in the laboratory, these curves would be applied to test an actual UV system to validate the actual Log I. The important thing is that the microbes used to test the UV system are cultured (albeit in much higher volumes) under exactly the same conditions as were used in the laboratory. This microbe culture is presented to the UV system mixed with the water of known UVT and tested over the range of UVTs and flow rates of interest.
7) A sample of the water is taken at the entry and exit of the UV system and recultured to quantify the reduction of the microbe.
8) Using the observed log reduction of the microbe and comparing it with the microbe's dose–response curve, the UV dose delivered can be determined. This UV dose is termed the reduction equivalent dose (RED).

US EPA (2006) recommends test procedure using a challenge microbe such as MS2 for health, safety, and repeatability. To validate the full-scale UV reactor, biodosimetric-tested RED dose results from the real-life tests with actual microbes should be used to measure the reactor performance using the following three steps:

Step 1: Conduct experimental tests using a challenge microorganism.

A *challenge microorganism* whose sensitivity to UV light is similar to the target pathogen is recommended in all experiments by US EPA (2006). However, it introduces uncertainty into the testing results. This uncertainty is accounted for by applying a validation factor (VF) as described in step 3. Two types of experimental tests can be used for UV reactor validation: (i) bench-scale testing using a collimated beam apparatus and (ii) full-scale reactor testing.

Bench-scale testing using a collimated beam apparatus. Collimated beam testing characterizes the UV dose–response relationship of the challenge microorganism. UV light is directed through a collimating tube to dose a sample of challenge microorganisms of a known concentration. After a specified exposure time, the sample is analyzed to determine the log inactivation (where log inactivation in this situation equals the log concentration prior to UV light exposure minus the log concentration after UV light exposure) as a function of UV dose. The UV dose delivered to the microorganisms is calculated based on the UV intensity, exposure time, and other experimental factors. Collimated beam tests are performed at a range of doses to generate a UV dose–response curve for the specific challenge microorganism. The functional forms of the equations for UV dose–response curves can vary depending on the results. EPA provided guidance on developing UV dose–response curves (US EPA, 2006). A quadratic UV dose–response equation can be used as follows:

$$\text{UV dose} = A \times (\text{log inactivation}) + B \times (\text{log inactivation})^2 \tag{5.1}$$

The coefficients A and B would be solved for using the collimated beam testing data by Example 5.2

using Matlab. UVT is typically reported at 254 nm because UV manufacturers, WTPs and WWTPs widely use A_{254}.

Full-scale reactor testing. The challenge microorganisms are injected upstream of the UV reactor. Samples are analyzed to determine the log inactivation at the test conditions of flow rate, UVT, lamp status, and UV intensity as measured by UV sensors. Full-scale reactor testing can be performed on-site at a water or a wastewater treatment plant or off-site at a validation test center.

Step 2: Estimate the RED.

The results from the two experimental tests in step 1 are combined. The log inactivation of the challenge microorganism measured during the full-scale testing is entered into the UV dose–response equation (if the relationship is quadratic) to calculate the **RED** of the reactor. Another way to conceptualize this step is to consider the RED to be "back calculated" using the field-measured log inactivation as the input variable. This approach is the opposite of most applications in which UV dose is the independent variable and log inactivation is the dependent variable. RED values are always specific to challenge microorganism used during experimental testing and the validation test conditions during full-scale reactor tests (flow rate, UVT, lamp status, and UV intensity as measured by the UV sensor).

Step 3: Adjust for uncertainty to calculate the validated dose.

RED is divided by a *VF* to produce the validated dose. The VF accounts for biases associated with using a challenge microorganism instead of the target pathogen, and experimental uncertainty provides a detailed description of how the VF is derived. The validated dose is associated with the validation test conditions of flow rate, lamp status, UV intensity as measured by a UV sensor, and, in some cases, UVT. The validated dose is compared with the required dose to determine the inactivation credit for the target pathogen.

5.3.2 Correlation of UV Sensitivity of Different Challenge Microorganisms with Target Microorganisms

Male-specific-2 bacteriophage (MS2) and *Bacillus subtilis* spores have historically been used for validation testing to receive treatment credit for *Cryptosporidium* and *Giardia*. Of importance is how much the UV sensitivity of each microorganism is related to the target microorganisms. The UV sensitivity of some commonly used and some candidate bioassay microorganisms is shown in Table 5.9.

Since the UVT of water has to be monitored during operations, it is critical to validate the UV reactor to confirm dose delivery.

Example 5.2

Given: The National Water Research Institute (NWRI) developed a standard dose–response curve and its 90% confident interval as follows:

At mean: Log inactivation = 0.5464 + 0.0368 ∗ (UV dose).

For 90% upper bound: Log inactivation = 0.8 + 0.0462 ∗ (UV dose).

For 90% lower bound: Log inactivation = 0.1 + 0.0295 ∗ (UV dose).

Find: If the UV dose is 40 mW/cm^2, what is the mean log inactivation?

Table 5.9 UV sensitivity of challenge microorganisms (US EPA, 2006).

	Reported delivered UV dose(mJ/cm^2) to achieve indicated log inactivation				
Microorganism	**1-log**	**2-log**	**3-log**	**4-log**	**Reference**
Bacillus subtilis	28	39	50	62	Sommer et al. (1998a)
MS2 phage	16	34	52	71	Wilson et al. (1992
Qß phage	10.9	22.5	34.6	47.6	Mackey et al.(2006)
PRD-1 phage	9.9	17	24	30	Mengand Gerba (1996)
B40-8 phage	12	18	23	28	Sommer et al. (1998b)
φx174 phage	2.2	5.3	7.3	11	Sommer et al. (1998b)
E. coli	3.0	4.8	6.7	8.4	Chang et al. (1985)
T7	3.6	7.5	11.8	16.6	Mackey et al. (2006)
T1	~5	~10	~15	~20	Wright (2006)

Solution:

The Matlab code is as follows:

```
clear
clc
x = 20:10:140;
y1 = 0.5464 + 0.0368*x;
yup = 0.8 + 3/65*x;
ylo = 0.1 + 2.8/95*x;
hold on
plot(x,y1,'k-o','MarkerSize',3,
  'MarkerFaceColor',[0 0 0],
  'MarkerEdgeColor',
  [0 0 0],'linewidth',1);
hold on
plot(x,yup,'k--','linewidth',1)
hold on
plot(x,ylo,'k-','linewidth',1)
set(gca,'ylim',[0 8],
  'FontWeight','bold','FontName',
  'Times New Roman');
xlabel('UV dose,
  mJ/cm^2','FontSize',14,'FontWeight',
  'bold','FontName', 'Times New Roman')
ylabel('Log Inactivation, -log(N/N_o)',
  'FontSize',14,'FontWeight','bold',
  'FontName', 'Times New Roman')
legend ('Mean', 'Upper bound',
  'Lower bound', 'location', 'Northwest')
box on
grid on
```

The previous dataset is plotted in the following graph (Figure 5.7):

Answer: If the UV dose is 40 mW/cm^2, the mean log inactivation is 2.

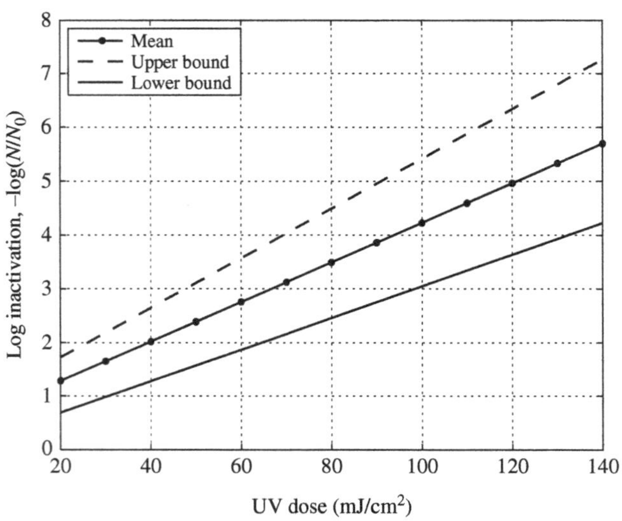

Figure 5.7 Dose–response standard relationship by NWRI.

5.3.2.1 Sampling Ports

The sampling points for microorganisms should be placed far enough from the UV reactor that the germicidal UV intensity at the point of sampling is less than 0.1% of the germicidal intensity within the UV reactor. If the outlet sample port is located downstream of a 90° bend (or the inlet sample port is upstream of a 90° bend), incident light is not a concern. To estimate intensity at a certain distance from the reactor, the following equation can be used:

$$I(r) = \frac{P_L}{2\pi r} e^{-\alpha_e r} \tag{5.2}$$

where

P_L = UV power emitted per unit arc length of the line source (mW/cm)
r = radial distance from the line source (cm)
α_e = Naperian (base e) absorption coefficient of the media for water (~0.015 cm^{-1})
$I(r)$ = UV intensity (mW/cm^2) at a distance r from the line source

US EPA (2006) provided ample examples in validating UV disinfection technology and process. Matlab codes were developed for EEIS designers to comply with the EPA requirements.

Example 5.3
Given: If the outlet sample port at a hypothetical UV test facility is located 10 ft downstream of the last UV lamp in a reactor, the lamp's maximum power per unit arc length is 100 watts per centimeter (W/cm).
Find: Is the sample port location acceptable?

Solution:

Using Equation 5.2, Matlab code of calculating the intensity at a different radial distance is developed as follows:
The Matlab code

```
clear
clc
PL=100;
r=305;
ae=0.015;
Ir=PL/(2*3.14*r)*exp(-ae*r)
```

Answer: Ir = 5.3808e-04.
Comments: When r = of 10 ft (305 cm), the calculated UV light intensity is 5.4 × 10^{-4} mW/cm. Because this intensity is less than 0.1% of the intensity within the reactor, the sample port location is acceptable.

The calculated dose approach uses a dose-monitoring equation to estimate the UV dose based on the parameters measured during reactor operations. The most common operational parameters in dose-monitoring equation are (i) flow rate, (ii) UV intensity, and (iii) UVT. Since different configurations of reactor have different UV distributions, the RED calculation becomes very complicated. As a result, Matlab codes will greatly facilitate the process.

5.3.3 Calculating the RED

The RED should be calculated for all full-scale reactor test conditions, individually for each replicate. For each test condition replicate (i.e. influent and effluent sample pairs), calculate the log inactivation (log *I*) using the following equation:

$$\text{log inactivation} = \log_{10}\frac{N_0}{N} \tag{5.3}$$

where

N_0 = concentration of infectious microorganisms before exposure to UV light
N = concentration of infectious microorganisms after exposure to UV light

Determine the RED in mJ/cm^2 for each test condition replicate pair using the measured log inactivation (log *I*) and the UV dose–response curve developed through collimated beam testing. If individual UV dose–response curves cannot be combined, the curve for a given day of testing should be used to determine RED for full-scale reactor testing data collected that day. If individual dose–response curves developed on the same day of testing cannot be combined, the curve resulting in the most conservative (lowest) RED values should be used. Replicates are evaluated separately to develop the UV dose-monitoring equation.

Example 5.4
Given: Collimated beam testing using a challenge microorganism produces the following UV dose–response curve:

$$\text{UV dose}\left(\text{mJ/cm}^2\right) = 2.18 \times (\log I)^2 + 15.3 \times (\log I) \tag{5.4}$$

Find: What would be the RED for the full-scale reactor if the full-scale testing produces the following data for each test condition and replicate test? (Tables 5.10 and 5.11)

Solution:

Matlab code is used for the equation.
The Matlab code

```
clear
clc
LogN0=[5.94 6.00 5.84 6.01
  5.99 6.04];
LogN=[4.57 4.54 4.56 4.10
  4.09 4.06];
```

Table 5.10 Dose-monitoring approaches: Key characteristics (US EPA 2006).

Dose-monitoring strategy	Parameter used as the operational set point	Parameters monitored during operations to confirm dose delivery
UV intensity set point approach	UV intensity	Flow rate lamp status UV intensity
Calculated dose approach	Calculated or validated dose	Flow rate lamp status UV intensity UVT

Table 5.11 Log N_0 and log N.

Test condition	Replicate	N_0 (pfu/ml)	N (pfu/ml)
1	1	5.94	4.57
1	2	6.00	4.54
1	3	5.84	4.56
2	1	6.01	4.10
2	2	5.99	4.09
2	3	6.04	4.06

```
LogI=LogN0-LogN
RED=2.18.*(LogI).^2+15.3.*LogI
```

Result:

```
LogI = 1.3700   1.4600   1.2800
  1.9100   1.9000   1.9800
RED = 25.0526  26.9849  23.1557
  37.1759  36.9398  38.8405
```

According to the test and replicate data, the log *I* is calculated first.

According to the log *I*, the RED values are calculated for each test using the equation and are shown in the last column (Table 5.12).

Table 5.12 Calculated RED according to the full-scale tests.

Test condition	Replicate	N_0 (pfu/ml)	N (pfu/ml)	log *I*	RED (mJ/cm^2)
1	1	5.94	4.57	1.37	25.1
1	2	6.00	4.54	1.46	27.0
1	3	5.84	4.56	1.28	23.2
2	1	6.01	4.10	1.91	37.2
2	2	5.99	4.09	1.9	36.9
2	3	6.04	4.06	1.98	38.8

5.3.3.1 Flow Rate for Validation

The water system's maximum and minimum flow rates, as well as one or more intermediate flow rates, are recommended by the US EPA to be selected as test conditions. To select intermediate flow rates, geometric progression in the following equation should be used because the relationship between UV dose and flow rate is nonlinear:

$$Q_n = Q_{MAX}\,\beta^{1-n} \tag{5.5}$$

where

Q_n = nth flow rate to be tested
Q_{MAX} = maximum flow rate to be tested
B = constant with a recommended value between 1.5 and 2.0 to achieve good separation of flow
n = flow rate test# to be evaluated (must be ≥3, if interpolating results)

The value of β should be sufficient to obtain at least three measured data points for developing the dose-monitoring equation. The value of n should be selected to span the range of flow rates.

Example 5.5 **Given**: A UV reactor using the calculated dose approach and operating within the range of 5–20 MGD is to be validated. The test engineer selects a β value of 1.6.

Find: What would be the four different flow rates for the validation test?

Solution:

The Matlab code is as follows:
The Matlab code

```
clear
clc
n=[1 2 3 4];
Qmax=20;
Constant=1.6;
Qn=Qmax*Constant.^(1-n)
```

Result:

```
Qn = 20.0000 12.5000 7.8125 4.8828
```

If n = 1, 2, 3, and 4, the following test flow rates should be used (Table 5.13):

Table 5.13 Flow rates for the validation test.

N	1	2	3	4
Q (MGD)	20	12.5	7.8	4.9

Comments: If the UV reactor uses MP lamps and validation testing is performed using a challenge microorganism other than MS2 phage or *B. subtilis*, an action spectra correction factor (CF) may need to be applied to the RED values to account for differences in the action spectra of the target pathogen and challenge microorganism. CF (US EPA, 2006) should be used for the RED adjustment (i.e. divide RED by the CF).

Example 5.6
Given: Flow effect on RED is described in the following table (Table 5.14).

Table 5.14 RED (mW/cm²) vs. flow rates.

	Flow rate (US gallon per minute (GPM))				
RED (mW/cm²)	RED5	RED8	RED11	RED15	RED20
3400	4.8	5.7	6.44	7.3	8
3150	5	6	6.8	7.7	8.5
2810	5.5	6.6	7.5	8.5	9.5
2460	6	7.3	8.2	9.5	10.5
2110	6.8	8.2	9.3	10.6	11.8
1760	7.9	9.5	10.8	12	13.4
1210	10.7	12.7	14.2	16	17.7
710	15.9	18.7	21.1	23.5	26
315	28.7	33.5	37.6	41.5	45.5
175	42.2	49	55	60.5	66.5

Find: Please develop a Matlab code to plot the RED versus flow rates.

Solution:

Matlab code is as follows:

```
clc
clear
y=[3400 3150  2810  2460  2110  1760  1210
  710 315 175];
ymax=max(y);
ymin=min(y);
yaxis=[ymin:0.01:ymax];
differenceymax=10^(floor(log10(abs
  (ymax)))-1);
differenceymin=10^(floor(log10(abs
  (ymin)))-1);
differencey=max(differenceymax,
  differenceymin);
x1 = [4.8  5  5.5 6  6.8 7.9 10.7  15.9
  28.7  42.2];
x2=[5.7 6  6.6 7.3 8.2 9.5 12.7  18.7
  33.5  49];
```

```
x3=[6.44  6.8 7.5 8.2 9.3 10.8  14.2  21.1
  37.6 55];
x4=[7.3 7.7 8.5 9.5 10.6  12 16 23.5
  41.5  60.5];
x5=[8  8.5 9.5 10.5  11.8  13.4  17.7  26
  45.5  66.5];
x=[x1 x2 x3 x4 x5];
xmax=max(x);
xmin=min(x);
differencexmax=10^(floor(log10
  (abs(xmax)))-1);
differencexmin=10^(floor(log10
  (abs(xmin)))-1);
differencex=max(differencexmax,
  differencexmin);
plot(x1,y,'ko-','MarkerSize',3,'Marker
  FaceColor',[0 0 0],'MarkerEdgeColor',
  [0 0 0]);
hold on
plot(x2,y,'ko-','MarkerSize',3,
  'MarkerFaceColor',[0.5 0.5 0.5],
  'MarkerEdgeColor',[0 0 0]);
hold on
plot(x3,y,'ko-','MarkerSize',3,'Marker
  FaceColor',[1 1 1],'MarkerEdgeColor',
  [0 0 0]);
hold on
plot(x4,y,'ks-','MarkerSize',3,'Marker
  FaceColor',[0 0 0],'MarkerEdgeColor',
  [0 0 0]);
hold on
plot(x5,y,'ks-','MarkerSize',3,'Marker
  FaceColor',[0.5 0.5 0.5],'MarkerEdge
  Color',[0 0 0]);
hold on
set(gca, 'FontName', 'Times New Roman');
set(gca,'xlim',[xmin-differencex,xmax
  +differencex],'FontSize',10,
  'FontWeight','bold','Font
  Name', 'Times New Roman');
set(gca,'ylim',[ymin-differencey,ymax
  +differencey],'FontSize',
  10,'FontWeight','bold','FontName',
  'Times New Roman');
xlabel('Q (GPM)','FontSize',14,'Font
  Weight','bold','FontName', 'Times
  New Roman')
ylabel({'RED (mW/cm^2)'},
  'FontSize',14,'FontWeight','bold',
  'FontName', 'Times New Roman')
legend('RED5','RED8',
  'RED11','RED15','RED20',
  'Location','NorthEast');
grid on
```

The plot of RED versus flow rate is presented in Figure 5.8.

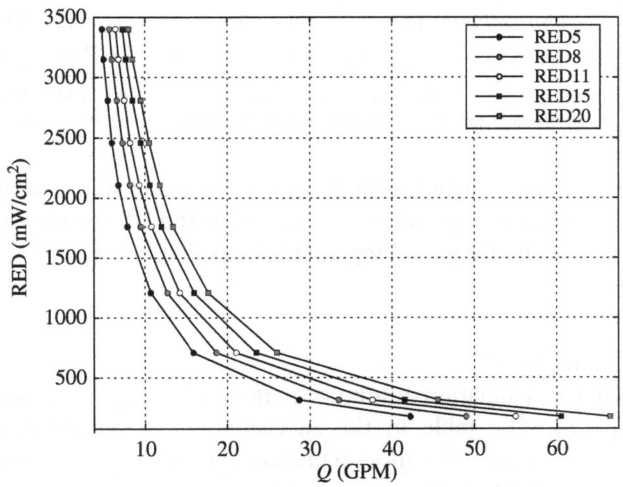

Figure 5.8 RED (mW/cm^2) versus flow rates.

The main advantage of testing two challenge microorganisms whose UV sensitivities of the target pathogen is that the RED bias factor for challenge microorganism can be set to 1.0 as described on the RED bias factor in the LT2ESWTR (US EPA, 2006). If validation testing is done with two challenge microorganisms with UV sensitivities of the target pathogen, i.e. one microorganism is more resistant and one is less resistant, the following approach can be used to estimate the RED of the target pathogen for each test condition:

For each test condition, calculate the UV sensitivity (mJ/cm^2 per log *I*) of the challenge microorganism using the following equation:

$$\text{UV sensitivity} = \frac{\text{RED}}{\log I} \tag{5.6}$$

where

RED = the RED for the test replicate as derived by inputting log *I* into the UV dose–response equation

log *I* = log inactivation for the test replicate as calculated

Create a graph of UV sensitivity on the *x*-axis and RED (mJ/cm^2) on the *y*-axis for each test condition.

For each challenge microorganism, plot paired UV sensitivity and RED values on the graph (2 values).

Draw a straight line between the two points.

Determine the UV sensitivity for the target pathogen by selecting the UV dose from Table 5.15 for 1 log inactivation (log *I* = 1).

Table 5.15 UV dose (in mJ/cm^2) required by LT2ESWTR to inactivate target pathogens (US EPA, 2006).

Log inactivation	0.5	1	1.5	2	2.5	3	3.5	4
Cryptosporidium	1.6	2.5	3.9	5.8	8.5	12	15	22
Giardia	1.5	2.1	3.0	5.2	7.7	11	15	22
Virus	39	58	79	100	121	143	163	186

Using the straight line in the graph created in step 4, read the corresponding RED value for the UV sensitivity of the target pathogen (as determined in step 5).

Example 5.7

Given: Validation testing is performed using MS2 and φx174. The table in the succeeding text summarizes average results for three replicates for one test condition (high UVT) (Table 5.16).

Find: Please develop a Matlab code to quantify the sensitivity of RED tests.

Solution:

Matlab code is developed for this problem:

The Matlab code

```
clc
x = [2 20];
y = [8 40];
p=polyfit(x, y, 1)
plot(y,x,'k-o','MarkerSize',3,
  'MarkerFaceColor', [0 0 0],
  'MarkerEdgeColor', [0 0 0],
  'linewidth',1);
xlabel('UV Sensitivity mW/cm^2 per
  log I','FontSize',14,
  'FontWeight','bold','FontName',
  'Times New Roman')
ylabel('RED, (mW/cm^2)','FontSize',
  18,'FontSize',14,'FontWeight',
  'bold','FontName',
  'Times New Roman')
set(gca,'ylim', [2 20],'Font
  Weight','bold','FontName',
  'Times New Roman');
set(gca, 'GridLineStyle' ,':',
  'GridAlpha',1)
grid on
```

Result:

```
p =1.7778   4.4444
```

The graph is the straight line between the two points (Figure 5.9).

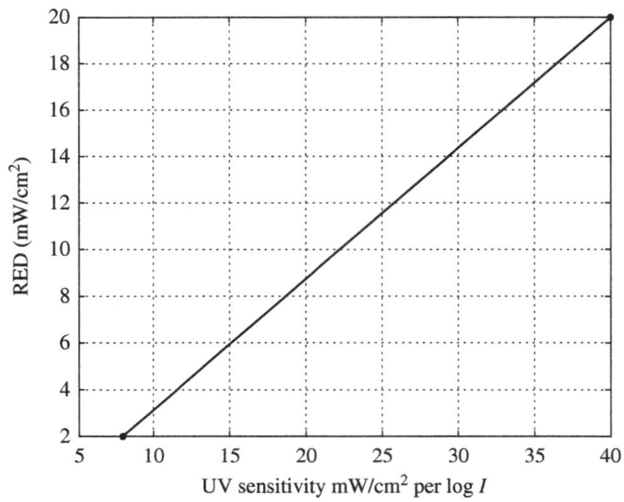

Figure 5.9 RED versus UV sensitivity.

Straight-line interpolation between the two points yields the following equation using the Matlab:

$$\text{RED} = 1.78 \times \text{UV sensitivity} + 4.44 \tag{5.7}$$

The equation earlier predicts that the RED delivered to *Cryptosporidium*, defined with a UV sensitivity of 3.9 mJ/cm^2 per log inactivation, is

$$\text{RED} = 1.78 \times 3.9 + 4.44 = 11.4\,\text{mJ/cm}^2 \tag{5.8}$$

Because the RED represents the dose delivered to *Cryptosporidium*, the RED bias factor is equal to 1.0.

Table 5.16 Results of validation testing.

Challenge microorganism	Influent conc. (pfu/ml)	Effluent conc. (pfu/ml)	UV sensitivity (mJ/cm^2 per log *I*)[1]	log*I*[2]	RED (mJ/cm^2)[3]
MS2	1×10^6	1×10^4	20	2.0	40
φ × 174	1×10^4	0	2	≥4.0	≥8.0

5.3.4 Uncertainty in Validation

The UV intensity set point approach has intrinsic uncertainty due to systematic or random errors. The uncertainty in the set point value is based on a prediction interval at a 95% confidence level using the following procedure:

1) Calculate the average and standard deviation of RED values for each test condition (typically at least 3–5 replicate pairs are generated for each test condition).
2) Calculate the uncertainty of the set point RED using:

$$U_{SP} = \frac{t \times SD_{RED}}{RED} \times 100\% \tag{5.9}$$

where RED = average RED value measured for each test condition

SD_{RED} = standard deviation of the RED values measured for each test condition

t = t-statistic for a 95% confidence level defined as a function of the number of replicate samples by using the following data (Table 5.17).

Table 5.17 *t*-Statistic for a 95% confidence level.

Number of samples	*T*
3	3.18
4	2.78
5	2.57

3) Select the highest U_{SP} from all test conditions for calculating the VF.

5.3.4.1 Calculating U_{IN} for the Calculated Dose Approach

For reactors using the calculated dose approach, the uncertainty of interpolation (U_{IN}) is calculated as the lower boundary of the 95% prediction interval for the dose-monitoring equation. This prediction interval reflects the noise in the data about that fit. In nonstatistical terms, the U_{IN} represents the difference between (i) the RED value as derived using measured log inactivation and the UV dose–response curve and (ii) the RED value as calculated using the dose-monitoring equation. The value of U_{IN} depends on the calculated RED (or calculated dose), increasing at low calculated RED values. EPA recommends that one U_{IN} be selected that represents the most conservative (largest) uncertainty value calculated for the validated dose-operating range (for the lowest calculated RED). Alternatively, U_{IN} can be expressed as a function of the calculated RED. U_{IN} is calculated using the following equation:

$$U_{IN} = \frac{t \times SD_{RED}}{RED} \times 100\% \tag{5.10}$$

where

SD = standard deviation of the differences between the test RED (based on the observed log inactivation and UV dose–response curve) and the RED calculated using the dose-monitoring equation for each replicate

RED = the RED as calculated using the dose-monitoring equation

t = t-statistic at a 95% confidence level for a sample size equal to the number of test conditions used to define the interpolation (Table 5.18).

Table 5.18 *t*-Statistic at a 95% confidence level for a sample size equal to the number of test conditions.

Number of data points used to develop the dose-monitoring equation	*T*	Number of data points used to develop the dose-monitoring equation	*t*
3	3.18	14	2.14
4	2.78	15	2.13
5	2.57	16	2.12
6	2.45	17	2.11
7	2.36	18	2.10
8	2.31	19–20	2.09
9	2.26	21	2.08
10	2.23	22–23	2.07
11	2.20	24–26	2.06
12	2.18	27–29	2.05
13	2.16	≥30	2.04

5.3.4.2 Determining the Validated Dose and Validated Operating Conditions

The last step is to adjust the RED results by the VF to determine the validated dose for the UV reactor using the following equation:

$$\text{Validated dose} = \frac{RED}{VF} \tag{5.11}$$

UV intensity set point approach can be used to validate the validated dose through the following example:

Example 5.8

Given: A water treatment plant plans to add UV disinfection to its treatment process to earn 2.5-log *Cryptosporidium* inactivation credit. Based on the LT2ESWTR dose requirements, the system needs to deliver a minimum required dose of 8.5 mJ/cm^2 to receive this level of inactivation credit. The hypothetical proposed installation has the following design specifications (Table 5.19):

Table 5.19 Design specifications.

Design flow rate	**400 gpm**
Minimum UVT	90%
Lamp aging factor	80%
Fouling factor	85%
Fouling/aging factor	68% (80% × 85%)
Disinfection goal	2.5-log *Cryptosporidium* inactivation credit

The water system's engineer selected a UV reactor with the following characteristics (Table 5.20):

Table 5.20 Characteristics of UV reactor.

UV reactor	• 2 banks of lamps • 4 300-W LPHO lamps per bank • Rated for flow rates of 50–500 gpm • 1 UV sensor/bank positioned equidistant from lamps 2 and 3
UV dose-monitoring approach	UV intensity set point approach with one alarm set point

Validation Test Plan

The validation test plan includes the following factors:

1) The UV manufacturer provided three reference UV sensors for calibrating the duty sensors during validation. These reference sensors had been calibrated by an independent, qualified sensor testing laboratory before validation and had a documented measurement uncertainty of 10%.
2) Because the number of UV sensors was less than the number of lamps, the highest output lamps were identified prior to testing and were positioned closest to the sensors at lamp positions 2 and 3. New lamps were used during validation testing (after a 100 h burn-in period).
3) The validation testing was conducted over a one-day period. The UV dose–response of the challenge microorganism (measured via a laboratory collimated beam test) was evaluated with 1 l influent water samples collected at high and low UVT values.
4) All recommended testing protocols were followed.

The UV manufacturer had already identified a target set point at 11.7 mW/cm^2 using numerical modeling. System X confirmed with the manufacturer that this set point was low enough to account for their combined conditions of minimum UVT and maximum lamp fouling and aging. The following two UVT–lamp power operating conditions were tested:

1) The UVT was lowered to produce the target UV sensor set point (in this case, the resultant UVT was 89.9%), while the lamp power was kept at 100%.
2) The UVT was raised back to its maximum value (no UV-absorbing chemical added), and the lamp power was reduced to produce the same UV sensor value (in this case, the resultant lamp power was 66%).

Test Data

The data collected during validation testing are presented as follows:

- Table 5.21 presents the UV dose–response data measured on the influent water collected during field validation testing for the laboratory collimated beam test.
- Tables 5.22, 5.23, and 5.24 present the data from full-scale reactor testing.
  - Table 5.22 presents the flow rate, UVT, lamp power, and UV sensor readings measured for each test condition.
  - Table 5.23 presents the measured challenge microorganism concentrations for the influent and effluent samples collected (in triplicate) from the UV reactor for each test condition.
- Table 5.24 presents the UV output of the eight lamps used during validation testing measured at the same lamp location (lamp 2 in row 1) adjacent to the same UV sensor (1), which was used to identify the highest output lamps.
- Table 5.25 presents data comparing the three reference UV sensor measurements with the duty UV sensors used during validation.

Find: Are these validation tests satisfying the EPA QA/QC criteria?

Solution: The following are the calculations to show how the data will be used to determine whether QA/QC

Table 5.21 Challenge microorganism UV dose–response measured using a collimated beam apparatus.

90% UVT					97% UVT				
	Replicate 1		Replicate 2			Replicate 1		Replicate 2	
UV dose (mJ/cm²)	*N* (pfu/ml)	Log *N*	*N* (pfu/ml)	Log *N*	UV dose (mJ/cm²)	*N* (pfu/ml)	Log *N*	*N* (pfu/ml)	Log *N*
0	882 329	5.95	944 980	5.98	0	1 148 154	6.06	1 300 460	6.11
10	180 120	5.26	198 394	5.30	10	316 328	5.50	257 749	5.41
20	64 217	4.81	69 438	4.84	20	113 644	5.06	74 396	4.87
30	20 622	4.31	20 100	4.30	30	34 679	4.54	25 189	4.40
40	7 257	3.86	8 145	3.91	40	12 624	4.10	9 226	3.97
60	1 274	3.11	1 399	3.15	60	1 980	3.30	1 722	3.24
80	188	2.27	261	2.42	80	387	2.59	211	2.32
100	80	1.90	90	1.95	100	80	1.90	100	2.00

Table 5.22 Flow rate, UVT, lamp power, and UV sensor data measured during validation testing.

Test ID	Banks on	Flow rate (gpm)	UVT (%)	Relative lamp output (%)	$S_{duty,\,1}$ (mW/cm²)	$S_{duty,\,2}$ (mW/cm²)
1	1, 2	394	89.9	100	11.7	11.7
2	1, 2	403	97.0	66	11.6	11.7

Table 5.23 Measured influent and effluent challenge microorganism concentrations.

	Influent challenge microorganism log concentration			Effluent challenge microorganism log concentration		
	Replicate			Replicate		
Test ID	1	2	3	1	2	3
1	5.94	6.00	5.84	4.57	4.54	4.56
2	6.01	5.99	6.04	4.10	4.09	4.06

Table 5.24 Sensor 1 measurements with lamp 2 operated at 100% ballast power.

Lamp ID	$S_{duty\ 1}$ (mW/cm²)	Lamp ID	$S_{duty\ 1}$ (mW/cm²)
1	13.6	5	13.9
2	14.6	6	13.3
3	14.2	7	14.5
4	13.4	8	14.3

criteria are met, to calculate the necessary CFs, and to determine the validated operating conditions for the target log inactivation.

Develop the UV Dose–Response Curve from the Collimated Beam Data

Figures 5.10 and 5.11 present the UV dose–response data from Table 5.21 at UVT values of 90 and 97%, respectively. The data have been fitted to quadratic equations that show log N as a function of UV dose. The fits were used to identify log N_0 values of 5.91 and 6.05 (i.e. log N_0 where the curves intersect the y-axes) from Figures 5.10 and 5.11, respectively. Using these values, Table 5.26 presents the UV dose–response data defined as UV dose versus log I, where log I = log(N_0/N).

The UV dose–response data described in Table 5.26 were analyzed to determine if the two datasets could be combined using the method referred to by Draper and

Table 5.25 Reference UV sensor checks.

Before/after validation testing	UVT(%)	Relative lamp power (%)	Sensor ID	S_{duty} (mW/cm^2)	$S_{ref,\ 1}$ (mW/cm^2)	$S_{ref,\ 2}$ (mW/cm^2)	$S_{ref,\ 3}$ (mW/cm^2)
Before	97	100	1	11.3	11.7	12.1	11.4
Before	97	68	1	5.1	5.5	5.7	5.3
Before	90	100	2	3.7	4.0	4.1	3.8
Before	90	68	2	2.0	1.9	1.9	1.8
After	97	100	1	11.6	11.8	12.2	11.4
After	97	68	1	5.1	5.4	5.6	5.3
After	90	100	2	3.9	4.0	4.1	3.9
After	90	68	2	1.9	1.8	2.0	1.8

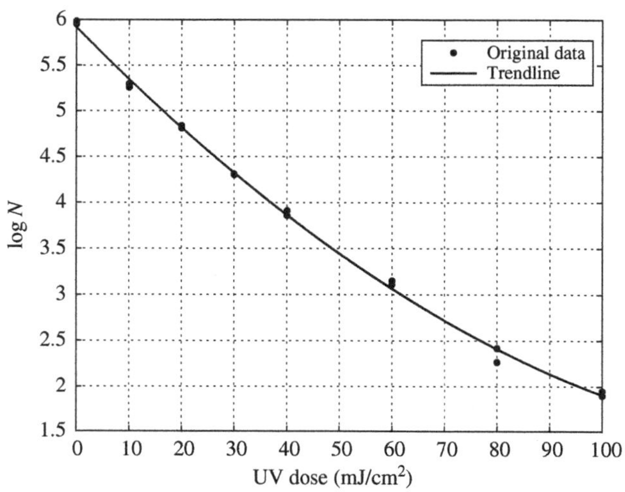

Figure 5.10 Log *N* versus UV dose using the data in Table 5.21 at 90% UVT.

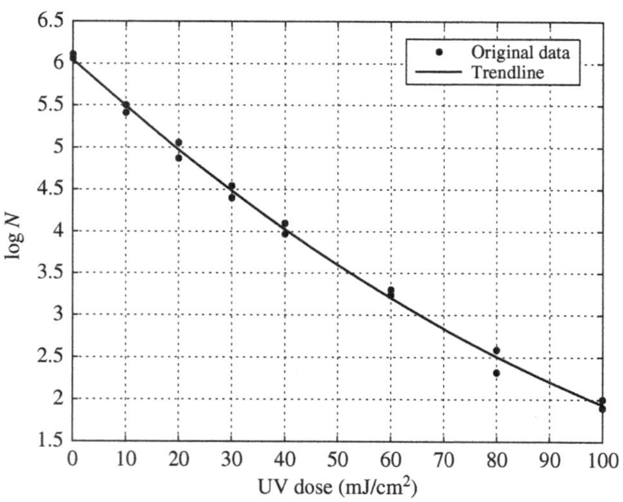

Figure 5.11 Log *I* versus UV dose using the data in Table 5.21 at 97% UVT.

Smith (1998). A statistical analysis of the collimated beam data is recommended to determine which terms are significant, p-value ≤ 0.05, using a standard regression tool. The process is iterative, and each time the regression tool is used, one term is dropped until all of the coefficients are deemed significant (p-values ≤ 0.05). In this example, three iterations were needed. The regression analysis showed that the two measured UV dose–response curves were statistically similar ($p < 0.05$) and could be combined. Figure 5.12 presents the plot of UV dose as a function of log inactivation for the combined dataset and the resultant UV dose–response equation.

The equation of trend line:

$$\log N = 0.0002(\text{UV dose})^2 - 0.059(\text{UV dose}) + 5.92 \quad r^2 = 0.9982 \tag{5.12}$$

The equation of trend line:

$$\log N = 0.0002(\text{UV dose})^2 - 0.057(\text{UV dose}) + 6.05 \quad r^2 = 0.9964 \tag{5.13}$$

The equation of trend line:

$$\text{RED}\left(\text{mJ/cm}^2\right) = 2.18(\log \text{I})^2 + 15.30(\log I) \quad r^2 = 0.9956 \tag{5.14}$$

5.3.5 Collimated Beam Data Uncertainty

The uncertainty in the UV dose calculation using the Equation (5.15):

$$U_{\text{DR}} = t\frac{\text{SD}}{\text{UV dose}_{\text{CB}}} \times 100\% \tag{5.15}$$

where

U_{DR} = uncertainty of the UV dose–response fit at a 95% confidence level

UV $dose_{CB}$ = UV dose calculated from the UV dose–response curve in Figure 5.12
SD = standard deviation of the difference between the calculated UV dose–response and the measured values from Table 5.21
t = t-statistic at a 95% confidence level for a sample size equal to the number of test condition replicates used to define the dose–response

In this case, SD = 2.2 at 1-log inactivation and $t = 2.04$ for 32 test condition replicate as shown in Table 5.26. Equation (5.15) can then be used to determine U_{DR} at various log inactivation values from Table 5.21. The graph in the succeeding text shows the relationship between log inactivation and UDR. The value of UDR should not exceed 30% at the UV dose corresponding to 1-log inactivation of the challenge organism. In this case, U_{DR} = 25% at 1.0-log inactivation, which is less than the recommended limit of 30% (Figure 5.13).

5.3.6 Electrical Energy per Order (EE/O)

The electrical energy (in kilowatt hours) required to reduce the concentration of a pollutant by one order of magnitude for 1000 US gallons of of water is defined as electrical energy per order (EE/O). For batch reactor, the equation is

$$\text{EEO} = \frac{1000\,P}{60V \times \log\frac{C_i}{C_f}} \tag{5.16}$$

$$\text{EEO} = \frac{1000P}{60V \times \log C_{in}/C_{eff}}$$

where

EEO = electrical energy per order (kWh)
P = lamp power (kW)
t = exposure time (min)

Table 5.26 Challenge microorganism UV dose–response defined as UV dose versus [log(N_o/N)] (i.e. log I).

90% UVT			97% UVT		
UV dose (mJ/cm²)	Replicate 1 5.92 – log N	Replicate 2	UV dose (mJ/cm²)	Replicate 1 6.05 – log N	Replicate 2
0	−0.03	−0.06	0	−0.01	−0.06
10	0.66	0.62	10	0.55	0.64
20	1.11	1.08	20	0.99	1.18
30	1.61	1.62	30	1.51	1.65
40	2.06	2.01	40	1.95	2.08
60	2.81	2.77	60	2.75	2.81
80	3.65	3.5	80	3.46	3.73
100	4.02	3.97	100	4.15	4.05

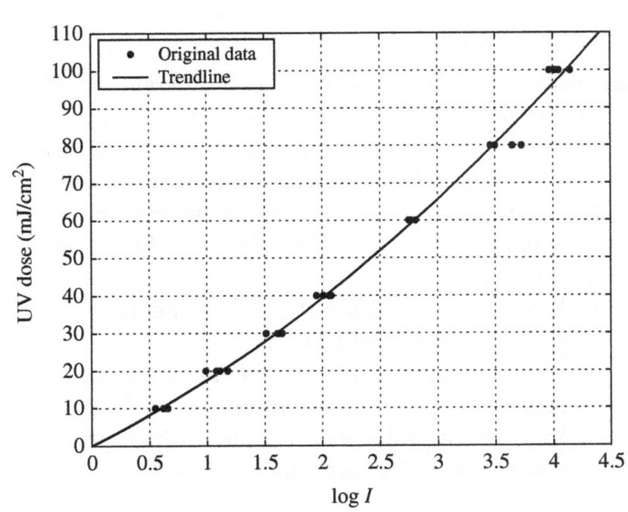

Figure 5.12 Log I versus UV dose using the data in Table 5.26.

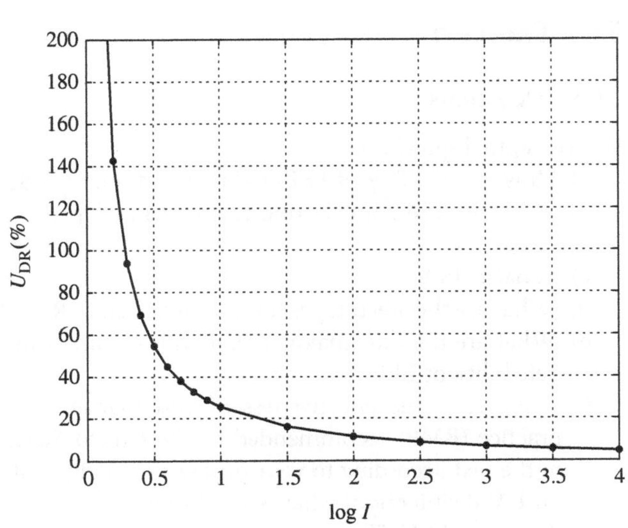

Figure 5.13 Uncertainty at different log I levels.

V = reactor volume (l)
C_i = initial concentration
C_f = final concentration

For a continuous flow reactor, the equation is

$$\text{EEO} = \frac{1000P}{60Q \times \log\frac{C_i}{C_f}} \tag{5.17}$$

where

P = lamp power (kW)
t = exposure time (min)
Q = flow rate (gpm)
C_i = initial concentration
C_f = final concentration

Power requirements can be estimated from manufacturer-supplied information regarding the number of lamps in a given system, the kilowatt draw of each lamp, the warranty power setting, and the average number of UV reactors needed. The total kilowatt draw from each manufacturer is then determined, and the average power consumption (kW) is calculated. The average power consumption can be used to calculate the total power costs by multiplying the total power requirements by the assumed power rate of 0.076$/kWh. One of the most important concepts to comparing power consumptions for different UV disinfection systems is electrical energy per order (EE/O). Usually, plug flow reactor will require less EE/O than batch reactors because the distribution of UV intensity in the plug flow reactor (PFR) is more uniform than batch reactor. In addition, the equations may not hold at extreme high inactivation orders such as greater than 6 or 7 because most UV dose will be wasted when *Giardia* concentration is extremely low.

5.4 Exercise

5.4.1 Questions

1) Conceptual questions:
   a) How is reliability of EEIS defined and quantified?
   b) What are the contributing factors to unit process failure?
   c) What is ISA?
   d) What are the advantages for decentralized WRRFs?
   e) What are the four major spatial scales in assessing reliability of EEIS?
   f) Please rank the effectiveness of best management practice (BMP) recommended by the United States in the text according to your professional judgment.
   g) In UV disinfection, what is the RED?
   h) How would UVT of a water sample affect the UV disinfection efficiency?
   a) Please rank fluence and cost required by UV disinfection of one cubic meter of drinking water, wastewater effluent, and reclaimed waste water? Why are there different requirements?
   b) Please rank what are the major factors contributing to the uncertainty in estimating RED?
   c) Please rank how is the uncertainty of interpolation (U_{IN}) calculated?
   d) Please rank what are the differences in calculation of the electrical energy per order for batch or CSTR reactors?
2) Reliability questions related to WTPs or WTWPs in your own city.
   In assessing the reliability of local utilities, the US EPA recommended the following questions on reliability level of service:
   a) What is plant useful life and value?
   b) What level of service do my stakeholders and customers demand?
   c) What do the regulators require and what is actual performance?
   d) How could unit processes of a plant fail and why do they fail?
   e) What are the likelihoods (probabilities) and consequences of unit process failure?
   f) What does it cost to repair the unit process and what are the other costs (social, environmental, etc.) that are associated with the failure?
   g) What alternative strategies exist for managing O&M, personnel, and capital budget to ensure reliability of a plant?
   h) What are the costs of rehabilitation, repair, and replacement for critical unit processes?
   i) Do utility companies have enough funding to maintain their plants for the required level of service?

5.4.2 Calculation

Eisenberg et al. (2001) reported two different criteria in quantifying the reliability of a WWTP. One is the expected time between failure (ETBF) in unit process based on chi-square distribution. Another is the fraction of the study period that all components in the unit were operating as shown in Table 5.27. Please answer the following:

1) Which unit process in the table is the most reliable process in terms of ETBF in days?
2) How would the reliability of all the unit processes be ranked?
3) In terms of probability of failure, which process is most likely to fail and which one is the least?

UV doses required to achieve required log inactivation are shown in this chapter. Please develop a Matlab

Table 5.27 Reliability of a WWTP (Eisenberg et al., 2001).

Unit processes	ETBF (days)	Fraction of all components are operating
Headworks	26	0.9953
Primary	41	0.9985
Secondary	9	0.9757
Tertiary	13	0.9994
UV	212	0.9991
Reverse osmosis	10	0.9990

program to do the following using the regression equations presented in this chapter:

1) What is the ratio of UV dose required if 4-, 3-, and 2-log inactivation are required for viruses, *Giardia*, and *Crypto*?
2) What are the strategies you may develop to achieve the three objectives at the lowest cost?

If a water sample has a UV absorption coefficient at 254 nm of 0.1 and I_0 of 120 mJ/cm, please develop a Matlab code to answer the following:

1) What would be the UV light intensity reading using a UV absorption photospectroscopy?
2) How often would you expect this UVT to occur if the sample was filtered water 2 as described in the example cumulative frequency diagram for three filtered waters by US EPA (2006)?

According to EPA's requirement, at least 3-log inactivation is required for *Giardia*, if the lamp power is 3 kW. Please answer the following by using Matlab code:

1) What would be the EE/O for a batch reactor if the log of inactivation is 3, 4, 5, and 6?
2) What would be the EE/O for a flow reactor if the log of inactivation is 3, 4, 5, and 6?
3) Please comment on the EE/O values if the log of inactivation is greater than 6.

5.4.3 Projects

5.4.3.1 Xiongan Design Project

Xiongan is located in the heart of Hebei province and next to Baiyangdian Lake. The Chinese government is relocating many nongovernmental related businesses and some government entities from Beijing to Xiongan such as science and technology, education, manufacturing, and certain corporations. It could require 350 billion USD in fixed asset investment over the next 20 years. Xiongan district is to expand to 10, 100, and 1000 km^2 in 5, 10, and 20 years, respectively. According to this plan, please do the following:

1) What is the historical air, water, and solid waste produced in Beijing in the past ten years?
2) What would be the required fixed investment and GDP in USD for 5, 10, 15, and 20 years according to the historical data?
3) What would be the air, water, and solid wastes in tons for 5, 10, 15, and 20 years if the district is to expand to 10, 100, and 1000 km^2 in 5, 10, and 20 years, respectively, according to the historical data?
4) How would you incorporate UV disinfection to WWTP and WTP in new Xiongan district? Why?

5.4.3.2 Community Proposal Project

The following topics may be considered:

1) After you collected air, water, and wastewater production data on your hometown, you calculated the current air, water, and soil quality indices. You identified that water pollution is the most serious environmental issue in your town. Please develop a proposal to design a sustainable wastewater engineering system for your hometown to achieve water quality indices of 10, 20, and 50% improvement in 5, 10, and 20 years, respectively.
2) Can your engineering design be directly applied to your state and your country? If not, what would be different at city, regional, and country levels? How would you improve your design?
3) For disinfecting wastewater effluent, the best way is UV. If your community did not use UV disinfection of wastewater effluent, please conduct throughout research on the following:
   a) Commercially available UV products that are economic for your community to adopt.
   b) What are the capital costs of the UV equipment?
   c) What are the validation tests that have to be carried out to satisfy the US EPA UV disinfection credit requirement test to ensure the reliability of the UV system?
   d) why UV disinfection of treated wastewater is not practiced? Is leachate in wastewater treatment plant a problem preventing UV to be adopted?
   e) What are the alternatives to co-treatment of leachate with wastewater?

References

Australia, the Natural Edge Project (TNEP) (2017). Worked example 5-domestic water systems. http://www.naturaledgeproject.net/Whole_System_Design.aspx (accessed 9 December 2017).

American Water Works Association and American Society of Civil Engineers (2012). *Water Treatment Plant Design*, 5e. New York: McGraw-Hill.

Chang, J.C.H., Ossoff, S.F., Lobe, D.C. et al.(1985). UV inactivation of pathogenic and indicator micro-organisms. *Applied Environmental Microbiology* 49 (6): 1361–1365.

Draper, N. and Smith, H. (1998). *Applied Regression Analysis*, 3e. New York: Wiley.

Eisenberg, D., Soller, J., Sakaji, S., and Olivieri, A. (2001) A methodology to evaluate water and wastewater treatment plant reliability. *Water Science & Technology* 43 (10): 91–99.

Sedlak, D. (2014). *Water 4.0: The Past, Present, and Future of the World's Most Vital Resource*. New Haven: Yale Press.

Mackey, E.D., Wright, H.B., Hargy, T. et al.(2006). *Optimization of UV Reactor Validation*. Denver: American Water Works Association Research Foundation.

Meng, Q.S. and Gerba, C.P.(1996). Comparative inactivation of enteric adenoviruses, poliovirus and coliphages by ultraviolet irradiation. *Water Research* 30 (11): 2665–2668.

Sommer, R., Haider, T., Cabaj, A. et al. (1998a). Time dose reciprocity in UV disinfection of water. *Water Science and Technology* 38: 145–150.

Sommer, R., Haider, T., Cabaj, A. et al. (1998b). Time UV dose reciprocity in UV disinfection of water. International Association of Water Quality (IAWQ), Vancouver, 1998 Poster.

Stasinopoulos, P., Smith, M., Hargroves, K., and Desha, C. (2008) *Whole System Design: An Integrated Approach to Sustainable Engineering*. Earthscan: The Natural Edge Project.

USEPA (2006). Ultraviolet disinfection guidance manual for the final long term 2 enhanced surface water treatment rule, 2006. http://www.epa.gov/safewater/disinfection/lt2/pdfs/guide_lt2_uvguidance.pdf (accessed 13 December 2017).

Wilson, B.R., Roessler, P.F., van Dellen, E. et al.(1992). Coliphage MS2 as a UV water disinfection efficacy test surrogate for bacterial and viral pathogens. Proceedings of the American Water Works Association Water Quality Technology Conference, Toronto, Canada (November 15–19).

Wright, H.B. (2006). Carollo Engineers. Internal communication with Erin Mackey, Carollo Engineers.

6

Resiliency on Temporal Scale

Minimal interruption of infrastructure service ensures the continuity of economic prosperity.

OBJECTIVES
To define and quantify the resiliency of EEIS To carry out mass balance for steady versus unsteady systems To formulate different mathematic models in predicting population growth To design an equalization tank to reduce unsteady effect To obtain rate constants using derivative versus integrated kinetic analysis To establish reaction kinetics and mechanisms of catalysts that reduce reaction time To estimate reaction kinetic rate constant of organic pollutants with aqueous hydroxyl radicals by group theory or QSAR models To establish the temperature effect on SEE system performance and resilience under temperature variation To illustrate how Hammett analysis could be used to explore reaction mechanism in heterogeneous catalytic system such as UV/TiO_2 To explore relationship between spatial and temporal scales in designing EEIS

6.1 Principle 3

Resiliency is the ability of EEIS to cope, recover, or adapt to chronic or acute stressors due to system properties such as robustness, adaptive capacity/redundancy, rapidity, and resourcefulness. In a simple term, resiliency refers to the ability to absorb shock such as temperature, flooding, or droughts. A simple quantification of resiliency is the time that an EEIS recovers from a stressor event and loss of function. On temporal scale, EEIS should be resilient over 20, 50, and 100 years with quantified uncertainty and the sensitivity of the EEIS should be quantified on different temporal scales.

There are two types of stressors: chronic stressors include temperature increases and sea level rise, while acute stressors include extreme hurricane, earthquake, storm, tornados, cold weather, power outage, and wildfire. Resiliency was applied to ecology and engineering by Holling (1973, 1996); to socioecological systems by Walker et al. (2004); to WRRFs by Currie et al. (2014), Schoen et al. (2015), and Xue et al. (2015); to water supply system by Hwang et al. (2014); and to drainage systems by Mugume et al. (2014) and Goldbloom-Helzner et al. (2015). Performance of EEIS includes effluent quality, level of service, unit energy consumption, waste production, and monetary loss. To quantitatively link the properties of EEIS to its performance, Cuppens et al. (2012) quantified resiliency as follows:

Principle 3 Design alternatives should be resilient with quantified uncertainty and sensitivity for different life expectancies

1) Steady versus unsteady system should be clearly identified to avoid overcapacity due to using safety factor in designing EEIS.
2) Design alternatives should be resilient on specific temporal scales such as 10-, 50-, and 100-year life expectancy.
3) EEIS should be flexible from atomic, molecular, nanoparticles, microcolloidal, reactor, process to plant on temporal scales.

Sustainable Environmental Engineering, First Edition. Walter Z. Tang and Mika Sillanpää.

$$M_{\text{resiliency}} = \int_{t_0}^{t_n} (M_0 - M_t)\mathrm{d}t \tag{6.1}$$

where

t_n is the total duration of the perturbation until recovery
t_0 is the initial time when the perturbation occurs
M_0 is the initial state of the chosen metric (a state variable representative of the state of the system)
M_t is the value of the chosen metric at a measured time t

Francis and Bekera (2014) quantified resiliency in terms of function recovery of performance level using an equation that reflects three properties: absorptive (robustness), adaptive, and restorative (rapidity) as follows:

$$\rho_i = S_p \times \frac{F_r}{F_o} \times \frac{F_d}{F_o} \tag{6.2}$$

where

S_p is the speed recovery factor
F_o is the original stable system performance level
F_d is the performance level immediately post-disruption
F_r is the performance at a new stable level after recovery efforts have been exhausted

Regardless of how to quantify the resiliency, it is critical to consider increase in resiliency as an opportunity and not an extra cost or effort. Indeed, Sweetapple et al. (2016) assessed the trade-off between resilience and reliability. Resiliency is maximum failure time under different levels of stressors for WRRFs. Multiobjective optimization could be used to evaluate the cost function of an intervention to build resiliency. Optimal conditions could then be determined by balancing cost/benefit between resiliency implementation and reliability. For Miami-Dade Water and Sewer Department (WASD), deep well injection does not increase resiliency of the system because it dispose 200 MGD treated wastewater effluent as deep as 10 000 ft. As a result, a reliable water resource from the reclaimed water would not be utilized. To decrease time to recover after major stress, Miami-Dade WASD will spend 13.6 billions to increase resiliency of it EEISs in next decade.

One of the most effective ways to quantify resiliency is to formulate different stressor scenarios and calculate the corresponding performance under different stressor scenarios. For example, Mabrouk et al. (2010) used mass balance of reactors and quantified the effect of influent variability on optimal design parameters of the reactors. Lempert et al. (2015) built a model of a system, developed sets of scenarios using Monte Carlo (MC) techniques, and assessed resiliency of the system in terms of specific performance metrics. To quantify the resiliency of WRRFs, Schoen et al. (2015) formulated failure profile for robustness, redundancy and recovery profile for resourcefulness, rapidity with consideration of lifespan of WRRFs. Weirich et al. (2015) applied a statistical method to avoid mathematical model of the system's components. The WRRF performance was predicted using effluent concentration in generalized linear models (GLMs). Hwang et al. (2014) assessed the resiliency of water supply system under different scenarios such as centralized versus decentralized plants using linear programming (LP). However, quantitative benchmarks of metrics are needed to increase the resiliency of EEIS (Juan-García et al., 2017). The metrics should include robustness, adaptive capacity/redundancy, rapidity, and resourcefulness of EEIS. Nevertheless, these methods provide quantitative way to assess resiliency of EEIS to future climate change and its resultant severe droughts and flooding on temporal scales.

Temporal scales have to be considered to reflect the economic development stage for current and projected time frame. For example, the recycling rate of solid waste is directly proportional to the GDP of a country. The higher the GDP, the higher the degree of separation and recycling. To increase the resiliency of stormwater management, green infrastructure should be implemented as much as possible at household, community, city, state, and country levels. For water, wastewater, and stormwater management, four dynamic processes need to be taken into consideration: population growth, urbanization rate, increasing affluence, and ecological demand that change with time. There are three major strategies to increase resiliency of EEIS: (i) Water micronets such as water supply, wastewater treatment, and stormwater management should be decentralized and designed for specific purpose.

In Miami-Dade County, it is unnecessary to pump wastewater for 20 miles to be treated at the central wastewater treatment plant (WWTP) such as WWTP in Virginia key of Miami; decentralized WWTPs could be more resilient than the centralized WWTPs in the long run by partnering with local city governments in Miami-Dade County. As a result, wastewater could be treated at the locations close to its reuse costumers. (ii) Water management has to be color coded: blue, green, grey, brown, and black water management should be resilient at 20-, 50-, and 100-year climate change scenarios. (iii) Fit-for-purpose should replace fit-for-all design. With different classifications of water, wastewater could be treated according to its intended purpose such as recharge aquifer or irrigation lawn to reduce consumption of tap water. As a result, sewer mains could be significantly reduced, and sewer pipe outbreaks could be reduced. All the EEIS should be adaptive to climate change in 20, 50, and 100 years in EEI design (Table 6.1).

Table 6.1 Resilience on temporal scales.

Design	Problems	Temporal	Scale
Air	Coal power plant	Winter seasons as worst design scenarios	Household Community City State Country
Water	WWTPs	Winter as the lowest biomass production season	Same as the previous
	WTPs	Algae bloom seasons as the worst time	Same as the previous
Land	Landfill	Mobile-integrated leachate treatment system	Same as the previous
	Leachate	BOD–COD ratios for young, medium, and mature landfill leachates	Same as the previous
Ecological	River delta hypoxic zone	Summer as the worst season due to algae bloom	Same as the previous
	Wildlife extinction	Winter as most vulnerable season for wildlife	Same as the previous

6.2 Challenges and Opportunities

Climate change in the next 100 years will cause more intensive flooding, droughts, hurricanes, and tornados. These extreme weather events will pose challenges to flooding and drought-related water shortage in major cities all over the world. Therefore, the resiliency of water, wastewater, and stormwater infrastructure will determine agricultural, industrial, and social productivities of a region. Reclamation water, desalination, and stormwater are part of the water supply portfolio of Israel and Singapore; their water supply and wastewater management set excellent examples in terms of the resiliency of water management system. In Australia, a decade-long millennium drought from 2000 to 2010 imposed grand challenges to the country (Radcliffe, 2015). The country was forced to adopt alternative water resources such as reclamation and desalination and recycle water for indirect potable purpose such as groundwater recharge. However, after a decade-long drought, intensive storms occur frequently and cause flooding everywhere. As a result, 450 million m^3/year desalination plants went to stand by due to high operation cost. Three potable recycling plants with a total capacity of 90 million m^3/year were closed. From Australia's experience in coping with prolonged drought and flooding, the challenges are clear: (i) How to optimize the water supply portfolio in terms of reclamation, desalination, and stormwater harvesting to achieve resiliency of water supply and (ii) how to set up the design benchmarks of EEIS in terms of energy consumption, unit cost, and unit resource utilization to achieve specific resiliency targets in 20, 50, and 100 years.

For Miami-Dade, decentralized WRRFs could increase resiliency of the EEISs by serving many purposes (i) If total treatment capacity reaches 120 MGD by decentralized WRRFs in different cities, one of three current WWTPs could be closed and reclaimed to commercial property by 2025. As a result, it would save the county \$2 billion for ending the ocean outfall using deep well injection and \$2 billion for upgrading the aged WWTP in addition to the reclaimed land value. (ii) The decentralized WRRFs would recharge the local Biscayne aquifer to provide additional water resource, prevent saltwater intrusion, and irrigate parks or golf courses. (iii) The decentralized WRRFs would reduce the vulnerability of the three coastal WWTPs to avoid catastrophic damage of future hurricanes. (iv) The WRRFs could also save hundred million dollars of energy bill for the county over their life expectancy because the innovative technologies can ensure that the new WRRFs be designed as energy positive.

Many cities in Florida are facing the regulatory mandates of ending ocean outfall; SEE designers should search for new paradigms in terms of planning, designing, and operating and maintenance of SEEI systems to meet this requirement. Opportunities are abundant. In the United States, it is estimated that \$322–600 billion investment is needed to bring the EEIS to be resilient. In China, this number could be easily doubled or even tripled in the next 20 years.

6.3 Discharge Standards

Since the total pollutant loading is proportional to population equivalent (PE), WWTPs have to meet different discharge standards to control the total pollutant loading of their nearby water bodies. As a result, it is not surprising that large cities may impose more stringent discharge standards than smaller cities. Precipitation, flow rate, and temperature due to climate change over time will impact the performance of WWTPs. Under extreme circumstance, an EEIS could lose functionality to varying degrees. Therefore, the resilience of WWTP is usually expressed in terms of time to recover after failure. Since the total amount of pollutant increases with flow rate, the larger the flow rate, the larger the loading. Population growth will impact plant design, operational capacity such as flow rate, and other treatment process constraints. A WWTP's total hydraulic flow or organic loading could be used to calculate PE. Common discharge standards for biochemical oxygen demand (BOD), suspended solids, nitrogen, and phosphorus are 30, 30, 10, and 0.2 mg/l, respectively. A biological WWTP could be designed based upon a PE of 0.2 lbs. of BOD and 100 gal/person/day. The treatment plant design capacity is the major classification criteria in determining such PE (Table 6.2).

Table 6.2 Treatment technology for different population equivalents.

Size (PE)	Secondary: biological	Advanced treatment[a]	Other defined treatment[b]
Up to 5 000	I	I	I
	I	I	
		I	
5 000–15 000	I	I	I
	I	V	I
	I		
15 000–50 000	I	V	I
	V		I
			I
50 000 and over	V	V	I
			V

[a] Advanced treatment includes one or more of the following:

1) Advanced biological/chemical methods.
2) Ion exchange, reverse osmosis, or electrodialysis.
3) Chemical recovery or carbon regeneration.

[b] Other defined treatment includes one or more of the following:

1) Separate sludge digestion with gas collection.
2) Mechanical sludge dewatering or sludge incineration.

6.4 Population Growth

One of major drivers causing the unsteady conditions is the population growth and urbanization. The human footprints on water and energy are directly proportional to the population. To reduce such footprints, different technologies serve different scale of PE. Therefore it is important to accurately predict population growth so that EEIS could accommodate human needs and economic development with minimal footprints.

Example 6.1

Given: The population data for a city were obtained from the US Census Bureau (Table 6.3). The historical population data are used to predict the future population at the end of a WRRF design life expectancy of 50 years.

Table 6.3 Population data.

Year	Population
1968	32 927
1978	34 666
1988	39 096
1998	55 820
2008	67 867
2018	79 662

Find:

1) Prepare the appropriate graphs for calculating each of the following four models for the population for this city:
   - Linear model.
   - Exponential model.
   - Decreasing rate of increase model (assume 100 000 as a maximum population).
   - Decreasing rate of increase model (assume 200 000 as a maximum population).

   Show the calibrated model equation for each of these four models.
2) Show the predicted population based on all four of these models on the same graph. The x-axis should cover the time period from 1968 to 2018. Also this graph shows the city population from the 6 census years.

Solution (Table 6.4):

Linear model

$$K_L = \frac{\Delta P}{\Delta Y}$$

$$\overline{K_L} = \frac{\sum_{i=1}^{x} K_{Li}}{x} = 934.70$$

$$\text{Model equation}: P = P_0 + \overline{K_L}(t - t_0) = 32927 + 934.70(t - 1940)$$

Table 6.4 Calculation.

Year	Population	ΔY	ΔP/ΔY	LN(P)	ΔLN(P)/ΔY	S–P (S = 100 000)	LN(S–P)	–LN(S–P)/ΔY	S–P (S = 200 000)	LN(S–P)	–LN (S–P)/ΔY
1968	32 927			10.402		67 073	11.113		167 073	12.026	
1978	34 666	10	173.90	10.453	0.005	65 334	11.087	0.003	165 334	12.016	0.001
1988	39 096	10	443.00	10.573	0.012	60 904	11.017	0.007	160 904	11.989	0.005
1998	55 820	10	1672.40	10.929	0.035	44 180	10.696	0.032	144 180	11.879	0.011
2008	67 867	10	1204.70	11.125	0.019	32 133	10.377	0.032	132 133	11.792	0.009
2018	79 662	10	1179.50	11.285	0.016	20 338	9.920	0.046	120 338	11.698	0.009
Ave			934.70		0.017			0.024			0.006

Exponential model

$$K_E = \frac{\Delta \text{LN}(P)}{\Delta Y}$$

$$\overline{K_E} = \frac{\sum_{i=1}^{x} K_{Ei}}{x} = 0.0177$$

$$\text{Model equation}: \text{LNP} = \text{LNP}_0 + \overline{K_E}(t - t_0)$$
$$= 10.4020 + 0.0177(t - 1968)$$

$$\text{Model equation}: P = e^{10.4020 + 0.0177(t-1968)}$$

Decreasing rate of increase model (assume 100 000 as the maximum population)

$$K_D = \frac{-\Delta \text{LN}(S-P)}{\Delta Y}$$

$$\overline{K_D} = \frac{\sum_{i=1}^{x} K_{Di}}{x} = 0.023866$$

$$\text{Model equation}: P = P_0 + (S - P_0)\left(1 - e^{-\overline{K_d}(t-t_0)}\right)$$

$$\text{Model equation}: P = 32927 + (67073)\left(1 - e^{-0.023866(t-1968)}\right)$$

Decreasing rate of increase model (assume 200 000 as the maximum population)

$$K_D = \frac{-\Delta \text{LN}(S-P)}{\Delta Y}$$

$$\overline{K_D} = \frac{\sum_{i=1}^{x} K_{Di}}{x} = 0.006563$$

$$\text{Model equation}: P = P_0 + (S - P_0)\left(1 - e^{-\overline{K_d}(t-t_0)}\right)$$

$$\text{Model equation}: P = 32927 + (167073)\left(1 - e^{-0.006563(t-1968)}\right)$$

Answer: Results are displayed in Table 6.5 and Figure 6.1.

Table 6.5 Result of the predicted population.

Year	Population (linear model)	Population (exponential model)	Population (decreasing rate of increase model, S = 200 000)	Population (decreasing rate of increase model, S = 100 000)
1968	32 927	32 927	32 927	32 927
1978	42 274	39 291	47 168	43 539
1988	51 621	46 885	58 385	53 477
1998	60 968	55 946	67 220	62 784
2008	70 315	66 759	74 180	71 500
2018	79 662	79 662	79 662	79 662
2028	89 009	95 059	83 980	87 306
2038	98 356	113 431	87 381	94 464
2048	107 703	135 354	90 060	101 167
2058	117 050	161 514	92 171	107 445
2068	126 397	192 730	93 833	113 324

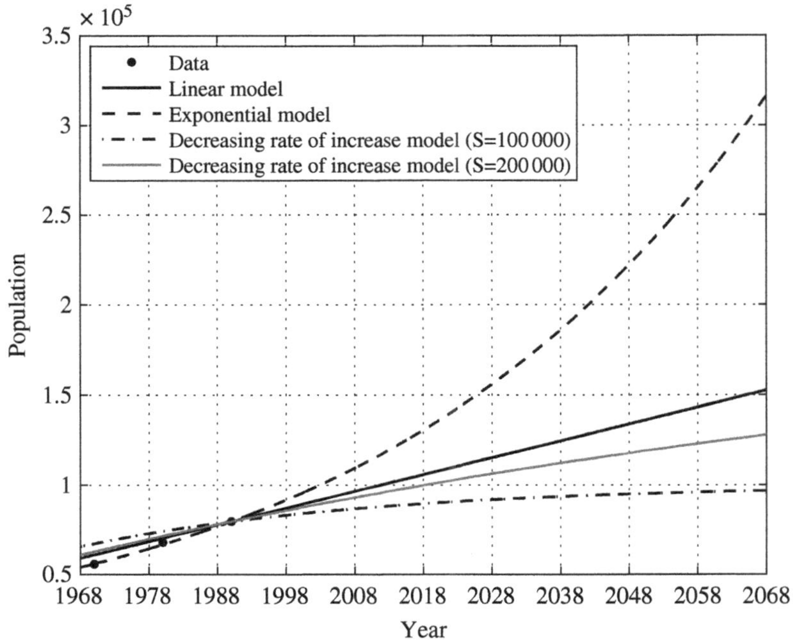

Figure 6.1 Population versus year.

6.5 Steady Versus Unsteady

For a specific EEIS, if the net accumulation within the system changes with time, it is referred as unsteady state. As a result, any equation should contain time as an independent variable. To reduce the uneven loading, equalization tank is designed to reduce the unsteady effect on the process efficiency (Davis, 2010). Equalization tanks can be implemented as offline or online. For offline equalization, if the unsteady flow rate Q_U is greater than the steady flow rate Q_S, the difference, i.e. $Q_U - Q_S$, is stored in the equalization tank. If Q_U is less than Q_S, the difference, i.e. $Q_U - Q_{SV}$, is supplied from the equalization tank (Figure 6.2).

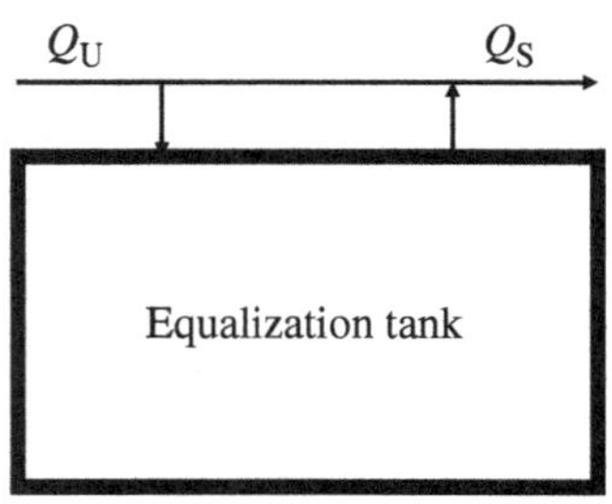

Figure 6.2 Offline equalization tank.

For online equalization, Q_U is influent to the equalization tank, while a steady flow Q_S is effluent from this tank. If Q_U is greater than Q_S, excess wastewater is stored in the tank. If Q_U is less than Q_S, the deficit is building up in the tank. The two types of equalization tanks serve the same purpose of providing steady flow rate, Q_S, for the rest of treatment process because of the constant pump rate out of the equalization tank to the rest of treatment processes. As a result, the system would work under relative constant loading rate. All the WTPs and WWTPs are designed in such a way to reduce the effect of hydraulic shocking load. However, if the influent water quality changes dramatically in WWTPs, the shocking load could significantly reduce the performance of the system. Therefore, equalization tank has to be designed in EEIS such as a WRRF.

6.5.1 Equalization Basin

One way to increase the resiliency of EEIS is to use equalization tank to reduce the impact of unsteady flow on the performance of the WRRFs.

Example 6.2

Given: A wastewater treatment plant (WWTP) requires an equalization basin to even out flow and BOD variations. The following flows and BOD_5 have been found to be typical of the average variation over a day.

Find: What size of equalization basin, in cubic meters, is required to provide a uniform outflow equal to the average daily flow, if flow rates are hourly averages? Compute and plot the unequalized and equalized hourly BOD mass loadings to the WWTP. Using the plot and calculations, determine the following ratios for BOD mass loading:

1) Peak to average
2) Minimum to average
3) Peak to minimum

Solution:

$$\text{Average daily flow}: 10\,\text{MGD} = 37850\,\text{m}^3/\text{day} = 0.438\,\text{m}^3/\text{s}$$

If the initial value of the first flow rate is greater than the average flow rate after the sequence of nighttime low flows, then the last row of the computation would result in a storage value of zero for a perfect sinusoidal flow pattern.

In Table 6.6, the flow that comes first is 0.4881 m^3/s at 900 h. Table 6.7 shows the tabular flow rates versus time. Calculations for each column are given in the following steps.

Table 6.6 Flow rate and BOD_5 variation in a day.

Time	Flow (m^3/s)	BOD_5 (mg/l)
0	0.3721	123
100	0.2781	118
200	0.175	95
300	0.145	80
400	0.145	85
500	0.153	95
600	0.175	100
700	0.2781	118
800	0.3941	136
900	0.4881	170
1000	0.5191	220
1100	0.5281	250
1200	0.5562	268
1300	0.5762	282
1400	0.5802	280
1500	0.6052	268
1600	0.6242	250
1700	0.6532	205
1800	0.6612	168
1900	0.6242	140
2000	0.6052	130
2100	0.5191	146
2200	0.4511	158
2300	0.4071	154

In the third column, the flows are converted to volumes using the time interval between flow measurements:

$$V_{\text{in}} = (0.4881\,\text{m}^3/\text{s})(1\,\text{h})(3600\,\text{s/h}) = 1757.16\,\text{m}^3$$

The average volume that leaves the equalization basin is calculated in the fourth column. It is the average flow rate computed on an hourly basis:

$$V_{\text{out}} = (0.438\,\text{m}^3/\text{s})(1\,\text{h})(3600\,\text{s/h}) = 1577.083\,\text{m}^3$$

The fifth column is the difference between the inflow volume and the outflow volume:

$$dS = V_{\text{in}} - V_{\text{out}} = 1577.16 - 1577.083 = 180.0767\,\text{m}^3$$

The required storage is computed in the sixth column. It is the cumulative sum of the difference between the inflow and outflow. For the second time interval, it is

$$\text{Storage} = \sum dS = 180.08 + 291.68 = 471.75\,\text{m}^3$$

Here, the last value for the cumulative storage is −0.32 m^3. It is not zero because of roundoff truncation in the calculations. At this point the equalization basin is empty and ready to begin the next day's cycle.

The required volume for the equalization basin is the maximum cumulative storage. Assume that an addition of 25% to the estimated volume will be provided to account for contingencies. With the requirement for 25% excess, the volume would then be

$$\text{Storage} = (6689.51\,\text{m}^3)(1.25) = 8361.89\,\text{m}^3$$

The size of the equalization basin is $\boxed{8361.89\,\text{m}^3}$.

The mass of BOD_5 entering the equalization basin is the product of the inflow (Q), the concentration of BOD_5 (S_0), and the integration time (Δt):

$$M_{\text{BOD-in}} = (Q)(S_0)(\Delta t)$$

The mass of BOD_5 leaving the equalization basin is the product of the average outflow (Q_{avg}), the average concentration (S_{avg}) in the basin, and the integration time (Δt):

$$M_{\text{BOD-out}} = (Q_{\text{avg}})(S_{\text{avg}})(\Delta t)$$

The average concentration is determined as

$$S_{\text{avg}} = \frac{(V_i)(S_0) + (V_s)(S_{\text{prev}})}{V_i + V_s}$$

where

V_i is volume of inflow during time interval Δt (m^3)

S_0 is average BOD_5 concentration during time interval Δt (g/m^3)

V_s is volume of wastewater in the basin at the end of the previous time interval Δt (m^3)

Table 6.7 Spreadsheet calculations for equalization basin.

Time	Flow (m^3/s)	V_{in} (m^3)	V_{out} (m^3)	ds (m^3)	$\sum$ds (m^3)	$M_{BOD\text{-}in}$ (kg)	S (mg/l)	$M_{BOD\text{-}out}$ (kg)
900	0.4881	1757.16	1577.08	180.08	180.08	298.72	170.00	268.10
1000	0.5191	1868.76	1577.08	291.68	471.75	411.13	215.61	340.03
1100	0.5281	1901.16	1577.08	324.08	795.83	475.29	243.16	383.49
1200	0.5562	2002.32	1577.08	425.24	1221.07	536.62	260.94	411.52
1300	0.5762	2074.32	1577.08	497.24	1718.30	584.96	274.19	432.43
1400	0.5802	2088.72	1577.08	511.64	2229.94	584.84	277.38	437.45
1500	0.6052	2178.72	1577.08	601.64	2831.58	583.90	272.74	430.14
1600	0.6242	2247.12	1577.08	670.04	3501.61	561.78	262.68	414.27
1700	0.6532	2351.52	1577.08	774.44	4276.05	482.06	239.51	377.72
1800	0.6612	2380.32	1577.08	803.24	5079.29	399.89	213.94	337.40
1900	0.6242	2247.12	1577.08	670.04	5749.32	314.60	191.26	301.63
2000	0.6052	2178.72	1577.08	601.64	6350.96	283.23	174.42	275.08
2100	0.5191	1868.76	1577.08	291.68	6642.64	272.84	167.96	264.89
2200	0.4511	1623.96	1577.08	46.88	6689.51	256.59	166.00	261.80
2300	0.4071	1465.56	1577.08	−111.52	6577.99	225.70	163.85	258.40
0	0.3721	1339.56	1577.08	−237.52	6340.47	164.77	156.94	247.50
100	0.2781	1001.16	1577.08	−575.92	5764.54	118.14	151.63	239.13
200	0.175	630.00	1577.08	−947.08	4817.46	59.85	146.05	230.33
300	0.145	522.00	1577.08	−1055.08	3762.38	41.76	139.59	220.15
400	0.145	522.00	1577.08	−1055.08	2707.29	44.37	132.94	209.66
500	0.153	550.80	1577.08	−1026.28	1681.01	52.33	126.53	199.54
600	0.175	630.00	1577.08	−947.08	733.93	63.00	119.29	188.14
700	0.2781	1001.16	1577.08	−575.92	158.00	118.14	118.55	186.96
800	0.3941	1418.76	1577.08	−158.32	−0.32	192.95	134.25	211.73
Ave						296.98		296.98
Max (peak)					6689.51	584.96		437.45
Min						41.76		186.96

S_{prev} is concentration of BOD_5 in the tank at the end of the previous time interval Δt (g/m^3)

Noting that 1 mg/l = 1 g/m^3, at time 900 h,

$$M_{BOD-in} = (0.4881\,m^3/s)(170\,g/m^3)(1\,h)(3600\,s/h)(10^{-3}\,kg/g) = 298.72\,kg$$

$$S_{avg} = \frac{(1757.16)(170)+0}{1757.16+0} = 170\,g/m^3 = 170\,mg/l$$

$$M_{BOD-out} = (0.4838\,m^3/s)(170\,g/m^3)(1\,h)(3600\,s/h)(10^{-3}\,kg/g) = 268.10\,kg$$

Here, the zero values in the computation of S_{avg} are valid only at start-up of an empty basin. Also note that in this case, $M_{BOD\text{-}in}$ and $M_{BOD\text{-}out}$ differ only because of the difference in flow rates.

For the second row (1000 h), the calculations are

$$M_{BOD-in} = (0.5191\,m^3/s)(220\,g/m^3)(1\,h)(3600\,s/h)(10^{-3}\,kg/g) = 411.13\,kg$$

$$S_{avg} = \frac{(1868.76)(220)+(180.08)(170)}{1868.76+180.08} = 215.61\,g/m^3 = 215.61\,mg/l$$

$$M_{BOD-out} = (0.4838\,m^3/s)(215.61\,g/m^3)(1\,h)(3600\,s/h)(10^{-3}\,kg/g) = 340.03\,kg$$

Here, V_s is the volume of wastewater in the tank at the end of the previous time interval. Therefore, it is equal to the accumulated dS. The concentration of BOD_5 (S_{prev}) is the average concentration at the end of previous interval (S_{avg}) and not the influent concentration for the previous interval (S_O).

For the third row (1100 h), the concentration of BOD_5 is

$$S_{avg} = \frac{(1901.16)(250) + (471.75)(215.61)}{1901.16 + 471.75}$$
$$= 243.16\,\text{g/m}^3 = 243.16\,\text{mg/l}$$

Using data from Table 6.7 to plot Figure 6.3, Figure 6.3 shows the unequalized and equalized hourly BOD mass loadings.

Using data from Table 6.7 and Figure 6.3 to make Table 6.8, Table 6.8 shows the ratios for BOD mass loading: (i) peak to average, (ii) minimum to average, and (iii) peak to minimum.

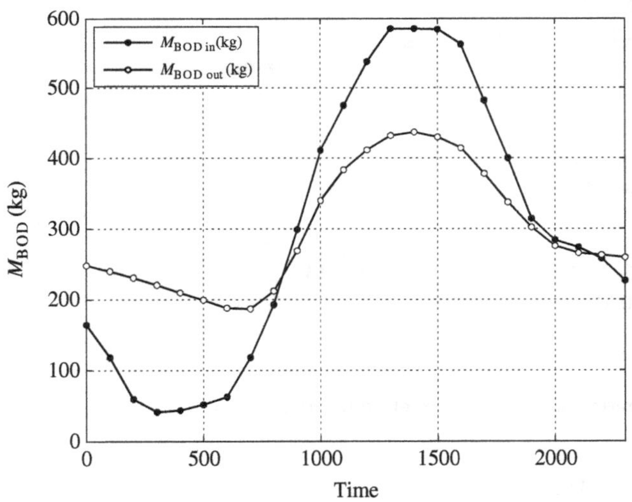

Figure 6.3 BOD loading.

Table 6.8 Ratios for BOD mass loading.

Ratios for BOD mass loading	Unequalized	Equalized
Peak to average	584.96/296.98 = 1.97	437.45/296.98 = 1.47
Minimum to average	41.76/296.98 = 0.14	186.96/296.98 = 0.63
Peak to minimum	584.96/41.76 = 14.01	437.45/41.76 = 2.34

Example 6.3

Given: Figure 6.4 shows the ratio of extreme flows to average daily flow.

Find:

1) Use Excel to determine the correlation equation for the ratio of maximum (peak) day flow to average daily flow versus average daily flow (m^3/day).
2) Calculate peak day flow rate (MGD) for different average daily flow rates (1, 2, 4, 6, 8, 10, 20, 40, 60, 80, 100 MGD).
3) Plot maximum (peak) day flow to average daily flow versus average daily flow (MGD).
4) Determine correlation equation for ratio of maximum (peak) day flow to average daily flow versus average daily flow (MGD).

Solution:

1) From Figure 6.5, the equation for the ratio of maximum (peak) day flow to average daily flow versus average daily flow (m^3/day):

$$\text{Ratio of peak day flow to average daily flow}$$
$$= 5.8176\ (\text{average daily flow})^{-0.109} \quad r^2 = 1$$

2) When $Q = 1$ MGD (Table 6.9),

$$Q = 1\ \text{MGD} = 3785\,\text{m}^3/\text{day}$$
$$\text{Ratio of peak day flows to average daily flow}$$
$$= 5.8176\left(3785\ (\text{m}^3/\text{day})\right)^{-0.109} = 2.3699$$
$$\text{Peak day flow} = 3785 \times 2.3699 = 8970.17\,\text{m}^3/\text{day}$$
$$= 2.3699\,\text{MGD}$$

3) From Figure 6.6, the equation for the ratio of maximum (peak) day flow/average daily flow versus average daily flow (MGD):

The regression equation of the trend line:

$$\log(\text{ratio of peak day flow to average daily flow})$$
$$= -0.1090\log(\text{average daily flow}) - 0.3747 \quad r^2 = 1$$

or the equation of the regression:

$$\text{Ratio of peak day flow to average daily flow}$$
$$= 2.3699\ (\text{average daily flow})^{-0.109} \quad r^2 = 1$$

Matlab Code:

```
clear all
close all
%data%
QrangeMGD= [1  2 4 6 8 10 20 40 60 80 100];
%function%
```

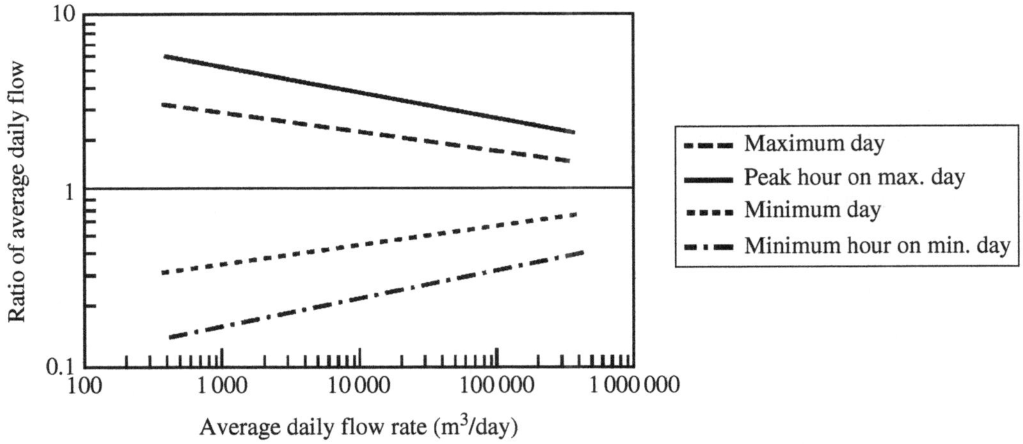

Figure 6.4 Ratio of extreme flows to average daily flow. (*Source:* Davis (2010).)

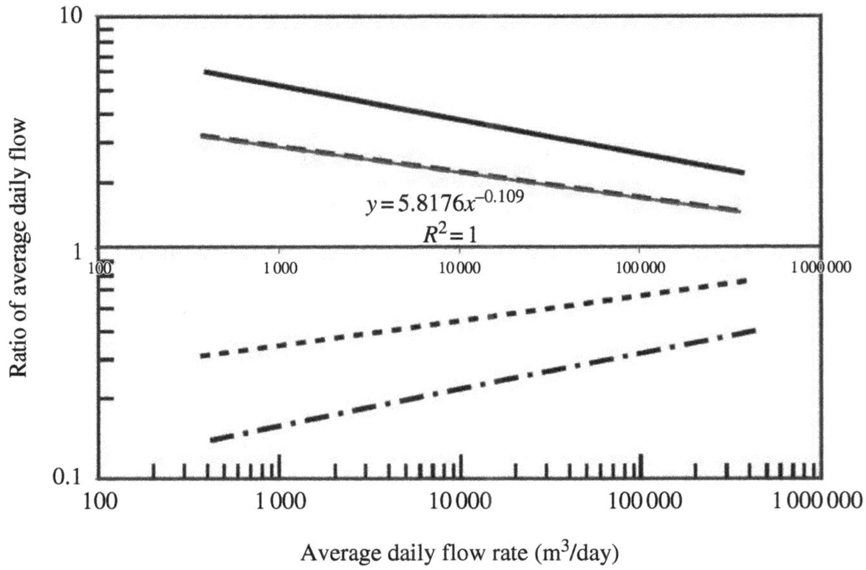

Figure 6.5 Ratio of extreme flows to average daily flow and excel simulation.

Table 6.9 Ratio of peak day flows to average daily flow at different flow rates.

Average daily flow (MGD)	Average daily flow (m^3/day)	Ratio of peak day flows to average daily flow	Peak day flow (m^3/day)	Peak day flow (MGD)
1	3 785	2.3699	8 970.17	2.3699
2	7 570	2.1975	16 634.83	4.3949
4	15 140	2.0376	30 848.64	8.1502
6	22 710	1.9495	44 272.43	11.6968
8	30 280	1.8893	57 207.60	15.1143
10	37 850	1.8439	69 791.18	18.4389
20	75 700	1.7097	129 425.03	34.1942
40	151 400	1.5853	240 013.69	63.4118
60	227 100	1.5168	344 455.66	91.0055
80	302 800	1.4699	445 096.05	117.5947
100	378 500	1.4346	543 000.91	143.4613

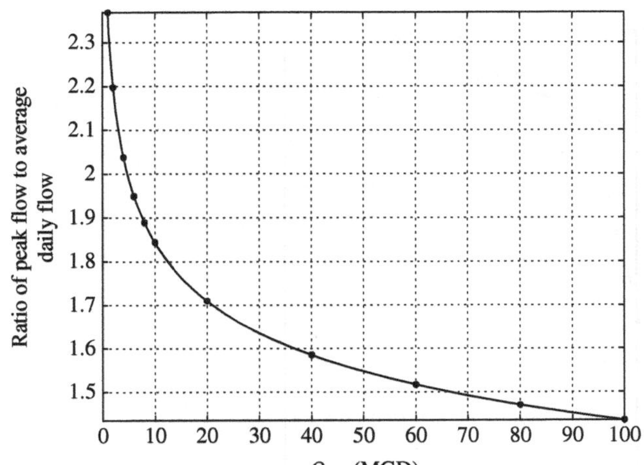

Figure 6.6 Ratio of maximum (peak) day flow/average daily flow versus average daily flow (MGD).

```
Qrangem3d=QrangeMGD.*3785;
Ratio=5.8176.*Qrangem3d.^(-0.109);
PeakQrangem3d=Ratio.*Qrangem3d;
PeakQrangeMGD=PeakQrangem3d./3785;
p=polyfit(log10(QrangeMGD),
  log10(Ratio), 1)
log10ratiofit=polyval(p, log10
  (QrangeMGD));
ratiofit=10.^log10ratiofit;
k=10^p(:,2)
R2 = 1 - sum((log10(Ratio) -
  log10ratiofit).^2)/sum((log10(Ratio) -
  mean(log10(Ratio))).^2)
xmax=max(QrangeMGD);
xmin=min(QrangeMGD);
xaxis=[xmin:0.1:xmax];
ymax=max(Ratio);
ymin=min(Ratio);
yaxis=[ymin:0.1:ymax];
log10ratiofitplot=polyval
  (p, log10(xaxis));
ratiofitplot=10.^log10ratiofitplot;
  %plot%
figure(1)
plot(QrangeMGD,Ratio,'ko','MarkerSize',
  3,'MarkerFaceColor',[0 0 0],
  'MarkerEdgeColor',[0 0 0],
  'linewidth',1);
hold on
plot(xaxis,ratiofitplot,'k-',
  'linewidth',1);
set(gca,'xtickmode','auto',
  'FontSize',10,'FontWeight','bold',
  'FontName', 'Times New Roman');
set(gca,'ylim',[ymin,
  ymax],'FontSize',10,'FontWeight',
  'bold','FontName', 'Times New Roman');
xlabel({'Q_a_v_e (MGD)'},'FontSize',
  14,'FontWeight','bold','FontName',
  'Times New Roman')
ylabel({'Ratio of Peak Flow to Average';
  'Daily Flow'},'FontSize',14,
  'FontWeight','bold','FontName',
  'Times New Roman')
grid on
hold on
```

Answer: Ratio of peak flow to average daily flow MGD:

$$p = -0.1090\,0.3747;\ k = 2.3699;\ R^2 = 1$$

6.6 Hydraulic Condition of Different Reactors

In designing WRRFs, hydraulic conditions of different reactors and biological or chemical reaction kinetics determine the resiliency of a WRRF. Batch reactors do not have flow in or flow out, and it is assumed that it is a perfectly mixed reactor. In continuous stirred tank reactors (CSTRs), there is flow in and out with a perfect mix at steady state. When it is at steady state, the change in the CSTR (dN_A) with time would be zero. Ideal plug flow reactor (PFR) assumes no radial gradients, only axial, and also exists in steady state. To characterize hydraulic profile in ideal CSTR or PFR, mathematical models do not consider the effects of nonideal flow pattern such as mixing and short circuits on conversion. Therefore, in an ideal PFR, the velocity profile is assumed to be flat, and the dispersion of reactants and products by turbulence and diffusion is negligible. In an ideal CSTR, all reactants and products are presumed to be perfectly mixed with the reactants. However, these conditions of idealized models are usually not true in practice. The back-mixing created by turbulence and diffusion in a tubular reactor produces a conversion that is lower than predicted by the mathematical plug flow model. Since back-mixing is not as extensive as in a stirred tank, conversion in a tubular reactor is often higher than conversion in a CSTR. One of the best ways to describe the nonideal flow pattern in different reactors is to use a tracer test. Figure 6.7 presents dimensionless trace concentration versus dimensionless residence time. CSTR would give the similar hydraulic characteristic as complete back-mixing, while PFR would give a spike at dimensionless residence time.

To improve mixing efficiency is to reduce reaction time. As a result, better mixing will improve reaction efficiency. Figure 6.7 shows that the ideal PFR is the best, static mixer is the second, empty pipe is third, and complete back-mixing is the worst in terms of tracer concentration profile. For this reason, pipe flow reactors should perform better than container reactors, which would outperform tank reactors as designed in traditional WTPs and WWTPs. Indeed, dead zones and short circuits in

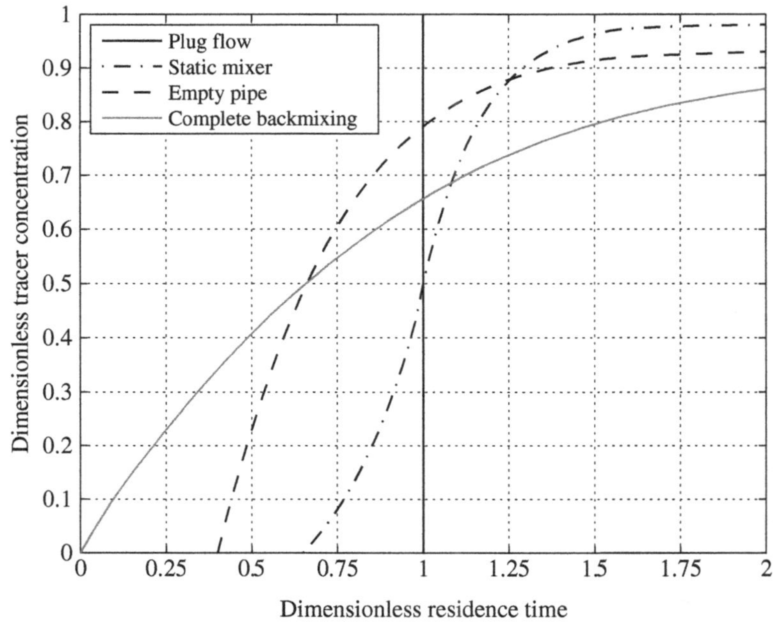

Figure 6.7 Cumulative residence time distributions for various flow systems.

the huge concrete tanks of traditional WTPs and WWTPs significantly increase required hydraulic retention time and reduce overall treatment efficacy. As a result, traditional WWTPs should be much less resilient than next generation of WRRFs. (Tables 6.10, 6.11, and 6.12).

6.7 Chemical Kinetics

As one of three pillars of modern reactor design, kinetics quantifies the speed and how the reaction takes place by establishing the reaction mechanism of any reaction. Kinetics can be studied by plotting concentration versus time. Both average and instantaneous time can be used in derivative and integration form. The *rate constant* identifies these trends based upon different orders of the slope. *Rate law* is used to determine the different rates based on different starting points and different trends of the reaction rate. Reaction order is found by summation of all the orders to the concentration of each reactant. The derivative of concentration over time equals to the rate constant to the "*n*th" power (Table 6.13).

To unveil reaction mechanism, elementary reaction refers to one step in chemical reactions where a bond of a molecule breaks, a bond between atoms is formed, or an electron is transferred. Elementary reactions refer to atoms or molecules. The formation of carbon dioxide through carbon and oxygen gas, for example, is not a elementary reaction. The split of the bond between oxygen gas and its individual addition to carbon is an elementary reaction. Reaction mechanism describes how the bonds are split and created or how the electrons are transferred in the reaction. In most chemical reactions, active intermediates are produced by elementary reactions during a chemical reaction. This transition compound of bonds that have yet to form but is proceeding toward the products is termed as *activated complex*. For a chemical bond to break or form, *activation energy* (E_a) should be overcome by the kinetic energy necessary for a reaction to occur. The kinetic energy a molecular possesses is proportional to temperature. *Rate determination steps* can be understood by the reactants consumed and product produced, where slow reactions have limited steps in reactor design. *Pseudo-first order* refers to reactions where the changes in the concentration of other reactant except one are so minimal that they do not represent a real first-order reaction.

The reaction orders are determined by the rate of reactant disappeared. Reaction orders can be determined by the linear plots according to Table 6.14.

Half-life could either be independent of the reactant as a first-order reaction and only proportional to the rate coefficient or be dependent on the reactant and the rate coefficient as shown in the second-order reaction. The rate coefficient is dependent on the initial concentration temperature.

Table 6.10 Complete mixing reactor performance at different reaction orders.

Reaction order, n ($r = -k_n c^n$)	C_{out}	$\frac{C_{out}}{C_{in}}$	t_d
0	$C_{in} - k_0 t_d$	$1 - \frac{k_0 t_d}{C_{in}}$	$\frac{C_{out}}{k_0}\left(\frac{C_{in}}{C_{out}} - 1\right)$
1	$\frac{C_{in}}{1 + k_1 t_d}$	$\frac{1}{1 + k_1 t_d}$	$\frac{1}{k_1}\left(\frac{C_{in}}{C_{out}} - 1\right)$
2	$\frac{-1 + \sqrt{1 + 4k_2 t_d C_{in}}}{2k_2 t_d}$	$\frac{1}{1 + k_2 t_d C_{out}}$	$\frac{1}{k_2 C_{out}}\left(\frac{C_{in}}{C_{out}} - 1\right)$
Any n	Numerical solution	$\frac{1}{1 + k_n t_d\, C_{out}^{n-1}}$	$\frac{1}{k_n\, C_{out}^{n-1}}\left(\frac{C_{in}}{C_{out}} - 1\right)$

Table 6.11 PFR performance at batch or steady state.

Reaction order, n ($r = -k_n c^n$)	C_{out}	$\frac{C_{out}}{C_{in}}$	t_d
0	$C_{in} - k_0 t_d$	$1 - \frac{k_0 t_d}{C_{in}}$	$\frac{1}{k_0}(C_{in} - C_{out})$
1	$C_{in}\exp(-k_1 t_d)$	$\exp(-k_1 t_d)$	$\frac{1}{k_1}\ln\frac{C_{in}}{C_{out}}$
2	$\frac{C_{in}}{1 + k_2 t_d C_{in}}$	$\frac{1}{1 + k_2 t_d C_{in}}$	$\frac{1}{k_2 C_{in}}\left(\frac{C_{in}}{C_{out}} - 1\right)$
Any $n \neq 1$	$\left[(n-1)k_n t_d + C_{in}^{1-n}\right]^{\frac{1}{1-n}}$	$\left(1 + (n-1)k_n t_d C_{in}^{n-1}\right)^{\frac{1}{1-n}}$	$\frac{1}{(n-1)k_n t_d\, C_{in}^{n-1}}\left[\left(\frac{C_{in}}{C_{out}}\right)^{n-1} - 1\right]$

Table 6.12 Tracer behavior in a complete mixing reactor and PFR.

Reactor type	Equation	Notes
CMR	$C(t) = C_{in} - (C_{in} - C(0))\exp\left(-\frac{t}{t_d}\right)$	C_{out} at t for system with initial concentration $C(0)$ and steady input of C_{in} thereafter
CMR	$C(t) = C(0)\exp\left(-\frac{t}{t_d}\right)$	C_{out} at t for system with initial concentration $C(0)$ and $C_{in} = 0$ thereafter[a]
PFR	C_{out} at $t = C_{in}$ at $(t - t_d)$	Regardless if input concentration, output is identical, but delayed by t_d

[a] For example, for a spike input at $t = 0$ and input of clean water thereafter.

Table 6.13 Rate constants and units for n-th order of chemical reactions

*n*th power	*k*-function	Rate units
0	C/t	Moles/seconds
1	$1/t$	1/seconds
2	$1/C/t$	1/moles/seconds

In general kinetic analysis, a linear straight line is plotted between properly arranged concentration data against time by either differential or integral methods. Differential method (nth-order reaction) can be expressed as follows:

$$\underbrace{\log\left(-\frac{dC_A}{dt}\right)}_{y} = \log k + \underbrace{n\log C_A}_{x}$$

Table 6.14 Linearized kinetic equations.

Reaction order	Zeroth order	First order	Second order
Derivative rate law $r = (-dC/dt)$	k	kC	kC^2
Integrated rate law	$C = -kt + C_0$	$C = C_0 e^{-kt}$	$C = 1(kt + 1/C_0)$
Linear equation	$C = -kt + C_0$	$\ln C = -kt + \ln C_0$	$1/C = kt + 1/C_0$
Linear plots	C vs. t	$\ln C$ vs. t	$1/C$ vs. t
Half-life	$C_0/(2k)$	$\ln 2/k$	$1/(kC_0)$
Units of rate constant k	M time^{-1}	time^{-1}	M^{-1} time^{-1}

Integral method (nth-order reaction):

$$F(C_{A1}) = F(C_{A0}) = \text{ktf}(n)$$

$$\underbrace{\frac{1}{C_A^{n+1}}}_{y} = \frac{1}{C_{A0}^{n+1}} + k(n+1)\underbrace{t}_{x}$$

For rate forms other than nth-order, straight-line relationships are also sought. For example, for Michaelis–Menten rate of a biological process,

$$-\frac{dC_A}{dt} = -r_A = \frac{kC_A}{K + C_A}$$

The equation could be arranged in a differential form:

$$\underbrace{\left[\frac{1}{\left(-\frac{dC_A}{dt}\right)}\right]}_{y} = \frac{1}{k} + \frac{K}{k}\underbrace{\left(\frac{1}{C_A}\right)}_{x} \quad \text{or} \quad \underbrace{\left[\frac{C_A}{\left(-\frac{dC_A}{dt}\right)}\right]}_{y} = \frac{K}{k} + \frac{1}{k}\underbrace{(C_A)}_{x}$$

and an integral form:

$$\underbrace{\left[\frac{\ln(C_{A0}/C_A)}{(C_{A0} - C_{A0})}\right]}_{y} = -\frac{1}{K} + \frac{k}{K}\underbrace{\left(\frac{1}{C_{A0} - C_{A0}}\right)}_{x}$$

Either forms would yield a straight-line relationship and could be used to determine reaction kinetics. Elementary reactions are the fundamental steps on how reactions break or form chemical bonds. Collision models consider that the atoms collide with each other to form molecule bonds and to also break old bonds with the correct orientation and sufficient energy. Reaction coordinate diagrams show how the energy changes are processed through a step-by-step timeline to see how rearrangement of atoms and molecules occurs. Kinetic energy distribution graphs show that temperature plays a role in how kinetic energy is distributed and used. For example, Maxwell–Boltzmann distribution is used to describe kinetic energy needed for the reaction of the compound. The Arrhenius equation describes the effect of temperature and catalyst on reaction rate for a reaction:

$$aA + bB = cC$$

$$\text{Rate} = k[A]^x[B]^y$$

This equation explicitly shows the effect of reactant concentrations on the rate. The effect of temperature and a catalyst on the rate of a reaction is described as the Arrhenius equation, which links the collisions with temperature and energy. The Arrhenius equation indicates that rate constant k depends on temperature. It can be used to determine the value of the activation energy, E_a.

The effect of temperature on rate constants is given quantitatively by the Arrhenius equation:

$$k = Ae^{-\frac{E_a}{RT}}$$

where

A is the "frequency factor"
E_a is the activation energy
R is the ideal gas constant
T is the absolute temperature
$e^{-E_a/RT}$ is the fraction of molecules whose kinetic energy is equal to or greater than E_a

The value of k increases as the temperature increases and in the presence of a catalyst. The Arrhenius equation can be linearly plotted according to the following equation. Activation energy and concentration could be used to experimentally obtain the activation energy of a reaction, E_a:

$$\ln(k) = -\frac{E_a}{RT} + \ln A$$

where

T must be in Kelvin
A is the Arrhenius constant (frequency of successful collisions based on collision geometry and where energy)

$R = 8.314$ J/K/mol
When $\ln k$ is plotted against $1/T$, the slope $= -E_a/R$

Example 6.4
Given: Kinetic rate constants at different temperature are given as follows (Table 6.15):

Table 6.15 Rate constant k and temperature T.

Rate constant k (s^{-1})	Temperature T (°C)
2.88×10^{-4}	320
4.87×10^{-4}	340
7.96×10^{-4}	360
1.26×10^{-3}	380
1.94×10^{-3}	400

Find: Please determine the value for E_a.

Solution:

Matlab code is as follows:

```
clear
clc
load Rateconstantandtemperaturedata
RateConstant=z(:,1);
TemperatureC=z(:,2);
TemperatureK=TemperatureC+273.15;
TemperatureK2=1./TemperatureK;
LNRateConstant=log(RateConstant);
hold on
n=1;
p=polyfit(TemperatureK2,
LNRateConstant,n)
x4=[0:0.01:100];
z3=polyval(p,TemperatureK2);
z4=polyval(p,x4);
plot(TemperatureK2,LNRateConstant,
  'ko','MarkerSize',3,
  'MarkerFaceColor',[0 0 0],
  'MarkerEdgeColor',[0 0 0],
  'linewidth',1);
hold on
plot(x4,z4,'k-','LineWidth',1);
hold on
plot(TemperatureK2,
  LNRateConstant,'ko',
  'MarkerSize',3,'MarkerFaceColor',
  [0 0 0], 'MarkerEdgeColor',[0 0 0],
  'linewidth',1);
hold on
R2 = 1 - sum((LNRateConstant - z3).^2)/sum
  ((LNRateConstant - mean
  (LNRateConstant)).^2)
set(gca,'xlim',[0.00145
  0.0017],'FontWeight','bold',
  'FontName', 'Times New Roman');
set(gca,'ylim',[-8.5 -6],
  'FontWeight','bold','FontName',
  'Times New Roman');
xlabel('1/T, K^-^1',
  'FontSize',14,'FontWeight',
  'bold','FontName',
  'Times New Roman')
ylabel('ln(k)','FontSize',14,
  'FontWeight','bold',
  'FontName', 'Times New Roman')
legend ('Original Data','Trendline',
  'location', 'NorthEast')
box on
grid on
```

The plot of $\ln(k)$ versus $1/T$ is as follows (Figure 6.8);

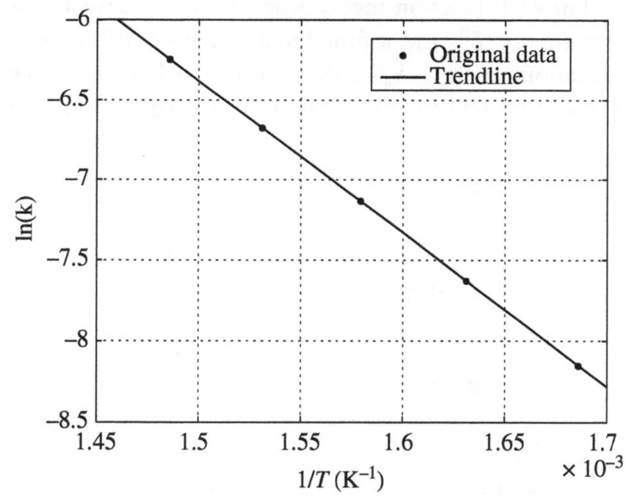

Figure 6.8 ln(k) versus 1/T.

$$\text{The equation of trend line: } \ln(k) = -9520.2\left(\frac{1}{T}\right) + 7.8989 \quad r^2 = 1$$

Determine E_a:

$$\ln(k) = -\frac{E_a}{RT} + \ln A$$

Therefore,

$$\ln(k) = -\frac{9520.2}{T} + 7.8989$$

$$\frac{E_a}{R} = 9520.2\text{K}$$

$$R = 8.314\text{J/K/mol}$$

$$E_a = 8.314 \times 9520.2$$

$$E_a = 79151\text{J} = 79.151\text{kJ}$$

6.8 Group Theory Predicting Hydroxyl Radical Kinetic Constants

In an advanced oxidation process (AOP), hydroxyl radical reaction holds the key to all oxidation of different organic pollutants (Li and Crittenden, 2009). Since kinetic experiments are costly and time consuming, thermokinetics could be used to theoretically calculate the reaction rate constant k according to the group theory (Minakata and Crittenden, 2011; Minakata et al., 2009, 2011). Benson's thermochemical group additivity theory hypothesizes that an observed experimental rate constant for a given organic compound is the combined rate of all elementary reactions involving HO•, which can be estimated using E_a. For each reaction mechanism, there is a base activation energy, E_a^0, and a functional group contribution of activation energy, E_aR_i, due to the neighboring (i.e. α-position) and/or the next-nearest neighboring (i.e. β-position) functional group (i.e. R_i). These contributions to the rate constant can be parameterized and determined empirically when sufficient rate constant data are available. Four reaction mechanisms that HO• initiates in the aqueous phase are (i) H-atom abstraction, (ii) HO• addition to alkenes, (iii) HO• addition to aromatic compounds, and (iv) HO• interaction with sulfur (S)-, nitrogen (N)-, or phosphorus (P)-atom-containing compounds. Accordingly, an overall reaction rate constant, $k_{overall}$, may be given in Equation (6.1), where k_{abs}, $k_{add\text{-}alkene}$, $k_{add\text{-}aromatic}$, and k_{int} are the rate constants for the aforementioned reaction mechanisms. Figure 6.9 shows that linear relationship exists between the Taft constant of a substituent group σ^* and the group contribution factor X. Electrophilic substituent σ^+ would contribute linearly to the group contribution factor Z (Figure 6.10);

$$k_{overall} = k_{abs} + k_{add-alkene} + k_{add-aromatic} + k_{int}$$

6.9 Photocatalytic Oxidation of Halogen-substituted Meta-phenols by UV/TiO$_2$

One of the most important ways to reduce reaction time is through a catalyst. TiO_2 has wide applications in oxidation and disinfection of water, wastewater, and volatile

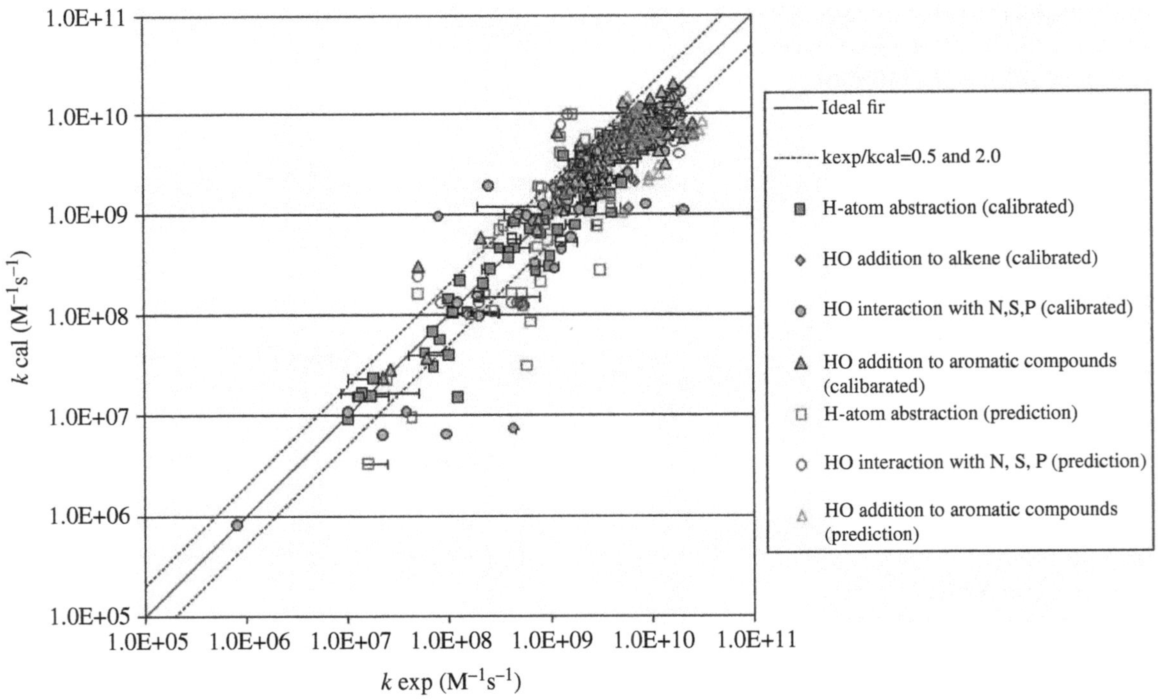

Figure 6.9 Total of 434 HO reaction rate constant from calibrations and predictions versus experimental rate constants for four reaction mechanism. Error bars represent the range of experimentally reported values.

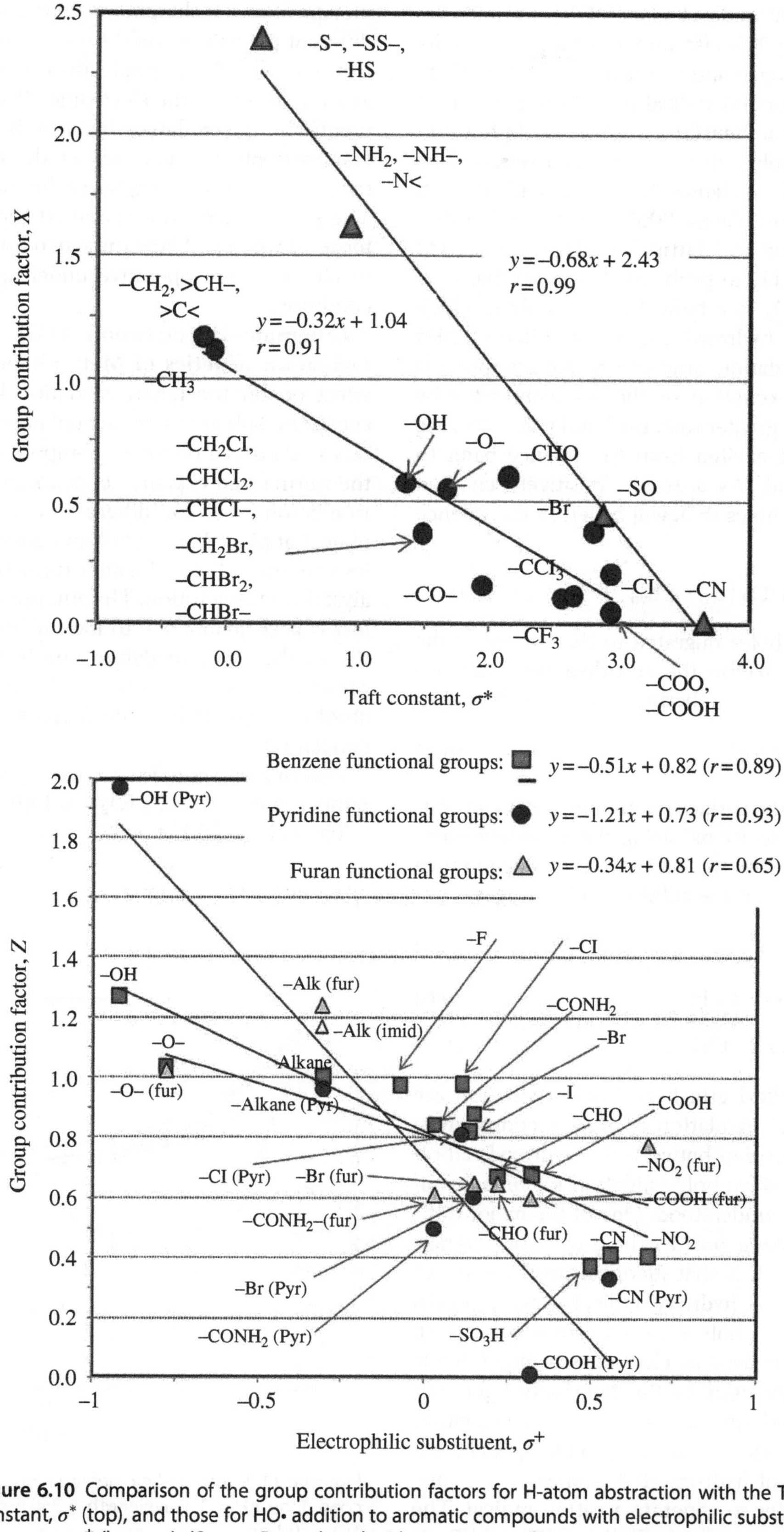

Figure 6.10 Comparison of the group contribution factors for H-atom abstraction with the Taft constant, σ^* (top), and those for HO• addition to aromatic compounds with electrophilic substituent parameter, σ^+ (bottom). (*Source:* Reproduced with permission of American Chemical Society.)

organic in air (Hammouda et al., 2017; Tang and An, 1995). In addition, TiO_2 is also important for water splitting to produce hydrogen and medical cancer phototherapy. Hole versus hydroxyl radical oxidation is a crucial step toward full comprehension and control of photooxidation processes. Hole oxidation has been reported by much experimental evidence (Meekins and Kamat, 2011; Thompson and Yates, 2005; Zhang and Yates, 2010; and Di Valentin and Fittipaldi, 2013). QSAR was applied by Tang (2011) to probe oxidation mechanisms using UV/TiO_2. TiO_2 is a typical photocatalyst, which generates hole and hydroxyl radical at different pHs and reduces the oxidation time of organic compounds significantly. Photoexcitation of the semiconductor by photons with energy greater than the band gap will cause electrons (e^-) to get excited from the valence band to the conduction band. As a result, positively charged "vacancies," positive holes (h^+), will be left in the valence band:

$$TiO_2 + h\nu \rightarrow TiO_2\,(e_{cb}^{\;-} + h_{vb}^{\;+}) \tag{6.3}$$

After the positive holes migrated to the surface of the catalyst, they could oxidize the adsorbed substrates via direct electron transfer if it is thermodynamically favorable:

$$h_{cb}^{\;+} + Org \rightarrow Org^{+} \tag{6.4}$$

At the same time, the surface containing holes can generate hydroxyl radicals by oxidizing the adsorbed water and/or hydroxide ions on the TiO_2 surface, which in turn can oxidize pollutants at the solid–liquid interface:

$$h_{vb}^{\;+} + OH^{-} \rightarrow OH \tag{6.5}$$

$$h_{vb}^{\;+} + H_2O \rightarrow OH + H^{+} \tag{6.6}$$

$$OH + Org \rightarrow Org - OH \tag{6.7}$$

Both hydroxyl radical attack and hole oxidation are operative during TiO_2 oxidation of organic compounds. However, the competition between the hydroxyl radical attack and direct positive hole oxidation for organics in the process is poorly understood. One of the major difficulties in distinguishing positive hole oxidation versus hydroxyl radical attack is that identical products might be expected from either hydroxyl radical attack or initial hydration of a primary hole-oxidized cation radicals. In most cases, oxygen is used as electron scavenger, while methanol is typically used as the hole scavenger. For example, the hydroxylation of phenol can proceed either by direct attack by hydroxyl radical, producing a characteristic distribution of hydroxylated isomers, or by primary electron transfer to generate a cation radical. The effect of halogen substituents including F, Cl, and Br at metaposition on the photocatalytic oxidation kinetics at different pH values could shine some light into this controversy issue. For example, the Hammett analysis is used as a tool to probe the electronic effect of halogen substituents on the oxidation kinetics, because the slopes of Hammett plots usually reflect the electronic nature of the transition state complex of the species being oxidized. The pH effect on the Hammett plots is interpreted in terms of when and how these two fundamental oxidation mechanisms are operative under specific experimental conditions.

Langmuir–Hinshelwood (LH) Modeling of the Oxidation Kinetics of Meta-phenols: The substituent effect on photocatalytic oxidation kinetics of halogen-substituted phenols is presented here using phenol, which was studied as a reference compound. Figure 6.11 shows the normalized organic concentration remaining versus irradiation time for different initial concentrations of phenol at pH 9. Without illumination, the initial recoveries vary from 92 to 74% after the separation of TiO_2 catalyst from the solution. The amount of organic pollutants lost is presumably due to loss by initial adsorption onto the catalyst. The oxidation rate in terms of percentage removal decreases with initial concentration, but the absolute degradation rate increases with higher initial concentrations.

Experimental conditions: pH = 3, wavelength = 350 nm, temperature = 42°C, TiO_2 = 0.1 g/l.

The LH model is used:

$$r_0 = k_{deg}\frac{K_{LH}C_0}{1 + K_{LH}C_0}$$

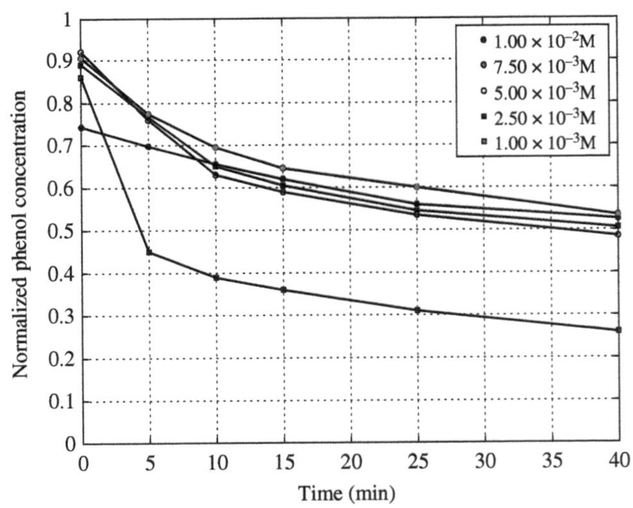

Figure 6.11 Photocatalytic degradation of phenol. Experimental conditions: pH = 3, wavelength = 350 nm, temperature = 42°C, TiO_2 = 0.1 g/l.

where

r_o is the initial oxidation rate $dC/dt|_{t=0}$ (mM/min)
C_o is the initial concentration of the pollutant (mM)
k is the oxidation rate constant (mM/min)
K is the adsorption equilibrium constant onto the catalyst surface (mM^{-1})

Based upon the LH model, the inverse initial oxidation rate ($1/r_0$) is linearly proportional to the inverse initial concentration ($1/C_0$). The y-intercept and the slope of the line give $1/k$ and $1/kK$, respectively.

pH Effect on Photocatalytic Oxidation of Halogen-substituted Meta-phenols: Figures 6.12 and 6.13 illustrate that the LH kinetics is maintained at pH 3 and 9 with the regression coefficient above 0.95 for each correlation line. From these figures, the adsorption constants and the oxidation constants are obtained at each pH level.

Experimental conditions: pH = 3, wavelength = 350 nm, temperature = 42°C, TiO_2 = 0.1 g/l.

Experimental conditions: pH = 9, wavelength = 350 nm, temperature = 42 C, TiO_2 = 0.1 g/l.

pH Effects on the Adsorption Constants: Table 6.16 presents the LH adsorption constants of phenol, m-chlorophenol, m-bromophenol, and m-fluorophenol as affected by pH. The adsorption constants for all the substituents except m-fluorophenol decrease with increasing pH and have the maximum values at pH 3.

Since the isopotential pH for titanium dioxide is about 6.6 and pK_a for substituted phenols is about 8.8, the pH effect on the adsorption typically results from the electrical repulsion between the negatively charged TiO_2 surface and the phenolate anions as pH increases. The increasing pH will increase not only the negative charge of TiO_2 surface but also the hydroxyl ion concentration competing with the substrates for adsorption sites. As a result, the competitive adsorption between hydroxide ions and phenolate ions may result in minimum oxidation rates at pH near 10, assuming the generation of hydroxyl radicals by adsorbed hydroxyl ions as the rate-limiting step.

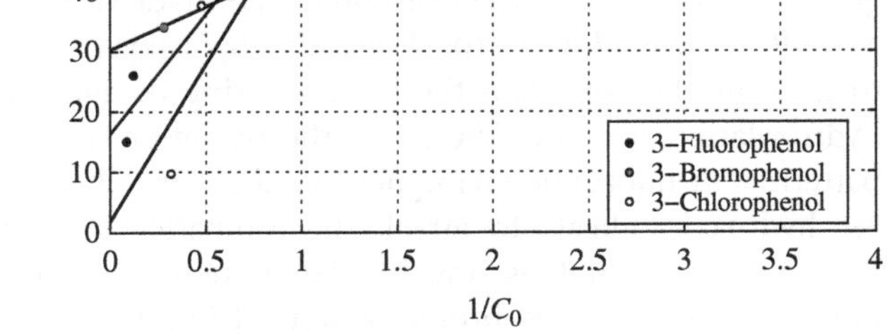

Figure 6.12 Reciprocal initial rate versus reciprocal initial concentration for halogen-substituted phenols (pH = 3). Experimental conditions: pH = 3, wavelength = 350 nm, temperature = 42°C, TiO_2 = 0.1 g/l.

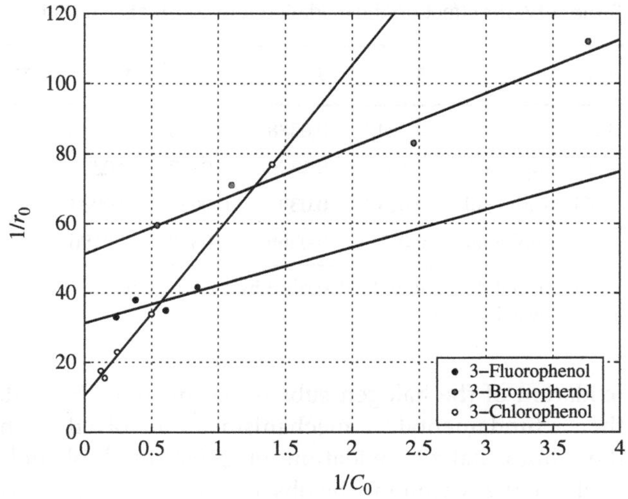

Figure 6.13 Reciprocal initial rate versus reciprocal initial concentration for halogen-substituted phenols (pH = 9). Experimental conditions: pH = 9, wavelength = 350 nm, temperature = 42°C, TiO_2 = 0.1 g/l.

pH Effects on the Oxidation Constants: The pH effect on the oxidation constants of the halogen-substituted phenols is presented in Table 6.17. When Table 6.17 is compared with Table 6.16, pH has the exactly opposite effect on the oxidation and adsorption constants. In other words, whenever K values reach a maximum for a compound, the k values at that pH level reach a minimum, and vice versa. This suggests that adsorption may not be the determining factor in terms of the

Table 6.16 pH effect on the adsorption constants K (mM^{-1}).

pH	3	5	7	9	11
Phenol	2.997	1.242	0.485	0.850	[a]
3-Fluorophenol	0.0172	0.084	2.430	2.890	[a]
3-Chlorophenol	2.557	0.510	0.229	0.263	[a]
3-Bromophenol	2.1089	0.346	0.296	3.300	[a]

[a] Experimental data cannot be described by the Langmuir–Hinshelwood model.

Table 6.17 pH effect on the oxidation rate constants *k* (mM/min).

pH	3	5	7	9	11
Phenol	0.0247	0.0388	0.0385	0.0685	[a]
3-Fluorophenol	0.1097	0.1396	0.0237	0.032	[a]
3-Chlorophenol	0.0247	0.033	0.062	0.0984	[a]
3-Bromophenol	0.033	0.0369	0.0858	0.0196	[a]

[a] Experimental data cannot be described by the Langmuir–Hinshelwood model,

oxidation of the halogen-substituted phenols. Instead, the fundamental mechanisms involved in the photocatalytic oxidation of substituted phenols could be the reason for the observed decoupling trends of the pH effect on oxidation and adsorption constants.

These results could be understood in terms of the thermodynamic requirements in photocatalytic oxidation of halogen-substituted phenols. The thermodynamic requirement for direct hole oxidation is that the redox potential of the electron donor, e.g. oxidizable species, should be lower than the valence band gap. For an electron acceptor, e.g. reducible species, the redox potential is greater than that of the conduction band gap. These reactions compete with electron–hole recombination, and their efficiency depends mainly on how fast the charge carriers can reach the surface. The thermodynamic feasibility of these redox reactions can be assessed by considering that the typical redox potentials for many compounds found in an electrolyte like water indeed lie within the band potential of TiO_2. If the solution pH decreases, the band gap between valence band and the conduction band could change from 2.2 up to 3.3 eV for anatase TiO_2. Therefore, direct oxidation of organic substrates by positive hole will be favored. On the other hand, more hydroxide ion in basic solutions will become the predominant species on catalyst's surface, and more hydroxyl radicals will be generated. This would favor the hydroxyl radical oxidation of organic compounds.

Different concentrations of potential-determining adsorbable ions can change the potential drop across the semiconductor/electrolyte interface, thus altering the flat band positions. H^+ and OH^- ions are potential-determining ions for metal oxides; thus the electrode potential of the semiconductor changes accord with the following equation:

$$E_f = E_f\,(\mathrm{pH}=0) - 0.059\ \mathrm{V/pH} \tag{6.8}$$

Therefore, electrons in the conduction band are better reducing agents under basic conditions, while positive holes in the valence band are stronger oxidizers at low pH. Since the conduction band of anatase TiO_2 lies at about −3 V (vs. NHE) at pH 0 and roughly −2.2 V at pH 14, the oxidation potentials for water and hydroxide ions remain thermodynamically favorable within all the pH range. The generation of $OH^{\cdot}$ is expected through the following reactions when the potential of a positive hole in a valence band of TiO_2 is greater than the following values at pH 7:

$$h_{vb}{}^{+} + OH^{-} \rightarrow OH \tag{6.9}$$

$$h^{+} + H_2O \rightarrow OH + H^{+} \quad E_0 = 2.8\ \mathrm{V\,(NHE)} \tag{6.10}$$

$$h^{+} + OH^{-} \rightarrow OH \quad E_0 = 2.0\ \mathrm{V\,(NHE)} \tag{6.11}$$

OH^- is easier to be oxidized by hydroxyl radical than water thermodynamically. However, in acidic pH range, the amount of hydroxide ions is several times less than in the basic solution. As a result, the formation of hydroxyl radicals may be kinetically slow due to significantly less hydroxide ions available in the acidic solution. When all the above factors are taken into consideration, it could be reasonable that positive holes are mostly consumed through the direct oxidation of organic compounds when pH is less than 5. Therefore, the inverse relationship between the adsorption constants and the oxidation rate constants suggests that the reactions at high pH are more dependent upon hydroxyl radical attack, because hydroxyl radicals are generated at the catalyst surface and then diffuse into the bulk solution within a few nanometers to subsequently react with the substrates. This would obviate the need for strong substrate adsorption at higher pH. The faster reaction rates at high pH suggest that reaction via hydroxyl radical attack is faster than via adsorption and subsequent direct electron transfer to positive holes.

The oxidation rate constants of *m*-fluorophenol significantly decreased at pH 9 compared with *m*-chlorophenol and *m*-bromophenol as shown in Table 6.17. Since the fluorine atom is the strongest electron-withdrawing group, the fluorine substituent will greatly delocalize the π electrons on the aromatic ring. As a result, there is a lower probability for a hydroxyl radical to gain an electron from the aromatic ring during the hydroxylation process. Therefore, the oxidation kinetic pattern of *m*-fluorophenol implies that the energy barrier for hydroxyl radicals to attack the aromatic rings of the *m*-fluorophenol is much higher than that for *m*-chlorophenol and *m*-bromophenol. At lower pH, the TiO_2 surface is less hydroxylated, so hydroxyl production may be limited. This would make the reaction rely more on the oxidation induced by positive holes, which are also better oxidizers at low pH since the band gap potential is higher. The overall results are the increasing oxidation constants of *m*-fluorophenol at pH of 3 and 5.

Hammett Correlations for Halo-substituted Meta-phenols: Based on the apparent relationship of the LH kinetic forms, Hammett correlation analysis may further explore the different reaction mechanisms; Table 6.18 lists all the Hammett constants of the related organic compounds.

Table 6.18 Hammett constants for the substituents.

Substituents	σ_m	σ_m^+	F	R
F	0.35	0.34	0.43	−0.34
Cl	0.4	0.37	0.41	−0.15
Br	0.41	0.37	0.44	−0.17

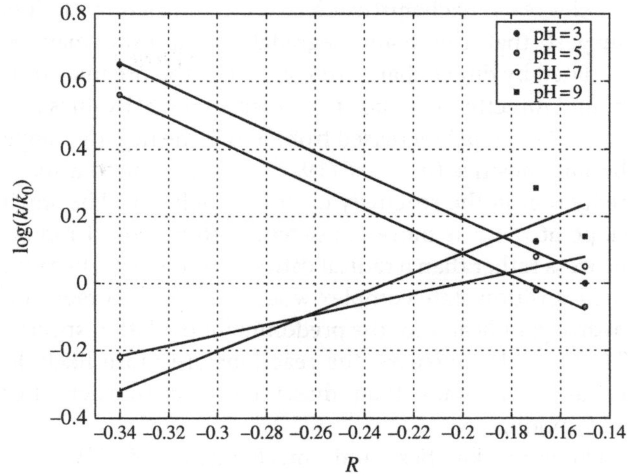

Figure 6.14 Hammett plot using Swain–Lupton aromatic substituent constant (*R*) for halogen-substituted meta-phenols. Experimental conditions: wavelength = 350 nm, temperature = 42°C, TiO_2 = 0.1 g/l.

Electronic Resonance Effect on the Oxidation Rates: At first, $\log(k/k_H)$ was plotted against σ_m, but no linear conformity was observed (r^2 = 0.5952). Since σ_m is the Hammett constant that reflects the electronic inductive effect and excludes the resonance effect, the nonlinearity suggests that the electronic inductive effect is minor. Construction of a plot with σ_m^+ constants yielded a good correlation (r^2 = 0.947). Therefore, this implies that there is indeed a direct conjugation of the substituent with the reaction center. Since reactions that involve interaction between a substituent and a developing positive charge in the transition state are well correlated with σ^+, the correlation suggests that the transition state complex formed during the oxidation very likely has a positive charge. The electronic requirements of photocatalytic oxidation of the metasubstituted phenols can be satisfied by the conjugative interaction between the halogen substituents and the π electrons on the aromatic ring.

The resonance effect has been further verified by even better correlations between $\log(k/k_H)$ and Swain–Lupton *R* parameters (r^2 = 0.9966), as presented in Figure 6.14. The rather successful correlation with *R* values implies that there is extensive resonance interaction of the substituents with the π electrons on the aromatic ring. As a result, the resonance effect will stabilize the radical cation by charge delocalization after adsorption and electron transfer to a photo-generated hole at the TiO_2 surface. The sign and magnitude of the slope ρ at pH 3 are −0.4623, which is consistent with the radical cation formation in the rate-determining step. The negative sign indicates strong electron demand at the reaction site. The data obtained show better linearity (r = 0.97) with σ^+ than with σ values (r = 0.83), indicating that direct electron transfer to the photo-generated hole gives rise to a radical cation. The magnitude of the ρ value (−0.56, r^2 = 0.97) was assumed to reflect a near diffusion-controlled electron transfer to a positive hole, a process that must be preceded by adsorption of the substrate onto the active TiO_2.

pH Effects on Hammett Correlations: Figure 6.14 gives further insight into the variations exhibited in the Hammett plots due to increasing pH. By inspection of Figure 6.14, the slopes of Hammett plots at pH 3 and 5 are negative and parallel. This implies that the same reaction process and pathway may be operative at both pH levels, respectively. The negative slopes of the Hammett plots at pH 3 and 5 imply that the transition state complex formed between the substrate and the oxidation species at this pH range is positively charged at the reaction center.

Experimental conditions: Wavelength = 350 nm, temperature = 42°C, TiO_2 = 0.1 g/l.

Several factors favor the formation of a positive-charged transition complex in acidic condition. First, all the halogen-substituted phenols have higher adsorption density at low pH than at high pH. This favors the direct electron transfer reaction pathway. Second, a positive hole has higher valence band potential in an acidic solution than in a basic solution. Third, a positive hole is a full positive charge while hydroxyl radicals are neutral species. As a mechanistic inference, these Hammett plots suggest that the main degradation pathway may be initiated by direct charge transfer from the organic compound directly to the positive holes at low pH values at 3 and 5. When pH increased from 5 to 7, the negative slope became positive (ρ = +0.6719, r^2 = 0.883) due to a sharp reduction in the reactivity of *m*-fluorophenol. The small slope at pH 7 is more consistent with a neutral radical process rather than a radical cation intermediate, because the transition state complex would be a neutral radical if hydroxyl radical were the predominant oxidation species. Thus, as pH increases, the reactions are more likely to be an $^{\cdot}OH$ attack than direct electron transfer after adsorption.

Different kinetics and mechanisms of UV/TiO_2 suggests that reaction time could change significantly under different reaction conditions.

6.10 Environmental Issues on Different Temporal Scales

Temporal scale could span from product life cycle, human lifetime, to civilization. Within production use at its disposal, life cycle assessment (LCA) is a major tool in quantifying its environmental impacts using software such as LCA2.1 by the US EPA (Bare, 2011). Figure 6.15 illustrates different scopes of environmental concerns with respect to different time horizons such as product, human lifetime, or civilization. Indeed, sustainable development aims to protect the capital of future generations so that mankind can thrive on planet Earth and is in harmony with nature for generations to come.

6.10.1 Correlation Between Temporal and Spatial Scales in the Sustainable Design of WTPs and WWTPs

In designing EEIS, there is correlation between spatial and temporal scales. For example, at the molecular level of nanometer, chemical reactions take place within nanoseconds in terms of bond breaking or forming. For particles of less than one millimeter such as colloidal particles, several seconds might be required. A typical example is reaction mechanism of Fenton's reagent. In 1895, Henry Fenton (Fenton and Jones 1895) reported the color change during hydrogen peroxide oxidation of tartaric acids catalyzed by ferrous

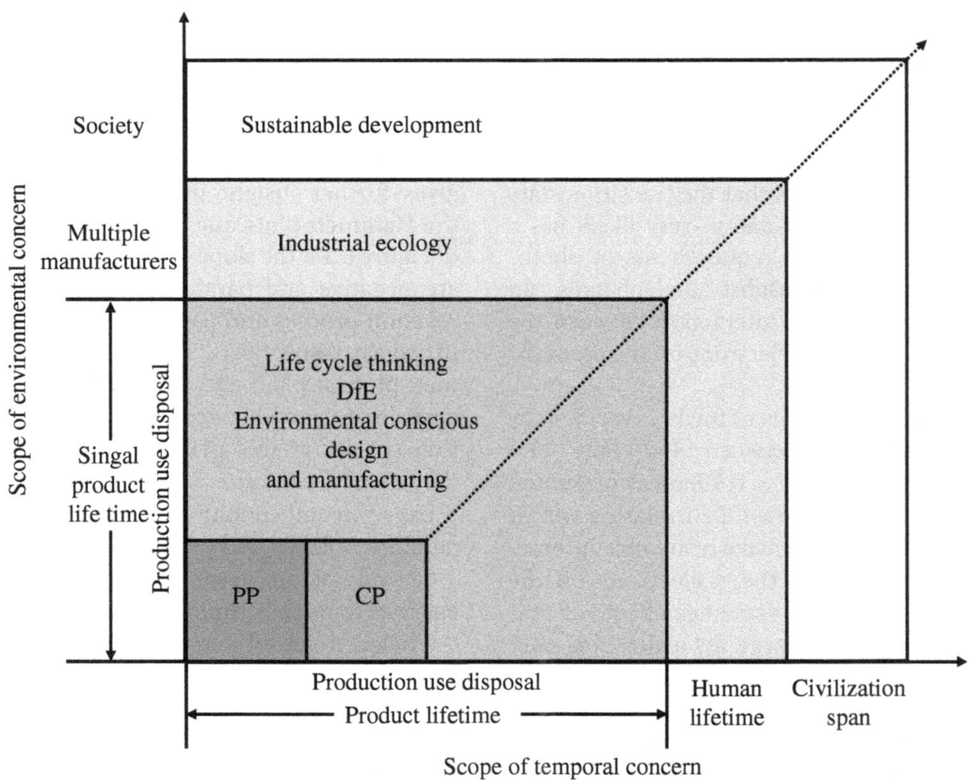

Figure 6.15 Scopes of environmental versus temporal concerns.

ion. Oxygen was thought as the oxidant because the kinetic studies were carried out according to the weight of oxygen releasing from the system. Fifty years later, Haber and Weiss (1934) proposed hydroxyl radical as the dominant oxidative species which initiated the vigorous radical chemistry. While Bray and Gorin (1932) proposed ferrate Fe(IV) as active intermediate in Fenton oxidation. Only after Kremer (2003) confirmed that the formation rate of hydroxyl radical versus ferryl ion is pH dependence. Yamamoto et al. (2012) confirmed that the ferryl–oxo species is the primary oxidizing agent in reactions of Fenton's reagent because the ferryl–oxo species is more energetically favorable than that yielding a free hydroxyl radical, by 24.4 kcal/mol. For these reasons, chemical reaction mechanisms should be fully understood at different time scales so that the reactions could be accurately scaled up to unit process at several meters scale such as a reactor. Since hydraulic retention time is usually measured in hours, and EEIS requires reliable design on its corresponding temporal scales (Figure 6.16).

Figure 6.17 shows that the correlation of spatial and temporal scales could be further divided into 10^{-8} to 10^{-16} m on a spatial scale and 10^{-8} to 10^{-16} s on a temporal scale at the molecular/electronic level for a reaction mechanism, which is adopted from Charpentier (2016). How to scale up kinetics to unit process and full plant poses true challenges for the next generation of SEE designers.

Figure 6.18 illustrates different levels of active research being conducted. Quantum mechanics requests electron, photon, and magnetic effects on innovative technologies, WTPs and WWTPs. For example, vacuum ultraviolet disinfection and oxidation, magnetic destabilization of colloidal particles in water or wastewater, and photocatalytic oxidation of organic pollutants all belong to this group. On a mesoscale, nanotechnologies such as nanocarbon tubes, nano-TiO_2, and oxidized graphene for water treatment have also shown as promising technologies.

Figure 6.19 presents different computation tools for the multiscale modeling of the sustainable design of WTPs and WWTPs. Process system modeling (PSM) can be carried out by specific software such as activated sludge model 3 (ASM3). Computational fluid dynamics (CFD) includes finite volume (FV), finite element (FE), and the lattice Boltzmann (LB) approach. CFD is used extensively in simulating reactor, process, and plants for water and wastewater treatment. Computational chemistry (CCH) includes mesoscale models (MM), microfinite element (microFE), molecular dynamics (MD), and quantum chemistry (QCH).

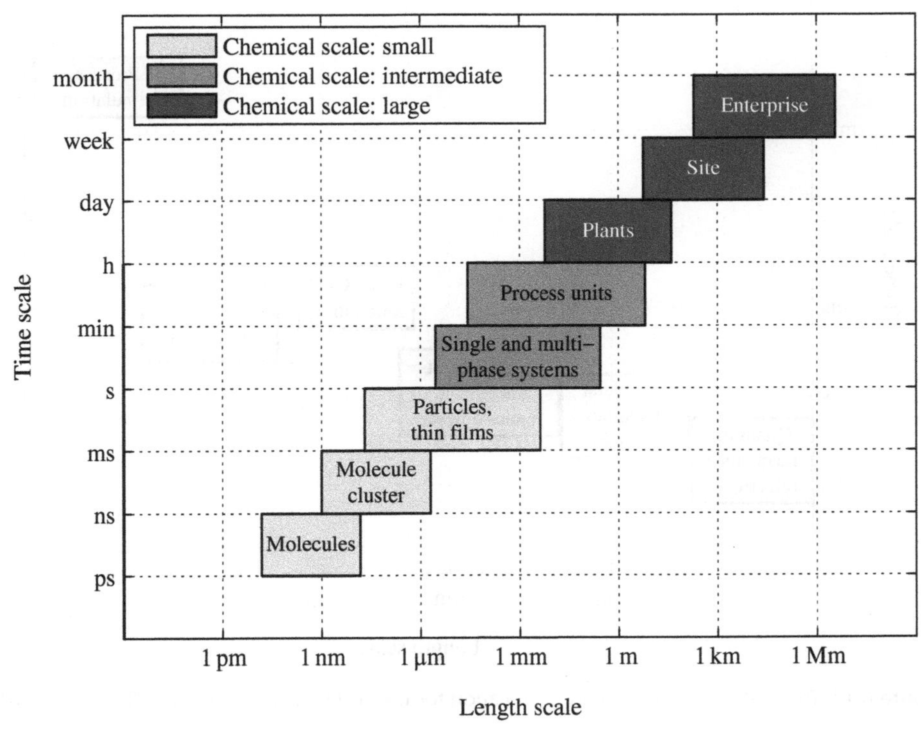

Figure 6.16 Time and length multiscales for the sustainable design of WTPs and WWTPs.

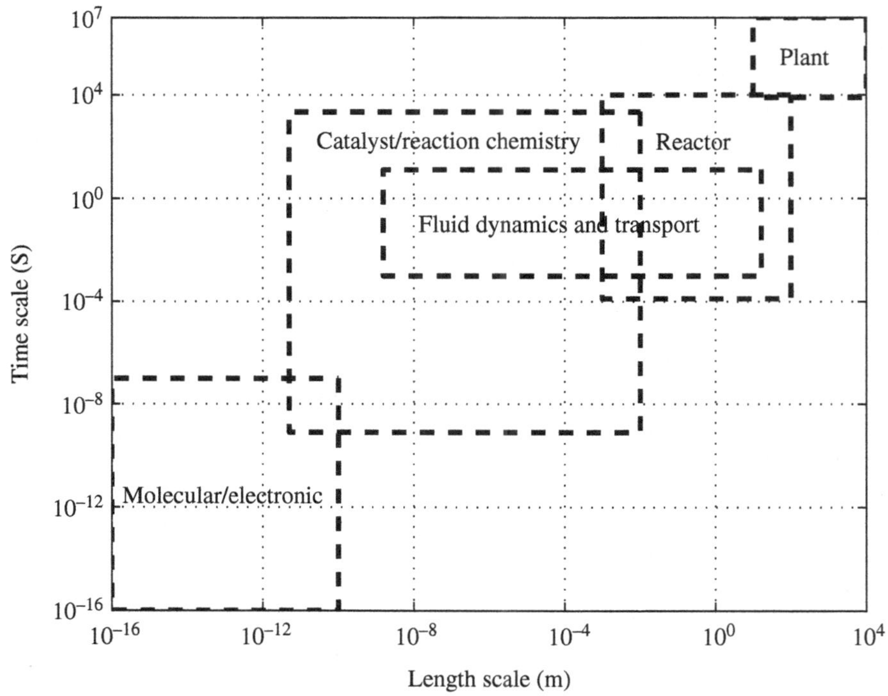

Figure 6.17 Organizing the complexity levels on the time and length scales by the integrated multiscale approach for sustainable design of WTPs and WWTPs.

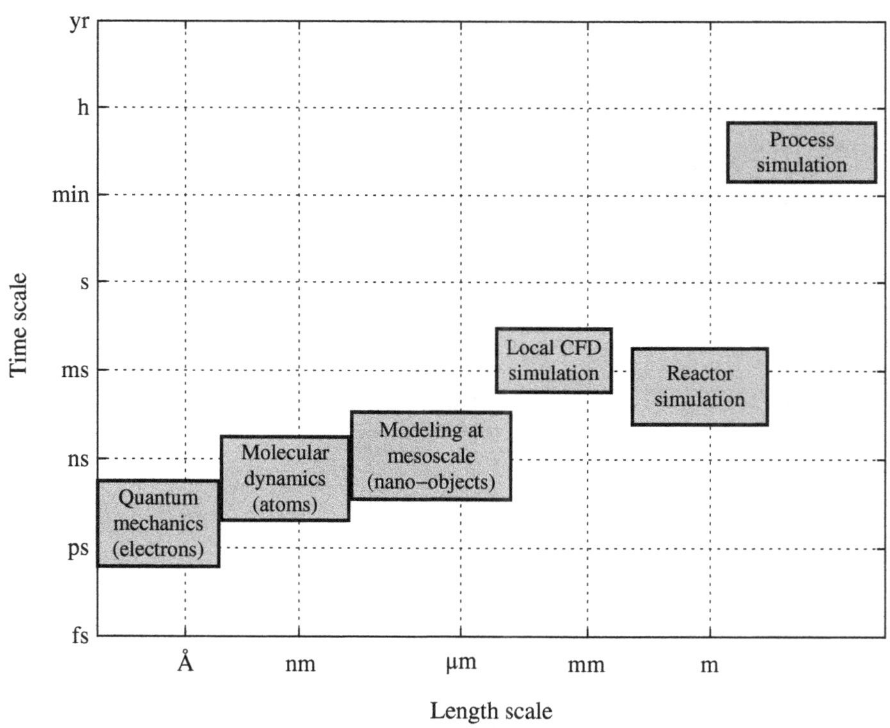

Figure 6.18 The scales of modeling and simulation for the sustainable design of WTPs and WWTPs.

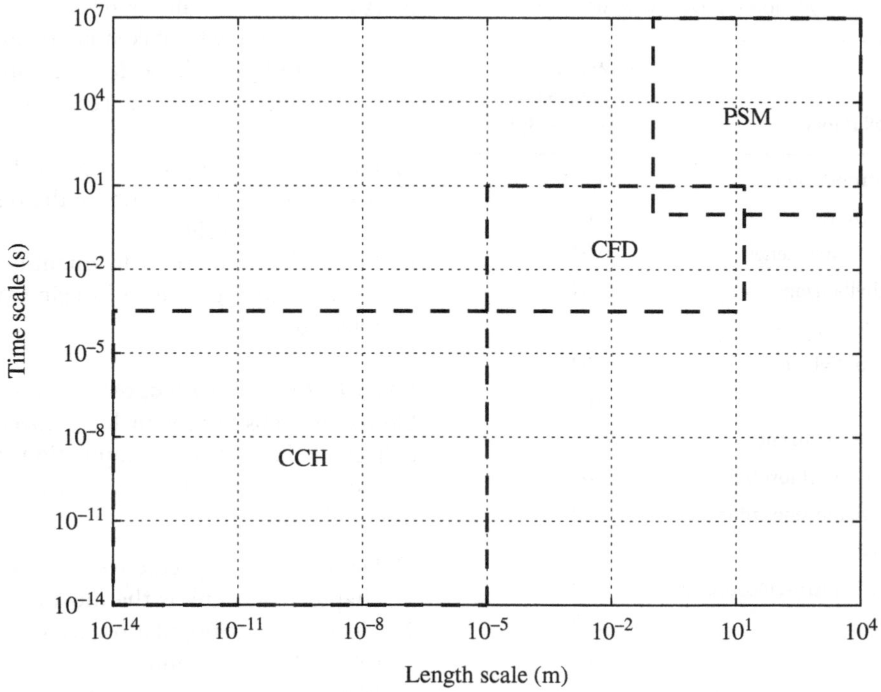

Figure 6.19 Computational tools in multiscale modeling for sustainable design of WTPs and WWTPs.

6.11 Exercise

6.11.1 Questions

1) How is resiliency of EEIS defined and quantified?
2) What is an unsteady system, and what is the characteristic of an equation describing an unsteady system?
3) What is elementary reaction? Please give one example in water or wastewater treatment for bond breaking, bond forming, or electron transfer reaction add, respectively.
4) What is transition state complex?
5) What is activation energy?
6) How does reaction rate increase with temperature?
7) Why are traditional WTPs and WWTPs not efficient?
8) Please search and list the major advantages of a static mixer.

6.11.2 Calculation

1) Please use group theory and estimate the hydroxylation rate constants of phenol, m-chlorophenol, m-bromophenol, and m-fluorophenol by hydroxyl radical in aqueous solution by using the method developed in the paper by Minakata and Crittenden (2011). Please answer the following:
   i) Are these the hydroxylation rate constants of metasubstituted chlorophenols by aqueous hydroxyl radicals agreed with literature data? Why or why not?
   ii) How would you rank the degradation rate of these metasubstituted chlorophenols by aqueous hydroxyl radicals?
   iii) Do these ranking orders also agreed with the experimental degradation kinetic rate k presented in Table 6.17 at pH 3 and 9, respectively? Why or why not?
2) The following pollutants are listed. Please answer the following questions about their reaction kinetic constants with hydroxyl radical according to Table 6.19:
   a) Which group has a high-rate constant reacting with hydroxyl radical? Substituted benzene versus substituted phenol?
   b) For substituted benzene, please rank the order of rate constant reacting with hydroxyl radical.

6.11.3 Project

6.11.3.1 Xiongan Project

Since Xiongan will have a population of 1, 2, and 5 million in 5, 10, and 20 years, respectively, what would be the ratio of peak daily flow/average daily flow of your design for 5, 10, or 20 years? How would these ratios affect the design of WRRFs? Population equivalency determines which technology is the best. According to

Table 6.19 Pollutions and pollution equivalent values.

	Pollutants	Pollution equivalent values (kg)
Substituted benzene	Nitrobenzene	0.2
	Benzene	0.02
	Methylbenzene	0.02
	Ethylbenzene	0.02
	ortho-Xylene	0.02
	para-Xylene	0.02
	meta-Xylene	0.02
	Chlorobenzene	0.02
	ortho-Dichlorobenzene	0.02
	p-Dichlorobenzene	0.02
	p-Nitrochlorobenzene	0.02
	2.4-Dinitro-chlorobenzene	0.02
Substituted phenol	Phenol	0.02
	m-Cresol	0.02
	2.4-Dichlorophenol	0.02
	2.4.6-Trichlorophenol	0.02

the population growth, please develop an integrated urban water management plan according to the following principles:

1) Water resource, WTP, WWTP, and urban stormwater should be designed with consideration of water cycle and human and ecological demands.
2) All the EEIS should be designed for specific purpose not fit for all with stakeholders' input.
3) All the EEIS should be designed, integrated, and interconnected to increase resiliency over 20, 50, and 100 years.
4) Please identify top five major factors that would contribute to the uncertainty of the resiliency of different EEIS, respectively.
5) Please find a method to quantify the uncertainty of the resiliency of the EEIS using the metric that you selected.

6.11.3.2 Community Proposal Project

Go to the website of your local governmental agency to collect the last 10 years' population statistics. Is the city of your hometown in a steady or unsteady state in terms of population?

1) Discuss which predictive models presented in this chapter you think is the most reliable and why.
2) Please predict population growth of the city that you selected in 5, 10, and 20 years.
3) Identify three major stressors on water supply, WRRFs, and drainage systems, and rank the resiliency of each EEIS under these stressors.
4) Propose design strategies to increase resiliency of each EEIS under each stressor.

Contact your local WWTP manager to obtain the daily flow rate of wastewater, design an equalization tank according to the example, and compare the equalization tank in the local WWTP and make recommendations for improvement, if any.

References

Bare, J.C., (2011). TRACI 2.0 – the tool for the reduction and assessment of chemical and other environmental impacts. *Clean Technologies and Environmental Policy* 13 (5): 687–696.

Bray, W.C. and Gorin, M.H. (1932). Ferryl ion, a compound of tetravalent iron. *Journal of American Chemical Society* 54: 2124–2125.

Charpentier, J.C., (2016). What kind of modern "green" chemical engineering is required for the design of the "Factory of Future"? *Procedia Engineering* 138: 445–458.

Cuppens, A., Smets, I., and Wyseure, G. (2012). Definition of realistic disturbances as a crucial step during the assessment of resilience of natural wastewater treatment systems. *Water Science and Technology* 65: 1506–1513.

Currie, J., Wragg, N., Roberts, C. et al. (2014). Transforming "value engineering" from an art form into a science – process resilience modelling. *Water Practice and Technology* 9: 104–114.

Davis, M.L. (2010). *Water and Wastewater Engineering: Design Principles and Practice*, 1e. New York: McGraw-Hill Education.

Di Valentin, C. and Fittipaldi, D. (2013). Hole scavenging by organic adsorbates on the TiO2 surface: a DFT model study. *Journal of Physical Chemistry Letters* 4 (11): 1901–1906.

Fenton, H.J. and Jones, H.O. (1895). The oxidation of organic acids in the presence of ferrous iron, Part I. *Proceeding of Chemical Society* 899–910.

Francis, R. and Bekera, B. (2014). A metric and frameworks for resilience analysis of engineered and infrastructure systems. *Reliability Engineering and System Safety* 121: 90–103.

Goldbloom-Helzner, D., Opie, J., Pickard, B., and Mikko, M. (2015). Flood resilience: a basic guide for water and wastewater utilities. *WEFTEC2015 proceedings*, Chicago (26–30 September), pp. 2029–2032.

Haber, F. and Weiss, J. (1934). The catalytic decomposition of hydrogen peroxide by iron salts. *Proceedings of the Royal Society of London, A* 147: 332–351.

Hammouda, S.B., Zhao, F.P., Safaei, Z. et al. (2017). Degradation and mineralization of phenol in aqueous medium by heterogeneous monopersulfate activation on nanostructured cobalt based-perovskite catalysts ACoO3 (A = La, Ba, Sr and Ce): Characterization, kinetics and mechanism study. *Applied Catalysis B: Environmental* 215: 60–73.

Holling, C.S. (1973). Resilience and stability of ecological systems. *Annual Review of Ecology, Evolution, and Systematics* 4: 1–23.

Holling, C.S. (1996). Engineering resilience versus ecological resilience. Engineering within Ecological Constraints, pp. 31e43. https://www.nap.edu/catalog/4919/engineering-within-ecological-constraints (accessed 1 March 2017).

Hwang, H., Forrester, A., and Lansey, K. (2014). Decentralized water reuse: regional water supply system resilience benefits. *Procedia Engineering* 70: 853–856.

Juan-García, P., Butler, D., Comas J.C., Darch, G. et al. (2017). Resilience theory incorporated into urban wastewater systems management: state of the art. *Water Research* 115: 149–161.

Kremer, M.L. (2003). The Fenton reaction. Dependence of the rate on pH. *Journal of the Physical Chemistry A* 107: 1734–1741.

Lempert, R.J., Groves, D.G., Popper, S.W. et al. (2015). A general, analytic strategies for generating and narrative scenarios. *Method Robust* 52: 514–528.

Li, K. and Crittenden, J.C (2009). Computerized pathway elucidation for hydroxyl radical-induced chain reaction mechanisms in aqueous phase advanced oxidation processes. *Environmental Science and Technology* 43 (8): 2831–2837.

Mabrouk, N., Mathias, J.D., and Deffuant, G. (2010). Computing the resilience of a wastewater treatment bioreactor. Proceedings of the 5th International Multi-conference Computing in the Global Information Technology ICCGI 20e25 (September 2010), pp. 185–188.

Meekins, B. and Kamat P. (2011). Role of water oxidation catalyst IrO2 in shuttling photogenerated holes across TiO2 interface. *Journal of Physical Chemistry Letters* 2: 2304–2310.

Minakata, D. and Crittenden, J. (2011). Linear free energy relationships between aqueous phase hydroxyl radical reaction rate constants and free energy of activation. *Environmental Science and Technology* 45: 3479–3486.

Minakata, D., Li, K., Westerhoff, P., and Crittenden, J. (2009). Development of a group contribution method to predict aqueous phase hydroxyl radical (HO•) reaction rate constants. *Environmental Science and Technology* 43: 6220–6227.

Minakata, D., Song, W.H., and Crittenden, J. (2011). Reactivity of aqueous phase hydroxyl radical with halogenated carboxylate anions: experimental and theoretical studies. *Environmental Science and Technology* 45: 6057–6065.

Mugume, S., Gomez, D., and Butler, D. (2014). Quantifying the resilience of urban drainage systems using a hydraulic performance assessment approach. 13th International Conference on Urban Drainage, Sarawak, Malaysia (7–12 September).

Radcliffe, J.C. (2015). Water recycling in Australia – during and after the drought. *Environmental Science: Water Research and Technology* 1: 554–562.

Schoen, M., Hawkins, T., Xue, X. et al. (2015). Technologic resilience assessment of coastal community water and wastewater service options. *Sustainability of Water Quality and Ecology* 6: 1–13.

Sweetapple, C., Fu, G., and Butler, D. (2016). Reliable, robust, and resilient system design framework with application to wastewater-treatment plant control. *American Society of Civil Engineering* 1–10. 10.1061/(ASCE)EE.1943-7870.0001171, 04016086.

Tang, W.Z. (2011). Photocatalytic oxidation of halogen substituted meta-phenols by UV/TiO_2 in acidic and basic aqueous solutions. *Finish Conference of Environmental Science*, Turku, Finland (5–6 May 2011).

Tang, W.Z. and An, H. (1995). UV/TiO2 photocatalytic oxidation of commercial dyes in aqueous solutions. *Chemosphere* 31 (9): 4157–4170.

Thompson, T. and Yates, J. (2005). Monitoring hole trapping in photo-excited TiO2 using a surface photoreaction. *Journal of Physical Chemistry B* 109: 18230–18236.

Walker, B., Holling, C.S., Carpenter, S.R., and Kinzig, A. (2004). Resilience, adaptability and transformability in social–ecological systems. *Ecology and Society* 9 (2): 5.

Weirich, S.R., Silverstein, J., and Rajagopalan, B. (2015). Resilience of secondary wastewater treatment plants: prior performance is predictive of future process failure and recovery time. *Environmental Engineering Science* 32: 222–231.

Xue, X., Schoen, M.E., Ma, X.C. et al. (2015). Critical insights for a sustainability framework to address integrated community water services: technical metrics and approaches. *Water Research* 77: 155–169.

Yamamoto, N., Koga, N., and Nagaoka, M. (2012). Ferryl –oxo species produced from Fenton's reagent via a two-step pathway: minimum free-energy path analysis. *Journal of Physical Chemistry B* 116: 14178–14182.

Zhang, Z. and Yates, J. (2010). Direct observation of surface-mediated electron-hole pair recombination in TiO_2. *Journal of Physical Chemistry C* 114: 3098–3101.

7

Efficiency of Renewable Materials

All the materials should be utilized to their bes advantages.

OBJECTIVES
To develop master equation for reactor design using mass balance analysis To avoid chemicals to reduce both capital and operation and maintenance costs To compare capital and operation and maintenance costs of UV disinfection vs. chlorination To design the most efficient batch, continuous stirring tank reactor (CSTR), and plug flow reactor (PFR) using experimental kinetic data To size different reactors to achieve a specific percentage of conversion To quantify molecular structure effect on reactor volume required to achieve a specific conversion factor To provide correlation between cost and reactor volume

7.1 Principle 4

Biomasses are considered as renewable materials. Therefore, there is a major driver to replace fossils such as crude oil with renewable biomass as feeding stock in chemical engineering. For the same reason, environmental engineering (EE) should also select biomass as feeding stock for anaerobic digestion whenever it is feasible. For example, in designing a WRRF organic compounds in wastewater could be considered as the feeding stock to produce energy. In designing EEIS, if chemicals or materials are not avoidable, each unit process should be designed with maximal efficiency to reduce footprint on the environments.

Principle 4 Mass efficient alternatives should be regenerative or neutral to avoid negative design in terms of footprints on air, water, and soil.

1) Mass efficiency requires the avoidance of the addition of chemicals.
2) Maximal renewable mass should be incorporated into the final products such as nutrients.
3) Components, units, processes, and systems should be designed with maximal material efficiency.

Mass balance, stoichiometry, and reaction kinetics are the three pillars in effective design of reactors such as batch, continuous stirring tank reactor (CSTR), plug-flow reactor (PFR), and packed bed reactor (PBR). Each reactor should achieve its maximum efficiency at minimal volume to avoid short circulation and dead zone. From the mass transfer efficiency perspective, the mass transfer efficiency of PFR, CSTR, and batch reactors decreases sequentially.

7.2 Stoichiometry

A balanced chemical equation gives stoichiometric information on what reactants are reacting and what products are formed. It also gives quantitative information on how much reacts and how much is formed. To show stoichiometric relationship, mole is used in the chemical reaction equation. A mole is a fixed number of molecules that could be expressed as gram-mole (g-mol) or pound-mole (lb-mol). One g-mol contains 6.022×10^{23} molecules, while one lb-mol contains 2.7×10^{26} molecules.

$$\text{Number of moles} = \frac{\text{mass}}{\text{molecular weight}} \tag{7.1}$$

Sustainable Environmental Engineering, First Edition. Walter Z. Tang and Mika Sillanpää.

The combustion of 1 mol of methane requires 2 mol of oxygen and produces 1 mol of carbon dioxide and 2 mol of water. Therefore, the stoichiometric equation is

$$\underset{1}{CH_4} + \underset{2}{2O_2} \rightarrow \underset{1}{CO_2} + \underset{2}{2H_2O} \tag{7.2}$$

Example 7.1
Given: Combustion of propane in the air.
Find:

1) How many moles of O_2 are required for each mole of propane?
2) How many grams of O_2 are required to burn 100 g of propane?
3) At STP, what volume of air at STP is required to burn 100 g of propane?

Solution:

1)
$$\begin{array}{lccccccc} & C_3H_8 & + & 5O_2 & \rightarrow & 3CO_2 & + & 4H_2O \\ C & 3 & & & & 3 & & \\ H & 8 & & & & & & 8 \\ O & & & 10 & & 6 & & 4 \end{array} \tag{7.3}$$

Each mole of propane requires 5 mol of O_2.

2) Under standard pressure and temperature, 1 mol of gas occupies 22.4 l; therefore,

$$\text{MW propane} = 3(12.01) + 8(1.008) + 44.09\ \text{g/mol}$$

$$\text{Moles propane} = \frac{100\ \text{g}}{44.09\ \text{g/mol}} = 2.27\ \text{mol}$$

$$\text{Moles oxygen required} = 2.27 \times 5 = 11.34\ \text{mol}$$

$$\text{MW oxygen} = 2(16.00) = 32.00\ \text{g/mol}$$

$$\begin{aligned} \text{Mass oxygen} &= \text{moles} \times \text{MW} \\ &= (11.34\,\text{mol})(32.00\,\text{g/mol}) \\ &= 363\,\text{g}\,O_2 \end{aligned}$$

$$(11.34\,\text{mol})(22.4\ \text{l/mol}) = 254\ \text{l}\ O_2$$

3) Since oxygen occupies 21% of air by volume,

$$254\ \text{l}O_2/(0.21\ \text{l}O_2/\text{l air}) = 1210\ \text{l air}$$

Combustion of propane gives energy; therefore, it reflects the intrinsic heat value of the organic compound. In traditional EE, the amount of organic compounds is expressed as the amount of oxygen needed to oxidize them to CO_2. Historically, degradation of organic compounds in wastewater is achieved through activated sludge (AS). To facilitate design, *theoretical oxygen demand* is the amount of O_2 required to completely oxidize a chemical substance to CO_2 and H_2O based on stoichiometry. *Biochemical oxygen demand* (BOD) is the amount of O_2 consumed by a substance in a standard test using microorganisms for oxidation. Chemical oxygen demand (COD) is the amount of O_2 consumed by a substance in a standard test using strong chemical oxidant such as potassium dichromate. As a result, wastewater treatment plants (WWTPs) are designed to provide sufficient oxygen to degrade organic compounds in wastewater. Unfortunately, aeration is extremely energy intensive and consumes about one-half electricity in a typical WWTP. In the last hundred years, the intrinsic energy of wastewater was not considered as energy resource. Even worse, the AS process incorporates water that should have been the product into sludge as wastes. As a result, sludge disposal costs another one-third of electricity needed by the WWTP.

Example 7.2
Given: For BOD test, 1 l of glucose stock solution of a BOD concentration of 10 g/l is to be prepared.
Find:

1) How many grams of glucose is needed?
2) How many liters of this stock solution is needed to prepare 100 ml of 1 mg/l BOD?

Solution:

$$\begin{array}{lccccccc} & C_6H_{12}O_6 & + & 6O_2 & \rightarrow & 6CO_2 & + & 6H_2O \\ C & 6 & & & & 2 & & \\ H & 12 & & & & & & 12 \\ O & 6 & & 6 & & 12 & & 6 \end{array} \tag{7.4}$$

$$\begin{aligned} MW_{\text{Glucose}} &= 6 \times (12.01) + 12 \times (1.008) + 6 \times (16.00) \\ &= 180\ \text{g/mol} \end{aligned}$$

1)
$$\frac{10\ \text{g}O_2}{\text{l}} \times \frac{\text{mole}\,O_2}{\text{g}} \times \frac{1\,\text{mol}\,C_6H_{12}O_6}{6\,\text{mol}\,O_2} \times \frac{180\,\text{g}\,C_6H_{12}O_6}{1\,\text{mol}\,C_6H_{12}O_6} = 9.375\,\text{g}\,C_6H_{12}O_6\ \text{in}\ 1\ \text{l}$$

2) Glucose dilutions

$$\frac{1\,\text{mg BOD}}{\text{l}} \times \frac{\text{l}}{10\,000\,\text{mg}\,O_2} \times 100\,\text{ml} = 10\,\mu\text{l}$$

According to the aforementioned calculations, 10 μl of stock solution diluted into 100 ml provides 1 mg BOD/l.

7.3 Avoid the Addition of Chemicals

7.3.1 Avoid Acid Addition

In WTP or WWTP design, whenever it is possible chemicals should not be added to water or wastewater for many reasons: (i) Addition of chemicals is extremely expensive over the life span of an EEIS. Chemical dosing equipment entails capital cost, while operation and maintenance (O&M) add cost. (ii) Reaction by-products such as DBPs during chlorination could be formed. (iii) Sludge is produced as the waste which has to be separated and disposed. Indeed, sludge disposal could be as high as one third to one half of the operation and maintenance costs. The following example shows the capital and O&M costs at different flow rates:

Example 7.3
Given: Table 7.1 shows the capital cost and O&M cost of acid feed (96% H_2SO_4) at different flow rates.
Find:

1) Calculate unit capital and O&M costs.
2) Plot unit cost/unit cost$_0$ vs. Q/Q_0.
3) Determine correlation equations between unit cost/unit cost$_0$ for capital and O&M vs. Q/Q_0, respectively (Tables 7.2 and 7.3).

Solution:

1) When $Q = 1$ MGD

$$\text{Unit capital cost} = 23\,868.26\$ \div 1\,\text{MGD}$$
$$\times\, 1\,\text{million gal}/1\,000\,\text{kgal} \div 1\,\text{day} = 23.87\$/\text{kgal}$$
$$\text{Unit O\&M costs} = 4988.82\$/\text{year} \div 1\,\text{MGD}$$
$$\times\, 1\,\text{million gal}/1000\,\text{kgal} \div 365\,\text{day/year}$$
$$\text{Unit O\&M costs} = 0.0137\$/\text{kgal}$$

Table 7.1 Capital cost and operation and maintenance cost of acid feed (96% H_2SO_4) at different flow rates.

Q (MGD)	2006 capital cost ($)	2006 operation and maintenance costs ($/year)
1	23 868.26	4 988.82
2	24 714.14	6 466.51
4	26 090.77	9 421.88
6	27 267.50	12 377.26
8	28 330.61	15 332.63
10	29 316.48	18 287.99
20	33 577.31	33 064.77
40	40 505.34	62 618.07
60	46 422.95	92 171.04
80	51 766.51	121 723.68
100	56 719.83	151 275.98

When $Q = 1$ MGD

$$Q_0 = 1\,\text{MGD}$$
$$\text{Unit capital cost}_0 = 23.87\$/\text{kgal}$$
$$\text{Unit O\&M costs}_0 = 0.0137\$/\text{kgal}$$
$$\frac{Q}{Q_0} = 1$$
$$\frac{\text{Unit capital cost}}{\text{Unit capital cost}_0} = 1$$
$$\frac{\text{Unit O\&M costs}}{\text{Unit O\&M costs}_0} = 1$$

Table 7.2 Unit capital cost and unit operation and maintenance cost of acid feed (96% H_2SO_4) at different flow rates.

Q (MGD)	Unit capital cost ($/kgal)	Unit O&M costs ($/kgal)
1	23.87	0.0137
2	12.36	0.0089
4	6.52	0.0065
6	4.54	0.0057
8	3.54	0.0053
10	2.93	0.0050
20	1.68	0.0045
40	1.01	0.0043
60	0.77	0.0042
80	0.65	0.0042
100	0.57	0.0041

Table 7.3 Unit cost ratio of acid feed (96% H_2SO_4) at different flow ratios.

Q/Q_0	Unit capital cost/unit capital cost$_0$	Unit O&M costs/unit O&M costs$_0$
1	1.0000	1.0000
2	0.5177	0.6481
4	0.2733	0.4722
6	0.1904	0.4135
8	0.1484	0.3842
10	0.1228	0.3666
20	0.0703	0.3314
40	0.0424	0.3138
60	0.0324	0.3079
80	0.0271	0.3050
100	0.0238	0.3032

When $Q = 20$ MGD

$$\frac{Q}{Q_0} = 20$$

$$\frac{\text{Unit capital cost}}{\text{Unit capital cost}_0} = 0.0703*$$

$$\frac{\text{Unit O\&M costs}}{\text{Unit O\&M costs}_0} = 0.3314$$

2) Plot unit cost/unit $cost_0$ vs. Q/Q_0 for capital and OM, respectively (Figure 7.1).
3)
   a) Equation for unit cost/unit $cost_0$ vs. Q/Q_0 for capital:
      The regression equation of the trend line:

$$\log(\text{unit cost/unitcost}_0) = -0.8057\log(Q/Q_0) - 0.0645 \quad r^2 = 0.9951$$

   or the equation of the regression:

$$\text{unit cost/unit cost}_0 = 0.8620(Q/Q_0)^{-0.8057} \quad r^2 = 0.9951$$

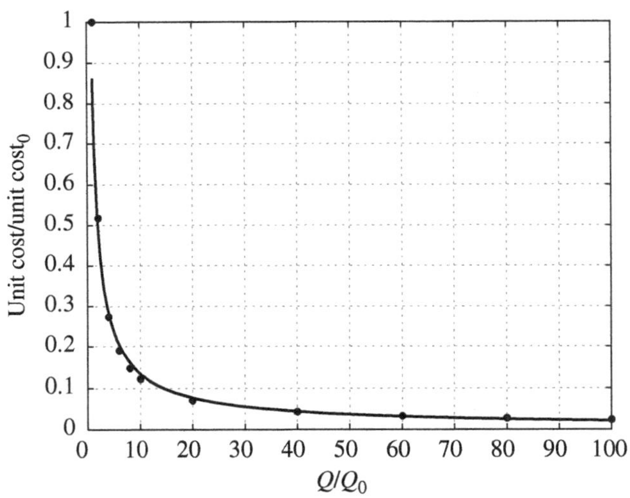

Figure 7.1 Unit capital cost/unit capital $cost_0$ vs. Q/Q_0 for acid feed (96% H_2SO_4), unit capital $cost_0$ = 23.87$/kgal, Q_0 = 1 MGD.

Matlab code:

```
clear all
close all
%data%
Q=[1];
Qrange=[1 2 4 6 8 10 20 40 60 80 100];
totalcostrange=[23868.26 24714.14
  26090.77 27267.50 28330.61 29316.48
  33577.31 40505.34 46422.95 51766.51
  56719.83];
totalcost=totalcostrange(1);
%function%
unitcost=totalcost./Q./1000;
unitcostrange=totalcostrange./
  Qrange./1000;
p=polyfit(log10(Qrange/Q), log10
  (unitcostrange /unitcost), 1)
log10ratiounitcostfit=polyval
  (p, log10(Qrange/Q));
ratiounitcostfit=10.
  ^log10ratiounitcostfit;
k=10^p(:,2)
R2=1 - sum((log10(unitcostrange
  /unitcost) - log10ratiounitcostfit).
  ^2)/sum((log10(unitcostrange
  /unitcost) - mean(log10
  (unitcostrange /unitcost))).^2)
xmax=max(Qrange/Q);
xmin=min(Qrange/Q);
xaxis=[xmin:0.1:xmax];
log10ratiounitcostfitplot=polyval(p,
  log10(xaxis));
ratiounitcostfitplot=10.
  ^log10ratiounitcostfitplot;
%plot%
figure(1)
  plot(Qrange/Q,unitcostrange
  /unitcost,'ko','MarkerSize',3,
  'MarkerFaceColor',[0 0 0],
  'MarkerEdgeColor',[0 0 0],
  'linewidth',1);
hold on
plot(xaxis,ratiounitcostfitplot,
  'k-','linewidth',1);
  set(gca,'xtickmode','auto',
  'FontSize',10,'FontWeight','bold',
  'FontName', 'Times New Roman');
  xlabel({'Q/Q_0'},'FontSize',14,
  'FontWeight','bold','FontName',
  'Times New Roman')
  ylabel({'Unit Cost/Unit Cost_0'},
  'FontSize',14,'FontWeight','bold',
  'FontName', 'Times New Roman')
  grid on
  hold on
```

Answer: H_2SO_4Capitalcostnormal

$p = -0.8057 - 0.0645$; $k = 0.8620$; $R^2 = 0.9951$

The aforementioned figure shows that the unit capital cost at different flow rates decreases exponentially from $23.87/kgal at Q_0 = 1 MGD for acid feed (96% H_2SO_4) to 50 MGD. After this flow rate, the unit cost ratio slightly decreases from 0.3 to 0.2

despite the fact that the total cost will continue to increase by \$56 719 at 100 MGD (Figure 7.2).

b) Equation for unit cost/unit $cost_0$ vs. Q/Q_0 for O&M:

The regression equation of the trend line ($1 \le Q/Q_0 \le 10$):

$$\log(\text{unit cost/unitcost}_0) = -0.4332 \log(Q/Q_0) - 0.0327 \quad r^2 = 0.9723$$

or the equation of the regression:

$$\text{unit cost/unitcost}_0 = 0.9274(Q/Q_0)^{-0.4332} \quad r^2 = 0.9723$$

The regression equation of the trend line ($10 \le Q/Q_0 \le 100$):

$$\text{unit cost/unitcost}_0 = -0.0006Q/Q_0 + 0.3516 \quad r^2 = 0.7013$$

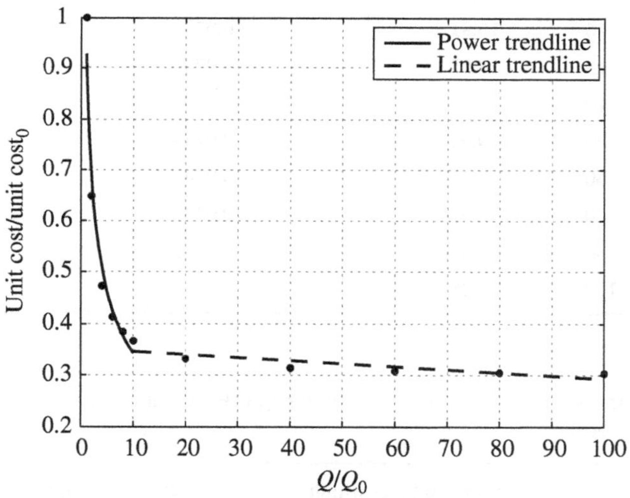

Figure 7.2 Unit O&M cost/unit O&M $cost_0$ vs. Q/Q_0 for acid feed (96% H_2SO_4), unit O&M $cost_0$ = 0.0137\$/kgal, Q_0 = 1 MGD.

Matlab code:

```
clear all
close all
%data%
Qrange=[1 2 4 6 8 10 20 40 60 80 100];
Q=Qrange(1);
totalcostrange=[4988.82 6466.51
  9421.88 12377.26 15332.63 18287.99
  33064.77 62618.07 92171.04 121723.68
  151275.98];
totalcost=totalcostrange(1);
xmid=6;
%function%
unitcost=totalcost./Q./1000./365;
unitcostrange=totalcostrange./
  Qrange./1000./365;
x=Qrange/Q;
y=unitcostrange /unitcost;
x1=x(1:xmid);
y1=y(1:xmid);
x2=x(xmid:11);
y2=y(xmid:11);
%power
p=polyfit(log10(x1), log10(y1), 1)
log10ratiounitcostfit=polyval
  (p, log10(x1));
ratiounitcostfit=10.
  ^log10ratiounitcostfit;
k=10^p(:,2)
R2power=1 - sum((log10(y1) -
  log10ratiounitcostfit).^2)/sum
  ((log10(y1) - mean(log10(y1))).^2)
xmax=max(x1);
xmin=min(x1);
xaxis=[xmin:0.1:xmax];
log10ratiounitcostfitplot=polyval(p,
  log10(xaxis));
ratiounitcostfitplot=10.
  ^log10ratiounitcostfitplot;
%line
p2=polyfit((x2),(y2), 1)
z=polyval(p2, (x2));
R2linear=1 - sum(((y2) - z).^2)/
  sum(((y2) - mean((y2))).^2)
xmax2=max(x2);
xmin2=min(x2);
xaxis2=[xmin2:0.1:xmax2];
z2=polyval(p2, xaxis2);
%plot%
figure(1)
plot(Qrange/Q,unitcostrange
/unitcost,'ko','MarkerSize',3,
  'MarkerFaceColor',[0 0 0],
  'MarkerEdgeColor',[0 0 0],
  'linewidth',1);
hold on
line1=plot(xaxis,
  ratiounitcostfitplot,'k-',
  'linewidth',1);
hold on
line2=plot(xaxis2,z2,'k--',
  'linewidth',1);
set(gca,'xtickmode','auto',
  'FontSize',10,'FontWeight',
```

```
    'bold','FontName', 'Times
    New Roman');
legend([line1,line2],'Power
    trendline','Linear trendline',
    'Location','Northeast');
xlabel({'Q/Q_0'},'FontSize',14,
    'FontWeight','bold','FontName',
    'Times New Roman')
ylabel({'Unit Cost/Unit Cost_0'},
    'FontSize',14,'FontWeight','bold',
    'FontName', 'Times New Roman')
grid on
hold on
```

Answer: H2SO4OMcostnormaltwoline

$$p = -0.4332\ \ -0.0327;\ k = 0.9274;\ R^2\text{power} = 0.9723$$

$$p^2 = -0.0006\ \ 0.3516;\ R^2\text{linear} = 0.7013$$

The O&M cost clearly shows that, when flow rate is $1 \le Q/Q_0 \le 10$, the Unit O&M Cost/Unit O&M Cost$_0 = 0.9274(Q/Q_0)^{-0.4332}$ with $r^2 = 0.9723$. When flow rate is $10 \le Q/Q_0 \le 100$, the Unit O&M Cost/Unit O&M Cost$_0 = -0.0006Q/Q_0 + 0.3516$, which shows a linear decrease trend.

Comment: This is true for the addition of HCl. If the acid concentration is diluted from 96% H_2SO_4 to 37% HCl, the volume of the feeding tank will increase. As a result, the unit capital and O&M costs will increase as shown in the following example.

Example 7.4
Given: Table 7.4 shows the capital cost and O&M cost of acid feed (37% HCl) at different flow rates.

Table 7.4 Capital cost and operation and maintenance cost of acid feed (37% HCl) at different flow rates.

Q (MGD)	2006 capital cost ($)	2006 operation and maintenance costs ($/year)
1	43 212.16	30 501.52
2	56 211.88	57 491.08
4	77 368.33	111 467.76
6	95 452.67	165 441.17
8	111 790.97	219 411.34
10	126 942.03	273 378.29
20	192 423.93	543 165.23
40	298 896.17	1 082 506.81
60	389 839.84	1 621 552.01
80	471 961.32	2 160 316.17
100	548 085.60	2 698 813.80

Find:

1) Calculate unit capital and O&M costs.
2) Plot unit cost/unit cost$_0$ vs. Q/Q_0.
3) Determine correlation equations between unit cost/unit cost$_0$ for capital and O&M vs. Q/Q_0, respectively (Tables 7.5 and 7.6).

Solution:

1) When $Q = 1$ MGD

$$\text{Unit capital cost} = 43212.16\$ \div 1\,\text{MGD} \times 1\,\text{million gal}/1000\,\text{kgal} \div 1\,\text{day} = 43.21\$/\text{kgal}$$

Table 7.5 Unit capital cost and unit operation and maintenance cost of acid feed (37% HCl) at different flow rates.

Q (MGD)	Unit capital cost ($/kgal)	Unit O&M costs ($/kgal)
1	43.21	0.0836
2	28.11	0.0788
4	19.34	0.0763
6	15.91	0.0755
8	13.97	0.0751
10	12.69	0.0749
20	9.62	0.0744
40	7.47	0.0741
60	6.50	0.0740
80	5.90	0.0740
100	5.48	0.0739

Table 7.6 Unit cost ratio of acid feed (37% HCl) at different flow ratios.

Q/Q_0	Unit capital cost/unit capital cost$_0$	Unit O&M costs/unit O&M costs$_0$
1	1.0000	1.0000
2	0.6504	0.9424
4	0.4476	0.9136
6	0.3682	0.9040
8	0.3234	0.8992
10	0.2938	0.8963
20	0.2227	0.8904
40	0.1729	0.8873
60	0.1504	0.8860
80	0.1365	0.8853
100	0.1268	0.8848

$$\text{Unit O\&M costs} = 30501.52\$/\text{year} \div 1\,\text{MGD} \times 1\,\text{million gal}/1000\,\text{kgal} \div 365\,\text{day/year}$$

$$\text{Unit O\&M costs} = 0.0836\$/\text{kgal}$$

When $Q = 1$ MGD

$$Q_0 = 1\ \text{MGD}$$

$$\text{Unit capital cost}_0 = 43.21\$/\text{kgal}$$

$$\text{Unit O\&M costs}_0 = 0.0836\$/\text{kgal}$$

$$\frac{Q}{Q_0} = 1$$

$$\frac{\text{Unit capital cost}}{\text{Unit capital cost}_0} = 1$$

$$\frac{\text{Unit O\&M costs}}{\text{Unit O\&M costs}_0} = 1$$

When $Q = 20$ MGD

$$\frac{Q}{Q_0} = 20$$

$$\frac{\text{Unit capital cost}}{\text{Unit capital cost}_0} = 0.2227$$

$$\frac{\text{Unit O\&M costs}}{\text{Unit O\&M costs}_0} = 0.8904$$

2) Plot unit cost/unit cost$_0$ vs. Q/Q_0 for capital and OM, respectively (Figure 7.3):

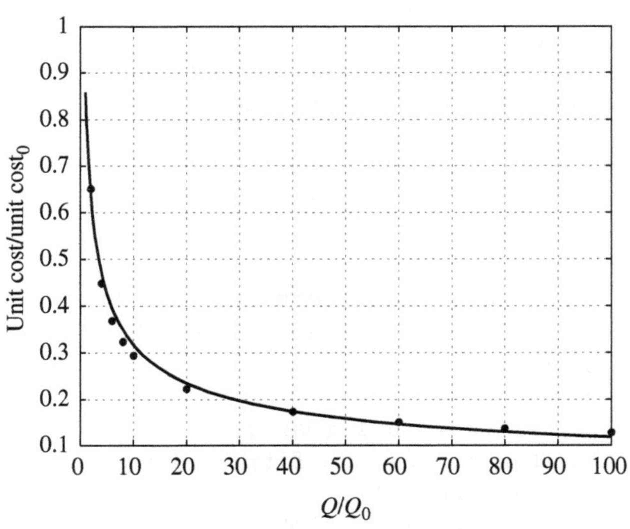

Figure 7.3 Unit capital cost/unit capital cost$_0$ vs. Q/Q_0 for acid feed (37% HCl), unit capital cost$_0$ = 43.21$/kgal, Q_0 = 1 MGD.

3)

a) Equation for unit cost/unit cost$_0$ vs. Q/Q_0 for capital:

The regression equation of the trend line:

$$\log(\text{Unit Cost/Unit Cost}_0) = -0.4317\log(Q/Q_0) - 0.0671 \qquad r^2 = 0.9873$$

or the equation of the regression:

$$\text{Unit Cost/Unit Cost}_0 = 0.8569(Q/Q_0)^{-0.4317} \qquad r^2 = 0.9873$$

Matlab code:

```
clear all
close all
%data%
Q=[1];
Qrange=[1 2 4 6 8 10 20 40 60 80 100];
totalcostrange=[43212.16 56211.88
  77368.33 95452.67 111790.97 126942.03
  192423.93 298896.17 389839.84
  471961.32 548085.60];
totalcost=totalcostrange(1);
%function%
unitcost=totalcost./Q./1000;
unitcostrange=totalcostrange./
  Qrange./1000;
p=polyfit(log10(Qrange/Q), log10
  (unitcostrange /unitcost), 1)
log10ratiounitcostfit=polyval
  (p, log10(Qrange/Q));
ratiounitcostfit=10.
  ^log10ratiounitcostfit;
k=10^p(:,2)
R2=1 - sum((log10(unitcostrange
  /unitcost) - log10ratiounitcostfit).
  ^2)/sum((log10(unitcostrange
  /unitcost) - mean(log10
  (unitcostrange /unitcost))).^2)
xmax=max(Qrange/Q);
xmin=min(Qrange/Q);
xaxis=[xmin:0.1:xmax];
log10ratiounitcostfitplot=polyval(p,
  log10(xaxis));
ratiounitcostfitplot=10.
  ^log10ratiounitcostfitplot;
%plot%
figure(1)
plot(Qrange/Q,unitcostrange
  /unitcost,'ko','MarkerSize',3,
  'MarkerFaceColor',[0 0 0],
  'MarkerEdgeColor',[0 0 0],
  'linewidth',1);
```

```
hold on
plot(xaxis,ratiounitcostfitplot,'k-',
   'linewidth',1);
set(gca,'xtickmode','auto',
   'FontSize',10,'FontWeight',
   'bold','FontName',
   'Times New Roman');
xlabel({'Q/Q_0'},'FontSize',14,
   'FontWeight','bold','FontName',
   'Times New Roman')
ylabel({'Unit Cost/Unit Cost_0'},
   'FontSize',14,'FontWeight',
   'bold','FontName', 'Times New Roman')
grid on
hold on
```

Answer: HClCapitalcostnormal

$p = -0.4317 \quad -0.0671; \; k = 0.8569; \; R^2 = 0.9873$

The unit capital cost at different flow rates decreases exponentially. For unit capital cost_0 is \$43.21/kgal at $Q_0 = 1$ MGD as follows:

$$\text{Unit Cost/Unit Cost}_0 = 0.8569(Q/Q_0)^{-0.4317} \qquad r^2 = 0.9873$$

The capital cost at the reference flow rate of 1 MGD would double to \$43.21/kgal of HCl from \$23.87/kgal of acid feed (96% H_2SO_4) (Figure 7.4).

b) Equation for unit cost/unit cost_0 vs. Q/Q_0 for O&M:

The regression equation of the trend line ($1 \le Q/Q_0 \le 10$):

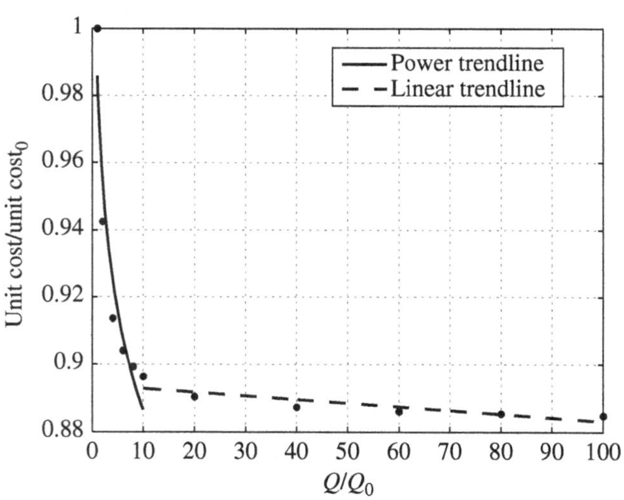

Figure 7.4 Unit O&M cost/unit O&M cost_0 vs. Q/Q_0 for acid feed (37% HCl), unit O&M $\text{cost}_0 = 0.0836\$/\text{kgal}$, $Q_0 = 1$ MGD.

$$\log(\text{Unit Cost/Unit Cost}_0) = -0.0460\log(Q/Q_0) - 0.0063 \qquad r^2 = 0.9244$$

or the equation of the regression:

$$\text{Unit Cost/Unit Cost}_0 = 0.9857(Q/Q_0)^{-0.0460} \qquad r^2 = 0.9244$$

The regression equation of the trend line ($10 \le Q/Q_0 \le 100$):

$$\text{Unit Cost/Unit Cost}_0 = -0.0001Q/Q_0 + 0.8939 \qquad r^2 = 0.7478$$

Matlab code:

```
clear all
close all
%data%
Qrange=[1 2 4 6 8 10 20 40 60 80 100];
Q=Qrange(1);
totalcostrange=[30501.52 57491.08
   111467.76 165441.17 219411.34
   273378.29 543165.23 1082506.81
   1621552.01 2160316.17 2698813.80];
totalcost=totalcostrange(1);
xmid=6;
%function%
unitcost=totalcost./Q./1000./365;
unitcostrange=totalcostrange./
   Qrange./1000./365;
x=Qrange/Q;
y=unitcostrange /unitcost;
x1=x(1:xmid);
y1=y(1:xmid);
x2=x(xmid:11);
y2=y(xmid:11);
%power
p=polyfit(log10(x1), log10(y1), 1)
log10ratiounitcostfit=polyval
   (p, log10(x1));
ratiounitcostfit=10.
   ^log10ratiounitcostfit;
k=10^p(:,2)
R2power=1 - sum((log10(y1) -
   log10ratiounitcostfit).^2)/sum
   ((log10(y1) - mean(log10(y1))).^2)
xmax=max(x1);
xmin=min(x1);
xaxis=[xmin:0.1:xmax];
log10ratiounitcostfitplot=polyval(p,
   log10(xaxis));
ratiounitcostfitplot=10.
   ^log10ratiounitcostfitplot;
%line
```

```
p2=polyfit((x2),(y2), 1)
z=polyval(p2, (x2));
R2linear=1 - sum(((y2) - z).^2)/sum
   (((y2) - mean((y2))).^2)
xmax2=max(x2);
xmin2=min(x2);
xaxis2=[xmin2:0.1:xmax2];
z2=polyval(p2, xaxis2);
%plot%
figure(1)
plot(Qrange/Q,unitcostrange
   /unitcost,'ko','MarkerSize',3,
   'MarkerFaceColor',[0 0 0],
   'MarkerEdgeColor',[0 0 0],
   'linewidth',1);
hold on
line1=plot(xaxis,
   ratiounitcostfitplot,'k-',
   'linewidth',1);
hold on
line2=plot(xaxis2,z2,'k--',
   'linewidth',1);
set(gca,'xtickmode','auto',
   'FontSize',10,'FontWeight','bold',
   'FontName', 'Times New Roman');
legend([line1,line2],'Power
   trendline','Linear trendline',
   'Location','Northeast');
xlabel({'Q/Q_0'},'FontSize',14,
   'FontWeight','bold',
   'FontName', 'Times New Roman')
ylabel({'Unit Cost/Unit Cost_0'},
   'FontSize',14,'FontWeight','bold',
   'FontName', 'Times New Roman')
grid on
hold on
```

Answer: HClOMcostnormaltwoline

$p = -0.0460 \quad -0.0063;\ k = 0.9857;$

$R^2\text{power} = 0.9244$

$p^2 = -0.0001 \quad 0.8939;\ R^2\text{linear} = 0.7478$

The O&M cost clearly shows that when flow rate is $1 \leq Q/Q_0 \leq 10$, the Unit O&M Cost/Unit O&M $\text{Cost}_0 = 0.9857(Q/Q_0)^{-0.0460}$ with $r^2 = 0.9244$. When flow rate is $10 \leq Q/Q_0 \leq 100$, the Unit O&M Cost/Unit O&M $\text{Cost}_0 = -0.0001Q/Q_0 + 0.8939$, which shows a linear decrease trend.

7.3.2 Replacing Chlorination with UV Disinfection

Traditionally, chlorination is used in disinfection for economic reasons. For example, Table 7.7 shows that the annual capital and O&M costs of chlorination at 1 MGD are \$39 831.77 and \$16 551.66, respectively, while the annual capital and O&M costs of UV disinfection at 1 MGD are \$21 617.18 and \$4 014.69, respectively. Clearly, the O&M cost of UV disinfection is significantly lower than chlorination. At 100 MGD, however, the annual capital and O&M costs of chlorination are \$371 543.92 and \$77 792.03, while the annual capital and O&M costs of UV disinfection are \$2 009 248.43 and \$410 206.17, respectively. More importantly, Figure 7.5 shows that the unit cost ratio to baseline flow rate at 1 MGD slightly increased. Apparently, this trend is contradictory to the economy of scale. The major reason for

Table 7.7 Capital cost and operation and maintenance cost of disinfection with chlorine at different flow rates.

Q (MGD)	2006 capital cost (\$)	2006 operation and maintenance costs (\$/year)
1	39 831.77	16 551.66
2	46 929.77	17 402.64
4	58 975.19	18 990.26
6	69 608.12	20 496.97
8	79 416.46	21 953.66
10	88 653.99	23 374.36
20	129 781.85	30 141.88
40	199 576.47	42 789.50
60	261 186.67	54 808.99
80	318 019.05	66 440.17
100	371 543.92	77 792.03

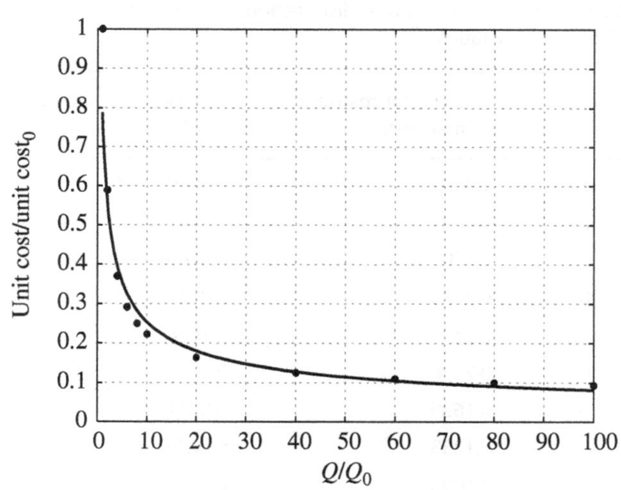

Figure 7.5 Unit capital cost/unit capital cost_0 vs. Q/Q_0 for disinfection with chlorine, unit capital $\text{cost}_0 = 39.83\$/\text{kgal}$, $Q_0 = 1$ MGD.

this is that the probability of the failure lamp and maintenance is much higher at large flow rate than that at a smaller flow rate.

Example 7.5

Given: Table 7.7 shows the capital cost and O&M cost of disinfection with chlorine at different flow rates.

Find:

1) Calculate unit capital and O&M costs.
2) Plot unit cost/unit cost$_0$ vs. Q/Q_0.
3) Determine correlation equations between unit cost/unit cost$_0$ for capital and O&M vs. Q/Q_0, respectively (Tables 7.8 and 7.9).

Table 7.8 Unit capital cost and unit operation and maintenance cost of disinfection with chlorine at different flow rates.

Q (MGD)	Unit capital cost ($/kgal)	Unit O&M costs ($/kgal)
1	39.83	0.0453
2	23.46	0.0238
4	14.74	0.0130
6	11.60	0.0094
8	9.93	0.0075
10	8.87	0.0064
20	6.49	0.0041
40	4.99	0.0029
60	4.35	0.0025
80	3.98	0.0023
100	3.72	0.0021

Table 7.9 Unit cost ratio of disinfection with chlorine at different flow ratios.

Q/Q_0	Unit capital cost/unit capital cost$_0$	Unit O&M costs/unit O&M costs$_0$
1	1.0000	1.0000
2	0.5891	0.5257
4	0.3702	0.2868
6	0.2913	0.2064
8	0.2492	0.1658
10	0.2226	0.1412
20	0.1629	0.0911
40	0.1253	0.0646
60	0.1093	0.0552
80	0.0998	0.0502
100	0.0933	0.0470

Solution:

1) When $Q = 1$ MGD

$$\text{Unit capital cost} = 39831.77\$ \div 1\,\text{MGD} \times 1\,\text{million gal}/1\,000\,\text{kgal} \div 1\,\text{day} = 39.83\$/\text{kgal}$$

$$\text{Unit O\&M costs} = 16551.66\$/\text{year} \div 1\,\text{MGD} \times 1\,\text{million gal}/1\,000\,\text{kgal} \div 365\,\text{day/year}$$

$$\text{Unit O\&M costs} = 0.0453\$/\text{kgal}$$

When $Q = 1$ MGD

$$Q_0 = 1\,\text{MGD}$$

$$\text{Unit capital cost}_0 = 39.83\$/\text{kgal}$$

$$\text{Unit O\&M costs}_0 = 0.0453\$/\text{kgal}$$

$$\frac{Q}{Q_0} = 1$$

$$\frac{\text{Unit capital cost}}{\text{Unit capital cost}_0} = 1$$

$$\frac{\text{Unit O\&M costs}}{\text{Unit O\&M costs}_0} = 1$$

When $Q = 20$ MGD

$$\frac{Q}{Q_0} = 20$$

$$\frac{\text{Unit capital cost}}{\text{Unit capital cost}_0} = 0.1629$$

$$\frac{\text{Unit O\&M costs}}{\text{Unit O\&M costs}_0} = 0.0911$$

2) Plot unit cost/unit cost$_0$ versus Q/Q_0 for capital and OM, respectively (Figure 7.6):

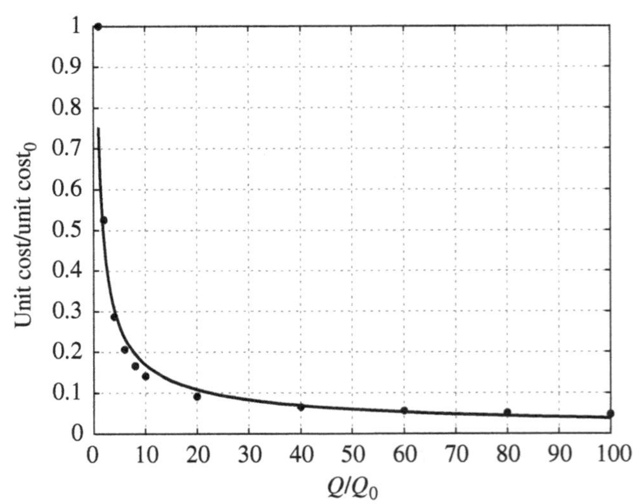

Figure 7.6 Unit O&M cost/unit O&M cost$_0$ vs. Q/Q_0 for disinfection with chlorine, unit O&M cost$_0$ = 0.0453$/kgal, Q_0 = 1 MGD.

a) Equation for unit cost/unit $cost_0$ vs. Q/Q_0 for capital (Figure 7.7):
The regression equation of the trend line:

$$\log(\text{Unit Cost/Unit Cost}_0) = -0.4926\log(Q/Q_0) - 0.1038 \quad r^2 = 0.9746$$

or the equation of the regression:

$$\text{Unit Cost/Unit Cost}_0 = 0.7874(Q/Q_0)^{-0.4926} \quad r^2 = 0.9746$$

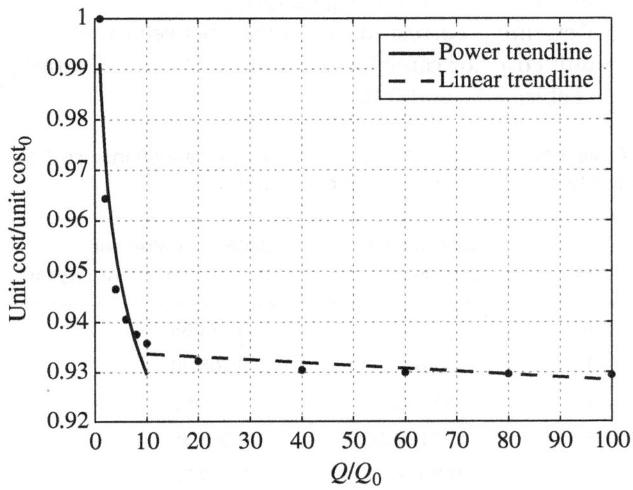

Figure 7.7 Unit capital cost/unit capital $cost_0$ vs. Q/Q_0 for disinfection with ultraviolet, unit capital $cost_0$ = 21.6172$/kgal, Q_0 = 1 MGD.

Matlab code:

```
clear all
close all
%data%
Q=[1];
Qrange=[1 2 4 6 8 10 20 40 60 80 100];
totalcost=[39831.77];
totalcostrange=[39831.
  77 46929.77419 58975.19491
  69608.11675 79416.46422 88653.98862
  129781.85 199576.47 261186.67
  318019.05 371543.92];
%function%
unitcost=totalcost./Q./1000;
unitcostrange=totalcostrange./
  Qrange./1000;
p=polyfit(log10(Qrange/Q), log10
  (unitcostrange /unitcost), 1)
log10ratiounitcostfit=polyval
  (p, log10(Qrange/Q));
ratiounitcostfit=10.
  ^log10ratiounitcostfit;
k=10^p(:,2)
R2=1 - sum((log10(unitcostrange
  /unitcost) - log10ratiounitcostfit).
  ^2)/sum((log10(unitcostrange
  /unitcost) - mean(log10
  (unitcostrange /unitcost))).^2)
xmax=max(Qrange/Q);
xmin=min(Qrange/Q);
xaxis=[xmin:0.1:xmax];
log10ratiounitcostfitplot=polyval(p,
  log10(xaxis));
ratiounitcostfitplot=10.
  ^log10ratiounitcostfitplot;
%plot%
figure(1)
plot(Qrange/Q,unitcostrange
  /unitcost,'ko','MarkerSize',3,
  'MarkerFaceColor',[0 0 0],
  'MarkerEdgeColor',[0 0 0],
  'linewidth',1);
hold on
plot(xaxis,ratiounitcostfitplot,'k-',
  'linewidth',1);
set(gca,'xtickmode','auto',
  'FontSize',10,'FontWeight',
  'bold','FontName',
  'Times New Roman');
xlabel({'Q/Q_0'},'FontSize',14,
  'FontWeight','bold','FontName',
  'Times New Roman')
ylabel({'Unit Cost/Unit Cost_0'},
  'FontSize',14,'FontWeight','bold',
  'FontName', 'Times New Roman')
grid on
hold on
```

Answer: Cl2ConstructionCostnormal

$p = -0.4926 \quad -0.1038;\ k = 0.7874;\ R^2 = 0.9746;$

b) Equation for unit cost/unit $cost_0$ vs. Q/Q_0 for OM:
The regression of the trend line:

$$\log(\text{Unit Cost/Unit Cost}_0) = -0.6482\log(Q/Q_0) - 0.1241 \quad r^2 = 0.9737$$

or the regression equation:

$$\text{Unit Cost/Unit Cost}_0 = 0.7514(Q/Q_0)^{-0.6482} \quad r^2 = 0.9737$$

Matlab code:

```
clear all
close all
%data%
```

```
Q=[1];
Qrange=[1 2 4 6 8 10 20 40 60 80 100];
totalcost=[16551.66];
totalcostrange=[16551.66 17402.63537
  18990.26237 20496.97108 21953.65622
  23374.35647 30141.88 42789.50
  54808.99 66440.17 77792.03];
%function%
unitcost=totalcost./Q./1000./365;
unitcostrange=totalcostrange./
  Qrange./1000./365;
p=polyfit(log10(Qrange/Q), log10
  (unitcostrange /unitcost), 1)
log10ratiounitcostfit=polyval
  (p, log10(Qrange/Q));
ratiounitcostfit=10.
  ^log10ratiounitcostfit;
k=10^p(:,2)
R2=1 - sum((log10(unitcostrange
  /unitcost) - log10ratiounitcostfit).
  ^2)/sum((log10(unitcostrange
  /unitcost) - mean(log10
  (unitcostrange /unitcost))).^2)
xmax=max(Qrange/Q);
xmin=min(Qrange/Q);
xaxis=[xmin:0.1:xmax];
log10ratiounitcostfitplot=polyval(p,
  log10(xaxis));
ratiounitcostfitplot=10.
  ^log10ratiounitcostfitplot;
%plot%
figure(1)
plot(Qrange/Q,unitcostrange
  /unitcost,'ko','MarkerSize',3,
  'MarkerFaceColor',
  [0 0 0],'MarkerEdgeColor',
  [0 0 0],'linewidth',1);
hold on
plot(xaxis,ratiounitcostfitplot,
  'k-','linewidth',1);
set(gca,'xtickmode','auto',
  'FontSize',10,'FontWeight',
  'bold','FontName', 'Times
  New Roman');
xlabel({'Q/Q_0'},'FontSize',14,
  'FontWeight','bold','FontName',
  'Times New Roman')
ylabel({'Unit Cost/Unit Cost_0'},
  'FontSize',14,'FontWeight',
  'bold','FontName',
  'Times New Roman')
grid on
hold on
```

Answer: Cl2OMCostnormal

$p = -0.6482 \quad -0.1241;\ k = 0.7514;\ R^2 = 0.9737$

Example 7.6

Given: Table 7.10 shows capital cost and O&M cost of disinfection with ultraviolet at different flow rates.

Find:

1) Calculate unit capital and O&M costs.
2) Plot unit cost/unit cost$_0$ vs. Q/Q_0.
3) Determine correlation equations between unit cost/unit cost$_0$ for capital and O&M vs. Q/Q_0, respectively (Tables 7.11 and 7.12).

Table 7.10 Capital cost and operation and maintenance cost of disinfection with ultraviolet at different flow rates.

Q (MGD)	2006 capital cost ($)	2006 operation and maintenance costs ($/year)
1	21 617.18	4 014.69
2	41 694.27	8 108.90
4	81 848.43	16 375.97
6	122 002.60	24 564.38
8	162 156.77	32 752.78
10	202 310.93	40 941.19
20	403 081.77	81 961.90
40	804 623.43	164 081.96
60	1 206 165.10	246 123.36
80	1 607 706.77	328 164.77
100	2 009 248.43	410 206.17

Table 7.11 Unit capital cost and unit operation and maintenance cost of disinfection with ultraviolet at different flow rates.

Q (MGD)	Unit capital cost ($/kgal)	Unit O&M costs ($/kgal)
1	21.6172	0.010999
2	20.8471	0.011108
4	20.4621	0.011216
6	20.3338	0.011217
8	20.2696	0.011217
10	20.2311	0.011217
20	20.1541	0.011228
40	20.1156	0.011238
60	20.1028	0.011239
80	20.0963	0.011239
100	20.0925	0.011239

Table 7.12 Unit cost ratio of disinfection with ultraviolet at different flow ratios.

Q/Q_0	Unit capital cost/unit capital cost$_0$	Unit O&M costs/unit O&M costs$_0$
1	1.0000	1.0000
2	0.9644	1.0099
4	0.9466	1.0198
6	0.9406	1.0198
8	0.9377	1.0198
10	0.9359	1.0198
20	0.9323	1.0208
40	0.9305	1.0218
60	0.9299	1.0218
80	0.9296	1.0218
100	0.9295	1.0218

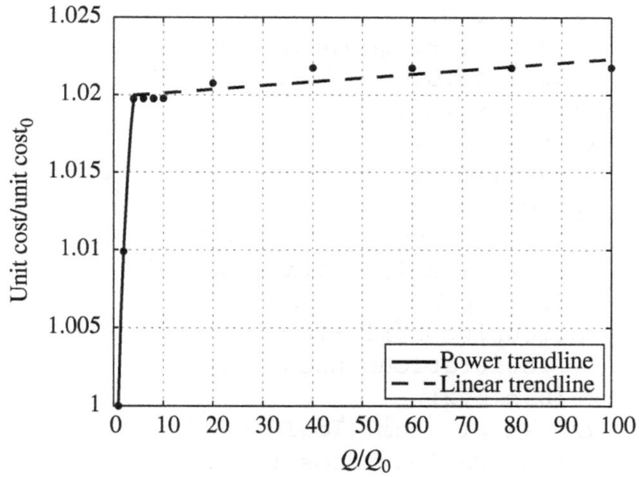

Figure 7.8 Unit O&M cost/unit O&M cost$_0$ vs. Q/Q_0 for disinfection with ultraviolet, unit O&M cost$_0$ = 0.010999$/kgal, Q_0 = 1 MGD.

Solution:

1) When $Q = 1$ MGD

$$\text{Unit capital cost} = 21617.18\$ \div 1\,\text{MGD} \times 1\,\text{million gal}/1000\,\text{kgal} \div 1\,\text{day} = 21.6172\$/\text{kgal}$$

$$\text{Unit O\&M costs} = 4014.67\$/\text{year} \div 1\,\text{MGD} \times 1\,\text{million gal}/1000\,\text{kgal} \div 365\,\text{day/year}$$

$$\text{Unit O\&M costs} = 0.010999\$/\text{kgal}$$

When $Q = 1$ MGD

$$Q_0 = 1\,\text{MGD}$$

$$\text{Unit capital cost}_0 = 21.6172\$/\text{kgal}$$

$$\text{Unit O\&M costs}_0 = 0.010999\$/\text{kgal}$$

$$\frac{Q}{Q_0} = 1$$

$$\frac{\text{Unit capital cost}}{\text{Unit capital cost}_0} = 1$$

$$\frac{\text{Unit O\&M costs}}{\text{Unit O\&M costs}_0} = 1$$

When $Q = 20$ MGD

$$\frac{Q}{Q_0} = 20$$

$$\frac{\text{Unit capital cost}}{\text{Unit capital cost}_0} = 0.9323$$

$$\frac{\text{Unit O\&M costs}}{\text{Unit O\&M costs}_0} = 1.0208$$

2) Plot unit cost/unit cost$_0$ vs. Q/Q_0 for capital and OM, respectively (Figure 7.8).

a) Equation for unit cost/unit cost$_0$ vs. Q/Q_0 for capital:

When $1 < Q/Q_0 < 10$, the regression equation is

$$\log(\text{Unit Cost/Unit Cost}_0) = -0.0278 \log(Q/Q_0) - 0.0039 \quad r^2 = 0.9213$$

or the equation of the regression:

$$\text{Unit Cost/Unit Cost}_0 = 0.9911(Q/Q_0)^{-0.0278} \quad r^2 = 0.9213$$

When $10 < Q/Q_0 < 100$, the regression equation is (Table 7.8)

$$\text{Unit Cost/Unit Cost}_0 = -0.0001(Q/Q_0) + 0.9344 \quad r^2 = 0.7012$$

Matlab code:

```
clear all
close all
%data%
Qrange=[1 2 4 6 8 10 20 40 60 80 100];
Q=Qrange(1);
totalcostrange=[21617.18
  41694.26667 81848.43333 122002.6
  162156.7667 202310.9333 403081.77
  804623.43 1206165.10 1607706.77
  2009248.43];
totalcost=totalcostrange(1);
xmid=6;
%function%
unitcost=totalcost./Q./1000;
unitcostrange=totalcostrange./
  Qrange./1000;
```

```
x=Qrange/Q;
y=unitcostrange /unitcost;
x1=x(1:xmid);
y1=y(1:xmid);
x2=x(xmid:11);
y2=y(xmid:11);
%power
p=polyfit(log10(x1), log10(y1), 1)
log10ratiounitcostfit=polyval
  (p, log10(x1));
ratiounitcostfit=10.
  ^log10ratiounitcostfit;
k=10^p(:,2)
R2power=1 - sum((log10(y1) -
  log10ratiounitcostfit).^2)/sum
  ((log10(y1) - mean(log10(y1))).^2)
xmax=max(x1);
xmin=min(x1);
xaxis=[xmin:0.1:xmax];
log10ratiounitcostfitplot=polyval(p,
  log10(xaxis));
ratiounitcostfitplot=10.
  ^log10ratiounitcostfitplot;
%line
p2=polyfit((x2),(y2), 1)
z=polyval(p2, (x2));
R2linear=1 - sum(((y2) - z).^2)/
  sum(((y2) - mean((y2))).^2)
xmax2=max(x2);
xmin2=min(x2);
xaxis2=[xmin2:0.1:xmax2];
z2=polyval(p2, xaxis2);
%plot%
figure(1)
plot(Qrange/Q,unitcostrange
  /unitcost,'ko','MarkerSize',3,
  'MarkerFaceColor',[0 0 0],
  'MarkerEdgeColor',[0 0 0],
  'linewidth',1);
hold on
line1=plot(xaxis,
  ratiounitcostfitplot,'k-',
  'linewidth',1);
hold on
line2=plot(xaxis2,z2,'k--',
  'linewidth',1);
set(gca,'xtickmode','auto',
  'FontSize',10,'FontWeight',
  'bold','FontName',
  'Times New Roman');
legend([line1,line2],'Power
  trendline','Linear
  trendline','Location','Northeast');
xlabel({'Q/Q_0'},'FontSize',14,
  'FontWeight','bold','FontName',
  'Times New Roman')
ylabel({'Unit Cost/Unit
  Cost_0'},'FontSize',14,
  'FontWeight','bold','FontName',
  'Times New Roman')
grid on
hold on
```

Answer: UVConstructionCosttwoline

$p = -0.0278 \quad -0.0039;\ k = 0.9911;$

$R2power = 0.9213$

$p2 = -0.0001 \quad 0.9344;\ R2linear = 0.7012$

b) Equation for unit cost/unit cost$_0$ vs. Q/Q_0 for OM:
The regression equation is $(0 < Q/Q_0 < 4)$:

$$\log(\text{Unit Cost/Unit Cost}_0) = -0.0141 \log(Q/Q_0) \qquad r^2 = 1.0000$$

Or the regression equation:

$$\text{Unit Cost/Unit Cost}_0 = (Q/Q_0)^{0.0141} \qquad r^2 = 1.0000$$

The regression equation is $(4 < Q/Q_0 < 100)$:

$$\text{Unit Cost/Unit Cost}_0 = 0.00002(Q/Q_0) + 1.0199 \qquad r^2 = 0.7739$$

Matlab code:

```
clear all
close all
%data%
Qrange=[1 2 4 6 8 10 20 40 60 80 100];
Q=Qrange(1);
totalcostrange=[4014.69 8108.899172
  16375.96764 24564.37612 32752.78459
  40941.19306 81961.90 164081.96
  246123.36 328164.77 410206.17];
totalcost=totalcostrange(1);
xmid=3;
%function%
unitcost=totalcost./Q./1000./365;
unitcostrange=totalcostrange./
  Qrange./1000./365;
x=Qrange/Q;
y=unitcostrange /unitcost;
x1=x(1:xmid);
y1=y(1:xmid);
x2=x(xmid:11);
y2=y(xmid:11);
```

```
%power
p=polyfit(log10(x1), log10(y1), 1)
log10ratiounitcostfit=polyval
  (p, log10(x1));
ratiounitcostfit=10.
  ^log10ratiounitcostfit;
k=10^p(:,2)
R2power=1 - sum((log10(y1) -
  log10ratiounitcostfit).^2)/sum
  ((log10(y1) - mean(log10(y1))).^2)
xmax=max(x1);
xmin=min(x1);
xaxis=[xmin:0.1:xmax];
log10ratiounitcostfitplot=polyval(p,
  log10(xaxis));
ratiounitcostfitplot=10.
  ^log10ratiounitcostfitplot;
%line
p2=polyfit((x2),(y2), 1)
z=polyval(p2, (x2));
R2linear=1 - sum(((y2) - z).^2)/sum
  (((y2) - mean((y2))).^2)
xmax2=max(x2);
xmin2=min(x2);
xaxis2=[xmin2:0.1:xmax2];
z2=polyval(p2, xaxis2);
%plot%
figure(1)
plot(Qrange/Q,unitcostrange
  /unitcost,'ko','MarkerSize',3,
  'MarkerFaceColor',[0 0 0],
  'MarkerEdgeColor',[0 0 0],
  'linewidth',1);
hold on
line1=plot(xaxis,
  ratiounitcostfitplot,'k-',
  'linewidth',1);
hold on
line2=plot(xaxis2,z2,'k--',
  'linewidth',1);
set(gca,'xtickmode','auto',
  'FontSize',10,
  'FontWeight','bold','FontName',
  'Times New Roman');
legend([line1,line2],'Power
  trendline','Linear
  trendline','Location','Southeast');
xlabel({'Q/Q_0'},'FontSize',14,
  'FontWeight','bold','FontName',
  'Times New Roman')
ylabel({'Unit Cost/Unit Cost_0'},
  'FontSize',14,'FontWeight','bold',
  'FontName', 'Times New Roman')
grid on
hold on
```

Answer: UVOMCostnormaltwoline

$p = 0.0141\ \ 0.0000;\ k = 1.0000;\ R2power = 1.0000$

$p2 = 0.0000\ \ 1.0199;\ R2linear = 0.7739$

Comments: Comparing the capital and O&M costs of chlorination with those of UV disinfection, it is clear that the unit capital cost of UV is much higher than chlorination. However, the O&M cost of UV disinfection is much smaller than that of chlorination when flow rate is smaller than 6 MGD. After the flow rate exceeds 6 MGD, the O&M cost of UV disinfection is greater than that of chlorination.

7.3.3 Anammox to Replace Nitrification/Denitrification

7.3.3.1 Nitrogen Forms

Nitrogen in municipal wastewater originates from human urine and excreta. Of 100% influent nitrogen, 85% is in the treated wastewater, and 15% is in the digested sludge solids. About 40% is ammonia and 60% is bound in organic matter. The annual per capita nitrogen ranges from 4 to 6 kg/cap/year that results in 35 mg/l total nitrogen in wastewater stream (Figure 7.9).

Total nitrogen is the sum of organic, ammonia, nitrate, nitrite, and gaseous nitrogen. Decomposition of nitrogenous organic matter releases ammonia to solution:

$$\text{Organic nitrogen compounds decomposition} \xrightarrow[\text{bacterial}]{----} NH_3 \tag{7.5}$$

Nitrifying bacteria will oxidize ammonia to nitrite and nitrate under aerobic conditions as shown in (7.6):

$$NH_3 + O_2 \xrightarrow[\text{aerobic nitrification}]{--------} NO_3^- \tag{7.6}$$

Denitrification will take place under anaerobic or anoxic conditions. Organic matter (AH_2) is oxidized by nitrate which serves as the hydrogen acceptor. The end product is nitrogen gas:

$$NO_3^- + AH_2 \xrightarrow[\text{anaerobic denitrification}]{----------} A + H_2O + N_2\uparrow \tag{7.7}$$

Primary sedimentation removes less than 15% of total nitrogen, while AS removes another 10%. As a result, only 4% BOD contributed by N could be removed. In anaerobic digester, 40% of the organic nitrogen in the sludge could be converted to ammonia, and 10% of the original

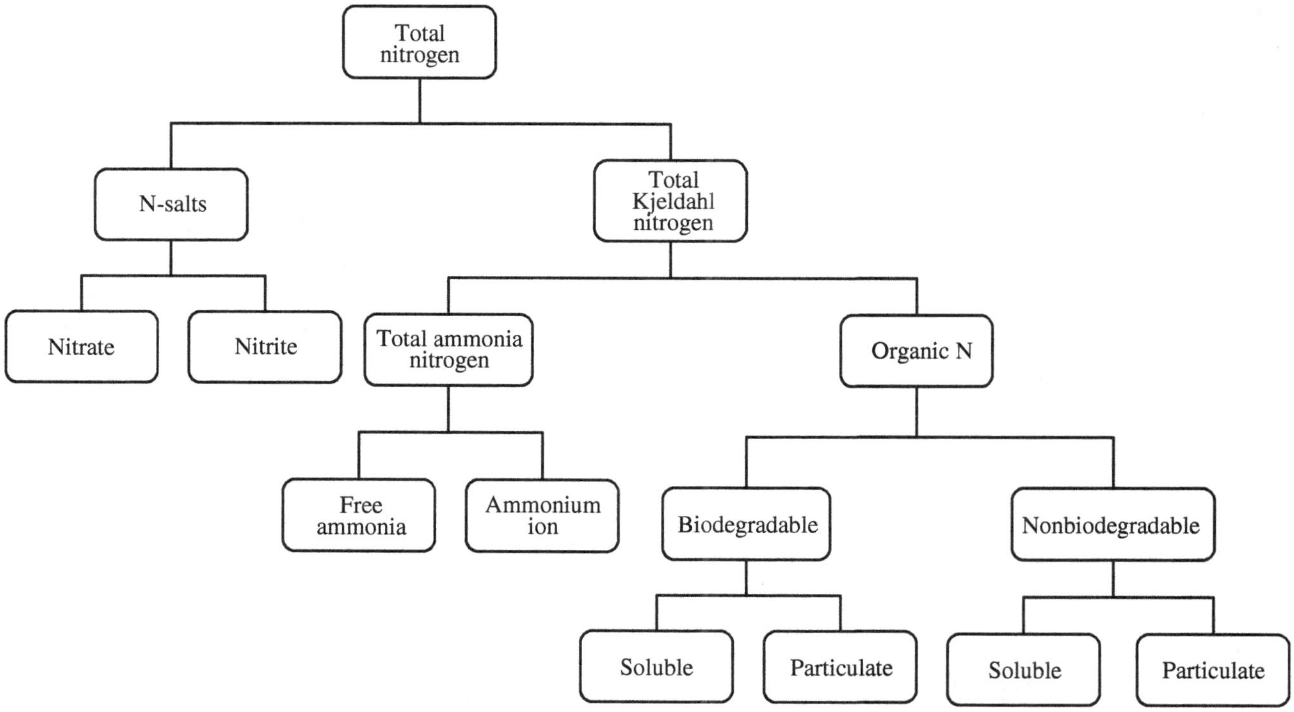

Figure 7.9 Nitrogen forms in domestic wastewater.

25% is recycled to the treatment plant and appears in the effluent, which then contains 85% of the influent nitrogen. Typically, nitrogen removal in conventional biological treatment systems is less than 40%. Therefore, tertiary treatment referred as nitrification/denitrification is required to achieve 1 mg/l discharge standard in most states of the United States.

7.3.3.2 Nitrification

Conventional activated sludge (CAS) is designed to remove carbonaceous BOD with minimal removal of ammonia nitrogen. The high ammonia content and low BOD provide greater growth potential for the nitrifiers relative to the heterotrophs. Therefore, the nitrification process at an increased sludge age compensates for lower operating temperature to ensure the growth rate of nitrifying bacteria. In effluent, some of these bacteria is lost and would be supplemented by newly produced bacteria. Temperature, pH, and dissolved oxygen concentration are the important parameters. The dissolved oxygen level should be greater than 1.0 mg/l at optimal pH near 8.4. Lime or soda ash may be needed to raise the pH to the optimum level in the nitrification tank to supplement the depleting alkalinity. Ammonia nitrogen loadings should vary from 160 to 320 g/m^3/day at 10–20 °C, respectively, with aeration period from 4 to 6 h.

7.3.3.3 Denitrification

Nitrate can be reduced to nitrogen gas by facultative heterotrophic bacteria in an anoxic environment. An organic carbon source such as methanol is needed to act as a hydrogen donor and to supply carbon for biological synthesis. Methanol is commonly used due to its high biodegradability. Since methanol is expensive, the cost of methanol is the major limit of denitrification process. Biodegradable organic matter can be used as an oxygen acceptor (hydrogen donor) for the conversion of nitrate to nitrogen gas. Therefore, a portion of raw wastewater could supply the carbonaceous BOD required:

$$\begin{array}{c}\text{Unoxidized organic matter} \\ \text{(BOD)}\end{array} + NO_3^- \rightarrow \begin{array}{c}\text{Oxidized organic matter} \\ \text{(reduced BOD)}\end{array} + N_2\uparrow \quad (7.8)$$

In the past engineering design, a biological nitrification/denitrification process requires the mixing of raw organic matter with nitrified wastewater. As a result, an aerobic zone is needed for nitrification and an anoxic zone for denitrification (Figure 7.10).

Methanol reacts with nitrate as follows:

$$5CH_3OH + 6NO_3^- \rightarrow 3N_2\uparrow + 5CO_2\uparrow + 7H_2O + 6OH^- \quad (7.9)$$

Since the effluent from nitrification also contains dissolved oxygen and nitrite, the total methanol required

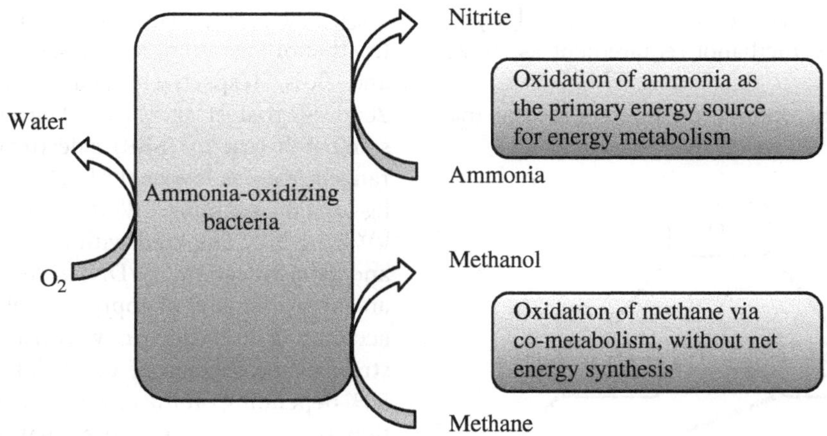

Figure 7.10 Schematic of cometabolism in ammonia-oxidizing bacteria using ammonia and methane, respectively.

as a hydrogen donor in denitrification can be estimated using Equation 7.10:

$$CH_3OH = 0.7DO + 1.1NO_2 - N + 2.0NO_3 - N \quad (7.10)$$

In addition, methanol is a carbon source in bacterial synthesis (Figure 7.11). Assuming that 30–50% excess methanol is needed for bacteria synthesis, the total methanol demand can be calculated using Equation 7.11:

$$CH_3OH = 0.9DO + 1.5NO_2 - N + 2.5NO_3 - N \quad (7.11)$$

where

CH_3OH = methanol, mg/l
DO = dissolved oxygen, mg/l
NO_2–N = nitrite nitrogen, mg/l
NO_3–N = nitrate nitrogen, mg/l

A plug-flow tank with underwater mixers is usually followed by a clarifier for sludge separation and return in designing the denitrification system. The retention time from 2 to 4 h is required for denitrification of a domestic wastewater depending on nitrate loading and temperature. Since methanol is expensive, traditional denitrification following nitrification is gradually being replaced by anaerobic ammonium oxidation (Anammox).

7.3.3.4 Anammox

The conventional biological nitrification and denitrification requires costly methanol. Therefore, a huge energy unit consumption up to 4.0 kWh/kg-N is required. As a result, this denitrification process is considered as unsustainable design in wastewater treatment. To reduce this energy consumption and to eliminate the addition of methanol, anaerobic ammonia oxidation (anammox) was discovered in 1990 (Van Loosdrecht and Salem, 2006). In anammox, ammonium is the electron donor, and nitrite is the electron acceptor that is converted anaerobically into mainly nitrogen gas and some nitrate by anammox bacteria (Strous et al., 1998). Figure 7.12 shows that anammox is the shortcut between ammonia and nitrate under biological degradation by planctomycetes. Since NO_2^- donates

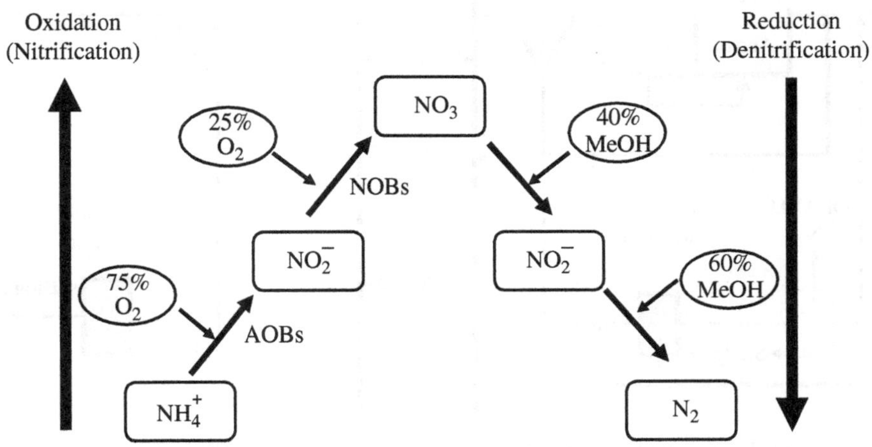

Figure 7.11 Nitrification/denitrification redox pyramid.

electron to ammonia, it reduces aeration and sludge production and eliminates methanol requirement as shown in Figure 7.12.

Table 7.13 shows the optimal parameters for the anammox process.

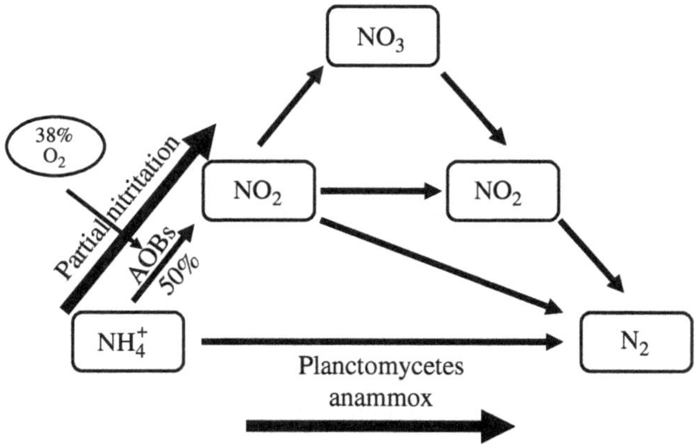

Figure 7.12 The biological N cycle and nitrate reduction to ammonium.

Table 7.13 Optimum anammox parameters.

Anammox parameters	Optimum data
Temperature (°C)	30–35
pH	7.0–8.0
SRT (days)	11+
HRT (h)	8
Nitrite to ammonia ratio	1–1.3

Quantitatively, anammox could reduce organic carbon (methanol), aeration, and sludge production by 100, 60, and 90%, respectively (Van Loosdrecht and Salem, 2006; Siegrist et al., 2008). The energy demand of the semibatch reactor (SBR) sidestream treatment systems ranged from as low as 0.8 kWh/kg-N to around 2 kWh/kg-N. For example, Wett et al. (2010) reported 1.2 kWh/kg-N. Compared with a conventional nitrification and denitrification (N/DN) sidestream treatment with an energy demand of approximately 4.0 kWh/kg-N (only accounting for electricity consumption in the sidestream), the savings of PN/A SBR systems are at least 50% depending on the oxygen transfer efficiency of aeration type. Therefore, more than 100 plants have anammox worldwide in 2014. SBR consists of 50% of all the plants, while granular systems include moving bed biofilm reactors (Rosenwinkel and Cornelius, 2005).

Nitrite is both an essential substrate and an inhibitor to the reaction. Strous et al. (1999) suggested that nitrite concentrations greater than 100 mg-N/l will completely stop the reaction. Egli et al. (2001) showed that anammox was only inhibited at nitrite concentrations higher than 182 mg-N/l. However, as high as 350 mg-N/l barely caused a 50% inhibition of the anammox process (Dapena-Mora et al. 2007). Inhibition frequency and time are all important for the performance of anammox. For these reasons, more than 50% of operation problems of anammox are due to the buildup of either nitrite or nitrate (Lackner et al., 2014).

Figure 7.13 illustrates the patented OLAND scheme and the mainstream line of WWTPs, respectively.

Table 7.14 shows that the chemical costs of nitrification/denitrification of $1.839 million/year due to addition

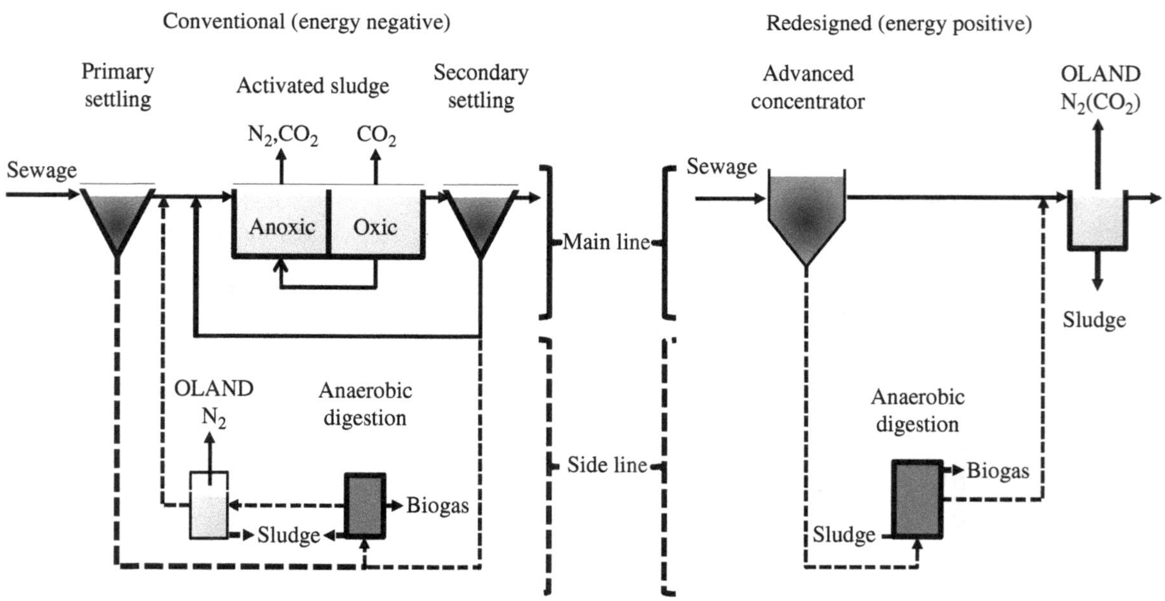

Figure 7.13 Conventional and redesigned sewage treatment schemes with OLAND in the side and mainstream line.

Table 7.14 Eliminate addition of methanol using sidestream reactor.

Process parameters	Unit	Nitrification/denitrification	Nitritation/denitritation	PN/anammox
Aerobic reactor volume	m^3	8 559	6 587	6 587
Anoxic reactor volume	m^3	4 431	4 431	N/A
Oxygen demand SOTR	kg/h	545	580	663
Aeration power consumption	kWh/day	1.70	1.81	2.07
Methanol demand	kg/day	11 298	6 998	0
Energy cost	\$/year	61 999	65 964	\$75 448
Chemical cost[a]	\$/year	1 839 188	1 139 254	0

[a] Assume blower efficiency is 18 Wh/m^3: fine bubble diffusers 25%.
- 2012 electricity price assumed to be \$0.10/kWh.
- 2012 methanol price assumed to be \$446/metric ton.

of methanol could be completely eliminated with the anammox process. Indeed, anammox is the key technology in retrofitting traditional WWTP to energy-positive WWTP in Austria.

7.4 Design Efficient Reactors

Efficiency of material could be achieved through the optimal design of reactors for many reasons: (i) the optimal reactor could reduce reactor volume that in turn reduce the embedded material and energy; (ii) smaller reactor would have better hydraulic conditions for reactions to take place and therefore reduce the operation cost; and (iii) the life cycle capital and O&M costs could be reduced. To design the most efficient reactor, mass balance can be described by Equation 7.13, which shows that the flux accumulation in the reactor equals feeding flux minus outgoing flux plus the generation of that component in the reactor:

$$\text{In} - \text{Out} + \text{Generation} = \text{Accumulation}$$

$$\text{For A}: F_{A0} - F_A + \int_0^V r_A \, dV = \frac{dN_A}{dt} \tag{7.12}$$

Therefore, the general mole balance equation of accumulation is

$$\text{Accumulation} = F_{A0} - F_A + G_A = \frac{dN_A}{dt} \tag{7.13}$$

where

F_A = rate of flow of A into the system
F_{A0} = rate of flow of A out the system
G_A = the generation of other components.

X_A is the equivalent of moles of A reacted over moles fed and is usually used as a master variable:

$$X_A(\text{conversion}) = \frac{\text{No. of moles of A reacted}}{\text{No. of moles of A fed in}} = \frac{N_{A0} - N_A}{N_{A0}} \tag{7.14}$$

X_A is the conversion factor between 0 and 1. At 0, A did not react. At 1, all of A has reacted. X_A, if the right hand is divided by volume at both nominator and denominator, could also be defined as follows:

$$X_A = \frac{(F_{A0} - F_A)}{F_{A0}} \tag{7.15}$$

Therefore, the master equation could be expressed as follows:

$$F_{A0} - F_A + \int_0^V r_A \, dV = \frac{dN_A}{dt} \tag{7.16}$$

$$\int_0^V r_A dV = \frac{dN_A}{dt} \tag{7.17}$$

$$\text{Ideal reactor}: r_A dV = \frac{dN_A}{dt} \tag{7.18}$$

$$\text{Constant volume reactor}: r_A = \frac{d(N_A/dt)}{dt} = \frac{dC_A}{dt} \tag{7.19}$$

Variables to consider in reactor design are the limiting reagent that is the reagent with the highest coefficient value and the time to achieve a certain conversion from mole balance in a batch reactor. Conversion factor X_A applies to all of the components in the system: N_A, F_A, and C_A. The main differences among batch reactor,

CSTR, and PFR are that (i) batch reactors rely on time, not volume to dictate the results required, (ii) CSTR relies on total volume required regardless of rate due to steady-state concentration, and PFR relies on volume required due to reaction rate. A chemical reactor volume for given experimental kinetic data can be found through the Levenspiel plot (Levenspiel, 1999), which uses conversion factor as the master variable. CSTR always operates at the lowest reaction rate:

$$r_A \mathrm{d}V = \frac{\mathrm{d}(N_{A0} - N_{A0}X_A)}{\mathrm{d}t} = -N_{A0}\frac{\mathrm{d}(X_A)}{\mathrm{d}t} \tag{7.20}$$

$$\mathrm{d}t = N_{A0}\frac{\mathrm{d}(X_A)}{-r_A V} \tag{7.21}$$

Example 7.7
Given: In a tubular reactor, the first-order reaction A → B is taking placing with a constant volumetric flow rate, v.
Find:

1) Derive an equation relating the reactor volume to the entering and exiting concentrations of A, the rate constant k, and the volumetric flow rate v.
2) Determine the reactor volume V necessary to reduce the exiting concentration to 0.1% of the entering concentration when the volumetric flow rate is 10 dm^3/min (i.e. l/min) and the specific reaction rate, k, is 0.23 min^{-1}.

Solution:

For a tubular reactor, the mole balance on species A was shown to be

$$\frac{\mathrm{d}F_A}{\mathrm{d}V} = r_A$$

For a first-order reaction, the rate law is

$$-r_A = kC_A$$

Since the volumetric flow rate, v_0, is constant,

$$\frac{\mathrm{d}F_A}{\mathrm{d}V} = \frac{\mathrm{d}(C_A v_0)}{\mathrm{d}V} = v_0\frac{\mathrm{d}C_A}{\mathrm{d}V} = r_A$$

Substituting for r_A yields

$$-v_0\frac{\mathrm{d}C_A}{\mathrm{d}V} = -r_A = kC_A$$

Rearranging gives

$$-\frac{v_0}{k}\frac{\mathrm{d}C_A}{C_A} = \mathrm{d}V$$

Using the conditions at the entrance of the reactor that when $V = 0$, then $C_A = C_{A0}$,

$$-\frac{v_0}{k}\int_{C_{A0}}^{C_A}\frac{\mathrm{d}C_A}{C_A} = \int_0^V \mathrm{d}V$$

This equation gives

$$V = \frac{v_0}{k}\ln\frac{C_{A0}}{C_A}$$

$$V = \frac{10}{0.23}\ln\frac{1}{0.001} = 300\,\mathrm{l}$$

There is no accumulation at steady state; mass flux has a relationship as shown by the following equations for CSTR:

$$0 = F_A - F_{A0} + r_A V$$

where

F_A = rate of flow of A into the system (moles/time)
F_{A0} = rate of flow of A out the system (moles/time)
r_A = rate of formation of species A (moles/time/volume)
V = system volume (volume)

To design a PFR, the master equation is

$$0 = F_A - F_{A0} + r'_A W \tag{7.22}$$

where

W = the weight of the catalyst (mass) = volume × density of catalyst
r'_A = rate of the catalyst (moles/time/mass)

Mole balance can be converted to mass balance as follows:

$$F_A \times M_A = \frac{\text{(moles)}}{\text{(time)}} \times \frac{\text{(mass)}}{\text{(mole)}} = \frac{\text{(mass)}}{\text{(time)}} = W_A$$

$$N_A \times M_A = \text{(moles)} \times \frac{\text{(mass)}}{\text{(mole)}} = \text{(mass)} = m_A$$

where

W_A = the mass flow rate of A (mass/time)
M_A = molar mass (mass or moles)
F_A = rate of flow of A into the system (moles/time)
N_A = number of moles of A (moles)

General mole balance equation:

$$F_A - F_{A0} + \int_0^V r_A \mathrm{d}V = \frac{\mathrm{d}N_A}{\mathrm{d}t} \tag{7.23}$$

While the general mass balance is

$$W_A - W_{A0} + M_A\int_0^V r_A \mathrm{d}V = \frac{\mathrm{d}m_A}{\mathrm{d}t} \tag{7.24}$$

Example 7.8
Given: Reaction A → B is to be carried out isothermally in a continuous-flow reactor. The following information are

also known: $C_A = 0.01C_{A0}$, $F_{A0} = 5$ mol/h, $v = 10$ l/h, and $C_{A0} = 0.5$ mol/l. The entering volumetric flow rate v is 10 l/h. For a constant volumetric flow rate, $v = v_0 =$ 10 l/h, $F_{A0} = C_{A0}\ v_0$, and $C_{A0} = F_{A0}/v_0 = (5\ \text{mol/h})/(10\ \text{l/h}) = 0.5$ mol/l. If the reaction has zero, first, and second order,

1) $-r_A = k$ with $k = 0.05$ mol l/h.
2) $-r_A = kC$ with $k = 0.0001\ \text{s}^{-1}$.
3) $-r_A = kC^2$ with $k = 3$ l/mol/h.

Find: The required volumes for CSTR and PFR for each reaction kinetics, respectively.

Solution:

1) $-r_A = k$ with $k = 0.05$ mol l/h.
   CSTR equation:

$$V = \frac{F_{A0} - F_A}{-r_A} = \frac{v(C_{A0} - 0.01C_{A0})}{k}$$

$$= \frac{10(0.5 - 0.01 \times 0.5)}{0.05} = 99\text{l}$$

   PFR equation:

$$\int_0^V dV = \frac{1}{r_A}\int_{F_{A0}}^{F_A} dF = -\frac{v}{k}\int_{C_{A0}}^{C_A} dC$$

$$V = -\frac{10}{0.05}(0.5 \times 0.01 - 0.5) = 99\ \text{l}$$

2) $-r_A = kC$ with $k = 0.0001\ \text{s}^{-1}$.
   CSTR equation (using $-r_A = kC_A$):

$$V = \frac{F_{A0} - F_A}{-r_A} = \frac{v(C_{A0} - 0.01C_{A0})}{k0.01C_{A0}}$$

$$= \frac{10(0.5 - 0.01 \times 0.5)}{0.0001 \times \frac{3600\text{s}}{1\text{h}} \times 0.01 \times 0.5} = 2750\text{l}$$

   PFR equation (using $-r_A = kC$):

$$\int_0^V dV = \frac{1}{r_A}\int_{F_{A0}}^{F_A} dF = -\frac{v}{kC}\int_{C_{A0}}^{C_A} dC$$

$$V = -\frac{v}{kC}\int_{C_{A0}}^{C_A} dC = -\frac{v}{k}\int_{C_{A0}}^{C_A} \frac{dC}{C}$$

$$= -\frac{10}{0.0001 \times \frac{3600\text{s}}{1\text{h}}}\ln\left(\frac{0.01C_{A0}}{C_{A0}}\right) = 128\text{l}$$

3) $-r_A = kC^2$ with $k = 3\ \text{dm}^3\ \text{mol}^{-1}\ \text{h}^{-1}$.
   CSTR equation (using $-r_A = kC_A$):

$$V = \frac{F_{A0} - F_A}{-r_A} = \frac{v(C_{A0} - 0.01C_{A0})}{k(0.01C_{A0})^2}$$

$$= \frac{10(0.5 - 0.01 \times 0.5)}{3 \times (0.01 \times 0.5)^2} = 66\ 000\ \text{l}$$

   PFR equation (using $-r_A = kC$):

$$\int_0^V dV = \frac{1}{r_A}\int_{F_{A0}}^{F_A} dF = -\frac{v}{kC^2}\int_{C_{A0}}^{C_A} dC$$

$$V = -\frac{v}{kC^2}\int_{C_{A0}}^{C_A} dC = -\frac{v}{k}\int_{C_{A0}}^{C_A}\frac{dC}{C^2} = 660\text{l}$$

Comment: There is no difference in reactor volume for both CSTR and PFR if the reaction has zero order, while there is a huge difference in reactor volumes if the reaction kinetics is the second order with magnitude less of reactor volume of PFR to achieve the same conversion.

Example 7.9
Given: For the aforementioned example, assuming $F_{A0} = 0.867$ mol/s.

Find: The reactor volume necessary to achieve 98% conversion using data in Table 7.15.

Table 7.15 Reaction rate vs. conversion factor.

Raw data	
X	$-r_A$
0.00	0.00530
0.10	0.00520
0.20	0.00500
0.30	0.00450
0.40	0.00400
0.50	0.00330
0.60	0.00250
0.70	0.00180
0.80	0.00125
0.85	0.00100

Solution:

Using Matlab plot $1/(-r_A)$ vs. $1/(1-X)$

Matlab code:

```
clear
clc
load RawData2
a=rawdata2(:,1);
b=rawdata2(:,2);
x=1./(1-a);
y=1./b;
hold on
n=1;
p=polyfit(x,y,n)
x4=[0:0.01:100];
z3=polyval(p,x);
z4=polyval(p,x4);
plot(x,y,'ko',
  'MarkerSize',3,'MarkerFaceColor',
  [0 0 0],'MarkerEdgeColor',
  [0 0 0],'linewidth',1);
hold on
plot(x4,z4,'k-','LineWidth',1);
hold on
R2=1 - sum((y - z3).^2)/sum((y - mean(y)).
  ^2)
set(gca,'xlim',[1 7],'FontWeight',
  'bold','FontName', 'Times New Roman');
xlabel('1/(1-X)','FontSize',14,
  'FontWeight','bold','FontName',
  'Times New Roman')
ylabel({'1/-r_A'},'FontSize',14,
  'FontWeight','bold','FontName',
  'Times New Roman')
legend('Original Data',
  'Trendline','Location','NorthWest');
grid on
```

The equation of trend line is $1/(-r_A) = 150.97 \times 1/(1-X) + 19.312$ $r^2 = 0.9948$

Therefore, at $X = 0.98$ $-r_A = 0.000132$

$F_{A0} = 0.867$ mol/s, then

$$V = \frac{F_{A0}X}{-r_A} = 64301$$

Example 7.10

Given: The rate of chemical reaction of 2-monochlorophenol (2-MCP) oxidation by Fenton process is determined according to kinetic data reported by Tang and Huang (1996a, b). The oxidation rate and its reciprocal values vs. conversion factors are listed in Table 7.16.

If 2-MCP enters a reactor with a volumetric flow rate, v_0, of 2 l/s, C_{A0} of 2-MCP is 5×10^{-4}M.

Solution:

To size a number of reactors for the reaction,

$$F_{A0} = v_0 \times C_{A0} = \frac{2\text{l}}{\text{s}} \times 5 \times \frac{10^{-4}\text{mol}}{\text{l}} = 0.001\,\text{mol/s}$$

$$F_{A0} = 0.001\,\text{mol/s} = 0.06\,\text{mol/min},$$

$F_{A0}/-r_A$ vs. conversion factor, X, is listed in Table 7.17.

Table 7.16 Oxidation rate vs. conversion factor.

X	0	0.1	0.2	0.3	0.4	0.6	0.7	0.8
$-r_A\left(\frac{\text{mol}}{\text{m}^3\cdot\text{min}}\right)$	0.832	0.74876	0.665531	0.582459	0.499083	0.333148	0.249977	0.166368
$(-1/r_A)\left(\frac{\text{m}^3\cdot\text{min}}{\text{mol}}\right)$	1.201923	1.335542	1.50256	1.71686	2.003675	3.001673	4.000361	6.010766

Table 7.17 $F_{A0}/-r_A$ vs. conversion.

X	0	0.1	0.2	0.3	0.4	0.6	0.7	0.8
$-r_A\left(\frac{\text{mol}}{\text{m}^3\cdot\text{min}}\right)$	0.832	0.74876	0.665531	0.582459	0.499083	0.333148	0.249977	0.166368
$(-1/r_A)\left(\frac{\text{m}^3\cdot\text{min}}{\text{mol}}\right)$	1.201923	1.335542	1.50256	1.71686	2.003675	3.001673	4.000361	6.010766
$(F_{A0}/-r_A)(\text{m}^3)$	0.072115	0.080133	0.090154	0.103012	0.12022	0.1801	0.240022	0.360646

1) Using the data in Table 7.16 to plot $(-1/r_A)$ vs. X.
2) Using the data in Table 7.17 to plot $(F_{A0}/-r_A)$ vs. X.

Solution:

1) Using the data in Table 7.16 to plot $(-1/r_A)$ vs. X (Figure 7.14).
2) Using the data in Table 7.17 to plot $(F_{A0}/-r_A)$ vs. X (Figure 7.15).

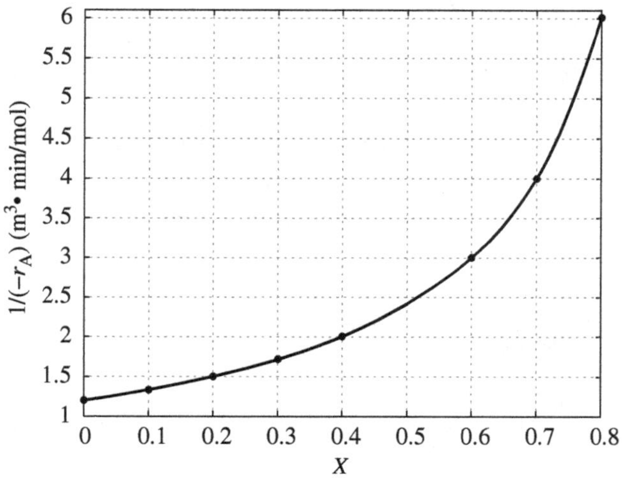

Figure 7.14 $(-1/r_A)$ vs. X.

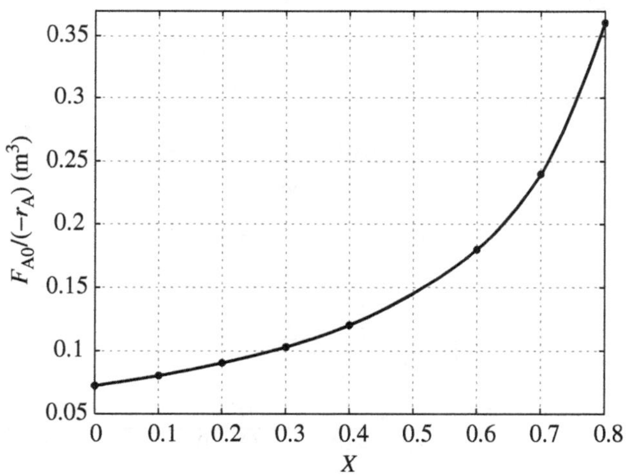

Figure 7.15 $(F_{A0}/-r_A)$ vs. X.

Example 7.11
Given:

1) Using the data in Tables 7.16 and 7.17, or Figure 7.14, to calculate the volume necessary to achieve 80% conversion in a CSTR.
2) Show the area in Figure 7.14 that would give the CSTR volume necessary to achieve 80% conversion.

Solution:

1)

$$V = \frac{F_{A0}X}{(-r_A)_{exit}}$$

$$\left(\frac{1}{-r_A}\right)_{X=0.8} = 6.01\ m^3 \cdot min/mol$$

$$V = (6.01\ m^3 \cdot min/mol) \times 0.06\ mol/min \times 0.8 = 0.28848\ m^3$$

2) Figure 7.16

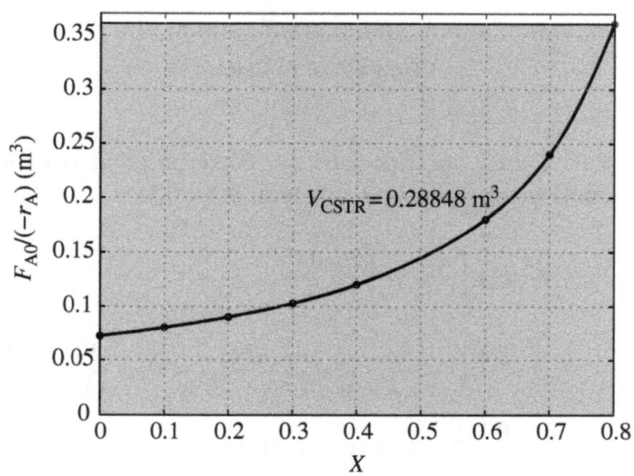

Figure 7.16 CSTR plot.

Example 7.12
Given: Using the data in Tables 7.16 and 7.17 or Figure 7.14 to calculate the volume necessary to achieve 80% conversion in a PFR.
Find: The shaded area in Figure 7.14 that would give the PFR the volume necessary to achieve 80% conversion. Also make a qualitative sketch of the conversion X and the rate of reaction, $-r_A$, down the length (volume) of the reactor (Figure 7.17).

Solution:

$$V = F_{A0}\int_0^{0.8}\frac{dX}{-r_A} = \int_0^{0.8}\frac{F_{A0}dX}{-r_A}$$

$$= \frac{\Delta X}{3}\left(\frac{F_{A0}}{-r_A(X=0)} + \frac{4F_{A0}}{-r_A(X=0.2)} + \frac{2F_{A0}}{-r_A(X=0.4)} + \frac{4F_{A0}}{-r_A(X=0.6)} + \frac{F_{A0}}{-r_A(X=0.8)}\right)$$

$$= \frac{0.2}{3}(0.072 + 4(0.090) + 2(0.120) + 4(0.180) + 0.361)$$

$$= 0.1169\ m^3$$

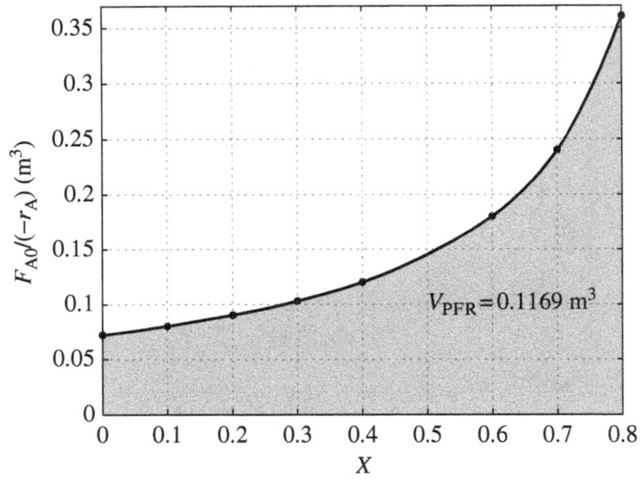

Figure 7.17 PFR plot.

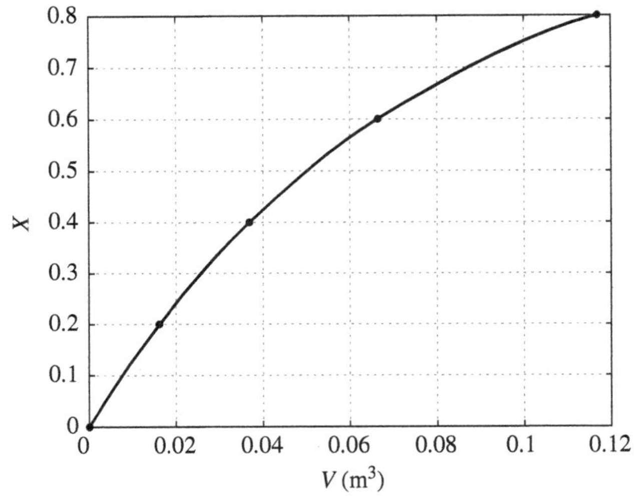

Figure 7.18 Conversion profile.

For $X = 0.2$, we calculate the corresponding reactor volume using Simpson's rule with $\Delta X = 0.1$:

$$V = F_{A0}\int_0^{0.2}\frac{dX}{-r_A} = \int_0^{0.2}\frac{F_{A0}dX}{-r_A}$$

$$= \frac{\Delta X}{3}\left(\frac{F_{A0}}{-r_A(X=0)} + \frac{4F_{A0}}{-r_A(X=0.1)} + \frac{F_{A0}}{-r_A(X=0.2)}\right)$$

$$= \frac{0.1}{3}(0.072 + 4(0.080) + 0.090)$$

$$= 0.0161\,\text{m}^3$$

For $X = 0.4$, we calculate the corresponding reactor volume using Simpson's rule with $\Delta X = 0.2$:

$$V = F_{A0}\int_0^{0.4}\frac{dX}{-r_A} = \int_0^{0.4}\frac{F_{A0}dX}{-r_A}$$

$$= \frac{\Delta X}{3}\left(\frac{F_{A0}}{-r_A(X=0)} + \frac{4F_{A0}}{-r_A(X=0.2)} + \frac{F_{A0}}{-r_A(X=0.4)}\right)$$

$$= \frac{0.2}{3}(0.072 + 4(0.090) + 0.120)$$

$$= 0.0369\,\text{m}^3$$

For $X = 0.6$, we calculate the corresponding reactor volume using Simpson's rule with $\Delta X = 0.3$ (Figures 7.18 and 7.19; Table 7.18):

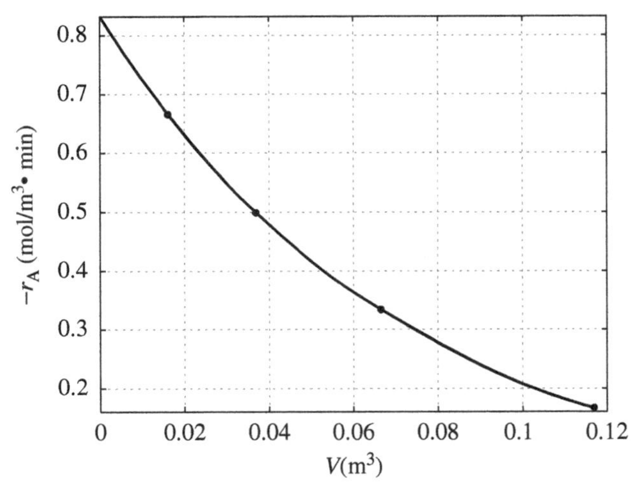

Figure 7.19 Reaction rate profile.

$$V = F_{A0}\int_0^{0.6}\frac{dX}{-r_A} = \int_0^{0.6}\frac{F_{A0}dX}{-r_A}$$

$$= \frac{\Delta X}{3}\left(\frac{F_{A0}}{-r_A(X=0)} + \frac{4F_{A0}}{-r_A(X=0.3)} + \frac{F_{A0}}{-r_A(X=0.6)}\right)$$

$$= \frac{0.3}{3}(0.072 + 4(0.103) + 0.180)$$

$$= 0.0664\,\text{m}^3$$

Table 7.18 Conversion and reaction rate profiles.

X	0	0.2	0.4	0.6	0.8
$-r_A\left(\frac{\text{mol}}{\text{m}^3\cdot\text{min}}\right)$	0.832	0.665531	0.499083	0.333148	0.166368
$V(\text{m}^3)$	0	0.016093	0.036863	0.066426	0.116948

The reason the isothermal CSTR volume is usually greater than the PFR volume is that the CSTR is always operating at the lowest reaction rate (e.g. $-r_A = 0.1664\,\text{mol/m}^3\cdot\text{min}$). The PFR on the other hand starts at a high rate at the entrance and gradually decreases to the exit rate, thereby requiring less volume because the volume is inversely proportional to the rate (Figures 7.20 and 7.21). However, for autocatalytic reactions, product-inhibited reactions, and nonisothermal exothermic reactions, these trends will not always be the same.

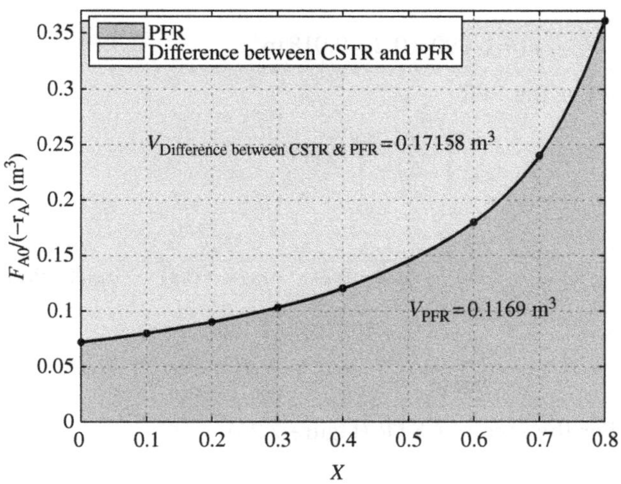

Figure 7.20 Comparison of CSTR and PFR reactor size.

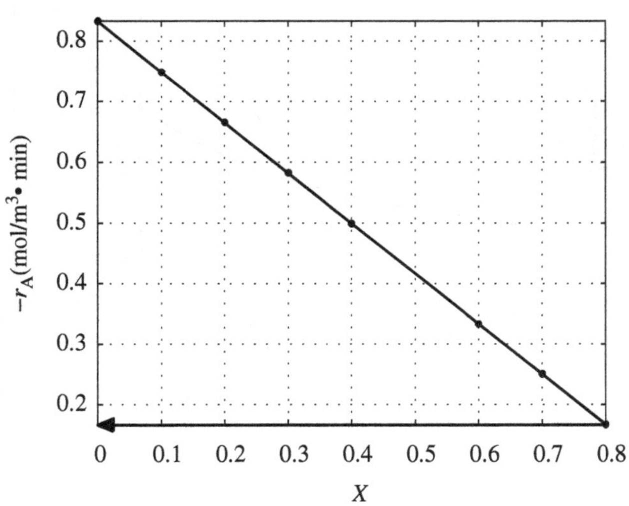

Figure 7.21 $-r_A$ vs. X.

Example 7.13

Given: 40% conversion is to be achieved in the first reactor for two CSTRs in series (Figure 7.22).

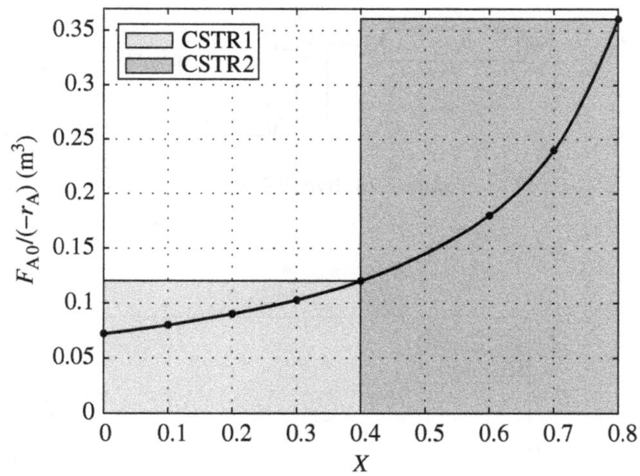

Figure 7.22 Two CSTRs in series.

Find: The volume of each of the two reactors necessary to achieve 80% overall conversion of the entering species A.

Solution:

$$V_1 = \frac{F_{A0}X_1}{(-r_A)_{X=0.4}}$$

$$V_1 = 0.120 \times 0.4 = 0.048\,\text{m}^3$$

$$V_2 = \frac{F_{A0}(X_2 - X_1)}{(-r_A)_{X=0.8}}$$

$$V_2 = 0.361 \times 0.4 = 0.1444\,\text{m}^3$$

$$V = V_1 + V_2 = 0.048\text{m}^3 + 0.1444\text{m}^3 = 0.1924\text{m}^3$$

Example 7.14

Given: Two PFRs are to be designed in series. If the data in Table 7.16 is used to achieve the intermediate conversion of 40% and the final conversion of 80%, the entering molar flow rate is the same as in the previous example, $F_{A0} = 0.001\,\text{mol/s} = 0.06\,\text{mol/min}$.

Find: The reactor volumes V_1 and V_2 for the PFR in sequence as shown in Figures 7.23 and 7.24.

Solution:

$$V_1 = F_{A0}\int_0^{0.4} \frac{dX}{-r_A} = \int_0^{0.4} \frac{F_{A0}dX}{-r_A}$$

$$= \frac{\Delta X}{3}\left(\frac{F_{A0}}{-r_A(X=0)} + \frac{4F_{A0}}{-r_A(X=0.2)} + \frac{F_{A0}}{-r_A(X=0.4)}\right)$$

$$= \frac{0.2}{3}(0.072 + 4(0.090) + 0.120)$$

$$= 0.0369\,\text{m}^3$$

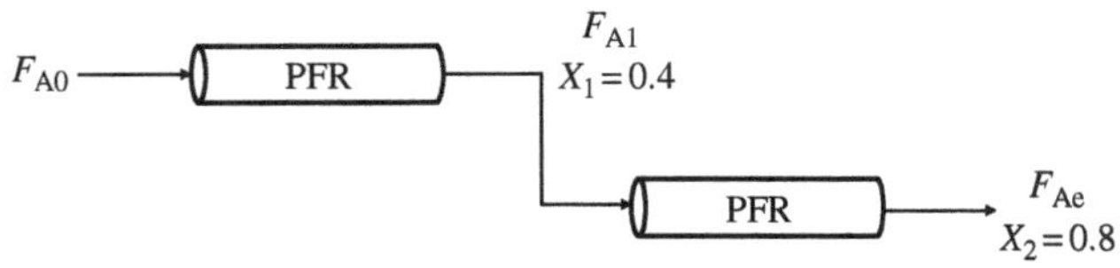

Figure 7.23 Two PFRs in series.

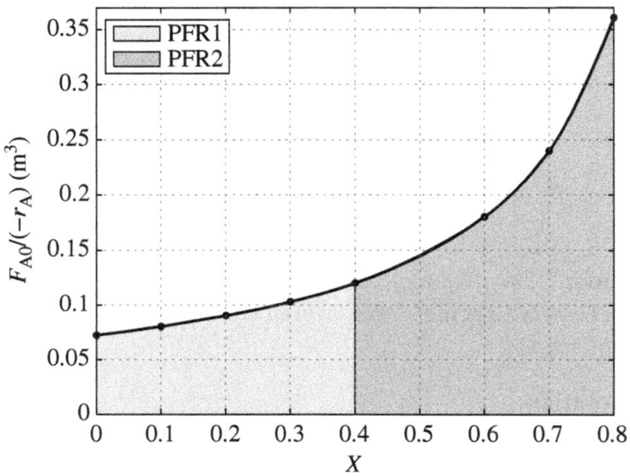

Figure 7.24 Two PFRs in series.

$$V_2 = F_{A0}\int_{0.4}^{0.8}\frac{dX}{-r_A} = \int_{0.4}^{0.8}\frac{F_{A0}dX}{-r_A}$$

$$= \frac{\Delta X}{3}\left(\frac{F_{A0}}{-r_A(X=0.4)} + \frac{4F_{A0}}{-r_A(X=0.6)} + \frac{F_{A0}}{-r_A(X=0.8)}\right)$$

$$= \frac{0.2}{3}(0.120 + 4(0.180) + 0.361)$$

$$= 0.0801\,\text{m}^3$$

$$V = V_1 + V_2 = 0.0369\,\text{m}^3 + 0.0801\,\text{m}^3 = 0.117\,\text{m}^3$$

Example 7.15
Given: The entering molar flow rate is the same as in the previous example, $F_{A0} = 0.001\,\text{mol/s} = 0.06\,\text{mol/min}$.
Find: The reactor volumes of V_1, V_2, and V_3 as shown in Figure 7.25 by using the data in Table 7.16.

Solution:

For the first CSTR

$$V_1 = \frac{F_{A0}X_1}{(-r_A)_{X=0.2}}$$

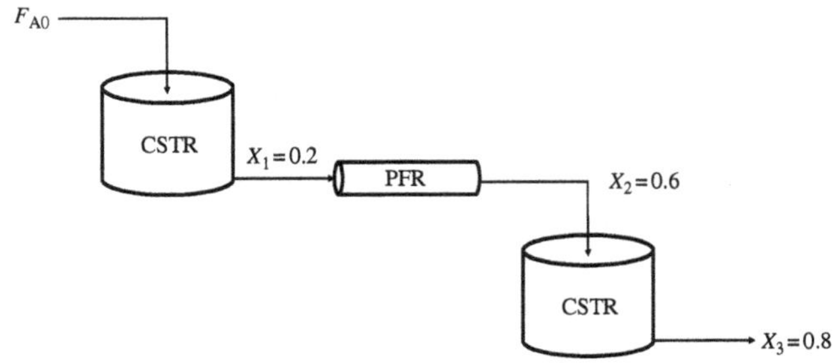

Figure 7.25 Reactors in series.

$$V_1 = 0.090 \times 0.2 = 0.018\,\text{m}^3$$

For the PFR

$$V_2 = F_{A0}\int_{0.2}^{0.6}\frac{dX}{-r_A} = \int_{0.2}^{0.6}\frac{F_{A0}dX}{-r_A}$$

$$= \frac{\Delta X}{3}\left(\frac{F_{A0}}{-r_A(X=0.2)} + \frac{4F_{A0}}{-r_A(X=0.4)} + \frac{F_{A0}}{-r_A(X=0.6)}\right)$$

$$= \frac{0.2}{3}(0.090 + 4(0.120) + 0.180)$$

$$= 0.05\,\text{m}^3$$

For the second CSTR (Figure 7.26),

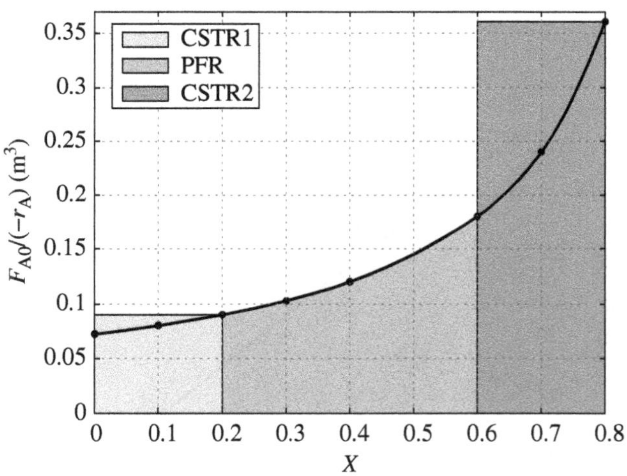

Figure 7.26 Reactors in series.

$$V_3 = \frac{F_{A0}(X_3 - X_2)}{(-r_A)_{X=0.8}}$$

$$V_3 = 0.361 \times 0.2 = 0.0722\,\text{m}^3$$

$$V = V_1 + V_2 + V_3 = 0.018\,\text{m}^3 + 0.05\,\text{m}^3 + 0.0722\,\text{m}^3$$

$$= 0.1402\,\text{m}^3$$

Example 7.16

Given: Consider reactor design data in Examples 7.13 and 7.14.

Find: The space time, τ, and space velocities, SV, for each reactor.

Solution:

From Example 7.13, the CSTR volume is 0.28848 m^3 and the corresponding space time and space velocity are

$$\tau = \frac{V}{v_0} = \frac{0.28848\,m^3}{0.002 m^3/s} = 144.24 s = 0.04007 h$$

$$SV = \frac{1}{\tau} = \frac{1}{0.04007 h} = 24.96 h^{-1}$$

From Example 7.14, the PFR volume is 0.1169 m^3, and the corresponding space time and space velocity are

$$\tau = \frac{V}{v_0} = \frac{0.1169\,m^3}{0.002 m^3/s} = 58.45\,s = 0.01624 h$$

$$SV = \frac{1}{\tau} = \frac{1}{0.01624\,h} = 61.58 h^{-1}$$

Molecular structure determines its reactivity. As more chlorine substitutes the phenol ring, the reactivity toward hydroxyl radical attack decreases. As a result, the reaction kinetics of Fenton oxidation decreases. In reactor design, it can be clearly seen through the Levenspiel plots. Figures 7.27, 7.28, and 7.29 illustrate how these plots can be used to evaluate the molecular structure such as degree of chlorine substituent on the aromatic ring of phenol increase the required reactor volume to achieve a specific oxidation target.

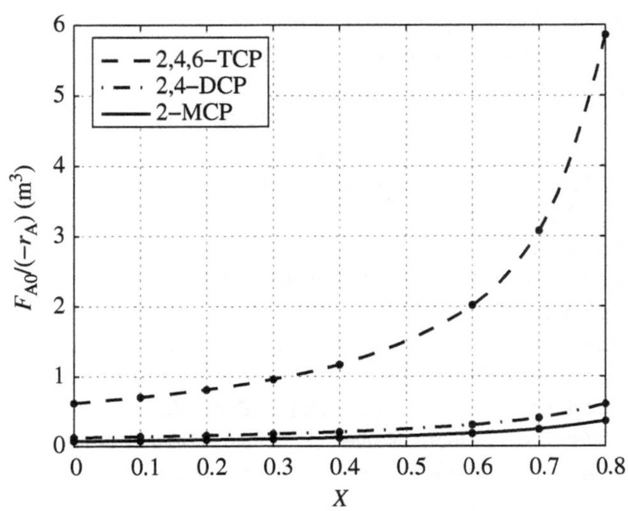

Figure 7.27 Volume comparison of PFR for 2-MCP, 2,4-DCP, and 2,4,6-TCP.

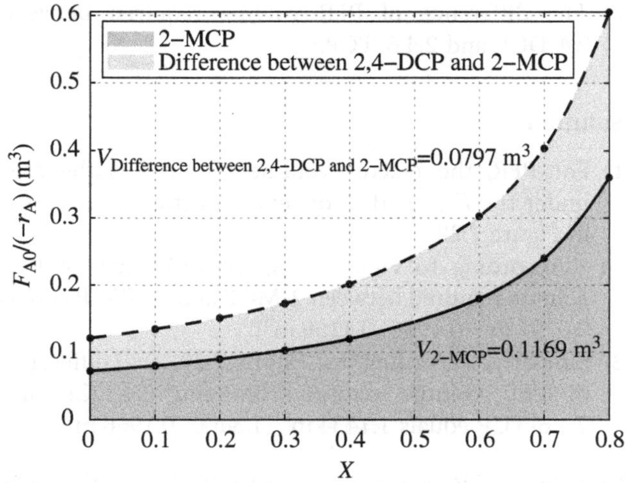

Figure 7.28 Volume comparison of PFR for 2-MCP and 2,4-DCP.

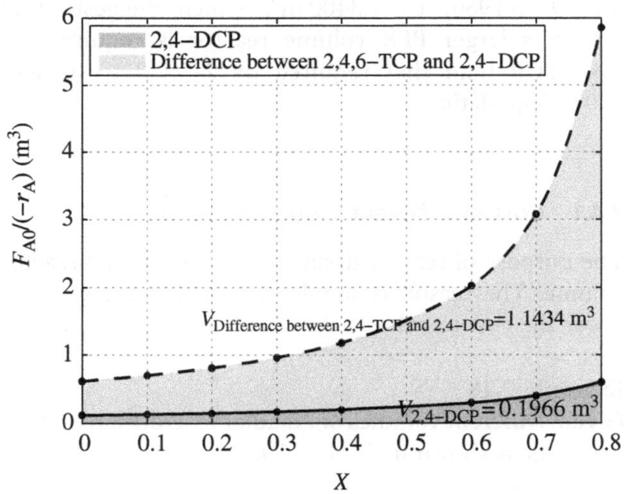

Figure 7.29 Volume comparison of PFR for 2,4-DCP and 2,4,6-TCP.

Example 7.17

Given: According to the Fenton oxidation kinetic data of 2-monochlorphenol (2-MTP), 2,4-dichlorophenol (2,4-DCP), and 2,4,6-trichlorphenol (2,4,6-TCP) by Tang and Huang (1996a, b), the following F_{A0} vs. the conversion factor, X, is plotted. PFR is to be designed to achieve 100% removal of chlorophenols.

Figure 7.27 can be used to determine the PFR volume for a specific conversion factor.

Find:

1) The difference of PFR volume required between 2-MCP and 2,4-DCP.

2) The difference of PFR volume required between 2,4-DCP and 2,4,6-TCP.

Solution:

1) For PFR, the reactor volume required is the area under the F_{A0} vs. the conversion factor, X, as shown in Figure 7.28.
   According to Figure 7.28, the difference of PFR volume required between 2-MCP and 2,4-DCP equals 0.0797 m^3 (0.1966–0.1169 m^3).
2) Similarly, according to Figure 7.29, the difference of PFR volume required between 2,4-DCP and 2,4,6-TCP equals 1.1434 m^3 (1.3400–0.1966 m^3).

Comments: Molecular structure significantly influences the oxidation kinetics of chlorophenols. As chlorination degree increases, the carbon of the aromatic ring is more oxidized. As a result, it has less reactivity toward hydroxyl radicals generated by Fenton process. Therefore, the PFR volume increased significantly from 0.1169, 0.1966, to 1.3400 m^3, which suggests that 10 times larger PFR volume reactor is required for 2,4,6-TCP than that required by 2-MCP to achieve 100% degradation.

7.4.1 Cost of Different Volume Reactors

The purpose of reactor design is to optimize the reactor volume. The reason is simple: the smaller the size, the lower the cost to achieve the same conversion factor.

Example 7.18
Given: Different volumes of Pfaudler reactors have different costs as shown in Table 7.19.

Table 7.19 Price and volume of Pfaudler reactors.

Pfaudler reactors			
Reactor capacity (gal)	Price (thousands of dollars)	ln(volume)	ln(price)
5	27	1.61	3.30
50	35	3.91	3.56
500	67	6.21	4.20
1000	80	6.91	4.38
4000	143	8.29	4.96
8000	253	8.99	5.53

Find: The cost of a 6 000-gal and a 15 000-gal Pfaudler reactor.

Solution:

Using Matlab plot Figure 7.30 ln(price) vs. ln(volume)

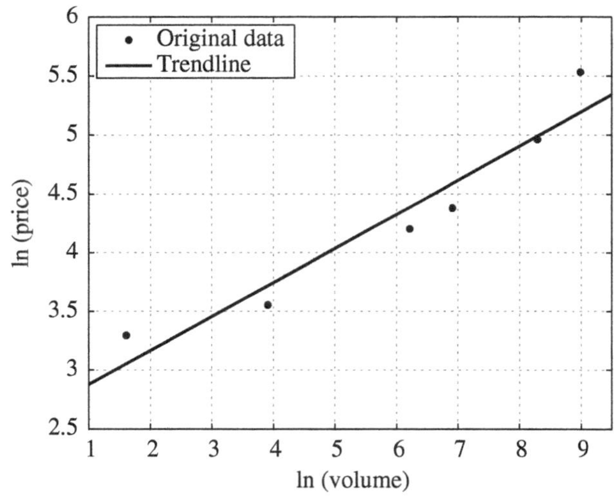

Figure 7.30 ln (price) vs. ln (volume).

Matlab code:

```
clear
clc
load RawData1ofPfaudlerReactors
ReactorCapacity=rawdata1(:,1);
Price=rawdata1(:,2);
LNvolume=log(ReactorCapacity);
LNprice=log(Price);
hold on
n=1;
p=polyfit(LNvolume,LNprice,n)
x4=[0:0.01:100];
z3=polyval(p,LNvolume);
z4=polyval(p,x4);
plot(LNvolume,
  LNprice,'ko','MarkerSize',3,
  'MarkerFaceColor',[0 0 0],
  'MarkerEdgeColor',[0 0 0],
  'linewidth',1);
hold on
plot(x4,z4,'k-','LineWidth',1);
hold on
R2=1 - sum((LNprice - z3).^2)/sum((LNprice
  - mean(LNprice)).^2)
set(gca,'xlim',[1 9.5],'FontWeight',
  'bold','FontName', 'Times New Roman');
xlabel('ln (volume)','FontSize',14,
  'FontWeight','bold','FontWeight',
  'bold','FontName', 'Times New Roman')
```

```
ylabel({'ln (price)'},'FontSize',14,
  'FontWeight','bold')
legend('Original Data','Trendline',
  'Location','NorthWest','FontWeight',
  'bold','FontName', 'Times New Roman');
box on
grid on
```

The equation of trend line: ln (price) = 0.2901ln (volume) + 2.5854: $r^2 = 0.9208$

$$\ln(\text{price}) = 0.2901\ln(\text{volume}) + 2.5854$$

$$\text{price} = e^{0.2901\ln(\text{volume}) + 2.5854}$$

For 6000-gal reactor

$$\text{price} = e^{0.2901\ln(\text{volume}) + 2.5854}$$

$$\text{pri} = e^{0.2901\ln(6000) + 2.5854}$$

$$\text{pri} = \$165526$$

For 15 000-gal reactor

$$\text{price} = e^{0.2901\ln(\text{volume}) + 2.5854}$$

$$\text{pri} = e^{0.2901\ln(15000) + 2.5854}$$

$$\text{pri} = \$215928$$

Therefore,

$$6000\,\text{gal. reactor} = \$165526$$

$$15000\,\text{gal. reactor} = \$215928$$

7.5 Exercise

7.5.1 Questions

1) What is the mass balance master equation for unsteady vs. steady reactors, respectively?
2) What are the design equations for batch, CSTR, and PFR, respectively?
3) Why reactor volume for PFR is much smaller than that of CSTR for a given conversion factor?
4) Explain the principle involved in selecting different combinations of CSTR and PFR.
5) Mass balance, stoichiometry, and kinetics are considered as three pillars of reactor design in chemical reactors. In water and WWTP, the product is water. How would you apply these three pillars to design a reactor to produce the maximum amount of treated water?
6) For pentachlorophenol, what is the PFR volume required to achieve 100 conversions if all the other reaction conditions remain the same as Tang and Huang (1996a)? Why or why not?

7.5.2 Calculation

1) A continuous-flow reactor will convert A to B at constant temperature. If $C_A = 0.02C_{A0}$, $F_{A0} = 5$ mol/h, $v =$ 10 l/h, $C_{A0} = 0.5$ mol/l, and the entering volumetric flow rate v is 10 l/h, for a constant volumetric flow rate, $v = v_0 = 50$ l/h, $F_{A0} = C_{A0}\ v_0$, and $C_{A0} = F_{A0}/v_0 =$ (5 mol/h)/(10 l/h) = 0.5 mol/l. If the required CSTR volumes are 99, 2 750, and 66 000, respectively, when the reaction has zero, first, and second order, (i) find the corresponding reaction rate constants. And what is the corresponding reactor volume of PFRs?
2) Fenton process requires optimal pH of 3.5 in oxidizing organic pollutants such as chlorophenols. As a result, strong acid such as HCl or H_2SO_4 has to be added. In addition, the Fenton process requires oxidant H_2O_2 and catalyst such as $FeSO_4$, if a vacuum UV reactor is used to oxidize chlorophenols, please answer the following design questions:
   a) How large is the reactor volume required to oxidize 1 mg/l mono-chlorophenol to 0.01 mg/l at a flow rate of 1 MGD?
   b) How much capital and O&M costs could be saved if VUV replaces the Fenton process for the same flow rate?
3) If a Pfaudler reactor is designed for the aforementioned problem, what is the capital cost of the Pfaudler reactor?

7.5.3 Project

7.5.3.1 Xiongan Project

1) On-site WWTP could play a significant role since Xiongan has many villages or satellite cities. For on-site WWT, what type of wastewater treatment reactor would you recommend?
2) As the Xiongan district expands with WW generated from one million people in 1 year and two million people in 5 years, would you still recommend on-site WWTP only? What are the advantages of centralized WWTP?
3) How much sludge and methanol could be saved if your new WWTP adopted anammox annually when OLAND was used in the side and mainstream line, respectively?
4) What would be the O&M cost saving over a life cycle of 50 years when OLAND is used in the side and mainstream line, respectively?
5) What would be the O&M cost saving over a life cycle of 50 years when UV disinfection is to replace chlorination?

7.5.3.2 Proposal Project

1) Make an appointment with the WWTP manager of your local town, and get data on methanol consumption and sludge produced annually. After you have analyzed the WWTP processes, please do the following:
   a) Develop an alternative using anammox and quantify monetary saving by eliminating methanol for nitrification/denitrification for over 50 years. Assuming flow rate is 120 MGD average daily flow; inflow total is 40 mg/l as nitrogen. Forty percent of the load would be removed after AS process. The side stream anammox is being proposed to retrofit the plant. Quantify the monetary saving by eliminating methanol for nitrification/denitrification by using side stream anammox over 50 years. Assume that 0.49 gal of methanol is needed for every pound of nitrogen removed in the conventional nitrification/denitrification process.
   b) Design an alternative using UV disinfection to replace chlorination for this 120 MGD. Following the examples used in the chapter, estimate the capital and O&M costs for both UV disinfection and chlorination over 50 year design life expectancy.

References

Dapena-Mora, A., Fernandez, I., Campos, J.L. et al. (2007). Evaluation of activity and inhibition effects on anammox process by batch tests based on the nitrogen gas production. *Enzyme and Microbial Technology* 40: 859–865.

Egli, K., Fanger, U., Alvarez, P. et al. (2001). Enrichment and characterization of an anammox bacterium from a rotating biological contactor treating ammonium-rich leachate. *Archives of Microbiology* 175: 198–207.

Lackner, S., Gilbert, E.M., Vlaeminck, S.E. et al. (2014). Full-scale partial nitritation/anammox experiences – An application survey. *Water Research*, 55: 292–303.

Levenspiel, O. (1999). *Chemical Reaction Engineering*, 3e. New York: Wiley.

Rosenwinkel, K.-H. and Cornelius, A. (2005). Deammonification in the moving-bed process for the treatment of wastewater with high ammonia content. *Chemical Engineering and Technology* 28: 49–52.

Siegrist, H., Salzgeber, D., Eugster, J., and Joss, A. (2008). Anammox brings WWTP closer to energy autarky due to increased biogas production and reduced aeration energy for N-removal. *Water Science and Technology* 57: 383–388.

Strous, M., Heijnen, J.J., Kuenen, J.G., and Jetten, M.S.M. (1998). The sequencing batch reactor as a powerful tool for the study of slowly growing anaerobic ammonium-oxidizing microorganisms. *Applied Microbiology and Biotechnology* 50: 589–596.

Strous, M., Kuenen, J.G., and Jetten, M.S.M. (1999). Key physiology of anaerobic ammonium oxidation. *Applied Environmental Microbiology* 65: 3248–3250.

Tang, W.Z. and Huang, C.P. (1996a). Effect of chlorine content of chlorinated phenols on their oxidation kinetics by Fenton's reagent. *Chemosphere* 8: 1621–1635.

Tang, W.Z. and Huang, C.P. (1996b). Oxidation kinetics and mechanisms of 2,4-dichlorophenol by Fenton's reagent. *Environmental Technology* 17: 1371–1378.

Van Loosdrecht, M.C.M. and Salem, S. (2006). Biological treatment of sludge digester liquids. *Water Science and Technology* 53: 11.

Wett, B., Hell, M., Nyhuis, G. et al. (2010). Syntrophy of aerobic and anaerobic ammonia oxidisers. *Water Science and Technology* 61: 1915–1922.

8

Efficiency of Renewable Energy

The solar energy harnessed and used measures human civilization.

OBJECTIVES
To show that coal-powered plants are inefficient and cause smog To demonstrate why renewable energy such as solar, biomass, and wind should be used to power the EEIS To calculate entropy, enthalpy, and exergy of organic pollutants in wastewater To present energy auditing procedure of WWTP in terms of unit energy consumption To estimate solar panels required in designing zero energy residential house To quantify energy saving by optimizing pump operation efficiency

8.1 Principle 5

Principle Energy-efficient alternatives should be regenerative or neutral to avoid negative energy design:

1) Renewable energy such as solar, biomass, or wind energy should have priority in designing EEISs.
2) Energy efficiency demands each unit or process to have maximum energy efficiency with renewable energy as the best choice.
3) An integrated system approach should be used to optimize overall efficiency to reduce operation costs of energy consumption.

For EE to improve energy efficiency, Table 8.1 lists design issues and the targets in designing energy-efficient EEIS.

The efficiency of renewable energy such as biomass, solar, and wind should be maximized for all the EEIS. In gauging a civilization, mankind is at its infancy because only 0.75% of solar energy is currently harnessed for human activities. If the efficiency of solar cell were to double, the amount of renewable energy harvested could increase significantly. For example, if 2% of solar energy that reached the Earth could be efficiently utilized, human civilization could reach to the next level. If 2% of energy in the Milky Way galaxy reached the Earth and were harnessed by humans, mankind could colonize most nearby planets. However, it is estimated that about 100 years is needed to harvest 2% of the solar power.

To reduce operation cost of WWTP, the best way is to generate renewable energy such as electricity and heat while increasing energy efficiency of the WWTP. Therefore, energy efficiency is a critical step to retrofit WWTP to energy positive or design and build new WRRF.

Before major improvements of EEIS energy efficiency were started across the United States in 2008, design standards of WWTP and WTP have not changed for the last 100 years since the introduction of the activated sludge process in 1916. Major design theory in air, water, and solid waste management was taught without data and tools critical to design sustainable EEIS. The traditional design philosophy of one size fits all resulted in urgent environmental issues and challenges. For example, to calculate the loading factors, chemical oxygen demand (COD) is used to quantify organic matter in terms of oxygen, while the concept of biological oxygen demand (BOD) was specifically designed for biological treatment of wastewater using activated sludge. As a result, COD and BOD intrinsically represent pollutants in terms of oxygen needed to oxidize and are not meant to represent how much energy is embedded in the organic

Sustainable Environmental Engineering, First Edition. Walter Z. Tang and Mika Sillanpää.

Table 8.1 SEE design metrics.

Design	Problems	Energy
Air	Incomplete combustion	Efficiency of any conversion process is less than 100%
Water	AS in WWTP	Energy intensive and not productive
Land	Pollutants in runoff	Regenerative design for biomass production
Ecological	Energy efficiency in food chain	Higher energy efficiency process

compounds. Only recently was the intrinsic energy contained in the organic pollutants considered as energy resource to be recovered in the design. The paradigm is shifting from organic pollutants as an energy liability required by aeration to energy source to be recovered. In theory, 1.7 kWh energy is contained in one cubic meter of a typical domestic wastewater in the United States. If only 25% of the energy were to be recovered, it could reach energy neutral to run the WWTP at 0.3 kWh/m^3. However, no design criteria in terms of energy efficiency were established to ensure the sustainability of the EEIS. No ranking system exists to assess the sustainability of WWTP or recovery of materials. For these reasons, many challenging questions beg answers: How did the EEIS in the past get it so wrong in designing energy negative WWTPs? How can we treat our wastewater better in terms of minimal footprint on water, energy, and nutrients? What are the regenerative ways to design WWTP? Conceptual, technology, cost, and professional isolation may all contribute to the current challenges faced by the United States and all over the world. Fundamental sciences of thermodynamics may offer us a starting point to shine light into this paradigm shift through creative thinking.

8.2 Challenges and Opportunities

Solar energy has a unit radiation intensity of 1366 W/m^2 and is clean, free, abundant, and reliable renewable energy. If all the solar energy were captured and stored, it could provide all the energy in one day consumed by the mankind for 1 year. Due to low efficiency of energy conversion, however, only 0.1% of 100 000 terawatts (TW) were captured by biomass through photosynthesis. These 100 TW biomasses would be consumed by herbivores to store 10 TW, primary carnivores to store 1 TW, and secondary carnivores to store 0.1 TW. Considering that the total human demand is 15 TW, 15% of primary biomass would meet human demand. Therefore, biomass is a natural candidate to harvest the solar energy in nature. For this reason, algae farming through wastewater is an important emerging technology for future renewable energy. In addition, radiant heat and light from the sun could be harnessed through photovoltaics, solar heating, artificial photosynthesis, solar architecture, and solar thermal electricity.

To achieve zero energy design, solar energy will play critical role to provide electricity or hot water as unit cost of solar photovoltaic technology decreases. To appreciate solar energy, it is important to point out that it is also the sun that produced the fossil fuels such as coal, oil, and natural gas in billion years of Earth's evolution. The National Academy of Engineering proposed the following grand plan for replacing fossil fuel with renewable energy as shown in Figure 8.1. First, in the next 20 years (2005–2025) of continued use of fossil fuels (especially oil) as the predominant source of energy and chemical feedstocks, managing carbon, reducing the intense use of energy resources, and promoting sustainability through educational efforts will be critical. Second, in the next 20–100 years (2025–2105), the use of fossil fuels will be phased out, and the ability to carry out green chemistry and engineering (built on fundamental understanding of the full life cycle impacts and toxicology of chemicals) and having access to alternative renewable sources of fuels and feedstocks will then be critical. In designing EEISs, SEE needs to synchronize the pace of the paradigm shift and contribute to this grand vision. For example, by 2025 all the WWTPs in the United States should be retrofitted to energy positive as energy-production centers, recovery of nutrients by 2035, and recover all the materials by 2045.

In the past, EE designers failed to harvest large potential energy and resource from wastewater because they were encouraged to optimize only parts of the system such as pumps in WWTPs. Components are optimized in isolation from other components; optimization typically considers single rather than multiple benefits. The optimal sequence of design steps is not usually considered because the designers have become highly specialized in their own professions. Integrated engineering design (IED) optimization has not been fully utilized to optimize the whole system. However, IED could enable designers to achieve efficiency improvements from 75 to 95% and to reduce environmental impacts due to the increased efficiency.

8.2.1 Inefficient Combustion of Fossil Fuels

The combustion of fossil fuels results in pollutant emissions that can damage human health and the environment on a number of levels. Most environmental concern has recently focused on global warming induced

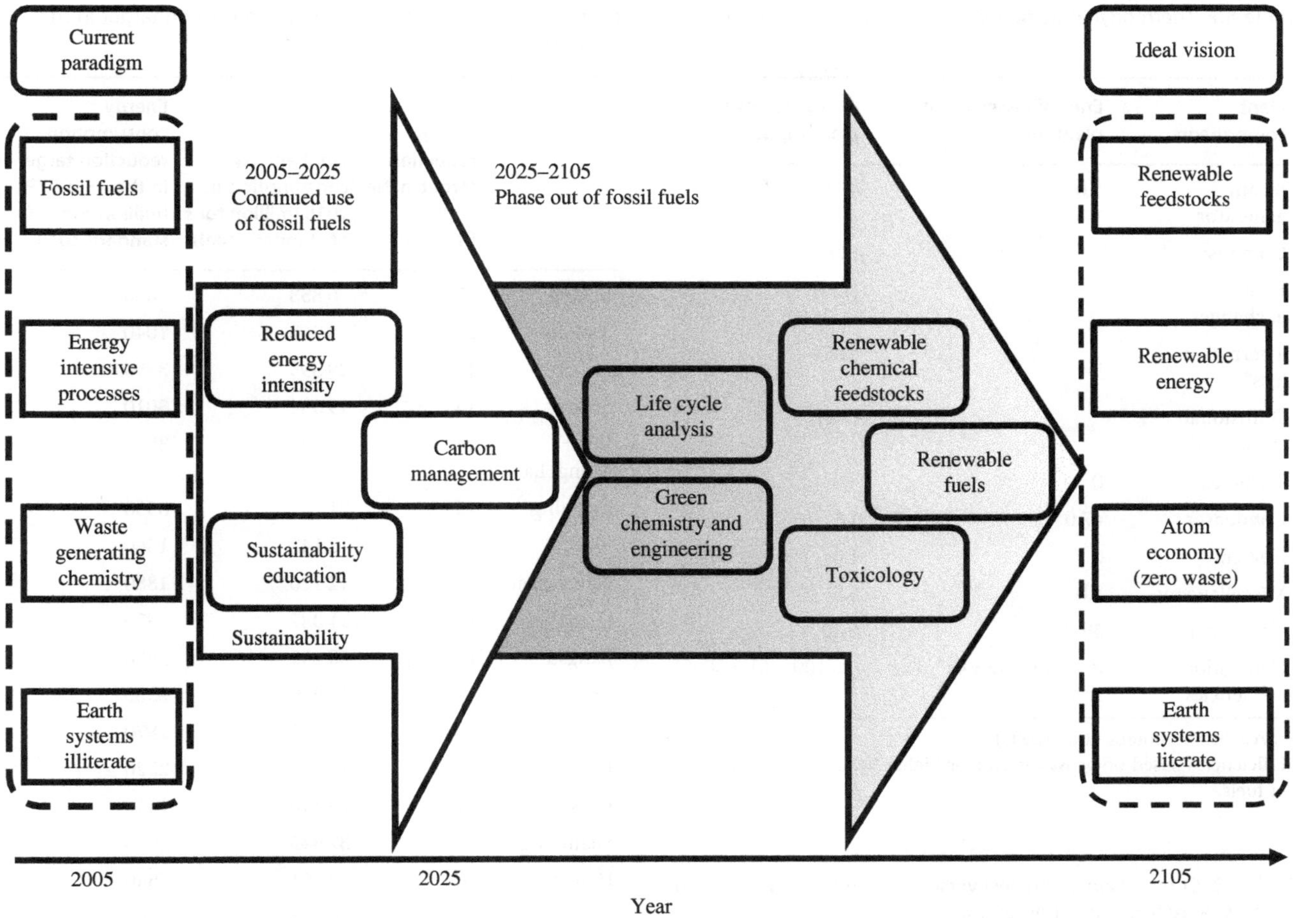

Figure 8.1 The grand challenges (box) for sustainability (arrow) address the transition from current thinking to the ideal vision for the chemical industry over the next 100 years. (*Source:* Reprint with permission from the National Academy of Science.)

by the emission of so-called "greenhouse" gases (GHGs), principally CO_2 emissions from fossil fuel combustion. However, combustion has very low efficiency. Table 8.2 lists the energy efficiency of a coal-powered electrical station. To design the most energy-efficient EEIS, the overall energy efficiency can be estimated using an IED approach. For example, an electric motor driving a pump that transports wastewater in WWTP comprises:

1) An electric motor (sizing and efficiency rating)
2) Motor controls (switching, speed or torque control)
3) Motor drive system (belts, gearboxes, etc.)
4) Pump
5) Pipework
6) Demand for fluid.

The efficiencies of these elements interact in complex ways. For example, if each element in the chain is improved in efficiency by 10%, then the overall level of energy use is $0.9 \times 0.9 \times 0.9 \times 0.9 \times 0.9 \times 0.9 = 0.53$, which results in a saving of 47%. Therefore, an IED approach could yield over 50% improvements. Corresponding to the energy efficiency improvement is the correlated reductions in negative environmental impacts. If the most efficient component is chosen for each part of a motor system, overall the efficiency of the whole system would be about seven times greater as shown in Table 8.3.

8.2.2 Challenges in China

In China, intensive fossil fuel consumption causes severe smog, which results in damage to human health and the environment. As a result, the Chinese government (2017) plans to reduce energy consumption as shown in Tables 8.4, 8.5, 8.6, 8.7, 8.8, and 8.9.

Table 8.2 Thermodynamic performance of coal-fired power stations.

Plant components	Energy losses (% of plant input)	Exergy losses (% of plant input)
Steam generator	9.0	49.0
Combustion[a]		(29.7)
Heat exchanger[a]		(14.9)
Thermal stack loss[a]		(0.6)
Diffusional stack loss[a]		(3.8)
Turbines	0	4.0
Condenser	47.0	1.5
Heaters	0	1.0
Miscellaneous	3.0	5.5
Plant totals	59.0	61.0
Generation efficiencies	$Z = 100 - 59 = 41$	$\eta = 100 - 61 = 39$

Source: Stasinopoulos et al. (2008).
[a] Efficiencies based on gross calorific or higher heating value (HHV) of fuels.

Table 8.3 Comparison of the best versus worst efficiency motoring systems (Stasinopoulos et al., 2008).

System component	Best efficiency	Worst efficiency
Electrical wiring	0.98	0.9
Motor	0.92	0.75
Drive (e.g. gearbox or belt)	1.0	0.7
Pump	0.85	0.4
Pipes	0.9	0.5
Overall efficiency of system	0.69	0.095

8.3 Energy Conservation Laws

Sustainable design alternatives are constrained by basic physical laws and costs. We start with the thermodynamic laws, then exergy, and lastly, technical benchmark data of different processes of WWTPs.

8.3.1 Thermodynamics Laws

Thermodynamic constraints and principles are applicable across both ecological and economic systems. Economic activity is entropic, while entropy increases as order decreases. Entropy leads the tendency of matter and energy to degrade or disperse into less useful forms during economic activity, although waste could be converted back to resources. Recycling of materials could close the loop of materials within society. The second law of thermodynamics requires that materials used in the economy tend to be used entropically. They get

Table 8.4 China energy consumption reduction target in the 13th FYP.

Regions	Energy reduction target in the 13th FYP (%)	Total energy consumption in 2015 (million tons of standard coal)	Energy consumption reduction target in the 13th FYP (million tons of standard coal)
Beijing	17	6 853	800
Tianjin	17	8 260	1040
Hebei	17	29 395	3390
Shanxi	15	19 384	3010
Inner Mongolia	14	18 927	3570
Liaoning	15	21 667	3550
Jilin	15	8 142	1360
Heilongjiang	15	12 126	1880
Shanghai	17	11 387	970
Jiangsu	17	30 235	3480
Zhejiang	17	19 610	2380
Anhui	16	12 332	1870
Fujian	16	12 180	2320
Jiangxi	16	8 440	1510
Shandong	17	37 945	4070
Henan	16	23 161	3540
Hubei	16	16 404	2500
Hunan	16	15 469	2380
Guangdong	17	30 145	3650
Guangxi	14	9 761	1840
Hainan	10	1 938	660
Chongqing	16	8 934	1660
Sichuan	16	19 888	3020
Guizhou	14	9 948	1850
Yunnan	14	10 357	1940
Tibet	10	—	—
Shaanxi	15	11 716	2170
Gansu	14	7 523	1430
Qinghai	10	4 134	1120
Ningxia	14	5 405	1500
Xinjiang	10	15 651	3540

Table 8.5 China emission reduction target of chemical oxygen demand (COD) in the 13th FYP.

Regions	Emission in 2015 (million tons)	Emission reductions in 2020 (%)	Emission reduction of key projects in 2020 (million tons)
Beijing	16.2	14.4	2.33
Tianjin	20.9	14.4	2.47
Hebei	120.8	19	16.14
Shanxi	40.5	17.6	4.75
Inner Mongolia	83.6	7.1	5.19
Liaoning	116.7	13.4	8.41
Jilin	72.4	4.8	2.32
Heilongjiang	139.3	6	7.33
Shanghai	19.9	14.5	2.72
Jiangsu	105.5	13.5	10.39
Zhejiang	68.3	19.2	7.64
Anhui	87.1	9.9	7.7
Fujian	60.9	4.1	2.14
Jiangxi	71.6	4.3	2.73
Shandong	175.8	11.7	13.3
Henan	128.7	18.4	16.98
Hubei	98.6	9.9	8.25
Hunan	120.8	10.1	10.49
Guangdong	160.7	10.4	11.06
Guangxi	71.1	1	0.35
Hainan	18.8	1.2	0.16
Chongqing	38	7.4	2.36
Sichuan	118.6	12.8	14.09
Guizhou	31.8	8.5	2.77
Yunnan	51	14.1	5.85
Tibet	2.9	—	—
Shaanxi	48.9	10	2.63
Gansu	36.6	8.2	2.4
Qinghai	10.4	1.1	0.07
Ningxia	21.1	1.2	0.1
Xinjiang	56	1.6	0.71
Xinjiang production and construction corps	10	1.6	0.04

Table 8.6 China emission reduction target of ammonia nitrogen in the 13th FYP.

Regions	Emission in 2015 (million tons)	Emission reductions in 2020 (%)	Emission reduction of key projects in 2020 (million tons)
Beijing	1.6	16.1	0.24
Tianjin	2.4	16.1	0.38
Hebei	9.7	20	1.59
Shanxi	5	18	0.61
Inner Mongolia	4.7	7	0.28
Liaoning	9.6	8.8	0.85
Jilin	5.1	6.4	0.2
Heilongjiang	8.1	7	0.48
Shanghai	4.3	13.4	0.53
Jiangsu	13.8	13.4	1.25
Zhejiang	9.8	17.6	0.85
Anhui	9.7	14.3	1.07
Fujian	8.5	3.5	0.3
Jiangxi	8.5	3.8	0.32
Shandong	15.3	13.4	1.49
Henan	13.4	16.6	1.93
Hubei	11.4	10.2	1.02
Hunan	15.1	10.1	1.41
Guangdong	20	11.3	1.54
Guangxi	7.7	1	0.08
Hainan	2.1	1.9	0.04
Chongqing	5	6.3	0.32
Sichuan	13.1	13.9	1.74
Guizhou	3.6	11.2	0.41
Yunnan	5.5	12.9	0.67
Tibet	0.3	—	—
Shaanxi	5.6	10	0.38
Gansu	3.7	8	0.28
Qinghai	1	1.4	0.01
Ningxia	1.6	0.7	0.01
Xinjiang	4	2.8	0.09
Xinjiang production and construction corps	0.5	2.8	

dissipated within the economic system unless reused or recovered. However, energy resource is usually degraded and difficult to be recycled depending upon the energy degradation rate. Typically, low-entropy matter/energy comes in, while high-entropy matter/energy goes out. The irreversibility of the real processes implies exergy destruction and waste discharged to the environment. Ecological economics starts from a recognition of the

Table 8.7 China emission reduction target of SO_2 in the 13th FYP.

Regions	Emission in 2015 (million tons)	Emission reductions in 2020 (%)	Emission reduction of key projects in 2020 (million tons)
Beijing	7.1	35	1.8
Tianjin	18.6	25	2.8
Hebei	110.8	28	18.4
Shanxi	112.1	20	22.4
Inner Mongolia	123.1	11	13.5
Liaoning	96.9	20	14.4
Jilin	36.3	18	5.2
Heilongjiang	45.6	11	4.3
Shanghai	17.1	20	3.4
Jiangsu	83.5	20	13.3
Zhejiang	53.8	17	9.1
Anhui	48	16	5.2
Fujian	33.8	—	3.5
Jiangxi	52.8	12	6.3
Shandong	152.6	27	35
Henan	114.4	28	20.5
Hubei	55.1	20	10.9
Hunan	59.6	21	8.5
Guangdong	67.8	3	2
Guangxi	42.1	13	4.5
Hainan	3.2	—	0.4
Chongqing	49.6	18	8.1
Sichuan	71.8	16	11.2
Guizhou	85.3	7	6
Yunnan	58.4	1	0.6
Tibet	0.5	—	—
Shaanxi	73.5	15	11
Gansu	57.1	8	4.6
Qinghai	15.1	6	0.9
Ningxia	35.8	12	4.3
Xinjiang	66.8	3	2
Xinjiang production and construction corps	11	13	0.9

Table 8.8 China emission reduction target of nitrogen oxides in the 13th FYP.

Regions	Emission in 2015 (million ton)	Emission reductions in 2020 (%)	Emission reduction of key projects in 2020 (million tons)
Beijing	13.8	25	0.7
Tianjin	24.7	25	3.5
Hebei	135.1	28	19.9
Shanxi	93.1	20	16.3
Inner Mongolia	113.9	11	12.5
Liaoning	82.8	20	14.9
Jilin	50.2	18	9
Heilongjiang	64.5	11	7.1
Shanghai	30.1	20	5.2
Jiangsu	106.8	20	18.7
Zhejiang	60.7	17	10.3
Anhui	72.1	16	9
Fujian	37.9	—	4.6
Jiangxi	49.3	12	5.9
Shandong	142.4	27	31
Henan	126.2	28	15.8
Hubei	51.5	20	5.9
Hunan	49.7	15	6.3
Guangdong	99.7	3	3
Guangxi	37.3	13	3.3
Hainan	9	—	1.2
Chongqing	32.1	18	2.8
Sichuan	53.4	16	3.7
Guizhou	41.9	7	2.9
Yunnan	44.9	1	0.4
Tibet	5.3	—	—
Shaanxi	62.7	15	9.4
Gansu	38.7	8	3.1
Qinghai	11.8	6	0.7
Ningxia	36.8	12	4.4
Xinjiang	63.7	3	1.9
Xinjiang production and construction corps	9.9	13	1.3

Table 8.9 China emission reduction target of volatile organic carbon in the 13th FYP.

Regions	Emission in 2015 (million tons)	Emission reductions in 2020 (%)	Emission reduction of key projects in 2020 (million tons)
Beijing	23.4	25	3.5
Tianjin	33.9	20	4.6
Hebei	154.6	20	19.5
Liaoning	105.4	10	10.5
Shanghai	42.1	20	8.4
Jiangsu	187	20	31.2
Zhejiang	139.2	20	25.5
Anhui	95.9	10	9.2
Shandong	192.1	20	38.4
Henan	167.5	10	16.6
Hubei	98.7	10	9.9
Hunan	98.3	10	7.9
Guangdong	137.8	18	20.7
Chongqing	40.2	10	4
Sichuan	111.3	5	5.6
Shaanxi	67.5	5	3.4

environmental, technological, individual, and social sources and support systems of productivity. Both natural and economic capitals should be maximally utilized. Table 8.10 is a brief history of sustainability by applying thermodynamic laws.

Table 8.10 Brief history of applying thermodynamic laws to sustainability limit of economic activities.

Year	Author	Research work	References
1969	R. U. Ayres and A. V. Kneese	Introduced mass and energy balances into static input–output analysis	Ayres and Kneese (1969)
1971	N. Georgescu-Roegen	Consequences of thermodynamic laws for economic analysis	Georgescu-Roegen (1971)
1986	Perrings	Extended input–output analysis to linear dynamic models that showed instabilities of economic activity	Perrings (1986)
1992	Van Gool W.	Exergy analysis of industrial processes	Van Gool (1992)
2006	F. C. Krysiak	Mass and energy conservation with the second law of thermodynamics may challenge the fundamental concepts of economic analysis	Krysiak (2006)
2009	Geoffrey P. Hammond and Adrian B. Winnett	Thermodynamic ideas on ecological economics	Hammond and Winnett (2009)

8.3.2 The First Thermodynamic Law

The first thermodynamic law defines conservation of materials and energy, which are not to be created or destroyed. Pearce and Turner (1990) analyzed the production process waste in the environment–economy interaction. The processing of resources creates waste W_R, as with overburden tips at coal mines; production creates waste W_P in the form of industrial effluent and air pollution and solid waste; final consumers create waste W_C by generating sewage and municipal solid wastes. The amount of waste in any period equals the amount of natural resources used, R:

$$R = W = W_R + W_P + W_C \tag{8.1}$$

The first law of thermodynamics states that energy and matter cannot be created or destroyed, but only transformed. Due to insatiable human demand, mining depletes natural resource, while consumers discard waste into the environment. As a result, mounting tasks are in front of the new generation of SEE designers to sustainably manage air, water, and solid wastes produced by modern societies.

8.3.3 The Second Thermodynamic Law

The second thermodynamic law states that work input into a system can be fully converted to heat and internal energy (via dissipative processes) without perfect efficiency. The maximum amount of work achievable in a given system with different energy sources in reference to its surrounding is referred as exergy. Exergy is the available energy for conversion from a donating source (or reservoir) with reference to a specified datum under ambient environmental conditions. It predicts the theoretical maximum energy available. Usually, high exergy degrades to low exergy continuously and irrevocably.

8.3.3.1 Energy Conversion

Example 8.1

Given: COD is traditionally referring to the amount of organic matter in wastewater in terms of amount of oxygen required to stabilize the organic matter chemically.

Find: How much energy in kJ is there in 1 kg COD of organic matter?

Solution:

Energy released from organic matter depends upon the final reduction forms of the carbon.

Step 1: Theoretical mole of electron can be accepted by 1 kg oxygen:

$$\frac{1}{4}O_2 + H^+ + e^- \rightarrow \frac{1}{2}H_2O$$

1 kg COD is equivalent to 125 mol electrons when oxygen is completely reduced.

Step 2: Depending upon different electron acceptors, the above 125 mole of electrons contains the following kJ energy:

1) During reduction of CO_2 to CH_4:

$$\frac{1}{8}CO_2 + H^+ + e^- \rightarrow \frac{1}{8}CH_4 + \frac{1}{4}H_2O$$

According to thermodynamics, 1 kg COD will be equivalent to 12 500 kJ.

2) Reduction to hydrogen:

$$H^+ + e^- \rightarrow \frac{1}{2}H_2$$

Thermodynamically, 1 kg COD would be equal to 17 800 kJ.

3) Production of electricity:

$$\Delta G = -\text{emf} \times n \times F$$

Then, 1 kg COD would produce 8300 kJ.

8.3.3.2 Enthalpy

Enthalpy is the heat content of gases, which depends not only on the internal energy (*U*) but also on the pressure (*P*) and volume (*V*). Thermodynamically, enthalpy is defined as

$$H = U + PV \tag{8.2}$$

where

H = enthalpy (kJ)
U = internal energy (kJ)
P = pressure (kPa)
V = volume (m^3)

The reaction heat is the enthalpy change of a reaction:

$$H_{\text{products}} - H_{\text{reactants}} = \Delta H$$

If there is no pressure change, then this is the heat absorbed:

ΔH positive → heat absorbed → *endothermic*
ΔH negative → heat released → *exothermic*

Example 8.2
Given: Combustion of 1 mol of propane and the enthalpy as follows:

$$C_3H_8(g) = -104.7\ \text{kJ/mol}$$
$$O_2(g) = 0\ \text{kJ/mol}$$
$$CO_2(g) = -393.5\ \text{kJ/mol}$$
$$H_2O(8) = -241.8\ \text{kJ/mol}$$

Find:

1) The heat of combustion of 1 mol propane to form liquid water
2) The net heat of combustion of 1 mole propane to form water vapor

Solution: Heat of combustion is the gross heat of combustion or higher heating value (HHV) and assumes liquid water as the product. The net heat of combustion or lower heating value (LHV) assumes gaseous water as the product, and high-efficiency (>90%) furnaces recover this heat by condensing water vapor.

1) From thermodynamic tables *H* (kJ/mol) of different reactants and products, the HHV released should be

$$\underset{(-104.7)}{C_3H_8(g)} + \underset{5(0)}{5O_2(g)} \rightarrow \underset{3(-393.5)}{3CO_2(g)} + \underset{4(-285.8)}{4H_2O(l)}$$

$$\Delta H = [3(-393.5) + 4(-285.8)] - [(-104.7) + 5(0)]$$
$$= -2219\ \text{kJ/mol} \rightarrow \text{heat released}$$

2) Net heat of combustion of propane can be found for water vapor, *H* (kJ/mol):

$$\underset{(-104.7)}{C_3H_8(g)} + \underset{5(0)}{5O_2(g)} \rightarrow \underset{3(-393.5)}{3CO_2(g)} + \underset{4(-241.8)}{4H_2O(g)}$$

$$\Delta H = [3(-393.5) + 4(-241.8)] - [(-104.7) + 5(0)]$$
$$= -2043\ \text{kJ/mol} \rightarrow \text{heat released}$$

8.3.3.3 Conservation of Energy

For nonbiodegradable materials in wastewater such as cellular fiber from papers, incineration could be the best process to recover heat energy (Figure 8.2). However, for a thermal power plant, about one-fourth or one-third of exergy is wasted due to the inefficiency of mixing and heat loss. On the other hand, chemical efficiency is usually much higher at 94–97%. For this reason, a fuel cell is a

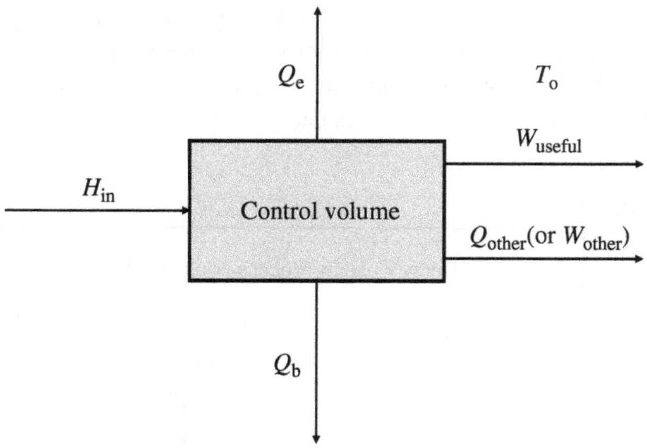

Figure 8.2 An energy balance for a simple control volume or unit operation.

much more efficient electricity generation process than a thermal power plant.

8.4 Energy Balances

Energy balances are analogous to mass balances because energy can be neither created nor destroyed. A system can be defined by the boundary separating the system from its surroundings. Open systems refer to mass and energy transfer across system boundaries, while a closed system is only energy transfer across system boundaries. *Internal energy* (U) is associated with mass by virtue of its state (phase) and temperature. *Enthalpy* (H) is associated with mass by virtue of its state, temperature, and pressure:

$$H = U + PV \tag{8.3}$$

where

H = enthalpy (kJ)
U = internal energy (kJ)
P = pressure (kPa)
V = volume (m^3)

Energy changes in a system are a function of mass and energy inputs and outputs. For systems involving only solids and liquids, the PV term is essentially constant, so $\Delta U = \Delta H$:

$$\Delta H = Mc_V \Delta T \tag{8.4}$$

where

M = mass
C_V = specific heat (J/kg/°C, BTU/lb/°F, cal/g /°C)
ΔT = change in temperature

When a substance changes phases, there are changes in internal energy and enthalpy not directly related to temperature change. Energy required from solid to liquid is referred as the enthalpy of fusion, while energy required from liquid to gas is called the enthalpy of vaporization.

Example 8.3
Given: 1.00 kg of +20°C water needs to be cooled by –20°C ice to 0°C ice in a container.
Find: How much ice at –20°C must be added to a well-insulated container with 0.25 kg of +20°C water in order for the final temperature to be 0°C and all the ice to be melted (ignore the heat capacity of the container)?

Solution:

1) Energy needed to cool 1.00 kg of +20°C water equals the energy released from x gram of –20°C ice (x_{ice}).
2) Heat loss by initial water $= M_{H_2O} c_V \Delta T$
   $= (1.00\,\text{kg})(4.184\,\text{kJ/kg/°C})(20 - 0°\text{C}) = 83.68\,\text{kJ}$
3) Heat need to raise ice temp. $= x_{ice} c_V \Delta T$
   $= x_{ice}(2.302\ \text{kJ/kg/°C})(0°\text{C} - (-20°\text{C}))$
   $= x_{ice}(46.05)\text{kJ/kg}$
   Heat needed to melt the ice $= x_{ice} H_{fusion}$
   $= x_{ice}(333\ \text{kJ/kg})$
4) Balance the terms:

$$\text{Heat gain by ice} = \text{Heat loss by water}$$

$$x_{ice}(46.05 + 333)\text{kJ/kg} = 83.68\ \text{kJ}$$

$$x_{ice} = 0.22\text{kg} = 220\,\text{g (about 8 ice cubes)}$$

Answer: 220 g of –20°C ice is required to cool 1.00 kg of +20°C water to 0°C.

Energy can be transferred through conduction, convection, and radiation. Conduction normally associated with solids, e.g., vibrational energy of hotter molecules translated to adjacent molecules:

$$q = h_{tc} A \frac{\Delta T}{x} \tag{8.5}$$

where

q = heat transfer rate (J/s)
h_{tc} = thermal conductivity (J/s/m/K)
A = cross-sectional area (m^2)
T = temperature (K)
ΔT = change in temperature (K)
x = linear distance corresponding to temperature differential (m)

Convection transfers heat from a fluid as a result of movement of the fluid and contact with another substance. For example, a pump or blower causes coolant in auto or hot air from a heat duct by forced convection, while natural convection is a result of density differences and plays an important role in designing energy positive residential houses.

Empirical equation for convective heat transfer between a fluid and surface can be expressed as

$$q = h_c A(T_s - T_f) \quad (8.6)$$

where

h_c = system-specific constant (J/s/m^2/K)
T_f = temperature of the fluid
T_s = temperature of the surface

In the second law of thermodynamics, there is always inefficiency when energy is converted to work because waste heat is always produced.

8.4.1 Physical Framework by Thermodynamics

Energy, entropy, and exergy have been applied in environmental sustainability. Energy transformations within society have been viewed as resource depletion and environmental degradation by ecology, economics, and engineering.

Figure 8.3 shows that there is a trade-off between capital cost, which is referred to as embodied energy, and OM cost.

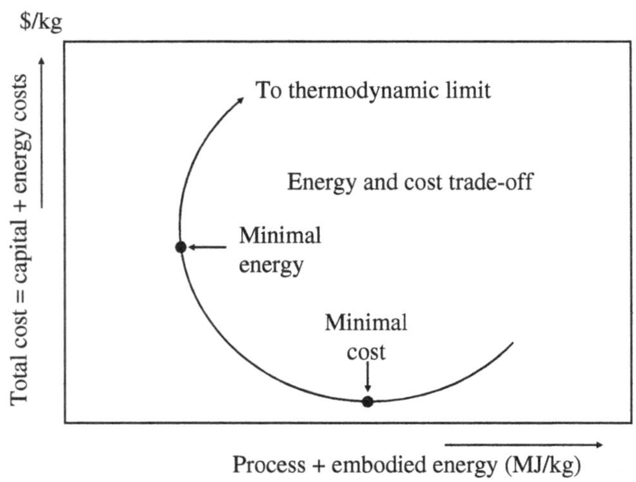

Figure 8.3 Trade-off between capital and operating costs. (*Source:* Adapted from Hammond and Winnett (2009).)

Figure 8.4 plots the energy consumption against equipment size. There is a minimal value associated with the total energy requirement at a specific equipment size, which is also referred to as capital cost.

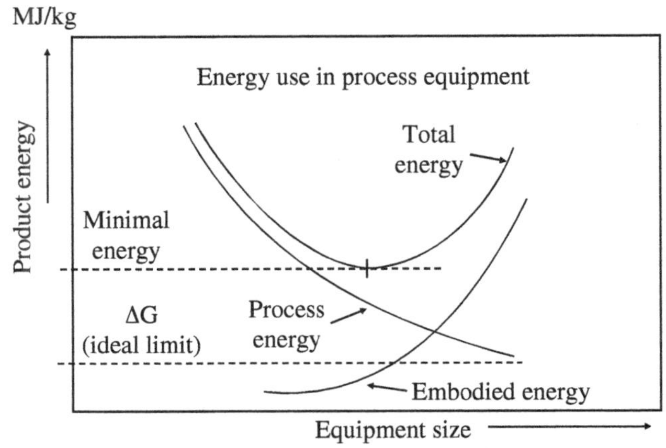

Figure 8.4 Trade-off between product energy and equipment size. (*Source:* Adapted from Hammond and Winnett (2009).)

Hammond and Winnett (2009) argued that although the thermodynamic (or exergetic) improvement potential is around 80%, only about 50% of energy currently used could be saved by technical means, and when economic barriers are taken into account, this is reduced to perhaps some 30%. Nevertheless, there is significant potential for improvement in energy efficiency as shown in Figure 8.5.

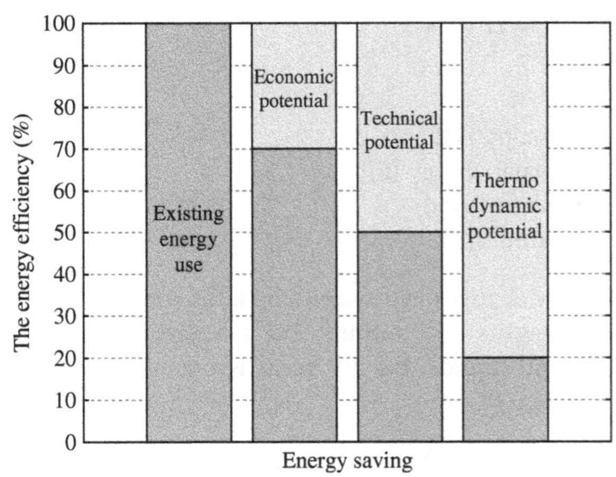

Figure 8.5 The energy efficiency gap between theory and practice. (*Source:* Adapted from Hammond and Winnett (2009).)

8.4.2 Exergy

Exergy is the maximum theoretical work that a system can do before being brought into equilibrium with its environment. It is not conserved due to irreversibility, friction, or heat loss due to temperature difference. Exergy represents a theoretical thermodynamic value, while energy efficiency reflects an actual system. It measures the thermodynamic efficiency of a system by reflecting the quality of an energy carrier. In other words, exergy is simply a measure of the maximum theoretical useful work that is obtainable from a thermal system (as it is brought into equilibrium with its surrounding environment). Therefore, exergy analysis should be employed as a tool among several quantitative approaches to study energy systems, in addition to the traditional first law energy analysis:

$$E = (U - U_o) + P(V - V_o) - T_o(S - S_o) + \Delta PE + \Delta KE \tag{8.7}$$

where

E = exergy
U = internal energy
P = pressure
V = volume
T = thermodynamic temperature
S = entropy
ΔKE = change in kinetic energy
ΔPE = change in potential energy
Subscript "$_o$" = reference environmental state

When a system is not moving nor has potential energy, it can be simplified as

$$E = (H - H_o) - T_o(S - S_o) \tag{8.8}$$

Material resources are viewed as having chemical exergy, because of their disequilibrium with surrounding conditions. An increase in process efficiency will reduce the amount of waste and exergy degradation. Thus, exergy is seen as a measure of the value of the material, whereas process wastes have a potential to cause adverse changes in the environment. The exergy can be used as a potential tool for resource and/or emission accounting.

Example 8.4
Given: 1 kg of steam at 500°C and 1 MP.
Find: How much exergy is there in 1.0 kg of steam at 500°C and 1 MP?

Solution: Since the temperature and pressure are constant, therefore, $\Delta PE = 0$ and $\Delta KE = 0$:

$$\text{Internal energy } \Delta U = 1.0\,\text{kg}\,(3124.34 - 104.86)\,\text{kJ/kg} = 3019.60\,\text{kJ}$$

$$P\Delta V = 101\,\text{kPa}\,(1.0\,\text{kg})(0.354 - 0.001)\,\text{m}^3/\text{kg} = 35.60\,\text{kJ}$$

$$T_o\Delta S = 298\,\text{K}\,(1.0\,\text{kg})\,(7.762 - 0.367)\,\text{kJ/kg/K} = 2203.60\,\text{kJ}$$

$$E = (U - U_o) + P(V - V_o) - T_o(S - S_o) + \Delta PE + \Delta KE$$

$$E = 3019.60 + 35.60 - 2203.60 = 851.60\,\text{kJ}$$

Exergy is a positive value, which measures the departure from the environment. If the end state of a system is in both thermal and mechanical equilibriums with its environment, exergy is the minimal theoretical work to bring the system from dead state to the given state. Human economic activity increases material entropy and decreases the exergy of energy.

8.5 Benchmarks for Unit Energy Consumption in WTP and WWTP

8.5.1 Unit Energy Consumption Values in WTP

For energy auditing in water treatment plants, typical unit value of energy consumption for each unit process of a WTP is very important. Table 8.11 compiles these unit values for WTP from Howe et al. (2012). It could be used as a benchmark in conducting energy auditing of any WTP.

8.5.2 Unit Energy Consumption Values in WWTP

For WWTPs, unit energy consumption data are extremely important when energy auditing is carried out in a WWTP. More importantly, the benchmark data for each unit process are critical to identify the energy gaps where energy efficiency could be improved by retrofitting the current WWTPs. Table 8.12 lists unit energy consumption data of each unit process of WWTP.

In the net zero guideline for WWTP, best practice values are significantly less than the typical value among 24 WWTPs surveyed in New York (Trarallo et al., 2015), so there is great potential to improve the performance of energy inefficient processes so that WWTPs can be retrofitted to be energy positive. In the last decade, major research efforts have been carried out all over the world such as in Singapore and Beijing, China (Gans et al., 2007) (Table 8.13). It shows that the overall energy efficiencies of WWTPs in Jurong and Gaobeidian are 35.9 and 31%, respectively. Since aeration and dewatering are the most

Table 8.11 Unit energy consumption and retrofitting measures for water treatment plants.

Unit process	Operation condition	Unit energy consumption (kWh/m^3)	Retrofitting measures
Coagulation	$G = 1200$ (1/s)	0.0004 (theoretical)	Motionless mixer
	$t = 1$ (s)	0.0014 (in practice)	
Flocculation	$G = 50$ (1/s)	0.0013	Motionless mixer
	$t = 30$ (min)		
Sedimentation	$\Delta h = 0.6$ m (2 ft)	0.0016	Plate or tube settler
Granular filtration	$\Delta h = 5–4$ m during filtration	0.01–0.014	Increasing filtration rate from 10 to 15 m/h (4–6 gpm/ft^2), total filter area decreases 33%
	$\Delta h = 1$ m (4 gpm/ft^2) every 24 h interval for backwashing	0.0007–0.0031	Reduce solid concentration before filtration
Membrane filtration	Low flux 6 gal/ft^2/day (or 10 l/m^2/h)	0.022	Impacts: global warming, human, aquatic, marine, and terrestrial toxicity
			Multistage stage membrane filtration
	Medium flux 18–36 gal/ft^2/day (or 30–60 l/m^2/h)	Photooxidant formation potential	Multistage stage membrane filtration
		Eutrophication	Dead-end flow is better than cross-flow membrane filtration
Reverse osmosis	TDS 35 000 mg/l	5.6	Better membrane
	50% recovery		Two or more stages with booster pumps between stages
	85 bar (1230 psi)		Water recovery rate varies from 50 to 85%
			Brine concentrators and crystallizers
Adsorption	$\Delta P = 1.7$ bar (25)	0.06	Regeneration of 1 kg GAC releases 2–5 kg CO_2
Ion exchange	$\Delta P = 1.7$ bar (25)	0.06	Brine management by evaporation ponds and falling film evaporator
Air stripping	Countercurrent packed towers with height from 10 to 15 m	From 0.051 to pump water to 15 m	Optimal air/water = 3.5 for minimal pressure drop from 50 to 100%
		0.025–0.051 for air blower	
Advanced oxidation process	O_3: 6–10 kWh/kg O_3	0.04–0.08	Fine bubble diffuser to increase mass transfer of O_3
	4–6 mg/l		
	H_2O_2: 2–4 kWh/kg H_2O_2	0.003–0.01	Optimal pH
	H_2O_2/O_3	0.043–0.09	Optimal pH and fine bubble diffuser to increase mass transfer of O_3
	H_2O_2 1.4–2.8 mg/l		
	UV		
	40 mJ/cm^2 for disinfection	0.003–0.025	Reduce turbidity of water
	100–1000 mJ/cm^2 for oxidation	0.0075–0.63	Better UV reactor design
	UV/H_2O_2		
	H_2O_2 3–5 mg/l	0.01–0.9	Optimal pH
	H_2O_2 3–100 mg/l for remediation	0.004–0.28	Better UV reactor design

Source: Data from Howe et al. (2012).

Table 8.12 Unit values of benchmark parameters.

Benchmark parameters	Unit values
Energetic content of methane	803 kJ/mol
Energetic content of hydrogen gas	286 kJ/mol
Biomass caloric content	20.4–25.6 MJ/kg fuel
Biomass heat content by combustion with 0.7 efficiency	14.2–21.4 MJ/kg fuel
Crude oil heat content by hydrothermal	33.2 MJ/kg fuel
Liquefaction with efficiency from	
Methane by anaerobic digestion with efficiency from 0.4 to 0.6	35.8 MJ/kg fuel
Biodiesel by transesterification with efficiency from 0.1 to 0.3	37.2 MJ/kg fuel
N/P limited determination (Redfield ratio)	<16 is N limited
Energy content of COD	3.86 kWh/kg COD
Combustion energy yield	14.2–21.4 MJ/kg
COD production per capita per day	180
N production per capita per day	13
P production per capita per day	2.1
Lipid storage compound	Stearic acid ($C_{18}H_{36}O_2$)
Carbohydrate storage compound	Glucose ($C_6H_{12}O_6$)
Protein compound	$C_{16}H_{24}O_5N_4$

energy-intensive processes, typical and best practice values in terms of per kg pollutant removed, and unit energy consumptions in WWTPs all over the world have been compared in Table 8.14.

In the United States, the energy consumption data of 24 WWTPs in New York were reported (Tarallo et al., 2015). Here, technical benchmarks are statistically analyzed. Table 8.15 compares typical values and best practice values in terms of per kg pollutant removed and unit energy consumption in 24 WWTPs in the state of New York.

According to these data, the following histogram and cumulative plots are produced to establish the benchmark unit energy consumption values at percentiles of 10, 25, 50, and 90 to facilitate energy auditing and identify opportunities to retrofit existing WWTPs (Figures 8.6, 8.7, 8.8, 8.9, 8.10, 8.11, 8.12, 8.13, 8.14, 8.15, 8.16, 8.17, 8.18, 8.19, 8.20, and 8.21).

To provide quantitative unit energy consumption benchmark data, Table 8.15 listed values at normal and best practice of the 24 WWTPs in New York in terms of MJ/MG for normalized 10 MGD WWTP. In terms of unit

Table 8.13 Comparison of overall plant-specific energy consumption between Jurong (Singapore) and Gaobeidian, Beijing (China) (Gans et al., 2007).

	Jurong	Gaobeidian, Beijing
Population served (PE)	0.99 million	2.4 million
Flow (m^3/day)	0.154 million	1 million
BOD_5 (mg/l)	329	173
COD (mg/l)	924	346
Total electricity consumption (kWh/m^3)	0.45	0.258
Total electricity consumption (kWh/kgCOD)	0.58	0.746
Total electricity consumption (kWh/kgO_2D)	0.58	0.443
Preliminary (kWh/m^3)	0.002 Fine screen-perforated plate, horizontal flow grit chamber	0.083 Coarse screen (6), fine screen (4), and grit chamber
Primary treatment (kWh/m^3)	0.0012 Gravity sedimentation	0.015 Traveling bridge, mixer, shredder, sludge pump
Aeration (kWh/m^3)	0.169 Surface-fixed and variable speed aerators	0.1484 Centrifugal blowers, fine bubble aerators, mixer
Sludge thickening (kWh/m^3)	0.003 Dissolved air flotation	0.002 Gravity thickening
Sludge dewatering (kWh/m^3)	0.021 Centrifuge, belt filter press	0.003 Belt filter press
Digester (kWh/m^3)	0.043 Draft tube system, gas injection system, mesophilic, no external heating	0.007 Mesophilic, external heating: spiral-plate heat exchange
Pump system (kWh/m^3)	0.151	—
Total electricity generation(kWh/m^3)	0.160 Gas collection, genset electricity generation	0.081 Gas holder for collected gas, methane booster compressor (8)
Energy efficiency (%)	35.9	31

Table 8.14 Unit energy consumption of WWTP at various locations.

Energy consumption (kWh/m^3)	Austria	Austria (Strass)	Sweden	China (Beijing)	Japan	Iran
Total energy consumption	0.304	0.317	0.475	0.258	0.320	0.300
Preliminary	0.039	0.068	0.136	0.083	0.013	0.013
Primary	—			0.015		0.001
Aeration	0.212	0.181	0.226	0.148	0.148	0.231
Sludge thickening	0.040	0.040	0.068	0.002	0.100	0.021
Sludge dewatering				0.003		
Sludge digester				0.007		
Pumping	0.013	0.028	0.045	—	0.059	0.034
Total energy generation	—	0.346	—	0.081	0.170	0.182
Energy efficiency (%)	—	109	—	31	50	60.67

(1) Austria: Jonasson (2007); (2) Austria (Strass): Jonasson (2007); (3) Sweden: Jonasson (2007); (4) Iran: Nouri et al. (2006); (5) China (Beijing): Gans et al. (2007); (6) Japan: Mizuta and Shimada (2010).

energy consumption, influent pump station, biological treatment, and sludge handling are three major components. For each unit process, best minimal values are significantly less than the normal minimal values while the best maximal values are much smaller than that of the normal maximal values. Therefore, it could be used as an excellent technical benchmark in identifying energy efficiency project for retrofitting purpose of any WWTP. To visualize these data, histogram and cumulative curves for each unit process would help SEE designer to identify feasible technical targets in retrofitting projects at different quartiles in terms of the unit energy consumption performance. For example, 50% of typical bar screen and grit removal would consume 13.5 MJ/MG as shown in Figure 8.8 for side stream pump. If the WWTP is to be retrofit to achieve 10% in terms of unit energy consumption, the unit energy consumption for bar screen and grit removal should be less than 9 MJ/MG as shown in Figure 8.8.

Table 8.15 Summary of the unit energy consumption for best and typical practice in 24 WWTPs in New York.

MJ/MG Activity	Normal 10%	Best 10%	Normal 25%	Best 25%	Normal 50%	Best 50%	Normal 75%	Best 75%	Normal 90%	Best 90%	Normal Min	best Min	Normal Max	Best Max
Influent pump station									210	148	210	148	257	181
Screening and grit removal									188	188			188	188
Screening and grit removal – sidestream pump	9	4	11	6	14	9.5	18.5	14	21	17	8.32	3.8	26.45	17.25
Primary clarifiers	84.5	95.5	85.5	96.3	88	98	89.5	100.5	91.5	102.5	84.4	95.8	93.2	102.9
Biological reactors and final clarifiers	46	36	49	39	54.5	45	80.5	51	125.1	106	45.9	37.8	126.9	119.9
Disinfection – discharge	11.5	11.5	15.4	15	16.75	16.6	24.1	24	24.5	24.5	9.8	10.4	24.7	24.7
Gravity thickener	76	92	77	94	82	96	85.5	101	123	136	79.1	90.4	148.8	153.2
Gravity thickener (sidestream pump)	4	4.4	4.57	5.1	5	6.2	6.1	12	20.1	13.4	3.8	4.5	20.1	14.75
Mechanical thickener	43	33	46	36	50	47	79	100	108	125	44	37	132	132
Mechanical thickener (sidestream pump)	2.1	0.6	2.6	0.8	2.8	1.25	4	1.8	7	3.6	1.9	0.7	7.9	3.6
Lime stabilization – cake	72	72	75	75	80	80	75	105	105	108	70.7	71.8	101.9	107.1
Anaerobic digestion	67	55	63	68	65	72	70	74	74	77.5	49.3	36.5	76.3	79
Anaerobic digestion – dewatering	54	51	58	52	61	56	66.5	58	73	64	54.9	51.3	74.4	66.4

Table 8.15 (Continued)

MJ/MG Activity	Normal 10%	Best 10%	Normal 25%	Best 25%	Normal 50%	Best 50%	Normal 75%	Best 75%	Normal 90%	Best 90%	Normal Min	best Min	Normal Max	Best Max
Flare	22.5	36	24	38	28	44	38	50	40	51	20.8	33.6	42.6	51.3
Boiler	17.4	15.2	17.8	15.4	19.7	16	24	19.5	24	20	17.5	14.2	25.4	20.7
Dewatering – cake	49	48.5	52	50.5	56	54	62	60.5	68	65.5	49.6	49.4	69.8	69.5
Dewatering – sidestream pump	4.53	2.1	4.3	2.2	5.1	2.4	5.8	2.7	11	4	17.5	2.14	11.1	5.9
Disinfection – discharge	11.5	11.5	15.4	15	16.75	16.6	24.1	24	24.5	24.5	9.8	10.4	24.7	24.7
Gravity thickener	76	92	77	94	82	96	85.5	101	123	136	79.1	90.4	148.8	153.2
Gravity thickener (sidestream pump)	4	4.4	4.57	5.1	5	6.2	6.1	12	20.1	13.4	3.8	4.5	20.1	14.75
Mechanical thickener	43	33	46	36	50	47	79	100	108	125	44	37	132	132
Mechanical thickener (sidestream pump)	2.1	0.6	2.6	0.8	2.8	1.25	4	1.8	7	3.6	1.9	0.7	7.9	3.6
Lime stabilization – cake	72	72	75	75	80	80	75	105	105	108	70.7	71.8	101.9	107.1

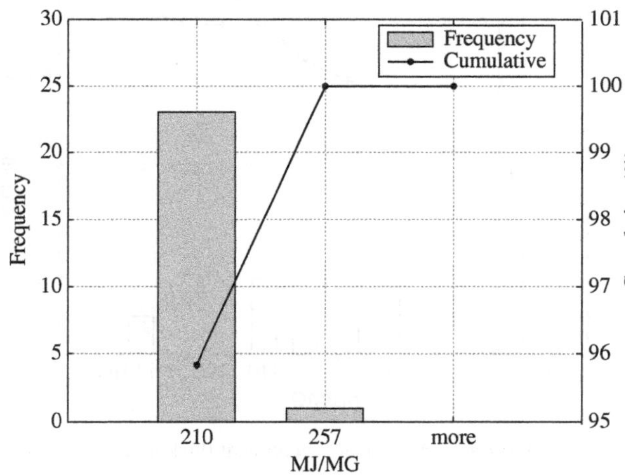

Figure 8.6 Typical influent pump station.

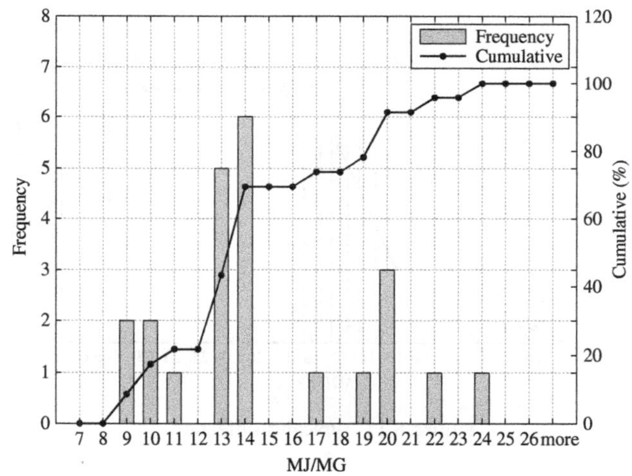

Figure 8.8 Typical screening and grit removal – sidestream pump.

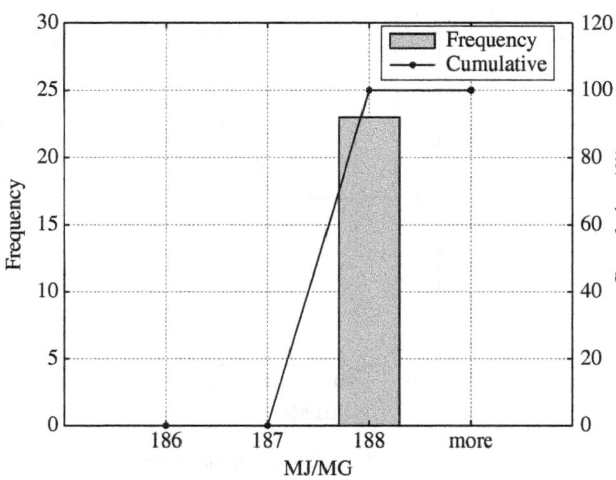

Figure 8.7 Typical screening and grit removal.

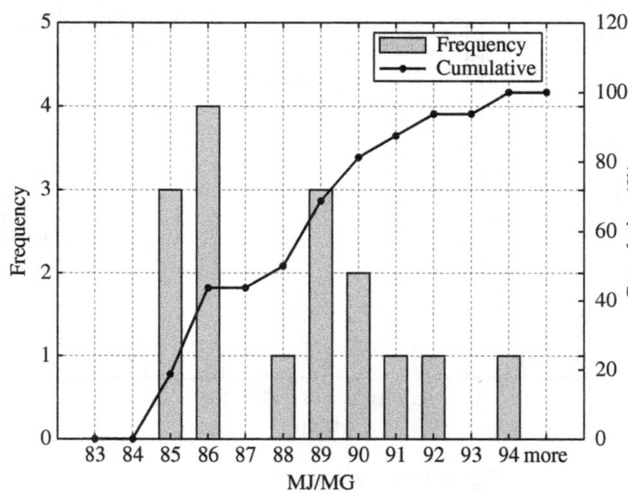

Figure 8.9 Typical primary clarifier.

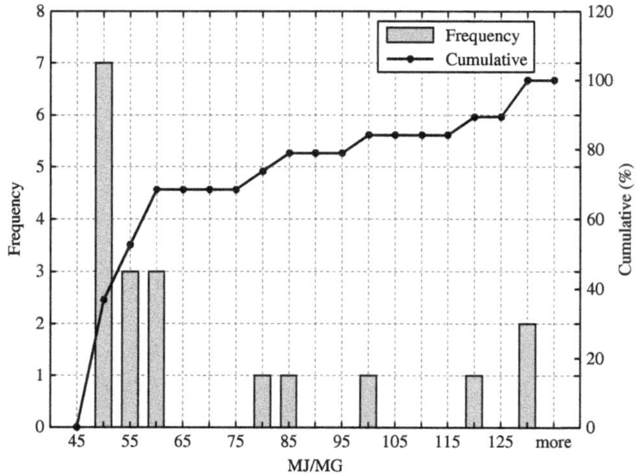

Figure 8.10 Typical biological reactor and final clarifiers.

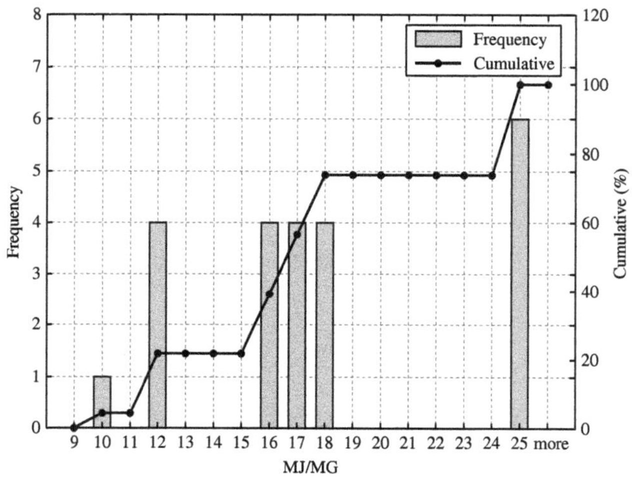

Figure 8.11 Typical disinfection.

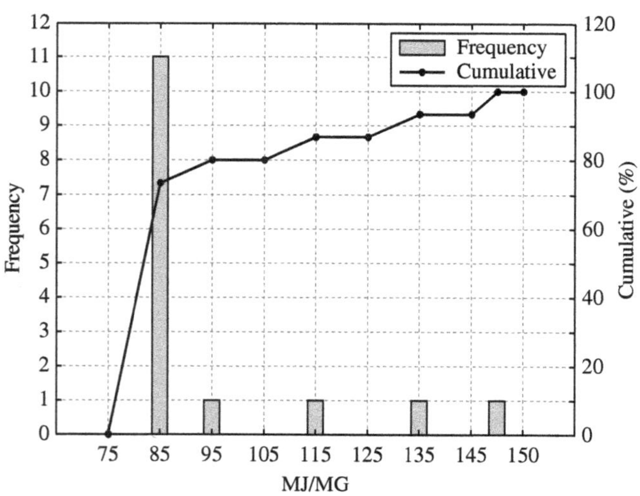

Figure 8.12 Typical gravity thickener.

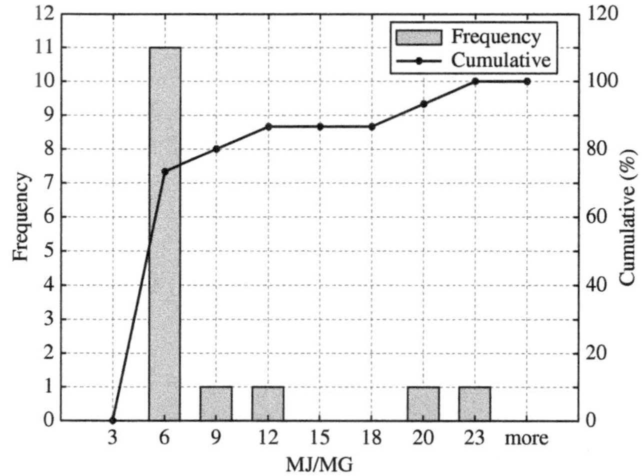

Figure 8.13 Typical gravity thickener (sidestream pump).

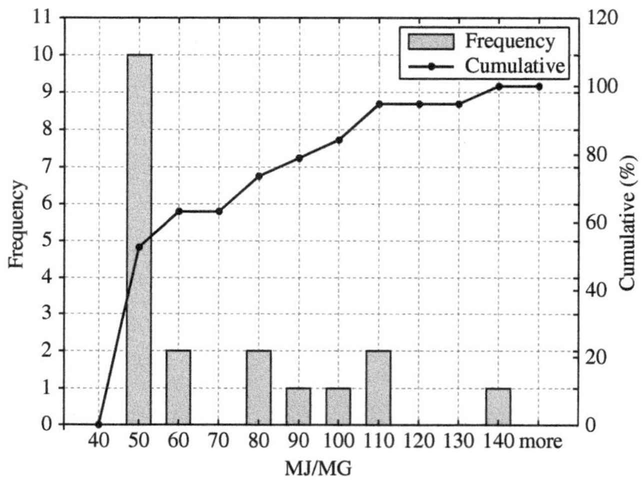

Figure 8.14 Typical mechanical thickener.

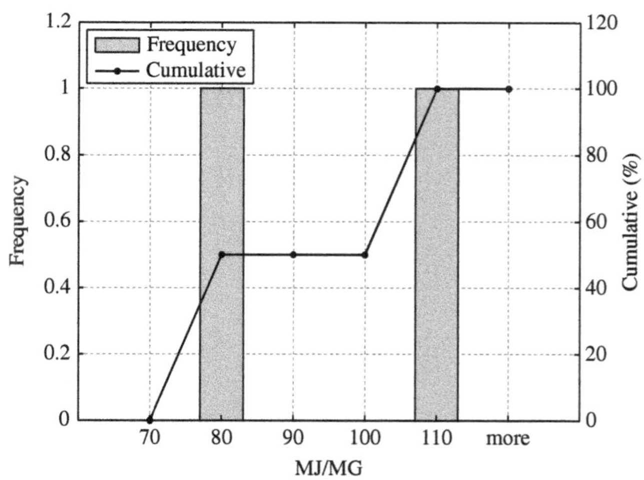

Figure 8.15 Typical lime stabilization.

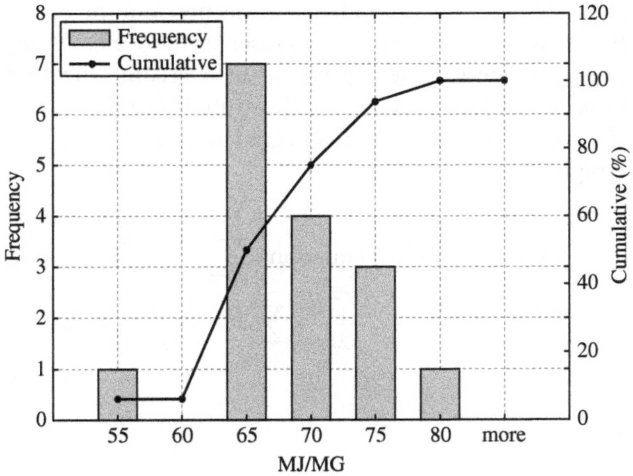

Figure 8.16 Typical anaerobic digestion.

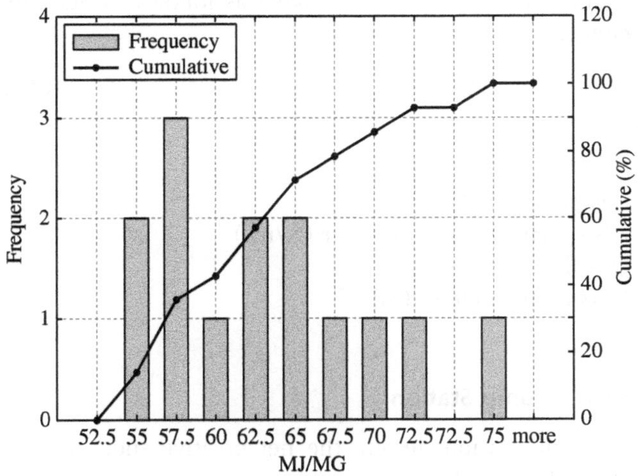

Figure 8.17 Typical anaerobic digestion dewatering.

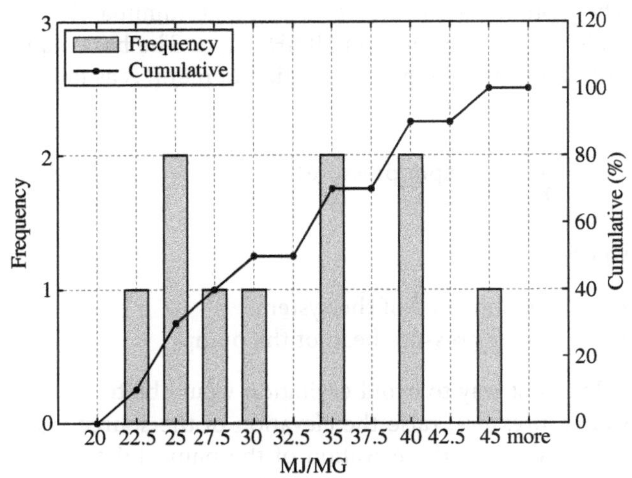

Figure 8.18 Typical flare.

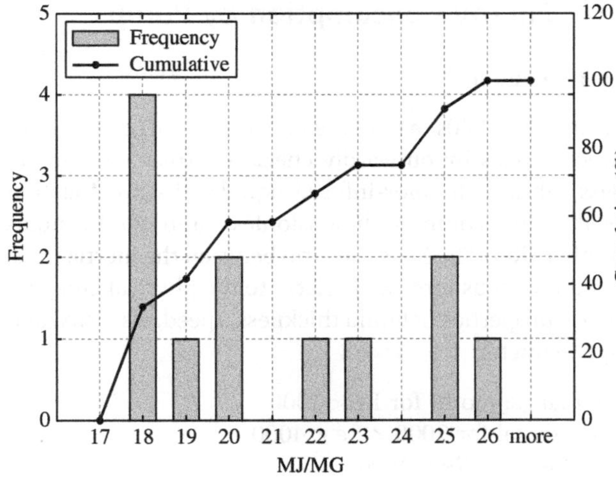

Figure 8.19 Typical boiler.

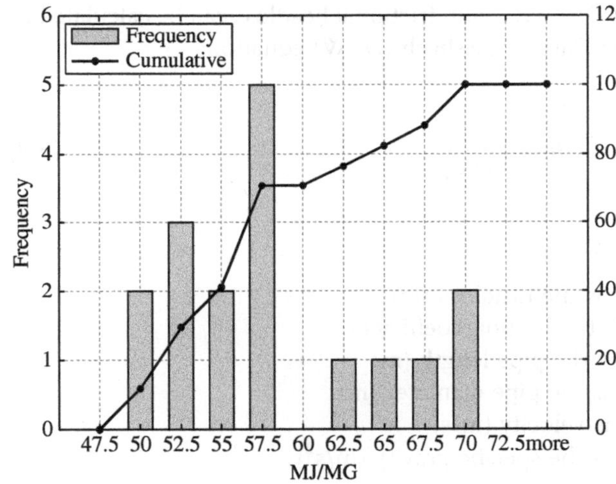

Figure 8.20 Typical dewatering cake.

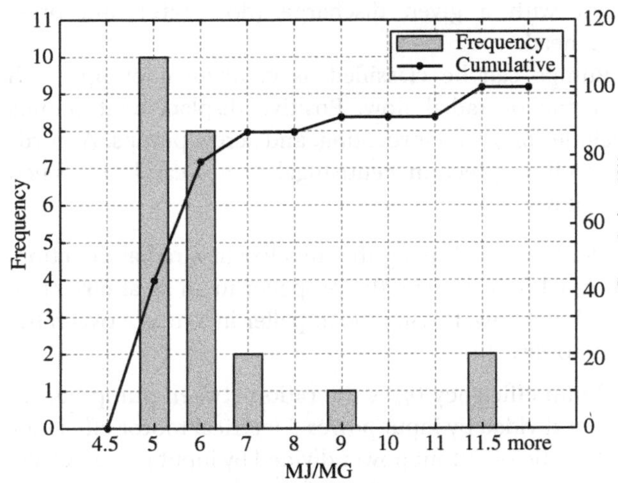

Figure 8.21 Typical dewatering sidestream pump.

8.6 Energy Consumption by Pump

8.6.1 Flow in Pipe

During the 1880s, an Irish engineer, Osborne Reynolds, directed tests by outwardly checking stream designs in glass tubes with color-infused liquids. His investigation resulted in a common dimensionless ratio, the Reynolds number (Re). The Re is the ratio between the inertia force and the viscous force and relates to the physical and geometric properties of liquid thickness, speed, viscosity, and pipe diameter:

Laminar (smooth) for Re < 2000
Transitional for 2000 < Re < 4000
Turbulent for Re > 4000

Moody diagram summarizes friction in pipe experimental results and establishes the relationship between the friction efficiency and Re and pipe relative roughness (ε/D). As a result, frictional headloss can be calculated by the Darcy–Weisbach (D–W) equation:

$$h_L = f\frac{l}{D}\frac{V^2}{2g} \tag{8.9}$$

where

h_L is the headloss (m)
f is the friction coefficient
L is the pipe length (m)
D is the pipe diameter (m)
V is velocity (m/s)
G is the specific gravity (m/s^2)

Water pumps are devices designed to convert mechanical energy into hydraulic energy, which is used to change the water level from a lower point to a higher point with a given discharge (flow rate) and pressure head.

Pumps can be classified as centrifugal pumps with dynamic or radial flow. Positive displacement pumps include rotary, reciprocating, and screw pumps. According to flow direction, centrifugal pumps can be classified as follows:

1) Radial flow, leaving the impeller in a radial direction.
2) Axial flow, leaving the propeller in an axial direction.
3) Mixed flow, leaving the impeller in a mix between the two directions.

Pump efficiency (η) is the ratio between pump output power divided by input power. Electrical motor efficiency is the ratio of output power divided by input power. Overall efficiency is the product of the efficiency of the pump and the efficiency of the motor. Over time, the pump characteristic curves will become steeper, and the operation point will get closer and closer to the y-axis. Pump must provide sufficient pressure to overcome the total headloss of different ratios of flow rate (Q) and diameter (D) and deliver pressure at an equal or greater value than design standards:

$$\text{Total head-loss equation} = \sum h_f + \sum h_{\text{minor}} = \sum f_i \times \frac{L_i}{D_i} \times \frac{V_{\text{pipe}}^2}{2g} + \sum k_i \times \frac{V_{\text{pipe}}^2}{2g} = \sum h_f \tag{8.10}$$

where

h_f is friction headloss (m)
h_{minor} is the component headloss (m)

Pump power can be calculated as follows to quantify energy footprint of a pump:

$$P = \gamma Q h_L \tag{8.11}$$

where

P is power (W)
γ is specific weight of water (N/m^3)
Q is flow rate (m^3/s)
h_L is total headloss (m)

8.6.2 Pump Station

In pump station design, the net positive suction head available ($NPSH_{\text{available}}$) must be greater than the net positive suction head required ($NPSH_{\text{required}}$) to prevent cavitation of the pump. Otherwise the pressure will create implosions in the pump and decrease pump efficiency. The optimal guideline is to have at minimal 5–10% $NPSH_{\text{available}}$ greater than $NPSH_{\text{required}}$. The basic equation to satisfy to prevent cavitation of pump is as follows:

$$\frac{P_s}{\gamma} < \frac{P_\nu}{\gamma}\text{(vapor pressure)} \tag{8.12}$$

where

P_s/γ = suction head of the system
P_v/γ = vapor pressure head of the pump

The best way to avoid cavitation would be to increase source pressure, raise the elevation of the source reservoir, and lower the elevation of the pump inlet. Otherwise, different pumps have to be selected to prevent cavitation of the pump.

For the pump and system to operate under steady conditions, pump total delivery head TDH_{tot} will have to equal the headloss in the system, as characterized by the system curve. That equality is satisfied under the conditions where the TDH_{tot} curve intersects the system curve. The value of Q at the operating point indicates the flow rate through both pumps and the piping. The head added by the original pump is the value of TDH_{orig} at that Q, and the head added by the pump which would have the value of TDH_{boost} at that Q.

8.7 Solar Energy

Solar energy is the only natural force that can reverse the entropy from disorder to order through photosynthesis. For this reason, green infrastructure (GI) plays a critical role in designing and building sustainable communities in regenerative way through harvest of solar energy using photosynthesis. Fundamentally, GI could increase production of primary biomass critical in the ecological system. It prevents pollution, adsorbs wastes, assimilates nutrients, and produces primary biomass by solar energy (Figure 8.22).

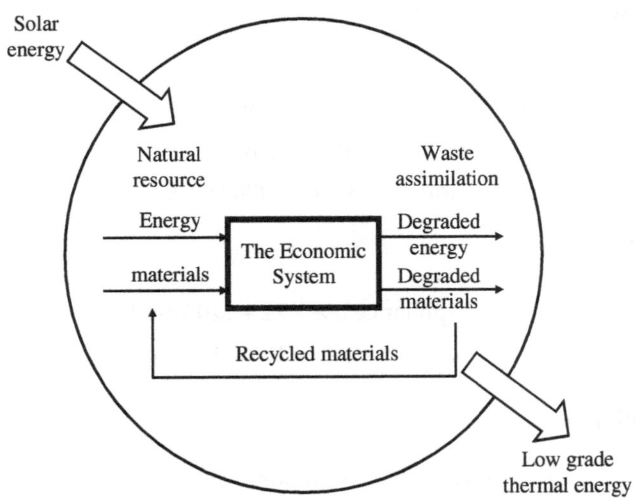

Figure 8.22 Simplified representation of energy and material flows across the biosphere and the economic system.

8.7.1 Calculation Solar Energy

In designing zero energy residential houses, solar energy and biogas are two major sources. Therefore, how much energy is needed and how much solar energy could be harvested are the critical design problems that are illustrated by the following examples as provided in Solar Photovoltaic Systems Technical Training Manual developed by the UNESCO (Wade, 2003).

Example 8.5
Given: The manufacturer states that a refrigerator takes 325 Wh/day to cool to 8°C when the outside air is 32°C. It will be used to make two 0.5-l ice packs per day, and about 1 l of water and vaccines will be cooled each day. It will be opened about six times a day.
Find: About how many watt-hours per day will the battery need to supply to operate the refrigerator?

Solution:

$$\text{Base load} = 325\ \text{Wh/day}$$
$$\text{Ice-pack load} = 1 \times 200 = 200\ \text{Wh/day}$$
$$\text{Cooling load} = 1 \times 45 = 45\ \text{Wh/day}$$
$$\text{Opening load} = 6 \times 1 = 6\ \text{Wh/day}$$
$$\text{Total load} = 325 + 200 + 45 + 6 = 576\ \text{Wh/day}$$

Answer: Total load = 576 Wh/day

Example 8.6
Given: A house has the following appliances in use:

1) One 18-W light used 4 h/day.
2) One 60-W fan used 2 h/day.
3) One 75-W refrigerator that runs 12 h/day.

The system will be powered by 110 Wp panels.
Find: How many panels will be needed if the climate is that in a typical tropical coastal area (class 2)?

Solution:

Step 1:

$$\text{Light}: 18 \times 4 = 72\ \text{Wh/day}$$
$$\text{Fan}: 60 \times 2 = 120\ \text{Wh/day}$$
$$\text{Refrigerator}: 75 \times 12 = 900\ \text{Wh/day}$$

Step 2:

$$\text{Total appliance use} = 72 + 120 + 900 = 1092\ \text{Wh/day}$$

Step 3:

$$\text{Panel energy needed} = 1.3 \times 1092 = 1419.6\ \text{Wh/day}$$

Step 4:

$$\text{Wp of panel capacity needed} = \frac{1419.6}{3.43} = 413.9\ \text{Wp}$$

Step 5:

$$\text{Number of panels needed} = \frac{413.9}{110} = 3.76\ \text{panels}$$
$$\text{Actual requirement} = 4\ \text{panels}$$

Answer: Number of panels needed = 413.9/110 = 3.76 panels; actual requirement = 4 panels

Example 8.7

Given: A video system that needs 200 W to operate is used for 4 h a week.

Find: What Wp of panel capacity will be needed to provide enough energy in a typical class 2 tropical coastal climate?

Solution:

$$200\ \text{W} \times 4\ \text{h} = 800\ \text{Wh/week}$$

$$\frac{800}{7} = 114.29\ \text{Wh/day}$$

If 114.29 Wh/day is required by the appliances,

$$114.29 \times 1.3 = 148.58\ \text{Wh/day}$$

must be provided by the panels.

1 Wp of panel can be expected to produce 3.43 Wh/day.

So at least

$$\frac{148.58}{3.43} = 43.3\ \text{Wp}$$

of panel capacity will be needed.

Answer: At least 43.3 Wp of panel capacity will be needed.

Example 8.8

Given: Appliances in a house need 200 Wh/day to operate.

Find: How many 40 Wp panels will be needed if the climate is that in a typical Pacific island?

Solution:

The panels must produce 200 Wh/day plus the energy lost in the system:

$$200 \times 1.3 = 260\ \text{Wh/day}$$

The total Wp needed will be

$$\frac{260}{3.43} = 75.8\ \text{Wp}$$

Therefore it will take

$$\frac{75.8}{40} = 1.89\ \text{panels to do the job}$$

Therefore at least two panels will have to be installed.

Answer: At least two panels will have to be installed.

Example 8.9

Given: Appliances in a house need 200 Wh/day to operate.

Find: How many 40 Wp panels will be needed for the conditions in Example 8.9?

Solution:

Panel watt-hours per day must still be 260 Wh/day.

The total peak watts must be

$$\frac{260}{3.0} = 86.7\ \text{Wp in a class 3 climate}$$

Therefore the number of panels will be

$$\frac{86.7}{40} = 2.16\ \text{panels}$$

So three panels will be needed.

Answer: Three panels will be needed.

Example 8.10

Given: A house with a 12-V solar system has the following appliances in use:

One 18-W light used 4 h/day.
One 60-W fan used 2 h/day.
One 75-W refrigerator that runs 12 h/day.

Find: What battery capacity will be needed?

Solution:

Step 1:

$$\text{Light}: 18 \times 4 = 72\ \text{Wh/day}$$
$$\text{Fan}: 60 \times 2 = 120\ \text{Wh/day}$$
$$\text{Refrigerator}: 75 \times 12 = 900\ \text{Wh/day}$$

Step 2:

$$\text{Total appliance use} = 72 + 120 + 900 = 1092\ \text{Wh/day}$$

Step 3:

$$\text{Total appliance watt-hours} \times 5 = 1092 \times 5 = 5460\ \text{Wh}$$

Step 4:

Divide watt-hours by battery voltage:

$$\frac{5460}{12} = 455\ \text{Ah}$$

So for the house in the example, a deep-discharge battery of at least 455 Ah should be used.

Answer: A deep-discharge battery of at least 455 Ah should be used.

Example 8.11
Given:

1) Charge controller:

A 24-V PV system includes 300 Wp of panels.

2) Discharge controller:

The above PV system operates:
Four 18-W fluorescent lights
One 60-W refrigerator
One 35-W television

Find:

1) What size of charge controller is needed?
2) What size of discharge controller is needed?

Solution:

1) Divide the panel peak watts by the system voltage to find the charge controller amperes:

$$\frac{300}{24} = 12.5 \text{ A}$$

So the charge controller should be rated at least.
2) What size of discharge controller is needed?

$$\frac{18\text{ W}}{24\text{ V}} = 0.75 \text{ A per light}$$

$$\frac{35\text{ W}}{24\text{ V}} = 1.46 \text{ A for the television}$$

$$4 \times 0.75 + 1.46 = 4.46 \text{ A running current}$$

$$4.46 \times 1.5 = 6.7 \text{ A for starting}$$

For the refrigerator:

$$\frac{60\text{ W}}{24\text{ V}} = 2.5 \text{ A per light}$$

$$2.5 \text{ A} \times 3 = 7.5 \text{ A for starting}$$

$$4 \times 0.75 + 1.46 = 4.46 \text{ A running current}$$

$$4.46 \times 1.5 = 6.7 \text{ A for starting}$$

So the discharge controller should be rated at least

$$6.7 + 7.5 = 14.2 \text{ A}$$

Answer:

1) The charge controller should be rated at least 12.5 A.
2) The discharge controller should be rated at least 14.2 A.

8.7.2 Solar-powered WWTP

100% natural treatment of wastewater at a treatment plant by John Todd at Ecological Design known as the Eco-Machine. The OCSL building is powered by solar and geothermal power. Unlike other WWTPs, the OCSL does not use chemicals to treat the water, but a combination of microorganisms, algae, plants, and gravel and sand filtration to clean sewage water and return clean drinkable water back to the aquifer. Environmental sustainability, green energy, and regenerative design are the key. All the wastewater from toilets, sinks, and showers on the Omega campus feeds into storage tanks that collect human waste. The wastewater is then sent to the Eco-Machine and is fed to "microscopic algae, fungi, bacteria, plants, and snails." The first stage is two 5000-gal anoxic tanks located underground and microbial organisms use the wastewater digest carbon, ammonia, phosphorus, nitrogen, potassium, and many other substances in the water. Water then flows to four man-made wetlands behind the OCSL building. The wetlands use microorganisms and native plants such as cocktails and bulrushes to reduce biochemical oxygen demand, odorous gases, and nutrients such as ammonia and phosphorus. As the wastewater flows through the wetlands, the microorganisms and plants are fed. After that, there is a 75% increase in water's clarity and a 90% reduction in its odor. The aerated lagoons are divided into four cells, each 10 feet deep. Without soil, beautiful tropical plants thrive here. The plants live on metal racks and their roots extend up to 5 ft into the water. The roots of the plants act as a habitat for the organisms in the lagoon and are sustained by them. After the lagoons, the water moves back outdoors to a sand filter; after which it reaches the standards for treated water. The treated water irrigates Omega's parking lot where it infiltrates into the groundwater table. As it trickles down to the aquifer that sits 250–300 ft below, campus water is further purified as it is withdrawn from deep wells that tap the aquifer; the water is used in sinks, toilets, and showers. Naturally the used water is reclaimed with the Eco-Machine at OCSL, and the purified water is released back to the aquifer, where the process can begin again. This full-circle process makes the Eco-Machine incredible and is an inspiration to interconnectedness utilizing processes to meet human's needs.

8.8 Exercise

8.8.1 Questions

1) In a typical WWTP, how is high-grade energy such as electricity degraded to mechanical, chemical, and biological energy? Please quantify these energy in kWh/m^3.
2) What is embodied energy? Why there is an optimality between the embodied energy and operation energy consumption?

3) How do thermodynamic laws dictate energy recovery in WRRF? Please analyze how energy could be recovered from chemical and biological to electrical energy. What is the unit energy production rate in kWh/m^3 during each major process?
4) In a typical WRRF, how much energy could be recovered in terms of theoretical energy contained in a typical wastewater sample of COD 500 mg/l and BOD 350 mg/l?
5) Where is the optimal point for operation cost versus equipment size?
6) What is the 10-percentile unit energy benchmark for each process in WWTP?
7) How large is the roof area needed for a typical house in Miami, Florida, to achieve net zero energy design by using 40-W solar panels?

8.8.2 Calculation

1) For three Miami-Dade WWTPs of 300 MGD average flow rate, how much energy contained in average concentration of 330 mg/l COD and 223 mg/l BOD energy in theory if CO_2 is reduced to CH_4 or H_2, respectively? If only biological fraction is able to be reduced to CH_4, what are the maximal amount of energy that could be recovered in theory? If a typical theoretical value of energy content is about 1.75 kWh/m^3, how is your estimate compared with this value?
2) If all the WWTPs are to recover energy BOD from anaerobic digestion and the biogas produced is to generate electricity with 25% efficiency through combustion engine or to generate electricity through fuel cell with 40% efficiency, what are the percentages of self-sustaining electricity if the plant needs 0.39 kw/m^3 to operate? Please list any assumptions that you would in terms of conversion efficiency.
3) If electricity costs $ 0.1/kWh, what would be the annual electricity bill for the operation of three WWTPs with the above energy recovery systems?

8.8.3 Project

8.8.3.1 Xiongan Project

1) Xiongan has very ambitious plan to use 100% renewable energy. One of the unique energy in Xiongan is geothermal energy. Please collect necessary data, and estimate how many billion watts of energy could be extracted annually from geothermal energy with commercially available technologies.
2) Energy efficiency is critical for energy-positive plants. For 1, 2, and 5 million populations in 5, 10, and 20 years, respectively, please develop the most energy-efficient WWTP process alternatives.
3) Please estimate the total energy saving in billion watts if all the WWTPs in Xiongan is at minimal energy neutral.

8.8.3.2 Community Project

Please contact the plant manager of WWTP of your local city, and request energy consumption for each process of the plant and actual flow rate associated with energy consumption:

1) Please conduct energy auditing at the WWTP: The energy consumption of each unit process should be divided by the actual flow rate.
2) Please compare the unit energy consumption rate in MJ/MGD with the benchmark data provided in this chapter.
3) Rank the performance of each unit process by percentage.
4) Identify the biggest potential in energy efficiency improvement.
5) Recommend different retrofitting methods to improve unit process energy efficiency.

References

Ayres, R.U. and Kneese, A.V. (1969). Production, consumption and externalities. *The American Economic Review* 59: 282–297.

Gans, N., Mobini, S., and Zhang, X. (2007). Mass and energy balances at the Gaobeidian wastewater treatment plant in Beijing, China. http://www.chemeng.lth.se/exjobb/E458.pdf (accessed 16 January 2018).

Georgescu-Roegen, N. (1971). *The Entropy Law and the Economic Process*. Cambridge, MA: Harvard University Press.

Hammond, G.P. and Winnett, A.B. (2009). The influence of thermodynamic ideas on ecological economics: an interdisciplinary critique. *Sustainability* 1: 1195–1225.

Howe, K.J., Hand, D.W., Crittenden, J.C. et al. (2012). *Principles of Water Treatment*. New York: Wiley Blackwell.

Jonasson, M. (2007). Energy benchmark for wastewater treatment processes – a comparison between Sweden and Austria. Master thesis, Lund University.

Krysiak, F. (2006). Entropy, limits to growth, and the prospects for weak sustainability. *Ecological Economics* 58: 182–191.

Mizuta, K. and Shimada, M. (2010). Benchmarking energy consumption in municipal WWTPs in Japan. *Water Science and Technology* 62 (10): 2256–2262.

Nouri, J., Jafarinia, M., Naddafi, K. et al. (2006). Energy recovery from wastewater treatment plant. *Pakistan Journal of Biological Science* 9 (1): 3–6.

Pearce, D.W. and Turner, R.K. (1990). *Economics of Natural Resources and the Environment*. Baltimore, MD: Johns Hopkins University Press.

Perrings, C. (1986). Conservation of mass and instability in a dynamic economy-environment system. *Journal of Environmental Economics and Management* 13(3): 199–211.

Stasinopoulos, P., Smith, M., Hargroves, K. and Desha, C. (2008). *Whole System Design: An Integrated Approach to Sustainable Engineering*. Earthscan: The Natural Edge Project.

Tarallo, S., Shaw, A., Kohl, P., and Eschborn, R. (2015). *A Guide to Net-Zero Energy Solutions for Water Resource Recovery Facilities*. Alexandria: Water Environment Research Foundation.

Van Gool, W. (1992). Exergy analysis of industrial processes. *Energy* 17: 791–803.

Wade, H.A. (2003). *Solar Photovoltaic Systems Technical Training Manual, the UNESCO the United Nations Educational*. Bangalore: Scientific and Cultural Organization.

9
Prevention

One ounce prevention is better than a pound of medicine.

OBJECTIVES
To design green infrastructure (GI) to prevent nonpoint water pollution To define the integrated urban water management (IUWM) by considering precipitation, runoff, evaporation, infiltration, purification, and storage as a whole in water management To develop flowcharts in designing rain harvest systems using Matlab To present Matlab codes in designing a rain harvest system as key technology of IUWM for the development of sustainable communities To achieve effective zero water design by using blue, green, gray, brown, and black water concepts To design green roofs and rain gardens To illustrate how life cycle assessment (LCA), cost and benefit analysis, and footprint could be used in assessing GI technologies

9.1 Principle 6

Pollution prevention requires SEE designers to reuse, reduce, and recycle materials and energy to protect the natural environment. The aim of the Pollution Prevention Act of 1990 in the US Environmental Protection Agency (EPA) is to reduce the amount of pollution through cost-effective changes in production, operation, and use of raw materials in industrial and domestic sectors.

> **Principle 6** Prevention, reduction, and minimization strategies should be implemented as early as possible:
>
> 1) The ultimate goals of prevention should be zero water, waste, and energy design.
> 2) Source reduction, recycling, and minimization on- or off-site should be maximized in management of air, water, and solid wastes.
> 3) Blue, green, gray, brown, and black water need to be separately managed and treated according to specific application.

Source reduction is more desirable than waste management or pollution control. Reducing waste at the source minimizes the cost of treatment and the transfer of pollution. It increases economic competitiveness due to the efficient use of raw materials. For example, one gallon of gasoline is not much fuel but would cost thousands of dollars to clean up if it leaked into groundwater. Volume reduction can be used to reduce treatment cost and to reduce handling and disposal costs for residues after treatment. Volume reduction can be accomplished using a variety of methods:

1) Reuse of treated wastewater and wastes.
2) Treatment modifications to reduce solid residues.
3) Segregated treatments to reduce hazardous waste mixtures.
4) Incineration to reduce waste volume and to render a hazardous waste nonhazardous.

Waste requires capital, energy, and resources to be sustainably managed. Prevention is the best strategy to reduce waste production in the first place. Therefore, on every design scale such as molecule, process, product, and system, opportunities exist to prevent waste rather than treat it after it is generated. Among all the wastes, hazardous wastes should be considered as the most important wastes to be recycled, reused, or reduced before treatment or disposal. To avoid toxic chemicals, green chemistry (GC) provided many successful solutions in terms of renewable biofeedstock and nontoxic solvent. To design plants efficiently and economically, green

Sustainable Environmental Engineering, First Edition. Walter Z. Tang and Mika Sillanpää.

engineering (GE) developed 12 principles for chemical engineering (Table 9.1).

Table 9.1 Green engineering and chemistry principles related to pollution prevention.

Green chemistry	Green engineering
Preventing the production of waste is better than cleaning or treating the waste	Ensure designs are as efficient and nonhazardous as possible
Maximize incorporation of raw materials in to product	Preventing the production of waste is better than cleaning or treating the waste
Use less hazardous chemicals	Any process should be designed to its maximum efficiency in terms of mass, energy, space, and time
Design safer chemicals	
Use safer solvents and auxiliaries	

9.2 Challenges and Opportunities

Human population growth, loss of biodiversity, fragility of nutrient cycles, air and water pollution, waste disposal, dwindling energy resources, and other environmental and sustainability problems demand prevention. According to the US EPA, urban pollutants are the most important contributors to the contamination of the nation waters (NRC and NAP, 2008) because urban development typically involves converting undeveloped areas to impervious, often paved surfaces, thereby transforming surfaces. In a natural landscape, less than 10% of the rainfall volume converts to runoff. When a land is developed, the water cycle and water quality of the area will deteriorate during the construction phases. Stormwater runoff has a significant impact on natural water bodies. During storm events, impervious surfaces cause runoff to rapidly flow toward receiving pipes or water bodies, picking up pollutants along its path. These nonpoint pollutants include sediment, metals, bacteria, nutrients, pesticides, trash, and polycyclic aromatic hydrocarbons. They either infiltrate into the ground, potentially degrading the groundwater, or flow into surface waters, depositing pollutants, sediments, and debris, while also eroding stream channels, altering sediment loads, and affecting stream temperature (Paul and Meyer, 2001; Roy et al., 2008). Furthermore, impervious surfaces, especially roads and parking lots, increase the risk of flooding for downstream regions by increasing water volume and decreasing travel time, resulting in rapid flood peaks (NRC, 2008; Jacob and Lopez, 2009). In urban landscapes, roads, parking lots, rooftops, and compacted soils cover 45–90% of land, preventing the infiltration of rainwater and altering the hydrology. Urban cement surface results in no room for woodland, grassland, lakes, and wetlands for naturally detained water. Since rainwater can only be discharged as wastewater instead of being reused, there is a direct relationship between land cover and downstream water quality. In addition, the cumulative impacts of urbanization on hydrology, water quality, and habitats should be addressed (NRC, 2008). When agriculture is fertilized at a constant rate, the nutrients will run into nearby water bodies, causing nonpoint pollution. Urban stormwater runoff is an important contributor to water pollution; it both contaminates and physically harms aquatic environments. When an urban drainage system is not designed properly, frequent flooding could take place, which disrupts social economic activities and damage natural environments.

To prevent flooding, integrated urban water management (IUWM) for smart cities have been given different names. For example, the United States refers to low impact development (US EPA, 1999), the United Kingdom promotes sustainable drainage systems (Lehmann, 2010), and Australia and New Zealand emphasize on water-sensitive urban design (Sharma et al., 2016). The key concept is to manage whole water cycles of stormwater, wastewater, irrigation, and water supply naturally by retaining 70–90% of average annual precipitation on-site through green infrastructure (GI). As a result, urban runoff could be cleaned and stored to reduce flooding and alleviate the impacts on natural ecosystems and urban heat island (China General Office of the State Council (CGOSC), 2017). Taking China as example, Chinese urban population expanded from 20% in 1980 to 58% in 2018, which implies that more than 400 million people had migrated to a city from the countryside in the past four decades. The increasing population causes lots of problems on the Chinese cities. Urban flooding is one of the most frequent disasters and causes enormous impacts on the economy. Direct economic losses from 2011 to 2014 are estimated up to \$100 billion (NBSA, 2015) because more than 62% of Chinese cities experienced urban flooding. The Chinese State Council initiated sponge city to reuse at least 70% of rainwater in 80% of urban areas by 2020. Sponge city is defined as a city of an infrastructure that collects excess rainfall and integrates flood control in urban planning to alleviate flooding impact on runoff quantity and quality. A sponge city aims for sustainable urban development including flood control, water conservation, water quality improvement, and ecosystem protection. As a result, a city water management system would operate like a sponge to absorb, store, infiltrate, and purify runoff and release it for reuse when needed. However, Li et al. (2017) surveyed 30 pilot sponge cities in China and identified challenges from technical, physical, regulatory, financial, and institutional perspectives. Technically, one standard definition could not cover all the functions of a sponge city. Currently, no GI industry exists to monitor, operate, and maintain the GI. In addition, physical limitations such as land scarcity, differences in climate, and soil properties exist. Uncertainty of life cycle costs and benefits prevents public–private partnership (PPP) and poses major financial challenges. Lastly local ordinances; design code of building, plumbing, and health and drainage may impose restrictions on the use of reclaimed stormwater.

Despite the aforementioned challenges, the sponge city program is creating investment opportunities in infrastructure upgrading, engineering products, and GI technologies. Opportunities are open to PPP for safer, greener, more holistic urban environments by leveraging the governmental investments. For example, 60 pilot cities received 400 and 600 million Chinese yuan (CNY) each year for three consecutive years since 2016.

To prevent water contamination, GI is critical to sustainable communities because it filters, purifies, and stores stormwater and reduces runoff and contamination. GI could significantly increase the resiliency of city by sustaining water needs for both humans and ecosystems. One of the objectives of GI is to delay peak flow and remove nutrients by absorbing nutrients. In designing GI system, stormwater quantity is usually estimated as follows:

$$Q = C \times I \times A \tag{9.1}$$

where

Q = flow rate
C = runoff coefficient
I = rainfall rate in inches
A = surface area

Sediment control is important during land developing so that the impact on the water quality is minimized. Runoff is filtered through a natural means to maintain ambient nitrogen concentration of 1 mg/l and phosphorus of 0.1 mg/l. To reduce water footprint on the natural water body, most WWTPs need a tertiary wastewater treatment process to achieve this discharge standards of nitrogen and phosphorus.

9.3 Green Infrastructure

9.3.1 Integrated Urban Water Management Paradigm

GI is one of the key technologies for IUWM. When water is managed as a whole, precipitation, runoff, evaporation, infiltration, purification, and storage are considered together on community, city, and regional scales. The aim of IUWM is to reduce water footprint so that the economy, society, and environment could be in harmony. The IUWM emphasizes the paradigm shifts as shown in Table 9.2.

Table 9.2 Paradigm shifts.

High tech	Smart	Low tech	Natural
Large scale	Centralized	Small scale	Decentralized
Linear	Separate	Cyclical	Networked
Wastewater	Pollution	Resource water	Recovery
Potable water	Highest quality	Uniform	Fit for purpose
Robust	Capacity	Resilient	Flexible
Customers	Bill payers	Partners	Stakeholders

In terms of prevention, GI provides green spaces to other physical features in a terrain such as coastal and marine areas. It increases the permeable area for groundwater recharge, attenuates peak runoff flows to reduce flooding, improves runoff water quality to protect ecosystem, and reduces operational energy consumption and greenhouse gas (GHG) emissions. In addition, GI reduces costs; improves local air quality, biodiversity, habitat, and connectivity; benefits human health; and provides potential amenity and recreation benefits. Therefore, GI should be a strategically planned network of natural and seminatural areas with other environmental features designed and managed to deliver a wide range of ecosystem services. The design principle is to prevent contamination or shortage of water supply through color-coded design such as blue, green, gray, brown, and black water to reduce blue or gray water footprints. Green water footprint refers to the amount of precipitation volume required. Blue water is defined as freshwater volume such as surface water and groundwater required. Gray water footprint, however, is defined as the freshwater volume required to assimilate pollutants at background level (Hoekstra et al., 2011). To prevent water contamination, two of the most effective ways are using rainwater and gray water to irrigate gardens and laws. Typically, water demands for irrigation, shower, laundry, kitchen, and toilet for a residential house are 40, 20, 16, 11, and 11%, respectively.

Since water footprints are measured by the amount of water to restore the original water quality due to the discharge of gray or black water, strict requirements for water efficiency have been established by the Leadership in Energy and Environmental Design (LEED). LEED is a rating system devised by the US Green Building Council (USGBC) to evaluate the environmental performance of a building and encourage market transformation toward sustainable design. For a new construction and renovation of commercial and institutional projects, performance is evaluated in five environmental categories, one of which is water efficiency. LEED is a point-based system with points awarded for meeting the specific requirements of credits within each category. Five of the 69 possible points in LEED are directly associated with water efficiency. These five points are among three LEED water efficiency credits as follows:

1) Credit 1 – Water-efficient landscaping, two points.
2) Credit 2 – Innovative wastewater technologies, one point.
3) Credit 3 – Water use reduction, two points.

The design strategies for meeting those requirements and the planning process are very important to successfully develop, incorporate, and optimize water efficiency on LEED projects. For example, water-efficient landscaping credit is to "limit or eliminate the use of potable water, or other natural surface or subsurface water resources available on or near the project site, for landscape irrigation." One point is awarded for a 50% reduction in water consumption for irrigation from a calculated midsummer baseline case, and a total of two points is awarded for a 100% reduction. High-efficiency drip and micro- and subsurface systems can reduce the amount of water required to irrigate a given landscape. The USGBC reports that drip systems alone can reduce water use by 30–50%. Climate-based controls, such as moisture sensors with rain shutoffs and weather-based evapotranspiration controllers, can further reduce demands by allowing naturally occurring rainfall to meet a portion of irrigation needs.

Since rainwater runoff volume management is directly linked to infiltration/inflow of sewer system, stream erosion and water quality, rainwater harvest is the most important key technology in using water efficiently. Rainwater collection involves collecting and holding on-site rainfall in cisterns, underground tanks, or ponds during rainfall and can be used for the irrigation systems. Rainwater collection and wastewater treatment systems span multiple conventional project disciplines; an experienced team including the architect, landscape architect, civil and plumbing engineers, and rainwater system designer should be assembled to achieve the targeted goals efficiently and effectively.

9.3.2 Green Infrastructure Design Tools

GI tools can be used to assess the potential economic and monetary value of multiple benefits, including reduced water treatment needs, increased groundwater recharge, and improved neighborhood aesthetics. The Center for Neighborhood Technology (CNT, 2010) developed a National Green Values™ calculator to quickly estimate the performance, costs, and benefits of GI compared with conventional stormwater management practices. Other resources for estimating the costs and benefits of alternative approaches to stormwater management are available at the EPA's GI website (EPA, 2014).

9.3.3 Green Infrastructure Modeling Tools

Modeling tools support planning and design decisions on a range of scales from setting a GI target for an entire watershed to designing a GI practice for a particular site. Outputs that are particularly helpful include runoff volume, runoff rate, pollutant loading, and cost. Some models can predict the water quality and water quantity impacts of GI. Models can be used to present or simulate a site or watershed and apply various environmental data to quantify the possible impacts. In addition, they can be used to predict the environmental outcomes of different design and management approaches.

Models can guide site designers in meeting mandatory or voluntary performance standards. For example, a site's land cover and stormwater controls can be linked to the volume of stormwater discharged by the site and the pollutant loads exported by the discharges. Models could compare the water quantity and quality outcomes associated with different design scenarios. Many sites offer design tools, for example, USGBC's LEED program and Sustainable Sites Initiative (USGBC, 2018). Land cover and stormwater controls can be implemented throughout a watershed to achieve specific targets of hydrological, chemical, and ecological outcomes in receiving waters. All activities occurring in a watershed and the pattern of precipitation in a given year impact receiving waters, making models particularly valuable at the watershed scale. The receiving water impacts associated with stormwater management approaches can be compared with the environmental outcomes of alternative management scenario.

However, no model can accurately predict all environmental outcomes on all scales, but a limited range of environmental outcomes within a limited range of scales. Environmental parameters such as water quality, streamflow rates, and groundwater recharge rates should be identified. A watershed scale such as a site, headwater stream, or lake should be defined. For each model, the amount of data and the spatial and temporal resolution should be determined. The level of accuracy required to meet your objective could be used to efficiently allocate staff and budget resources by weighing a simple model's level of accuracy and cost. Table 9.3 and 9.4 list initial and operation cost of GI.

Table 9.3 Potential operating costs.

Materials	Regulatory compliance
Direct product materials	Monitoring
Catalysts and solvents	Manifesting
Wasted raw materials	Reporting
Transport	Notification
Storage	Recordkeeping
	Training (right to know, safety, etc.)
Waste management (materials and labor)	Training materials
Pretreatment	Inspections
On-site handling	Protective equipment
Storage	Labeling
Hauling	Penalties/fines
Insurance	Lab fees
Disposal	Insurance
	R&D to comply with regulations
Utilities	Handling (raw materials and waste)
Electricity	Closure and post-closure care
Steam	
Cooling and process water	**Revenues**
Refrigeration	Sale of product
Fuel (gas or oil)	Marketable by-product
Plant air and inert gas	Manufacturing throughput change
Sewerage	Change in sales from
	Increased market share
Direct labor	Improved corporate image
Operating labor and supervision	
Manufacturing clerical labor	**Future liability**
Inspection (QA and QC)	Fines and penalties
Worker productivity changes	Personal injury
Indirect labor	
Maintenance (materials and labor)	
Miscellaneous (housekeeping)	
Medical surveillance	

Table 9.4 Potential initial costs.

Purchased equipment	Materials
Equipment	Piping
Sales tax	Electrical
Price for initial spare parts	Instruments
Process equipment	Structural
Monitoring equipment	Insulation
Preparedness/ protective equipment	Building construction materials
Safety equipment	Painting materials
Storage and material handling equipment	Ducting materials
Laboratory/analytical equipment	
Freight, insurance	
Utility connections and new systems	**Site preparation**
Electricity	Demolition, clearing, etc.
Steam	Disposal of old equipment, rubbish
Cooling and process water	Walkways, roads, and fencing
Refrigeration	Grading, landscaping
Fuel (gas or oil)	
Plant air	**Engineering/contractor (in-house and ext)**
Inert gas	Planning
General plumbing	Engineering
Sewerage	Procurement
	Consultants
Installation	Design
Vendor	Drafting
Contractor	Accounting
In-house staff	Supervision
Construction/ installation	
Labor and supervision	**Contingency**
Taxes and insurance	
Equipment rental	**Permitting – fees and in-house staff**
Start-up and training	**Initial charge for catalysts and chemicals**
Vendor/contractor	
In-house	**Working capital** (funds for raw materials, inventory, materials/ supplies)
Trials/manufacturing variances	**Salvage value** of replaced equipment

9.4 Design Tools of Rain Harvest

Rain harvest is the most effective way to reduce runoff, prevent soil erosion, nutrient runoff, and delay peak flow for flooding. To illustrate the design process, the design tools for a rain harvest system are illustrated through five examples. The tools were developed to meet the water demand for a public bathroom in the ecocorridor of the Indian Creek in the South Miami Beach as a case study by using Matlab. However, the design tools should be applicable to design rain harvest system for all the public spaces. Figure 9.1 presents a design flowchart that shows the relationship between source water, wastewater, and water user.

For the public bathroom in Indian Creek, the rainwater demand is for handwashing and the toilet flush. To calculate the water demand for the public bathroom, Figure 9.2 shows the design flowchart of estimating water demand.

9.4.1 Determine the Water Demand of a Public Bathroom

Example 9.1

Given: Total number of visitors = 500, ratio of men to women = 1.

Find: Calculate flow rate of wastewater treatment system and rainwater demand for public bathrooms (Tables 9.5 and 9.6).

Solution (Tables 9.7 and 9.8):

$$\text{Number of men} = \frac{500 \times (1)}{(1)+1} = 250$$

$$\text{Number of women} = \frac{500}{(1)+1} = 250$$

Water demand of toilets for men

$= 250 \times 0.25 \times 1.28 = 80\ \text{gal/day}$

Water demand of toilets for women

$= 250 \times 2 \times 1.28 = 640\ \text{gal/day}$

Water demand of urinals for men

$= 250 \times 1.75 \times 0.5 = 218.75\ \text{gal/day}$

Water demand of urinals for men

$= 250 \times 2 \times 0.25 = 125\ \text{gal/day}$

Water demand of urinals for men

$= 250 \times 2 \times 0.25 = 125\ \text{gal/day}$

Flow rate of wastewater treatment system

$=$ total daily water demand $= 1188.75\ \text{gal/day}$

Daily rainwater demand

$=$ daily water demand for handwashing

$= 250\ \text{gal/day}$

For January (Table 9.10),

Monthly water demand $= 250 \times 31 = 7750\ \text{gal}$

For February,

Monthly water demand $= 1188.75 \times 31 = 36\,851.25\ \text{gal}$

Matlab code:

```
clear all
close all
% input
ratioofmentowomen=1;
totalnumberofpeople=500;
```

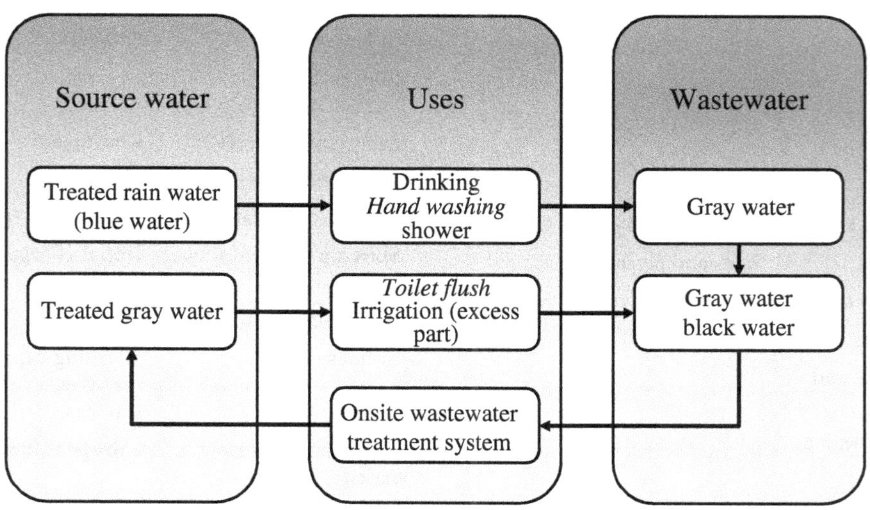

Figure 9.1 Design flowchart.

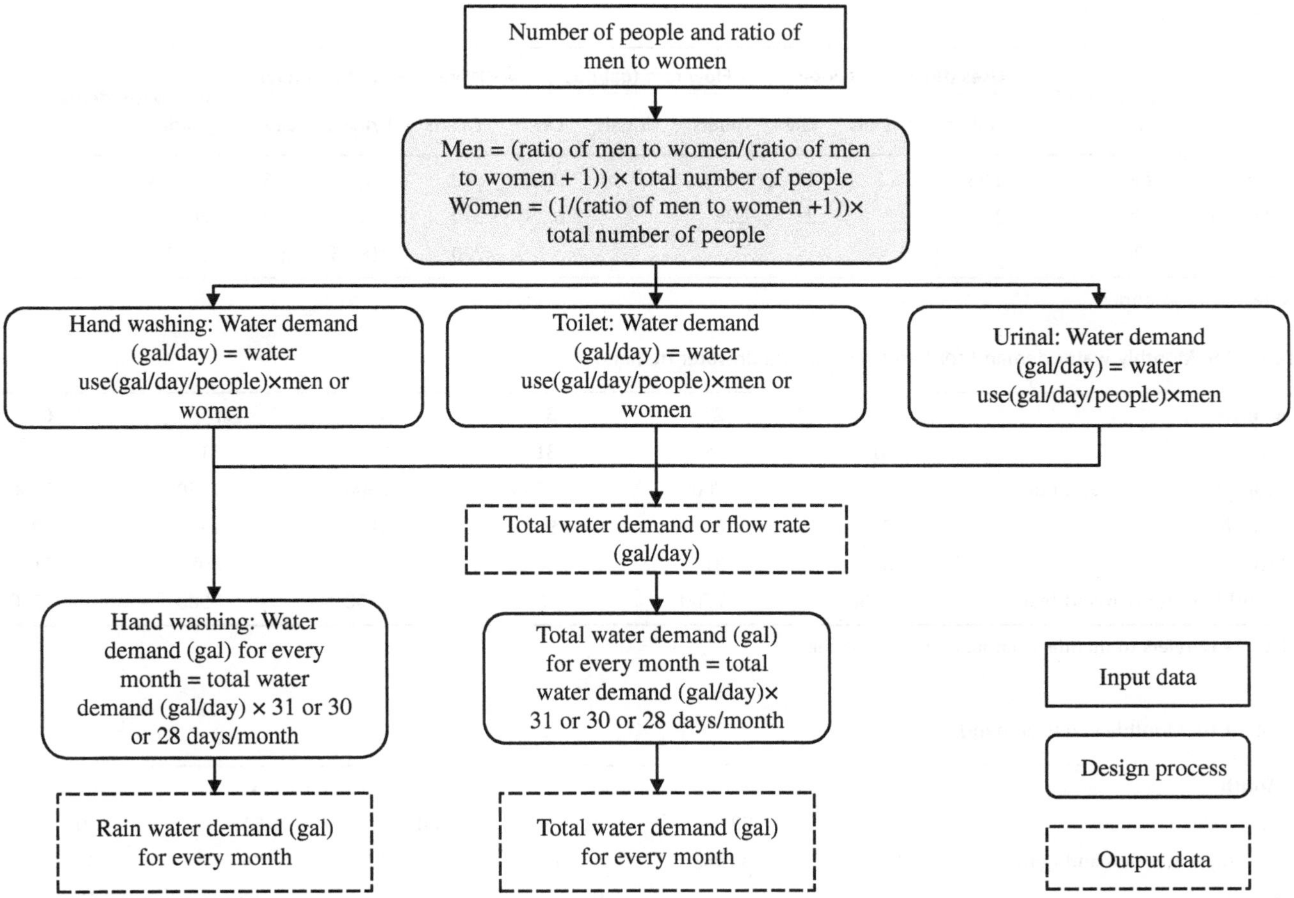

Figure 9.2 Flowchart for water demand.

Table 9.5 Uses and flow rate of bathroom.

	Uses (flush/day/people)			Flow rate (gal/flush)		
	Toilets	**Urinals**	**Lavatories**	**Toilets**	**Urinals**	**Lavatories**
Men	0.25	1.75	2	1.28	0.5	0.25
Women	2		2	1.28	0.5	0.25

Source: http://www.allianceforwaterefficiency.org/commercial_restroom_audit.aspx

Table 9.6 Day of month.

Month	**1**	**2**	**3**	**4**	**5**	**6**	**7**	**8**	**9**	**10**	**11**	**12**
Day	31	28	31	30	31	30	31	31	30	31	30	31

Table 9.7 Input data.

Ratio of men to women	**Total number of people**
1	500

```
% function
% number of people
men=ratioofmentowomen/
  (ratioofmentowomen+1)
  *totalnumberofpeople;
women=1/(ratioofmentowomen+1)
  *totalnumberofpeople;
```

Table 9.8 Calculated results.

	Number of people	Uses (flush/day/people)			Flow rate (gal/flush)			Water demand (gal/day)			Total water demand (gal/day)
		Toilets	Urinals	Lav	Toilets	Urinals	Lav	Toilets	Urinals	Lav	
Men	250	0.25	1.75	2	1.28	0.5	0.25	80	218.75	125	423.75
Women	250	2		2	1.28		0.25	640	0	125	765
Total	500							720	218.75	250	1188.75

Note: Lav is lavatories.

Table 9.9 Monthly water demand for handwashing or rainwater demand.

Month	**1**	**2**	**3**	**4**	**5**	**6**
Day	31	28	31	30	31	30
Monthly water demand (gal)	7750	7000	7750	7500	7750	7500
Month	**7**	**8**	**9**	**10**	**11**	**12**
Day	31	31	30	31	30	31
Monthly water demand (gal)	7750	7750	7500	7750	7500	7750

Note: 1–12 refers to months from January to December.

Table 9.10 Monthly water demand.

Month	**1**	**2**	**3**	**4**	**5**	**6**
Day	31	28	31	30	31	30
Monthly water demand (gal)	36 851.25	33 285	36 851.3	35 662.5	36 851	35 662.5
Month	**7**	**8**	**9**	**10**	**11**	**12**
Day	31	31	30	31	30	31
Monthly water demand (gal)	36 851.25	36 851	35 662.5	36 851.25	35 662.5	36 851.25

Note: 1–12 refers to months from January to December.

```
%uses
usestoiletmen=0.25;
usestoiletwomen=2;
usesurinalmen=1.75;
useslavatoriesmen=2;
useslavatorieswomen=2;
% flow rate
flowratetoiletmen=1.28;
flowratetoiletwomen=1.28;
flowrateurinalmen=0.5;
flowratelavatoriesmen=0.25;
flowratelavatorieswomen=0.25;
% day for one year
dayforoneyear=[31  28 31 30 31 30 31 31 30
  31 30 31];
% day for three years
dayforthreeyears=[31 28 31 30 31 30 31 31
  30 31 30 31 31 28 31 30 31 30 31 31 30 31
  30 31 31 28 31 30  31 30 31 31 30 31
  30 31];
% water demand for handwashing
% water demand
waterdemandtoiletmen=men*usestoiletmen*
  flowratetoiletmen;
waterdemandtoiletwomen=women*
  usestoiletwomen*flowratetoiletwomen;
waterdemandurinalmen=men*
  usesurinalmen*flowrateurinalmen;
waterdemandlavatoriesmen=men*
  useslavatoriesmen*
  flowratelavatoriesmen;
waterdemandlavatorieswomen=women*
  useslavatorieswomen*
  flowratelavatorieswomen;
% output
% daily water demand
dailywaterdemandofbathroom=
  waterdemandtoiletmen+
  waterdemandtoiletwomen
  +waterdemandurinalmen
```

```
  +waterdemandlavatoriesmen
  +waterdemandlavatorieswomen
% monthly water demand for handwashing
monthlywaterdemandforhandwashing=
  dayforoneyear.*
  (waterdemandlavatoriesmen
  +waterdemandlavatorieswomen)
waterdemandforthreeyearhandwash=
  dayforthreeyears.*
  (waterdemandlavatoriesmen
  +waterdemandlavatorieswomen);
% monthly water demand
monthlywaterdemand=dayforoneyear.
  *dailywaterdemandofbathroom
waterdemandforthreeyear=
  dayforthreeyears.
  *dailywaterdemandofbathroom;
```

Answer:

```
waterdemandofbathroom
dailywaterdemandofbathroom = 1.1888e+03
monthlywaterdemandforhandwashing =

 Columns 1 through 8

   7750   7000   7750   7500   7750   7500
7750    7750

 Columns 9 through 12

    7500     7750     7500     7750

monthlywaterdemand =

  1.0e+04 *

 Columns 1 through 10

  3.6851  3.3285  3.6851  3.5663  3.6851
3.5663  3.6851  3.6851  3.5663  3.6851

 Columns 11 through 12

  3.5663  3.6851
```

9.4.2 Determine the Roof Area and the Tank Size

Example 9.2
Given: Use the same rainwater demand value in Example 9.1, rain collection efficiency = 0.8 (Table 9.11).

Table 9.11 Precipitation (inch).

Month	**1**	**2**	**3**	**4**	**5**	**6**
Precipitation	2.34	2.22	3.2	3.9	6.08	10.24
Month	**7**	**8**	**9**	**10**	**11**	**12**
Precipitation	7	9.2	8.88	6.56	3.83	2.59

Source: http://www.intellicast.com/Local/History.aspx?unit=F&month=6&location=USFL9856

Find: Calculate roof area and tank size that meet the requirement for zero water design of the public bathroom at the Greenway Corridor of Indian Creek.

Solution:

Assuming roof area = 4000 ft^2 and tank is big enough that it does not limit end-of-month water storage,

$$\text{Rainwater collected (gal)} = \text{precipitation (in)} \times \text{roof area}\left(\text{ft}^2\right) \times 0.625\left(\text{gal/ft}^2\text{/in}\right) \times \text{collection efficiency}$$

For January

$$\text{Rainwater collected} = 2.34\,\text{in} \times 4000\,\text{ft}^2 \times 0.623\left(\text{gal/ft}^2\text{/in}\right) \times 0.8 = 4665.024\,\text{gal}$$

$$\text{Rainwater waterdemand (gal)} = 7750\,\text{gal}$$

$$\text{Rainwater waterdemand} > \text{rainwater collected}$$

City water has to be added to the tank as water supplement to meet the required water demand.

To calculate cumulative water storage at the end of each month, the water supplement and cumulative water storage for the previous month should be equal to the rainwater collected, and subtract the monthly demand:

$$\text{Cumulative water storage (gal)} = \text{cumulative water storage (gal) for last month} + \text{rainwater collected (gal)} + \text{water supplement (gal)} - \text{rainwater demand (gal)}$$

For January

$$\text{Cumulative water storage} = 4665.024 + 3084.976 - 7750 = 0\,\text{gal}$$

For February

$$\text{Cumulative water storage} = 0 + 4425.792 + 2574.208 - 7000 = 0\,\text{gal}$$

For May (Table 9.12)

$$\text{Cumulative water storage} = 275.04 + 12121.088 + 0 - 7750 = 4646.128\ \text{gal}$$

$$\text{Maximum cumulative water storage} = 49943.84\ \text{gal}$$

49 943.84 gal is selected as tank size.
This tank can collect all rainwater during this year.

Matlab code:

```
clear all
close all
% input
run('waterdemandofbathroom.m');
waterdemand=
  monthlywaterdemandforhandwashing;
Averageprecipitationmonthin=[2.34 2.22
  3.2 3.9 6.08 10.2 7 9.2 8.88 6.56
  3.83 2.59];
AssumedCollectionEfficiency=0.8;
Catchmentarea=4000;
% function
Averageprecipitationmonthgalft2=
  Averageprecipitationmonthin*0.623;
Rainfallcollectedgal=
  Averageprecipitationmonthgalft2*
  AssumedCollectionEfficiency*
  Catchmentarea;
storage=Rainfallcollectedgal-
  waterdemand;
supply(1)=0;
endofmonthstorage(1)=0;
endofmonthstorage;
maxendofmonthstorage=max
  (endofmonthstorage);
monthsupply(1)=0;
for i=2:13
if endofmonthstorage(i-1)+storage
  (i-1)<=0
endofmonthstorage(i)=0;monthsupply(i)
  =-(endofmonthstorage(i-1)+storage
  (i-1));
else if endofmonthstorage(i-1)+storage
  (i-1)>0
endofmonthstorage(i)=endofmonthstorage
  (i-1)+storage(i-1);monthsupply(i)=0;
end
end
end
% output
tanksize=max(endofmonthstorage)
```

Answer: `tanksizebymethod1`

`tanksize = 4.9944e+04`

Table 9.13 shows that water supplement is 0 after the first year and that roof area and tank size meet the requirement.

Table 9.12 Water balance calculations (tank size not limited).

Month	A Rainwater demand (gal)	B Precipitation (inch)	C Rainwater collected (gal)	D Cumulative end-of-month water storage (gal)	E Water supplement (gal)
Jan	7750	2.34	4 665.024	0	3084.976
Feb	7000	2.22	4 425.792	0	2574.208
Mar	7750	3.2	6 379.520	0	1370.48
Apr	7500	3.9	7 775.040	275.04	0
May	7750	6.08	12 121.088	4646.128	0
Jun	7500	10.2	20 334.720	17 480.848	0
July	7750	7	13 955.200	23 686.048	0
Aug	7750	9.2	18 341.120	34 277.168	0
Sep	7500	8.88	17 703.168	44 480.336	0
Oct	7750	6.56	13 078.016	49 808.352	0
Nov	7500	3.83	7 635.488	49 943.84	0
Dec	7750	2.59	5 163.424	47 357.264	0

Table 9.13 Water balance calculations (tank size = 49 943.84 gal).

Month	A Rainwater demand (gal)	B Precipitation (inch)	C Rainwater collected (gal)	D Cumulative end-of-month water storage (gal)	E Water supplement (gal)
Jan	7750	2.34	4 665.024	0	3084.976
Feb	7000	2.22	4 425.792	0	2574.208
Mar	7750	3.2	6 379.520	0	1370.48
Apr	7500	3.9	7 775.040	275.04	0
May	7750	6.08	12 121.088	4 646.128	0
Jun	7500	10.2	20 334.720	17 480.848	0
July	7750	7	13 955.200	23 686.048	0
Aug	7750	9.2	18 341.120	34 277.168	0
Sep	7500	8.88	17 703.168	44 480.336	0
Oct	7750	6.56	13 078.016	49 808.352	0
Nov	7500	3.83	7 635.488	49 943.84	0
Dec	7750	2.59	5 163.424	47 357.264	0
Jan	7750	2.34	4 665.024	44 272.288	0
Feb	7000	2.22	4 425.792	41 698.08	0
Mar	7750	3.2	6 379.520	40 327.6	0
Apr	7500	3.9	7 775.040	40 602.64	0
May	7750	6.08	12 121.088	44 973.728	0
Jun	7500	10.2	20 334.720	49 943.84	0
July	7750	7	13 955.200	49 943.84	0
Aug	7750	9.2	18 341.120	49 943.84	0
Sep	7500	8.88	17 703.168	49 943.84	0
Oct	7750	6.56	13 078.016	49 943.84	0
Nov	7500	3.83	7 635.488	49 943.84	0
Dec	7750	2.59	5 163.424	47 357.264	0

Matlab code:

```
clear all
close all
% input
run('waterdemandofbathroom.m');
waterdemand=
  waterdemandforthreeyearhandwash;
Averageprecipitationmonthin=[2.34 2.22
  3.2 3.9 6.08 10.2 7 9.2 8.88 6.56
  3.83 2.59];
Averageprecipitationmonthin=
  [Averageprecipitationmonthin
  Averageprecipitationmonthin
  Averageprecipitationmonthin];
AssumedCollectionEfficiency=0.8;
Catchmentarea=4000;
stoagetankvolume=49943.84;
% function
Averageprecipitationmonthgalft2=
  Averageprecipitationmonthin*0.623;
Rainfallcollectedgal=
  Averageprecipitationmonthgalft2*
  AssumedCollectionEfficiency*
  Catchmentarea;
storage=Rainfallcollectedgal-
  waterdemand;
supply(1)=0;
endofmonthstorage(1)=0;
endofmonthstorage;
maxendofmonthstorage=max
  (endofmonthstorage);
monthsupply(1)=0;
for i=2:37
    if endofmonthstorage(i-1)+storage
      (i-1)<=0
```

```
        endofmonthstorage(i)=0;monthsupply
          (i)=-(endofmonthstorage(i-1)
          +storage(i-1));
      elseif endofmonthstorage(i-1)+
        storage(i-1)>=stoagetankvolume
        endofmonthstorage(i)
          =stoagetankvolume;monthsupply
          (i)=0;
      else
        endofmonthstorage(i)
          =endofmonthstorage(i-1)+storage
          (i-1);monthsupply(i)=0;
      end
  end
monthsupply;
totalmonthsupply=sum(monthsupply);
supplementafterthefirstyear=
  monthsupply';
supplementafterthefirstyear([1],:)=[];
supplementafterthefirstyear([1],:)=[];
supplementafterthefirstyear([1],:)=[];
supplementafterthefirstyear([1],:)=[];
supplementafterthefirstyear([1],:)=[];
supplementafterthefirstyear([1],:)=[];
supplementafterthefirstyear([1],:)=[];
supplementafterthefirstyear([1],:)=[];
supplementafterthefirstyear([1],:)=[];
supplementafterthefirstyear([1],:)=[];
supplementafterthefirstyear([1],:)=[];
supplementafterthefirstyear([1],:)=[];
supplementafterthefirstyear([1],:)=[];
% output
maxsupplementafterthefirstyear=max
  (supplementafterthefirstyear)
```

Answer: `watersupplementafterthefirstyear`

```
maxsupplementafterthefirstyear = 0
```

9.4.3 Design Rainwater System by Cumulative Plot Method

Example 9.3
Given: Use the same input value in Example 9.2, and design rainwater harvesting system to achieve zero water.
Find: Calculate roof area and tank size that achieve zero water design.

Solution (Table 9.14):

$$\text{Tank size} = \max - \min = 42914.18 - (-7029.66)$$
$$= 49943.84\,\text{gal}$$

Since roof area and tank size are the same as the values in Example 9.2, they meet the requirement (Figure 9.3).

Table 9.14 Water balance calculations by method 2.

Month	A Rainwater demand (gal)	B Rainwater collected (gal)	C Cumulative rainwater demand (gal)	D Cumulative rainwater collected (gal)	E Difference between cumulative rainwater collected and cumulative rainwater demand (gal)
Jan	7750	4 665.024	7 750	4 665.024	−3084.98
Feb	7000	4 425.792	14 750	9 090.816	−5 659.18
Mar	7750	6 379.520	22 500	15 470.336	−7 029.66
Apr	7500	7 775.040	30 000	23 245.376	−6 754.62
May	7750	12 121.088	37 750	35 366.464	−2 383.54
Jun	7500	20 334.720	45 250	55 701.184	10 451.18
July	7750	13 955.200	53 000	69 656.384	16 656.38
Aug	7750	18 341.120	60 750	87 997.504	27 247.5
Sep	7500	17 703.168	68 250	105 700.672	37 450.67
Oct	7750	13 078.016	76 000	118 778.688	42 778.69
Nov	7500	7 635.488	83 500	126 414.176	42 914.18
Dec	7750	5 163.424	91 250	131 577.6	40 327.6
Max					42 914.18
Min					−7 029.66

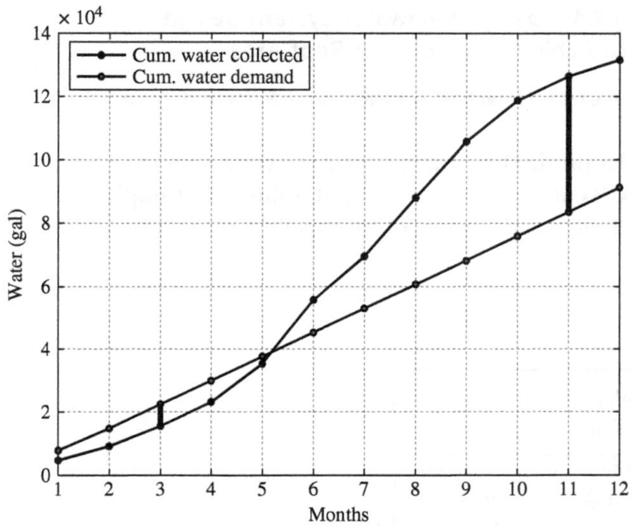

Figure 9.3 Mass curve for calculation of required tank size.

Matlab code:

```
clear all
close all
% input
run('waterdemandofbathroom.m');
waterdemand=
  monthlywaterdemandforhandwashing;
Averageprecipitationmonthin=[2.34 2.22
  3.2 3.9 6.08  10.2  7 9.2 8.88  6.56
  3.83  2.59];
AssumedCollectionEfficiency=0.8;
Catchmentarea=4000;
% function
Averageprecipitationmonthgalft2=
  Averageprecipitationmonthin*0.623;
Rainfallcollectedgal=
  Averageprecipitationmonthgalft2*
  AssumedCollectionEfficiency*
  Catchmentarea;
CumRainfallcollectedgal(1)=0;
Cumwaterdemand(1)=0;
for i=2:13
  CumRainfallcollectedgal(i)
    =CumRainfallcollectedgal(i-1)
    +Rainfallcollectedgal(i-1);
end
for i=2:13
  Cumwaterdemand(i)=Cumwaterdemand(i-1)
    +waterdemand(i-1);
end
Difference=CumRainfallcollectedgal-
  Cumwaterdemand;
Max=max(Difference);
Min=min(Difference);
% output
if Min<0
tanksize=Max-Min
else if Min>=0
tanksize=Max
  end
end

CumRainfallcollectedgal=
  CumRainfallcollectedgal';
CumRainfallcollectedgal([1],:)=[];
Cumwaterdemand=Cumwaterdemand';
Cumwaterdemand([1],:)=[];
months=[1 2 3 4 5 6 7 8 9 10 11 12];

plot(months,CumRainfallcollectedgal,
  'k-o','MarkerSize',3,
  'MarkerFaceColor',[0 0 0],
  'MarkerEdgeColor',[0 0 0],
  'linewidth',1);
hold on
plot(months,Cumwaterdemand,'k-o',
  'MarkerSize',3,'MarkerFaceColor',
  [0.5 0.5 0.5],'MarkerEdgeColor',
  [0 0 0],'linewidth',1);
grid on
xmax=max(months);
xmin=min(months);
set(gca,'xtickmode','auto','FontSize',
  10,'FontWeight','bold','FontName',
  'Times New Roman');
set(gca,'xlim',[xmin xmax],
  'FontSize',10,
  'FontWeight','bold','FontName',
  'Times New Roman');
xlabel({'Months'},
  'FontSize',14,'FontWeight',
  'bold','FontName', 'Times New Roman')
ylabel({'Water (gal)'},
  'FontSize',14,'FontWeight',
  'bold','FontName', 'Times New Roman')
legend('Cum. water collected','Cum. water
  demand','Location','NorthWest');
box on
grid on
hold on
Maxline=[CumRainfallcollectedgal(11)
  Cumwaterdemand(11)];
Minline=[Cumwaterdemand(3)
  CumRainfallcollectedgal(3)];
plot([11 11],Maxline,'k-',
  'linewidth',2.5);
```

```
hold on
plot([3 3],Minline,'k-',
  'linewidth',2.5);
grid on
```

Answer: `tanksizebymethod1`

`tanksize = 4.9944e+04`

9.4.4 Design Rainwater System Design to Achieve the Smallest Roof Area

9.4.4.1 Flowchart for Rainwater System (Figure 9.4)

Example 9.4 Rainwater harvesting system
Given: Use the same input value in Example 9.2, and design rain harvest of the smallest roof area.

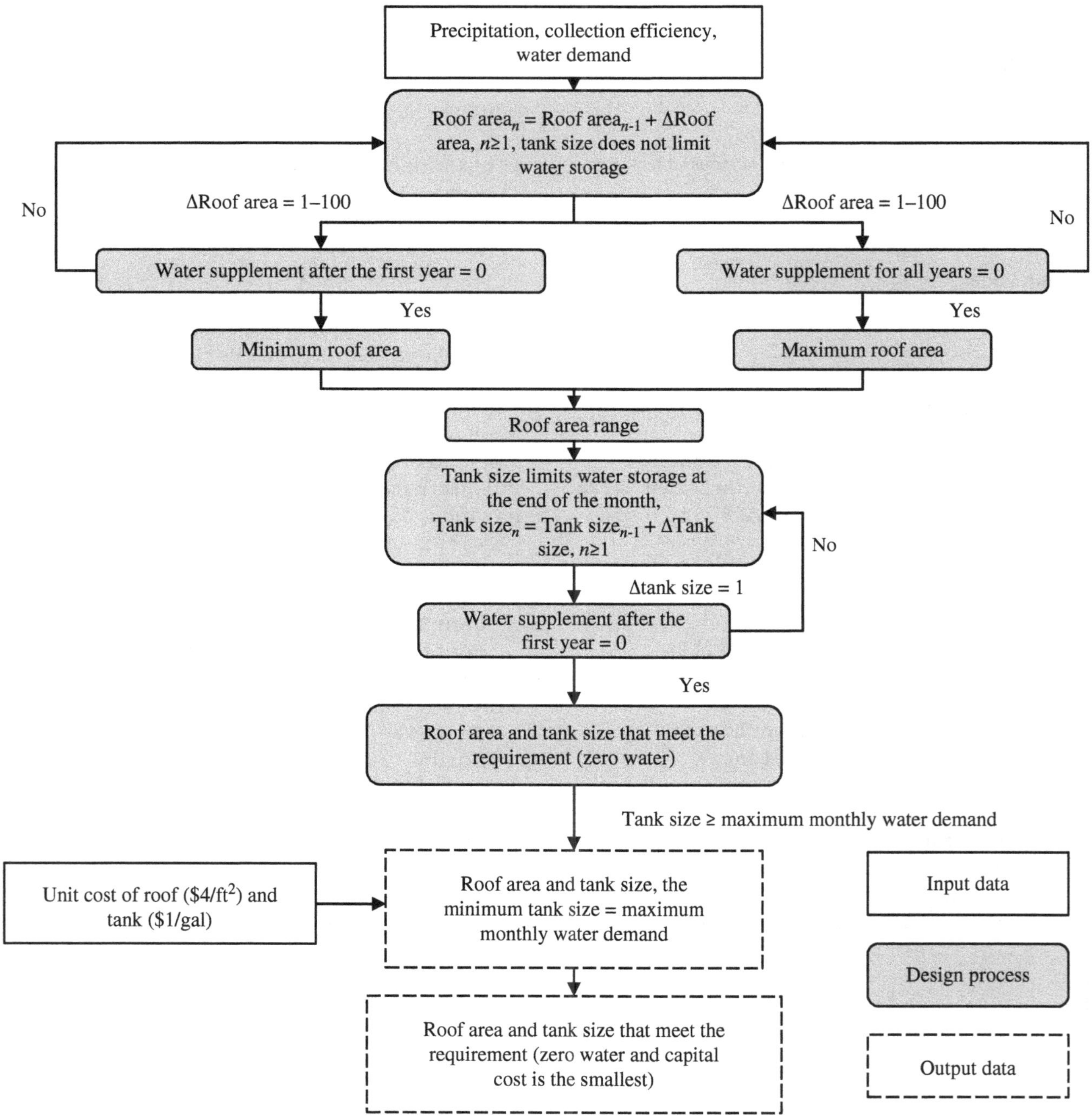

Figure 9.4 Flowchart for rainwater system.

Find: The smallest roof area and tank size that meet the requirement (zero water) after the first year.

Solution:

Find the smallest roof area that meets the requirement (zero water) by Matlab.

Matlab function code:

Myfunforindiancreek

```
function [ maxsecondyearmonthsupply,
  endofmonthstorage] =
  myfunforindiancreek
  (waterdemand,Catchmentarea,
  Averageprecipitationmonthin,
  AssumedCollectionEfficiency)
% function
Averageprecipitationmonthgalft2=
  Averageprecipitationmonthin*0.623;
Rainfallcollectedgal=
  Averageprecipitationmonthgalft2*
  AssumedCollectionEfficiency*
  Catchmentarea;
storage=Rainfallcollectedgal-
  waterdemand;
supply(1)=0;
endofmonthstorage(1)=0;
endofmonthstorage;
maxendofmonthstorage=max
  (endofmonthstorage);
monthsupply(1)=0;
for i=2:37
  if endofmonthstorage(i-1)+storage
  (i-1)<=0
endofmonthstorage(i)=0;monthsupply(i)=
  -(endofmonthstorage(i-1)+storage
  (i-1));
else if endofmonthstorage(i-1)+storage
  (i-1)>0
endofmonthstorage(i)=endofmonthstorage
  (i-1)+storage(i-1);monthsupply(i)=0;
  end
  end
end
monthsupply;
totalmonthsupply=sum(monthsupply);
secondyearmonthsupply=monthsupply';
secondyearmonthsupply([1],:)=[];
secondyearmonthsupply([1],:)=[];
secondyearmonthsupply([1],:)=[];
secondyearmonthsupply([1],:)=[];
secondyearmonthsupply([1],:)=[];
secondyearmonthsupply([1],:)=[];
secondyearmonthsupply([1],:)=[];
secondyearmonthsupply([1],:)=[];
secondyearmonthsupply([1],:)=[];
secondyearmonthsupply([1],:)=[];
secondyearmonthsupply([1],:)=[];
secondyearmonthsupply([1],:)=[];
secondyearmonthsupply([1],:)=[];
maxsecondyearmonthsupply=max
  (secondyearmonthsupply);
end
```

Myfunforindiancreeklim

```
function [ maxsecondyearmonthsupply,
  endofmonthstorage] =
  myfunforindiancreeklim
  (waterdemand,Catchmentarea,
  stoagetankvolume,
  Averageprecipitationmonthin,
  AssumedCollectionEfficiency)
Averageprecipitationmonthgalft2=
  Averageprecipitationmonthin*0.623;
Rainfallcollectedgal=
  Averageprecipitationmonthgalft2*
  AssumedCollectionEfficiency*
  Catchmentarea;
storage=Rainfallcollectedgal-
  waterdemand;
%
supply(1)=0;
endofmonthstorage(1)=0;
endofmonthstorage;
maxendofmonthstorage=max
  (endofmonthstorage);
monthsupply(1)=0;
for i=2:37
    if endofmonthstorage(i-1)+storage
      (i-1)<=0
      endofmonthstorage(i)=0;monthsupply
        (i)=-(endofmonthstorage(i-1)
        +storage(i-1));
   elseif endofmonthstorage(i-1)+storage
     (i-1)>=stoagetankvolume
      endofmonthstorage(i)=
        stoagetankvolume;monthsupply
        (i)=0;
    else
      endofmonthstorage(i)
        =endofmonthstorage(i-1)+storage
        (i-1);monthsupply(i)=0;
    end
  end
monthsupply;
totalmonthsupply=sum(monthsupply);
secondyearmonthsupply=monthsupply';
secondyearmonthsupply([1],:)=[];
```

```
secondyearmonthsupply([1],:)=[];
secondyearmonthsupply([1],:)=[];
secondyearmonthsupply([1],:)=[];
secondyearmonthsupply([1],:)=[];
secondyearmonthsupply([1],:)=[];
secondyearmonthsupply([1],:)=[];
secondyearmonthsupply([1],:)=[];
secondyearmonthsupply([1],:)=[];
secondyearmonthsupply([1],:)=[];
secondyearmonthsupply([1],:)=[];
secondyearmonthsupply([1],:)=[];
secondyearmonthsupply([1],:)=[];
maxsecondyearmonthsupply=max
  (secondyearmonthsupply);
end

smallstorageforminarea

function [minCatchmentarea,
  Smallstorageforminarea] =
  smallstorageforminarea
  (waterdemand,Catchmentarea,
  Averageprecipitationmonthin,
  AssumedCollectionEfficiency);
[ maxsecondyearmonthsupply,
  endofmonthstorage] =
  myfunforindiancreek
  (waterdemand,Catchmentarea,
  Averageprecipitationmonthin,
  AssumedCollectionEfficiency);
while (maxsecondyearmonthsupply>0)
Catchmentarea=Catchmentarea+1;
[ maxsecondyearmonthsupply,
  endofmonthstorage] =
  myfunforindiancreek
  (waterdemand,Catchmentarea,
  Averageprecipitationmonthin,
  AssumedCollectionEfficiency);
end
Catchmentarea;
minCatchmentarea=Catchmentarea;
Catchmentarea=minCatchmentarea;
stoagetankvolume=0;
[ maxsecondyearmonthsupply,
  endofmonthstorage] =
  myfunforindiancreeklim
  (waterdemand,Catchmentarea,
  stoagetankvolume,
  Averageprecipitationmonthin,
  AssumedCollectionEfficiency);
while (maxsecondyearmonthsupply>0)
stoagetankvolume=stoagetankvolume+1;
[ maxsecondyearmonthsupply,
  endofmonthstorage] =
  myfunforindiancreeklim
  (waterdemand,Catchmentarea,
  stoagetankvolume,
  Averageprecipitationmonthin,
  AssumedCollectionEfficiency);
end
Smallstorageforminarea=
  stoagetankvolume;
end
```

Matlab main code:

```
clear all
close all
% input
run('waterdemandofbathroom.m');
waterdemand=
  waterdemandforthreeyearhandwash;
Averageprecipitationmonthin=[2.34 2.22
  3.2 3.9 6.08 10.2 7 9.2 8.88 6.56 3.83
  2.59 2.34 2.22 3.2 3.9 6.08 10.2 7 9.2
  8.88 6.56 3.83 2.59 2.34 2.22 3.2 3.9
  6.08 10.2 7 9.2 8.88 6.56 3.83 2.59];
AssumedCollectionEfficiency=0.8;
Catchmentarea=2700;
% function
[minCatchmentarea,
  Smallstorageforminarea] =
  smallstorageforminarea
  (waterdemand,Catchmentarea,
  Averageprecipitationmonthin,
  AssumedCollectionEfficiency);
% output
roofarea=minCatchmentarea
tanksize=Smallstorageforminarea
```

Answer: `smalllestroofareaandtanksize`

```
roofarea = 2775
tanksize = 20245
```

Roof area = 2775 ft^2

Tank size = 20245 gal

Table 9.15 shows that water supplement is 0 after the first year and that roof area and tank size meet the requirement.

9.4.5 Determine Roof Area for a Rainwater Harvest Tank Without Adding City Water in the First Year

Example 9.5

Given: Use the same input value in Example 9.2, and design rainwater harvesting system without city water in the first year.

Table 9.15 Water balance calculations (tank size = 20 245 gal and roof area = 2 775 ft²).

Month	A Rainwater demand (gal)	B Precipitation (in)	C Rainwater collected (gal)	D Cumulative end-of-month water storage (gal)	E Water supplement (gal)
Jan	7750	2.34	3 236.360	0	4513.6396
Feb	7000	2.22	3 070.393	0	3929.6068
Mar	7750	3.2	4 425.792	0	3324.208
Apr	7500	3.9	5 393.934	0	2106.066
May	7750	6.08	8 409.005	659.004	0
Jun	7500	10.2	14 107.212	7 266.216	0
July	7750	7	9 681.420	9 197.636	0
Aug	7750	9.2	12 724.152	14 171.788	0
Sep	7500	8.88	12 281.573	18 953.361	0
Oct	7750	6.56	9 072.874	20 245	0
Nov	7500	3.83	5 297.120	18 042.119	0
Dec	7750	2.59	3 582.125	13 874.245	0
Jan	7750	2.34	3 236.3604	9 360.605	0
Feb	7000	2.22	3 070.3932	5 430.998	0
Mar	7750	3.2	4 425.792	2 106.790	0
Apr	7500	3.9	5 393.934	0.7248	0
May	7750	6.08	8 409.0048	659.729	0
Jun	7500	10.2	14 107.212	7 266.9416	0
July	7750	7	9 681.42	9 198.3616	0
Aug	7750	9.2	12 724.152	14 172.5136	0
Sep	7500	8.88	12 281.5728	18 954.0864	0
Oct	7750	6.56	9 072.8736	20 245	0
Nov	7500	3.83	5 297.1198	18 042.1198	0
Dec	7750	2.59	3 582.1254	13 874.2452	0

Find: The roof area that meets the requirement that water supplement is 0 in the first year.

Solution:

Find the roof area that meets the requirement that water supplement is 0 in the first year by Matlab.

Matlab function code:
Myfunforindiancreekmax

```
function
  [ stoagetankvolume,maxyearmonthsupply,
  endofmonthstorage] =
  myfunforindiancreekmax
  (waterdemand,Catchmentarea,
  Averageprecipitationmonthin,
  AssumedCollectionEfficiency)
Averageprecipitationmonthgalft2=
  Averageprecipitationmonthin*0.623;
Rainfallcollectedgal=
  Averageprecipitationmonthgalft2*
  AssumedCollectionEfficiency*
  Catchmentarea;
storage=Rainfallcollectedgal-
  waterdemand;
supply(1)=0;
endofmonthstorage(1)=0;
endofmonthstorage;
maxendofmonthstorage=max
  (endofmonthstorage);
monthsupply(1)=0;
for i=2:37
  if endofmonthstorage(i-1)+storage
   (i-1)<=0
endofmonthstorage(i)=0;monthsupply(i)
    =-(endofmonthstorage(i-1)+storage
    (i-1));
else if endofmonthstorage(i-1)+storage
  (i-1)>0
endofmonthstorage(i)=endofmonthstorage
  (i-1)+storage(i-1);monthsupply(i)=0;
```

```
    end
  end
end
monthsupply;
totalmonthsupply=sum(monthsupply);
stoagetankvolume=maxendofmonthstorage;
secondyearmonthsupply=monthsupply';
secondyearmonthsupply([1],:)=[];
secondyearmonthsupply([1],:)=[];
secondyearmonthsupply([1],:)=[];
secondyearmonthsupply([1],:)=[];
secondyearmonthsupply([1],:)=[];
secondyearmonthsupply([1],:)=[];
secondyearmonthsupply([1],:)=[];
secondyearmonthsupply([1],:)=[];
secondyearmonthsupply([1],:)=[];
secondyearmonthsupply([1],:)=[];
secondyearmonthsupply([1],:)=[];
secondyearmonthsupply([1],:)=[];
secondyearmonthsupply([1],:)=[];
maxyearmonthsupply=max(monthsupply');
end
```

Smallstorageformaxarea

```
function [maxCatchmentarea,
  Smallstorageformaxarea] =
  smallstorageformaxarea
  (waterdemand,Catchmentarea,
  Averageprecipitationmonthin,
  AssumedCollectionEfficiency);
[ stoagetankvolume,maxyearmonthsupply,
  endofmonthstorage] =
  myfunforindiancreekmax
  (waterdemand,Catchmentarea,
  Averageprecipitationmonthin,
  AssumedCollectionEfficiency);
while (maxyearmonthsupply>0)
Catchmentarea=Catchmentarea+1;
[ stoagetankvolume,maxyearmonthsupply,
  endofmonthstorage] =
  myfunforindiancreekmax
  (waterdemand,Catchmentarea,
  Averageprecipitationmonthin,
  AssumedCollectionEfficiency);
end
Catchmentarea;
maxCatchmentarea=Catchmentarea;
Catchmentarea=maxCatchmentarea;
stoagetankvolume=0;
[ maxsecondyearmonthsupply,
  endofmonthstorage] =
  myfunforindiancreeklim
  (waterdemand,Catchmentarea,
  stoagetankvolume,
  Averageprecipitationmonthin,
  AssumedCollectionEfficiency);
while (maxsecondyearmonthsupply>0)
stoagetankvolume=
  stoagetankvolume+1;
[ maxsecondyearmonthsupply,
  endofmonthstorage] =
  myfunforindiancreeklim
  (waterdemand,Catchmentarea,
  stoagetankvolume,
  Averageprecipitationmonthin,
  AssumedCollectionEfficiency);
end
Smallstorageformaxarea=
  stoagetankvolume;
end
```

Matlab main code:

```
clear all
close all
% input
run('waterdemandofbathroom.m');
waterdemand=
  waterdemandforthreeyearhandwash;
Averageprecipitationmonthin=[2.34  2.22
  3.2 3.9 6.08  10.2  7  9.2 8.88  6.56
  3.83  2.59 2.34  2.22  3.2 3.9 6.08  10.2
  7  9.2 8.88  6.56  3.83  2.59 2.34  2.22
  3.2 3.9 6.08  10.2  7  9.2 8.88  6.56
  3.83  2.59];
AssumedCollectionEfficiency=0.8;
Catchmentarea=2700;
% function
[maxCatchmentarea,
  Smallstorageformaxarea] =
  smallstorageformaxarea
  (waterdemand,Catchmentarea,
  Averageprecipitationmonthin,
  AssumedCollectionEfficiency);
% output
roofarea=maxCatchmentarea
tanksize=Smallstorageformaxarea
```

Answer: `tanksizeandroofareathatmeet requirementforallyears`

```
roofarea = 6646
tanksize = 0
```

Roof area = 6646 ft^2

Tank size = 0 gal

Table 9.16 shows that water supplement is 0 for all years and that roof area and tank size meet the requirement.

Table 9.16 Water balance calculations (tank size = 0 gal and roof area = 6646 ft^2).

Month	A Rainwater demand (gal)	B Precipitation (in)	C Rainwater collected (gal)	D Cumulative end-of-month water storage (gal)	E Water supplement (gal)
Jan	7750	2.34	7 750.937	0	0
Feb	7000	2.22	7 353.453	0	0
Mar	7750	3.2	10 599.572	0	0
Apr	7500	3.9	12 918.228	0	0
May	7750	6.08	20 139.187	0	0
Jun	7500	10.2	33 786.137	0	0
July	7750	7	23 186.564	0	0
Aug	7750	9.2	30 473.770	0	0
Sep	7500	8.88	29 413.813	0	0
Oct	7750	6.56	21 729.123	0	0
Nov	7500	3.83	12 686.363	0	0
Dec	7750	2.59	8 579.028	0	0
Jan	7750	2.34	7 750.937	0	0
Feb	7000	2.22	7 353.453	0	0
Mar	7750	3.2	10 599.572	0	0
Apr	7500	3.9	12 918.228	0	0
May	7750	6.08	20 139.187	0	0
Jun	7500	10.2	33 786.137	0	0
July	7750	7	23 186.564	0	0
Aug	7750	9.2	30 473.770	0	0
Sep	7500	8.88	29 413.813	0	0
Oct	7750	6.56	21 729.123	0	0
Nov	7500	3.83	12 686.363	0	0
Dec	7750	2.59	8 579.028	0	0

$$\text{Tank size} = 0\,\text{gal}$$

Since rainwater collected is greater than rainwater demand, water storage at the end of month can be 0. In the design, the smallest tank size is equal to the maximum water demand.

$$\text{Tank size} = 7750\,\text{gal}$$

9.4.6 Design Rainwater Harvest Tank for Specific Roof Areas

Example 9.6 Rainwater harvesting system
Given: Use the same input value in Example 9.2, and design a rainwater harvest tank for specific roof areas.
Find: The corresponding tank size and roof area: $2775\,ft^2 : 100\,ft^2 : 6646\,ft^2$.

Solution:

Find the corresponding tank size and roof area: $2775\,ft^2 : 100\,ft^2 : 6646\,ft^2$ by Matlab.

Matlab main code:

```
clear all
close all
% input
run('waterdemandofbathroom.m');
waterdemand=
  waterdemandforthreeyearhandwash;
Averageprecipitationmonthin=
  [2.34  2.22  3.2 3.9 6.08  10.2  7
  9.2 8.88  6.56  3.83  2.59 2.34  2.22
  3.2 3.9 6.08  10.2  7  9.2 8.88  6.56
  3.83  2.59 2.34  2.22  3.2 3.9 6.08
```

```
  10.2   7  9.2 8.88   6.56
  3.83   2.59];
AssumedCollectionEfficiency=0.8;
Catchmentarea=2700;
% function
[minCatchmentarea,
  Smallstorageforminarea] =
  smallstorageforminarea
  (waterdemand,Catchmentarea,
  Averageprecipitationmonthin,
  AssumedCollectionEfficiency);
[maxCatchmentarea,
  Smallstorageformaxarea] =
  smallstorageformaxarea
  (waterdemand,Catchmentarea,
  Averageprecipitationmonthin,
  AssumedCollectionEfficiency);
differencearea=10^(floor(log10
  (minCatchmentarea))-1);
minCatchmentareaInteger=
  minCatchmentarea/differencearea;
minCatchmentareaInteger=ceil
  (minCatchmentareaInteger)
  *differencearea;
%first
i=1;
Catchmentarea=minCatchmentareaInteger;
stoagetankvolume=0;
[ maxsecondyearmonthsupply,
  endofmonthstorage] =
  myfunforindiancreeklim
  (waterdemand,Catchmentarea,
  stoagetankvolume,
  Averageprecipitationmonthin,
  AssumedCollectionEfficiency);
while (maxsecondyearmonthsupply>0)
stoagetankvolume=stoagetankvolume+1;
[ maxsecondyearmonthsupply,
  endofmonthstorage] =
  myfunforindiancreeklim
  (waterdemand,Catchmentarea,
  stoagetankvolume,
  Averageprecipitationmonthin,
  AssumedCollectionEfficiency);
end
stoagetankvolumesmall(i)
  =stoagetankvolume;
% second
while (Catchmentarea<maxCatchmentarea)
i=i+1;
Catchmentarea=Catchmentarea
  +differencearea;
% function
stoagetankvolume=0;
[ maxsecondyearmonthsupply,
  endofmonthstorage] =
  myfunforindiancreeklim
  (waterdemand,Catchmentarea,
  stoagetankvolume,
  Averageprecipitationmonthin,
  AssumedCollectionEfficiency);
while (maxsecondyearmonthsupply>0)
stoagetankvolume=stoagetankvolume+1;
[ maxsecondyearmonthsupply,
  endofmonthstorage] =
  myfunforindiancreeklim
  (waterdemand,Catchmentarea,
  stoagetankvolume,
  Averageprecipitationmonthin,
  AssumedCollectionEfficiency);
end
stoagetankvolumesmall(i)
  =stoagetankvolume;
end
maxCatchmentareaInteger=Catchmentarea-
  differencearea;
stoagetankvolumesmall(:,end)=[];
midstoagetankvolumesmall=
  stoagetankvolumesmall;
stoagetankvolumerange=
  [Smallstorageforminarea,
  midstoagetankvolumesmall,
  Smallstorageformaxarea];
catcharearange=[minCatchmentarea,
  minCatchmentareaInteger:
  differencearea:
  maxCatchmentareaInteger,
  maxCatchmentarea];
x=catcharearange;
y=stoagetankvolumerange;
xmax=max(x);
xmin=min(x);
n=2;
p=polyfit(x,y,n)
x4=[xmin:differencearea:xmax];
z3=polyval(p,x);
z4=polyval(p,x4);
figure (1)
plot(x,y,'ko','MarkerSize',3,
  'MarkerFaceColor',
  [0 0 0],'MarkerEdgeColor',
  [0 0 0],'linewidth',1);
hold on
plot(x4,z4,'k-','LineWidth',1);
hold on
R2 = 1 - sum((y - z3).^2)/sum
  ((y - mean(y)).^2)
grid on
```

```
xmax=max(x);
xmin=min(x);
set(gca,'xtickmode',
  'auto','FontSize',10,
  'FontWeight','bold','FontName',
  'Times New Roman');
set(gca,'xlim',[xmin xmax],
  'FontSize',10,'FontWeight',
  'bold','FontName', 'Times New Roman');
xlabel({'Catchment Area (ft^2)'},
  'FontSize',14,
  'FontWeight','bold','FontName',
  'Times New Roman')
ylabel({'Storage Tank Volume (gal)'},
  'FontSize',14,
  'FontWeight','bold','FontName',
  'Times New Roman')
grid on
hold on
figure (2)
waterdemandmax=max(waterdemand);
yrowsize=size(y,1);
ycolumnsize=size(y,2);
ysize=max(yrowsize,ycolumnsize);
for i=1:ysize;
  if y(i)<waterdemandmax
    y(i)=waterdemandmax;
  end
end
plot(x,y,'ko',
  'MarkerSize',3,'MarkerFaceColor',
  [0 0 0],'MarkerEdgeColor',
  [0 0 0],'linewidth',1);
grid on
xmax=max(x);
xmin=min(x);
set(gca,'xtickmode',
  'auto','FontSize',10,
  'FontWeight','bold','FontName',
  'Times New Roman');
set(gca,'xlim',[xmin xmax],
  'FontSize',10,
  'FontWeight','bold','FontName',
  'Times New Roman');
xlabel({'Catchment Area (ft^2)'},
  'FontSize',14,
  'FontWeight','bold','FontName',
  'Times New Roman')
ylabel({'Storage Tank Volume (gal)'},
  'FontSize',14,
  'FontWeight','bold','FontName',
  'Times New Roman')
grid on
hold on
```

Answer: `plotareavssizetankv2smalltank handwashing`

```
p = 1.0e+04 *
  0.0000  -0.0016  5.5189
R2 = 0.9985
```

Equation for tank size vs. roof area (Figures 9.5 and 9.6).

The regression equation of the trend line (2775 ≤ roof area ≤ 4400).

Tank size = (0.0011)(roof area)^2 − (15.7827) (roof area) + 55 189 $r^2 = 0.9985$.

The regression equation of the trend line (roof area ≥ 4400).

Tank size = 7750.

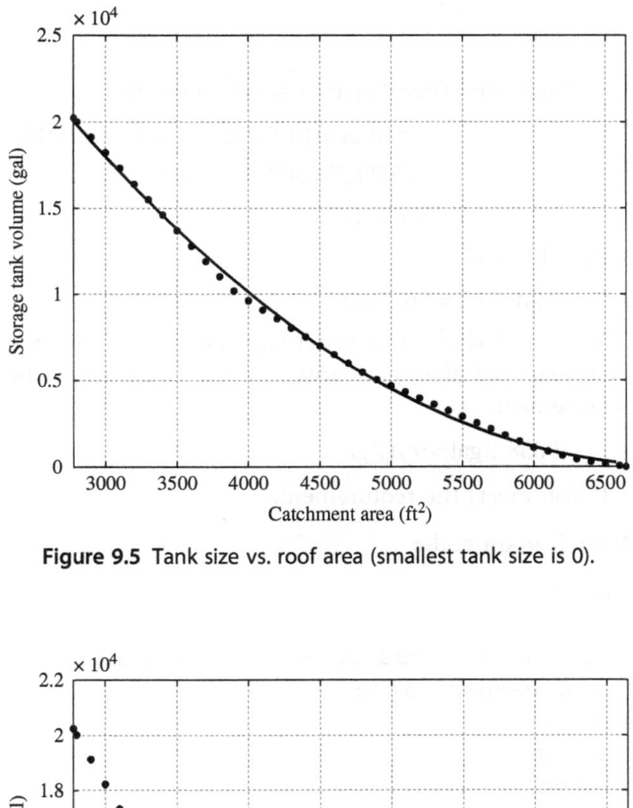

Figure 9.5 Tank size vs. roof area (smallest tank size is 0).

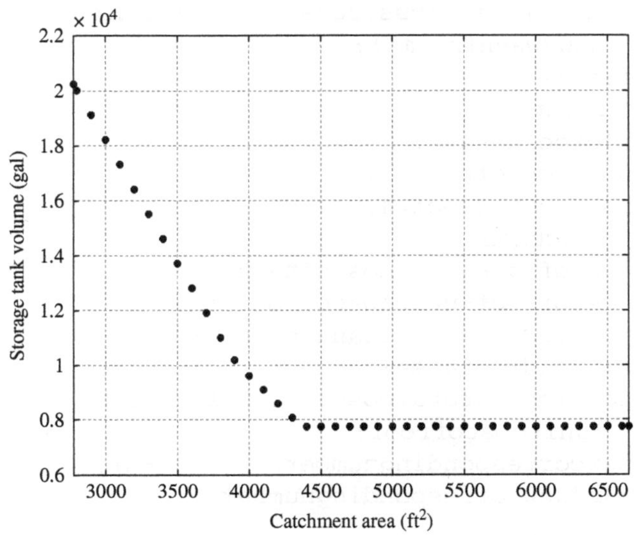

Figure 9.6 Tank size vs. roof area (smallest tank size is 7750 gal).

9.4.7 Design a Rainwater Harvest Tank of the Optimized Size

Example 9.7

Given: Use the same input value in Example 9.2, and design a rainwater harvest tank of the optimized size.

There are two methods to select tank size:

Method A: Select tank size based on roof area (given) and Equation or Table 9.17.

Method B: Select size of tank and roof based on cost (cost is the smallest).

Find:

a) Roof area = 4000 ft^2; select tank size by method A.
b) Unit cost of tank = 1$/gal; unit cost of roof = 4$/$ft^2$; select tank size by method B.

Solution:

$$\text{Tank size} = (0.0011)(\text{roof area})^2 - (15.7827) \times (\text{roof area}) + 55189 = (0.0011)(4000)^2 - (15.7827)(4000) + 55189 = 9658.2\ \text{gal}$$

In Table 9.17

$$\text{Tank size} = 9617\ \text{gal}$$

Table 9.18 shows that water supplement is 0 after the first year and that roof area and tank size meet the requirement.

$$9658.2\ \text{gal} > 9617\ \text{gal}$$

It also meets the requirement.

Matlab main code:

```
clear
close
run('plotareavssizetankv2smalltank
  handwashing.m');
figure
% input
oldroofarea=0;
unitcostoftank=1;
unitcostofroof=4;
% function
costoftank=unitcostoftank.*y;
costofroof=unitcostofroof.*x;
totalcost=costoftank+costofroof;
x=x-oldroofarea;
totalcost=totalcost-oldroofarea.
  *unitcostofroof;
newcorrespondingnumber=find(x>=0);
x=x(newcorrespondingnumber);
```

Table 9.17 Roof area and tank size.

Roof area (ft^2)	Tank size (gal)
2775	20 245
2800	20 019
2900	19 118
3000	18 217
3100	17 316
3200	16 415
3300	15 514
3400	14 613
3500	13 712
3600	12 811
3700	11 910
3800	11 008
3900	10 188
4000	9 617
4100	9 101
4200	8 585
4300	8 069
4400	7 750
4500	7 750
4600	7 750
4700	7 750
4800	7 750
4900	7 750
5000	7 750
5100	7 750
5200	7 750
5300	7 750
5400	7 750
5500	7 750
5600	7 750
5700	7 750
5800	7 750
5900	7 750
6000	7 750
6100	7 750
6200	7 750
6300	7 750
6400	7 750
6500	7 750
6600	7 750
6646	7 750

Table 9.18 Water balance calculations (tank size = 9617 gal).

Month	A Rainwater demand (gal)	B Precipitation (in)	C Rainwater collected (gal)	D Cumulative end-of-month water storage (gal)	E Water supplement (gal)
Jan	7750	2.34	4 665.024	0	3084.976
Feb	7000	2.22	4 425.792	0	2574.208
Mar	7750	3.2	6 379.520	0	1370.48
Apr	7500	3.9	7 775.040	275.04	0
May	7750	6.08	12 121.088	4646.128	0
Jun	7500	10.2	20 334.720	9617	0
July	7750	7	13 955.200	9617	0
Aug	7750	9.2	18 341.120	9617	0
Sep	7500	8.88	17 703.168	9617	0
Oct	7750	6.56	13 078.016	9617	0
Nov	7500	3.83	7 635.488	9617	0
Dec	7750	2.59	5 163.424	7030.424	0
Jan	7750	2.34	4 665.024	3945.448	0
Feb	7000	2.22	4 425.792	1371.24	0
Mar	7750	3.2	6 379.520	0.76	0
Apr	7500	3.9	7 775.040	275.8	0
May	7750	6.08	12 121.088	4646.888	0
Jun	7500	10.2	20 334.720	9617	0
July	7750	7	13 955.200	9617	0
Aug	7750	9.2	18 341.120	9617	0
Sep	7500	8.88	17 703.168	9617	0
Oct	7750	6.56	13 078.016	9617	0
Nov	7500	3.83	7 635.488	9617	0
Dec	7750	2.59	5 163.424	7030.424	0

```
totalcost=totalcost
  (newcorrespondingnumber);
y=y(newcorrespondingnumber);
mintotalcost=min(totalcost);
correspondingnumberoftotalcost=find
  (totalcost==min(totalcost));
correspondingroofadd=x
  (correspondingnumberoftotalcost);
correspondingtank=y
  (correspondingnumberoftotalcost);
plot(x,totalcost,'ko','MarkerSize',3,
  'MarkerFaceColor',[0 0 0],
  'MarkerEdgeColor',[0 0 0],
  'linewidth',1);
grid on
xmax=max(x);
xmin=min(x);
set(gca,'xtickmode','auto',
  'FontSize',10,'FontWeight','bold',
  'FontName','Times New Roman');
set(gca,'xlim',[xmin xmax],
  'FontSize',10,'FontWeight','bold',
  'FontName', 'Times New Roman');
xlabel({'Catchment Area (ft^2)'},
  'FontSize',14,'FontWeight',
  'bold','FontName', 'Times New Roman')
ylabel({'Total Cost ($)'}, 'FontSize',14,
  'FontWeight','bold','FontName',
  'Times New Roman')
grid on
hold on
%output
roofarea=correspondingroofadd
tanksize=correspondingtank
```

Answer: `costcomparehandwashing`

```
p = 1.0e+04 *
  0.0000  -0.0016  5.5189
R2 = 0.9985
roofarea = 4300
tanksize = 8069
```

Roof area = 4300 ft^2

Tank size = 8069 gal

In Table 9.17, tank size = 8069 gal when roof area = 4300 ft^2 (Figure 9.7).

In Table 9.17, total cost is the smallest when roof area = 4300 ft^2.

For Indian Creek, since roof area is not given,

The value selected for roof area = 4300 ft^2

The value selected for tank size = 8069 gal

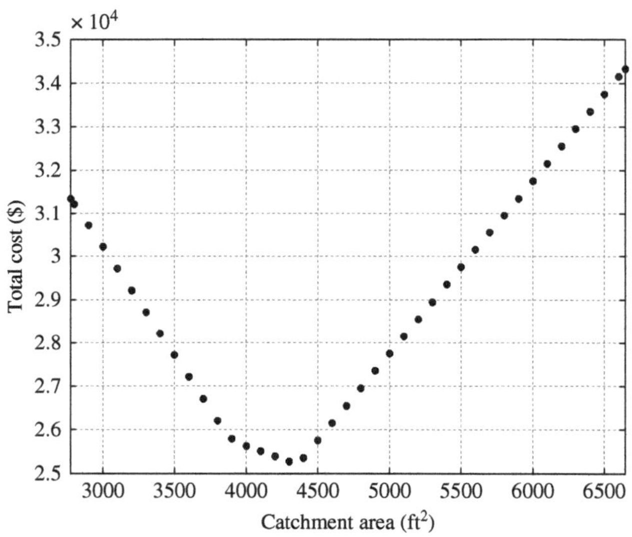

Figure 9.7 Total cost vs. roof area.

9.5 Design Anaerobic Digester Reactor

To design a septic tank to treat black water, Figure 9.8 presents the flow chart of the design process. The procedure is applied in Example 9.8.

Example 9.8
Given: Anaerobic digester can be used for wastewater treatment to produces biogas. Input data are shown in Tables 9.19 and 9.20.

Find: Volume of anaerobic tank and amount of biogas and electricity.

Solution:

Matlab code:

```
clear all
close all
% input
run('waterdemandofbathroom.m');
flowrate=dailywaterdemandofbathroom;
flowrate=flowrate/264.17;
BOD=300;
COD=450;
TSS=400;
VSS=270;
pH=7.7;
SO4=85;
% function
BODremovalefficiency=0.8;
Yieldcoefficient=0.1;
NewvolatilesuspendedsolidsNVSSproduced
  inBODremoval=Yieldcoefficient*BOD*
  BODremovalefficiency;
degradationrate=0.4;
Nondegradableresidue=VSS*
  (1-degradationrate);
Ashreceivedininflow=TSS-VSS;
Sludgeproduced=
  Newvolatilesuspendedsolids
  NVSSproducedinBODremoval
  +Nondegradableresidue
  +Ashreceivedininflow;
Totalsludgeproduced=
  flowrate*Sludgeproduced;
HRT=9.6;
Methanegasproducedat25OC=300/1000;
% output
Volumeofanaerobicdigester=HRT*
  flowrate/24
Volumetricorganicloading=COD*flowrate/
  Volumeofanaerobicdigester/1000
TotalCODremoved=BODremovalefficiency*
  flowrate*COD/1000;
Sulphateremoval=BODremovalefficiency*
  flowrate*SO4/1000;
CODavailableforMethanegasproduction=
  TotalCODremoved-Sulphateremoval;
totalMethanegasproducedperday=
  Methanegasproducedat25OC*
  CODavailableforMethanegasproduction;
Methanegasleaving=0.028*flowrate;
usableMethane=
  totalMethanegasproducedperday-
  Methanegasleaving
electricity=usableMethane*1.4
```

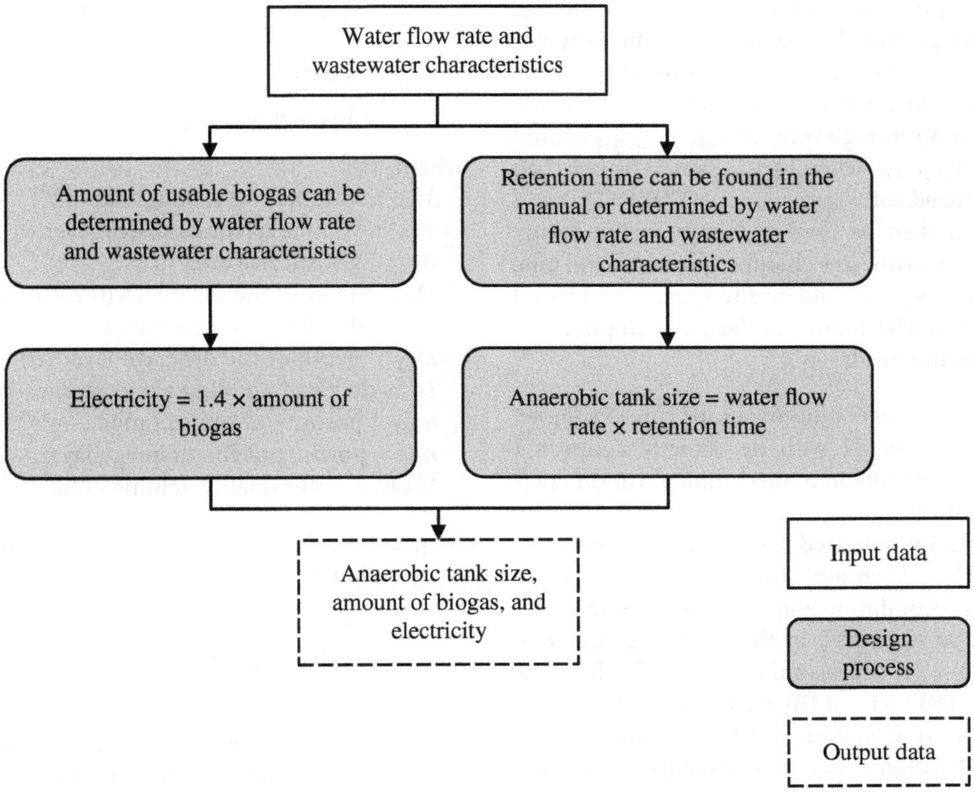

Figure 9.8 Flowchart for wastewater treatment system.

Table 9.19 Flow rate.

Flow rate (gal/day)	Flow rate (m^3/day)
1188.75	4.5

Table 9.20 Wastewater characteristics.

Wastewater characteristics		
BOD	300	mg/l
COD	450	mg/l
TSS	400	mg/l
VSS	270	mg/l
pH	7.7	
SO_4	85	mg/l

Answer: `anaerobicdigester` (Table 9.21)

```
Volumeofanaerobicdigester = 1.8000
Volumetricorganicloading = 1.1250
usableMethane = 0.2682
electricity = 0.3755
```

Table 9.21 Output data for wastewater treatment.

Output data for wastewater treatment		
Volume	1.8	m^3
Amount of biogas	0.2682	m^3/day
Electricity	0.37548	kWh

9.6 Green Roof Design

Green roof (GR) design is to determine the size of storage vaults/tanks, gravel beds, perforated pipes, stormwater chambers, blue roofs, and GRs. Rooftop systems include blue roofs and GRs, while subsurface systems include storage vaults/tanks, gravel beds, perforated pipes, and stormwater chamber, an impermeable membrane between the roof outer surface and the water, which protects the roof itself and also provides insulation. The slope should be less than 2% with 0% being optimal (DEPNY, 2015). Three inches of ponding depth is assumed for storage volume calculations. A pre-treatment section should be designed at the inlet of the system to sift out any sediment, hydrocarbons, and debris that may clog the system and an overflow to reduce the risk of damaging the system during large rainfall events (DEPNY, 2015).

In designing the green roof, the available rooftop storage volume must be greater than or equal to the required storage volume. Another requirement is for the volume of the subsurface stormwater to be less than the volume required for rooftop storage volume, e.g. $V_{A\ ROOF}$ should be greater than $V_{R\ ROOF}$, while $V_{R\ SUB}$ should be less than $V_{R\ ROOF}$. Specialized software, such as HYDROCAD and AUTOCAD, can then be used to fine-tune the design using a variety of stormwater chamber products and pipe modeling. A GR design should be site specific and based on the user needs. The following design example is for detached houses in Florida:

Step 1: Calculate the precondition annual runoff volume. A curve number of 82 with no directly connected impervious area results in a mean annual runoff coefficient (C) of 0.13.

Step 2: Calculate the required annual volume retention efficiency. The retention efficiency necessary to achieve the precondition annual runoff volume and nutrient load is expressed in the following equation. Efficiency (%) = (1 − precondition runoff volume) × 100 efficiency (%) = (1 − 0.13) × 100 = 87%. Therefore, the GR and cistern system must retain 87% of the annual rainfall volume. An alternative to considering the square footage of the landscape is to estimate the number of plants that can be irrigated with 1120 gallons of gray water per week. A standard look-up chart can be used to determine approximately how much water an individual tree or shrub will need for 1 week in July.

Step 3: Determine the necessary cistern volume to attain water with the 87% retention efficiency of the system. To achieve 85% removal, approximately 4.6 in of cistern volume over the GR area must be used.

Step 4: Convert the retention in inches to cubic feet and gallons. Cistern volume = 4.6 in/12 in/ft × 10 000 ft^2 = 3 833 ft^3, or 28 675 gal is required to accompany the GR system. This can be provided by a number of cisterns located at different sites near the roof drainage or by one central location.

Step 5: Determine classification of GR surface. There are two types of GRs. An extensive GR is one where the root zone (pollution control layer and growth media layer) is less than 6 in in depth, whereas intensive GRs have root zones greater than or equal to 6 in and are typically intended for public or private access. There are two distinct functions for GRs; one is passive and the other is active. Passive GRs are intended only for maintenance access and typically require less maintenance, while an active roof is used for public and private access. GRs can be built on any type of roof deck with a minimum slope of 1 in/ft:

$$\text{Storage volume} = V_{SM} + V_{DL} + (D_P \times A_{GR})$$

$$V_{SM} = A_{GR} \times D_{SM} \times n_{SM}$$

$$V_{DL} = A_{GR} \times D_{DL} \times n_{DL}$$

where

V_{SM} = volume of the soil media (ft^3)
V_{DL} = volume of the drainage layer (ft^3)
A_{GR} = GR surface area (ft^2)
D_{SM} = depth of the soil media (0.25–0.5 ft for extensive; 0.5–2.0 ft for intensive)
D_{DL} = depth of the drainage layer (ft)
D_P = depth of ponding above surface (ft)
n_{SM} = porosity of the soil media (~20%)
n_{DL} = porosity of the drainage layer (~25%)
WQv = water quality volume cubic (ft^3)

Step 6: Calculate the drainage layer and soil media storage volume:

$$V_{SM} = A_{GR} \times D_{SM} \times n_{SM}$$

$$V_{DL} = A_{GR} \times D_{DL} \times n_{DL}$$

1) Determine implications of roof slope:

   Reinforcement is needed to protect GRs from sliding on slopes steeper than 2 : 12. Even with reinforcement, slopes should be limited. The United States recommend that GRs should not be installed on slopes steeper than 40° (Figure 9.9).

   Note: EPA Comparison of roof slope is expressed as roof pitch vs. roof slope in degrees. Pitch and degrees on the same line express the same roof slope. For example, a 1 : 12 slope is a 4-degree roof slope.

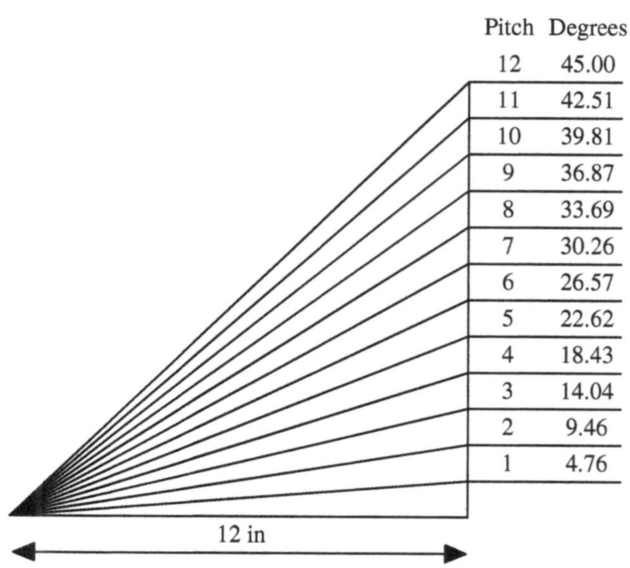

Figure 9.9 Roof pitch vs. roof slope.

2) Determining what areas of roof can be vegetated and what areas need to remain vegetation-free.
   GRs may include vegetation-free zones designed to:
   a) Resist wind uplift and scour.
   b) Reduce fire risk associated with air intakes or proximity to flammable materials and equipment.
   c) Provide access for roof maintenance-related issues.
   d) Provide enhanced flow path toward drains and scuppers for runoff sheeting off walls and parapets.
   e) In areas where exhausts onto the roof surface or the presence of condensate releases would negatively affect plant growth.
3) Determine GR size needed to meet stormwater:
   a) Calculate the stormwater runoff volume reduction performance goal for a site:

   $$R = D_r \times A \times 3630$$

   where
   D_r is the annual runoff depth in inches
   A is the total watershed area in acres
   3630 is a conversion factor to convert the final result to cubic feet

   b) Calculate annual stormwater runoff volumes and pollutant loads (total suspended sediment, dissolved phosphorus, and particulate phosphorus) from a site:

   $$D_r = P \times P_j \times R_v$$

   where
   P is the total annual rainfall depth in inches
   P_j is the fraction of annual rainfall events that product runoff
   R_V is the runoff coefficient, which is dimensionless

   c) Provide a method to add stormwater best management practices (BMPs), input BMP-specific parameters, and route overflow stormwater and pollutants to downstream practices (Table 9.22).

Table 9.22 Runoff coefficients by land use and soil type.

R_v coefficients	A soils	B soils	C soils	D soils
Forest/open space	0.02	0.03	0.04	0.05
Managed turf (disturbed soils)	0.15	0.20	0.22	0.25
Impervious cover	0.95	0.95	0.95	0.95

   d) Calculate stormwater runoff volume reduction achieved toward the performance goal and annual stormwater volume and pollutant load reductions for each individual BMP and the entire site:

   $$L = R \times C \times 6.243 \times 10^{-5}$$

   where
   R is the annual runoff volume (ft^3)
   C is an average annual pollutant concentration (mg/l)
   6.243×10^{-5} is a conversion factor to convert the final result to pounds

   e) Stormwater runoff volume could reduce annual pollutant load reductions for TSS, dissolved phosphorus, and particulate phosphorus:
      - If an overland path is used, a stabilized channel must be provided for erosive velocities (3.5–5.0 fps) for a 1-year storm event.
      - The GR system must safely convey runoff from a 100-year storm away from the building and into a downstream drainage system.

Step 7: Determine required treatment:

To account for the runoff reduction from GRs in hydraulic and hydrologic models, designers may use a reduced CN. Intensive GRs should use the runoff CNs for wood, brush, or grass, depending on the specific plant communities used (Table 9.23).

Table 9.23 Effective curve numbers for extensive green roof.

Growing media thickness (in)	2	3	4	6	8
Effective CN	94	92	88	85	77

Green Roof Cost/Benefit

GR capital costs vary widely. Examples of important factors that influence GR capital costs include:

Roof size: All other factors being equal (location, ease of access, etc.), per square foot cost, would typically decrease by a factor of at least 3 as size increases from a 1 000-ft^2 roof to a 20 000 ft^2-roof. Based on local projects, extensive GRs typically range from \$10 to \$30 per square foot for the components above the waterproofing assembly and a simple irrigation system (Table 9.24).

9.6.1 Life Cycle Assessment

The economic input–output life cycle assessment (EIO-LCA) model can be used for the material extraction, material processing, and manufacturing phases for each

Table 9.24 Benefit/cost analysis.

First cost	$1000
Project life	50
Annual saving	$170.33
Annual O&M costs	$20
Salvage value	$0
MARR	3%
B/C	3.86

material or component such as copper and plastic pipes (PVC). The model creates LCA per $1 million expenditure for the copper and PVC materials. The tables from 9.25 to 9.28 contain the results of the LCA. The tables represent the impacts on other sectors caused by a $1 million of PVC production such as plastic pipe and pipe fitting manufacturing).

9.6.2 Footprint

GRs can help prevent water resource adversely impacted by climate change by reducing electricity usage, improving air quality, and shrinking carbon footprint and can also greatly reduce the volume of stormwater runoff from rainfall events, helping to keep coastal and inland waters clean. In addition, GRs can save energy, reduce neighborhood temperatures, and protect human health. They have a strong regulating effect on the temperature of underlying roof surfaces and building interiors, reducing the energy needed for building cooling and the urban heat island effect. GRs can also protect our waters from pollution. They have substantial capacity to both absorb and delay rainfall runoff, reducing the volume of rainfall runoff and pollutants that flow to rivers, lakes, and beaches. A GR with a 3–4-in soil layer can generally absorb 1–1½ in of rainfall from a given storm event. Even when saturated, GRs can substantially delay runoff, reducing flooding and erosion.

Table 9.29 provides a list of types of energy contributed by GRs.

Table 9.25 Economic activity.

	Sector	Total economic ($mill)	Total value added ($mill)	Employee comp VA ($mill)	Net tax VA ($mill)	Profits VA ($mill)	Direct economic ($mill)	Direct economic (%)
	Total for all sectors	1.720	1.000	0.446	0.035	0.521	1.380	80.200
111400	Greenhouse and nursery production	1.110	0.690	0.290	0.010	0.390	1.100	99.000
324110	Petroleum refineries	0.051	0.004	0.001	0.000	0.002	0.036	72.400
420000	Wholesale trade	0.049	0.034	0.018	0.008	0.008	0.031	64.400
211000	Oil and gas extraction	0.047	0.024	0.003	0.004	0.017	0.003	6.010
115000	Agriculture and forestry support activities	0.043	0.025	0.022	0.001	0.001	0.038	88.700
531000	Real estate	0.040	0.032	0.003	0.004	0.025	0.027	66.700
52A000	Monetary authorities and depository credit intermediation	0.032	0.022	0.009	0.000	0.013	0.021	65.600
221100	Power generation and supply	0.021	0.014	0.004	0.002	0.007	0.013	63.100
221200	Natural gas distribution	0.017	0.006	0.001	0.001	0.003	0.011	66.700
550000	Management of companies and enterprises	0.017	0.010	0.009	0.000	0.001	0.000	0.000

Table 9.26 Conventional air pollutants.

	Sector	CO (t)	NH_3 (t)	NO_x (t)	PM 10 (t)	PM 2.5 (t)	SO_2 (t)	VOC (t)
	Total for all sectors	8.430	3.300	2.780	31.400	5.470	1.410	1.490
111400	Greenhouse and nursery production	6.360	3.060	1.740	31.000	5.320	0.320	1.030
115000	Agriculture and forestry support activities	0.784	0.000	0.011	0.002	0.002	0.000	0.064
221200	Natural gas distribution	0.190	0.000	0.008	0.000	0.000	0.003	0.009
211000	Oil and gas extraction	0.189	0.000	0.137	0.001	0.001	0.009	0.192
532400	Commercial and industrial machinery and equipment rental and leasing	0.125	0.000	0.002	0.000	0.000	0.000	0.010
1111B0	Grain farming	0.116	0.055	0.012	0.097	0.025	0.004	0.013
484000	Truck transportation	0.106	0.000	0.112	0.032	0.006	0.002	0.012
221100	Power generation and supply	0.047	0.002	0.340	0.047	0.038	0.756	0.003
324110	Petroleum refineries	0.035	0.002	0.048	0.007	0.006	0.083	0.035
113A00	Forest nurseries, forest products, and timber tracts	0.033	0.000	0.031	0.004	0.003	0.003	0.005

Table 9.27 Greenhouses gases.

	Sector	Total t CO_{2e}	CO_2 fossil t CO_{2e}	CO_2 process t CO_{2e}	CH_4 t CO_{2e}	N_2O t CO_{2e}	HFC/PFCs t CO_{2e}
	Total for all sectors	971.000	649.000	45.100	75.000	199.000	3.100
111400	Greenhouse and nursery production	453.000	307.000	0.000	0.000	146.000	0.000
221100	Power generation and supply	183.000	180.000	0.000	0.495	1.120	1.160
325310	Fertilizer manufacturing	76.800	19.000	25.700	0.000	32.000	0.000
211000	Oil and gas extraction	74.300	20.900	13.600	39.700	0.000	0.000
324110	Petroleum refineries	51.700	51.500	0.000	0.160	0.000	0.000
1121A0	Cattle ranching and farming	14.800	0.972	0.000	8.440	5.430	0.000
484000	Truck transportation	12.800	12.800	0.000	0.000	0.000	0.000
486000	Pipeline transportation	12.600	5.780	0.016	6.850	0.000	0.000
1111B0	Grain farming	9.400	1.390	0.000	0.768	7.250	0.000
221200	Natural gas distribution	6.340	0.573	0.000	5.770	0.000	0.000

Table 9.28 Transportation.

	Sector	Air (ton-km)	Oil pipe (ton-km)	Gas pipe (ton-km)	Rail (ton-km)	Truck (ton-km)	Water (ton-km)	Intl air (ton-km)	Intl water (ton-km)	Total (ton-km)
	Total for all sectors	1790	248 000	103 000	242 000	717 000	252 000	44 600	2 410 000	4 020 000
111400	Greenhouse and nursery production	1420	0	0	30 000	601 000	161 000	44 000	49 900	887 000
325190	Other basic organic chemical manufacturing	68	0	0	5 390	1 700	1 370	22.6	11 700	20 200
325320	Pesticide and other agricultural chemical manufacturing	67.6	0	0	983	3 370	1 630	9.91	833	6 900
316900	Other leather and allied product manufacturing	29.5	0	0	0.884	37	0	11.3	343	422
1111B0	Grain farming	16.7	0	0	25 100	3 910	18 300	2.65	1 010	48 300
316100	Leather and hide tanning and finishing	13.1	0	0	0.217	26.7	0	3.55	4.5	48.1
32619A	Other plastic product manufacturing	12.7	0	0	532	4 050	9.69	64.5	4 410	9 080
3259A0	All other chemical product and preparation manufacturing	12.2	0	0	181	381	32.4	7.62	263	877
325181	Alkali and chlorine manufacturing	10.2	0	0	1 030	127	77.3	0.556	233	1 480
335911	Storage battery manufacturing	8.4	0	0	4.84	1 150	18.8	24	582	1 790

Table 9.29 Energy.

	Sector	Total energy (TJ)	Coal (TJ)	NatGas (TJ)	Petrol (TJ)	Bio/ waste (TJ)	NonFossElec (TJ)
	Total of all sectors	11.600	1.720	4.270	4.010	0.159	1.470
111400	Greenhouse and nursery production	6.140	0.000	2.390	2.710	0.000	1.040
221100	Power generation and supply	2.230	1.620	0.475	0.079	0.000	0.052
324110	Petroleum refineries	0.866	0.000	0.231	0.561	0.043	0.031
211000	Oil and gas extraction	0.445	0.000	0.363	0.038	0.000	0.044
325310	Fertilizer manufacturing	0.407	0.002	0.364	0.009	0.005	0.027
484000	Truck transportation	0.174	0.000	0.000	0.172	0.000	0.002
486000	Pipeline transportation	0.151	0.000	0.115	0.000	0.000	0.036
325190	Other basic organic chemical manufacturing	0.144	0.018	0.055	0.020	0.043	0.008
331110	Iron and steel mills	0.062	0.037	0.017	0.000	0.000	0.007
325211	Plastic material and resin manufacturing	0.057	0.002	0.030	0.012	0.006	0.006

9.7 Rain Garden Design

Similar to GRs, rain gardens provide the similar beneficial effects such as reducing peak flow to sewer system, retaining and degrading nutrients. Example 9.9 presents a design procedure of rain gardens.

Example 9.9 Design a rain garden for a residential house located at 9240 SW 70th Street, Miami, FL, USA. Based on a Miami-Dade County property appraiser, the total lot size of the property is approximately 12 638 ft^2. Two rain gardens should be designed to treat rooftop runoff. The rain garden depth, d = rain garden depth (inches), can be obtained from Table 9.30 and Figure 9.10.

Table 9.30 Rain garden depth.

Slope (%)	Rain garden depth (in)
4	3–5
5–7	6–7
8–12	8

Solution:

Design steps

1) **Rain garden depth**

$$s = \frac{h}{L_s}$$

where

L_s = length of string (inches)
h = height (inches)
s = slope (%) In this case, L_s = 600 in (50 ft), h = 2 (in), and s = slope (%) should be

$$s = \frac{2}{600} \times 100 = 0.3\%$$

d = 4 in

2) **Drainage area**

$$A_d = A_r \times \frac{d_w}{100}$$

where

A_r = roof area (ft^2)
d_w = downspout collection of water (%)
A_d = drainage area (ft^2)
In this case, A_r = 2900 (ft^2),
d_w = 25%, and A_d = drainage area (ft^2):

$$A_d = 2900 \times \frac{25}{100} = 725\,ft^2$$

Figure 9.10 Project location.

3) **Surface area**

$$A = A_d \times c$$

where

S_t = soil type
d = rain garden depth (inches) obtained from Table 9.30
c = size factor obtained from Table 9.31

Table 9.31 Size factor area.

Soil type	3–5 in deep	6–7 in deep	8 in deep
Sandy	0.19	0.15	0.08
Silty	0.34	0.25	0.16
Clayey	0.43	0.32	0.20

A = rain garden surface area (ft^2)

In this case, S_t = sandy soil, d = 4 in = (1/3) ft, c = 0.19, and A = rain garden surface area (ft^2) should be

$$A = 725 \times 0.19 = 138\ \text{ft}^2$$

This project will include two rain gardens of the same size. Therefore, the total rain garden surface area is approximately 276 ft^2 (138 × 2).

4) **Rain garden length**

$$L_g = \frac{A}{W_g}$$

where

W_g = garden width (ft)
L_g = garden length (ft)

In this case, W_g = 8 ft and L_g = garden length (ft) should be

$$L_g = \frac{138}{8} = 17.25\ \text{ft} = 18\ \text{ft}$$

5) **Rain garden volume**

$$V = A \times d$$

where

A = surface area (ft^2)
d = depth (ft)

In this case: V = 138 ft^2 × (1/3) ft = 46 ft^3
V_t = 46 ft^3 × 2 rain gardens = 92 ft^3

Based on the Florida Drainage Manual IDF Curves for Zone 10, the intensity of the rainfall for the 5-year 24-h storm is 0.32 in/h. Therefore, the runoff that will occur in one drainage area (725 ft^2) in 24 h is approximately 7.68 in (0.32 in/h × 24 h).

9.7.1 Life Cycle Assessment

Stormwater treatment technologies managing runoff during rain events are primarily designed to reduce flood risks, settle suspended solids, and concurrently immobilize metals and nutrients. The US EPA TRACI 2.0 (Bare, 2011) can be used to conduct the LCA of the designed rain garden to evaluate the environmental, economic, and social performance of GI stormwater control measures (SCMs). The LCA modeling quantified the environmental impacts associated with the materials, construction, transport, operation, and maintenance of different stormwater management systems. For example, the US EPA TRACI 2.1 lists the eutrophication potential of air and water pollutants (Table 9.32).

Table 9.32 The eutrophication potential of air and water pollutants listed by the US EPA TRACI 2.1.

Substance name	Eutrophication air (kg N eq/kg substance)	Eutrophication water (kg N eq/kg substance)
Phosphorus	1.120	7.290
Phosphorus pentoxide	0.490	3.190
Phosphate	0.366	2.380
Phosphoric acid	0.355	2.310
Nitrogen	0.150	0.986
Ammonium	0.119	0.779
Ammonia	0.119	0.779
Nitric oxide	0.069	0.451
Nitrogen dioxide	0.044	0.291
Nitrogen oxides	0.044	0.291
Nitrate	0.036	0.237
Nitric acid	0.034	0.227
Biological oxygen demand	0.000	0.050
Chemical oxygen demand	0.000	0.050

The metrics to evaluate benefits and impacts include carbon footprint (global warming potential) in terms of GHG emission, energy utilization, and water withdrawal during the production of the material or systems. The results of this rain garden show that the construction phase is the main contributing life cycle phase for all adverse environmental impacts with total life cycle cost and labor impacts. An EIO-LCA can be used for the material refining, extraction, production, processing, manufacturing, and transportation phases for the aluminum downspout and gardening plants. The model assesses the impacts of $1 million expenditure on material. As the EIO-LCA model is linear, the results obtained from the simulation can be scaled to the cost of the materials of the rain gardening system and used to evaluate the impact.

9.7.2 Environmental Impacts of Aluminum

The impact of the aluminum downspout used in the project was accessed for the sectors of the economy via EIO-LCA modeling, and the results of the EIO-LCA modeling for the GHG emission, energy utilization, and water withdrawal sectors are shown in Tables 9.33, 9.34, 9.35, 9.36, and 9.37, respectively. Although the absolute values could be an good estimate, the relative change of the impacts under different design alternatives could be very important for sustainable EEIS design.

9.7.3 Cost and Benefit Analysis of Rain Garden

Rain gardens reduce the overall quantity of runoff by temporarily storing rainwater in a landscaped area before letting it infiltrate the ground. Plants and soils filter out pollutants from the water before returning them to the ground, where they restore groundwater supplies. Rain gardens often result in cost savings by reducing the size of sewer infrastructure. The use of rain gardens may also reduce irrigation costs. Unlike grass or nonnative plants, the native plants used in rain gardens require little water after they are established. Therefore, rain gardens can be designed to meet the necessary runoff volume reduction goal in a cost-effective way.

Table 9.33 Water withdrawals.

	Sector	Water withdrawals (kgal)
	Total of all sectors	19 200.0
221100	Power generation and supply	16 000.0
331314	Secondary smelting and alloying of aluminum	1 410.0
1111B0	Grain farming	356.0
2122A0	Gold, silver, and other metal ore mining	220.0
322130	Paperboard mills	173.0
325510	Paint and coating manufacturing	168.0
33131B	Aluminum product manufacturing from purchased aluminum	142.0
212230	Copper, nickel, lead, and zinc mining	71.0
325190	Other basic organic chemical manufacturing	58.1
325188	All other basic inorganic chemical manufacturing	54.5

Table 9.34 Energy.

	Sector	Total energy (TJ)	Coal (TJ)	NatGas (TJ)	Petrol (TJ)	Bio/ waste (TJ)	NonFossElec (TJ)
	Total of all sectors	24.300	5.390	9.330	1.840	1.700	6.020
33131A	Alumina refining and primary aluminum production	7.210	0.000	1.920	0.061	0.180	5.050
221100	Power generation and supply	6.860	5.000	1.460	0.243	0.000	0.160
33131B	Aluminum product manufacturing from purchased aluminum	5.250	0.000	4.090	0.123	1.040	0.000
331314	Secondary smelting and alloying of aluminum	0.885	0.000	0.586	0.018	0.118	0.163
484000	Truck transportation	0.419	0.000	0.000	0.415	0.000	0.004
331110	Iron and steel mills	0.415	0.246	0.113	0.004	0.002	0.050
322130	Paperboard mills	0.220	0.020	0.015	0.009	0.130	0.015
211000	Oil and gas extraction	0.211	0.000	0.172	0.018	0.000	0.021
324110	Petroleum refineries	0.204	0.000	0.054	0.132	0.010	0.007
325190	Other basic organic chemical manufacturing	0.193	0.024	0.073	0.027	0.058	0.010

Table 9.35 Greenhouse gases.

	Sector	Total $t\ CO_{2e}$	CO_2 fossil $t\ CO_{2e}$	CO_2 process $t\ CO_{2e}$	CH_4 $t\ CO_{2e}$	N_2O $t\ CO_{2e}$	HFC/PFCs $t\ CO_{2e}$
	Total for all sectors	1560.000	1080.000	195.000	69.700	7.040	206.000
221100	Power generation and supply	563.000	554.000	0.000	1.520	3.440	3.570
33131A	Alumina refining and primary aluminum production	448.000	102.000	159.000	0.000	0.000	187.000
33131B	Aluminum product manufacturing from purchased aluminum	215.000	315.000	0.000	0.000	0.000	0.000
331110	Iron and steel mills	35.800	13.500	22.000	0.218	0.000	0.000
211000	Oil and gas extraction	35.200	9.920	6.450	18.800	0.000	0.000
331314	Secondary smelting and alloying of aluminum	30.900	30.900	0.000	0.000	0.000	0.000
484000	Truck transportation	30.900	30.900	0.000	0.000	0.000	0.000
212100	Coal mining	29.000	3.270	0.000	25.700	0.000	0.000
486000	Pipeline transportation	14.100	6.460	0.018	7.640	0.000	0.000
482000	Rail transportation	13.000	13.000	0.000	0.000	0.000	0.000

Table 9.36 Greenhouse gases.

	Sector	Total $t\ CO_{2e}$	CO_2 fossil $t\ CO_{2e}$	CO_2 process $t\ CO_{2e}$	CH_4 $t\ CO_{2e}$	N_2O $t\ CO_{2e}$	HFC/PFCs $t\ CO_{2e}$
	Total for all sectors	971.000	649.000	45.100	75.000	199.000	3.100
111400	Greenhouse and nursery production	453.000	307.000	0.000	0.000	146.000	0.000
221100	Power generation and supply	183.000	180.000	0.000	0.495	1.120	1.160
325310	Fertilizer manufacturing	76.800	19.000	25.700	0.000	32.000	0.000
211000	Oil and gas extraction	74.300	20.900	13.600	39.700	0.000	0.000
324110	Petroleum refineries	51.700	51.500	0.000	0.160	0.000	0.000
1121A0	Cattle ranching and farming	14.800	0.972	0.000	8.440	5.430	0.000
484000	Truck transportation	12.800	12.800	0.000	0.000	0.000	0.000
486000	Pipeline transportation	12.600	5.780	0.016	6.850	0.000	0.000
1111B0	Grain farming	9.400	1.390	0.000	0.768	7.250	0.000
221200	Natural gas distribution	6.340	0.573	0.000	5.770	0.000	0.000

Table 9.37 Energy from all sectors.

	Sector	Total energy (TJ)	Coal (TJ)	NatGas (TJ)	Petrol (TJ)	Bio/ waste (TJ)	NonFossElec (TJ)
	Total of all sectors	11.600	1.720	4.270	4.010	0.159	1.470
111400	Greenhouse and nursery production	6.140	0.000	2.390	2.710	0.000	1.040
221100	Power generation and supply	2.230	1.620	0.475	0.079	0.000	0.052
324110	Petroleum refineries	0.866	0.000	0.231	0.561	0.043	0.031
211000	Oil and gas extraction	0.445	0.000	0.363	0.038	0.000	0.044
325310	Fertilizer manufacturing	0.407	0.002	0.364	0.009	0.005	0.027
484000	Truck transportation	0.174	0.000	0.000	0.172	0.000	0.002
486000	Pipeline transportation	0.151	0.000	0.115	0.000	0.000	0.036
325190	Other basic organic chemical manufacturing	0.144	0.018	0.055	0.020	0.043	0.008
331110	Iron and steel mills	0.062	0.037	0.017	0.000	0.000	0.007
325211	Plastic material and resin manufacturing	0.057	0.002	0.030	0.012	0.006	0.006

Example 9.10
Given:
1) Lot Information
Zip code: 33173
Annual rainfall∗ (in): **43.7**
Storm type (in): 90%
Storm rainfall (in): **1.42**
Size of lot (acres): 21 780 SF = 0.5 ac.
Soil type: B

2) Predevelopment

Impervious area (%) : 19.4

Lawn in good condition (%) : 78.9

Lawn in fair condition (%) : 0

Urban fill (%) : 0

Natural desert landscaping (%) : 0

Newly graded areas, no vegetation (%) : 0

Straight row crops in good condition (%) : 0

Pasture in fair condition (%) : 0

Meadow (%) : 0

Woods in good condition (%) : 0

Desert shrub in fair condition (%) : 0

Wetlands (%) : 0

Water (%) : 1.7

3) Runoff Reduction Goal

Precipitation depth capture (in) : 1

Volume captured over : whole site

Required volume to capture on-site : 1815 ft^3

4) Conventional Development

Roof size (ft^2) = 2900

Parking lot size (ft^2) = 1220

Sidewalk (ft^2) = 112

5) Green Improvements

Amount (ft^2) : 138 × 2 = 256

Rain gardens do not reduce impermeable area of the site but capture 27.1% of the runoff volume. Compared with conventional approaches, the green practices in this case will decrease the total life cycle construction and maintenance costs by 2% (in net present value).

Find: Cost and benefit of a rain garden which captures 1 inch precipitation at a 1815 ft^2 site as shown in Tables 9.38 to 9.41.

Solution: 1) Volume Control

Required volume capture from 1 in over whole site (ft^3) = 1815

Volume captured by current BMPs (ft^3) = 210

Rain garden (ft^3) = 210

Percentage of required volume captured by current BMPs (%) = 12

Decrease in impervious area (%) = 0

Runoff coefficients are listed in Table 9.38.
Land use pattern is listed in Table 9.39.
Costs are presented in Table 9.40.

2) Monetary Benefit
When a homeowner invests time and money, even modest amounts, in one or more rain gardens, there is usually a significant benefit to the family. A rain garden adds to the aesthetic value and even property value of a home. Rain garden and other landscaping can increase a home value by **3–12%**. Rain gardens improve environmental quality by attracting insects or birds. They reduce the potential for damaging or messy flooding of a basement or a sensitive part of the landscape. Reduced stormwater nuisance could be estimated by assuming that once a year, a resident avoids the necessity of cleaning up a mess caused by stormwater. If the cleanup takes an average of 8 at $10 per hour, the monetary benefits of rain gardens, in terms of reduced stormwater nuisance, would be $80 per year.

3) Cost Estimate
The cost of rain garden is estimated in Table 9.41.

4) Miami-Dade County Stormwater Fee
Reducing the quantity of stormwater entering public stormwater systems through the use of rain gardens not only benefits stormwater system performance but also creates considerable economic benefits for both public and private sectors. Stormwater utility fees are assigned to all developed residential and nonresidential properties and are determined as a function of equivalent residential units (ERUs) for residential properties and by determining the impervious area for nonresidential properties. For residential properties, the utility conducted a statistical analysis to establish an ERU. The value of one ERU is set at **1548** SF and incurs a

Table 9.38 Coefficient and runoff.

Description	Predevelopment	Conventional	Green	Predevelopment to conventional difference (%)	Conventional to green difference (%)
Average annual rainfall					
Total runoff (in)	38.69	38.7	38.69	−1	−0
Total runoff volume (ft^3)	70 230	70 234	70 215	−1	0
Cumulative abstractions (in)		4.1	4.11		0.21
90% storm					
Total runoff (in)	0.05	0.05	0.05	−27	−1
Total runoff volume (ft^3)	95	95	94	−27	−1
Cumulative abstractions (in)		0.46	0.46		−0.34
CN	69	69	69		
Initial abstractions (in)		0.91	0.91		0.23

Table 9.39 Land use pattern.

	Conventional area (ft^2)	Green (using BMPs) area (ft^2)
Conventional roof	2900	2900
Green roof	0	0
Parking lot	1220	1220
Permeable parking lot	0	0
Swales in parking lot	0	0
Streets	0	0
Reduced street width	0	0
Roadside swales	0	0
Driveway and alleys	0	0
Permeable driveway and alleys	0	0

charge of \$4 per month for residential homes. Commercial, industrial, and other nonresidential properties are calculated using the same ERU concept but account for the exact footage of a facility and all impervious surfaces when determining the total monthly stormwater fee.

To determine the stormwater utility fee for nonresidential properties, the total square footage of property's impervious surfaces needs to be divided by 1548 (the size of one ERU) and multiplied by \$4 (the cost of one ERU). As an example, a nonresidential property with a total impervious area of 4644 ft^2 would pay \$12 per month in stormwater utility fees (using this calculation: 4644/1548 × \$4 = \$12).

9.7.4 Water Footprint

Water pollution could be quantified as blue, green, and gray water footprint. The blue water footprint refers to the consumption of blue water resources, such as surface and groundwater, along with the supply chain of a product. Loss of water from the available ground–surface water body in a catchment area is consumed when water evaporates, returns to another catchment area or the sea, or is incorporated into a product. The green water footprint denotes the consumption of green water resources such as rainwater, which is not wasted as runoff. The gray water footprint refers to the volume of freshwater to dilute pollutant concentration to its natural background (Hoekstra et al., 2011). In terms of reducing water footprint rain harvest is the most effective way to reduce green water footprint while gray water reuse is the most effective way to reduce gray water footprint.

9.7.5 Nitrogen and Phosphorus Footprint

One of the most widespread, costly, and challenging environmental problems in America is nutrient pollution. Nutrient pollution is caused by excess nitrogen and phosphorus in the air and water. Immense quantities of pollution enter almost all streams without ever flowing through pipes, sewers, treatment plants, or stormwater outfall structures. Such wastewater sources have been characterized as nonpoint source pollution. Nonpoint sources include agricultural areas, abandoned and active mine sites, urban areas and highway facilities, suburban lawns, and natural areas. It occurs when precipitation falls on the urban environment and picks up pollutants and deposits them into surface waters or introduces them into groundwater. Nitrates and phosphates are nutrients that

Table 9.40 Cost.

	Construction cost ($)				Annual maintenance cost ($)				Life cycle cost ($, NPV)			
	Conventional	Green	Difference	%	Conventional	Green	Difference	%	Conventional	Green	Difference	%
Concrete sidewalk	581	581	0	0	3	3	0	0	598	598	0	0
Parking lot	6 722	6 722	0	0	183	183	0	0	7 637	7 637	0	0
Conventional stormwater storage	20 963	18 538	−2426	−12	54	48	-6	−12	21 236	19 779	−2 457	-12
Standard roof	21 750	21 750	0	0	145	145	0	0	22 475	22 475	0	0
Turf	3 685	3 664	−21	−1	1 579	1 570	-9	−1	11 582	11 516	−66	−1
Rain garden	0	700	700	0	0	34	34	0	0	870	870	0
Downspout disconnection	0	35	35	0	0	0	0	0	0	36	36	0
Total	53 702	51 990	−1 712	−3	1 965	1 984	19	1	63 528	61 911	−1 617	−3

Table 9.41 Rain garden cost estimate.

Item description	Unit	Unit price	Quantity		Total
3 in × 4 in × 10 ft Aluminum Downspout White	EA	$13.37	2	$26.74	
Amerimax Home Products 3 in × 4 in White Aluminum Downspout Clip	EA	$1.48	2	$2.96	
3 in × 4 in Aluminum Downspout A Elbow	EA	$3.97	2	$7.94	
Ames 16-Tine Double Play Bow Rake with Fiberglass Handle	EA	$15.97	1	$15.97	
Fiskars 3 in Softouch Hand Trowel	EA	$5.97	1	$5.97	
Grade Stakes-Pine (12-Pack) (0.562 in × 1.375 in × 23.5 in)	EA	$4.57	1	$4.57	
Goldilocks Lysimachia 4.25 in Grande	EA	$5.99	4	$23.96	
Pure Joy Sedum 4.5 in Quart	EA	$7.46	2		$14.92
Hocus Pocus Veronica 4.5 in Quart	EA	$7.46	2	$14.92	
Banana Cream Leucanthemum 4.5 in Quart	EA	$7.46	2	$14.92	
Surfinia Red Petunia 4.25 in Grande	EA	$5.99	4	$23.96	
ColorBlaze Lime Time Coleus 4.25 in Grande	EA	$5.99	4	$23.96	
White Album ColorChoice Euonymus 4.5 in Quart	EA	$17.45	2		$34.90
2.5 qt. Helleri Holly	EA	$16.25	2		$32.50
Total					**$250.00**

plants need to grow. In small amounts they are beneficial to many ecosystems. In excessive amounts, however, nutrients cause eutrophication. Eutrophication is the enrichment of an aquatic ecosystem with nutrients that accelerate biological productivity, such as growth of algae and weeds, and an undesirable accumulation of algal biomass. This explosive growth of algae reduces the oxygen in surface water during massive algae die-off. As a result, estuarine waters could have poor oxygen (hypoxia) or could be completely without oxygen (anoxia). Therefore, prevention using GI such as green roof, rain harvest, rain garden, bioswales, and constructed wetlands is the BMP to combat nonpoint pollution of natural water body and ecosystem.

9.8 Exercise

9.8.1 Questions

1) What are the major design rules in prevention?
2) What are the benefits of GI?
3) What are the major elements of GI?
4) Why is rain harvest so important for a sustainable community?
5) What are the different methods in sizing rain harvest tank size?
6) In your opinion, what is the order of GI in terms of benefit/cost ratio?

9.8.2 Calculations

1) In Miami-Dade, the annual average precipitation is 60 in. If 80% of the rain can be harvested, answer the following:
   a) How much water can be harvested for 1 in of rain over a 2000-ft^2 roof?
   b) What size of the rain harvest tank should be used using three different design methods?
   c) How much annual saving could be recognized if the irrigation area is 5000 ft^2?
2) What would be the annual improvement of gray water footprint if the average runoff has 2-mg/l N and 0.26-mg/l P due to the rain harvest system for a single-family house with a 2000-ft^2 roof in Miami-Dade County if the N and P natural standards are 0.2 and 0.01 mg/l, respectively? What would be the total annual gray water footprint saved if all the single-family houses of 200 000 installed such rain harvest system in Miami-Dade County?

9.8.3 Projects

9.8.3.1 Xiongan Project

1) Xiongan special economic zone plans to apply GI to maintain clean stormwater discharge to its receiving water body. It has ambitious target to maintain 70% green space and to achieve zero water design. One of the ambitious plans is to require vertical forest

house where each apartment should have sufficient large area for trees and grass to combat the smog and water pollution. Collect literature data, and estimate the area required for an apartment of five people to maintain zero water through using gray water for irrigation and rainwater for cooking, shower, and laundry purposes.

2) For vertical forest high-rise building, the challenges are to maximize precipitation areas for rain harvest. However, the high density of the apartment requires different layers of balcony with different arrangement so that each apartment could receive sufficient water for water demand by trees and grass of the balcony. For these reasons, it is critical to team with architectural, structural, and urban planning engineers to achieve zero water design. Form a team with these experts to design a 40-floor modern residential condo of 800 residences with 4 units on each floor and four people in one apartment.
3) From a related website, collect monthly precipitation data in Xiongan, and design the following GI for a sustainable Xiongan:
   a) Rain harvest system for individual houses.
   b) GR for high-rise buildings taller than 40 floors.
   c) Constructed wetlands throughout the city.
   d) Use the US EPA software package TRACI 2.0 to conduct the LCA of your design alternatives.
   e) Compare the corresponding impacts for each alternative, and select the best alternative with supporting quantitative impacts of different categories.
   f) State your reasons for your selected alternative.

9.8.3.2 Community Proposal Project

1) Work with a department of public works or community development, and get an inventory of stormwater management infrastructure in your city.
2) Conduct a review, and develop a summary of current literature related to the stormwater benefits of different GIs.
3) Identify opportunities to design and build different GIs such as rain harvest, rain gardens, GR, wetland, and on-site WWT for wherever is appropriate.
4) Develop design alternatives of your GI for stormwater management.
5) Develop guidelines for the use of rainwater harvesting for stormwater management.
6) Conduct cost and benefit analysis of your design alternatives.
7) Use the US EPA software package TRACI 2.0 to conduct the LCA of your design alternatives.
8) Compare the corresponding impacts for each alternative, and select the best alternative with supporting quantitative impacts of different categories.
9) State your reason for your selected alternative.
10) Plan and execute the comprehensive GI plan for the community to reduce infiltration/inflow to sewer system by 50%.

References

Bare, J.C. (2011). TRACI 2.0 – The Tool for the Reduction and Assessment of Chemical and other environmental Impacts. *Clean Technologies and Environmental Policies* 13 (5): 687–696.

Center for Neighborhood Technology and American Rivers (2010). The value of green infrastructure: a guide to recognizing its economic, environmental and social benefits. https://www.americanrivers.org/conservation-resource/value-green-infrastructure/ (accessed 22 January, 2018).

China General Office of the State Council (CGOSC) (2017). Guideline to promote building sponge cities. http://www.gov.cn/zhengce/content/2015-10/16/content_10228.htm (accessed 16 September 2017).

Department of Environmental Protection of New York State (DEPNY) (2015). Guidelines for the design and construction of stormwater management systems. Albany: Department of Environmental Conservation.

EPA (2014). Green infrastructure cost-benefit resources. http://water.epa.gov/infrastructure/greeninfrastructure/gi_costbenefits.cfm (accessed 12 January 2018).

Hoekstra, A.Y., Chapagain, A., Martinez-Aldaya, M., and Mekonnen, M. (2011). *The Water Footprint Assessment Manual: Setting the Global Standard*. London: Earthscan.

Jacob, J.S. and Lopez, R. (2009). Is denser greener? An evaluation of higher density development as an urban stormwater-quality best management practice. *Journal of the American Water Resources Association* 45 (3): 687–701.

Lehmann, S. (2010). *The Principles of Green Urbanism. UNESCO Chair in Sustainable Urban Development*. London: Earthscan.

Li, H., Ding, L.Q., Ren, M.L. et al. (2017). Sponge City Construction in China: a survey of the challenges and opportunities. *Water* 9: 594.

National Bureau of Statistics of China (NBSA) (2015). *China Statistical Yearbook 2015*. Beijing: China Statistics Press.

NRC (National Research Council) and NAP (National Academies Press) (2008). Urban Stormwater Management in the United States. Washington, DC: National Academies Press.

National Research Council (2008). Minerals, Critical Minerals, and the U.S. Economy. Washington, DC: The National Academies Press. https:/doi.org/10.17226/12034 (accessed 19 December 2017).

Paul, M.J. and Meyer, J.L. (2001). Streams in the urban landscape. *Annual Review of Ecology and Systematics* 32: 333–365.

Roy, A., Wenger, S., Fletcher, T. et al. (2008). Impediments and solutions to sustainable, watershed-scale urban stormwater management: lessons from Australia and the United States. *Environmental Management* 42 (2): 344–359.

Sharma, A.K., Pezzaniti, D., Myers, B. et al. (2016). Water Sensitive Urban Design: an investigation of current systems, implementation drivers, community perceptions and potential to supplement urban water services. *Water* 8: 272.

United States Environmental Protection Agency (US EPA) (1999). Low-impact development design strategies: an integrated design approach. *EPA 841-B-00003.* Washington, DC: US EPA.

The U.S. Green Building Council (USGBC) (2018). Leadership in Energy and Environmental Design (LEED). https://www.usgbc.org/education-at-usgbc (accessed January 2018).

10

Recovery

No material is wasted in a functioning ecosystem.

OBJECTIVES
To understand that water is the most valuable product to be recovered To explain that traditional tertiary treatment of removing N and P from wastewater is expensive and unsustainable To recover nutrients, energy, and water (NEW) from wastewater by designing WRRFs To understand the struvite precipitation mechanism and process To provide capital and operation and maintenance (O&M) cost equations for decision making on water, energy, and nutrient recovery To design a cost-effective system to reclaim water to the required discharge standards

10.1 Principle 7

The degree of recovering water, energy, nutrient, and materials from wastewater (WW) reflects the sustainable level in EEIS design. In theory, almost all the wastewater (WW) could be recovered to reusable water because contaminants in WW is less than 0.1% by weight. However, not all the water could be recovered because water used in economy tend to dissipate within the economic system due to the second law of thermodynamics. In addition, energy resources are usually not recyclable. Energy would deteriorate to lower grade energy in addition to dissipation in the environments. Indeed, entropy would put a limit in designing recovery processes of EEIS. This irreversible energy degradation rate depends upon not only the ratio between the intensities of output and input flows but also the energy quality. Therefore, the ultimate decision whether to recover water, energy, nutrient, and materials from WW depends upon the life cycle cost and benefit analysis (LCCBA) of different alternatives. The prevention principle is illustrated in Box 10.1.

In nature, almost all the material, nutrients, and water are recycled. In modern society, however, the traditional environmental engineering design focuses on treatment and disposal. However, when the accumulation rate is significantly greater than the treatment capacity, waste would end in the environments without adequate treatment. If cost is not a limited factor, all materials could be recovered through environmental engineering infrastructure system as shown in Table 10.1.

To recover methane, landfills can be designed in such a way that they produce methane and is referred to as bioreactor landfill. WW contains nutrients such as N and P, energy as methane, and treated WW. For land, green infrastructure aims to maximize primary biomass production and reduce pollution of water bodies. Algae could be recovered from eutrophic lakes. For WW, most of the nitrogen are in soluble and colloidal organic forms. The amount removed by primary sedimentation is limited to about 15%, and activated sludge (AS) removes another 10%. Twenty-five percentage of nitrogen removal equals about four percentage of the BOD applied. With a total reduction of only 25%, the effluent contains 26 mg/l nitrogen of the original 35 mg/l. Approximately 2 mg/l is organic nitrogen bound in the effluent suspended solids. The remaining 24 mg/l is in the form of ammonia, except when nitrification occurs during aeration. Figure 10.1 is a diagram tracing nitrogen through a hypothetical treatment plant. The nitrogen concentration in the influent WW is assigned a value of 100%. To recover energy, sludge is stabilized by anaerobic digestion, and the ammonia released from

Sustainable Environmental Engineering, First Edition. Walter Z. Tang and Mika Sillanpää.

Box 10.1

Principle 7 All waste should be recovered in terms of benefit and cost ratio within an allowable budget.

1) Water, nutrients, energy, and materials should be recovered in the investment return period of SEEI system.
2) Life cycle benefit and cost should be analyzed in determining what material can be recovered and to what extent.
3) Within the same investment return period, life cycle assessment (LCA) of environmental impacts should be quantified to determine design strategy.

Table 10.1 Recovery of materials and energy.

Design	Problems	Mass	Energy
Air	Landfill CH_4	Bioreactor landfills	CH_4
	Carbon dioxide from coal power plants	Carbon sequestration	Methanol
Water	Nutrients	N and P	Struvite
	Wastewater	Reclaimed water	CH_4 or electricity
Land	Soil erosion	Regenerative GI	Biomass
Ecological	Algae boom	Constructed wetlands	Blue algae for biodiesel

decomposition of the sludge solids is returned to the influent of the treatment plant in supernatant from the digester. Assuming 40% of the organic nitrogen in the sludge is converted to ammonia, 10% of the original 25% is recycled to the treatment plant and appears in the effluent, which then contains 85% of the influent nitrogen. Depending on variations in nitrogen content of the WW and methods of sludge processing, nitrogen removal in conventional biological treatment systems ranges from 20% up to 40%.

10.2 Phosphorus Removal from Wastewater

The common forms of phosphorus in WW are orthophosphate (PO_4^{3-}), polyphosphates (polymers of phosphoric acid), and organically bound phosphates. Polyphosphates such as hexametaphosphate gradually hydrolyze in water to the soluble orthoform, and bacterial decomposition of organic compounds also releases orthophosphate. With the majority of compounds in WW being soluble, phosphorus is removed only sparingly by sedimentation. Secondary biological treatment removes phosphorus by biological uptake; however, relative to the quantities of nitrogen and carbon, the amount of phosphorus is greater than necessary for biological synthesis. As a result, conventional treatment process removes only about 20–40% of the influent phosphorus.

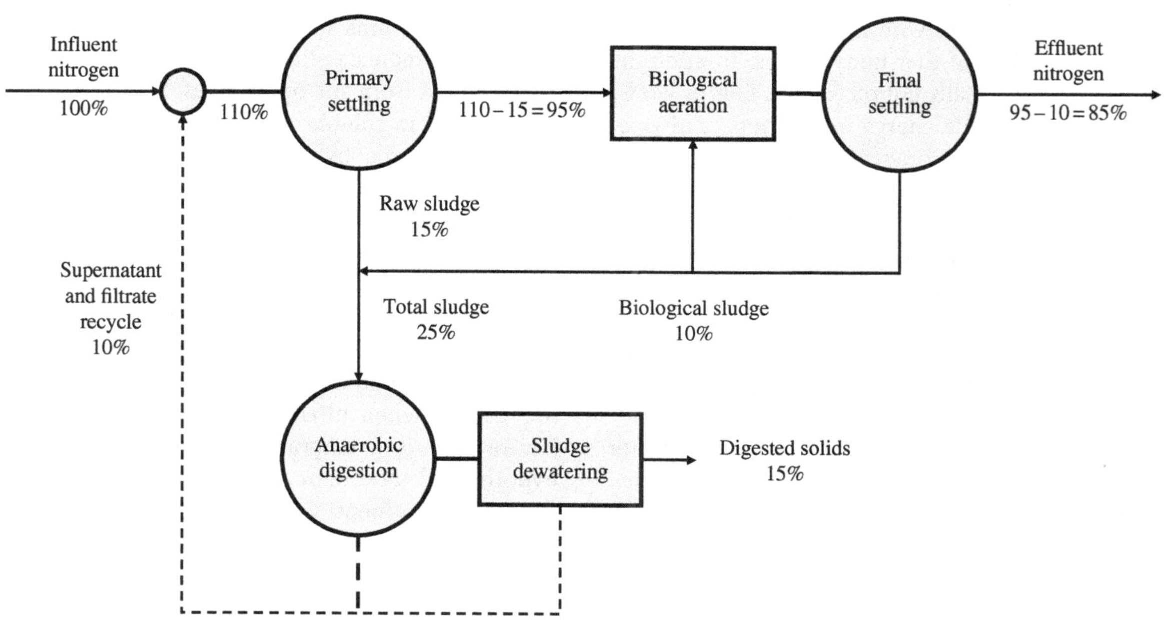

Figure 10.1 Diagram tracing wastewater nitrogen through a hypothetical treatment plant.

10.2.1 Phosphorus Removal in Conventional Treatment

Phosphorus enters the sewer in the form of soluble and organically bound phosphates. Biological activity in the sewer releases organically bound phosphates into solution that are not removed by sedimentation. The amount of organically bound phosphates released into a soluble form varies with the sewer age, WW temperature, and biological conditions. Table 10.2 shows typical N and P forms in raw, primary, and secondary WW treatment. The total phosphorus is reduced from 7 to 6 mg/l by sedimentation. Secondary biological treatment removes phosphorus by biological uptake; however, the amount of phosphorus is surplus relative to the quantity of nitrogen and carbon necessary for synthesis. In general, the amount of phosphorus in the excess biological floc produced in activated sludge treatment of a WW is equal to about 1% of the BOD applied. For this reason, the total phosphorus is further reduced from 6 mg/l to approximately 5 mg/l.

Table 10.2 Approximate nutrient composition of average domestic wastewater and effluents.

Determinant	Raw	Primary	Secondary
Organic content (mg/l)			
Suspended solids	240	120	30
Biochemical oxygen demand	200	130	30
Nitrogen content (mg/l as N)			
Inorganic nitrogen	22	22	24
Organic nitrogen	13	8	2
Total nitrogen	35	30	26
Phosphorus content (mg/l as P)			
Inorganic phosphorus	4	4	4
Organic phosphorus	3	2	1
Total phosphorus	7	6	5

10.2.2 Chemical Phosphorus Removal

Chemical precipitation using aluminum or iron coagulants is effective in phosphate removal. Aluminum ions precipitate with phosphate ions as follows:

$$\begin{aligned}&Al_2(SO_4)_3 14.3H_2O + 2PO_4^{3-} \rightarrow \\ &\quad 2AlPO_4 \downarrow + 3SO_4^{2-} + 14.3H_2O\end{aligned} \tag{10.1}$$

The molar ratio of Al to P is 1 : 1 and the mass ratio of commercial alum to phosphorus is 9.7 to 1.0. Coagulation studies have shown that greater than this alum dosage be necessary to precipitate phosphorus from WW. One of the competing reactions, which accounts in part for the excess alum requirement, is the alum reaction with natural alkalinity as follows:

$$\begin{aligned}&Al_2(SO_4)_3 14.3H_2O + 6HCO_3^- \rightarrow \\ &\quad 2Al(OH)_3 \downarrow + 3SO_4^{2-} + 6CO_2 + 14.3H_2O\end{aligned} \tag{10.2}$$

As a result, 75, 85, and 95% phosphorus reductions require alum to phosphorus mass ratios of about 13 to 1, 16 to 1, and 22 to 1, respectively. For example, to achieve 85% phosphorus removal from a WW containing 10 mg/l of P, the alum dosage needed is approximately 16 × 10 = 160 mg/l, which is substantially greater than the 9.7 × 10 = 97 mg/l stoichiometric quantity of alum according to Reaction (10.1).

Iron coagulants precipitate orthophosphate by combining with the ferric ion as shown in Reaction (10.3) at a molar ratio of 1 to 1:

$$FeCl_3 6H_2O + PO_4^{3-} \rightarrow FePO_4 \downarrow + 3Cl^- + 6H_2O \tag{10.3}$$

Similar to aluminum, a greater amount of iron is required in actual coagulation than this chemical reaction predicts. One of the competing reactions with natural alkalinity is

$$\begin{aligned}&FeCl_3 6H_2O + 3HCO_3^- \rightarrow \\ &\quad Fe(OH)_3 \downarrow + 3CO_2 + 3Cl^- + 6H_2O\end{aligned} \tag{10.4}$$

If WW has sufficient natural alkalinity, ferric salts applied without coagulant aids result in phosphorus removal at Fe to P dosages of 1.8 to 1.0 ratio or greater. This is equivalent to an application of approximately 150 mg/l of commercial ferric chloride for treatment of WW containing 10 mg/l of P. Since the reaction of ferric chloride with natural alkalinity is relatively slow, lime may be applied to raise the pH and supply the hydroxide ion for coagulation as follows:

$$\begin{aligned}&2FeCl_3 6H_2O + 3Ca(OH)_2 \rightarrow \\ &\quad 2Fe(OH)_3 \downarrow + 3CaCl_3 + 6H_2O\end{aligned} \tag{10.5}$$

Ferrous sulfate also forms a phosphate precipitate with an Fe to P molar ratio of 1 : 1, and the dosages for coagulation are similar to ferric salts. Commercially available iron salts are ferric chloride, ferric sulfate, ferrous sulfate, and waste pickle liquor from the steel industry. For the purpose of recycle and reuse, pickle liquors with either sulfuric or hydrochloric acid are the two common waste liquors from metal finishing with iron content from a low of 0.5% to a high of 15%. Chemical–biological treatment combines chemical precipitation of phosphorus with biological removal of organic matter. Alum or iron salts are added prior to primary clarification, directly to the biological process, or prior to final clarification. For all application points, the amount of chemical added is about the same to achieve a specific phosphorus removal. Addition of coagulants to the primary clarifier enhances both suspended solids and BOD removal, resulting in 75% solids and 50% BOD removal.

The current paradigm is shifting design by taking WW as resources. For example, Anaerobic processes can be used to generate methane for energy production, nutrients can be recovered through struvite precipitation, and reclaimed water can be used for irrigation. In terms of per capita, 13 and 2.1 g N and P will be produced, which will result in 36 and 5.8 g N and P/m^3/capita/day, respectively. Although more than 80 and 50% of N and P are contributed to by human urine to the WW flow, urine itself only contributes 1% of the WW flow rate of 0.35 m^3. Therefore, for successful recovery of N and P, it is critical to separate urine from feces. Nutrients could be then be subsequently recovered through struvite precipitation, which is the most effective and easiest way to recover N and P from WW.

Example 10.1

Given: Alum is applied in the aeration tank of a conventional AS system (diagram in Figure 10.2) to reduce the total phosphorus in the effluent to 0.5 mg/l. The dosage of commercial alum is 90 mg/l. Calculate the alum dosage in terms of the mass ratio of alum applied to the phosphorus content of the WW and the molar ratio of aluminum to phosphorus. Calculate anticipated sludge solids production assuming the following: raw and settled WW characteristics as given in Figure 10.2, an operating F/M ratio in biological aeration of 0.40, and an effluent SS concentration of 15 mg/l. The reduction in effluent SS from 30 to 15 mg/l is the result of the increased settleability of the chemical–biological floc as compared with the biological floc without alum addition.

Find: How much would be the increase of the sludge?

Solution:

Alum to phosphorus and aluminum to phosphorus ratios:

$$\text{Alum dosage} = 90\ \text{mg/l}$$

$$\text{Aluminum dosage} = (90)\left[\frac{(2)(27)}{(600)}\right] = 8.1\ \text{mg/l}$$

$$(\text{Molecular mass of } Al_2(SO_4)_3 \cdot 14.3H_2O = 2 \times 27 + 3(32 + 4 \times 16) + 14.3(16 + 2) = 600\ \text{g/mol})$$

$$\text{Alum applied/P in wastewater} = \frac{90}{5.9} = 15$$

$$\text{Al applied/P in wastewater} = \frac{8.1}{5.9} = 1.4$$

Sludge solids production per liter WW processed:

$$\text{SS removal in primary} = 0.5 \times 240 = 120\ \text{mg}$$

The constant K is 0.50 for an F/M of 0.40; therefore,

$$\text{SS in waste activated sludge} = 0.5 \times 130 = 65\ \text{mg}$$

Since K assumes an effluent SS of 30 mg/l and the actual effluent is 15 mg/l,

$$\text{SS from improved settleability} = 30 - 15 = 15\ \text{mg}$$

$$\text{Total biological solids removal} = 65 + 15 = 80\ \text{mg}$$

Organic phosphorus in biological solids is 2.0% by weight; therefore,

$$\text{P in biological solids removed} = 0.02 \times 80 = 1.6\ \text{mg}$$

Phosphorus removed by alum precipitation is the phosphorus in the WW after primary sedimentation minus the phosphorus in the biological solids removed minus the phosphorus in the process effluent, which is 5.9–1.6–0.5 = 3.8 mg:

$$AlPO_4 \text{ precipitate} = \frac{(\text{P precipitated})(\text{molecular mass of } AlPO_4)}{(\text{molecular mass of P})} = \frac{(3.8\ \text{mg})(122)}{(31)} = 15\ \text{mg/l}$$

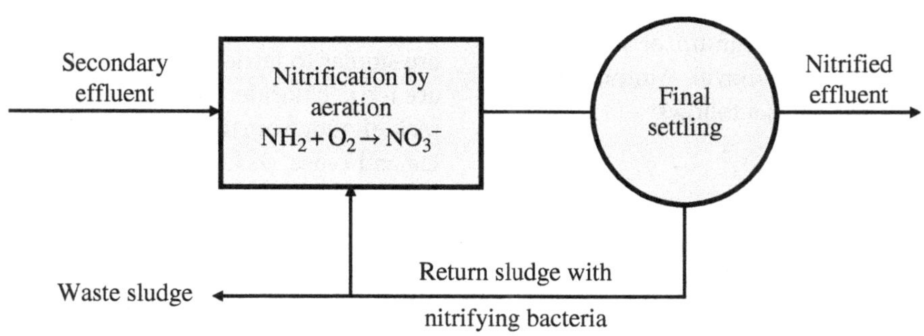

Figure 10.2 Flow diagram for nitrification by suspended growth aeration following conventional biological treatment.

Unused alum = Alum dosage

$$-\frac{(\text{P precipitated})(\text{molecular mass of alum})}{(2\times\text{molecular mass of P})}$$

$$= 90\ \text{mg/l} - \frac{(3.8)(600)}{(2\times 31)} = 53\ \text{mg/l}$$

The alum not used in phosphorus precipitation reacts with natural alkalinity to precipitate $Al(OH)_3$:

$Al(OH)_3$ precipitated

$$= \frac{(\text{Unused alum})(2\times\text{molecular mass of }Al(OH)_3)}{(\text{Molecular mass of alum})}$$

$$= \frac{(53)(2\times 78)}{(600)} = 14\ \text{mg/l}$$

The total sludge solids production equals the sum of the SS removal in primary sedimentation, SS in waste AS, SS from improved settleability in final clarification, $AlPO_4$ precipitate, and $Al(OH)_3$ precipitate. Therefore, solids production per liter of WW treated equals

$$120 + 65 + 15 + 15 + 14 = 229\ \text{mg}$$

increases in sludge solids production.

The solids production by chemical–biological processing relative to conventional AS processing, as given in Figure 10.2, is 229 – 185 = 44 mg/l of WW treated, which is an increase of 24%.

10.3 Phosphorus Recovery

Phosphorus can be recovered from either sewage sludge or sludge liquor of a wastewater treatment plant (WWTP). Since cost and benefit are the major concerns in decision making, LCA and LCCBA are good methods to assess the sustainability of different technologies in terms of process design, operational conditions, and the type of phosphorus product. Up to 90% of the phosphorus entering the WWTP becomes sewage sludge (Pinnekamp et al., 2013), so the recovery of phosphorus from the sewage sludge line offers a higher potential of phosphorus recovery in comparison with a recovery process from the effluent of the WWTP. Almost all processes developed for phosphorus recovery focus on this fraction of the phosphorus originally contained in the raw WW. Elimination of dissolved phosphate from WW and the transformation to sewage sludge take place partly through bacterial growth (38–45%). The phosphorus is used by the microorganisms in the AS as nutrient and included in the biomass. Tertiary WWTPs additionally remove around 30–52% of the phosphorus from WW into the solid sludge phase. There are two main concepts for phosphorus elimination: (i) Enhanced biological phosphorus removal (EBPR) through phosphorus-accumulating bacteria. Under specific process conditions, these bacteria are able to store large amounts of phosphorus. Under anaerobic conditions and in the presence of easily degradable organic compounds, the bacteria can use stored phosphorus as an energy source. Under aerobic conditions, the bacteria fill up the phosphorus storage again. The difference between the phosphorus uptake and the phosphorus released is referred as the net elimination in the form of polyphosphates (Seviour et al., 2003; Oehmen et al., 2007; Pinnekamp et al., 2007). (ii) Chemical phosphorus elimination applies aluminum, iron, or calcium salts. The phosphorus is removed as precipitation product of inorganic phosphates such as strengite, variscite, or apatite, respectively. As a result, larger amounts of sewage sludge are produced by the EBPR than WWTPs.

10.3.1 Enhanced Phosphorus Uptake

Some bacteria have the ability to accumulate phosphorus in the form of polyphosphates in excess of the phosphorus requirements by their growth. Conventional AS biomass typically contains 1–2% phosphorus on a dry weight basis, whereas biomass in an enhanced phosphate removal process is capable of accumulating phosphorus in excess of 3%; in some cases phosphorus contents up to 18% have been obtained with artificial, tailored substrates (Appeldoorn et al., 1992). The highest phosphorus concentration found in the biomass with domestic sewage as a substrate is about 7%. The essential features of the process are an anaerobic phase followed by an aerobic phase. Microorganisms are responsible for the phenomenon, but chemical precipitation of phosphorus may be a significant factor (Bark et al., 1992). The most commonly implicated species are from the genus *Acinetobacter*, but other related species may be involved. Operation of the process with anaerobic–aerobic sequencing provides favorable conditions for enrichment of the sludge with bio-P microorganisms. In the presence of short-chain fatty acids under aerobic conditions, the bio-P microorganisms are able to store polyphosphates as a phosphorus source for energy generation. The initial anaerobic phase is required to produce short-chain acids.

Phosphorus is released from the sludge during the anaerobic phase, but the released phosphorus is taken up later in the process. These acids are utilized by the bio-P microorganisms with concomitant phosphorus removal in an aerobic reactor. The phosphorus-rich sludge formed is settled and removed from the WW. AS that has aerobically accumulated phosphorus in an enhanced phosphorus uptake system will release phosphate when an oxygen deficiency occurs (Schön et al., 1993). Rasmussen et al. (1994) found that most of the release was accomplished in the short time of 4 h under anaerobic conditions. Designing clarifiers to minimize the residence time of the sludge biomass is essential for the proper operation of an enhanced phosphorus removal process.

10.3.2 Struvite Precipitation

Struvite crystallization is an important mineral phosphorus recycling product. If P is not recovered, struvite could also spontaneously precipitate within pipes, pumps, or centrifuges, and cause severe operational problems in sludge handling and treatment. Due to the degradation of the biomass, an important part of the phosphorus is remobilized as orthophosphate and therefore dissolved in the liquid phase of the sludge water typically with concentrations in the range of 80–300 mg/l and NH_4–N/L in the concentration range of 600–800 mg (Lahav et al., 2013). If magnesium is present and the pH is high enough (optimum above pH 8), struvite ($MgNH_4PO_4 \cdot 6H_2O$) precipitates according to the following reversible reaction:

$$Mg^{2+} + NH_4^+ + PO_4^{3-} + 6H_2O = MgNH_4PO_4 \cdot 6H_2O \tag{10.6}$$

The chemical balance is mainly influenced by stoichiometry (concentration) and matrix conditions (pH, temperature). A concentration of ortho-PO_4 above 50 mg/l and pH around 7.8–8.5 is the optimal conditions for struvite crystallization in the presence of sufficient magnesium and ammonia. The first step in crystallization is the nucleation, which forms the initial, smallest size crystals. The second step is the crystal growth, which is the mass transport of ions from the solution to the crystal surface and the incorporation of material into the crystal lattice. To produce fertilizer products, the kinetics of nucleation and growth are mainly determined by the supersaturation of the liquid phase in struvite, the pH, and the temperature. Oversaturation of struvite is defined as a state in which the solubility product is increased, which means that the product of the activities of phosphate, ammonium, and magnesium increases a certain value depending on temperature and ion strength. The pH influences the oversaturation indirectly by the concentrations of NH_4^+ and PO_4^{3-} with an optimal pH leading to the highest oversaturation. The optimal pH lies in the range of pH 8–10.7. Britton et al. (2005) determined the conditional solubility pP_S (product of magnesium, ammonium, and phosphate concentration) as a function of the pH: $pP_S = -0.203\ pH^2 + 4.09pH - 11.76$. As a consequence of struvite crystallization, proton is released and the pH is required to be kept stable during the crystallization process. Crossing the limit of the metastable region will result in a sudden increase of crystallization rate. Furthermore, the degree of oversaturation influences the growth rate of different crystal sizes. For higher supersaturation, smaller crystals grow faster than large ones because of diffusion mechanisms.

Solubility product of struvite, K_s, is defined as follows:

$$K_s = [Mg^{2+}][NH_4^+][PO_4^{3-}]$$

Struvite would precipitate according to following reaction:

$$Mg^{2+} + NH_4^+ + HPO_4^{2-} + 6H_2O \rightarrow MgNH_4PO_4 \cdot 6H_2O + H^+$$

pH determines the species of ammonia between NH_4^+ and NH_3 according to the following equilibrium reaction:

$$NH_4^+ + H_2O \leftrightarrow NH_3 + H_3O^+$$

The equilibrium constant, K_a, is as follows:

$$K_a = \frac{[NH_3][H_3O^+]}{[NH_4^+]}$$

Example 10.2

Given: The effluent of anaerobic digester treating molasses-based industrial WW has the following chemical composition as presented by Çelen and Türker (2001). At the optimal pH from 8.5 to 9, 90%, NH_3 removal is reported (Table 10.3).

Find: If pH is 9.0, what is the percentage lost to the atmosphere?

Solution:

Step 1: pH determines the species of ammonia between NH_4^+ and NH_3 according to the following equilibrium reaction:

$$NH_4^+ + H_2O \leftrightarrow NH_3 + H_3O^+$$

Table 10.3 Composition of wastewater.

Parameter	Concentration in water (mg/l)
NH_4^+	1400
Mg^{2+}	21.4
PO_4^{3-}	24
Ca^{2+}	21.2
K^+	2150
COD	3240
pH	7.9

Step 2: The equilibrium constant, K_a, is as follows:

$$K_a = \frac{[NH_3][H_3O^+]}{[NH_4^+]} = 5.7 \times 10^{-10}$$

Step 3: 17.9% of ammonia is lost to air, whereas at the same condition no loss of ammonia is determined at pH 7.9 (Table 10.4).

Table 10.4 pH and ammonia lost to air.

pH	NH_4^+ (mg/l)	Ammonia lost to air (%)
7.9	1354	—
8.5	1205	11
9.0	1112	17.9

The cost of the MAP recovering process was reported in the year 2000 as shown in Figure 10.3 and in Tables 10.5 and 10.6.

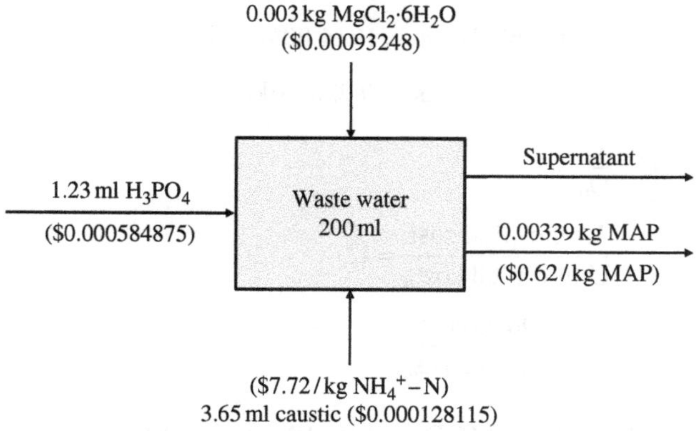

Figure 10.3 Purification and mass balances of struvite recovery in two-step process.

Table 10.5 The market prices of the chemicals in struvite precipitation.

Chemical	Price ($/kg)
H_3PO_4 (75%)	0.40
$MgCl_2{\cdot}6H_2O$	0.31
MgO (85%)	0.44
NaOH (100%)	0.12
NH_4^+	0.23

Table 10.6 Cost of $NH4^+$–N.

Unit cost ($/kg NH_4^+–N)	Authors
7.5–8.0	Çelen and Türker (2001)
9.1–11.38	Siegrist (1996)
4.55–9.92	Andrade and Schuiling (1999)
14.9	Webb et al. (1995)

In 2007, Turker and Celen (2007) were able to recover NH_3 at a cost of \$0.36/kg NH_4–N in their laboratory study as shown in Figure 10.3, while the market price of struvite in 2000 was \$8.75/kg NH_4^+–N.

Nutrients recovery from urine produced in a residential dorm at the University of Central Florida was studied by Ishii and Boyer (2015). Construction costs of urine source separation system, operation of drinking water treatment, decentralized urine treatment, and centralized WW treatment are estimated and compared.

Example 10.3

Given: A flow of 3 ML/day of settled sewage with 250 mg/l BOD and 33 mg/l organic and ammonia nitrogen is to be applied to a land irrigation area.

Find: If the limiting nitrogen loading is 300 kg/ha/year, what are the area of land required, the total annual equivalent depth of effluent applied, and the BOD annual loading rate? (note: 1 mg/l = 1 g/m^3 = 1 kg/ML.)

Solution:

Total daily nitrogen loading = 3 ML/day × 33 kg/ML = 99 kg/day

$$\text{Land area required} = \frac{\text{Total annual nitrogenload}}{\text{Annual application rate}}$$

$$\text{Land area required} = \frac{90\text{kg/day} \times 365\text{day/year}}{300\text{kg/ha/year}}$$

Land area required = 120.5 ha

$$\text{Annual effluent application rate} = \frac{3000\,\text{m}^3/\text{day} \times 365\,\text{day/year}}{120.5\,\text{ha}}$$

Annual BOD application rate = 2272 kg/ha/year

10.4 Capital and Operation Cost of Reclaiming Water for Reuse

The most valuable product to be recovered is the treated water due to huge volume of treated WW effluent. However, distribution infrastructure is required for irrigation or recharge of groundwater aquifer. To reduce pipeline for distribution of water reuse, clusters of WRRFs close to decentralized WWTPs are ideal to proximate reused water to its intended customers. In Miami-Dade County, for example, 300 MGD of treated WW could be reclaimed for irrigation, cooling, and agricultural irrigation. If the reclamation rates are 20, 40, and 60%, the Miami-Dade Water and Sewer Department (WASD) could increase its annual revenue to $65.7, 131, and 197 million/year, respectively, at a water price of $3/1000 gal. In reality, the actual charge for water is about $10/1000 gal due to the WW fee associated with water consumption. Miami-Dade WASD reclaims less than 7%. Therefore, there is great potential for Miami-Dade WASD to enhance its financial balance sheet through water reuse. The major barrier for Miami-Dade WASD is the capital investment for reclaiming the treated WW and building the infrastructure to reuse the water. Not far from Miami, Tampa sets up an exemplary model for cities over the world to follow. In 1995, the US EPA awarded Tampa–St. Petersburg $29.9 million for water reuse. Tampa has 2.5 million people, consumes 60 MGD, and had its share of $19.2 million dollars to build its water reclaim plant. Now, Tampa is able to provide an estimated 16 MGD of drinking water supply to the region through exchange of reclaimed water downstream for upstream surface water. It offsets potable water demands with reclaimed water usage for irrigation and other purposes. The water reclaim plant also benefits greatly from minimum environmental flows and the natural ecosystem in the Tampa Bay area. With additional financial support from the South Florida Water Management District (SWFWMD), Tampa is developing the South Tampa Area Reuse (STAR) project to provide reclaimed water for residents in the South Tampa area. In about 10 years, maximum STAR demand will be about 9 MGD in the dry season. The STAR project will recover about 10 MGD.

Due to economy of scale, unit capital and O&M costs of reclaiming water facility decreases with flow. To facilitate SEE design of water reclaiming plants, regression equations were developed using MatLab. These correlation equations can be used to estimate treatment project capital and O&M costs that can be estimated at different flow rates according to a California study (WateReuse Foundation, 2009). The regression equations use capital or O&M costs at 0.5 MGD as the reference cost. Dimensionless cost ratios between the cost at a specific flow rate and the corresponding cost at 0.5 MGD are plotted against the flow rate ratios. Most equations show that the unit capital and O&M cost decrease with flow rate exponentially before 10 MGD. When flow rate is greater than 10 MGD, the economy-of-scale effect decreases to a linear relationship.

10.4.1 Building

Example 10.4

Given: Table 10.7 shows the capital cost and O&M cost of the admin/lab/shop building at different flow rates.

Solution:

1) When $Q = 1$ MGD

$$\text{Unit capital cost} = 957000\$ \div 0.5\,\text{MGD} \times 1\ \text{million gal}/10^3\,\text{kgal} \div 1\ \text{day} = 1914.00\$/\text{kgal}$$

$$\text{Unit O\&M costs} = 43500\$/\text{year} \div 0.5\,\text{MGD} \times 1\,\text{million gal}/10^3\,\text{kgal} \div 365\,\text{day/year}$$

$$\text{Unit O\&M costs} = 0.2384\$/\text{kgal}$$

When $Q = 0.5$ MGD

$$Q_0 = 0.5\,\text{MGD}$$

$$\text{Unit capital cost}_0 = 1914.00\$/\text{kgal}$$

$$\text{Unit O\&M costs}_0 = 0.2384\$/\text{kgal}$$

$$\frac{Q}{Q_0} = 1$$

$$\frac{\text{Unit capital cost}}{\text{Unit capital cost}_0} = 1$$

$$\frac{\text{Unit O\&M costs}}{\text{Unit O\&M costs}_0} = 1$$

When $Q = 1$ MGD (Tables 10.8 and 10.9)

$$\frac{Q}{Q_0} = 2$$

Table 10.7 Admin/lab/shop building.

Expense ($)	Cost per amount of flow (MGD)						
	0.5	1.0	2.0	5.0	10	15	20
Concrete	0	0	0	0	0	0	0
Excavation and backfill (10%)	0	0	0	0	0	0	0
Miscellaneous metals (4%)	0	0	0	0	0	0	0
Yard piping (7%)	0	0	0	0	0	0	0
Total concrete	0	0	0	0	0	0	0
Equipment	0	0	0	0	0	0	0
Tax and delivery (11%)	0	0	0	0	0	0	0
Installation (20%)	0	0	0	0	0	0	0
Manufacturer services (4%)	0	0	0	0	0	0	0
Total mechanical	0	0	0	0	0	0	0
Protective coating (7%)	0	0	0	0	0	0	0
Electricity (10%)	0	0	0	0	0	0	0
Instrumentation (10%)	0	0	0	0	0	0	0
Housing[a]	600 000	600 000	750 000	1 800 000	3 000 000	3 600 000	4 200 000
Subtotal	600 000	600 000	750 000	1 800 000	3 000 000	3 600 000	4 200 000
Contractor overhead and profit (15%)	90 000	90 000	112 500	270 000	450 000	540 000	630 000
Scope-of-project contingency (30%)	180 000	180 000	225 000	540 000	900 000	1 080 000	1 260 000
Total construction cost	870 000	870 000	1 087 500	2 610 000	4 350 000	5 220 000	6 090 000
Engineering design (10%)	87 000	87 000	108 750	261 000	435 000	522 000	609 000
Total capital cost	957 000	957 000	1 196 250	2 871 000	4 785 000	5 742 000	6 699 000
Cost per gal ($)	1.91	0.96	0.60	0.57	0.48	0.38	0.33
O&M (5% construction cost)	43 500	43 500	54 375	130 500	217 500	261 000	304 500

[a] Based on an assumed $300/ft^2$ used for planning-level estimates:

1) Calculate unit capital and O&M costs.
2) Plot unit cost/unit cost$_0$ versus Q/Q_0.
3) Determine correlation equations between unit cost/unit cost$_0$ for capital and O&M versus Q/Q_0, respectively.

Table 10.8 Unit capital cost and unit operation and maintenance cost of admin/lab/shop building at different flow rates.

Q (MGD)	Unit capital cost ($/kgal)	Unit O&M costs ($/kgal)
0.5	1914.00	0.2384
1	957.00	0.1192
2	598.13	0.0745
5	574.20	0.0715
10	478.50	0.0596
15	382.80	0.0477
20	334.95	0.0417

Table 10.9 Unit cost ratio of admin/lab/shop building at different flow ratios.

Q/Q_0	Unit capital cost/unit capital cost$_0$	Unit O&M costs/unit O&M costs$_0$
1	1.0000	1.0000
2	0.5000	0.5000
4	0.3125	0.3125
10	0.3000	0.3000
20	0.2500	0.2500
30	0.2000	0.2000
40	0.1750	0.1750

$$\frac{\text{Unit capital cost}}{\text{Unit capital cost}_0} = 0.5$$

$$\frac{\text{Unit O\&M costs}}{\text{Unit O\&M costs}_0} = 0.5$$

2) Plot unit cost/unit cost$_0$ versus Q/Q_0 for capital and O&M, respectively (Figure 10.4).

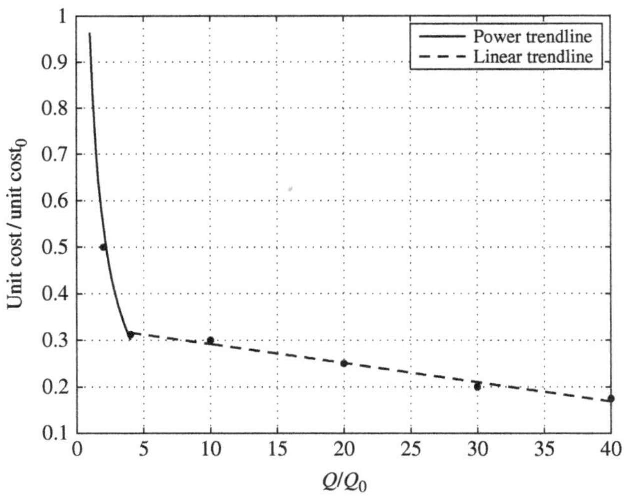

Figure 10.4 Unit capital cost/unit capital cost$_0$ versus Q/Q_0 for admin/lab/shop building, unit capital cost$_0$ = 1914.00$/kgal, Q_0 = 0.5 MGD.

3) a) Equation for unit cost/unit cost$_0$ versus Q/Q_0 for capital:

The regression equation of the trend line ($1 \le Q/Q_0 \le 4$):

$$\log\left(\frac{\text{unit cost}}{\text{unit cost}_0}\right) = -0.8390\log\left(\frac{Q}{Q_0}\right) - 0.0162$$

$$r^2 = 0.9879$$

or the equation of the regression:

$$\frac{\text{Unit cost}}{\text{Unit cost}_0} = 0.9635\left(\frac{Q}{Q_0}\right)^{-0.8390} \quad r^2 = 0.9879$$

The regression equation of the trend line ($4 \le Q/Q_0 \le 40$) (Figure 10.5) (note: Unit O&M cost was estimated by 5% of capital cost. Although the initial capital and O&M costs are not the same, the dimensionless plot for the unit costs of capital or O&M vs. flow rate ratios are the same for the rest of the examples.):

$$\frac{\text{Unit cost}}{\text{Unit cost}_0} = -0.0041Q/Q_0 + 0.3326 \quad r^2 = 0.9850$$

Matlab code:

```
clear all
close all
%data%
```

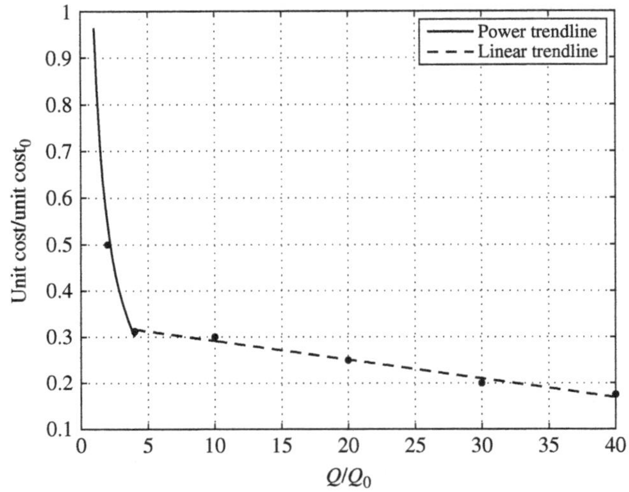

Figure 10.5 Unit O&M cost/unit O&M cost$_0$ versus Q/Q_0 for admin/lab/shop building, unit O&M cost$_0$ = 0.2384$/kgal, Q_0 = 0.5 MGD.

```
Qrange=[0.5 1 2 5 10 15 20];
Q=Qrange(1);
totalcostrange=[957000 957000 1196250
   2871000 4785000 5742000 6699000];
totalcost=totalcostrange(1);
xmid=3;
%function%
unitcost=totalcost./Q./10^3;
unitcostrange=totalcostrange./
   Qrange./10^3;
x=Qrange/Q;
y=unitcostrange /unitcost;
x1=x(1:xmid);
y1=y(1:xmid);
x2=x(xmid:end);
y2=y(xmid:end);
%power
p=polyfit(log10(x1), log10(y1), 1)
log10ratiounitcostfit=polyval(p,
   log10(x1));
ratiounitcostfit=10.
   ^log10ratiounitcostfit;
k=10^p(:,2)
R2power = 1 - sum((log10(y1) -
   log10ratiounitcostfit).^2)/sum((log10
(y1) - mean(log10(y1))).^2)
xmax=max(x1);
xmin=min(x1);
xaxis=[xmin:0.1:xmax];
log10ratiounitcostfitplot=polyval(p,
   log10(xaxis));
ratiounitcostfitplot=10.
   ^log10ratiounitcostfitplot;
```

```
%line
p2=polyfit((x2),(y2), 1)
z=polyval(p2, (x2));
R2linear = 1 - sum(((y2) - z).^2)/sum
   (((y2) - mean((y2))).^2)
xmax2=max(x2);
xmin2=min(x2);
xaxis2=[xmin2:0.1:xmax2];
z2=polyval(p2, xaxis2);
%plot%
figure(1)
plot(Qrange/Q,unitcostrange
   /unitcost,'ko',
   'MarkerSize',3,'MarkerFaceColor',
   [0 0 0],'MarkerEdgeColor',
   [0 0 0],'linewidth',1);
hold on
line1=plot(xaxis,ratiounitcostfitplot,
   'k-','linewidth',1);
hold on
line2=plot(xaxis2,z2,
   'k--','linewidth',1);
set(gca,'xtickmode','auto',
   'FontSize',10,'FontWeight',
   'bold','FontName', 'Times New Roman');
legend([line1,line2],'Power
trendline','Linear trendline',
   'Location','Northeast');
xlabel({'Q/Q_0'},'FontSize',14,
   'FontWeight','bold','FontName',
   'Times New Roman')
ylabel({'Unit Cost/Unit Cost_0'},
   'FontSize',14,'FontWeight',
   'bold','FontName', 'Times New Roman')
grid on
hold on
```

Answer: AdminLabShopBuildingCapital costtwoline

```
p = -0.8390 -0.0162; k = 0.9635;
  R^2power = 0.9879
p2 = -0.0041 0.3326; R^2linear = 0.9850
```

b) Equation for unit cost/unit cost_0 versus Q/Q_0 for O&M:

The regression equation of the trend line ($1 \le Q/Q_0 \le 4$):

$$\log\left(\frac{\text{unit cost}}{\text{unit cost}_0}\right) = -0.8390\log\left(\frac{Q}{Q_0}\right) - 0.0162$$

$$r^2 = 0.9879$$

or the equation of the regression:

$$\frac{\text{Unit cost}}{\text{Unit cost}_0} = 0.9635\left(\frac{Q}{Q_0}\right)^{-0.8390} \quad r^2 = 0.9879$$

The regression equation of the trend line ($4 \le Q/Q_0 \le 40$):

$$\frac{\text{Unit cost}}{\text{Unit cost}_0} = -0.0041Q/Q_0 + 0.3326 \quad r^2 = 0.9850$$

Matlab code:

```
clear all
close all
%data%
Qrange=[0.5 1 2 5 10 15 20];
Q=Qrange(1);
totalcostrange=[43500  43500  54375
  130500 217500 261000 304500];
totalcost=totalcostrange(1);
xmid=3;
%function%
unitcost=totalcost./Q./10^3./365;
unitcostrange=totalcostrange./Qrange./
  10^3./365;
x=Qrange/Q;
y=unitcostrange /unitcost;
x1=x(1:xmid);
y1=y(1:xmid);
x2=x(xmid:end);
y2=y(xmid:end);
%power
p=polyfit(log10(x1), log10(y1), 1)
log10ratiounitcostfit=polyval(p,
  log10(x1));
ratiounitcostfit=10.
  ^log10ratiounitcostfit;
k=10^p(:,2)
R2power = 1 - sum((log10(y1) -
  log10ratiounitcostfit).^2)/sum((log10
(y1) - mean(log10(y1))).^2)
xmax=max(x1);
xmin=min(x1);
xaxis=[xmin:0.1:xmax];
log10ratiounitcostfitplot=polyval(p,
  log10(xaxis));
ratiounitcostfitplot=10.
  ^log10ratiounitcostfitplot;
%line
p2=polyfit((x2),(y2), 1)
z=polyval(p2, (x2));
R2linear = 1 - sum(((y2) - z).^2)/sum(((y2)
  - mean((y2))).^2)
```

```
xmax2=max(x2);
xmin2=min(x2);
xaxis2=[xmin2:0.1:xmax2];
z2=polyval(p2, xaxis2);
%plot%
figure(1)
plot(Qrange/Q,unitcostrange
   /unitcost,'ko','MarkerSize',3,
   'MarkerFaceColor',
   [0 0 0],'MarkerEdgeColor',
   [0 0 0],'linewidth',1);
hold on
line1=plot(xaxis,
   ratiounitcostfitplot,'k-',
   'linewidth',1);
hold on
line2=plot(xaxis2,z2,'k--',
   'linewidth',1);
set(gca,'xtickmode',
   'auto','FontSize',10,
   'FontWeight','bold','FontName',
   'Times New Roman');
legend([line1,line2],'Power
   trendline','Linear trendline',
   'Location','Northeast');
xlabel({'Q/Q_0'},'FontSize',14,
   'FontWeight','bold','FontName',
   'Times New Roman')
ylabel({'Unit Cost/Unit
   Cost_0'},'FontSize',14,
   'FontWeight','bold','FontName',
   'Times New Roman')
grid on
hold on
```

Answer: `AdminLabShopBuildingOMcost twoline`

```
p = -0.8390 -0.0162; k = 0.9635; R2power
   = 0.9879
p2 = -0.0041 0.3326; R^2linear = 0.9850
```

10.4.2 Headwork

Example 10.5

Given: Table 10.10 shows capital cost and O&M cost of headworks at different flow rates.

Find:

1) Calculate unit capital and O&M costs.
2) Plot unit cost/unit cost$_0$ versus Q/Q_0.
3) Determine correlation equations between unit cost/unit cost$_0$ for capital and O&M versus Q/Q_0, respectively.

Solution:

1) When $Q = 0.5$ MGD

$$\text{Unit capital cost} = 473737\$ \div 0.5\,\text{MGD} \times 1\,\text{million gal}/10^3\,\text{kgal} \div 1\,\text{day} = 947.47\$/\text{kgal}$$

$$\text{Unit O\&M costs} = 21533\$/\text{year} \div 0.5\,\text{MGD} \times 1\,\text{million gal}/10^3\,\text{kgal} \div 365\,\text{day/year}$$

$$\text{Unit O \& M costs} = 0.1180\$/\text{kgal}$$

When $Q = 0.5$ MGD

$$Q_0 = 0.5\,\text{MGD}$$

$$\text{Unit capital cost}_0 = 947.47\$/\text{kgal}$$

$$\text{Unit O\&M costs}_0 = 0.1180\$/\text{kgal}$$

$$\frac{Q}{Q_0} = 1$$

$$\frac{\text{Unit capital cost}}{\text{Unit capital cost}_0} = 1$$

$$\frac{\text{Unit O\&M costs}}{\text{Unit O\&M costs}_0} = 1$$

When $Q = 1$ MGD (Tables 10.11 and 10.12)

$$\frac{Q}{Q_0} = 2$$

$$\frac{\text{Unit capital cost}}{\text{Unit capital cost}_0} = 0.5872$$

$$\frac{\text{Unit O\&M costs}}{\text{Unit O\&M costs}_0} = 0.5872$$

2) Plot unit cost/unit cost$_0$ versus Q/Q_0 for capital and O&M, respectively (Figure 10.6).
3) a) Equation for unit cost/unit cost$_0$ versus Q/Q_0 for capital:

The regression equation of the trend line:

$$\log\left(\frac{\text{unit cost}}{\text{unit cost}_0}\right) = -0.5472\log\left(\frac{Q}{Q_0}\right) - 0.0744$$

$$r^2 = 0.9694$$

Table 10.10 Headworks.

Expense ($)	Cost per amount of flow (MGD)						
	0.5	1.0	2.0	5.0	10	15	20
Concrete[a]	6 400	7 200	11 200	20 000	24 800	29 600	29 600
Excavation and backfill (10%)	640	720	1 120	2 000	2 480	2 960	2 960
Miscellaneous metals (4%)	256	288	448	800	992	1 184	1 184
Yard piping (7%)	448	504	784	1 400	1 736	2 072	2 072
Total concrete	7 744	8 712	13 552	24 200	30 008	35 816	35 816
Equipment[b,c]	167 500	197 000	206 500	349 000	545 000	695 000	835 000
Tax and delivery (11%)	18 425	21 670	22 715	38 390	59 950	76 450	91 850
Installation (20%)	33 500	39 400	41 300	69 800	109 000	139 000	167 000
Manufacturer services (4%)	6 700	7 880	8 260	13 960	21 800	27 800	33 400
Total mechanical	226 125	265 950	278 775	471 150	735 750	938 250	1 127 250
Protective coating (7%)	16 371	19 226	20 463	34 675	53 603	68 185	81 415
Electricity (10%)	23 387	27 466	29 233	49 535	76 576	97 407	116 307
Instrumentation (10%)	23 387	27 466	29 233	49 535	76 576	97 407	116 307
Housing	0	0	0	0	0	0	0
Subtotal	297 014	348 821	371 255	629 095	972 513	1 237 064	1 477 094
Contractor overhead and profit (15%)	44 552	52 323	55 688	94 364	145 877	185 560	221 564
Scope-of-project contingency (30%)	89 104	104 646	111 377	188 728	291 754	371 119	443 128
Total construction cost	430 670	505 790	538 320	912 187	1 410 143	1 793 743	2 141 786
Engineering design (10%)	43 067	50 579	53 832	91 219	141 014	179 374	214 179
Total capital cost	473 737	556 369	592 152	1 003 406	1 551 158	1 973 117	2 355 965
Cost per gal ($)	0.95	0.56	0.30	0.20	0.16	0.13	0.12
O&M (5% construction cost)	21 533	25 290	26 916	45 609	70 507	89 687	107 089

[a] Concrete was sized for grit chambers only.
[b] Equipment costs include both screens and Pista Grit.
[c] For screens, the average cost of 6-mm and 2-mm screens is used.

Table 10.11 Unit capital cost and unit operation and maintenance cost of headworks at different flow rates.

Q (MGD)	Unit capital cost ($/kgal)	Unit O&M costs ($/kgal)
0.5	947.47	0.1180
1	556.37	0.0693
2	296.08	0.0369
5	200.68	0.0250
10	155.12	0.0193
15	131.54	0.0164
20	117.80	0.0147

Table 10.12 Unit cost ratio of headworks at different flow ratios.

Q/Q_0	Unit capital cost/unit capital cost$_0$	Unit O&M costs/unit O&M costs$_0$
1	1.0000	1.0000
2	0.5872	0.5872
4	0.3125	0.3125
10	0.2118	0.2118
20	0.1637	0.1637
30	0.1388	0.1388
40	0.1243	0.1243

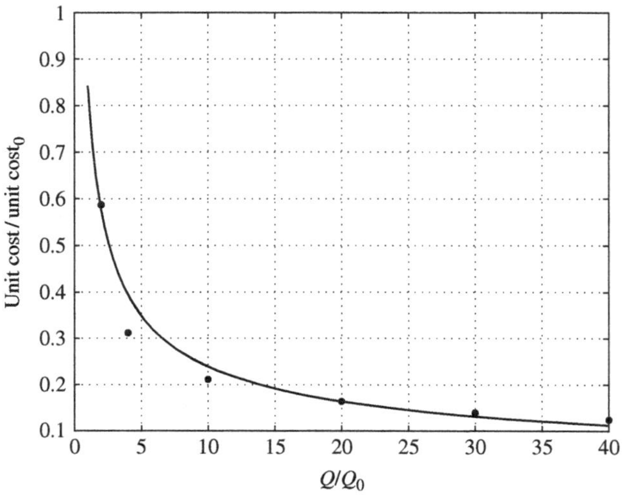

Figure 10.6 Unit capital cost/unit capital cost$_0$ versus Q/Q_0 for headworks, unit capital cost$_0$ = 947.47\$/kgal, Q_0 = 0.5 MGD.

or the equation of the regression (Figure 10.7):

$$\frac{\text{Unit cost}}{\text{Unit cost}_0} = 0.0744\left(\frac{Q}{Q_0}\right)^{-0.5472} \quad r^2 = 0.9694$$

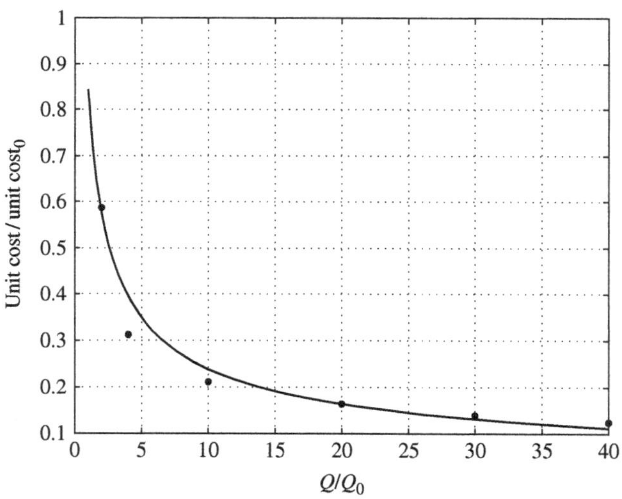

Figure 10.7 Unit O&M cost/unit O&M cost$_0$ versus Q/Q_0 for headworks, unit O&M cost$_0$ = 0.1180\$/kgal, Q_0 = 0.5 MGD.

Matlab code:

```
clear all
close all
%data%
Qrange=[0.5 1 2 5 10 15 20];
Q=Qrange(1);
totalcostrange=[473737 556369 592152
  1003406 1551158 1973117 2355965];
totalcost=totalcostrange(1);
unitcost=totalcost./Q./1000;
  unitcostrange=totalcostrange./
  Qrange./1000;
p=polyfit(log10(Qrange/Q), log10
  (unitcostrange /unitcost), 1)
log10ratiounitcostfit=polyval(p, log10
  (Qrange/Q));
ratiounitcostfit=10.
  ^log10ratiounitcostfit;
k=10^p(:,2)
R2 = 1 - sum((log10(unitcostrange
  /unitcost) - log10ratiounitcostfit).
  ^2)/sum((log10(unitcostrange
  /unitcost) - mean(log10(unitcostrange
  /unitcost))).^2)
xmax=max(Qrange/Q);
xmin=min(Qrange/Q);
xaxis=[xmin:0.1:xmax];
log10ratiounitcostfitplot=polyval(p,
  log10(xaxis));
ratiounitcostfitplot=10.
  ^log10ratiounitcostfitplot;
%plot%
figure(1)
plot(Qrange/Q,unitcostrange
  /unitcost,'ko',
  'MarkerSize',3,'MarkerFaceColor',
  [0 0 0],'MarkerEdgeColor',
  [0 0 0],'linewidth',1);
hold on
plot(xaxis,ratiounitcostfitplot,'k-',
  'linewidth',1);
set(gca,'xtickmode',
  'auto','FontSize',10,
  'FontWeight','bold','FontName',
  'Times New Roman');

xlabel({'Q/Q_0'},'FontSize',14,
  'FontWeight','bold','FontName',
  'Times New Roman')
ylabel({'Unit Cost/Unit
  Cost_0'},'FontSize',14,
  'FontWeight','bold','FontName',
  'Times New Roman')
grid on
hold on
```

Answer: Headworkcapitalcost

```
p = -0.5472 -0.0744; k = 0.8425;
R2 = 0.9694
```

b) Equation for unit cost/unit cost$_0$ versus Q/Q_0 for O&M:

The regression equation of the trend line:

$$\log\left(\frac{\text{unit cost}}{\text{unit cost}_0}\right) = -0.5472\log\left(\frac{Q}{Q_0}\right) - 0.0744$$
$$r^2 = 0.9694$$

or the equation of the regression:

$$\frac{\text{Unit cost}}{\text{Unit cost}_0} = 0.0744\left(\frac{Q}{Q_0}\right)^{-0.5472} \quad r^2 = 0.9694$$

Matlab code:

```
clear all
close all
%data%
Qrange=[0.5 1 2 5 10 15 20];
Q=Qrange(1);
totalcostrange=[21533 25290 26916 45609
   70507 89687 107089];
totalcost=totalcostrange(1);
unitcost=totalcost./Q./365./1000;
unitcostrange=totalcostrange./Qrange./
   365./1000;
p=polyfit(log10(Qrange/Q), log10
   (unitcostrange /unitcost), 1)
log10ratiounitcostfit=polyval(p, log10
   (Qrange/Q));
ratiounitcostfit=10.
   ^log10ratiounitcostfit;
k=10^p(:,2)
R2 = 1 - sum((log10(unitcostrange
   /unitcost) - log10ratiounitcostfit).
   ^2)/sum((log10(unitcostrange
   /unitcost) - mean(log10(unitcostrange
   /unitcost))).^2)
xmax=max(Qrange/Q);
xmin=min(Qrange/Q);
xaxis=[xmin:0.1:xmax];
log10ratiounitcostfitplot=polyval(p,
   log10(xaxis));
ratiounitcostfitplot=10.
   ^log10ratiounitcostfitplot;
%plot%
figure(1)
plot(Qrange/Q,unitcostrange
   /unitcost,'ko',
   'MarkerSize',3,'MarkerFaceColor',
   [0 0 0],'MarkerEdgeColor',
   [0 0 0],'linewidth',1);
hold on
plot(xaxis,ratiounitcostfitplot,'
   k-','linewidth',1);
set(gca,'xtickmode',
   'auto','FontSize',10,
   'FontWeight','bold','FontName',
   'Times New Roman');

xlabel({'Q/Q_0'},'FontSize',14,
   'FontWeight','bold','FontName',
   'Times New Roman')
ylabel({'Unit Cost/Unit
   Cost_0'},'FontSize',14,
   'FontWeight','bold','FontName',
   'Times New Roman')
grid on
hold on
```

Answer: `HeadworkOMcost`

```
p = -0.5472 -0.0744; k = 0.8425;
 R2 = 0.9694
```

10.4.3 Oxidation

Example 10.6

Given: Table 10.13 shows capital cost and O&M cost of oxidation process at different flow rates.

Find:

1) Calculate unit capital and O&M costs.
2) Plot unit cost/unit cost$_0$ versus Q/Q_0.
3) Determine correlation equations between unit cost/unit cost$_0$ for capital and O&M versus Q/Q_0, respectively.

Solution:

1) When Q = 0.5 MGD

$$\text{Unit capital cost} = 3396659\$ \div 0.5\,\text{MGD} \times 1\,\text{million gal}/10^3\,\text{kgal} \div 1\,\text{day} = 6793.32\$/\text{kgal}$$

$$\text{Unit O\&M costs} = 154394\$/\text{year} \div 0.5\,\text{MGD} \times 1\,\text{million gal}/10^3\,\text{kgal} \div 365\,\text{day/year}$$

$$\text{Unit O\&M costs} = 0.8460\$/\text{kgal}$$

When Q = 0.5 MGD

$$Q_0 = 0.5\,\text{MGD}$$

$$\text{Unit capital cost}_0 = 6793.32\$/\text{kgal}$$

Table 10.13 Ox (CA Title 22).

	Cost per amount of flow (MGD)						
Expense ($)	**0.5**	**1.0**	**2.0**	**5.0**	**10**	**15**	**20**
Concrete[a]	774 400	1 024 000	1 861 600	3 640 000	7 156 000	11 017 600	16 400 800
Excavation and backfill (10%)	77 440	102 400	186 160	364 000	715 600	1 101 760	1 640 080
Miscellaneous metals (4%)	30 976	40 960	74 464	145 600	286 240	440 704	656 032
Yard piping (7%)	54 208	71 680	130 312	254 800	500 920	771 232	1 148 056
Total concrete	937 024	1 239 040	2 252 536	4 404 400	8 658 760	13 331 296	19 844 968
Equipment[b,c]	548 000	598 000	676 000	900 000	1 800 000	2 580 000	3 360 000
Tax and delivery (11%)	60 280	65 780	74 360	99 000	198 000	283 800	369 600
Installation (20%)	109 600	119 600	135 200	180 000	360 000	516 000	672 000
Manufacturer services (4%)	21 920	23 920	27 040	36 000	72 000	103 200	134 400
Total mechanical	739 800	807 300	912 600	1 215 000	2 430 000	3 483 000	4 536 000
Protective coating (7%)	117 378	143 244	221 560	393 358	776 213	1 177 001	1 706 668
Electricity (10%)	167 682	204 634	316 514	561 940	1 108 876	1 681 430	2 438 097
Instrumentation (10%)	167 682	204 634	316 514	561 940	1 108 876	1 681 430	2 438 097
Housing	0	0	0	0	0	0	0
Subtotal	2 129 566	2 598 852	4 019 723	7 136 638	14 082 725	21 354 156	30 963 829
Contractor overhead and profit (15%)	319 435	389 828	602 958	1 070 496	2 112 409	3 203 123	4 644 574
Scope-of-project contingency (30%)	638 870	779 656	1 205 917	2 140 991	4 224 818	6 406 247	9 289 149
Total construction cost	3 087 871	3 768 335	5 828 598	10 348 125	20 419 952	30 963 526	44 897 553
Engineering design (10%)	308 787	376 834	582 860	1 034 813	2 041 995	3 096 353	4 489 755
Total capital cost	3 396 659	4 145 169	6 411 458	11 382 938	22 461 947	34 059 879	49 387 308
Cost per gal ($)	6.79	4.15	3.21	2.28	2.25	2.27	2.47
O&M (5% construction cost)	154 394	188 417	291 430	517 406	1 020 998	1 548 176	2 244 878

[a] Concrete cost is only for grit chambers.
[b] Equipment costs include both screens and Pista Grit (Western Environmental Equipment Co. and Siemens Technologies).
[c] For screens, the average cost of 6-mm and 2-mm screens was used.

$$\text{Unit O\&M costs}_0 = 0.8460\$/\text{kgal}$$

$$\frac{Q}{Q_0} = 1$$

$$\frac{\text{Unit capital cost}}{\text{Unit capital cost}_0} = 1$$

$$\frac{\text{Unit O\&M costs}}{\text{Unit O\&M costs}_0} = 1$$

When $Q = 1$ MGD (Tables 10.14 and 10.15)

$$\frac{Q}{Q_0} = 2$$

Table 10.14 Unit capital cost and unit operation and maintenance cost of Ox (CA Title 22) at different flow rates.

Q (MGD)	Unit capital cost ($/kgal)	Unit O&M costs ($/kgal)
0.5	6793.32	0.8460
1	4145.17	0.5162
2	3205.73	0.3992
5	2276.59	0.2835
10	2246.19	0.2797
15	2270.66	0.2828
20	2469.37	0.3075

Table 10.15 Unit cost ratio of Ox (CA Title 22) at different flow ratios.

Q/Q_0	Unit capital cost/unit capital $cost_0$	Unit O&M costs/unit O&M $costs_0$
1	1.0000	1.0000
2	0.6102	0.6102
4	0.4719	0.4719
10	0.3351	0.3351
20	0.3306	0.3306
30	0.3342	0.3342
40	0.3635	0.3635

$$\frac{\text{Unit capital cost}}{\text{Unit capital cost}_0} = 0.6102$$

$$\frac{\text{Unit O\&M costs}}{\text{Unit O\&M costs}_0} = 0.6102$$

2) Plot unit cost/unit $cost_0$ versus Q/Q_0 for capital and O&M, respectively (Figure 10.8).
3) a) Equation for unit cost/unit $cost_0$ versus Q/Q_0 for capital:

The regression equation of the trend line ($1 \le Q/Q_0 \le 4$):

$$\log\left(\frac{\text{unit cost}}{\text{unit cost}_0}\right) = -0.5417\log\left(\frac{Q}{Q_0}\right) - 0.0172$$

$$r^2 = 0.9679$$

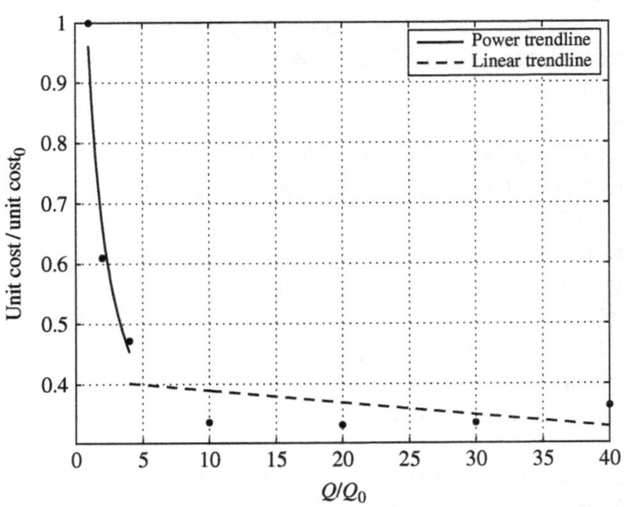

Figure 10.8 Unit capital cost/unit capital $cost_0$ versus Q/Q_0 for Ox (CA Title 22), unit capital $cost_0$ = 6793.32$/kgal, Q_0 = 0.5 MGD.

or the equation of the regression:

$$\frac{\text{Unit cost}}{\text{Unit cost}_0} = 0.9613\left(\frac{Q}{Q_0}\right)^{-0.5417} \quad r^2 = 0.9679$$

The regression equation of the trend line ($4 \le Q/Q_0 \le 40$) (Figure 10.9):

$$\frac{\text{Unit cost}}{\text{Unit cost}_0} = -0.0021Q/Q_0 + 0.4099 \quad r^2 = 0.2510$$

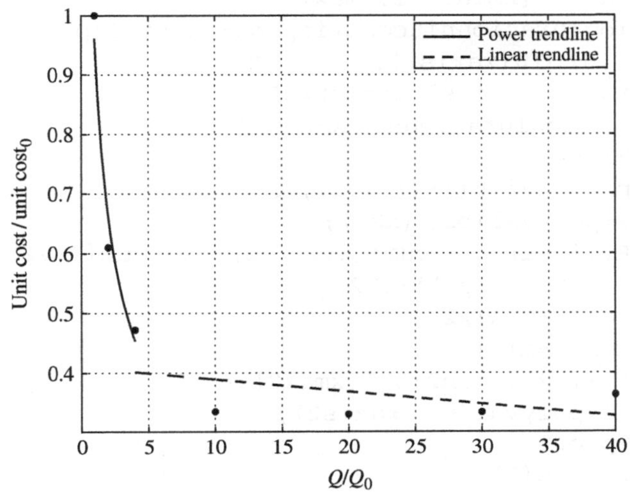

Figure 10.9 Unit O&M cost/unit O&M $cost_0$ versus Q/Q_0 for Ox (CA Title 22), unit O&M $cost_0$ = 0.8460$/kgal, Q_0 = 0.5 MGD.

Matlab code:

```
clear all
close all
%data%
Qrange=[0.5 1 2 5 10 15 20];
Q=Qrange(1);
totalcostrange=[3396659 4145169 6411458
  11382938  22461947  34059879
  49387308];
totalcost=totalcostrange(1);
xmid=3;
%function%
unitcost=totalcost./Q./10^3;
unitcostrange=totalcostrange./
  Qrange./10^3;
x=Qrange/Q;
y=unitcostrange /unitcost;
x1=x(1:xmid);
y1=y(1:xmid);
x2=x(xmid:end);
y2=y(xmid:end);
%power
p=polyfit(log10(x1), log10(y1), 1)
```

```
log10ratiounitcostfit=polyval(p,
  log10(x1));
ratiounitcostfit=10.
  ^log10ratiounitcostfit;
k=10^p(:,2)
R2power = 1 - sum((log10(y1) -
  log10ratiounitcostfit).^2)/sum((log10
  (y1) - mean(log10(y1))).^2)
xmax=max(x1);
xmin=min(x1);
xaxis=[xmin:0.1:xmax];
log10ratiounitcostfitplot=polyval(p,
  log10(xaxis));
ratiounitcostfitplot=10.
  ^log10ratiounitcostfitplot;
%line
p2=polyfit((x2),(y2), 1)
z=polyval(p2, (x2));
R2linear = 1 - sum(((y2) - z).^2)/sum(((y2)
  - mean((y2))).^2)
xmax2=max(x2);
xmin2=min(x2);
xaxis2=[xmin2:0.1:xmax2];
z2=polyval(p2, xaxis2);
%plot%
figure(1)
plot(Qrange/Q,unitcostrange
  /unitcost,'ko','MarkerSize',3,
  'MarkerFaceColor',
  [0 0 0],'MarkerEdgeColor',
  [0 0 0],'linewidth',1);
hold on
line1=plot(xaxis,
  ratiounitcostfitplot,'k-',
  'linewidth',1);
hold on
line2=plot(xaxis2,z2,'k--',
  'linewidth',1);
set(gca,'xtickmode',
  'auto','FontSize',10,
  'FontWeight','bold','FontName',
  'Times New Roman');
legend([line1,line2],'Power
  trendline','Linear trendline',
  'Location','Northeast');
xlabel({'Q/Q_0'},'FontSize',14,
  'FontWeight','bold','FontName',
  'Times New Roman')
ylabel({'Unit Cost/Unit Cost_0'},
  'FontSize',14,'FontWeight',
  'bold','FontName', 'Times New Roman')
grid on
hold on
```

Answer: `OxCATitle22CapitalCost`

```
p = -0.5417  -0.0172; k = 0.9613; R2power
  = 0.9679
p2 = -0.0021 0.4099; R2linear = 0.2510
```

b) Equation for unit cost/unit cost$_0$ versus Q/Q_0 for O&M:

The regression equation of the trend line ($1 \le Q/Q_0 \le 4$):

$$\log\left(\frac{\text{unit cost}}{\text{unit cost}_0}\right) = -0.5417\log\left(\frac{Q}{Q_0}\right) - 0.0172$$

$$r^2 = 0.9679$$

or the equation of the regression:

$$\frac{\text{Unit cost}}{\text{Unit cost}_0} = 0.9613\left(\frac{Q}{Q_0}\right)^{-0.5417} \quad r^2 = 0.9679$$

The regression equation of the trend line ($4 \le Q/Q_0 \le 40$):

$$\frac{\text{Unit cost}}{\text{unit cost}_0} = -0.0021Q/Q_0 + 0.4099 \quad r^2 = 0.2510$$

Matlab code:

```
clear all
close all
%data%
Qrange=[0.5 1 2 5 10 15 20];
Q=Qrange(1);
totalcostrange=[154394 188417 291430
  517406 1020998 1548176 2244878];
totalcost=totalcostrange(1);
xmid=3;
%function%
unitcost=totalcost./Q./10^3./365;
unitcostrange=totalcostrange./Qrange./
  10^3./365;
x=Qrange/Q;
y=unitcostrange /unitcost;
x1=x(1:xmid);
y1=y(1:xmid);
x2=x(xmid:end);
y2=y(xmid:end);
%power
p=polyfit(log10(x1), log10(y1), 1)
log10ratiounitcostfit=polyval(p,
  log10(x1));
```

```
ratiounitcostfit=10.
  ^log10ratiounitcostfit;
k=10^p(:,2)
R2power = 1 - sum((log10(y1) -
  log10ratiounitcostfit).^2)/sum
  ((log10(y1) - mean(log10(y1))).^2)
xmax=max(x1);
xmin=min(x1);
xaxis=[xmin:0.1:xmax];
log10ratiounitcostfitplot=polyval
  (p, log10(xaxis));
ratiounitcostfitplot=10.
  ^log10ratiounitcostfitplot;
%line
p2=polyfit((x2),(y2),1)
z=polyval(p2, (x2));
R2linear = 1 - sum(((y2) - z).^2)/sum(((y2)
  - mean((y2))).^2)
xmax2=max(x2);
xmin2=min(x2);
xaxis2=[xmin2:0.1:xmax2];
z2=polyval(p2, xaxis2);
%plot%
figure(1)
plot(Qrange/Q,unitcostrange
  /unitcost,'ko',
  'MarkerSize',3,'MarkerFaceColor',
  [0 0 0],'MarkerEdgeColor',
  [0 0 0],'linewidth',1);
hold on
line1=plot(xaxis,
  ratiounitcostfitplot,'k-
','linewidth',1);
hold on
line2=plot(xaxis2,z2,
  'k--','linewidth',1);
set(gca,'xtickmode',
  'auto','FontSize',10,
  'FontWeight','bold','FontName',
  'Times New Roman');
legend([line1,line2],'Power
  trendline','Linear trendline',
  'Location','Northeast');
xlabel({'Q/Q_0'},
  'FontSize',14,'FontWeight',
  'bold','FontName', 'Times New Roman')
ylabel({'Unit Cost/Unit
  Cost_0'},'FontSize',14,
  'FontWeight','bold','FontName',
  'Times New Roman')
grid on
hold on
```

Answer: `OxCATitle22CapitalCost`

```
p = -0.5417  -0.0172; k = 0.9613; R2power
  = 0.9679
p2 = -0.0021 0.4099; R2linear = 0.2510
```

10.4.4 Aerobic SBR

Example 10.7
Given: Table 10.16 shows capital cost and O&M cost of Aqua Aerobics SBR at different flow rates.
Find:

1) Calculate unit capital and O&M costs.
2) Plot unit cost/unit cost$_0$ versus Q/Q_0.
3) Determine correlation equations between unit cost/unit cost$_0$ for capital and O&M versus Q/Q_0, respectively.

Solution:

1) When Q = 0.5 MGD

$$\text{Unit capital cost} = 3150365\$ \div 0.5\,\text{MGD} \times 1\,\text{million gal}/10^3\,\text{kgal} \div 1\,\text{day} = 6930.80\$/\text{kgal}$$

$$\text{Unit O\&M costs} = 157518\$/\text{year} \div 0.5\,\text{MGD} \times 1\,\text{million gal}/10^3\,\text{kgal} \div 365\,\text{day/year}$$

$$\text{Unit O\&M costs} = 0.8631\$/\text{kgal}$$

When Q = 0.5 MGD

$$Q_0 = 0.5\,\text{MGD}$$

$$\text{Unit capital cost}_0 = 6930.80\$/\text{kgal}$$

$$\text{Unit O\&M costs}_0 = 0.8631\$/\text{kgal}$$

$$\frac{Q}{Q_0} = 1$$

$$\frac{\text{Unit capital cost}}{\text{Unit capital cost}_0} = 1$$

$$\frac{\text{Unit O\&M costs}}{\text{Unit O\&M costs}_0} = 1$$

When Q = 1 MGD (Tables 10.17 and 10.18)

$$\frac{Q}{Q_0} = 2$$

Table 10.16 Aqua Aerobics SBR (CA Title 22).

	Cost per amount of flow (MGD)						
Expense ($)	**0.5**	**1.0**	**2.0**	**5.0**	**10**	**15**	**20**
Concrete	856 000	1 524 800	1 680 000	2 872 800	3 537 600	4 330 400	5 434 400
Excavation and backfill (10%)	85 600	152 480	168 000	287 280	353 760	433 040	543 440
Miscellaneous metals (4%)	34 240	60 992	67 200	114 912	141 504	173 216	217 376
Yard piping (7%)	59 920	106 736	117 600	201 096	247 632	303 128	380 408
Total concrete	1 035 760	1 845 008	2 032 800	3 476 088	4 280 496	5 239 784	6 575 624
Equipment[a]	500 000	625 000	875 000	1 400 000	1 900 000	2 400 000	2 650 000
Tax and delivery (11%)	55 000	68 750	96 250	154 000	209 000	264 000	291 500
Installation (20%)	100 000	125 000	175 000	280 000	380 000	480 000	530 000
Manufacturer services (4%)	20 000	25 000	35 000	56 000	76 000	96 000	106 000
Total mechanical	675 000	843 750	1 181 250	1 890 000	2 565 000	3 240 000	3 577 500
Protective coating (7%)	119 753	188 213	224 984	375 626	479 185	593 585	710 719
Electricity (10%)	171 076	268 876	321 405	536 609	684 550	847 978	1 015 312
Instrumentation (10%)	171 076	268 876	321 405	536 609	684 550	847 978	1 015 312
Housing	0	0	0	0	0	0	0
Subtotal	2 172 665	3 414 723	4 081 844	6 814 932	8 693 780	10 769 326	12 894 467
Contractor overhead and profit (15%)	325 900	512 208	612 277	1 022 240	1 304 067	1 615 399	1 934 170
Scope-of-project contingency (30%)	651 800	1 024 417	1 224 553	2 044 480	2 608 134	3 230 798	3 868 340
Total construction cost	3 150 365	4 951 348	5 918 673	9 881 651	12 605 981	15 615 522	18 696 978
Engineering design (10%)	315 036	495 135	591 867	988 165	1 260 598	1 561 552	1 869 698
Total capital cost	3 465 401	5 446 483	6 510 540	10 869 816	13 866 579	17 177 074	20 566 676
Cost per gal ($)	6.93	5.45	3.26	2.17	1.39	1.15	1.03
O&M (5% construction cost)	157 518	247 567	295 934	494 083	630 299	780 776	934 849

[a] Equipment costs provided by Aqua Aerobics.

Table 10.17 Unit capital cost and unit operation and maintenance cost of Aqua Aerobics SBR (CA Title 22) at different flow rates.

Q (MGD)	Unit capital cost ($/kgal)	Unit O&M costs ($/kgal)
0.5	6930.80	0.8631
1	5446.48	0.6783
2	3255.27	0.4054
5	2173.96	0.2707
10	1386.66	0.1727
15	1145.14	0.1426
20	1028.33	0.1281

Table 10.18 Unit cost ratio of Aqua Aerobics SBR (CA Title 22) at different flow ratios.

Q/Q_0	Unit capital cost/unit capital $cost_0$	Unit O&M costs/unit O&M $costs_0$
1	1.0000	1.0000
2	0.7858	0.7858
4	0.4697	0.4697
10	0.3137	0.3137
20	0.2001	0.2001
30	0.1652	0.1652
40	0.1484	0.1484

$$\frac{\text{Unit capital cost}}{\text{Unit capital cost}_0} = 0.7858$$

$$\frac{\text{Unit O\&M costs}}{\text{Unit O\&M costs}_0} = 0.7858$$

2) Plot unit cost/unit cost$_0$ versus Q/Q_0 for capital and O&M, respectively (Figure 10.10).

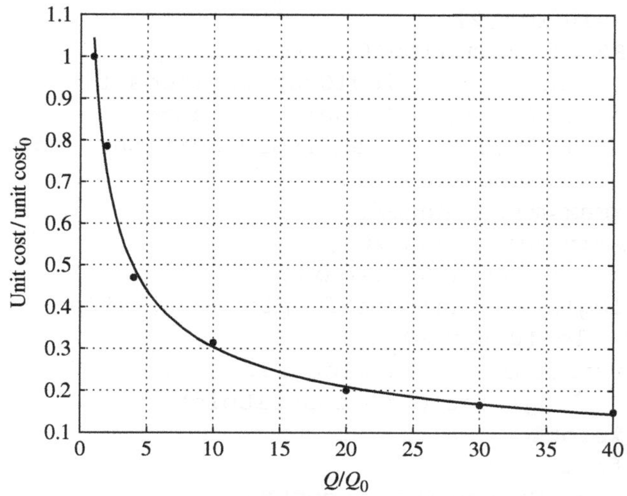

Figure 10.10 Unit capital cost/unit capital cost$_0$ versus Q/Q_0 for Aqua Aerobics SBR (CA Title 22), unit capital cost$_0$ = 6930.80\$/kgal, Q_0 = 0.5 MGD.

3) a) Equation for unit cost/unit cost$_0$ versus Q/Q_0 for capital:

The regression equation of the trend line:

$$\log\left(\frac{\text{unit cost}}{\text{unit cost}_0}\right) = -0.5372\log\left(\frac{Q}{Q_0}\right) + 0.0185$$

$$r^2 = 0.9952$$

or the equation of the regression (Figure 10.11):

$$\frac{\text{Unit cost}}{\text{Unit cost}_0} = 1.0435\left(\frac{Q}{Q_0}\right)^{-0.5372} \quad r^2 = 0.9952$$

Matlab code:

```
clear all
close all
%data%
Qrange=[0.5 1 2 5 10 15 20];
```

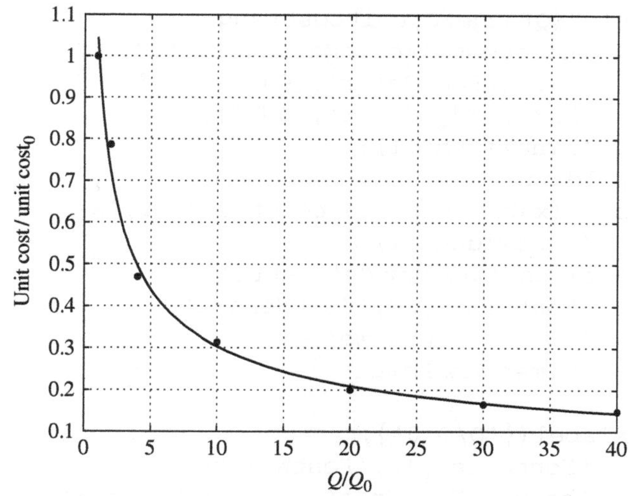

Figure 10.11 Unit O&M cost/unit O&M cost$_0$ versus Q/Q_0 for Aqua Aerobics SBR (CA Title 22), unit O&M cost$_0$ = 0.8631\$/kgal, Q_0 = 0.5 MGD.

```
Q=Qrange(1);
totalcostrange=[3465401 5446483 6510540
   10869816  13866579  17177074
   20566676];
totalcost=totalcostrange(1);
unitcost=totalcost./Q./1000;
unitcostrange=totalcostrange./
   Qrange./1000;
p=polyfit(log10(Qrange/Q), log10
   (unitcostrange /unitcost), 1)
log10ratiounitcostfit=polyval(p, log10
   (Qrange/Q));
ratiounitcostfit=10.
   ^log10ratiounitcostfit;
k=10^p(:,2)
R2 = 1 - sum((log10(unitcostrange
   /unitcost) - log10ratiounitcostfit).
   ^2)/sum((log10(unitcostrange
   /unitcost) - mean(log10(unitcostrange
   /unitcost))).^2)
xmax=max(Qrange/Q);
xmin=min(Qrange/Q);
xaxis=[xmin:0.1:xmax];
log10ratiounitcostfitplot=polyval
   (p, log10(xaxis));
ratiounitcostfitplot=10.
   ^log10ratiounitcostfitplot;
%plot%
figure(1)
```

```
plot(Qrange/Q,unitcostrange
   /unitcost,'ko','MarkerSize',3,
   'MarkerFaceColor',[0 0 0],
   'MarkerEdgeColor',[0 0 0],
   'linewidth',1);
hold on
plot(xaxis,ratiounitcostfitplot,'k-',
   'linewidth',1);
set(gca,'xtickmode','auto',
   'FontSize',10,'FontWeight',
   'bold','FontName',
   'Times New Roman');

xlabel({'Q/Q_0'},
   'FontSize',14,'FontWeight',
   'bold','FontName', 'Times New Roman')
ylabel({'Unit Cost/Unit Cost_0'},
   'FontSize',14,'FontWeight','bold',
   'FontName', 'Times New Roman')
grid on
hold on
```

Answer: AquaAerobicsSBRCATitle22Capitalcost

p = -0.5372 0.0185; k = 1.0435;
R^2 = 0.9952

b) Equation for unit cost/unit cost$_0$ versus Q/Q_0 for O&M:

The regression equation of the trend line:

$$\log\left(\frac{\text{unit cost}}{\text{unit cost}_0}\right) = -0.5372\log\left(\frac{Q}{Q_0}\right) + 0.0185$$
$$r^2 = 0.9952$$

or the equation of the regression:

$$\frac{\text{Unit cost}}{\text{Unit cost}_0} = 1.0435\left(\frac{Q}{Q_0}\right)^{-0.5372} \quad r^2 = 0.9952$$

Matlab code:

```
clear all
close all
%data%
Qrange=[0.5 1 2 5 10 15 20];
Q=Qrange(1);
totalcostrange=[157518 247567 295934
   494083 630299 780776 934849];
totalcost=totalcostrange(1);
unitcost=totalcost./Q./1000./365;
unitcostrange=totalcostrange./Qrange./
   1000./365;
p=polyfit(log10(Qrange/Q), log10
   (unitcostrange /unitcost), 1)
log10ratiounitcostfit=polyval(p, log10
   (Qrange/Q));
ratiounitcostfit=10.
   ^log10ratiounitcostfit;
k=10^p(:,2)
R2 = 1 - sum((log10(unitcostrange
   /unitcost) - log10ratiounitcostfit).
   ^2)/sum((log10(unitcostrange
   /unitcost) - mean(log10(unitcostrange
   /unitcost))).^2)
xmax=max(Qrange/Q);
xmin=min(Qrange/Q);
xaxis=[xmin:0.1:xmax];
log10ratiounitcostfitplot=polyval(p,
   log10(xaxis));
ratiounitcostfitplot=10.
   ^log10ratiounitcostfitplot;
%plot%
figure(1)
plot(Qrange/Q,unitcostrange
   /unitcost,'ko','MarkerSize',3,
   'MarkerFaceColor',[0 0 0],
   'MarkerEdgeColor',[0 0 0],
   'linewidth',1);
hold on
plot(xaxis,ratiounitcostfitplot,'k-',
   'linewidth',1);
set(gca,'xtickmode',
   'auto','FontSize',10,
   'FontWeight','bold','FontName',
   'Times New Roman');

xlabel({'Q/Q_0'},'FontSize',14,
   'FontWeight','bold','FontName',
   'Times New Roman')
ylabel({'Unit Cost/Unit
   Cost_0'},'FontSize',14,
   'FontWeight','bold','FontName',
   'Times New Roman')
grid on
hold on
```

Answer: AquaAerobicsSBRCATitle22OMcost

p = -0.5372 0.0185; k = 1.0435;
R^2 = 0.9952

10.4.5 MBR

Example 10.8
Given: Table 10.19 shows capital cost and O&M cost of Aqua Aerobics MBR (CA Title 22) at different flow rates:
Find:

1) Calculate unit capital and O&M costs.
2) Plot unit cost/unit cost$_0$ versus Q/Q_0.
3) Determine correlation equations between unit cost/unit cost$_0$ for capital and O&M versus Q/Q_0, respectively.

Solution:

1) When Q = 0.5 MGD

$$\text{Unit capital cost} = 4230740\$ \div 0.5\,\text{MGD} \times 1\,\text{million gal}/10^3\,\text{kgal} \div 1\,\text{day} = 8461.48\$/\text{kgal}$$

$$\text{Unit O\&M costs} = 192306\$/\text{year} \div 0.5\,\text{MGD} \times 1\,\text{million gal}/10^3\,\text{kgal} \div 365\,\text{day/year}$$

$$\text{Unit O\&M costs} = 1.0537\$/\text{kgal}$$

When Q = 0.5 MGD

$$Q_0 = 0.5\,\text{MGD}$$

$$\text{Unit capital cost}_0 = 8461.48\$/\text{kgal}$$

$$\text{Unit O\&M costs}_0 = 1.0537\$/\text{kgal}$$

$$\frac{Q}{Q_0} = 1$$

$$\frac{\text{Unit capital cost}}{\text{Unit capital cost}_0} = 1$$

Table 10.19 Aqua Aerobics MBR (CA Title 22).

	Cost per amount of flow (MGD)						
Expense ($)	**0.5**	**1.0**	**2.0**	**5.0**	**10**	**15**	**20**
Concrete	610 400	1 062 400	2 012 000	4 722 400	9 684 800	13 448 800	19 054 400
Excavation and backfill (10%)	61 040	106 240	201 200	472 240	968 480	1 344 880	1 905 440
Miscellaneous metals (4%)	24 416	42 496	80 480	188 896	387 392	537 952	762 176
Yard piping (7%)	42 728	74 368	140 840	330 568	677 936	941 416	1 333 808
Total concrete	738 584	1 285 504	2 434 520	5 714 104	11 718 608	16 273 048	23 055 824
Equipment[a]	1 000 000	2 000 000	2 900 000	8 000 000	14 500 000	20 500 000	26 000 000
Tax and delivery (11%)	110 000	220 000	319 000	880 000	1 595 000	2 255 000	2 860 000
Installation (20%)	200 000	400 000	580 000	1 600 000	2 900 000	4 100 000	5 200 000
Manufacturer services (4%)	40 000	80 000	116 000	320 000	580 000	820 000	1 040 000
Total mechanical	1 350 000	2 700 000	3 915 000	10 800 000	19 575 000	27 675 000	35 100 000
Protective coating (7%)	146 201	278 985	444 466	1 155 987	2 190 553	3 076 363	4 070 908
Electricity (10%)	208 858	398 550	634 952	1 651 410	3 129 361	4 394 805	5 815 582
Instrumentation (10%)	208 858	398 550	634 952	1 651 410	3 129 361	4 394 805	5 815 582
Housing	0	0	0	0	0	0	0
Subtotal	2 652 502	5 061 590	8 063 890	20 972 912	39 742 882	55 814 021	73 857 896
Contractor overhead and profit (15%)	397 875	759 239	1 209 584	3 145 937	5 961 432	8 372 103	11 078 684
Scope-of-project contingency (30%)	795 751	1 518 477	2 419 167	6 291 874	11 922 865	16 744 206	22 157 369
Total construction cost	3 846 127	7 339 306	11 692 641	30 410 723	57 627 179	80 930 330	107 093 950
Engineering design (10%)	384 613	733 931	1 169 264	3 041 072	5 762 718	8 093 033	10 709 395
Total capital cost	4 230 740	8 073 236	12 861 905	33 451 795	63 389 897	89 023 363	117 803 345
Cost per gal ($)	8.46	8.07	6.43	6.69	6.34	5.93	5.89
O&M (5% construction cost)	192 306	366 965	584 632	1 520 536	2 881 359	4 046 517	5 354 697

[a] Equipment costs provided by Aqua Aerobics.

$$\frac{\text{Unit O\&M costs}}{\text{Unit O\&M costs}_0} = 1$$

When $Q = 1$ MGD (Tables 10.20 and 10.21)

$$\frac{Q}{Q_0} = 2$$

$$\frac{\text{Unit capital cost}}{\text{Unit capital cost}_0} = 0.9541$$

$$\frac{\text{Unit O\&M costs}}{\text{Unit O\&M costs}_0} = 0.9541$$

2) Plot unit cost/unit cost$_0$ versus Q/Q_0 for capital and O&M, respectively (Figure 10.12).
3) a) Equation for unit cost/unit cost$_0$ versus Q/Q_0 for capital:
   The regression equation of the trend line ($1 \leq Q/Q_0 \leq 4$):

$$\log\left(\frac{\text{unit cost}}{\text{unit cost}_0}\right) = -0.1979\log\left(\frac{Q}{Q_0}\right) - 0.0131$$

$$r^2 = 0.8740$$

Table 10.20 Unit capital cost and unit operation and maintenance cost of Aqua Aerobics MBR (CA Title 22) at different flow rates.

Q (MGD)	Unit capital cost ($/kgal)	Unit O&M costs ($/kgal)
0.5	8461.48	1.0537
1	8073.24	1.0054
2	6430.95	0.8009
5	6690.36	0.8332
10	6338.99	0.7894
15	5934.89	0.7391
20	5890.17	0.7335

Table 10.21 Unit cost ratio of Aqua Aerobics MBR (CA Title 22) at different flow ratios.

Q/Q_0	Unit capital cost/unit capital cost$_0$	Unit O&M costs/unit O&M costs$_0$
1	1.0000	1.0000
2	0.9541	0.9541
4	0.7600	0.7600
10	0.7907	0.7907
20	0.7492	0.7492
30	0.7014	0.7014
40	0.6961	0.6961

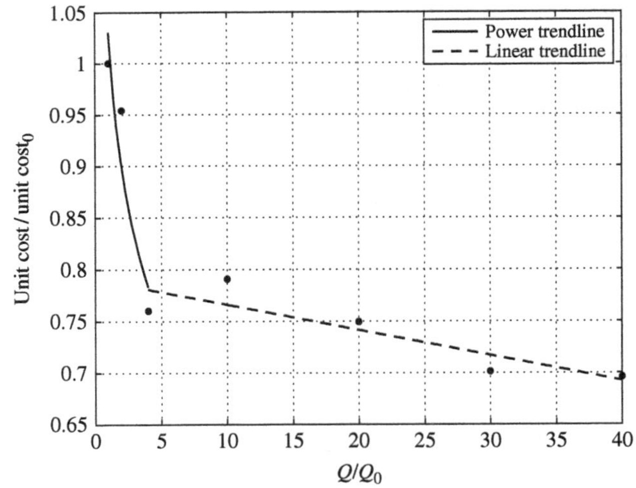

Figure 10.12 Unit capital cost/unit capital cost$_0$ versus Q/Q_0 for Aqua Aerobics MBR (CA Title 22), unit capital cost$_0$ = 8461.48 $/kgal, Q_0 = 0.5 MGD.

or the equation of the regression:

$$\frac{\text{Unit cost}}{\text{unit cost}_0} = 1.0305\left(\frac{Q}{Q_0}\right)^{-0.1979} \quad r^2 = 0.8740$$

The regression equation of the trend line ($4 \leq Q/Q_0 \leq 40$) (Figure 10.13):

$$\frac{\text{Unit cost}}{\text{Unit cost}_0} = -0.0024Q/Q_0 + 0.7904 \quad r^2 = 0.7910$$

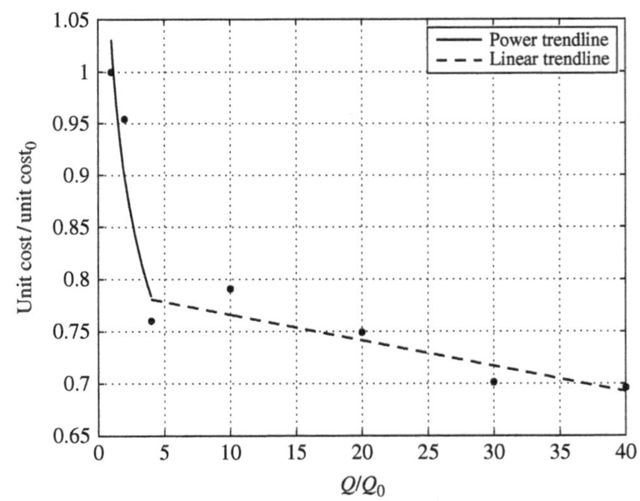

Figure 10.13 Unit O&M cost/unit O&M cost$_0$ versus Q/Q_0 for Aqua Aerobics MBR (CA Title 22), unit O&M cost$_0$ = 1.0537$/kgal, Q_0 = 0.5 MGD.

Matlab code:

```
clear all
close all
%data%
Qrange=[0.5 1 2 5 10 15 20];
Q=Qrange(1);
totalcostrange=[4230740 8073236 12861905
   33451795   63389897   89023363
   117803345];
totalcost=totalcostrange(1);
xmid=3;
%function%
unitcost=totalcost./Q./10^3;
unitcostrange=totalcostrange./
   Qrange./10^3;
x=Qrange/Q;
y=unitcostrange /unitcost;
x1=x(1:xmid);
y1=y(1:xmid);
x2=x(xmid:end);
y2=y(xmid:end);
%power
p=polyfit(log10(x1), log10(y1), 1)
log10ratiounitcostfit=polyval
   (p, log10(x1));
ratiounitcostfit=10.
   ^log10ratiounitcostfit;
k=10^p(:,2)
R2power = 1 - sum((log10(y1) -
   log10ratiounitcostfit).^2)/sum((log10
   (y1) - mean(log10(y1))).^2)
xmax=max(x1);
xmin=min(x1);
xaxis=[xmin:0.1:xmax];
log10ratiounitcostfitplot=polyval(p,
   log10(xaxis));
ratiounitcostfitplot=10.
   ^log10ratiounitcostfitplot;
%line
p2=polyfit((x2),(y2), 1)
z=polyval(p2, (x2));
R2linear = 1 - sum(((y2) - z).^2)/sum(((y2)
   - mean((y2))).^2)
xmax2=max(x2);
xmin2=min(x2);
xaxis2=[xmin2:0.1:xmax2];
z2=polyval(p2, xaxis2);
%plot%
figure(1)
plot(Qrange/Q,unitcostrange
   /unitcost,'ko',
   'MarkerSize',3,
   'MarkerFaceColor',
   [0 0 0],'MarkerEdgeColor',
   [0 0 0],'linewidth',1);
hold on
line1=plot(xaxis,ratiounitcostfitplot,
   'k-','linewidth',1);
hold on
line2=plot(xaxis2,z2,
   'k--','linewidth',1);
set(gca,'xtickmode','auto',
   'FontSize',10,
   'FontWeight','bold','FontName',
   'Times New Roman');
legend([line1,line2],'Power
   trendline','Linear trendline',
   'Location','Northeast');
xlabel({'Q/Q_0'},
   'FontSize',14,
   'FontWeight','bold','FontName',
   'Times New Roman')
ylabel({'Unit Cost/Unit
   Cost_0'},'FontSize',14,
   'FontWeight','bold','FontName',
   'Times New Roman')
grid on
hold on
```

Answer: AquaAerobicsMBRCATitle22 Capitalcost

p = -0.1979 0.0131; k = 1.0305;
R^2power = 0.8740
p2 = -0.0024 0.7904; R^2linear = 0.7910

b) Equation for unit cost/unit $cost_0$ versus Q/Q_0 for O&M:

The regression equation of the trend line ($1 \le Q/Q_0 \le 4$):

$$\log\left(\frac{\text{unit cost}}{\text{unit cost}_0}\right) = -0.1979\log\left(\frac{Q}{Q_0}\right) - 0.0131$$
$$r^2 = 0.8740$$

or the equation of the regression:

$$\frac{\text{Unit cost}}{\text{Unit cost}_0} = 1.0305\left(\frac{Q}{Q_0}\right)^{-0.1979} \quad r^2 = 0.8740$$

The regression equation of the trend line ($4 \le Q/Q_0 \le 40$):

$$\frac{\text{Unit cost}}{\text{Unit cost}_0} = -0.0024Q/Q_0 + 0.7904 \quad r^2 = 0.7910$$

Matlab code:

```
clear all
close all
%data%
Qrange=[0.5 1 2 5 10 15 20];
Q=Qrange(1);
totalcostrange=[192306 366965 584632
  1520536 2881359 4046517 5354697];
totalcost=totalcostrange(1);
xmid=3;
%function%
unitcost=totalcost./Q./10^3./365;
unitcostrange=totalcostrange./Qrange./
  10^3./365;
x=Qrange/Q;
y=unitcostrange /unitcost;
x1=x(1:xmid);
y1=y(1:xmid);
x2=x(xmid:end);
y2=y(xmid:end);
%power
p=polyfit(log10(x1), log10(y1), 1)
log10ratiounitcostfit=polyval(p,
  log10(x1));
ratiounitcostfit=10.
  ^log10ratiounitcostfit;
k=10^p(:,2)
R2power = 1 - sum((log10(y1) -
  log10ratiounitcostfit).^2)/sum
  ((log10(y1) - mean(log10(y1))).^2)
xmax=max(x1);
xmin=min(x1);
xaxis=[xmin:0.1:xmax];
log10ratiounitcostfitplot=polyval
  (p, log10(xaxis));
ratiounitcostfitplot=10.
  ^log10ratiounitcostfitplot;
%line
p2=polyfit((x2),(y2), 1)
z=polyval(p2, (x2));
R2linear = 1 - sum(((y2) - z).^2)/sum
  (((y2) - mean((y2))).^2)
xmax2=max(x2);
xmin2=min(x2);
xaxis2=[xmin2:0.1:xmax2];
z2=polyval(p2, xaxis2);
%plot%
figure(1)
plot(Qrange/Q,unitcostrange
  /unitcost,'ko',
  'MarkerSize',3,'MarkerFaceColor',
  [0 0 0],'MarkerEdgeColor',
  [0 0 0],'linewidth',1);
hold on
line1=plot(xaxis,
  ratiounitcostfitplot,'k-',
  'linewidth',1);
hold on
line2=plot(xaxis2,z2,'k--',
  'linewidth',1);
set(gca,'xtickmode',
  'auto','FontSize',10,
  'FontWeight','bold','FontName',
  'Times New Roman');
legend([line1,line2],'Power
  trendline','Linear trendline',
  'Location','Northeast');
xlabel({'Q/Q_0'},
  'FontSize',14,'FontWeight',
  'bold','FontName', 'Times New Roman')
ylabel({'Unit Cost/Unit
  Cost_0'},'FontSize',14,
  'FontWeight','bold','FontName',
  'Times New Roman')
grid on
hold on
```

Answer: `AquaAerobicsMBRCATitle22OMcost`

p = -0.1979 0.0131; k = 1.0305;
R^2power = 0.8740
p2 = -0.0024 0.7904; R^2linear = 0.7910

10.4.6 Microfiltration

Example 10.9
Given: Table 10.22 shows the capital cost and O&M cost of MF at different flow rates.
Find:

1) Calculate unit capital and O&M costs.
2) Plot unit cost/unit cost$_0$ versus Q/Q_0.
3) Determine correlation equations between unit cost/unit cost$_0$ for capital and O&M versus Q/Q_0, respectively.

Solution:

1) When Q = 0.5 MGD

$$\text{Unit capital cost} = 3230797\$ \div 0.5\,\text{MGD} \times 1\,\text{million gal}/10^3\,\text{kgal} \div 1\,\text{day} = 6461.59\$/\text{kgal}$$

Table 10.22 Annual cost of microfiltration.

Expense ($)	Cost per amount of flow (MGD)						
	0.5	1.0	2.0	5.0	10	15	20
Concrete	0	0	0	0	0	0	0
Excavation and backfill (10%)	0	0	0	0	0	0	0
Miscellaneous metals (4%)	0	0	0	0	0	0	0
Yard piping (7%)	0	0	0	0	0	0	0
Total concrete	0	0	0	0	0	0	0
Equipment	1 079 369	1 641 697	2 496 986	4 346 802	6 611 393	8 449 445	10 055 789
Tax and delivery (11%)	118 731	180 587	274 668	478 148	727 253	929 439	1 106 137
Installation (20%)	215 874	328 339	499 397	869 360	1 322 279	1 689 889	2 011 158
Manufacturer services (4%)	43 175	65 668	99 879	173 872	264 456	337 978	402 232
Total mechanical	1 457 148	2 216 291	3 370 931	5 868 183	8 925 381	11 406 751	13 575 315
Protective coating (7%)	102 000	155 140	235 965	410 773	624 777	798 473	950 272
Electricity (10%)	145 715	221 629	337 093	586 818	892 538	1 140 675	1 357 532
Instrumentation (10%)	145 715	221 629	337 093	586 818	892 538	1 140 675	1 357 532
Housing[a]	175 000	350 000	700 000	1 750 000	3 500 000	5 250 000	7 000 000
Subtotal	2 025 578	3 164 690	4 981 082	9 202 592	14 835 233	19 736 573	24 240 650
Contractor overhead and profit (15%)	303 837	474 703	747 162	1 380 389	2 225 285	2 960 486	3 636 098
Scope-of-project contingency (30%)	607 673	949 407	1 494 325	2 760 778	4 450 570	5 920 972	7 272 195
Total construction cost	2 937 088	4 588 800	7 222 570	13 343 758	21 511 088	28 618 032	35 148 943
Engineering design (10%)	293 709	458 880	722 257	1 334 376	2 151 109	2 861 803	3 514 894
Total capital cost	3 230 797	5 047 680	7 944 827	14 678 134	23 662 197	31 479 835	38 663 837
Cost per gal ($)	6.46	5.05	3.97	2.94	2.37	2.10	1.93
O&M (5% construction cost)	146 854	229 440	361 128	667 188	1 075 554	1 430 902	1 757 447

[a] Housing footprint based on general, unpublished rules of thumb. Assumed $250/ft^2.

$$\text{Unit O\&M costs} = 146854\$/\text{year} \div 0.5\,\text{MGD} \times 1\,\text{million gal}/10^3\,\text{kgal} \div 365\,\text{day/year}$$

$$\text{Unit O\&M costs} = 0.8047\$/\text{kgal}$$

When Q = 0.5 MGD

$$Q_0 = 0.5\,\text{MGD}$$

$$\text{Unit capital cost}_0 = 6461.59\$/\text{kgal}$$

$$\text{Unit O\&M costs}_0 = 0.8047\$/\text{kgal}$$

$$\frac{Q}{Q_0} = 1$$

$$\frac{\text{Unit capital cost}}{\text{Unit capital cost}_0} = 1$$

$$\frac{\text{Unit O\&M costs}}{\text{Unit O\&M costs}_0} = 1$$

When Q = 1 MGD (Tables 10.23 and 10.24)

$$\frac{Q}{Q_0} = 2$$

Table 10.23 Unit capital cost and unit operation and maintenance cost of MF at different flow rates.

Q (MGD)	Unit capital cost ($/kgal)	Unit O&M costs ($/kgal)
0.5	6461.59	0.8047
1	5047.68	0.6286
2	3972.41	0.4947
5	2935.63	0.3656
10	2366.22	0.2947
15	2098.66	0.2614
20	1933.19	0.2407

Table 10.24 Unit cost ratio of MF at different flow ratios.

Q/Q_0	Unit capital cost/unit capital cost$_0$	Unit O&M costs/unit O&M costs$_0$
1	1.0000	1.0000
2	0.7812	0.7812
4	0.6148	0.6148
10	0.4543	0.4543
20	0.3662	0.3662
30	0.3248	0.3248
40	0.2992	0.2992

$$\frac{\text{Unit capital cost}}{\text{Unit capital cost}_0} = 0.7812$$

$$\frac{\text{Unit O\&M costs}}{\text{Unit O\&M costs}_0} = 0.7812$$

2) Plot unit cost/unit cost$_0$ versus Q/Q_0 for capital and O&M, respectively (Figure 10.14).
3) a) Equation for unit cost/unit cost$_0$ versus Q/Q_0 for capital:
   The regression equation of the trend line:

$$\log\left(\frac{\text{unit cost}}{\text{unit cost}_0}\right) = -0.3269\log\left(\frac{Q}{Q_0}\right) - 0.0080$$

$$r^2 = 0.9990$$

or the equation of the regression (Figure 10.15):

$$\frac{\text{Unit cost}}{\text{Unit cost}_0} = 0.9817\left(\frac{Q}{Q_0}\right)^{-0.3269} \qquad r^2 = 0.9990$$

Matlab code:

```
clear all
close all
%data%
Qrange=[0.5 1 2 5 10 15 20];
Q=Qrange(1);
totalcostrange=[3230797 5047680 7944827
   14678134  23662197  31479835
   38663837];
totalcost=totalcostrange(1);
unitcost=totalcost./Q./1000;
unitcostrange=totalcostrange./
   Qrange./1000;
p=polyfit(log10(Qrange/Q), log10
   (unitcostrange /unitcost), 1)
log10ratiounitcostfit=polyval(p, log10
   (Qrange/Q));
ratiounitcostfit=10.
   ^log10ratiounitcostfit;
k=10^p(:,2)
R2 = 1 - sum((log10(unitcostrange
   /unitcost) - log10ratiounitcostfit).
   ^2)/sum((log10(unitcostrange
   /unitcost) - mean(log10(unitcostrange
   /unitcost))).^2)
xmax=max(Qrange/Q);
xmin=min(Qrange/Q);
xaxis=[xmin:0.1:xmax];
```

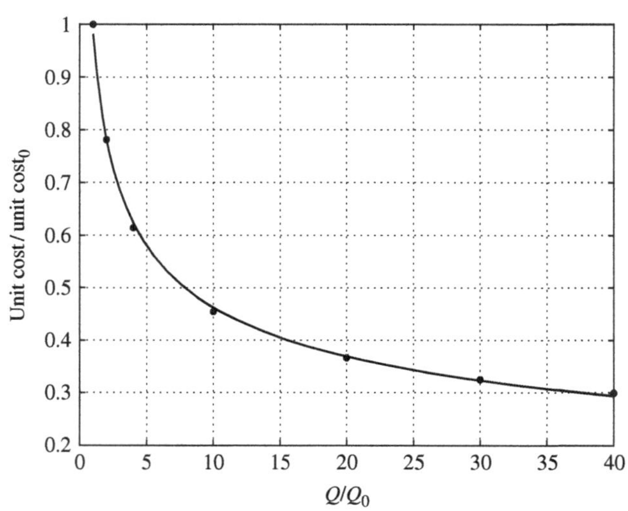

Figure 10.14 Unit capital cost/unit capital cost$_0$ versus Q/Q_0 for MF, unit capital cost$_0$ = 6461.59$/kgal, Q_0 = 0.5 MGD.

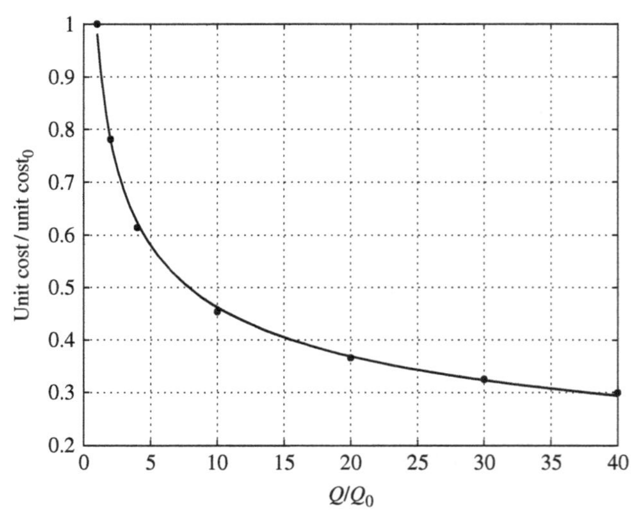

Figure 10.15 Unit O&M cost/unit O&M cost$_0$ versus Q/Q_0 for MF, unit O&M cost$_0$ = 0.8047$/kgal, Q_0 = 0.5 MGD.

```
log10ratiounitcostfitplot=polyval(p,
  log10(xaxis));
ratiounitcostfitplot=10.
  ^log10ratiounitcostfitplot;
%plot%
figure(1)
plot(Qrange/Q,unitcostrange
  /unitcost,'ko','MarkerSize',3,
  'MarkerFaceColor',
  [0 0 0],'MarkerEdgeColor',
  [0 0 0],'linewidth',1);
hold on
plot(xaxis,ratiounitcostfitplot,'k-',
  'linewidth',1);
set(gca,'xtickmode','auto',
  'FontSize',10,'FontWeight','bold',
  'FontName', 'Times New Roman');

xlabel({'Q/Q_0'},
  'FontSize',14,'FontWeight',
  'bold','FontName', 'Times New Roman')
ylabel({'Unit Cost/Unit Cost_0'},
  'FontSize',14,'FontWeight',
  'bold','FontName', 'Times New Roman')
grid on
hold on
```

Answer: MFCapitalcost

p = -0.3269 -0.0080; k = 0.9817;
R^2 = 0.9990

b) Equation for unit cost/unit cost$_0$ versus Q/Q_0 for O&M:

The regression equation of the trend line:

$$\log\left(\frac{\text{unit cost}}{\text{unit cost}_0}\right) = -0.3269\log\left(\frac{Q}{Q_0}\right) - 0.0080$$

$$r^2 = 0.9990$$

or the equation of the regression:

$$\frac{\text{Unit cost}}{\text{Unit cost}_0} = 0.9817\left(\frac{Q}{Q_0}\right)^{-0.3269} \quad r^2 = 0.9990$$

Matlab code:

```
clear all
close all
%data%
Qrange=[0.5 1 2 5 10 15 20];
Q=Qrange(1);
totalcostrange=[146854 229440 361128
  667188 1075554 1430902 1757447];
totalcost=totalcostrange(1);
unitcost=totalcost./Q./1000./365;
unitcostrange=totalcostrange./Qrange./
  1000./365;
p=polyfit(log10(Qrange/Q), log10
  (unitcostrange/unitcost), 1)
log10ratiounitcostfit=polyval(p, log10
  (Qrange/Q));
ratiounitcostfit=10.
  ^log10ratiounitcostfit;
k=10^p(:,2)
R2 = 1 - sum((log10(unitcostrange
  /unitcost) - log10ratiounitcostfit).
  ^2)/sum((log10(unitcostrange
  /unitcost) - mean(log10(unitcostrange
  /unitcost))).^2)
xmax=max(Qrange/Q);
xmin=min(Qrange/Q);
xaxis=[xmin:0.1:xmax];
log10ratiounitcostfitplot=polyval(p,
  log10(xaxis));
ratiounitcostfitplot=10.
  ^log10ratiounitcostfitplot;
%plot%
figure(1)
plot(Qrange/Q,unitcostrange
  /unitcost,'ko','MarkerSize',3,
  'MarkerFaceColor',[0 0 0],
  'MarkerEdgeColor',[0 0 0],
  'linewidth',1);
hold on
plot(xaxis,ratiounitcostfitplot,'k-',
  'linewidth',1);
set(gca,'xtickmode','auto',
  'FontSize',10,'FontWeight','bold',
  'FontName', 'Times New Roman');

xlabel({'Q/Q_0'},'FontSize',14,
  'FontWeight','bold','FontName',
  'Times New Roman')
ylabel({'Unit Cost/Unit Cost_0'},
  'FontSize',14,'FontWeight',
  'bold','FontName', 'Times New Roman')
grid on
hold on
```

Answer: MFOMcost

p = -0.3269 -0.0080; k = 0.9817;
R^2 = 0.9990

10.4.7 Reverse Osmosis

Example 10.10

Given: Table 10.25 shows the capital cost and O&M cost of RO at different flow rates.

Find:

1) Calculate unit capital and O&M costs.
2) Plot unit cost/unit cost$_0$ versus Q/Q_0.
3) Determine correlation equations between unit cost/unit cost$_0$ for capital and O&M versus Q/Q_0, respectively.

Solution:

1) When Q = 0.5 MGD

$$\text{Unit capital cost} = 2440613\$ \div 0.5\text{MGD} \times 1\,\text{million gal}/10^3\,\text{kgal} \div 1\,\text{day} = 4881.23\$/\text{kgal}$$

$$\text{Unit O\&M costs} = 110937\$/\text{year} \div 0.5\text{MGD} \times 1\,\text{million gal}/10^3\,\text{kgal} \div 365\,\text{day/year}$$

$$\text{Unit O\&M costs} = 0.6079\$/\text{kgal}$$

When Q = 0.5 MGD

$$Q_0 = 0.5\text{MGD}$$

$$\text{Unit capital cost}_0 = 4881.23\$/\text{kgal}$$

$$\text{Unit O\&M costs}_0 = 0.6079\$/\text{kgal}$$

$$\frac{Q}{Q_0} = 1$$

$$\frac{\text{Unit capital cost}}{\text{Unit capital cost}_0} = 1$$

Table 10.25 Annual cost of reverse osmosis.

	Cost per amount of flow (MGD)						
Expense ($)	**0.5**	**1.0**	**2.0**	**5.0**	**10**	**15**	**20**
Concrete	0	0	0	0	0	0	0
Excavation and backfill (10%)	0	0	0	0	0	0	0
Miscellaneous metals (4%)	0	0	0	0	0	0	0
Yard piping (7%)	0	0	0	0	0	0	0
Total concrete	0	0	0	0	0	0	0
Equipment	770 000	1 110 000	1 610 000	2 610 000	3 780 000	4 680 000	5 450 000
Tax and delivery (11%)	84 700	122 100	177 100	287 100	415 800	514 800	599 500
Installation (20%)	154 000	222 000	322 000	522 000	756 000	936 000	1 090 000
Manufacturer services (4%)	30 800	44 400	64 400	104 400	151 200	187 200	218 000
Total mechanical	1 039 500	1 498 500	2 173 500	3 523 500	5 103 000	6 318 000	7 357 500
Protective coating (7%)	72 765	104 895	152 145	246 645	357 210	442 260	515 025
Electricity (10%)	103 950	149 850	217 350	352 350	510 300	631 800	735 750
Instrumentation (10%)	103 950	149 850	217 350	352 350	510 300	631 800	735 750
Housing[a]	210 000	420 000	840 000	2 100 000	4 200 000	6 300 000	8 400 000
Subtotal	1 530 165	2 323 095	3 600 345	6 574 845	10 680 810	14 323 860	17 744 025
Contractor overhead and profit (15%)	229 525	348 464	540 052	986 227	1 602 122	2 148 579	2 661 604
Scope-of-project contingency (30%)	459 050	696 929	1 080 104	1 972 454	3 204 243	4 297 158	5 323 208
Total construction cost	2 218 739	3 368 488	5 220 500	9 533 525	15 487 175	20 769 597	25 728 836
Engineering design (10%)	221 874	336 849	522 050	953 353	1 548 717	2 076 960	2 572 884
Total capital cost	2 440 613	3 705 337	5 742 550	10 486 878	17 035 892	22 846 557	28 301 720
Cost per gal ($)	4.88	3.71	2.87	2.10	1.70	1.52	1.42
O&M (5% construction cost)	110 937	168 424	261 025	476 676	774 359	1 038 480	1 286 442

[a] Housing footprint based on general, unpublished rules of thumb. Assumed $250/ft^2.

$$\frac{\text{Unit O\&M costs}}{\text{Unit O\&M costs}_0} = 1$$

When $Q = 1$ MGD (Tables 10.26 and 10.27)

$$\frac{Q}{Q_0} = 2$$

$$\frac{\text{Unit capital cost}}{\text{Unit capital cost}_0} = 0.7591$$

$$\frac{\text{Unit O\&M costs}}{\text{Unit O\&M costs}_0} = 0.7591$$

2) Plot unit cost/unit cost_0 versus Q/Q_0 for capital and O&M, respectively (Figure 10.16).
3) a) Equation for unit cost/unit cost_0 versus Q/Q_0 for capital:
   The regression equation of the trend line:

$$\log\left(\frac{\text{unit cost}}{\text{unit cost}_0}\right) = -0.3346\log\left(\frac{Q}{Q_0}\right) - 0.0164$$

$$r^2 = 0.9962$$

Table 10.26 Unit capital cost and unit operation and maintenance cost of RO at different flow rates.

Q (MGD)	Unit capital cost ($/kgal)	Unit O&M costs ($/kgal)
0.5	4881.23	0.6079
1	3705.34	0.4614
2	2871.28	0.3576
5	2097.38	0.2612
10	1703.59	0.2122
15	1523.10	0.1897
20	1415.09	0.1762

Table 10.27 Unit cost ratio of RO at different flow ratios.

Q/Q_0	Unit capital cost/unit capital cost_0	Unit O&M costs/unit O&M costs_0
1	1.0000	1.0000
2	0.7591	0.7591
4	0.5882	0.5882
10	0.4297	0.4297
20	0.3490	0.3490
30	0.3120	0.3120
40	0.2899	0.2899

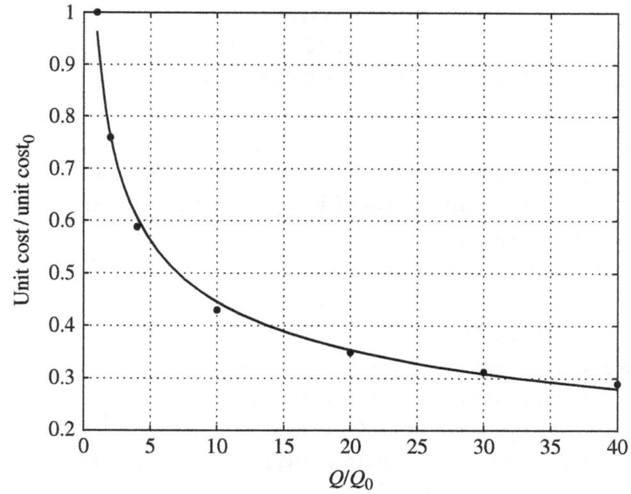

Figure 10.16 Unit capital cost/unit capital cost_0 versus Q/Q_0 for RO, unit capital cost_0 = 4881.23$/kgal, Q_0 = 0.5 MGD.

or the equation of the regression (Figure 10.17):

$$\frac{\text{Unit cost}}{\text{Unit cost}_0} = 0.9629\left(\frac{Q}{Q_0}\right)^{-0.3346} \quad r^2 = 0.9962$$

Matlab code:

```
clear all
close all
%data%
Qrange=[0.5 1 2 5 10 15 20];
Q=Qrange(1);
totalcostrange=[2440613 3705337 5742550
   10486878  17035892  22846557
   28301720];
```

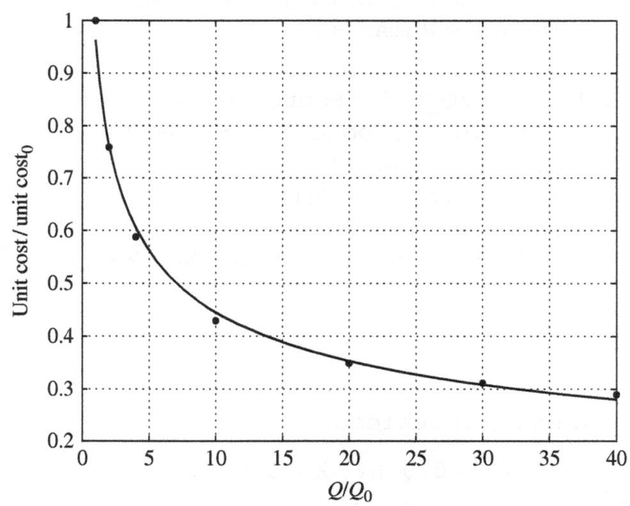

Figure 10.17 Unit O&M cost/unit O&M cost_0 versus Q/Q_0 for RO, unit O&M cost_0 = 0.6079$/kgal, Q_0 = 0.5 MGD.

```
totalcost=totalcostrange(1);
unitcost=totalcost./Q./1000;
unitcostrange=totalcostrange./
  Qrange./1000;
p=polyfit(log10(Qrange/Q), log10
  (unitcostrange /unitcost), 1)
log10ratiounitcostfit=polyval(p, log10
  (Qrange/Q));
ratiounitcostfit=10.
  ^log10ratiounitcostfit;
k=10^p(:,2)
R2 = 1 - sum((log10(unitcostrange
  /unitcost) - log10ratiounitcostfit).
  ^2)/sum((log10(unitcostrange
  /unitcost) - mean(log10(unitcostrange
  /unitcost))).^2)
xmax=max(Qrange/Q);
xmin=min(Qrange/Q);
xaxis=[xmin:0.1:xmax];
log10ratiounitcostfitplot
  =polyval(p, log10(xaxis));
ratiounitcostfitplot=10.
  ^log10ratiounitcostfitplot;
%plot%
figure(1)
plot(Qrange/Q,unitcostrange
  /unitcost,'ko','MarkerSize',3,
  'MarkerFaceColor',
  [0 0 0],'MarkerEdgeColor',
  [0 0 0],'linewidth',1);
hold on
plot(xaxis,ratiounitcostfitplot,'k-',
  'linewidth',1);
set(gca,'xtickmode',
  'auto','FontSize',10,
  'FontWeight','bold','FontName',
  'Times New Roman');

xlabel({'Q/Q_0'},'FontSize',14,
  'FontWeight','bold','FontName',
  'Times New Roman')
ylabel({'Unit Cost/Unit Cost_0'},
  'FontSize',14,'FontWeight',
  'bold','FontName', 'Times New Roman')
grid on
hold on
```

Answer: ROCapitalcost

p = -0.3346 -0.0164; k = 0.9629;
R^2 = 0.9962

b) Equation for unit cost/unit $cost_0$ versus Q/Q_0 for O&M:

The regression equation of the trend line:

$$\log\left(\frac{\text{unit cost}}{\text{unit cost}_0}\right) = -0.3346\log\left(\frac{Q}{Q_0}\right) - 0.0164$$

$$r^2 = 0.9962$$

or the equation of the regression:

$$\frac{\text{Unit cost}}{\text{Unit cost}_0} = 0.9629\left(\frac{Q}{Q_0}\right)^{-0.3346} \quad r^2 = 0.9962$$

Matlab code:

```
clear all
close all
%data%
Qrange=[0.5 1 2 5 10 15 20];
Q=Qrange(1);
totalcostrange=[110937 168424 261025
  476676 774359 1038480 1286442];
totalcost=totalcostrange(1);
unitcost=totalcost./Q./1000./365;
unitcostrange=totalcostrange./Qrange./
  1000./365;
p=polyfit(log10(Qrange/Q), log10
  (unitcostrange /unitcost), 1)
log10ratiounitcostfit=polyval(p, log10
  (Qrange/Q));
ratiounitcostfit=10.
  ^log10ratiounitcostfit;
k=10^p(:,2)
R2 = 1 - sum((log10(unitcostrange
  /unitcost) - log10ratiounitcostfit).
  ^2)/sum((log10(unitcostrange
  /unitcost) - mean(log10(unitcostrange
  /unitcost))).^2)
xmax=max(Qrange/Q);
xmin=min(Qrange/Q);
xaxis=[xmin:0.1:xmax];
log10ratiounitcostfitplot=polyval(p,
  log10(xaxis));
ratiounitcostfitplot=10.
  ^log10ratiounitcostfitplot;
%plot%
figure(1)
plot(Qrange/Q,unitcostrange
  /unitcost,'ko','MarkerSize',3,
  'MarkerFaceColor',[0 0 0],
  'MarkerEdgeColor',[0 0 0],
  'linewidth',1);
hold on
```

```
plot(xaxis,ratiounitcostfitplot,'k-',
  'linewidth',1);
set(gca,'xtickmode',
  'auto','FontSize'
  ,10,'FontWeight','bold','FontName',
  'Times New Roman');


xlabel({'Q/Q_0'},
  'FontSize',14,'FontWeight',
  'bold','FontName', 'Times New Roman')
ylabel({'Unit Cost/Unit Cost_0'},
  'FontSize',14,'FontWeight',
  'bold','FontName', 'Times New Roman')
grid on
hold on
```

Answer: ROOMcost

```
p = -0.3346 -0.0164; k = 0.9629;
R2 = 0.9962
```

10.4.8 Filtration

Example 10.11

Given: Table 10.28 shows the capital cost and O&M cost of filtration at different flow rates.

Find:

1) Calculate unit capital and O&M costs.
2) Plot unit cost/unit cost$_0$ versus Q/Q_0.
3) Determine correlation equations between unit cost/unit cost$_0$ for capital and O&M versus Q/Q_0, respectively.

Table 10.28 Filtration.

	Cost per amount of flow (MGD)						
Expense ($)	**0.5**	**1.0**	**2.0**	**5.0**	**10**	**15**	**20**
Concrete	88 800	120 376	171 406	282 643	466 206	558 743	743 428
Excavation and backfill (10%)	8 880	12 038	17 141	28 264	46 621	55 874	74 343
Miscellaneous metals (4%)	3 552	4 815	6 856	11 306	18 648	22 350	29 737
Yard piping (7%)	6 216	8 426	11 998	19 785	32 634	39 112	52 040
Total concrete	107 448	145 655	207 401	341 998	564 109	676 079	899 548
Equipment[a,b]	181 303	193 498	214 085	263 876	394 212	431 061	565 437
Tax and delivery (11%)	19 943	21 285	23 549	29 026	43 363	47 417	62 198
Installation (20%)	36 261	38 700	42 817	52 775	78 842	86 212	113 087
Manufacturer services (4%)	7 252	7 740	8 563	10 555	15 768	17 242	22 617
Total mechanical	244 759	261 222	289 015	356 233	532 186	581 932	763 340
Protective coating (7%)	24 654	28 481	34 749	48 876	76 741	88 061	116 402
Electricity (10%)	35 221	40 688	49 642	69 823	109 630	125 801	166 289
Instrumentation (10%)	35 221	40 688	49 642	69 823	109 630	125 801	166 289
Housing	0	0	0	0	0	0	0
Subtotal	447 303	516 734	630 448	886 753	1 392 295	1 597 674	2 111 868
Contractor overhead and profit (15%)	67 095	77 510	94 567	133 013	208 844	239 651	316 780
Scope-of-project contingency (30%)	134 191	155 020	189 134	266 026	417 689	479 302	633 560
Total construction cost	648 589	749 264	914 150	1 285 792	2 018 828	2 316 628	3 062 208
Engineering design (10%)	64 859	74 926	91 415	128 579	201 883	231 663	306 221
Total capital cost	713 448	824 191	1 005 565	1 414 371	2 220 711	2 548 291	3 368 429
Cost per gal ($)	1.43	0.82	0.50	0.28	0.22	0.17	0.17
O&M (5% construction cost)	32 429	37 463	45 708	64 290	100 941	115 831	153 110

[a] Average cost used for filtration equipment [Ox and SBR for Title 22 (California) and Class A+ (Arizona) requirements].

[b] Cost of equipment includes the cost of fiberglass reinforced plastic (FRP) tanks and metering pump skid for alum and cationic polymer (Augusta Fiberglass budget).

Solution:

1) When $Q = 0.5$ MGD

$$\text{Unit capital cost} = 713448\$ \div 0.5\,\text{MGD} \times 1\,\text{million gal}/10^3\,\text{kgal} \div 1\,\text{day} = 1426.90\$/\text{kgal}$$

$$\text{Unit O\&M costs} = 32429\$/\text{year} \div 0.5\,\text{MGD} \times 1\,\text{million gal}/10^3\,\text{kgal} \div 365\,\text{day/year}$$

$$\text{Unit O\&M costs} = 0.1777\$/\text{kgal}$$

When $Q = 0.5$ MGD

$Q_0 = 0.5\,\text{MGD}$

$\text{Unit capital cost}_0 = 1426.90\$/\text{kgal}$

$\text{Unit O\&M costs}_0 = 0.1777\$/\text{kgal}$

$$\frac{Q}{Q_0} = 1$$

$$\frac{\text{Unit capital cost}}{\text{Unit capital cost}_0} = 1$$

$$\frac{\text{Unit O\&M costs}}{\text{Unit O\&M costs}_0} = 1$$

When $Q = 1$ MGD (Tables 10.29 and 10.30)

$$\frac{Q}{Q_0} = 2$$

$$\frac{\text{Unit capital cost}}{\text{Unit capital cost}_0} = 0.5776$$

$$\frac{\text{Unit O\&M costs}}{\text{Unit O\&M costs}_0} = 0.5776$$

Table 10.29 Unit capital cost and unit operation and maintenance cost of filtration at different flow rates.

Q (MGD)	Unit capital cost ($/kgal)	Unit O&M costs ($/kgal)
0.5	1426.90	0.1777
1	824.19	0.1026
2	502.78	0.0626
5	282.87	0.0352
10	222.07	0.0277
15	169.89	0.0212
20	168.42	0.0210

Table 10.30 Unit cost ratio of filtration at different flow ratios.

Q/Q_0	Unit capital cost/unit capital cost_0	Unit O&M costs/unit O&M costs_0
1	1.0000	1.0000
2	0.5776	0.5776
4	0.3524	0.3524
10	0.1982	0.1982
20	0.1556	0.1556
30	0.1191	0.1191
40	0.1180	0.1180

2) Plot unit cost/unit cost_0 versus Q/Q_0 for capital and O&M, respectively (Figure 10.18).
3) a) Equation for unit cost/unit cost_0 versus Q/Q_0 for capital:
The regression equation of the trend line:

$$\log\left(\frac{\text{unit cost}}{\text{unit cost}_0}\right) = -0.5816\log\left(\frac{Q}{Q_0}\right) - 0.0571$$
$$r^2 = 0.9829$$

or the equation of the regression (Figure 10.19):

$$\frac{\text{Unit cost}}{\text{Unit cost}_0} = 0.8767\left(\frac{Q}{Q_0}\right)^{-0.5816} \quad r^2 = 0.9829$$

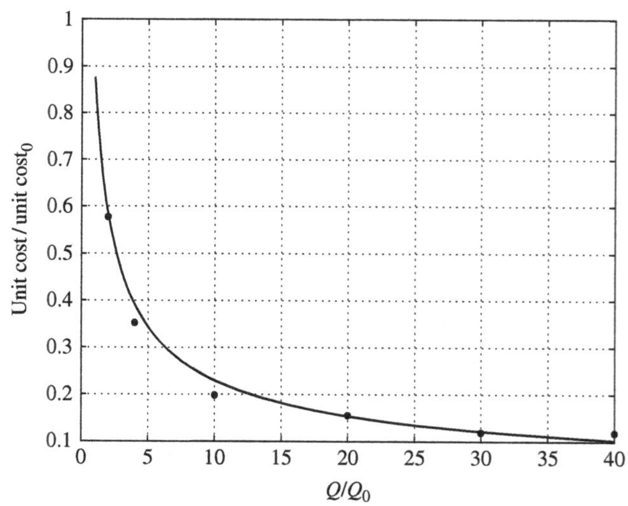

Figure 10.18 Unit capital cost/unit capital cost_0 versus Q/Q_0 for filtration, unit capital cost_0 = 1426.90$/kgal, Q_0 = 0.5 MGD.

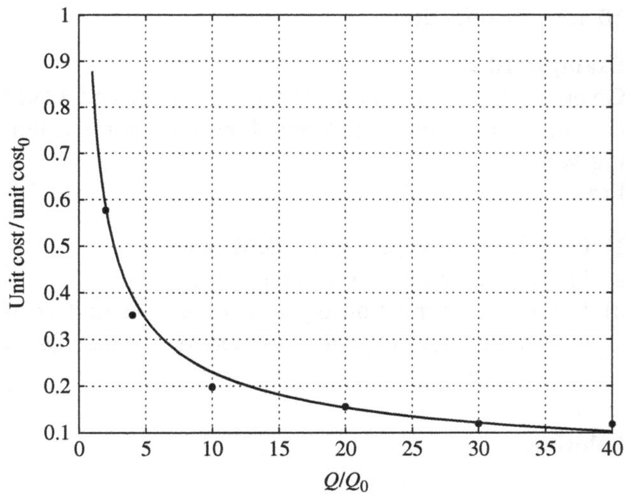

Figure 10.19 Unit O&M cost/unit O&M $cost_0$ versus Q/Q_0 for filtration, unit O&M $cost_0$ = 0.1777$/kgal, Q_0 = 0.5 MGD.

Matlab code:

```
clear all
close all
%data%
Qrange=[0.5 1 2 5 10 15 20];
Q=Qrange(1);
totalcostrange=[713448 824191 1005565
   1414371 2220711 2548291 3368429];
totalcost=totalcostrange(1);
unitcost=totalcost./Q./1000;
unitcostrange=totalcostrange./
   Qrange./1000;
p=polyfit(log10(Qrange/Q), log10
   (unitcostrange /unitcost), 1)
log10ratiounitcostfit=polyval(p, log10
   (Qrange/Q));
ratiounitcostfit=10.
   ^log10ratiounitcostfit;
k=10^p(:,2)
R2 = 1 - sum((log10(unitcostrange
   /unitcost) - log10ratiounitcostfit).
   ^2)/sum((log10(unitcostrange
   /unitcost) - mean(log10(unitcostrange
   /unitcost))).^2)
xmax=max(Qrange/Q);
xmin=min(Qrange/Q);
xaxis=[xmin:0.1:xmax];
log10ratiounitcostfitplot=polyval(p,
   log10(xaxis));
ratiounitcostfitplot=10.
   ^log10ratiounitcostfitplot;
%plot%
figure(1)
plot(Qrange/Q,unitcostrange
   /unitcost,'ko','MarkerSize',3,
   'MarkerFaceColor',[0 0 0],
   'MarkerEdgeColor',[0 0 0],
   'linewidth',1);
hold on
plot(xaxis,ratiounitcostfitplot,'k-',
   'linewidth',1);
set(gca,'xtickmode','auto',
   'FontSize',10,
   'FontWeight','bold','FontName',
   'Times New Roman');

xlabel({'Q/Q_0'},
   'FontSize',14,
   'FontWeight','bold','FontName',
   'Times New Roman')
ylabel({'Unit Cost/Unit Cost_0'},
   'FontSize',14,'FontWeight',
   'bold','FontName', 'Times New Roman')
grid on
hold on
```

Answer: `FiltrationCapitalcost`

`p = -0.5816 -0.0571; k = 0.8767;`
R^2 `= 0.9829`

b) Equation for unit cost/unit $cost_0$ versus Q/Q_0 for O&M:

The regression equation of the trend line:

$$\log\left(\frac{\text{unit cost}}{\text{unit cost}_0}\right) = -0.5816\log\left(\frac{Q}{Q_0}\right) - 0.0571$$

$$r^2 = 0.9829$$

or the equation of the regression:

$$\frac{\text{Unit cost}}{\text{Unit cost}_0} = 0.8767\left(\frac{Q}{Q_0}\right)^{-0.5816} \quad r^2 = 0.9829$$

Matlab code:

```
clear all
close all
%data%
Qrange=[0.5 1 2 5 10 15 20];
Q=Qrange(1);
totalcostrange=[32429 37463 45708 64290
   100941 115831 153110];
totalcost=totalcostrange(1);
```

```
unitcost=totalcost./Q./1000./365;
unitcostrange=totalcostrange./Qrange./
   1000./365;
p=polyfit(log10(Qrange/Q), log10
   (unitcostrange /unitcost), 1)
log10ratiounitcostfit=polyval(p, log10
   (Qrange/Q));
ratiounitcostfit=10.
   ^log10ratiounitcostfit;
k=10^p(:,2)
R2 = 1 - sum((log10(unitcostrange
   /unitcost) - log10ratiounitcostfit).
   ^2)/sum((log10(unitcostrange
   /unitcost) - mean(log10(unitcostrange
   /unitcost))).^2)
xmax=max(Qrange/Q);
xmin=min(Qrange/Q);
xaxis=[xmin:0.1:xmax];
log10ratiounitcostfitplot=polyval(p,
   log10(xaxis));
ratiounitcostfitplot=10.
   ^log10ratiounitcostfitplot;
%plot%
figure(1)
plot(Qrange/Q,unitcostrange
   /unitcost,'ko',
   'MarkerSize',3,'MarkerFaceColor',
   [0 0 0],'MarkerEdgeColor',
   [0 0 0],'linewidth',1);
hold on
plot(xaxis,ratiounitcostfitplot,'k-',
   'linewidth',1);
set(gca,'xtickmode',
   'auto',
   'FontSize',10,
   'FontWeight','bold','FontName',
   'Times New Roman');

xlabel({'Q/Q_0'},'FontSize',14,
   'FontWeight','bold','FontName',
   'Times New Roman')
ylabel({'Unit Cost/Unit
   Cost_0'},'FontSize',14,
   'FontWeight','bold','FontName',
   'Times New Roman')
grid on
hold on
```

Answer: `FiltrationOMcost`

`p = -0.5816 -0.0571; k = 0.8767;`
R^2 `= 0.9829`

10.4.9 Disinfection

Example 10.12

Given: Table 10.31 shows the capital cost and O&M cost of disinfection at different flow rates for reclaiming WW.

Find:

1) Calculate unit capital and O&M costs.
2) Plot unit cost/unit cost_0 versus Q/Q_0.
3) Determine correlation equations between unit cost/unit cost_0 for capital and O&M versus Q/Q_0, respectively.

Solution:

1) When $Q = 0.5$ MGD

$$\text{Unit capital cost} = 334772\$ \div 0.5\,\text{MGD} \times 1\,\text{million gal}/10^3\,\text{kgal} \div 1\,\text{day} = 669.54\$/\text{kgal}$$

$$\text{Unit O\&M costs} = 15217\$/\text{year} \div 0.5\,\text{MGD} \times 1\,\text{million gal}/10^3\,\text{kgal} \div 365\,\text{day/year}$$

$$\text{Unit O\&M costs} = 0.0834\$/\text{kgal}$$

When $Q = 0.5$ MGD

$$Q_0 = 0.5\,\text{MGD}$$

$$\text{Unit capital cost}_0 = 669.54\$/\text{kgal}$$

$$\text{Unit O\&M costs}_0 = 0.0834\$/\text{kgal}$$

$$\frac{Q}{Q_0} = 1$$

$$\frac{\text{Unit capital cost}}{\text{Unit capital cost}_0} = 1$$

$$\frac{\text{Unit O\&M costs}}{\text{Unit O\&M costs}_0} = 1$$

When $Q = 1$ MGD (Tables 10.32 and 10.33)

$$\frac{Q}{Q_0} = 2$$

$$\frac{\text{Unit capital cost}}{\text{Unit capital cost}_0} = 0.8984$$

$$\frac{\text{Unit O\&M costs}}{\text{Unit O\&M costs}_0} = 0.8984$$

Table 10.31 Disinfection (CA).

Expense ($)	Cost per amount of flow (MGD)						
	0.5	1.0	2.0	5.0	10	15	20
Concrete[a]	119 872	227 916	295 200	491 674	772 030	856 583	955 531
Excavation and backfill (10%)	11 987	22 792	29 520	49 167	77 203	85 658	95 553
Miscellaneous metals (4%)	4 795	9 117	11 808	19 667	30 881	34 263	38 221
Yard piping (7%)	8 391	15 954	20 664	34 417	54 042	59 961	66 887
Total concrete	145 045	275 778	357 192	594 926	934 156	1 036 465	1 156 192
Equipment[b]	14 979	15 682	17 034	19 887	23 126	28 347	44 118
Tax and delivery (11%)	1 648	1 725	1 874	2 188	2 544	3 118	4 853
Installation (20%)	2 996	3 136	3 407	3 977	4 625	5 669	8 824
Manufacturer services (4%)	599	627	681	795	925	1 134	1 765
Total mechanical	20 222	21 171	22 996	26 847	31 220	38 268	59 559
Protective coating (7%)	11 569	20 786	26 613	43 524	67 576	75 231	85 103
Electricity (10%)	16 527	29 695	38 019	62 177	96 538	107 473	121 575
Instrumentation (10%)	16 527	29 695	38 019	62 177	96 538	107 473	121 575
Housing	0	0	0	0	0	0	0
Subtotal	209 888	377 125	482 839	789 652	1 226 027	1 364 912	1 544 005
Contractor overhead and profit (15%)	31 483	56 569	72 426	118 448	183 904	204 737	231 601
Scope-of-project contingency (30%)	62 966	113 138	144 852	236 896	367 808	409 473	463 201
Total construction cost	304 338	546 832	700 116	1 144 995	1 777 740	1 979 122	2 238 807
Engineering design (10%)	30 434	54 683	70 012	114 500	177 774	197 912	223 881
Total capital cost	334 772	601 515	770 128	1 259 495	1 955 514	2 177 034	2 462 687
Cost per gal ($)	0.67	0.60	0.39	0.25	0.20	0.15	0.12
O&M (5% construction cost)	15 217	27 342	35 006	57 250	88 887	98 956	111 940

[a] Concrete cost is based on average contact time for Arizona and California SBR, Ox, and MBR.
[b] Equipment cost includes the cost of the chemical system in which only FRP tanks and metering pump skid system are estimated for (Augusta Fiberglass).

Table 10.32 Unit capital cost and unit operation and maintenance cost of disinfection at different flow rates.

Q (MGD)	Unit capital cost ($/kgal)	Unit O&M costs ($/kgal)
0.5	669.54	0.0834
1	601.52	0.0749
2	385.06	0.0480
5	251.90	0.0314
10	195.55	0.0244
15	145.14	0.0181
20	123.13	0.0153

Table 10.33 Unit cost ratio of disinfection at different flow ratios.

Q/Q_0	Unit capital cost/unit capital $cost_0$	Unit O&M costs/unit O&M $costs_0$
1	1.0000	1.0000
2	0.8984	0.8984
4	0.5751	0.5751
10	0.3762	0.3762
20	0.2921	0.2921
30	0.2168	0.2168
40	0.1839	0.1839

2) Plot unit cost/unit cost$_0$ versus Q/Q_0 for capital and O&M, respectively (Figure 10.20).
3) a) Equation for unit cost/unit cost$_0$ versus Q/Q_0 for capital:

The regression equation of the trend line:

$$\log\left(\frac{\text{unit cost}}{\text{unit cost}_0}\right) = -0.4716\log\left(\frac{Q}{Q_0}\right) - 0.0454$$

$$r^2 = 0.9872$$

or the equation of the regression (Figure 10.21):

$$\frac{\text{Unit cost}}{\text{Unit cost}_0} = 1.1103\left(\frac{Q}{Q_0}\right)^{-0.4716} \quad r^2 = 0.9872$$

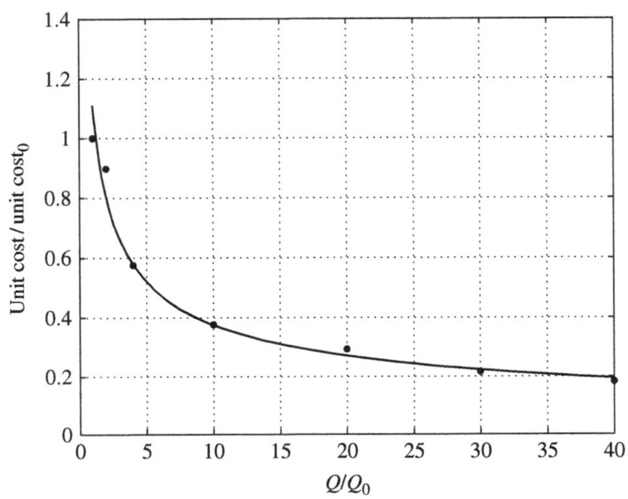

Figure 10.20 Unit capital cost/unit capital cost$_0$ versus Q/Q_0 for disinfection, unit capital cost$_0$ = 669.54$/kgal, Q_0 = 0.5 MGD.

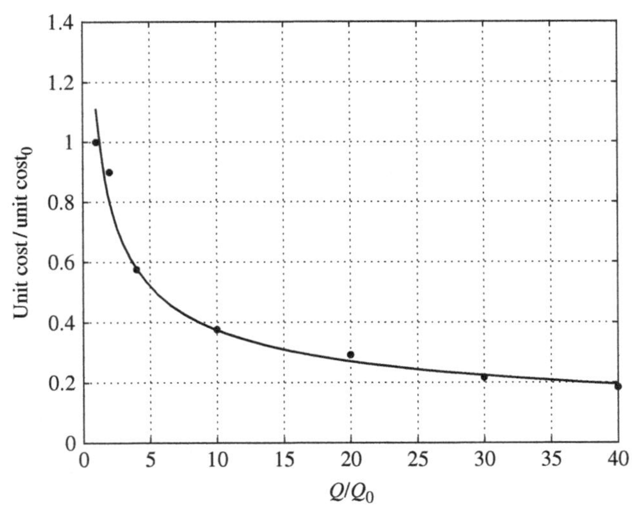

Figure 10.21 Unit O&M cost/unit O&M cost$_0$ versus Q/Q_0 for disinfection, unit O&M cost$_0$ = 0.0834$/kgal, Q_0 = 0.5 MGD.

Matlab code:

```
clear all
close all
%data%
Qrange=[0.5 1 2 5 10 15 20];
Q=Qrange(1);
totalcostrange=[334772 601515 770128
  1259495 1955514 2177034 2462687];
totalcost=totalcostrange(1);
unitcost=totalcost./Q./1000./365;
unitcostrange=totalcostrange./Qrange./
  1000./365;
p=polyfit(log10(Qrange/Q), log10
  (unitcostrange /unitcost), 1)
log10ratiounitcostfit=polyval
  (p, log10(Qrange/Q));
ratiounitcostfit=10.
  ^log10ratiounitcostfit;
k=10^p(:,2)
R2 = 1 - sum((log10(unitcostrange
  /unitcost) - log10ratiounitcostfit).
  ^2)/sum((log10(unitcostrange
  /unitcost) - mean(log10
  (unitcostrange /unitcost))).^2)
xmax=max(Qrange/Q);
xmin=min(Qrange/Q);
xaxis=[xmin:0.1:xmax];
log10ratiounitcostfitplot=polyval
  (p, log10(xaxis));
ratiounitcostfitplot=10.
  ^log10ratiounitcostfitplot;
%plot%
figure(1)
plot(Qrange/Q,unitcostrange
  /unitcost,'ko','MarkerSize',3,
  'MarkerFaceColor',[0 0 0],
  'MarkerEdgeColor',[0 0 0],
  'linewidth',1);
hold on
plot(xaxis,ratiounitcostfitplot,
  'k-','linewidth',1);
set(gca,'xtickmode','auto',
  'FontSize',10,'FontWeight','bold',
  'FontName', 'Times New Roman');


xlabel({'Q/Q_0'},'FontSize',14,
  'FontWeight','bold','FontName',
  'Times New Roman')
ylabel({'Unit Cost/Unit
  Cost_0'},'FontSize',14,
  'FontWeight','bold','FontName',
  'Times New Roman')
```

```
grid on
hold on
```

Answer: `DisinfectionCACapitalcost`

```
p = -0.4716 0.0454; k = 1.1103;
R^2 = 0.9872
```

b) Equation for unit cost/unit cost$_0$ versus Q/Q_0 for O&M:

The regression equation of the trend line:

$$\log\left(\frac{\text{unit cost}}{\text{unit cost}_0}\right) = -0.4716\log\left(\frac{Q}{Q_0}\right) - 0.0454$$

$$r^2 = 0.9872$$

or the equation of the regression:

$$\frac{\text{Unit cost}}{\text{Unit cost}_0} = 1.1103\left(\frac{Q}{Q_0}\right)^{-0.4716} \quad r^2 = 0.9872$$

Matlab code:

```
clear all
close all
%data%
Qrange=[0.5 1 2 5 10 15 20];
Q=Qrange(1);
totalcostrange=[15217 27342 35006 57250
  88887 98956 111940];
totalcost=totalcostrange(1);
unitcost=totalcost./Q./1000./365;
unitcostrange=totalcostrange./Qrange./
  1000./365;
p=polyfit(log10(Qrange/Q), log10
  (unitcostrange /unitcost), 1)
log10ratiounitcostfit=polyval(p, log10
  (Qrange/Q));
ratiounitcostfit=10.
  ^log10ratiounitcostfit;
k=10^p(:,2)
R2 = 1 - sum((log10(unitcostrange
  /unitcost) - log10ratiounitcostfit).
  ^2)/sum((log10(unitcostrange
  /unitcost) - mean(log10(unitcostrange
  /unitcost))).^2)
xmax=max(Qrange/Q);
xmin=min(Qrange/Q);
xaxis=[xmin:0.1:xmax];
log10ratiounitcostfitplot=polyval(p,
  log10(xaxis));
ratiounitcostfitplot=10.
  ^log10ratiounitcostfitplot;
%plot%
figure(1)
plot(Qrange/Q,unitcostrange
  /unitcost,'ko','MarkerSize',3,
  'MarkerFaceColor',[0 0 0],
  'MarkerEdgeColor',[0 0 0],
  'linewidth',1);
hold on
plot(xaxis,ratiounitcostfitplot,
  'k-','linewidth',1);
set(gca,'xtickmode',
  'auto','FontSize',10,
  'FontWeight','bold','FontName',
  'Times New Roman');

xlabel({'Q/Q_0'},'FontSize',14,
  'FontWeight','bold','FontName',
  'Times New Roman')
ylabel({'Unit Cost/Unit
  Cost_0'},'FontSize',14,
  'FontWeight','bold','FontName',
  'Times New Roman')
grid on
hold on
```

Answer: `DisinfectionCAOMcost`

```
p = -0.4716 0.0454; k = 1.1103;
R^2 = 0.9872
```

10.5 Exercise

10.5.1 Questions

1) What is the most valuable material in WWTP to be recovered?
2) Please rank economic values of nutrient, energy, and water recovery.
3) Water could be recovered in a large quantity, what prevents the reuse of the treated effluent from WWTPs?
4) If nutrients such as N and P in the form of struvite are so valuable, why is nutrient recovery not commonly practiced in WWTPs?
5) What are the common treatment processes for organic, toxic metals, and virus or bacteria to achieve wastewater quality standards?
6) Which unit process is the most expensive in reclaiming treated wastewater?

10.5.2 Calculations

1) Table 10.34 shows the capital cost and operation and maintenance (O&M) cost of Zenon MBR (CA Title 22) at different flow rates:
   a) Calculate unit capital and O&M costs.
   b) Plot unit cost/unit $cost_0$ versus Q/Q_0.
   c) Determine correlation equations between unit cost/unit $cost_0$ for capital and O&M versus Q/Q_0, respectively.
2) A water utility in Southern Florida conducted feasibility study of reclaiming water for reuse. The treatment train of the pilot WRRF is to nitrify the secondary effluent from WWTP effluent using UF, RO, and UV disinfection. The target is to achieve total nitrogen < 1.5 mg/l, total phosphorus < 0.02 mg/l, BOD_5 < 5 mg/l, and TSS < 5 mg/l. A pilot scale was tested with flow rate of 20 GPD. The influent has concentration of total nitrogen of 19.7 mg/l and total phosphorus of 2.2 mg/l, while the effluent has concentration of total nitrogen of 1.20 mg/l and total phosphorus of 0.007 mg/l. If the capital and O&M costs that were valid from the California tests were also applicable to Southern Florida, please answer the following if the WRRF is to be built to have a flow rate of 10 MGD:
   1) What would be the unit capital and O&M costs according to the equations present in this chapter?
   2) What would be the total capital? If the design life expectancy is 50 years, what would be the total O&M cost?
   3) If all the 10 MGD reclaimed water were sold for irrigation purpose at a price of $5/1000 gal, how many years the capital costs could be recovered after the WRRF is built and in operation?

Table 10.34 Zenon MBR (CA Title 22).

	Cost per amount of flow (MGD)						
Expense ($)	**0.5**	**1.0**	**2.0**	**5.0**	**10**	**15**	**20**
Concrete	610 400	1 062 400	2 012 000	4 722 400	9 684 800	13 448 800	19 054 400
Excavation and backfill (10%)	61 040	106 240	201 200	472 240	968 480	1 344 880	1 905 440
Miscellaneous metals (4%)	24 416	42 496	80 480	188 896	387 392	537 952	762 176
Yard piping (7%)	42 728	74 368	140 840	330 568	677 936	941 416	1 333 808
Total concrete	738 584	1 285 504	2 434 520	5 714 104	11 718 608	16 273 048	23 055 824
Equipment[a]	2 963 000	3 704 000	7 037 000	14 815 000	24 306 000	27 778 000	29 630 000
Tax and delivery (11%)	325 930	407 440	774 070	1 629 650	2 673 660	3 055 580	3 259 300
Installation (20%)	592 600	740 800	1 407 400	2 963 000	4 861 200	5 555 600	5 926 000
Manufacturer services (4%)	118 520	148 160	281 480	592 600	972 240	1 111 120	1 185 200
Total mechanical	4 000 050	5 000 400	9 499 950	20 000 250	32 813 100	37 500 300	40 000 500
Protective coating (7%)	331 704	440 013	835 413	1 800 005	3 117 220	3 764 134	4 413 943
Electricity (10%)	473 863	628 590	1 193 447	2 571 435	4 453 171	5 377 335	6 305 632
Instrumentation (10%)	473 863	628 590	1 193 447	2 571 435	4 453 171	5 377 335	6 305 632
Housing	0	0	0	0	0	0	0
Subtotal	6 018 065	7 983 098	15 156 777	32 657 230	56 555 269	68 292 152	80 081 531
Contractor overhead and profit (15%)	902 710	1 197 465	2 273 517	4 898 584	8 483 290	10 243 823	12 012 230
Scope-of-project contingency (30%)	1 805 420	2 394 929	4 547 033	9 797 169	16 966 581	20 487 646	24 024 459
Total construction cost	8 726 195	11 575 492	21 977 327	47 352 983	82 005 140	99 023 620	116 118 221
Engineering design (10%)	872 619	1 157 549	2 197 733	4 735 298	8 200 514	9 902 362	11 611 822
Total capital cost	9 598 814	12 733 041	24 175 059	52 088 281	90 205 654	108 925 982	127 730 043
Cost per gal ($)	19.20	12.73	12.09	10.42	9.02	7.26	6.39
O&M (5% construction cost)	436 310	578 775	1 098 866	2 367 649	4 100 257	4 951 181	5 805 911

[a] Equipment costs provided by Zenon Technologies.

4) In your option, what would be the major factors that would affect your cost and benefit analysis of this project?

10.5.3 Projects

10.5.3.1 Xiongan Project

If the Xiongan is planned for populations of 1, 2, and 5 million in 5, 10, and 20 years, respectively, what percentage of wastewater should be reclaimed? Please design a system that could be applied to reclaim wastewater using the similar treatment train as in the previous problem to achieve the target percentage. Please use the capital and O&M cost diagram to estimate the capital and O&M costs for 50 years.

10.5.3.2 Community Proposal Project

Please make an appointment with officials in your hometown and collect waste reuse data. According to the data, please do the following:

1) Calculate the percentage of water reuse rates.
2) Determine whether the city is sustainable in terms of water resources.
3) Please design a water reuse system so that the city becomes sustainable in terms of water resources.
4) Please estimate the capital and O&M costs of your proposed project for 20 and 50 years.
5) What is the return-on-investment period for the water reuse system?

References

Andrade, A and Schuling, O. (1999). *Literature search, digest and assessment concerning struvite recovery for recycling from waste waters.* Final Report for CEEP (CEFIC). Holland: Geochem Research BV/Utrecht University Holland.

Appeldoorn, K.J., Kortstee, G.J.J.K., and Zehnder, A.B. (1992). Biological phosphate removal by activated sludge under defined conditions. *Water Research* 26 (4): 453–460.

Bark, K., Sponner, A., Kämpfer, P. et al. (1992). Differences in polyphosphate adsorption by *Acinetobacter* isolates from wastewater producing polyphosphate: AMP phosphotransferase. *Water Research* 26 (10): 1379–1388.

Britton, A., Koch, F.A., Mavinic, D.S. et al. (2005). Pilot-scale struvite recovery from anaerobic digester supernatant at an enhanced biological phosphorus removal wastewater treatment plant. *Journal of Environmental Science* 4: 265–277.

Celen, I. and Türker, M. (2001). Recovery of ammonia as struvite from anaerobic digester effluents. *Environmental Technology* 22: 1263–1272.

Ishii, S.K.L. and Boyer, T.H. (2015). Life cycle comparison of centralized wastewater treatment and urine source separation with struvite precipitation: focus on urine nutrient management. *Water Research* 79: 88–103.

Lahav, O., Telzhensky, M., Zewuhn, A. et al. (2013). Struvite recovery from municipal-wastewater sludge centrifuge supernatant using seawater NF concentrate as a cheap Mg (II) source. *Separation and Purification Technology* 108: 103–110.

Oehmen, A., Lemos, P.C., Carvalho, G. et al. (2007). Advances in enhanced biological phosphorus removal: from micro to macro scale. *Water Research* 41: 2271–2300.

Pinnekamp, J., Montag, D., Gethke, K., and Herbst, H. (2007). Rückgewinnung eines schadstofffreien, mineralischen Kombinationsdüngers Magnesiumammoniumphosphat – MAP aus Abwasser und Klärschlamm. *UBA-Texte 25/07.* Dessau-Rosslau.

Pinnekamp, J., Baumann, P., Cornel, P. et al. (2013). Stand und Perspektiven der Phosphorrückgewinnung aus Abwasser und Klärschlamm – Teil 1, KA Korrespondenz Abwasser, Abfall, 60/10, 837–844.

Rasmussen, H., Bruus, J.H., Keiding, K., and Nielsen, P.H. (1994). Observations of dewaterability and physical, chemical and microbiological changes in anaerobically stored activated sludge from a nutrient removal plant. *Water Research* 28 (2): 417.

Schön, G.S., Geywitz, S., and Mertens, F. (1993). Influence of dissolved oxygen and oxidation reduction potential on phosphate release and uptake by activated sludge from sewage plants with enhanced biological phosphorus removal. *Water Research* 27 (3): 349–354.

Seviour, R.J., Mino, T., and Onuki, M. (2003). The microbiology of biological phosphorus removal in activated sludge system. *FEMS Microbiology Reviews* 27: 99–127.

Siegrist, H. (1996). Nitrogen removal from digester supernatant-comparison of chemical and biological methods. *Water Science and Technology* 34: 399–406.

Turker, M. and Celen, I. (2007). Removal of ammonia as struvite from anaerobic digester effluents and recycling of magnesium and phosphate. *Bioresource Technology* 98 (8): 1529–1534.

WateReuse Foundation (2009). Decision Support System for Selection of Satellite vs. Regional Treatment for Reuse Systems. *Project: 04-14.*

Webb K.M., Bhargava S.K., Priestley A.J. et al. (1995). Struvite ($MgNH_4PO_4.6H_2O$) precipitation: potential for nutrient removal and re-use from wastewaters. *Chemistry in Australia* 62: 42–44.

11

Separation

Dispersion of wastes is ignorant of their value.

OBJECTIVES
To understand why there are fundamental design flaws in activated sludge process To carry out balances of energy, C, N, P, solids, and water in WWTP To design energy-positive WWTPs with or without N and P recovery To develop energy diagram and calculate the energy balance of design alternatives To diagnose net energy-positive alternatives of WWTPs To develop retrofitting strategies of WWTP To design WRRFs as energy positive, energy neutral by avoiding energy negative

11.1 Principle 8

Separation may serve different purposes for reuse, recycle, and treatment. For solid waste, food waste (FW) should be separated from other waste because energy could be recovered from the FW. Papers and plastic separated from the solid waste stream could be reused or recycled. For zero water design, it is critical to separate blue (surface and groundwater), green (rainwater), gray (bath and laundry), and black (toilet water). Black water can be further separated into brown (urine) and black (feces). In developing countries, separated urine and feces toilets are promoted by many United Nations programs. N and P could be recovered from the concentrated urine, while anaerobic digestion of feces could generate methane and good organic fertilizer. In developed countries, waterless urinals could save significant amounts of tap water and reduce discharge to the sewer system. As a result, the design criteria of flow rate could be reduced due to water saving by waterless urinals. Separation also holds the key to sludge dewatering and disposal. Water as the final product in WWTP should be separated from any waste produced during the processes such as physicochemical treatment (coagulation) or biological treatment (activated sludge). The former needs to avoid adding chemicals, and the latter needs to separate organic content from water at the front end of a treatment systems. Enhanced separation of organic from wastewater (WW) using a cloth microsieving before WW enters the aeration tank prevents free water from being fixed as cellular water as part of activated sludge. The cloth microsieved solids could be directly diverted to the anaerobic digester to generate more methane to achieve energy-positive WWTP. As a result, organic loading to the aeration tank would be reduced, and major secondary waste such as activated sludge would also be reduced significantly.

Principle 8 Separation should be implemented as early as possible on every scale to minimize energy consumption and materials:

1) Waste should be separated according to its nature such as recyclability or biodegradability.
2) Separation technologies such as ultra-, micro-, and nanofiltrations should be selected according to the size of pollutants.
3) Phase-transferring technologies such as activated carbon should be secondary choices after destructive technologies of advanced oxidation process.

Sustainable Environmental Engineering, First Edition. Walter Z. Tang and Mika Sillanpää.

Separation plays important role in reuse, recycling, or treatment. For example, cardboard separated from domestic solid waste (DSW) stream could be used for cardboard reproduction. Fat–oil–grease (FOG) separated from kitchen waste could be codigested with sludge to generate more methane as shown in Table 11.1.

As a treatment strategy, separation technologies are largely dependent on particle size. The particle size is an important factor in the process of water and wastewater treatment (WWT) in the past. Figure 11.1 shows the approximate range of particle size for each specific separation technology.

Figure 11.1 lists sizes for typical particulates in water: viruses, humic acids, clay/silt, bacteria, algae, and sand. Nutrients can be considered to be in the molecules category, with 10^{-10} m of atomic size. Reverse osmosis can essentially treat all different elements in the figure by allowing water filter through its membrane. One of the major design faults is that tank processes were specifically designed for specific range of particles, coagulation of

Table 11.1 Separation for different media and purposes.

Media	Problems	For reuse	For recycle	For treatment
Air	Hazardous waste incineration	CO_2 sequestration	Hazardous waste such as acid and base recycle	Catalyst
Water	Leachate combined with WW	Dripping irrigation	Nutrients	BMR, MF, NF, RO
Land	Food wastes	Kitchen oil for biodiesel	Fat, oil, and grease (FOG) codigestion with sludge	Compositing
Ecological	Invasive plants and aquatic species		Isolate and destroy the invasive plants and aquatic species	

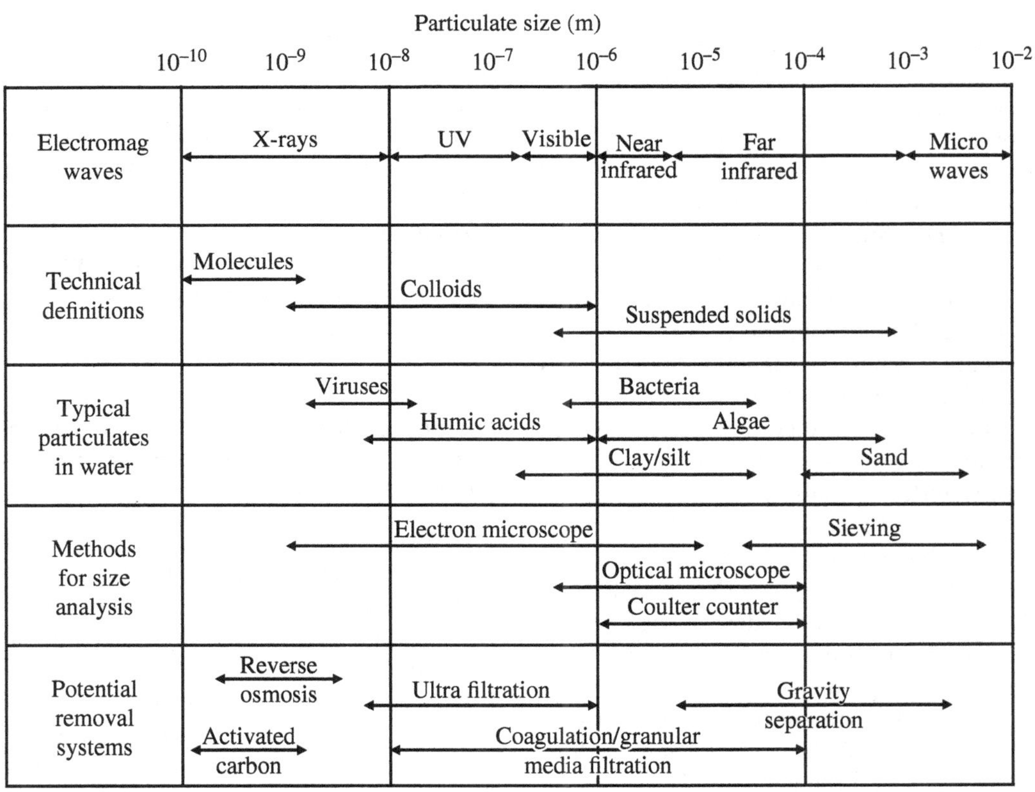

Figure 11.1 Particle size versus treatment technologies.

colloidal particles, and sedimentation for flocculated matters. In WWT, however, different sizes of organic matters ranging from 10^{-3} to 1 mm could be separated in one unit process such as cloth microsieving. Depending upon the opening of the microsieves, 40–60% of organic pollutants could be diverted to anaerobic digester to produce biogas instead of producing sludge. The process would reduce aeration and sludge as secondary waste, and produces more methane for energy-positive design.

11.2 Challenges and Opportunities

Domestic wastewater (DWW) constitutes 70% of WW in China. With an annual WW volume of 750 million m^3, cumulative pollution contributed 225 million tons of COD to Chinese water bodies from 1992 to 2007 (Yu et al., 2015). In rural China, the major pollution sources are DWW and agricultural runoff. From 2001 to 2010, annual discharge of TN was 0.446 and 2.2 million tons and TP 80 000 and 230 000 tons from DWW and the agriculture in China, respectively (Guan et al., 2014). The major challenges faced by rural WWTP are a long construction period, high operation and maintenance (O&M) cost, low technical reliability of a complicated process, high O&M costs, low recycling rates of plants, and geological conditions requiring different designs. More importantly, there are no sustainability design guidelines for rural WWTP and sludge disposal for increasing WW flow rate due to rapid urbanization and population growth.

To solve these urgent WWT challenges, separation is one of the most effective ways to remediate the deterioration of water quality in rural China. There are many opportunities for separation in China. WWT in rural areas of China could be the last big market where only 22% of WW has been treated. China has about 280 000 villages within 1 805 counties, but the WWT rate in rural China is only 13%. In Northern China, groundwater is the major drinking water resource. If WW is not properly treated, contamination of groundwater will spread widely. The 13th Five-Year Environmental Plan for rural China requires one-third of villages to have WWT facilities. Currently, 160 million farmers do not have WWT. If \$1400 of capital cost for WWT was required for each rural resident, the market capital to build WWT facilities alone would be \$240 billion. Indeed, the WWT capital market for rural China could increase exponentially to 3.5, 12, and 20 billion USD in 2016, 2020, and 2025, respectively. In the next decade, the market for WW treatment and the sewer pipe network in China is estimated at 300 billion USD. Separation of gray water from black water would decrease WW at minimal by half. The separated gray water could be used for irrigation of green infrastructures such as wetlands. Separation of urine from feces may offer a way to recover N and P. For example, Ishii and Boyer (2015) used the following unit values in their calculation of urine separation project for a dormitory at the University of Florida (Table 11.2).

According to the assumptions of Table 11.2, for 8 823 population equivalent (PE) annually, the total N and P that can be recovered from 1 920 m^3 of urine is 13 200 and 1 020 kg, respectively.

Table 11.2 Unit values for urine separation design.

Unit values for urine separation design			
Human consumption	Air	28	m^3/day
	Water	1.5	kg/day
	Food	1.5	kg/day
	Calories	2000	cal/day
Flash	Urine	0.05	l/day
	Feces	6	l/day
Total volume	Urine	0.2	l/day
	Feces	0.5	kg/day
Cost	Water from groundwater	0.917	\$/m^3
	WW treatment by central WWTP of 20 000 PE	1.336	kWh/m^3
	Electricity	0.1	\$/m^3
Energy composition in the United States	Coal	50	%
	Natural gas	20	%
	Nuclear	20	%
	Renewable	10	%
Nutrient removal efficiency	N	84	%
	P	49.8	%
Waterless urine toilet		700	\$
Regular urine toilet		200	\$

In the past, China designed and built conventional systems such as activated sludge as the major process for towns of more than 1000 people. However, high energy consumption and large amounts of sludge prevent the normal operation of the WWTPs built in recent decades. Two likely extremes exist in rural markets: the technologies were either too low tech and resulted in not enough protection of water resources or too high tech, such as membrane biological reactor (MBR). Unfortunately, a major difficulty also arises due to high capital cost and the intensive energy required to run. As a result, the unit operating cost might be as high as one USD per cubic meter of WW treated. With limited resources of rural WWTP, many of these plants will most likely run into financial difficulty. For these reasons, decentralized and energy positive WWTPs are critical to ensure sustainable WW management in rural China.

In urban China, activated sludge processes have been utilized in more than 95% of the WWTPs in the past four decades. However, most of the WWTPs do not have sludge treatment facility due to the cost of intensive energy. In 2015, cities and counties had a total of 3717 WWTPs with a treatment capacity of about 157 million cubic meters/day. The total treatment capacity of sewage is 48.06 billion cubic meters per year, with 84.1% operating load rate. However, sludge treatment and disposal facilities were only 43.4% at the end of 2013, because sludge treatment and disposal was not mandatory for WWTPs by law. The 12th Five-Year Plan (FYP) requires the disposal rate of sludge treatment to be 80% by the municipalities directly under the central government, the provincial capitals, and the cities to reach 70% while 30% is required for cities at county or township levels. With consideration of energy footprints and the energy-positive unit process in WWTP, only 60 of 3717 WWTPs in China have anaerobic digestion facilities of sludge, which is barely 1.7% of the total WWTPs in China. Therefore, it is estimated that sludge treatment and disposal market could be as high as 50 billion USD in the next 10 years. The major sludge disposal problem is the optimization of sludge digestion technology. For example, various composting technologies need to be improved. Sludge heat drying and incineration technology and equipment should be integrated. Land application of treated sludge needs to be monitored. Although incineration, carbon char, and landfill are different options, the land application of anaerobic digested sludge should be the future direction due to the energy production as biogas and pathogen-free organic fertilizer sludge. According to the US experience, treatment strategy depends upon the PE. To effectively and economically meet discharge standards, three scales are residential, community, and city as shown in Table 11.3, which lists the most suitable treatment technologies and the operation parameters and performance levels required in the United States.

Separation technologies are considered as phase transfer methods and do not destroy organic pollutants. Nevertheless, precipitation, coagulation, flocculation, and sedimentation are still very important unit processes in treating toxic wastewater containing metals. For water reuse, micro-filtration, nano-filtration, and reverse osmosis are critical to ensure the effluent water quality to meet the water reuse standards.

Table 11.3 Operating parameters and key performance metrics in the United States.

Parameter	Household	Community	City
Wastewater treatment standard	Secondary biological treatment for subsurface drip irrigation reuse	Advanced treatment with nitrogen removal for surface water discharge and reuse	Advanced biological treatment for reuse and deep well injection
BOD_5 in treated effluent(mg/l)	30 (20–40)[a]	1.8 (0.8–3.5)	2.1 (1.2–2.4)
Percentage of water reclaimed (%)	100	77	56
Effluent TN to soil from reclaimed water (mg/l)	16 (2–31)	0.23 (0.03–6.8)	2.3 (1.3–3.1)
Effluent TP to soil from reclaimed water (mg/l)	0.16 (0.12–0.20)	0.005 (0.004–0.04)	0.01 (0.004–0.03)
Total biosolids production (kg/year)	9.8[a]	60 000	2 894 136

Note: Numeric values presented are average values, where values in parentheses are minimum and maximum values.
[a] Asano et al. (2007).

11.3 Precipitation

Chemical precipitation of metal-contaminated water has long been used in treating WW from industrial applications, such as metal finishing and plating. Effective precipitation is usually accompanied by coagulation, flocculation, and clarification and/or filtration to separate the solid and liquid phases of the waste stream by raising the pH of the solution to lower the metal solubility. Alkali-precipitating agents such as caustic soda (NaOH), hydrated lime ($Ca(OH)_2$), and magnesium hydroxide ($Mg(OH)_2$) are common precipitating agents. The solubility of metal hydroxide precipitates decreases with increasing pH to a minimum value called the isoelectric point. Chemical precipitation depends on several variables:

1) Maintenance of a proper pH range throughout the precipitation reaction and subsequent settling time.
2) Addition of a sufficient excess of treatment ions (precipitating agent) to drive the precipitation reaction to completion.
3) Effective removal of precipitated solids.

Figure 11.2 is a typical schema of a metals treatment system.

After precipitation, coagulation and flocculation are required to destabilize colloidal particle and to grow the particle so that they can settle from WW.

11.4 Coagulation and Flocculation

Coagulation destabilizes colloidal particles, while flocculation refers to growth of destabilized colloidal particles. The stability of colloidal particles is defined by the size, area, and charge of the surface. Colloidal particles attract one another through van der Waals forces and have repulsion forces going against one another due to the zeta potential. When suspended in water, the charge on organic and inorganic colloids is typically negative. Flocculation is the process of bringing together the destabilized or "coagulated" particles to form a larger agglomeration of floc by physical mixing or addition of chemical coagulant aids or both. Electric double layer (EDL) refers to a fixed and a diffusing layer. The potential voltage from the shear of a plane is called the zeta potential. Destabilization reduces the zeta potential. Commonly used coagulants include $AlCl_3$ (alum), which might cause Alzheimer's disease due to residual alum. $FeCl_3$ and $FeSO_4$ (ferric ion) are a better choice than alum. The rate of flocculation depends upon several factors: (i) the nature of the particle surface, (ii) the presence of charges, (iii) the shape of the particles, and (iv) the density of the particles. Jar tests are used to determine the dose of coagulants through experiment (Figure 11.3).

Mixing provides greater uniformity of the WW feed and disperses precipitating agents, coagulants, and coagulant aids throughout reactor to ensure the most rapid precipitation reactions and subsequent settling of the precipitates. To quantify the degree of mixing, the

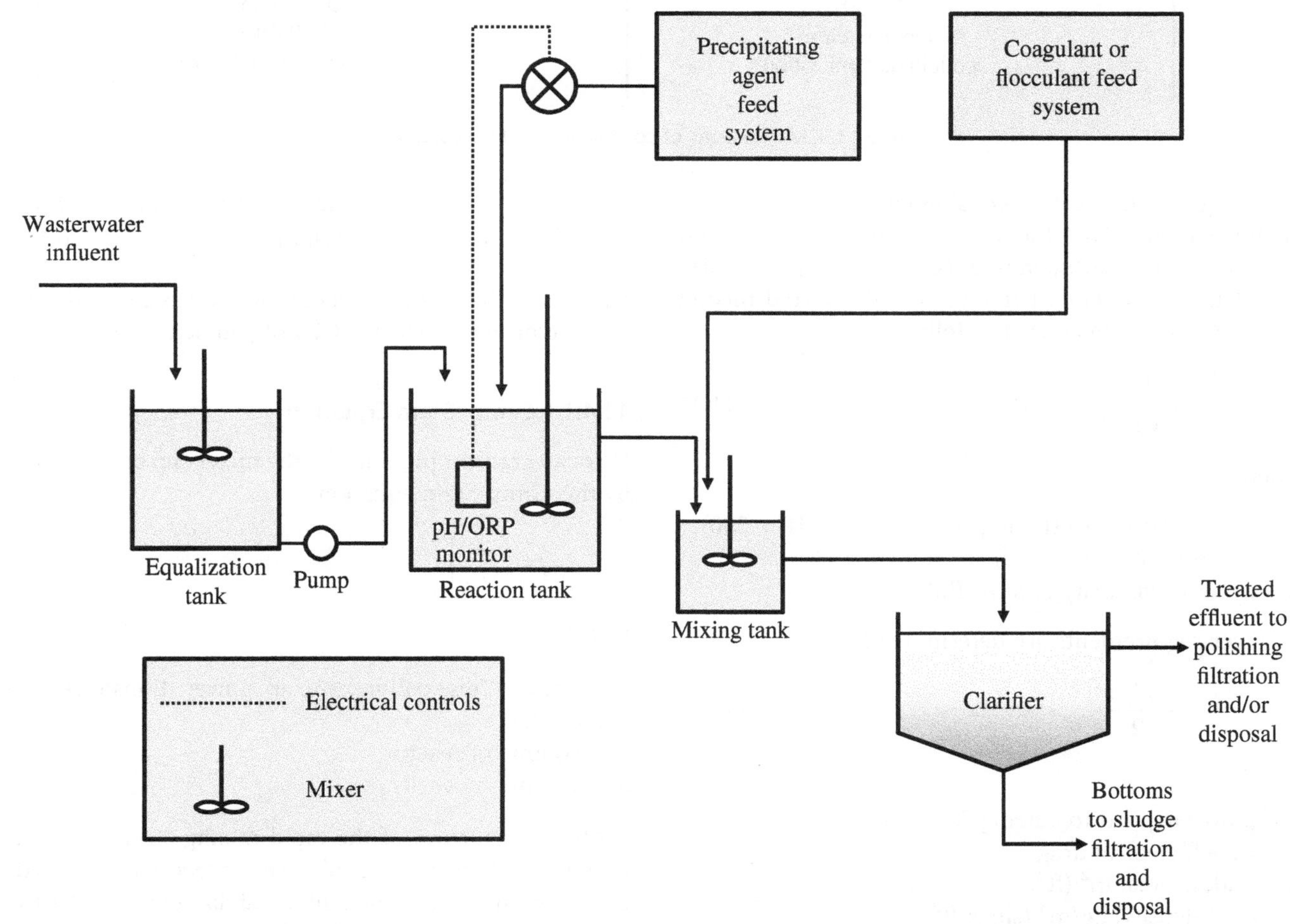

Figure 11.2 Continuous metal precipitation, coagulation, and flocculation (P/C/F) system.

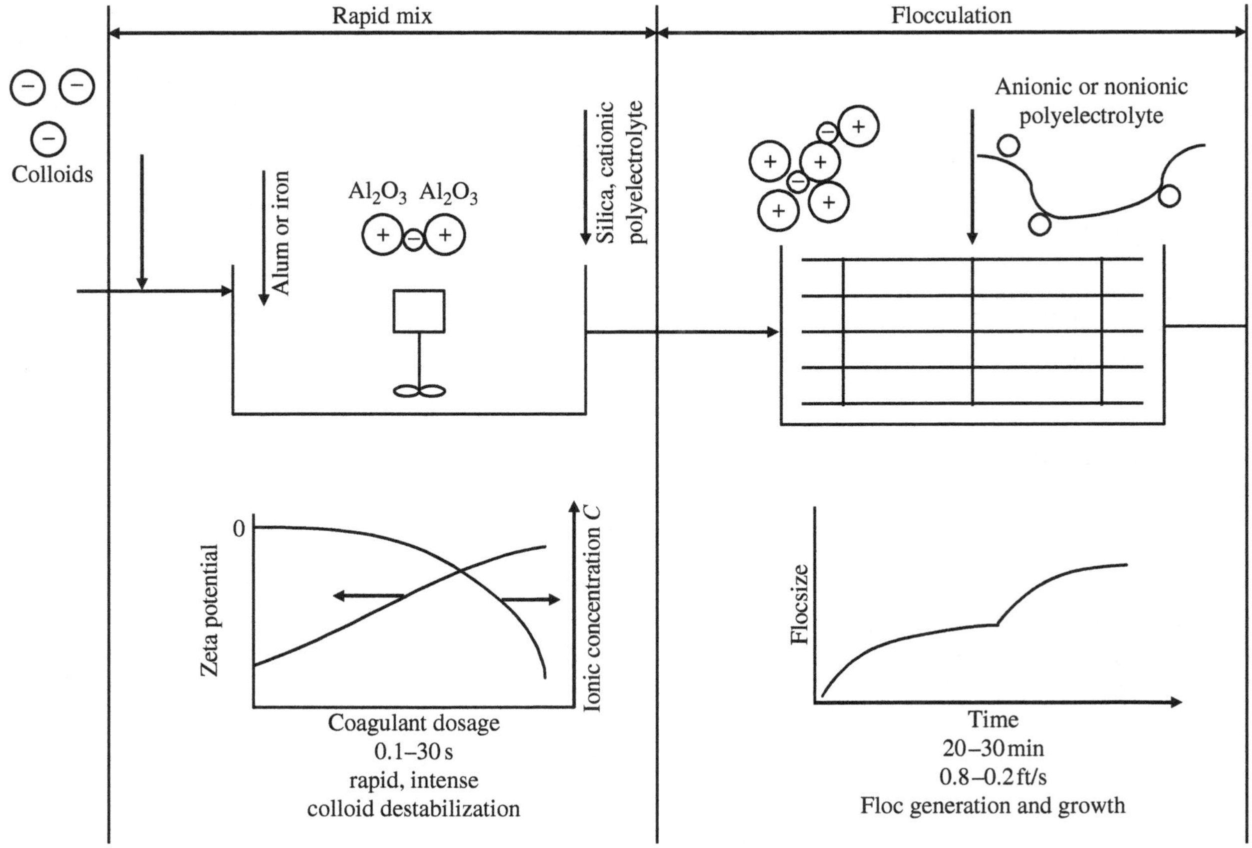

Figure 11.3 Mechanisms of coagulation and flocculation.

following factors must be considered: (i) the amount of energy supplied, (ii) the mixing residence time, and (iii) the related turbulence effects of the specific size and shape of the tank. For mechanically stirred mixing basins, G can be calculated as follows:

$$G = \left(\frac{P}{V\mu}\right)^{0.5} \tag{11.1}$$

where

P = power applied to stirring, W (ft-lbf/s = HP × 550)
V = reactor volume m^3 (ft^3)
μ = dynamic viscosity N-s/m^2 (lbf-s/ft^2)

Power requirements are determined from

$$P = \frac{C_D A \rho v^3}{2} \tag{11.2}$$

where

P = power applied to stirring, W (ft-lbf/s)
C_D = coefficient of drag
A = paddle area, m^2 (ft^2)
ρ = fluid density, kg/m^3 (slugs/ft^3)

For water at 20°C,

$$\rho = \frac{62.4\,\text{lbm}}{\text{ft}^3} \times \frac{\text{slug}}{32.124\,\text{lbm}} = \frac{1.945\,\text{slugs}}{\text{ft}^3} = \frac{1000\,\text{kg}}{\text{m}^3}$$

where v is the relative velocity of paddles in fluid (m/s [fps]), typically about 0.6–0.75 of paddle tip speed.

11.4.1 Camp-Stein Equation

Velocity gradient provided by the mixer can be estimated by the Camp-Stein equation:

$$\bar{G} = \left[\frac{\bar{P}}{V \cdot \mu}\right]^{0.5} \tag{11.3}$$

where

$\bar{P}$ = power dissipation (mixer power transferred to bulk fluid)
V = volume of reactor
μ = dynamic viscosity

After coagulation of the rapid mixing stage, flocculation is based on slow-to-enhance contact between particles to promote the growth of destabilized floc. Under the hydraulic force, G, the *orthokinetic flocculation* colloid number rate decreases according to the following equation:

$$J_{ok} = \frac{dN}{dt} = -\frac{2\alpha\bar{G}d^3N^2}{3} \quad (11.4)$$

In-line static mixers are a type of pipe reactors used for rapid mixing that can be done for any type of flow (laminar, transitional, and turbulent). When comparing both the static and tank mixed reactors, more energy is required in the tank than in the static mixer.

11.4.2 Static and Plug-flow Reactor Mixers

Figure 11.4 shows the hydraulic characters of PFR, Batch, and CSTR reactors.

11.4.3 Power, Pressure, and Pump in Reactors

Hydraulic power formula is used to calculate the power of each reactor:

$$\text{Hydraulic power} = P = \gamma \times Q \times h_L \quad (11.5)$$

where

γ = specific gravity
Q = flow rate
h_L = headloss

Tank reactors form dead zones and short-circuiting, whereas plug-flow conditions minimize short-circuiting and completely avoid dead zones. The US EPA Design Manual (2004) provided good design examples of separation technologies.

Example 11.1
Given: A landfill leachate and groundwater have the following characteristics:

1) pH: 6.5
2) Cr^{6+}: 15 mg/l
3) Total Cr: 75 mg/l
4) Zn^{2+}: 35 mg/l
5) Suspended solids: 25 mg/l
6) Dissolved oxygen: 2 mg/l

At the following leachate flow rates:

1) The feasibility study determined that the method of treatment would be chemical reduction of chromium (Cr^{6+} to Cr^{3+}) using sulfur dioxide, followed by

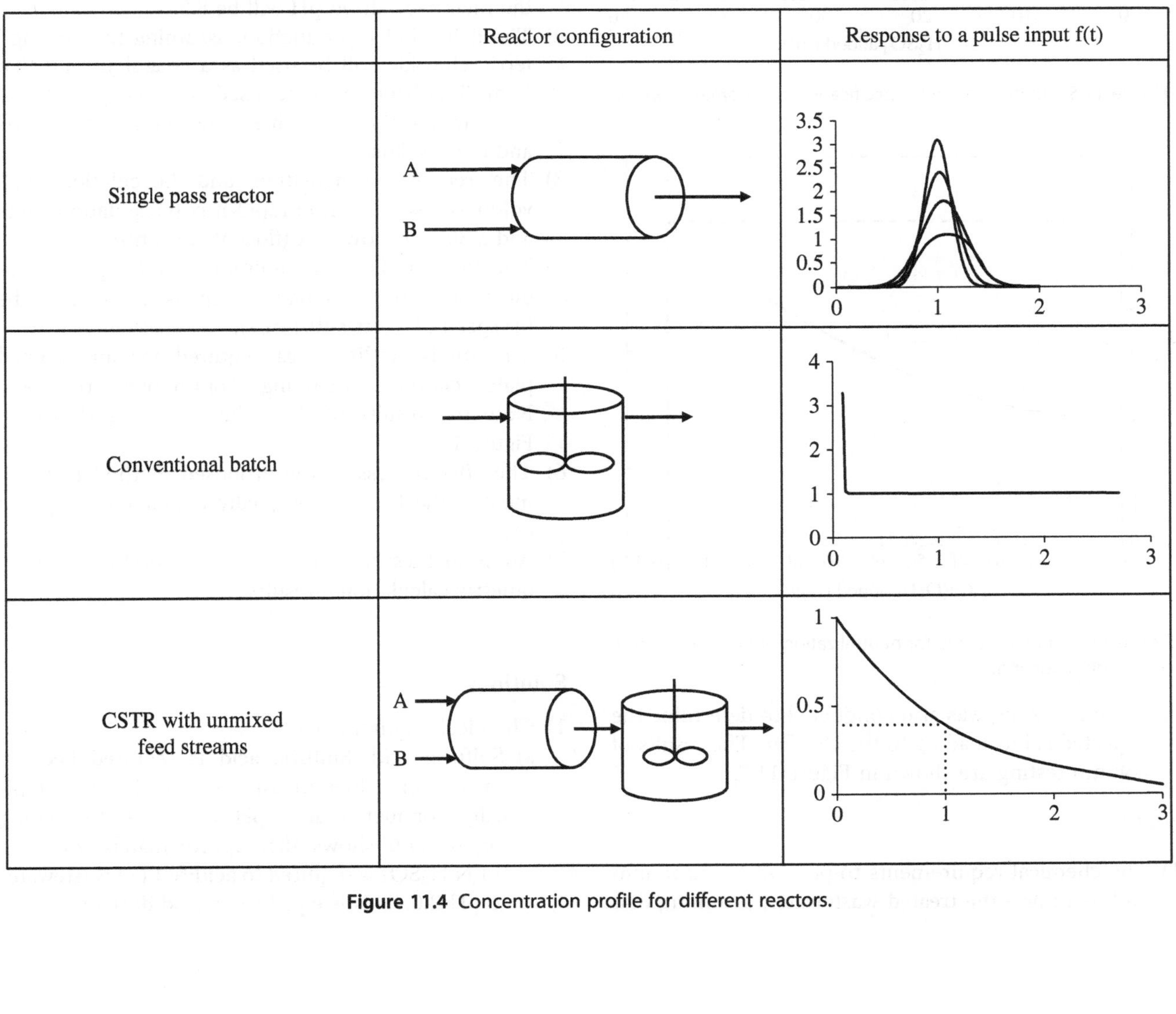

Figure 11.4 Concentration profile for different reactors.

hydroxide precipitation using lime, and then followed by coagulation, flocculation, and clarification. The full-scale treatment system will be designed to operate 24 h/day.

2) During initial bench-scale testing, a 1-l sample of raw wastewater was titrated with 0.1 N sulfuric acid and then neutralized with lime, yielding the results shown in Figures 11.5 and 11.6.

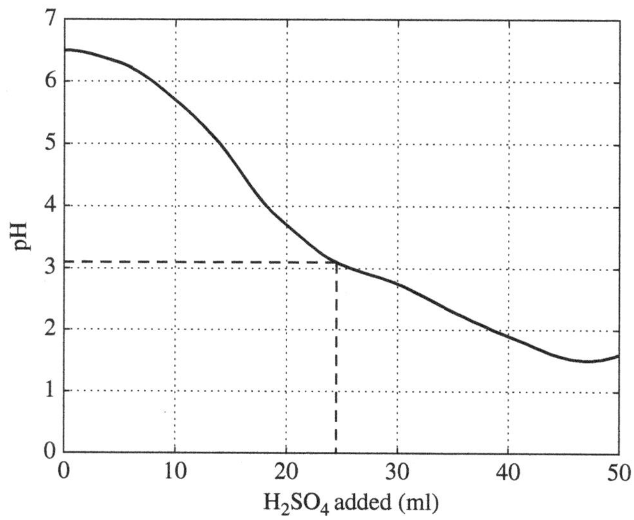

Figure 11.5 Titration curve for pacification of chromium waste.

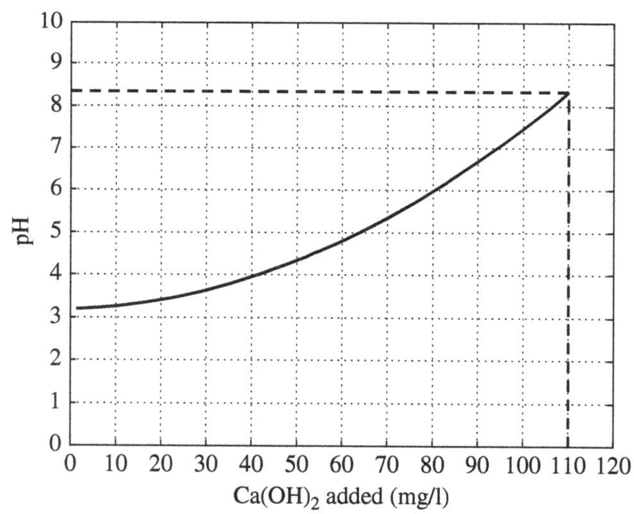

Figure 11.6 Titration curve for neutralization of waste following chromium reduction.

3) Column testing was also conducted to determine the expected solids loading to the clarifier. The results of column testing are shown in Figure 11.7.

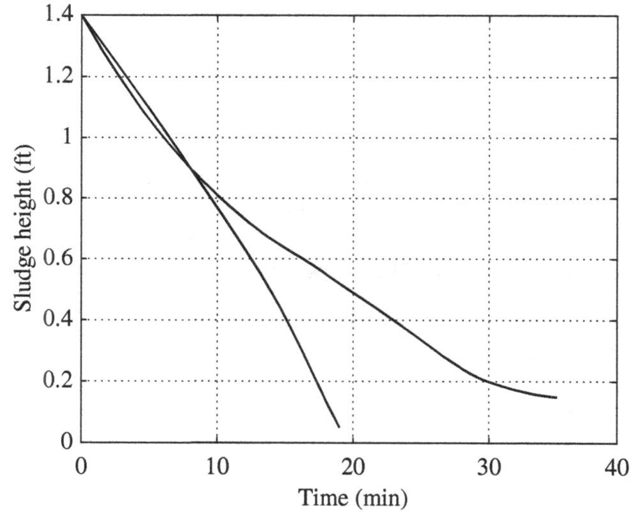

Figure 11.7 Column testing results.

Find:

1) The chemical requirements to precipitate chromium and neutralize the treated wastewater, assuming that sulfur dioxide will be used as the reducing agent and that lime will be used as the precipitant. Assume that the final neutralized pH will be 8.5.
2) The daily sludge production, assuming that 10 mg/l ferric chloride will be used as a coagulant and that 1 mg/l polymer will be used as a coagulant aid. Assume that the sludge has a specific gravity of 1.02 and is 3% solids.
3) The required coagulation and flocculation tank volumes, using a 1-min rapid-mix (coagulation) time and a 30-min slow-mix (flocculation) time.
4) The theoretical power requirement and required paddle area for the flocculation step, assuming a paddle tip speed of 1.2 ft/s (0.4 m/s).
5) The solids settling area required for an inclined plate clarifier, assuming continuous treatment. Note the results of the column testing shown in Figure 11.7.
6) The effectiveness of the proposed method of treatment if the leachate originally contained 5 mg/l of cadmium.
7) Additional testing to confirm the theoretical (stoichiometric calculations) results.

Solution:

1) Chemical requirements
   a) Sulfuric acid. Sulfuric acid is required because chromium reduction using sulfur dioxide is typically conducted at a pH of 2.5–3. Examining Figure 11.5 shows that approximately 25 ml of 0.1 N H_2SO_4 is required to acidify 1 l of wastewater to pH 3.0. This is equal to an acid dose of

$$(25\,\text{ml/l}) \times (0.1\,\text{N}) = 2.5\frac{\text{meqH}_2\text{SO}_4}{\text{l}}$$

The acid feed rate required to achieve this dose will depend on the normality of acid used for treatment. (Note: Typical required retention times for chromium reduction at a pH of 3 range from 2 to 20 min, depending on the initial hexavalent chromium concentration.)

b) Sulfur dioxide. The equations that govern the reduction of hexavalent chromium to trivalent chromium using sulfur dioxide are as follows:

$$SO_2 + H_2O \rightarrow H_2SO_3$$

$$2H_2CrO_4 + 3H_2SO_3 \rightarrow Cr_2(SO_4)_3 + 5H_2O$$

$$H_2SO_3 + \frac{1}{2}O_2 \rightarrow H_2SO_4$$

Using the aforementioned equations and molar ratios, and assuming that the reactions go to completion, gives the SO_2 requirements as follows for Cr^{6+}. Because 3 mol of SO_2 yields 3 mol of H_2SO_3, then

$$\frac{3 \times (32 + 2(16))}{2 \times 52} = 1.85\frac{\text{mg}}{\text{l}}SO_2 \text{ per}\frac{\text{mg}}{\text{l}}\,Cr^{6+}$$

(Note: The quantity (32 + 2(16)), or 64, is the gram atomic weight of sulfur dioxide. The quantity 52 is the gram atomic weight of chromium.) And the requirements for O_2 are

$$\frac{3 \times (32 + 2(16))}{(1/2) \times 2(16)} = 4\frac{\text{mg}}{\text{l}}SO_2 \text{ per}\frac{\text{mg}}{\text{l}}O_2$$

Therefore,

$$SO_2\text{required} = (1.85)15\frac{\text{mg}}{\text{l}}\,Cr^{6+} + 4(2.0)\frac{\text{mg}}{\text{l}}O_2 = 35.8\frac{\text{mg}}{\text{l}}$$

c) *Lime*. Lime is required to raise the pH of the wastewater to precipitate chromium. Assume that the lime is 90% pure.

i) The amount of lime required to neutralize the wastewater to pH 8.5, as indicated in Figure 11.6, is approximately 107 mg/l.

ii) The amount of lime required to precipitate the chromium can be calculated using the following equation:

$$Cr_2(SO_4)_3 + 3Ca(OH)_2 \rightarrow 2Cr(OH)_3 + 3CaSO_4$$

$$\frac{3 \times (40 + 2(16) + 2(1))}{3 \times 52} = 2.14\frac{\text{mg}}{\text{l}}\text{of } Ca(OH)_2 \text{ per}\frac{\text{mg}}{\text{l}}\text{ of } Cr^{3+}$$

The total lime requirement to precipitate trivalent chromium is based on the combined total of trivalent and hexavalent chromium present in the raw wastewater (75 mg/l):

$$Ca(OH)_2\text{required (90\% pure)} = \frac{2.14 \times 75}{0.90}\frac{\text{mg}}{\text{l}} = 178.3\frac{\text{mg}}{\text{l}}$$

iii) Lime is also required to precipitate zinc (35 mg/l). The minimum solubility of zinc hydroxide occurs at approximately pH 9.0 (relatively close to 8.5; see Figure 11.8). Therefore, for simplicity, assume that the lime reaction with zinc goes to completion at pH 8.5. The amount of lime can be calculated using the following equation:

$$Ca(OH)_2 + Zn^{2+} \rightarrow Zn(OH)_2 + Ca^{2+}$$

$$\frac{(40 + 2(16) + 2(1))}{65.4} = 1.13\frac{\text{mg}}{\text{l}}\text{of } Ca(OH)_2 \text{ per}\frac{\text{mg}}{\text{l}}\text{ of } Zn^{2+}$$

$$Ca(OH)_2\text{required (90\% pure)} = \frac{1.13 \times 35}{0.90}\frac{\text{mg}}{\text{l}} = 43.9\frac{\text{mg}}{\text{l}}$$

iv) The amount of lime required to neutralize the H_2SO_4 produced from dissolved oxygen initially present in the wastewater can be calculated from the following equations:

$$H_2SO_3 + \frac{1}{2}O_2 \rightarrow H_2SO_4$$

$$H_2SO_4 + Ca(OH)_2 \rightarrow CaSO_4 + 2H_2O$$

$$\frac{(2(2) + 32 + 16(4))}{(1/2)(2)(16)} = 6.12\frac{\text{mg}}{\text{l}}H_2SO_4 \text{ produced per}\frac{\text{mg}}{\text{l}}O_2 \text{ present}$$

$$H_2SO_4 \text{ produced} = 6.12 \times 2\frac{\text{mg}}{\text{l}}O_2 = 12.2\frac{\text{mg}}{\text{l}}$$

Therefore, the lime requirement, from the second equation, is

$$\frac{(40 + 2(16) + 2(1))}{(2(1) + 32 + 4(16))} = 0.76\frac{\text{mg}}{\text{l}}\text{lime per}\frac{\text{mg}}{\text{l}}H_2SO_4$$

$$Ca(OH)_2\text{required (assuming 90\% pure)} = \frac{0.76 \times 12.2}{0.90}\frac{\text{mg}}{\text{l}} = 10.3\frac{\text{mg}}{\text{l}}$$

v) Therefore,

The total amount of lime = 107 + 178.3 + 43.9 + 10.3 = 339.5 mg/l

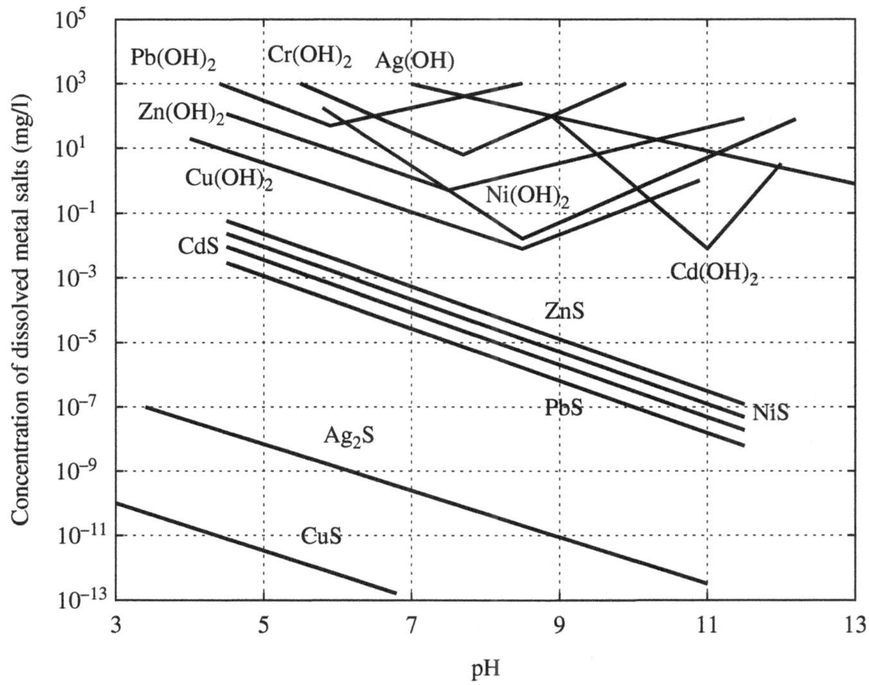

Figure 11.8 Solubility of metal hydroxides and sulfides as a function of pH (US EPA, 1980).

Or, in kg/day,

$$\left(339.5\frac{\text{mg}}{\text{l}}\right)\left(\frac{1000\text{l}}{\text{m}^3}\right)\left(\frac{8640\text{m}^3}{\text{day}}\right)\left(\frac{\text{kg}}{10^6\text{mg}}\right) = 2933.28\frac{\text{kg}}{\text{day}}$$

2) Daily sludge production.
   a) Chromium hydroxide sludge:

$$Ca(OH)_2 \text{ dosage (90\% pure)} = 178.3\frac{\text{mg}}{\text{l}}$$

$$178.3 \text{ mg} \times \frac{1\,\text{mol}}{74000\,\text{mg}} = 0.00241 \text{ mol}$$

3 mol of $Ca(OH)_2$ forms 2 mol of $Cr(OH)_3$ sludge

Therefore,

mg of $Cr(OH)_3$ formed per liter

$$= \frac{2}{3}(0.00241 \text{ mol})(103000 \text{ mg/mol}) = 165.5 \text{ mg}$$

   b) Zinc hydroxide sludge:

$$Ca(OH)_2 \text{ dosage} = 43.9\,\text{mg/l}$$

$$43.9 \text{ mg} \times \frac{1\,\text{mol}}{74000\,\text{mg}} = 0.000593 \text{ mol}$$

1 moles of $Ca(OH)_2$ forms 1 moles of $Zn(OH)_2$

Therefore,

mg of $Zn(OH)_2$ formed per liter

$$= (0.000593 \text{ mol})(99400 \text{ mg/mol}) = 58.9 \text{ mg}$$

   c) Suspended solids. Given in the problem statement (25 mg/l).
   d) Coagulant and coagulant aid. Assume that all coagulant and coagulant aid settle out of solution and subsequently contribute to the sludge volume. Therefore, coagulant at 10 mg/l + coagulant aid at 1 mg/l = 11 mg/l.
   e) Total sludge.

$$165.5\text{mg} + 58.9\text{mg} + 25\text{mg} + 11\,\text{mg/l} = 260.4\text{mg/l}$$

Or

$$\left(260.4\frac{\text{mg}}{\text{l}}\right)\left(1000\frac{\text{l}}{\text{m}^3}\right)\left(8640\frac{\text{m}^3}{\text{day}}\right)\left(\frac{\text{kg}}{10^6\text{mg}}\right)$$

$$= 2249.87856\frac{\text{kg}}{\text{day}}$$

As mentioned in the problem statement, on the assumption that the sludge is 3% solids and that the specific gravity of the sludge is 1.02, the volume that will require disposal each day can be calculated as follows:

$$\frac{2249\,\text{kg/day}}{(1000\text{kg/m}^3)(1.02)(0.03)} = 73524.71\frac{\text{m}^3}{\text{day}}$$

3) Required volumes of coagulation and flocculation units:
   a) Coagulation tank:

$$\left(8640\frac{\text{m}^3}{\text{day}}\right)(\text{day}/1440 \text{ min})(1 \text{ min}) = 6\text{m}^3$$

For a cubicle tank, approximately

$$(3 \times 3 \times 3)\text{ft or } (0.9 \times 0.9 \times 0.9)\text{m}$$

b) For flocculation tank:

$$\left(8640\frac{\text{m}^3}{\text{day}}\right)(\text{day}/1440\ \text{min})(30\ \text{min}) = 180\text{m}^3$$

Assume a length-to-width ratio of 2 : 1. A rectangular tank, 5 ft (2 m) deep, would approximately has reaction volume

$$(18 \times 9 \times 5)\text{ft} = 810\,\text{ft}^3 \text{ or } (5 \times 2 \times 2)\text{m} = 20\text{m}^3$$

4) Calculate the theoretical power requirement and required paddle area for the flocculation step
   a) Theoretical power requirement. Rearranging the equation,

$$G = \left(\frac{P}{V\mu}\right)^{0.5}$$

$$P = \mu G^2 V$$

where
μ = absolute fluid viscosity, lb-s/ft^2 or N-s/m^2 (for water at 20°C, $\mu = 0.21 \times 10^{-4}$ lb-s/ft^2 = 1×10^{-3} where N-s/m^2)
V = tank volume ft^3 or m^3
P = power, ft-lb/s or W or horsepower (hp)
G = mean velocity gradient, ft/s-ft or m/s-m (Note that typical G value for flocculation is 30 ft/s/ft (30/s).)
Therefore,

$$P = \left(0.21 \times 10^{-4}\text{lb-s/ft}^2\right)(30\text{ft/s/ft})^2\left(810\text{ft}^3\right)$$
$$= 15.3\text{ft-lb/s} = 15.3/550 = 0.03\text{hp}$$

Or

$$P = \left(1.00 \times 10^{-3}\text{N-s/m}^2\right)(30\text{m/s/m})^2\left(20\text{m}^3\right) = 18\,\text{W}$$

   b) Paddle area requirement. Rearranging the equation,

$$P = \frac{C_D A \rho v^3}{2}$$

$$A = \frac{2P}{C_D v^3 \rho}$$

Use a paddle tip speed of 1.2 fps (0.4 m/s).
where
A = paddle area
P = power requirement (ft-lb/s) or (W)
C_D = dimensionless coefficient of drag (typically taken as 1.8)
v = relative velocity of paddle in fluid, fps or m/s (Assume to be 0.75 times the paddle tip speed)
ρ = mass fluid density (at 20°C, 1.94 lb/ft^3, or slugs/ft^3, or kg/m^3)

Therefore,

$$A = \frac{2(15.3)}{(1.8 \times ((0.75)(1.2))^3(1.94))} = 12.0\text{ft}^2 = 0.74\text{m}^2$$

5) Calculate the solids settling area required for inclined plate clarifier. Figure 11.7 shows the results of the column testing. The column tests determined that the settling rate was 1.34 ft (0.41 m) in 18 min. Convert this to commonly used units. (Note: 1.34 ft (0.41 m) is the height of the solid/liquid interface at time zero, when the column is filled to 1000 ml.)

$$\frac{0.41\,\text{m}}{18\ \text{min}} = 0.023\frac{\text{m}^3/\,\text{min}}{\text{m}^3}$$

   a) For continuous rate divided by 2 (per manufacturer):

$$\frac{0.023}{2} = 0.011\frac{\text{m}^3/\,\text{min}}{\text{m}^3}$$

$$\left(8640\frac{\text{m}^3}{\text{day}}\right)\left(\frac{\text{day}}{1440\ \text{min}}\right) = 6\frac{\text{m}^3}{\text{min}}$$

$$\text{Solids settling volume required} = 6\frac{\text{m}^3}{\text{min}} \times \frac{\text{day}}{0.011\,\text{m}^3/\text{day}}$$
$$= 545.54\text{m}^3$$

(Note: Most manufacturers of inclined plate clarifiers recommend a rate of 0.25 gpm/ft^2 (0.010 (m^3/min)/m^2) for metal hydroxide precipitates. Rates should not exceed 1.0 gpm/ft^2 (0.04 (m^3/min)/m^2.)

Answer: The required dimensions of an inclined plate clarifier to accommodate 744 ft^2 of effective settling area would be 6 ft × 10 ft × 12 ft.

The minimum solubility of cadmium hydroxide occurs at pH > 11. At a pH of 8.5, the solubility of cadmium hydroxide is high (greater than 100 mg/l). Therefore, by raising the pH to only 8.5 or 9 (to remove chromium and zinc), the cadmium concentration would not be reduced below the original concentration of 5 mg/l. To effectively remove cadmium, a second-stage precipitation/clarification step would be required where the pH would have to be raised to 11. Results of jar testing could be conducted to quantify coprecipitation of cadmium hydroxide.

Comments: From SEE design principles, it is clear that precipitation of metal violates several SEE design rules:

1) Precipitation usually hydrolyzes toxic metal to form the corresponding metal oxides. This process

incorporates free water, a product, into sludge, which is the secondary waste and could be more difficult to dispose.

2) In precipitation, pH adjustment requires alkaline such as lime, which consumes chemicals and produces additional sludge.
3) To separate the fine precipitates, it requires coagulation and flocculation, which requires velocity gradient and therefore energy.

For the aforementioned reasons, metals should be separated using ion exchange so that they can be recovered and concentrated for recycling or reuse to reduce the footprint on materials and the energy required by the treatment systems.

After precipitation, sludge handling is an energy-intensive process. Typically, sludge from water treatment plants is pressure filtrated since it is often difficult to dewater, particularly alum sludge and softening precipitates containing magnesium hydroxide. Gravity-thickened alum waste is conditioned by the addition of lime slurry. The conditioned sludge is then fed continuously to the pressure filter until filtrate ceases, and the cake is consolidated under high pressure. Filtrate is measured through a weir tank and recycled to the inlet of the water treatment plant. The cake is transported by truck to a disposal site. Alum sludge is conditioned using lime and/or fly ash. Lime dosage is in the range 10–15% of the sludge solids. Ash from an incinerator, or fly ash from a power plant, is applied at a much higher dosage, approximately 100% of dry sludge solids. Polyelectrolytes may also be added to aid coagulation. Fly ash and diatomaceous earth are used for precoating; the latter requires about 50.2 kg/m^2 of filter area. Under normal operation, cake density is 40–50% solids and has a dense, dry, textured appearance. Alum sludge from surface water treatment is amenable to centrifuge dewatering to about 20% solids. The removal efficiency in a scroll centrifuge ranges from 50 to 95% based on operating conditions and polymer dosage. Alum sludge from surface water coagulation settles to a density in the range of 2–6% solids. Coagulation-softening mixtures from the treatment of turbid river waters gravity-thicken approximately as follows: alum–lime sludge, 4–10%; iron–lime settlings, 10–20%; alum–lime wash water, about 4%; and iron–lime backwash, up to 8%. The density achieved in gravity thickening relates to 8%. Lime-softening precipitates compact more readily than alum floc in a scroll centrifuge. A settled sludge input with 15–25% solids can be dewatered to a solidified cake of 65%. Suspended solids recovery is often 85–90% with polyelectrolyte flocculation. The performance of thickener handling water treatment plant waste varies with the character of the water being treated and the chemicals applied. Calcium carbonate residue from groundwater softening consolidates to 15–25% solids. Solids content and sludge volume are intrinsically linked as in Table 11.4 and Figures 11.9 and 11.10.

The regression equation between the log(kg sludge/kg dry solids) and the log(sludge solids content) is

$$\log(\text{kg sludge/kg dry solids}) = -1.0008 \log(\text{sludge solids content}) + 2.0005$$
$$r^2 = 1.0000$$

or

$$\text{kg sludge/kg dry solids} = 100.1117(\text{sludge solids content})^{-1.0008}$$
$$r^2 = 1.0000$$

Table 11.4 Density and volume of sludge.

Sludge solids content (%)	kg sludge/kg dry solids	m^3 sludge/ton dry solids
1	100	100
2	50	50
5	20	20
10	10	9.7
15	6.7	6.4
20	5.0	4.7
30	3.3	3.0
40	2.5	2.2

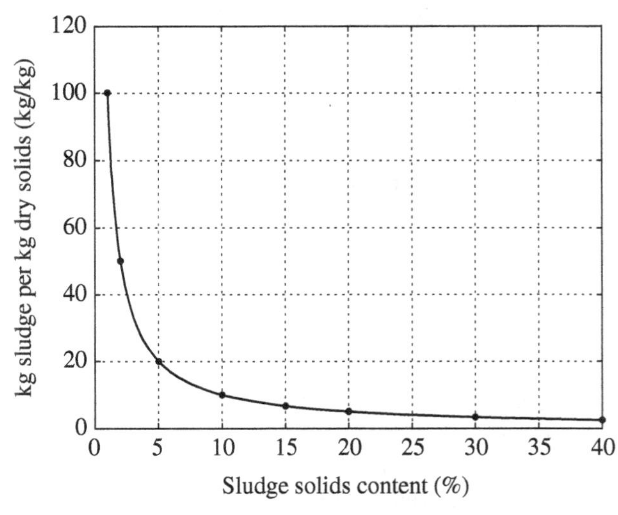

Figure 11.9 kg sludge/kg dry solids versus sludge solids content (%).

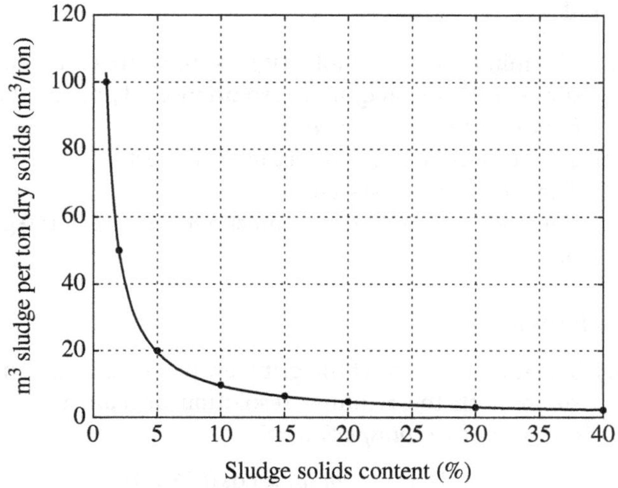

Figure 11.10 m^3 sludge/ton dry solids versus sludge solids content (%).

The regression equation of log(m^3 sludge/ton dry solids) and log(sludge solids content) is

$$\log(m^3 \text{ sludge/ton dry solids}) = -1.0332 \log(\text{sludge solids content}) + 2.0115$$
$$r^2 = 0.9997$$

or

$$m^3 \text{sludge/ton dry solids} = 102.6814(\text{sludge solids content})^{-1.0332}$$
$$r^2 = 0.9997$$

11.5 Membrane Filtration Systems

Water membrane filtration systems can be classified as (i) microfiltration, (ii) ultrafiltration, (iii) nanofiltration, (iv) reverse osmosis, and (v) biomembrane reactors. All of these separation technologies are designed to filter a specific range size particle with the corresponding pore size. For example, microfiltration focuses on the largest particle as shown in Figure 11.1. Reverse osmosis is to remove the smallest particles such as ion in the size ranging from several angstrom due to the small membrane pore sizes. It removes cations, acids, and salts at angstroms to nanometers. The filtration systems usually remove the large size particles first. Therefore, microfiltration and ultra-filtration are typically used for pre-treatment for nanofiltration or reverse osmosis to prevent fouling.

For water treatment or reclaiming treated wastewater, high pressure is required to force water passing through membrane as permeate and leave concentrate as waste by-product. Pressure through the pump to aim for permeate performance using hollow fibers is

$$\Delta P = \rho_L g h \tag{11.6}$$

This is the formula to determine the pressure of the flow.

$$\text{Resistance } (\text{kg/m}^2/\text{s}) : \bar{R} = \frac{\text{Driving force}}{\text{Flux}} = \frac{\Delta P}{J} \tag{11.7}$$

$$\text{Membrane resistance } (\text{m}^{-1}) : R = \frac{\bar{\bar{R}}}{\mu} = \frac{\Delta P}{\mu J} \tag{11.8}$$

Permeability for overall flow :

$$k_V = \frac{1}{\text{Resistivity}} = \frac{J_V}{\Delta P/\delta_{mem}} \tag{11.9}$$

$$\text{Permeability for individual species} : k_i = \frac{J_i}{\Delta P/\delta_{mem}} \tag{11.10}$$

where: ΔP = transmembrane differential pressure, psi or kPa

dynes/cm^2

J = flux, gfd or Lmh

kv = Permeability for overall flow

μ = viscosity constant of water

δ_{mem} = thickness of the membrane, ft or m

The Long Term 2 Enhanced Surface Water Treatment Rule (LT2ESWTR) requires state to report integrity test of membrane filtration system to earn credit for virus removal (US EPA 2005). The procedure is described in the Membrane Filtration Guidance Manual (US EPA, 2005). Example 11.2 illustrates how to calculate the required technical values to comply with the EPA's requirements for membrane filtration to earn 3 log virus removal credit. The system is a submerged vacuum-driven hollow-fiber membrane. The upper control limit (UCL) refers to the maximal pressure before the membrane is breached. Log removal value is defined as the log $C_{effluent}$ – log $C_{influent}$.

Example 11.2

Given: A submerged, vacuum-driven hollow-fiber membrane system that operates in a deposition (i.e. dead-end) mode hydraulic configuration is required to provide a total of at least 3 log removals for *Cryptosporidium* by the LT2ESWTR. A pressure decay test is applied to one of the membrane units in the system. Assume the plant is at sea level.

The parameters of the design project include operational parameters, direct integrity test parameters, and unit and membrane characteristics:

1) Operational parameters
   a) The permitted design capacity of the membrane unit is 1200 GPM.
   b) The maximum anticipated water temperature is 75°F (24°C).
   c) The minimum anticipated water temperature is 41°F (5°C).
   d) The maximum anticipated backpressure that might be exerted on the units during direct integrity testing is 75 in of water.
   e) The minimum anticipated backpressure that might be exerted on the units during direct integrity testing is 60 in of water.
   f) The backpressure measured prior to the most recent pressure decay test was 65 in of water.
   g) The filtrate flow measured prior to the most recent pressure decay test was 1000 GPM.
   h) The trans-membrane pressure (TMP) measured prior to the most recent pressure decay test was 10 psi.
2) Direct integrity test parameters
   a) The volume of pressurized piping during the test is 285 l.
   b) The initial applied test pressure is 16 psi.
   c) The duration of the pressure decay test is 10 min.
   d) Baseline (i.e. diffusive) decay is negligible over the duration of the test.
   e) The smallest verifiable rate of pressure decay under known compromised conditions is 0.10 psi/min.
   f) The most recent pressure decay test yielded a result of 0.13 psi/min.
   g) The temperature of both the water and the applied air was 68°F (20°C) during the most recent pressure decay test.
3) Unit and membrane characteristics
   a) The maximum rated TMP is 30 psi.
   b) The pore shape correction factor (κ) for the membrane material was not determined experimentally, and thus a conservative value of 1 is assumed.
   c) The membrane material is relatively hydrophilic and has a liquid-membrane contact angle (i.e. "wetting angle") of 300, as determined using a method acceptable to the state.
   d) The membrane surface variation (i.e. roughness) is 0.3 μm, as provided by the manufacturer.
   e) The hollow-fiber lumen diameter is 500 μm.
   f) The depth of the membrane into the potting material is 50 mm.
   g) All flow through any integrity breach that may be present is assumed to be turbulent.

Find:

1) The minimum direct integrity test pressure commensurate with the required resolution of 3 μm for the removal of *Cryptosporidium.*
2) The sensitivity of the direct integrity test.
3) The UCL for this system.
4) The LRV verified by the most recent direct integrity test.

Solution:

1) The minimum direct integrity test pressure commensurate with the required resolution of 3 μm for the removal of *Cryptosporidium*:

$$P_{test} = (0.193 \times \kappa \times \sigma_{water} \times \cos(\theta)) + BP_{max}$$
$$= \left(0.193\frac{\text{psi}\cdot\text{cm}}{\text{dyne}} \times 1.0 \times 74.9\frac{\text{dynes}}{\text{cm}} \times \cos(30°)\right) + \frac{75\,\text{in}\,H_2O}{\frac{27.7\,\text{in}\,H_2O}{\text{psi}}}$$
$$= 12.519\,\text{psi} + 2.708\,\text{psi} = 15.227\,\text{psi}$$

where

P_{test} = test pressure
κ = pore shape correction factor
σ_{water} = surface tension of water
BP_{max} = maximum anticipated backpressure over the conversion rate of inches of water to psi
θ = wetting angle

Since the scenario states that baseline diffusive losses are negligible, the pressure calculated earlier represents the lowest permissible initial applied test pressure.

2) The sensitivity of the direct integrity test:

$$LRV_{DIT} = \log\left(\frac{Q_p \times ALCR \times P_{atm}}{\Delta P_{test} \times V_{sys} \times VCF}\right)$$

$$ALCR = 170 \times Y \times \sqrt{\frac{(P_{test} - BP)(P_{test} + P_{atm})}{(460 + T) \times TMP_{max}}}$$

$$Y \propto \left[\frac{1}{\left(\frac{P_{test} - BP}{P_{test} + P_{atm}}\right)}, K\right]$$

$$K = f \times \frac{L}{d_{fiber}}$$

$$K = 0.037 \times \frac{50\,\text{mm}}{0.5\,\text{mm}} = 3.7$$

$$\left(\frac{P_{test} - BP}{P_{test} + P_{atm}}\right) = \left(\frac{16\,\text{psi} - \dfrac{60\,\text{in}\,H_2O}{\dfrac{27.7\,\text{in}\,H_2O}{\text{psi}}}}{16\,\text{psi} + 14.7\,\text{psi}}\right) = 0.45$$

$$Y \propto \left[\frac{1}{0.45}, 3.7\right] \approx 0.78$$

$$\text{ALCR} = 170 \times 0.78 \times \sqrt{\frac{\left(16\,\text{psi} - \dfrac{60\,\text{in}\,H_2O}{\dfrac{27.7\,\text{in}\,H_2O}{\text{psi}}}\right)(16\,\text{psi} + 14.7\,\text{psi})}{(460 + 75) \times 30\,\text{psi}}} = 21.57$$

$$\text{LRV}_{DIT} = \log\left(\frac{1200\,\text{GPM} \times 3.785\,\dfrac{1}{\text{gal}} \times 21.57 \times 14.7\,\text{psi}}{0.1\,\dfrac{\text{psi}}{\text{min}} \times 285\,1 \times 1}\right) = 4.7$$

where

Q_p = design capacity filtrate flow
ΔP_{test} = smallest verifiable decay rate
P_{test} = initial applied test pressure
P_{atm} = atmospheric pressure
V_{sys} = volume of the system
LRV_{DIT} = maximum removal value that a membrane filtration system is capable of verifying
ALCR = pipe flow model for a hollow-fiber membrane filtration system under conditions of turbulent flow
VCF = standard for deposition mode hydraulic configuration
BP = minimum backpressure over the conversion rate of inches of water to psi
T = maximum anticipated temperature
TMP_{max} = maximum allowable TMP value
Y = net expansion factor
K = flow resistance coefficient
f = friction factor
L = potting depth
d_{fiber} = fiber diameter

The maximum removal value that this membrane filtration system is capable of verifying is 4.7 log or 99.998%.

3) The UCL for this system:

$$\text{UCL} = \frac{Q_p \times \text{ALCR} \times P_{atm}}{10^{\text{LRC}} \times V_{sys} \times \text{VCF}}$$

$$= \frac{1200\,\text{GPM} \times 3.785\,\text{l/gal} \times 21.57 \times 14.7\,\text{psi}}{10^3 \times 285\,\text{l} \times 1}$$

$$= 5.05\,\text{psi/min}$$

The value of UCL is found to be 5.05 psi/min.
where

UCL = the pressure per minute needed for the membrane system
LRC = 3.0 based on the scenario
Q_p = design capacity filtrate flow
ALCR = pipe flow model for a hollow-fiber membrane filtration system under conditions of where turbulent flow
P_{atm} = atmospheric pressure
V_{sys} = volume of the system
VCF = standard for deposition mode hydraulic configuration

4) The LRV verified by the most recent direct integrity test:

$$\text{LRV}_{MRIT} = \log\left(\frac{Q_{pi} \times \text{ALCR} \times P_{atm}}{\Delta P_{test2} \times V_{sys} \times \text{VCF}}\right)$$

$$\text{LRV}_{MRIT} = \log\left(\frac{1000\,\text{GPM} \times 3.785\,\dfrac{1}{\text{gal}} \times 21.57 \times 14.7\,\text{psi}}{0.13\,\dfrac{\text{psi}}{\text{min}} \times 285\,1 \times 1}\right)$$

$$= 4.51$$

where

LRV_{MRIT} = removal value that a membrane filtration system is capable of verifying in the most recent direct integrity test
ΔP_{test2} = measured test decay rate
Q_{pi} = design capacity filtrate flow prior to testing
ALCR = pipe flow model for a hollow-fiber membrane filtration system under conditions of where turbulent flow
P_{atm} = atmospheric pressure
V_{sys} = volume of the system
VCF = standard for deposition mode hydraulic configuration

The maximum removal value that this membrane filtration system is capable of verifying is 4.51 log or 99.9969%.

11.6 Activated Carbon Adsorption

Activated carbon has large adsorption surface area and is easy to design and operate. Therefore, it has been widely used in industrial wastewater treatment. The design is based upon column test results as shown in Example 11.3.

Example 11.3
Given (Table 11.5):

Table 11.5 Design data.

Design data		
Flow rate	50	l/s
Contaminant concentration influent	69	μg
Contaminant concentration effluent	4	μg

Different types of activated carbons were evaluated on the basis of isotherm results, and the most promising carbon was selected for the pilot test; the pilot plant data is shown later, and the results of the test are graphically presented in Figure 11.11 (Table 11.6).

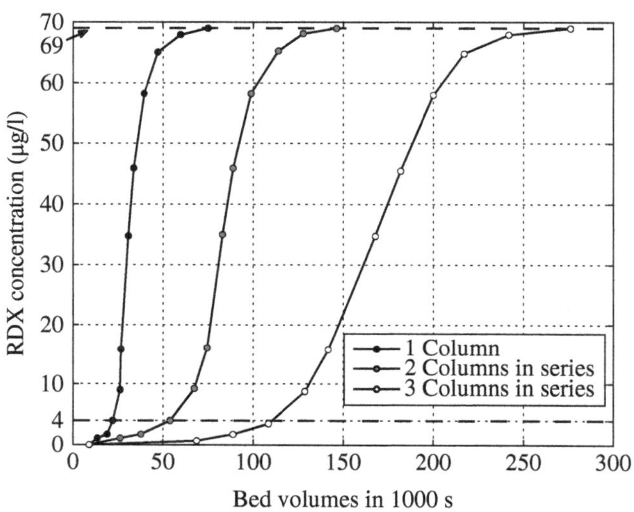

Figure 11.11 RDX concentration versus bed volumes.

Table 11.6 Pilot plant data.

Pilot plant data		
Carbon sample	XYZ	Carbon company
Column inside diameter	108	mm
Column area	9150	mm^2
Bed volume	5.56	l
Flow rate	2.84	l/min
Hydraulic loading	310	lpm/m^2
Bed depth	0.61	m
EBCT (each column)	2	min
RDX influent concentration	69	μg/l
Weight of GAC per column	2.32	kg
Weight of RDX per column	14.2	g
Run time	150	days

Find: Design a liquid phase granular activated carbon (GAC) adsorption system for a flow rate of 50 l/s to treat RDX-contaminated groundwater according to the given pilot study information.

Solution:

Surface Loading Rate
The surface loading rate is given by the pilot area, in this case 310 Lpm/m^2. As a general rule, the rate should be 80–400 Lpm/m^2. Since the range of the pilot is near the high side of the range, the hydraulic loading rate would be limited to 310 Lpm/m^2.

Empty Bed Contact Time (EBCT) and Number of Columns
The pilot data indicate a significant increase in the mass of RDX adsorbed per unit weight of carbon based on a minimum contact time of 4 minutes (if two columns in series are used, each having a contact time of 2 minutes) and an even greater increase for a contact time of 6 minutes. A minimum contact time of 6 minutes per bed (assuming two beds in series) will be used. *Actual EBCT would be found below.*

Nominal Column Diameter
Area:

$$A = \frac{\text{Flow rate}}{\text{Loading rate}} = \frac{Q}{v_0}$$

$$= \frac{3000 \text{l/min}}{310 \text{lpm/m}^2} = 9.68\,\text{m}^2$$

Diameter (Table 11.7):

Table 11.7 CalgonCarbon® manufacturer's data of Model 12-40.

CalgonCarbon® manufacturers		
Model 12-40		
Diameter	3.66	m
Depth	3.35	m
Carbon per adsorbent	18180	kg
GAC density	420	kg/m^3

$$d = \sqrt{\frac{4A}{\pi}} = \sqrt{\frac{4(9.68)}{\pi}} = 3.5\,\text{m}$$

Referring to CalgonCarbon® manufacturer's data, the nearest diameter unit is 3.66 m.

This unit would give us a new area of

$$A = 10.5\,\text{m}^2$$

Bed Depth

Pilot plant data indicates a depth of 0.61 m per column, giving a total depth of 1.22 m. However, according to manufacturer's data, the corresponding depth for the 3.66 m unit should be 3.35 m.

Actual EBCT Based on CalgonCarbon® Units

$$\text{Minimun carbon contact volume} = \text{pilot depth} \times \text{area}$$
$$= 12.81\ \text{m}^3$$

$$\text{Minimum flow rate} = \frac{\text{minimum volume}}{\text{EBCT}}$$
$$= \frac{12.81\ \text{m}^3}{6\ \text{min}} = 2.14\frac{\text{m}^3}{\text{min}}$$

The amount of carbon per bed is about 18 180 kg. The density of GAC is approximately 420 kg/m^3:

$$\text{Volume} = \frac{\text{Mass}}{\text{Density}}$$

$$\text{Volume} = \frac{2\ \text{beds} \times 18180\ \text{kg per bed}}{420\ \text{kg/m}^3}$$
$$= 86.6\ \text{m}^3\ \text{of GAC}$$

Actual EBCT is

$$\text{EBCT} = \frac{\text{Volume}}{\text{Minimum flow rate}}$$
$$= \frac{86.6\ \text{m}^3}{2.14\ \text{m}^3/\text{min}}$$
$$= 40\ \text{min}$$

Anticipated Volume of Water Treated

Contact time for the pilot was 6 min. Actual contact time is 40/6 × 100% = 666% longer. The replacing period for the full-scale system can be approximated by multiplying the ratio of the difference in the contact time by the pilot test run time to breakthrough. A vapor phase-activated carbon treatment unit consists of a blower, an air stripper, a heat exchanger, and carbon vessels to treat air from the air stripper. All the volatile organic compounds (VOCs) such as perchloroethylene (PCE), trichloroethylene (TCE), benzene, and toluene can be stripped from WW. The water containing the VOCs enters the top of the stripper. At the same time the clean air flows upward, and as the water and air have contact with each other, the VOCs are transferred from water phase to air phase. The now contaminated air flows through a heat exchanger to lower its humidity. Furthermore, it passes through the carbon vessels where the carbon absorbs the contaminants from the air. Finally, the air is free of VOCs and is released to the atmosphere (Figure 11.12).

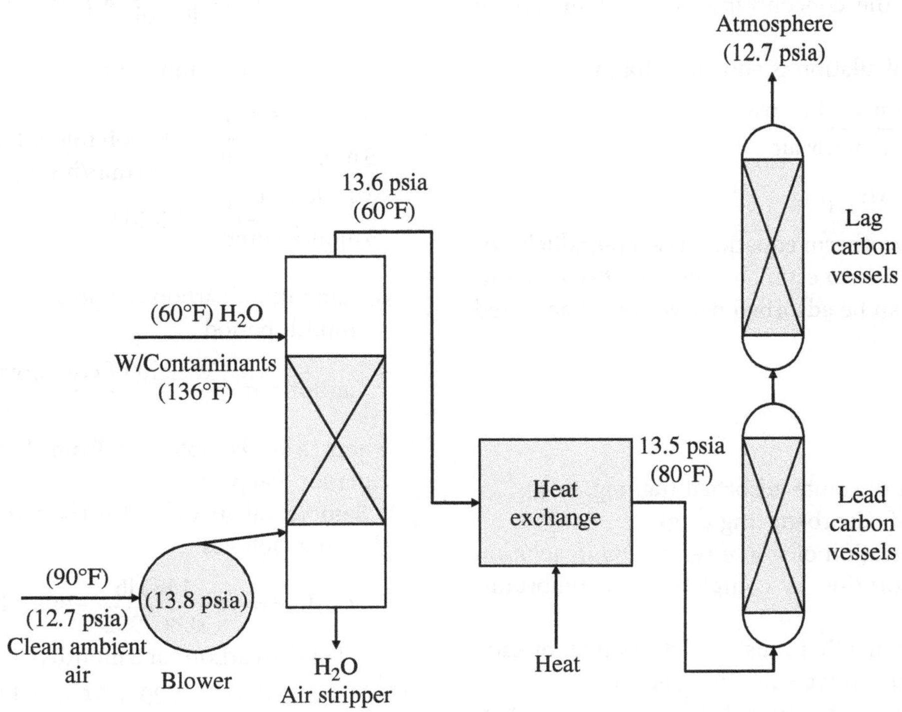

Figure 11.12 Air stripper system.

Example 11.4
Given:

1) Q (flow rate ft^3/min) = 2000 ft^3/min.
2) T_i (inlet temperature) = 90°F.
3) Run time between carbon changes: 3 months/vessel.
4) Number of carbon vessels: two (in series).
5) P_{atm} (atmospheric pressure): 12.7 psia (elevation of site approximately 1 mile above sea level).
6) T_f (outlet temperature) = 60°F.
7) Contaminants and their concentrations leaving the air stripper in the air phase:
   a) Perchloroethylene (PCE): 15 ppmv
   b) Trichloroethylene (TCE): 14 ppmv
   c) Benzene: 9 ppmv
   d) Toluene: 5 ppmv
8) 100% removal of all the organic contaminants.

Find:

1) Amount of carbon required
2) The size of the carbon adsorption vessel

Solution:

1) Determine the amount of carbon needed for 3 months.
   a) Partial pressure of each contaminant. Assuming the pressure in the carbon vessel is the same as the atmospheric discharge pressure of 12.7 psia,

$$P_{\text{contaminant}} = \frac{C \times P_{\text{atm}}}{10^6}$$

where C is the concentration of contaminant in ppmv.

Sample calculation is shown as follows:

$$P_{\text{PCE}} = \frac{15\,\text{ppmv} \times 12.7\,\text{psia}}{10^6\,\text{ppmv air}}$$

$$= 1.9 \times 10^{-4}\,\text{psia}$$

   b) Freundlich isotherm equation. Use Freundlich isotherm equation to estimate the weight of contaminant that can be adsorbed per weight of activated carbon:

$$\frac{x}{m} = KC^{\frac{1}{n}}$$

where
x = amount of solute adsorbed (μg, mg, or g)
m = mass of adsorbent (mg or g)
C = concentration of solute remaining in solution after adsorption is complete (at equilibrium) (mg/l)
K, n = constants that must be determined for each solute, carbon type, and temperature

   c) Freundlich isotherm table for this specific carbon with given K and $1/n$ values at 77°F is as follows (Table 11.8):

Table 11.8 Freundlich isotherm table for this specific carbon with given K and $1/n$ values at 77°F.

Contaminant	K	C (lb/in²)	$1/n$	x/m
PCE	1.4	0.0001905	0.156	0.367952
TCE	1.4	0.0001778	0.23	0.192138
Benzene	1.1	0.0001143	0.131	0.334963
Toluene	0.7	0.0000635	0.111	0.239445

Weight of each contaminant as a function of time (t):

$$n = \frac{PV}{RT}$$

where

P = partial pressure of contaminant
V = volume = 2000 ft^3/min
R = universal gas constant = 10.73 ft^3 psia/lb/mol°R
T = outlet temperature = 60°F = 520°R

$$\frac{\text{Weight of contaminant}}{\text{3 months time}} = n \times \text{MW} \times 3600\ \text{s/h} \times 24\text{h/day} \times 30\text{days} \times 3$$

MW = molecular weight of contaminant.
Sample calculation for PCE is shown as follows:

$$n = \frac{(1.9 \times 10^{-4}\text{psia}) \times (2000\text{ft}^3/\text{min})}{\left(10.73\frac{\text{ft}^3\text{psia}}{\text{lbmol}}{}^\circ\text{R}\right) \times (520\,^\circ\text{R})}$$

$$= 6.8 \times 10^{-5}\text{mol/min}$$

$$\frac{\text{Weight of PCE}}{\text{3 months time}} = 6.8\,\text{mol/min} \times 165.83\text{lb/lbmol} \times 60\ \text{min/h} \times 24\text{h/day} \times 30\text{days} \times 3$$

$$\frac{\text{Weight of PCE}}{\text{3 months time}} = 1465\,\text{lb}$$

   d) Amount of carbon required per contaminant per 3 month period:

$$\text{Carbon req} = \frac{\text{Weight of contaminant/3 months time}}{x/m}$$

where x/m is obtained from Freundlich isotherm equation (step 2).

Sample calculation of carbon required for PCE is shown as follows:

$$\text{Carbon req} = \frac{1465\,\text{lb}}{0.367952} = 3981\,\text{lb}\frac{\text{carbon}}{3\text{months}}$$

Total lb carbon for 3 months

$$= (3981 + 5720 + 2662 + 1138) = 13501\text{lb}$$

This calculation is only an estimate of the quantity of carbon required. Manufacturers recommend calculating the carbon needed for the three or four most prevalent constituents and then adding a safety factor. Safety factors may be as little as 20% more carbon than calculated for nonregenerable systems to as high as 100% for a very conservative design. Applying a conservative safety factor of 100%, we see that the estimated carbon total (M) becomes

$$M = 13\,501 \times 2 = (27\,002)\,\text{lb for 3 months}$$

2) Determine the size of the carbon adsorption vessels.
   Diameter of carbon vessel:

$$Q = A \times V$$

where

Q = flow rate
A = cross-sectional area of vessel = $\pi D^2/4$
V = assumed superficial air velocity = 50 ft/min

$$D = \left(\frac{4A}{3.14}\right)^{0.5} = \left(\frac{4Q}{3.14V}\right)^{0.5} = \left(\frac{4(2000\,\text{ft}^3/\text{min})}{3.14(50\,\text{ft/min})}\right)^{0.5} = 7.1\,\text{ft}$$

Depth of carbon in the vessels:

$$M = \rho \times \text{volume}$$

where

M = weight of carbon = 2.7002×10^4 lb
ρ = density of carbon = 30 lb /ft^3

$$\text{Volume} = A \times H$$

$$H = \frac{\text{Volume}}{A} = \frac{M}{\rho}\frac{4}{3.14D^2} = \frac{(27002)(4)}{(30)(3.14)(7.1)^2} = 23\,\text{ft}$$

where H is the depth of vessel.

3) Determine the total pressure drops through the units in the process train.
   a) Pressure drop from blower to air stripper = 5 in H_2O.
   b) Pressure drop from air stripper to heat exchanger = 1 in H_2O.
   c) Pressure drop through heat exchanger = 1 in H_2O.
   d) Pressure drop from heat exchanger to carbon vessels = 1 in H_2O.
   e) Pressure drop through carbon vessels=18 in H_2O.
   f) Pressure drop between carbon vessels = 1 in H_2O.
   g) Pressure drop from carbon vessels to discharge point = 1 in H_2O

$$\text{Total pressure drop} = \sum \text{Pressure drops of each component}$$
$$= (5 + 1 + 1 + 1 + \; + 18 + 1 + 1)\text{in}\,H_2O$$
$$= (28\,\text{in}\,H_2O)\left(0.036\frac{\text{psia}}{\text{in}\,H_2O}\right) = 1.008\,\text{psia}$$

4) Determine the size and type of blower.
   Size of blower
   Pressure at exit of blower:

$$P_{exit} = P_{atm} + \text{total pressure drop} = 12.7 + 1.008 = 13.708\,\text{psia}$$

Power of blower required:

$$\text{HP} = \left(\frac{(P_1)(Q)}{[33000]}\right) \times \left[\frac{k}{(k-1)}\right]\left[\left(\frac{P_2}{P_1}\right)^{\frac{k-1}{k}} - 1\right]$$

where

P_1 = inlet pressure = P_{atm} = (12.7 lb/in^2)(144 in^2/ft^2)= 1829 lb/ft^2
$P_2 = P_{exit}$ = 13.708 lb/in^2 = 1974 lb/ft^2
Q = flow rate
C_p = heat capacity for constant pressure
C_v = heat capacity for constant temperature
$k = C_p/C_v$ = 1.4 for air

$$\text{HP} = \left(\frac{(1829)(2000)}{[33000]}\right) \times \left[\frac{1.4}{(1.4-1)}\right]\left[\left(\frac{1974}{1829}\right)^{\frac{1.4-1}{1.4}} - 1\right]$$
$$= 8.55\,\text{HP (at 100\% efficiency)}$$

Type of blower: High-pressure centrifugal blowers are often used in this type of application.

11.7 Anaerobic Membrane Biological Reactor

Bioreactor eliminates BOD and nutrients through either aerobic or anaerobic processes. Anaerobic digestion is a key technology in recovering biogas from sludge. The primary treatment is to remove the suspended solids, the secondary treatment is to degrade BOD, and the tertiary treatment is to remove nutrients. MBR combines the secondary and tertiary treatment together. It could remove both BOD and significant portion of nutrients due to longer solid retention time of large biomass in the MBR. Table 11.9 lists the advantages of membrane bioreactors.

A typical layout of MF within a treatment plant is as follows (Figure 11.13).

According to the different arrangement of membrane bioreactors, they can be classified as submerged MBR (Figure 11.14) versus MBR sidestream (Figure 11.15).

Within the MBR, one of the most popular membranes is the hollow-fiber MBR (Figure 11.16).

According to loading rate to anaerobic reactors, low and high rates of AD are classified as follows (Table 11.10).

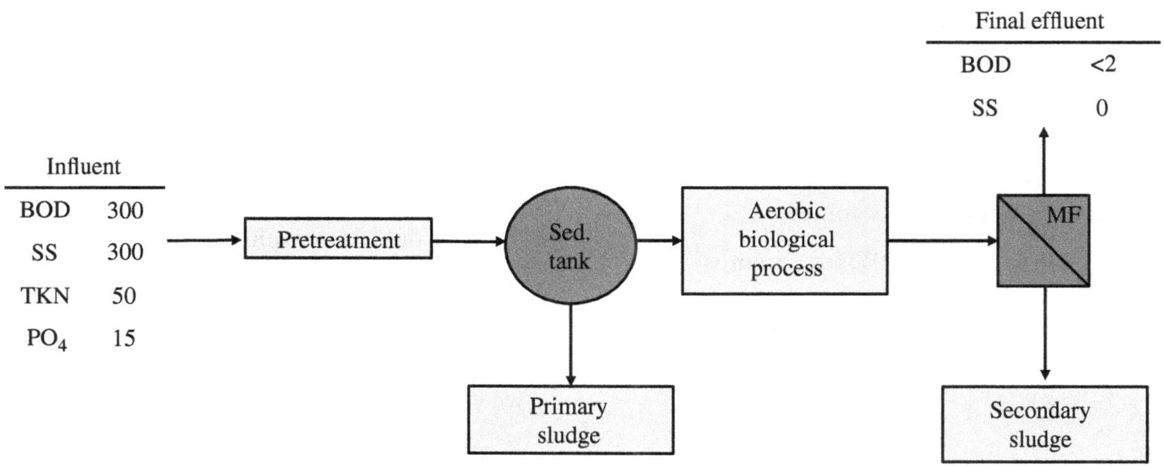

Figure 11.13 Aerobic biological treatment with membrane separation.

Table 11.9 Advantages and disadvantages of membrane bioreactors.

Advantages	Disadvantages
Small footprint	Aeration limitation
Complete solid removal	Membrane fouling
Effluent disinfection	Membrane costs
Removal of organic, nutrient, and solid	High operating costs
High loading rate capability	
Low/zero sludge production	
Rapid start-up	
Sludge bulking not a problem	
Modular/retrofit	

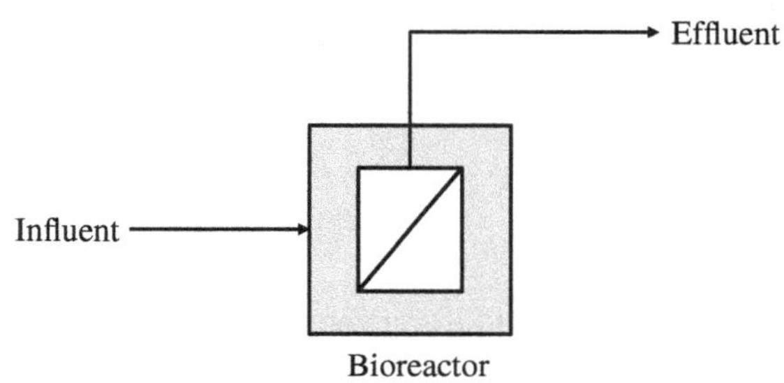

Figure 11.14 Submerged membrane bioreactor.

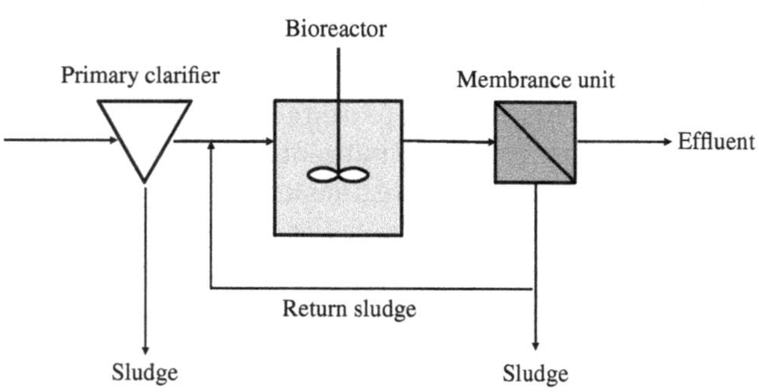

Figure 11.15 External membrane bioreactor.

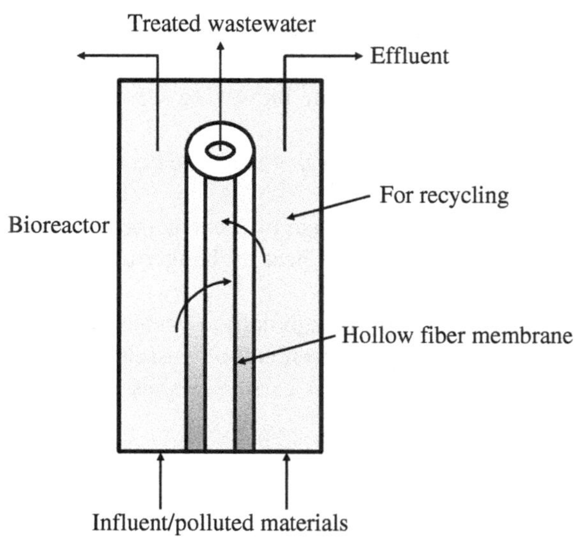

Figure 11.16 Hollow-fiber membrane bioreactor.

Table 11.10 Low- and high-rate anaerobic digesters.

	Low-rate anaerobic reactors	High-rate anaerobic reactors
Type	Anaerobic pond Septic tank Imhoff tank Standard rate anaerobic digester	Anaerobic contact process Anaerobic filter (AF) Upflow anaerobic sludge blanket (UASB) Fluidized bed reactor Hybrid reactor: UASB/AF Anaerobic sequencing batch reactor (ASBR)
Characteristics	Slurry-type bioreactor, temperature, mixing, SRT, or other environmental conditions are not regulated. Loading of 1–2 kg cod/m^3-day	Able to retain very high concentration of active biomass in the reactor Thus extremely high SRT could be maintained irrespective of HRT. Load 5–20 kg cod/m^3-day Cod removal efficiency: 80–90%

Using the volumetric organic loading rate, which is in kg COD/m^3-day, the volumetric organic loading rate can be calculated as follows:

$$\text{VOLR} = \frac{S_o \cdot Q}{V} \tag{11.11}$$

where

VOLR = volumetric organic loading rate (kg COD/m^3/day)
S_o = wastewater biodegradable COD (mg/l)
Q = wastewater flow rate (m^3/day)
V = bioreactor volume (m^3)

$$V = \theta_a \cdot Q \tag{11.12}$$

$$A = \frac{\theta_a \cdot Q}{H} \tag{11.13}$$

where

H= reactor height (m)
θ_a = allowable hydraulic retention time (h)
Q = wastewater flow rate (m^3/h)
A = surface area of the reactor (m^2)
For dilute WW with COD < 1000 mg/l,

The Long Term 2 Enhanced Surface Water Treatment Rule (LT2ESWTR) requires state to report integrate test to earn credit for virus removal (US EPA 2005). The procedure is described in the Membrane Filtration Guidance Manual (US EPA, 2005). Example 11.5 illustrate the calculation on how to calculate the required technical values to earn the 3 log virus removal credit for a submerged vacuum-driven hollow-fiber membrane.

11.8 Air Stripping

Air stripping is a common process used to remove VOCs from water. This is achieved by enabling contact between the liquid (in this case water) and gas (air). Gas is treated to remove the VOCs before it is released into the atmosphere. Air stripping involves the mass transfer of VOCs dissolved in water from the contaminated water to the air phase. This principle is known as Henry's law:

$$P_a = H_a X_a$$

where, at equilibrium, P_a is the partial pressure of a gas above a liquid and is proportional to the mole fraction of the gas dissolved in the liquid, X_a. Ha is the known Henry's constant for the gas.

Thermal oxidation and activated carbon are typically used to treat off-gas that exceeds regulation limits. Thermal oxidation eliminates the contaminant but is a costly, complex process. Low-profile air strippers can remove 99% of most VOC contaminants. The efficiency of the stripper can be improved, and additional stripping incorporated by increasing the number of trays. Efficiency can also be increased by increasing the airflow. Sieve tray air strippers operate in a similar way to packed-column air strippers. The difference is that the liquid (water) flows across trays that are perforated with small holes over a weir and through a downcomer to the next lower tray until the treated water flows from the bottom of the stripper. Gas (air) is bubbled through the holes in the trays, stopping the liquid from dripping through them. The VOCs are transferred from the liquid to the gas phase as the air is bubbled through the water on the trays. Figures 11.17 and 11.18 show a simplified diagram of a packed-column and a low-profile sieve air stripper.

The theoretical number of trays required is estimated using the following relationship:

$$N_{\text{theoretical}} = \frac{\log\left[\frac{X_0 - (Y_{n+1}/m)}{X_n - (Y_{n+1}/m)}\left(1 - \frac{1}{S}\right) + \frac{1}{S}\right]}{\log S}$$

where

X_0 = concentration of contaminant (TCE) in the inlet water phase: 10 mg/l
X_n = concentration of contaminant (TCE) in the treated water phase: 0.1 mg/l

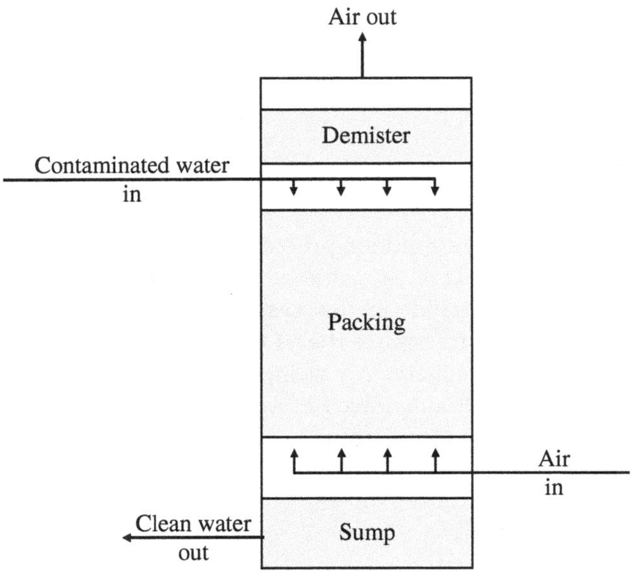

Figure 11.17 Simplified diagram of a packed-column air stripper.

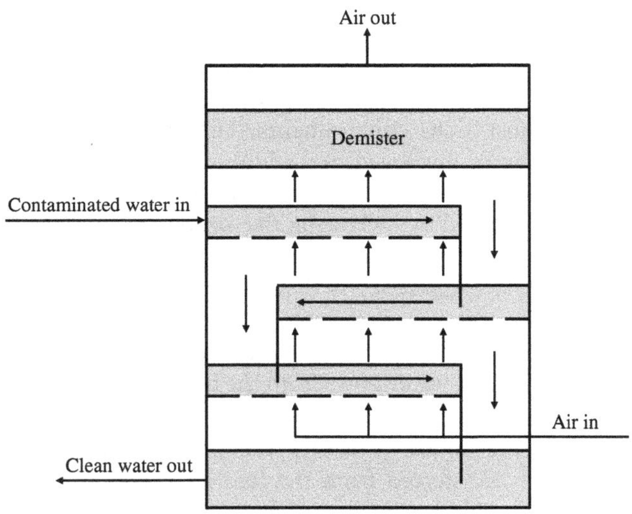

Figure 11.18 Simplified diagram of a low-profile sieve tray air stripper.

N = number of theoretical plates (assumes that the liquid on each plate is completely mixed and where that the vapor leaving the plates is in equilibrium with the liquid)
H = Henry's constant (kPa)
m = slope of equilibrium curve (H/P_t)
G = kg-moles air/min
L = kg-moles of water/min
P_t = ambient pressure (kPa)
S = stripping factor
Y_{n+1} = concentration of volatiles in the air entering the air

For air stripping $Y_{n+1} = 0$ (the concentration of TCE in the air entering the air stripper is zero), and the equation becomes

$$N_{\text{theoretical}} = \frac{\log\left[\frac{X_0}{X_n}\left(1-\frac{1}{S}\right)+\frac{1}{S}\right]}{\log S}$$

A packed-column air stripper consists of a cylindrical column that contains an open-structured packing material. Water containing the VOCs enters the top of the column and flows down through packing material. At the same time, air flows up through the column (countercurrent flow). As the water and air pass each other, the VOCs are transferred from the water phase to the air phase. The water phase leaves the bottom of the column with most of the VOCs removed. The VOCs that are in the air phase exit from the top of the column. For the packed-column air stripper, the mass balance equation for VOCs in the liquid versus gas phases is as follows:

$$(x_{ai} - x_{ae})L = (y_{ae} - y_{ai})\,G \tag{11.14}$$

where

L = molar liquid (water) flow per unit of stripper cross-sectional area $\left[\frac{\text{kg-mol}}{\text{m}^2\text{s}}\right]$
G = molar gas (air) flow per unit of stripper cross-sectional area $\left[\frac{\text{kg-mol}}{\text{m}^2\text{s}}\right]$
x_{ai} = mole fraction of contaminant in liquid (water) influent $\left[\frac{\text{kg-mol}}{\text{kg-molewater}}\right]$
x_{ae} = mole fraction of contaminant in liquid (water) effluent $\left[\frac{\text{kg-mol}}{\text{kg-molewater}}\right]$
y_{ai} = mole fraction of contaminant in gas (air) influent $\left[\frac{\text{kg-mol}}{\text{kg-moleair}}\right]$
y_{ae} = mole fraction of contaminant in gas (air) effluent $\left[\frac{\text{kg-mol}}{\text{kg-moleair}}\right]$

If influent air is not contaminated by VOC, the molar ratio of gas (air) to liquid (water) is

$$\left(\frac{x_{ai} - x_{ae}}{y_{ae} - y_{ai}}\right) = \frac{G}{L}\left[\frac{\text{kg-mol air}}{\text{kg-mol water}}\right]$$

$$y_{ai} = 0$$

$$\frac{G}{L} = \left(\frac{x_{ai} - x_{ae}}{y_{ae}}\right)$$

where

x_{ai} and L = field measurements and x_{ae} imposed by ARAR
p_{Te} = total pressure of gas (air) effluent (atm)
P_{ae} = partial pressure of contaminant a in gas (air) effluent (atm)

From Dalton's law of partial pressures,

$$y_{ae} = \frac{p_{ae}}{p_{Te}}\left[\frac{\text{mol}}{\text{mol}}\right] \text{ or } \left[\frac{\text{atm}}{\text{atm}}\right]$$

$$p_{ae} = y_{ae} p_{Te}[\text{atm}]$$

At equilibrium from Henry's law,

$$p_{ae} = H_a x_{ai}[\text{atm}]$$

Example 11.5
Given: The inlet water contains 10 mg/l of volatile organic compound (VOC), trichloroethylene (TCE). The desired TCE concentration in the charge is 0.1 mg/l. The wastewater flow rate is 0.6 m^3/min, while the air flow rate is 18 $m^3/min/m^2$ of tray area.
Find: Design a low-profile sieve tray air stripper (Figure 11.19).

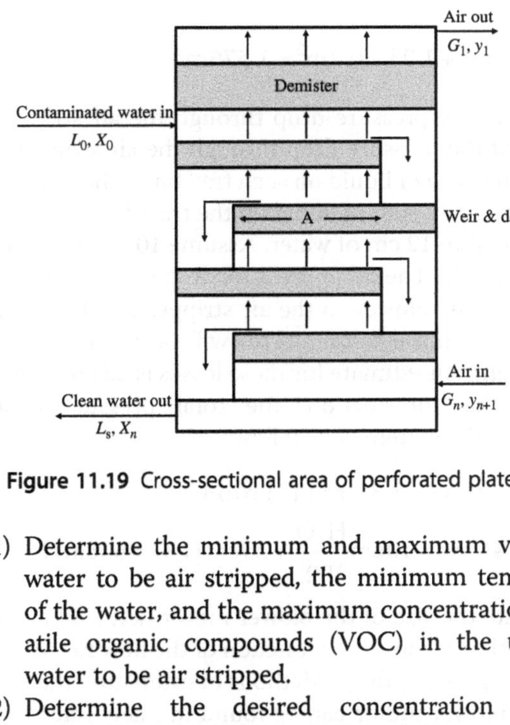

Figure 11.19 Cross-sectional area of perforated plate section.

1) Determine the minimum and maximum volume of water to be air stripped, the minimum temperature of the water, and the maximum concentration of volatile organic compounds (VOC) in the untreated water to be air stripped.
2) Determine the desired concentration (percent removed) of the VOC in the treated water.
3) Calculate the theoretical number of sieve trays needed to remove the VOC to the desired concentration.
4) Estimate the tray efficiency and the number of actual trays needed.
5) Estimate the size (cross-sectional area) of the perforated plate section of each tray.
6) Estimate the pressure drop through the air stripper.
7) Estimate the size of the air blower motor (kW).

Solution:

1) Determine the volume of water to be air stripped, the minimum temperature of the water, and the concentration of all the volatile organic compounds (VOC) in the untreated water. The inlet water contains 10 mg/l of the VOC trichloroethylene (TCE). (Note: If the inlet water contains more than one VOC, repeat the process for each to estimate the number of trays needed for each VOC. Use the largest number of trays for the estimated design.) The flow rate of water to be treated is 0.6 m^3/min. The minimum temperature of the water is 20°C.
2) Determine the desired concentration of the TCE in the treated water. The desired concentration of TCE in the discharge water is 0.1 mg/l (99% removal).
3) Calculate the theoretical number of sieve trays needed to remove the VOC to the desired concentration. The theoretical number of trays required is estimated using the following relationship (Treybal, 1980):

$$N_{\text{theoretical}} \frac{\log\left[\frac{X_0 - (Y_{n+1}/m)}{X_n - (Y_{n+1}/m)}\left(1 - \frac{1}{S}\right) + \frac{1}{S}\right]}{\log S} \tag{11.15}$$

where
X_0 = concentration of contaminant (TCE) in the inlet water phase: 10 mg/l
X_n = concentration of contaminant (TCE) in the treated water phase: 0.1 mg/l
N = number of theoretical plates (assume that the liquid on each plate is completely mixed and that the vapor leaving the plates is in equilibrium with the liquid)
H = Henry's constant (kPa)
m = slope of equilibrium curve (H/P_t)
G = kg-mol air/min
L = kg-mol of water/min
P_t = ambient pressure (kPa)
S = stripping factor
Y_{n+1} = concentration of volatiles in the air entering the air

For air stripping $Y_{n+1} = 0$ (the concentration of TCE in the air entering the air stripper is zero), and the equation becomes

$$N_{\text{theoretical}} = \frac{\log\left[\frac{X_0}{X_n}\left(1 - \frac{1}{S}\right) + \frac{1}{S}\right]}{\log S} \tag{11.16}$$

In this example, the inlet concentration of TCE (X_0), the desired outlet concentration of TCE (X_n), the liquid temperature, and the flow rate are known. The airflow rate (G) must be determined and is related to the perforated plate area of each tray. Several combinations of airflow rates and numbers of trays should be calculated to determine the best economic balance between having more trays and a lower airflow rate (higher capital costs vs. lower operating costs) and fewer trays and a higher airflow rate (lower capital costs vs. higher operating costs). For this example, use an air-to-water ratio of 37 m^3 of air to 1 m^3 of water. Substituting in the above equation yields

$$H = 5.57 \times 10^4 \text{ kPa (for TCE at 20°C)}$$

$$P_t = 101 \text{ kPa}$$

(101 kPa at sea level, 86 kPa at 1500 m elevation)

$$m = \frac{H}{P_t} = \left(\frac{5.57 \times 10^4 \text{kPa}}{101 \text{kPa}}\right) = 551 \frac{\text{mol } H_2O}{\text{mol air}}$$

$$= 551 \frac{\text{kg-mol } H_2O}{\text{kg-mol air}}$$

$$= \frac{37 \text{m}^3 \text{air}}{\text{min}} \times \frac{\text{kg-mol air}}{24 \text{ m}^3 \text{ air at 20°C}} = 1.54 \frac{\text{kg-mol air}}{\text{min}}$$

$$L = \frac{1 \text{m}^3 H_2O}{\text{min}} \times \frac{1000 \text{kg} H_2O}{\text{m}^3 H_2O} \times \frac{\text{kg-mol } H_2O}{18 \text{kg} H_2O}$$

$$= 55.6 \frac{\text{kg-mol } H_2O}{\text{min}}$$

$$S = \left(551 \frac{\text{mol } H_2O}{\text{mol air}}\right)\left(\frac{1.54 \frac{\text{kg-mol air}}{\text{min}}}{55.6 \frac{\text{kg-mol } H_2O}{\text{min}}}\right) = 1$$

$$N_{\text{theoretical}} = \frac{\log\left[\frac{10 \text{mg/l}}{0.1 \text{mg/l}}\left(1 - \frac{1}{1.53}\right) + \frac{1}{1.53}\right]}{\log 1.53} = 1.66$$

4) Estimate the tray efficiency and the number of actual trays needed. In practice, a condition of complete equilibrium does not exist. The overall plate efficiency is

$$E = \frac{N_{\text{theoretical}}}{N_{\text{actual}}} \tag{11.17}$$

Rearranging gives the number of actual trays as

$$N_{\text{actual}} = \frac{N_{\text{theoretical}}}{E} \tag{11.18}$$

The efficiency is highly dependent on the design of the trays and the vapor flow rate. From manufacturer's data, the appropriate range appears to be E = 0.4–0.6 (i.e. 40–60% efficient). Using the above relationship, assuming 50% tray efficiency, and substituting in the above equation gives the number of actual trays needed as

$$N_{\text{actual}} = \frac{1.66}{0.50} = 3.32 = 4$$

5) Estimate the size (cross-sectional area) of the perforated plate section of each tray. The cross-sectional area of the perforated plate section of each tray is related to the airflow rate; 9–18 $m^3/min/m^2$ of tray area is common. For this example, use 18 $m^3/min/m^3$ of tray area.

Since the design flow rate is 0.01 m^3/s,

$$\frac{0.01 \text{m}^3 H_2O}{\text{s}} \times \frac{60 \text{s}}{1 \text{ min}} = 0.6 \frac{\text{m}^3 H_2O}{\text{min}}$$

The area is

$$\frac{0.6 \text{m}^3 H_2O}{\text{min}} \times \frac{37 \text{m}^3 \text{air}}{\text{m}^3 H_2O} \times \frac{\text{m}^2 \text{plate area}}{\frac{18 \text{m}^3 \text{air}}{\text{min}}} = 1.23 \text{m}^2$$

Estimate that the downcomer and weir area are 20% of each plate. The total cross-sectional area of each plate is

$$1.23 \text{m}^2 + 1.23 \times 0.2 \text{m}^2 = 1.476 \text{m}^2$$

6) Estimate the pressure drop through the air stripper. Most of the pressure drop through the air stripper is from the head of liquid on each tray times the number of trays. The depth of liquid on the trays typically varies from 8 to 12 cm of water. Assume 10 cm water for this example. The other pressure drop is from the piping from the blower to the air stripper and inlet and exit losses in the column. This will vary from system to system. An estimate for these losses is 25 cm water. From this information, the total pressure drop through the system is as follows:

$$N \text{ trays} = 3.32; \text{ round up to 4}$$

$$4 \text{ trays} \times 10 \text{cm} \frac{H_2O}{\text{tray}} + 25 \text{cm} H_2O = 65 \text{cm} H_2O$$

7) Estimate the size of the blower motor (kW). The size of the blower motor is a function of the flow rate of air and the pressure drop. Methods of estimating the size of the blower motor can be found in many reference books (Perry, 1984; Avallone and Baumeister, 1987; McCabe et al., 1993):

$$\frac{0.01 \text{m}^3}{\text{s}} \times \frac{1000 \text{l}}{1 \text{m}^3} \times \frac{1 \text{gal}}{3.785 \text{l}} \times \frac{60 \text{s}}{1 \text{ min}} = 158.5 \text{gpm}$$

$$\Delta P_{\text{total}} = 65 \text{m} = 213.3 \text{ft}$$

Efficiency for a blower motor ranges anywhere from 40 to 60%. So we will assume 50% efficiency:

$$A = \frac{\pi d^2}{4}$$

$$A = \frac{\pi (2.5\text{ft})^2}{4} = \frac{\pi (0.6454\text{m})^2}{4}$$

$$A = 4.908\,\text{ft}^2 = 0.3271\,\text{m}^2$$

$$V_L = \frac{Q_L}{A} = \frac{220\,\text{gpm}}{4.908\,\text{ft}^2}\left(\frac{0.01388\ \text{m}^3/\text{s}}{0.456\,\text{m}^2}\right)\text{per stripper}$$
$$= 44.82\frac{\text{gpm}}{\text{ft}^2}\left(0.03044\frac{\text{m}}{\text{s}}\right)$$

$$W_{\text{pump}} = \frac{(158.5\,\text{gal/min})(213.3\,\text{ft-H}_2\text{O})}{3{,}960\,(\text{gal}\cdot\text{ft/min/hp}) \times 0.50} = 17.07\,\text{hp}$$

Air stripping by packed column is shown in Figure 11.20.

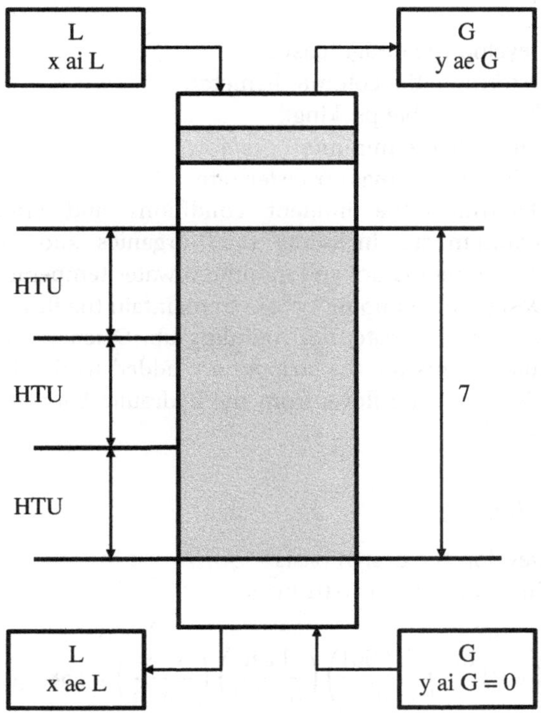

Figure 11.20 Random "dumped" packed tower.

Typical design parameters are as follows:

NTU = number of transfer units (m).

HTU = height of transfer unit (m).

Z = NTU × HTU packing depth.

L = molar liquid (water) flow per unit of stripper cross-sectional area $\left(\frac{\text{kg-mol}}{\text{m}^2\text{s}}\right)$.

G = molar gas (air) flow per unit of stripper cross-sectional area $\left(\frac{\text{kg-mol}}{\text{m}^2\text{s}}\right)$.

x_{ai} = mole fraction of contaminant in liquid (water) influent $\left(\frac{\text{kg-mol}}{\text{kg-molewater}}\right)$.

x_{ae} = mole fraction of contaminant in liquid (water) effluent $\left(\frac{\text{kg-mol}}{\text{kg-molewater}}\right)$.

y_{ai} = mole fraction of contaminant in gas (air) influent $\left(\frac{\text{kg-mol}}{\text{kg-moleair}}\right)$.

y_{ae} = mole fraction of contaminant in gas (air) effluent $\left(\frac{\text{kg-mol}}{\text{kg-moleair}}\right)$.

$$(x_{ai} - x_{ae})L = (y_{ae} - y_{ai})G \tag{11.19}$$

Molar ratio of contaminant transferred from liquid to gas per unit of stripper cross-sectional area per unit time (kg-mol/s) is:

$$\left(\frac{x_{ai} - x_{ae}}{y_{ae} - y_{ai}}\right) = \frac{G}{L}\left[\frac{\text{kg-mol air}}{\text{kg-mol water}}\right] \tag{11.20}$$

The molar ratio of gas (air) to liquid (water), assuming uncontaminated influent air $y_{ai} = 0$ is:

$$\frac{G}{L} = \left(\frac{x_{ai} - x_{ae}}{y_{ae}}\right) \tag{11.21}$$

where

x_{ai} and L = field measurements and x_{ae} = imposed by ARAR

p_{Te} = total pressure of gas (air) effluent (atm)

P_{ae} = partial pressure of contaminant a in gas (air) effluent (atm)

To design packed tower air stripping tower, Dalton's law of partial pressure and Henry's law of equilibrium between liquid and air phases are used. From Dalton's law of partial pressures,

$$y_{ae} = \frac{p_{ae}}{p_{Te}}\left[\frac{\text{mol}}{\text{mol}}\right] \text{or} \left[\frac{\text{atm}}{\text{atm}}\right] \tag{11.22}$$

$$p_{ae} = y_{ae}p_{Te}[\text{atm}]$$

At equilibrium from Henry's law,

$$p_{ae} = H'_a x_{ai}[\text{atm}]$$

Substituting yields

$$y_{ae}p_{Te} = H'_a x_{ai}[\text{atm}]$$

$$y_{ae} = \frac{H'_a x_{ai}}{p_{Te}}[\text{atm}]$$

and from the material balance

$$(x_{ai} - x_{ae})\left(\frac{L}{G}\right) = y_{ae}\left[\frac{\text{mol}}{\text{mol}}\right] \tag{11.23}$$

Again substituting gives

$$(x_{ai} - x_{ae})\left(\frac{L}{G}\right) = \frac{H'_a x_{ai}}{p_{Te}} \tag{11.24}$$

$$\frac{x_{ai} - x_{ae}}{x_{ai}} = = \frac{H'_a\left(\frac{G}{L}\right)}{p_{Te}} \tag{11.25}$$

The fraction of contaminant transferred from liquid (water) to gas (air) phase is

$$(x_{ai} - x_{ae})\left(\frac{L}{G}\right) = \frac{H'_a x_{ai}}{p_{Te}} \tag{11.26}$$

$$\frac{C_{ai} - C_{ae}}{C_{ai}} = \frac{x_{ai} - x_{ae}}{x_{ai}} \tag{11.27}$$

where

C_{ai} = concentration of contaminant a in liquid (water) influent (mg/l)

C_{ae} = concentration of contaminant a in liquid (water) effluent (mg/l)

For convenience, the flows of water and air are measured volumetrically:

$$C_{ai}(L)Q_L = C_{ae}(L)Q_L + C_{ae}(G)Q_G \tag{11.28}$$

$$\frac{C_{ai} - C_{ae}}{C_{ai}} = \left(\frac{H'_a}{p_{Te}}\right)\left(\frac{Q_G}{Q_L}\right) \tag{11.29}$$

where P_{Te} is measured as a fraction of the standard atmosphere (atm), H'_a is the dimensionless Henry's constant H_a/C_0RT, actually (volume/volume) Q_G/Q_L is reduced to common flow units (m^3/m^3), and C_0 is the molar density of water at 20°C, 55.41 kg mol/m^3. The theoretical minimum equilibrium moles of gas required G_{min}/L is calculated from the influent and effluent concentrations and the "dimensionless" Henry's constant (H'_a):

$$\text{The universal gas constant } R_u = 0.08205746\left[\frac{\text{m}^3\text{atm}}{\text{kg-mol}\cdot\text{K}}\right]$$

At 1 atm and 20°C, the molar density of water is C_0, 55.41 kg-mol/m^3. Q_G/Q_L (m^3/m^3) is the air-to-water ratio (ATW):

$$y_{ae} = \frac{H'_a x_{ai}}{p_{Te}}\left[\frac{\text{mol}}{\text{mol}}\right]$$

Substituting gives

$$(x_{ai} - x_{ae})\left(\frac{L}{G}\right) = \frac{H'_a x_{ai}}{p_{Te}} \tag{11.30}$$

and rearranging yields

$$\frac{G_{min}}{L} = \frac{(x_{ai} - x_{ae})p_{Te}}{H'_a x_{ai}} \tag{11.31}$$

which is the equilibrium molar ratio of gas (air) to liquid (water).

Example 11.6

Given: If flow rate is 0.01 m^3/s, contaminants are listed in Table 11.11.

Table 11.11 Contaminants.

Contaminant	Formula	GMW[a] (g/g-mol)	CAS number	H_a[b] (atm/mol/mol)
Benzene	C_6H_6	78.11	71-43-2	309.2
Toluene	$C_6H_5CH_3$	92.14	108-88-3	353.1
Trichloroethylene (TCE)	C_2HCl_3	131.50	79-01-6	506.1

[a] The [gram] molecular weight of the contaminant.
[b] H_a at 20°C (296.13 K).

Find:

1) Develop the design basis.
2) Determine the column diameter.
3) Find a suitable packing.
4) Calculate the minimum gas flow.
5) Calculate the mass transfer rate.
6) Determine the influent conditions and effluent requirements, including total organics and background inorganics and minimum water temperature.
7) Design the pumping system to maintain the flow. Use the real flow rate, not rounding up. Discharge head adjustments for the stripper are added to the TDH. The aggregate flows from the hydraulic barriers are as follows.

Solution:

1) Develop the design basis:
   Since flow rate is 0.01 m^3/s,

$$\left(0.01\frac{\text{m}^3}{\text{s}}\right)\left(\frac{1000\text{l}}{\text{m}^3}\right)\left(\frac{1\text{ gal}}{3.785\text{l}}\right)\left(\frac{60\text{s}}{1\text{ min}}\right) = 158.5\text{gpm}$$

   1) The pretreatment system to prevent scale/slime from clogging the stripper (if water is high in hardness, iron or manganese) (Table 11.12).
   2) A contaminant material balance for the stripping system (Table 11.13).
   3) The air pollution control requirements from the material balance and the regulations.

2) Determine the column diameter:
   1) Determine a preliminary stripper cross-sectional area for the sustained pumping rate, 158.5 gpm (0.01 m^3/s), using 45 gpm/ft^2 (0.03056 m/s) for the stripper surface loading:

Table 11.12 Background inorganic concentrations.

Ion	mgl	GMW	Valence	GEqW[a]	meq/l	mg/l as $CaCO_3$
CO_2	0	44	−2	22	0.00	0.00
Anions						
SO_4	60	96	−2	48	1.25	62.46
Cl	54	35	−1	35	1.52	76.15
HCO_3	30	61	−1	61	0.49	24.58
					Total	163.19
Cations						
Na	10	23	1	23	0.43	21.75
Ca	40	40	2	20	2.00	99.80
Fe	0.3	56	2	28	0.01	0.54
Mg	10	24	2	12	0.82	41.12
Mn	0.05	55	2	27	0.00	0.09
					Total	163.29

Note:
[a] GEqW is the [gram] equivalent weight of the inorganic ion.

$$A = \frac{Q\,[\mathrm{ft}^2]}{45\,\mathrm{gpm}} = \frac{Q[\mathrm{m}^2\mathrm{s}]}{0.03056[\mathrm{m}^3]}$$

$$A = 0.0222\frac{Q\,[\mathrm{ft}^2]}{\mathrm{gpm}} = 32.72Q\left[\frac{\mathrm{s}}{\mathrm{m}}\right]$$

$$A = 0.0222(158.5\,\mathrm{gpm})\frac{\mathrm{ft}^2}{\mathrm{gpm}} = 32.72\left[0.01\frac{\mathrm{m}^3}{\mathrm{s}}\right]\left[\frac{\mathrm{s}}{\mathrm{m}}\right]$$

$$= 3.5187\,\mathrm{ft}^2 = 0.3272\,\mathrm{m}^2$$

2) Divide a by the number of strippers:

$$a = \frac{3.5187}{2}\mathrm{ft}^2 = \frac{0.3272}{2}\mathrm{m}^2$$

$$= 1.75935\frac{\mathrm{ft}^2}{\mathrm{stripper}} = 0.1636\frac{\mathrm{m}^2}{\mathrm{stripper}}$$

3) Divide $a = \pi\,(d^2/4)$ the unit area by p, multiply by 4, and take the square root:

$$d \approx \sqrt{\frac{4a}{\pi}}$$

$$\approx \sqrt{\frac{4(1.75935)}{\pi}} = \sqrt{\frac{4(0.1636)}{\pi}}$$

$$\approx \sqrt{2.24007} = \sqrt{0.2084}$$

$$\approx 1.4966\,\mathrm{ft} = 0.4565\,\mathrm{m}$$

4) Bracket the calculated diameter with the nearest standard diameters. In this example, a 2.5-ft (0.762-m)-diameter column is standard for most manufacturers. The availability of standard metric sizes should be verified.

3) Find a suitable packing.
   1) Find packings in the diameter range of roughly 5–10% of the stripper diameter. The rule of thumb is 1 in of packing diameter per 1 ft of tower diameter; 2.5-in (0.0635-m) packing is not standard for most manufacturers.
   2) Reconsider the number of strippers if the packings and diameters do not correspond. Three 2-ft-diameter strippers with 2-in packing could be used in lieu of two 2.5-ft-diameter strippers.
   3) Find the area of the standard diameter strippers:

$$a = \pi\frac{d^2}{4}$$

$$= \pi\frac{(2.5\,\mathrm{ft})^2}{4} = \pi\frac{(0.762\,\mathrm{m})^2}{4}$$

$$= 4.908\,\mathrm{ft}^2 = 0.456\,\mathrm{m}^2$$

   4) Calculate the surface hydraulic loading Q/A and compare the loading with various packing manufacturers' recommendations:

$$V_L = \frac{Q_L}{A} = \frac{220\,\mathrm{gpm}}{4.908\,\mathrm{ft}^2}\left(\frac{0.01388\,\mathrm{m}^3/\mathrm{s}}{0.456\,\mathrm{m}^2}\right)\text{per stripper}$$

$$= 44.82\,\mathrm{gpm/ft}^2\;(0.03044\,\mathrm{m/s})$$

   5) Adjust the system configuration to get the hydraulics within the recommended range.

4) Calculate the minimum gas flow.

Table 11.13 Removal requirements.

Contaminant	Concentration (μg/l)			Mole fraction (mol/mol)	
	Influent, C_{ai}	Standard, C_{ae}	Removal requirement	X_{ai}	X_{ae}
Total VOCs	2500	NA	NA	NA	NA
Benzene	750	10	98.7%	0.17330	0.00231
Toluene	1000	100	90.0%	0.19588	0.01959
Trichloroethylene (TCE)	750	100	86.7%	0.10294	0.01373

Determine G_{min} and the critical contaminant from the following relationship (Table 11.14):

$$\frac{Q_{G_{min}}}{Q_L} = \frac{(C_{ai} - C_{ae})}{H'_a C_{ai}}$$

Table 11.14 Critical contaminants.

For P_{te} = 1 atm and 20°C (296.13 K) $H'_a = H_a/C_oRT$			
Contaminant	$\frac{(C_{ai} - C_{ae})}{C_{ai}}$	H'_a	$\frac{Q_{G_{min}}}{Q_L}$
Benzene	0.9867	0.2320	4.253 m^3/m^3
Toluene	0.9000	0.2649	3.397 m^3/m^3
Trichloroethylene (TCE)	0.8667	0.3797	2.283 m^3/m^3
Critical contaminant (benzene)			
$\frac{Q_{G_{min}}}{Q_L} = 4.253 m^3/m^3$ (maximum)			

5) Calculate the mass transfer rate.
Use a model, if available, to confirm the results:

$$\frac{a_w}{a_t} = 1 - e^{-1.45\left(\frac{S_c}{S}\right)^{0.75} N_{Re}^{0.01} N_{Fr}^{-0.05} N_{We}^{0.2}} \quad (11.32)$$

$$\frac{1}{K_{LA}} = \frac{1}{H'_a K_G a_w} + \frac{1}{K_L a_w} \quad (11.33)$$

where
a_w = wetted surface area of the packing (m^2/m^3)
a_t = total surface area of the packing (m^2/m^3)
K_{LA} = overall mass transfer rate (m/s)
K_L = liquid phase mass transfer rate (m/s)
K_G = gas phase mass transfer rate (m/s)

1) Calculate the dimensionless numbers (http://www.processassociates.com/process/dimen gives a comprehensive listing and definitions of dimensionless numbers):

$$\text{Reynolds number } N_{Re} = \left(\frac{1}{a_t}\right)\frac{V_L \rho_L}{\mu_L} \quad (11.34)$$

$$\text{Frounde number } N_{Fr} = a_t \frac{V_L^2}{g_c} \quad (11.35)$$

$$\text{Weber number } N_{We} = \left(\frac{1}{a_t}\right)\frac{V_L^2 \rho_L}{g_c S} \quad (11.36)$$

$$\text{Schmidt number } N_{Sc} = \frac{\mu_L}{D_L \rho_L} \quad (11.37)$$

$$\text{Gravitation constant } g_c = 9.807 \frac{m}{s^2}$$

2) Look up the properties of the liquid (water) at the minimum water temperature, T (Table 11.15).

Table 11.15 Water at 20°C (293.16 K).

The properties of the liquid (water) at 20°C	
$s = 0.072764 \frac{N}{m} = \frac{kg}{s^2}$	Liquid surface tension
$\mu_L = 0.0010042 \frac{kg}{m \cdot s}$	Liquid viscosity
$\rho_L = 998.20 \frac{kg}{m^3}$	Liquid density
$d_p = 0.0508 m$	Nominal diameter
$a_t = 157 \frac{m^2}{m^3}$	Total surface area
$s_c = 0.033 \frac{kg}{s^2}$	Critical surface tension for polyethylene packing
$C_f = 15$	Packing factor

3) Look up the properties of the critical contaminant, benzene, at the minimum water temperature, T:

Liquid diffusivity of benzene 20°C (296.13K) $D_L = 8.91 \times 10^{-10} m^2/s$

4) Obtain data from product literature (Table 11.15):

Nominal diameter: $d_p = 0.0508$ m

Total surface area: $a_t = 157 \frac{m^2}{m^3}$

Critical surface tension for polyethylene packing:

$$s_c = 0.033 \frac{kg}{s^2}$$

Packing factor: $C_f = 15$

5) Liquid mass velocity is as follows:

$$L = \rho_L \frac{Q_L}{A}\left[\frac{kg}{m^2 \cdot s}\right]$$

Liquid mass velocity at 0.01388 m^3/s with a nominal column diameter of 0.76 m is

$$L = 998.19 \times \left[\frac{0.014}{0.456}\right]$$

$$L = 30.38 \frac{kg}{m^2 \cdot s}$$

6) Calculate the Reynolds number, N_{Re}:

$$\text{Reynolds number } N_{Re} = \left(\frac{1}{a_t}\right)\frac{V_L \rho_L}{\mu_L}$$

$V_L = 0.304 m/s$

$$\rho_L = 998.19\,\text{kg/m}^3$$

$$a_t = 157\,\text{m}^2/\text{m}^3$$

$$\mu_L = 0.0010042\ \text{kg/m·s}$$

$$N_{Re} = \frac{0.3043 \times 998.19}{157 \times 0.001} = 192.7$$

$$N_{Re}^{0.1} = 1.692$$

7) Calculate the Froude number, N_{Fr}:

$$N_{Fr} = \frac{a_t\, V_L^2}{g_c} = \frac{157 \times (0.304)^2}{9.807} = 0.015$$

$$N_{Fr}^{-0.5} = 1.234$$

8) Calculate the Weber number, N_{We}:

$$N_{We} = \left(\frac{1}{a_t}\right)\frac{V_L^2 \rho_L}{g_c S} = \left(\frac{1}{157}\right)\frac{(30.39)^2 \times 998.19}{9.807 \times 0.073} = 0.081$$

$$N_{We}^{0.2} = 0.605$$

9) Calculate the wetted area of the packing, a_w, from the dimensionless relation:

$$\frac{a_w}{a_t} = 1 - e^{\left[-1.45\left(\frac{S_c}{S}\right)^{0.75} N_{Re}^{0.01} N_{Fr}^{-0.05} N_{We}^{0.2}\right]}$$

$$N_{Re}^{0.01} N_{Fr}^{-0.05} N_{We}^{0.2} = 1.692 \times 1.234 \times 0.6048 = 1.263$$

$$\left(\frac{S_c}{S}\right)^{0.75} = \left(\frac{0.033}{0.0728}\right)^{0.75} = (0.454)^{0.75} = 0.553$$

10) Calculate the wetted surface area:

$$\frac{a_w}{a_t} = 1 - e^{[-1.45 \times (0.553 \times 1.263)]}$$

$$= 1 - e^{-1.0125}$$

$$= 1 - 0.3633 = 63.37\%$$

$$a_t = 157\,\text{m}^2/\text{m}^3$$

$$a_w = 63.37\%(157)$$

$$= 99.96\,\text{m}^2/\text{m}^3$$

11) Calculate the liquid phase mass transfer coefficient, Onda K_L, from the following relationship:

$$K_L\left(\frac{\rho_L}{\mu_L g_c}\right)^{\frac{1}{3}} = 0.0051\left(\frac{V_L \rho_L}{a_w \mu_L}\right)^{\frac{2}{3}}\left(\frac{\mu_L}{\rho_L D_L}\right)^{-0.5}(a_t d_P)^{0.4}$$

$$\left(\frac{\rho_L}{\mu_L g_c}\right)^{\frac{1}{3}} = \left(\frac{998.19}{(0.0010042)(9.8066)}\right)^{\frac{1}{3}}$$

$$= (101361)^{\frac{1}{3}} = 46.63$$

$$\left(\frac{V_L \rho_L}{a_w \mu_L}\right)^{\frac{2}{3}} = \left(\frac{0.3043 \times 998.19}{99.96 \times 0.0010042}\right)^{\frac{2}{3}} = (302.7)^{\frac{2}{3}} = 45.08$$

$$\left(\frac{\mu_L}{\rho_L D_L}\right)^{-0.5} = \left[\frac{0.0010042}{(998.19)(8.91 \times 10^{-10})}\right]^{-0.5}$$

$$= (1129)^{-0.5} = 0.030$$

$$(a_t d_P)^{0.4} = (157 \times 0.0508)^{0.4} = (7.9756)^{0.4} = 2.295$$

$$K_L\left(\frac{\rho_L}{\mu_L g_c}\right)^{\frac{1}{3}} = 0.0051\left(\frac{V_L \rho_L}{a_w \mu_L}\right)^{\frac{2}{3}}\left(\frac{\mu_L}{\rho_L D_L}\right)^{-0.5}(a_t d_P)^{0.4}$$

$$K_L = \frac{0.0051(45.08)(0.030)(2.295)}{46.63} = 0.0003\,\text{m/s}$$

12) Calculate the gas phase mass transfer coefficient, Onda K_G, using a stripping factor (R) between 2 and 5. Try R = 2.5 if air pollution control is required or R = 4.5 if it is not:

$$\frac{K_G}{a_t D_G} = 5.23\left(\frac{G}{a_t \mu_G}\right)^{0.7}\left(\frac{\mu_G}{\rho_G D_G}\right)^{\frac{1}{3}}(a_t d_P)^{-2.0}$$

13) Look up the properties of the gas (air) at the minimum water temperature, T (Table 11.16).

Table 11.16 Air at 20°C (293.16 K) and 1 atm.

The properties of the gas (air) at 20°C (293.16 K) and 1 atm	
$\frac{A}{W} = \frac{0.4132\ (\text{m}^3/\text{s})\,\text{air}}{0.02776\ (\text{m}^3/\text{s})\text{water}} = 14.89$	Gas viscosity
$\mu_{G} = 1.773 \times 10^{-5}\,\frac{\text{kg}}{\text{m·s}}$	
$\rho_G = 1.2046\,\frac{\text{kg}}{\text{m}^3}$	Gas density

14) The properties of the critical contaminant, benzene, at the minimum water temperature, T, are:

Gas diffusivity $D_G = 9.37 \times 10^{-6}\,\text{m}^2/\text{s}$
(benzene in air at 20°C, 1 atm)

15) Calculate the gas flow rate from the relationship:

$$\frac{Q_{G_{min}}}{Q_L} = \frac{(C_{ai} - C_{ae})}{H_a' C_{ai}} = 4.253 \text{ from Table 11.14}$$

$$V_L = 0.031\,\text{m/s}$$

$$V_{G_{min}} = 4.2635 \times 0.031 = 0.130\ \text{m/s}$$

$$R = 3.5$$

$$V_G = R \times V_{G_{min}}$$

$$V_G = (3.5)(0.130) = 0.453\ \text{m/s}$$

$$G = V_G \rho_G = (0.4531\,\text{m/s})(1.2046\,\text{kg/m}^3)$$
$$= 0.546\,\text{kg/s·m}^2$$

16) See Table 11.15 for a_t and dp:

$$\left(\frac{G}{a_t \mu_G}\right)^{0.7} = \left(\frac{0.5458}{(157)(1.773 \times 10^{-5})}\right)^{0.7}$$
$$= (196.06)^{0.7} = 40.24$$

$$\left(\frac{\mu_G}{\rho_G D_G}\right)^{\frac{1}{3}} = \left(\frac{1.773 \times 10^{-5}}{(1.2046)(9.37 \times 10^{-6})}\right)^{\frac{1}{3}}$$
$$= (1.571)^{\frac{1}{3}} = 1.162$$

$$(a_t d_P)^{-2.0} = (157 \times 0.0508)^{-2.0} = 0.01572$$

$$a_t D_G = (157)(9.37 \times 10^{-6}) = 0.0015\,\text{m/s}$$

$$\frac{K_G}{a_t D_G} = (5.23)(40.24)(1.162)(0.157) = 3.846$$

$$K_G = (3.853)(0.0015) = 0.006\,\text{m/s}$$

17) Calculate the overall mass transfer coefficient, K_{LA}:

$$\frac{1}{K_{LA}} = \frac{1}{H_a' K_G a_w} + \frac{1}{K_L a_w}$$

$$\frac{1}{K_{LA}} = \frac{1}{(0.2320)(0.006)(99.96)} + \frac{1}{(0.003)(99.96)}$$
$$= 7.622 + 29.71 = 37.33$$

$$K_{LA} = 0.02679\,\text{s}^{-1}$$

$$\text{HTU} = \frac{V_L}{K_{LA}} = \frac{0.030}{0.027} = 1.136\,\text{m}$$

18) Determine NTU for the selected R:

$$R = \frac{G}{G_{min}} = \frac{H_a'}{P_{Te}} \times \frac{G}{L} = 3.5$$

$$\text{NTU} = \left(\frac{R}{R-1}\right) \ln \left[\frac{\left[\left(\frac{x_{ai}}{x_{ae}}\right)(R-1)\right] + 1}{R}\right]$$

$$= \left(\frac{3.5}{3.5-1}\right) \ln \left[\frac{\left[\left(\frac{750}{10}\right)(3.5-1)\right] + 1}{3.5}\right] = 5.88$$

$$Z = (\text{NTU})(\text{HTU}) = (5.88)(3.07) = 17.13\,\text{m}$$

$$\frac{A}{W} = \frac{0.4132\,\text{m}^3/\text{s air}}{0.02776\,\text{m}^3/\text{s water}} = 14.89$$

Design, emissions control, and operation and maintenance are factors that determine the cost of an air stripper. Capital costs include design considerations such as the size and number of towers, capacity, and the materials of construction. Equipment costs can include the tower shell, sieve trays, water distributor, mist eliminator, blower and motor, engineering, and contract overhead and profit. Operation and maintenance costs are influenced by operating labor, repair and maintanence, and the energy requirements of the blower motor and pumps. The cost of an air stripping system can double if air treatment is a requirement. The cost of air stripping may range from \$0.04 to \$0.17/gal. The following table lists the costs at the various flow rates (Table 11.17).

Table 11.17 Air stripping cost.

Flow rates (m^3/s)	Flow rates (Mgal/year)	Low estimate cost (M)	High estimate cost (M)
0.01	83.32	3.3	14.2
0.1	833.183	33.3	141.6
1	8 331.836	333.3	1 416.4

11.9 LCA Tools for WWTPs

The core criteria in SEE design are the footprints of EEIS on energy and materials such as water, carbon, and nutrients according to its life cycle of the EEIS. To assess these footprints, many LCA tools have developed as shown in Table 11.18 (Figure 11.21).

Databases that could be used for LCA of WWTP include:

1) E-PRTR: 28 000 industry pollutants by EU Pollutant Release and Transfer (Yoshida, Water Research, 56, 292–303, 2014).
2) International Reference Life Cycle Data System (ILCD) v2.2 USLCI (Eco-invent Center, 2007; NREL, 2012).
3) TRACI (Tool for Reduction and Assessment of Chemical and Other Environmental Impacts) (US EPA, 2014).
4) IMPACT World+.

Separation strategy could be effective on different spatial scales because treated WW should be disposed approximate to the reuse. From a sustainability point of view,

Table 11.18 Life cycle assessment tools for WWTPs.

Type	Description of methodology	Estimation tool	Tool format	Available	Applicable	Source
Process LCA-based tools	Use process-based inventory over life cycle[a]	SimaPro	Software	Commercial	Varies, any product or process	www.pre.nl
		GaBi	Software	Commercial		www.gabi-software.com
		SiSOSTAQUA	Software	Commercial		www.simpple.com
Hybrid LCA-based tools	Use both process-based and input–output-based inventory over life cycle[b]	WEST	MS-Excel	Upon request	Water, water reuse, desalination	Dr. Jennifer Stokes at ucbwaterlca@gmail.com
		WWEST	MS-Excel	Upon request	Wastewater	Dr. Jennifer Stokes at ucbwaterlca@gmail.com
		WESTWeb	Web based	Public	Water, water reuse, desalination wastewater	west.berkeley.edu
Specific tools	Uses input parameters specific to utility over O&M	Tampa Bay Water[c]	MS-Excel	Upon request	Water and desalination	www.tampabaywater.org
		Johnston Tool[d]	MS-Excel	Upon request	Water	Dr. Tanju Karafil at tkaranf@clemson.edu
Other related tools	Not specifically used to estimate emissions from water reuse or desalination facilities but contain aspects that are applicable	CHEApet[e]	Web based	Public	Wastewater	cheapet.werf.org
		UK Environment Agency tool[f]	MS-Excel	Upon request	Water supply, water reuse, desalination	Environment Agency at enquiries@environmentagency.gov.uk
		Bridle and BSM2G tool[g]	Software	Public	Wastewater	Author Lluis Corominas at lcorominas@icra.cat
		System dynamics[h]	Software	Commercial	Varies	www.iseesystems.com
		GPS-X[i]	Software	Commercial	Wastewater	www.hydromantis.com/GPS-X.html
		Carbon Accounting Workbook, 5th Version[j]	MS-Excel	Commercial	Water	www.ukwir.org
		mCO2[k]	Software	Commercial	Wastewater	www.mwhglobal.com
		BioWin 4.0[l]	Software	Commercial	Wastewater	www.envirosim.com

Sources:
[a] ISO (2006).
[b] Stokes and Horvath (2006, 2011a, b).
[c] Tampa Bay Water (2012).
[d] Johnston (2011).
[e] Crawford et al. (2011).
[f] Reffold et al. (2008).
[g] Corominas et al. (2012).
[h] Shrestha et al. (2011).
[i] Goel et al. (2012).
[j] UKWIR (2008).
[k] MWH (2012).
[l] EnviroSim Associates Ltd. (2014).

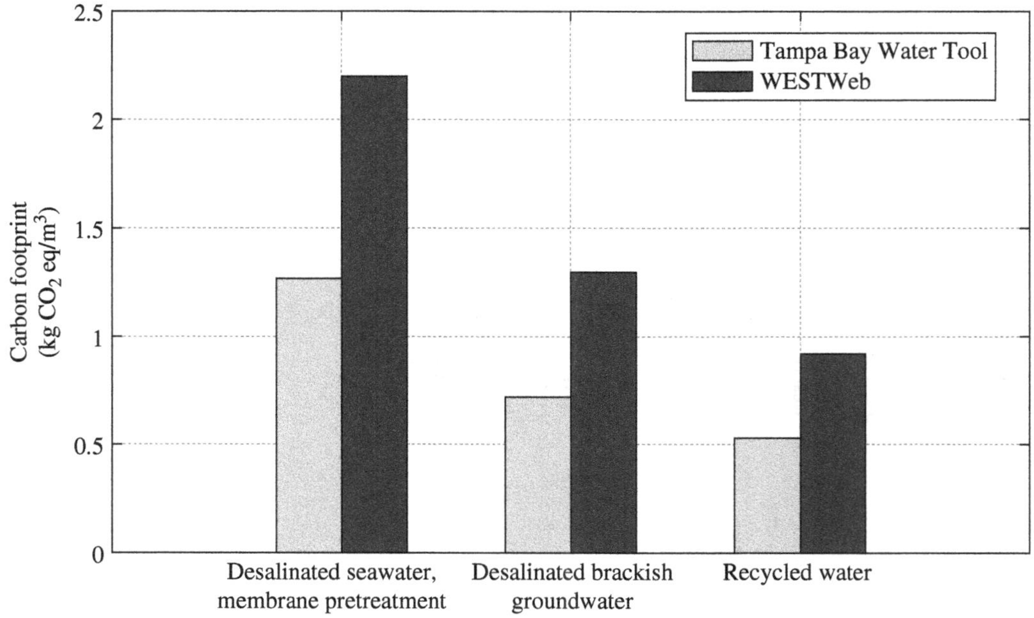

Figure 11.21 Comparison of carbon footprint estimate using Tampa Bay Water and WESTWeb tools. (*Source:* Cornejo et al. (2016). Reproduced with permission of American Chemical Society.)

the energy footprint of WW treatment per cubic meter of WW on different spatial scales such as household (250 GPD), community (0.31 MGD), and city (10.3 MGD) decreases as shown in Figure 11.22 due to economy of scale.

For all the treatment strategies, water reuse is the major positive offset of energy footprint regardless of scale. In conventional AS process, electricity is the major factor contributing to energy consumption. Table 11.19 illustrates the contribution of construction and O&M phases to energy footprint on different scales.

To achieve sustainable design in rural areas, solar energy is the best way to reduce energy footprint by using

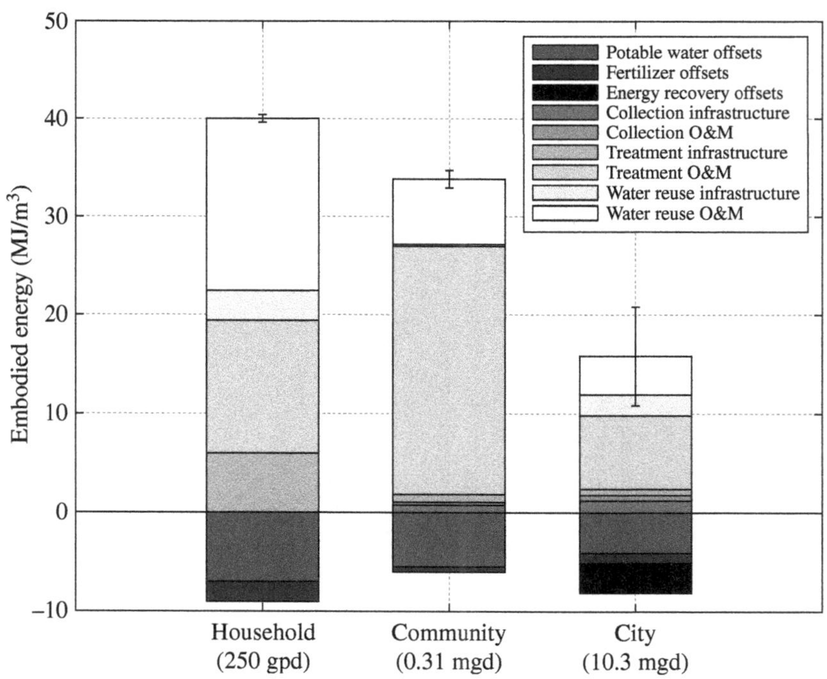

Figure 11.22 Embodied energy of wastewater systems with collection, treatment, water reuse, and resource recovery offsets on different scales. (*Source:* Cornejo et al. (2016). Reproduced with permission of American Chemical Society.)

Table 11.19 Percentage contribution from sewer collection, treatment, and water reuse to the total embodied energy of wastewater systems and resource recovery offsets on different scales.

Phase	Stage	Description	Household (250 GPD)	Community (0.31 MGD)	City (10.3 MGD)
Construction	Collection	Piping	—	2.5%	9.2%
	Treatment	Tanks	14.6%	2.3%	3.1%
	Reuse	Piping	7.2%	0.4%	12.5%
O&M	Collection	Electricity	—	1.5%	5.5%
	Treatment	Sludge removal	0.02%	0.02%	0.04%
		Chemicals	—	6.3%	17.5%
		Electricity	34.3%	62.8%	22.5%
		Diesel	—	2.4%	5.5%
	Reuse	Electricity	43.8%	17.9%	15.4%
		Diesel	—	3.9%	8.9%
	Resource recovery offsets	Potable water offsets	−17.9%	−15.4%	−25.2%
		Fertilizer offsets	−3.3%	−0.5%	−5.1%
		Energy offsets	—	—	−18.5%

Source: Cornejo et al. (2016). Reproduced with permission of American Chemical Society.

constructed wetland. A decentralized WWT system utilizing constructed wetlands will be the major technology due to the relatively large residential area for the photoremediation or treatment of WW. Solar energy would significantly reduce electricity consumption and treat WW efficiently and economically.

11.10 Capital and O&M Costs of Membrane Filtration

There are three major cost categories: (i) Capital cost includes cost of the system and process reactors. (ii) Operation includes cost of operating the system. (iii) Maintenance covers the cost of fixing and keeping the system up to date. Operation and maintenance costs are generally referred as O&M cost in terms of percentage of capital cost.

Example 11.7
Given: Table 11.20 shows capital cost and operation and maintenance (O&M) cost of microfiltration at different flow rates:
Find:

1) Calculate unit capital and O&M costs.
2) Plot unit cost/unit cost$_0$ versus Q/Q_0.
3) Determine correlation equations between unit cost/unit cost$_0$ for capital and O&M versus Q/Q_0, respectively.

Solution:

1) When Q = 1 MGD,

Table 11.20 Capital cost and operation and maintenance cost of microfiltration at different flow rates.

Q (MGD)	2006 capital cost ($)	2006 operation and maintenance costs ($/year)
1	1 145 772.56	290 645.63
2	1 751 658.86	315 291.25
4	2 967 431.47	361 582.50
6	4 181 204.09	410 873.75
8	5 395 976.70	457 165.00
10	6 611 749.31	503 456.25
20	12 685 612.37	742 912.50
40	25 372 224.74	1 230 825.00
60	37 520 950.86	1 706 737.51
80	50 208 563.23	2 194 650.01
100	62 356 289.35	2 670 562.51

$$\text{Unit capital cost} = 1145772.56\$ \div 1\,\text{MGD} \times 1\,\text{million gal}/1000\,\text{kgal} \div 1\,\text{day} = 1145.77\$/\text{kgal}$$

$$\text{Unit O\&M costs} = 290645.63\$/\text{year} \div 1\,\text{MGD} \times 1\,\text{million gal}/1000\,\text{kgal} \div 365\,\text{day/year}$$

$$\text{Unit O\&M costs} = 0.7963\,\$/\text{kgal}$$

When Q = 1 MGD,

$$Q_0 = 1\,\text{MGD}$$

$$\text{Unit capital cost}_0 = 1145.77\$/\text{kgal}$$

$$\text{Unit O\&M costs}_0 = 0.7963\$/\text{kgal}$$

$$\frac{Q}{Q_0} = 1$$

$$\frac{\text{Unit capital cost}}{\text{Unit capital cost}_0} = 1$$

$$\frac{\text{Unit O\&M costs}}{\text{Unit O\&M costs}_0} = 1$$

When Q = 20 MGD (Tables 11.21 and 11.22),

Table 11.21 Unit capital cost and unit operation and maintenance cost of microfiltration at different flow rates.

Q (MGD)	Unit capital cost ($/kgal)	Unit O&M costs ($/kgal)
1	1145.77	0.7963
2	875.83	0.4319
4	741.86	0.2477
6	696.87	0.1876
8	674.50	0.1566
10	661.17	0.1379
20	634.28	0.1018
40	634.31	0.0843
60	625.35	0.0779
80	627.61	0.0752
100	623.56	0.0732

Table 11.22 Unit cost ratio of microfiltration at different flow ratios.

Q/Q_0	Unit capital cost/ unit capital cost$_0$	Unit O&M costs/unit O&M costs$_0$
1	1.0000	1.0000
2	0.7644	0.5424
4	0.6475	0.3110
6	0.6082	0.2356
8	0.5887	0.1966
10	0.5771	0.1732
20	0.5536	0.1278
40	0.5536	0.1059
60	0.5458	0.0979
80	0.5478	0.0944
100	0.5442	0.0919

$$\frac{Q}{Q_0} = 20$$

$$\frac{\text{Unit capital cost}}{\text{Unit capital cost}_0} = 0.5536$$

$$\frac{\text{Unit O\&M costs}}{\text{Unit O\&M costs}_0} = 0.1278$$

2) Plot unit cost/unit cost$_0$ versus Q/Q_0 for capital and O&M, respectively (Figure 11.23).

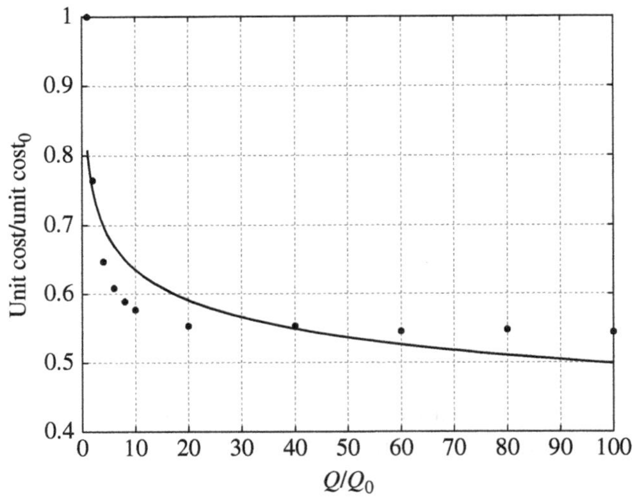

Figure 11.23 Unit capital cost/unit capital cost$_0$ versus Q/Q_0 for microfiltration, unit capital cost$_0$ = 1145.77$/kgal, Q_0 = 1 MGD.

3)

1) Equation for unit cost/unit cost$_0$ versus Q/Q_0 for capital:

The regression equation of the trend line:

$$\log\left(\frac{\text{unit cost}}{\text{unit cost}_0}\right) = -0.1047\log\left(\frac{Q}{Q_0}\right) - 0.0926$$

$$r^2 = 0.7258$$

or the equation of the regression (Figure 11.24):

$$\frac{\text{Unit cost}}{\text{Unit cost}_0} = 0.8080\left(\frac{Q}{Q_0}\right)^{-0.1047} \quad r^2 = 0.7258$$

Matlab code:

```
clear all
close all
%data%
Q=[1];
Qrange=[1 2 4 6 8 10 20 40 60 80 100];
totalcost=[1145772.56];
totalcostrange=[1145772.56 1751658.862
  2967431.474 4181204.086 5395976.698
  6611749.31 12685612.37 25372224.74
  37520950.86 50208563.23 62356289.35];
```

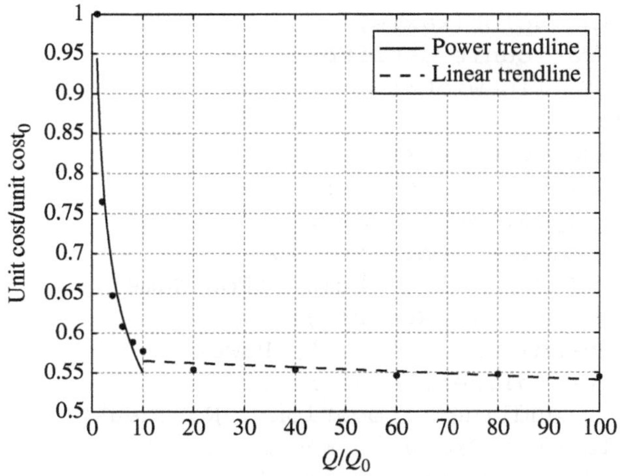

Figure 11.24 Unit capital cost/unit capital cost$_0$ versus Q/Q_0 for microfiltration, unit capital cost$_0$ = 1145.77$/kgal, Q_0 = 1 MGD.

```
%function%
unitcost=totalcost./Q./1000;
unitcostrange=totalcostrange./
  Qrange./1000;
p=polyfit(log10(Qrange/Q), log10
  (unitcostrange /unitcost), 1)
log10ratiounitcostfit=polyval(p, log10
  (Qrange/Q));
ratiounitcostfit=10.
  ^log10ratiounitcostfit;
k=10^p(:,2)
R2 = 1 - sum((log10(unitcostrange
  /unitcost) - log10ratiounitcostfit).
  ^2)/sum((log10(unitcostrange
  /unitcost) - mean(log10(unitcostrange
  /unitcost))).^2)
xmax=max(Qrange/Q);
xmin=min(Qrange/Q);
xaxis=[xmin:0.1:xmax];
log10ratiounitcostfitplot=polyval(p,
  log10(xaxis));
ratiounitcostfitplot=10.
  ^log10ratiounitcostfitplot;
%plot%
figure(1)
plot(Qrange/Q,unitcostrange
  /unitcost,'ko',
  'MarkerSize',3,'MarkerFaceColor',
  [0 0 0],'MarkerEdgeColor',
  [0 0 0],'linewidth',1);
hold on
plot(xaxis,ratiounitcostfitplot,
  'k-','linewidth',1);
set(gca,'xtickmode','auto',
  'FontSize',10,'FontWeight',
  'bold','FontName', 'Times New Roman');
xlabel({'Q/Q_0'},'FontSize',14,
  'FontWeight','bold','FontName',
  'Times New Roman')
ylabel({'Unit Cost/Unit
  Cost_0'},'FontSize',14,
  'FontWeight','bold','FontName',
  'Times New Roman')
grid on
hold on
```

Answer: `MFConstructionCostnormal`

```
p = -0.1047 -0.0926; k = 0.8080; R2 = 0.7258
```

2) Equation for unit cost/unit cost$_0$ versus Q/Q_0 for capital:

The regression equation of the trend line ($1 \le Q/Q_0 \le 10$):

$$\log\left(\frac{\text{unit cost}}{\text{unit cost}_0}\right) = -0.2344\log\left(\frac{Q}{Q_0}\right) - 0.025$$

$$r^2 = 0.9499$$

or the equation of the regression:

$$\frac{\text{Unit cost}}{\text{Unit cost}_0} = 0.9440\left(\frac{Q}{Q_0}\right)^{-0.2344} \quad r^2 = 0.9499$$

The regression equation of the trend line ($10 \le Q/Q_0 \le 100$) (Figure 11.25):

$$\frac{\text{Unit cost}}{\text{Unit cost}_0} = -0.0003Q/Q_0 + 0.5678 \quad r^2 = 0.6182$$

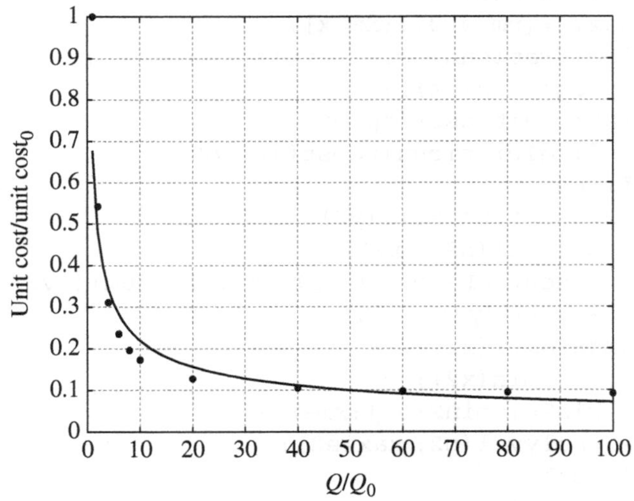

Figure 11.25 Unit O&M cost/unit O&M cost$_0$ versus Q/Q_0 for microfiltration, while unit O&M cost$_0$ = 0.7963 $/kgal, Q_0 = 1 MGD.

Matlab code:

```
clear all
close all
```

```
%data%
Qrange=[1  2  4  6  8  10 20 40 60 80 100];
Q=Qrange(1);
totalcostrange=[1145772.56 1751658.862
  2967431.474 4181204.086 5395976.698
  6611749.31 12685612.37 25372224.74
  37520950.86 50208563.23 62356289.35];
totalcost=totalcostrange(1);
xmid=6;
%function%
unitcost=totalcost./Q./1000;
unitcostrange=totalcostrange./
  Qrange./1000;
x=Qrange/Q;
y=unitcostrange /unitcost;
x1=x(1:xmid);
y1=y(1:xmid);
x2=x(xmid:11);
y2=y(xmid:11);
%power
p=polyfit(log10(x1), log10(y1), 1)
log10ratiounitcostfit=polyval(p,
  log10(x1));
ratiounitcostfit=10.
  ^log10ratiounitcostfit;
k=10^p(:,2)
R2power = 1 - sum((log10(y1) -
  log10ratiounitcostfit).^2)/sum((log10
  (y1) - mean(log10(y1))).^2)
xmax=max(x1);
xmin=min(x1);
xaxis=[xmin:0.1:xmax];
log10ratiounitcostfitplot=polyval(p,
  log10(xaxis));
ratiounitcostfitplot=10.
  ^log10ratiounitcostfitplot;
%line
p2=polyfit((x2),(y2), 1)
z=polyval(p2, (x2));
R2linear = 1 - sum(((y2) - z).^2)/sum(((y2)
  - mean((y2))).^2)
xmax2=max(x2);
xmin2=min(x2);
xaxis2=[xmin2:0.1:xmax2];
z2=polyval(p2, xaxis2);
%plot%
figure(1)
plot(Qrange/Q,unitcostrange
  /unitcost,'ko','MarkerSize',3,
  'MarkerFaceColor',[0 0
  0],'MarkerEdgeColor',[0 0
  0],'linewidth',1);
hold on
line1=plot(xaxis,
  ratiounitcostfitplot,'k-
  ','linewidth',1);
hold on
line2=plot(xaxis2,z2,'k--
  ','linewidth',1);
set(gca,'xtickmode',
  'auto','FontSize',10,
  'FontWeight','bold','FontName',
  'Times New Roman');
legend([line1,line2],'Power
  trendline','Linear
  trendline','Location','Northeast');
xlabel({'Q/Q_0'},'FontSize',14,
  'FontWeight','bold','FontName',
  'Times New Roman')
ylabel({'Unit Cost/Unit
  Cost_0'},'FontSize',14,
  'FontWeight','bold','FontName',
  'Times New Roman')
grid on
hold on
```

Answer: `MFConstructionCostnormaltwoline`

```
p = -0.2344 -0.0250; k = 0.9440; R2power
      = 0.9499
p2 = -0.0003 0.5678; R2linear = 0.6182
```

3) Equation for unit cost/unit cost$_0$ versus Q/Q_0 for O&M: The regression of the trend line:

$$\log\left(\frac{\text{unit cost}}{\text{unit cost}_0}\right) = -0.4891\log\left(\frac{Q}{Q_0}\right) - 0.1688$$
$$r^2 = 0.9254$$

or the regression equation (Figure 11.26):

$$\frac{\text{Unit cost}}{\text{Unit cost}_0} = 0.6780\left(\frac{Q}{Q_0}\right)^{-0.4891} \quad r^2 = 0.9254$$

Matlab code:

```
clear all
close all
%data%
Q=[1];
Qrange=[1  2  4  6  8  10 20 40 60 80 100];
totalcost=[290645.63];
totalcostrange=[290645.63  315291.2502
  361582.5003 410873.7505 457165.0007
  503456.2509 742912.50  1230825.00
  1706737.51 2194650.01 2670562.51];
```

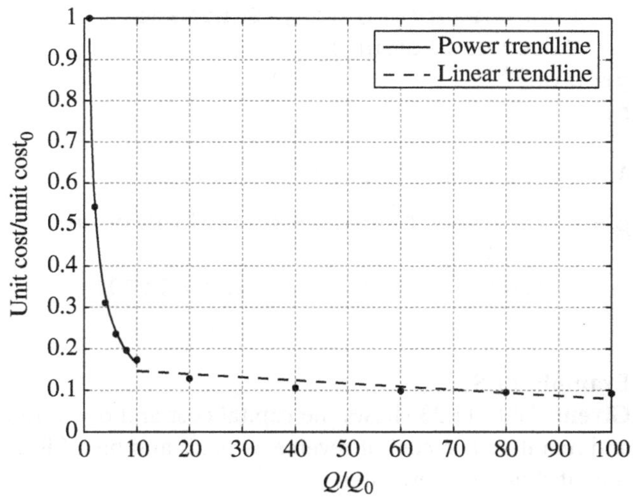

Figure 11.26 Unit O&M cost/unit O&M cost$_0$ versus Q/Q_0 for microfiltration, unit O&M cost$_0$ = 0.7963$/kgal, Q_0 = 1 MGD.

```
%function%
unitcost=totalcost./Q./1000./365;
unitcostrange=totalcostrange./Qrange./
  1000./365;
p=polyfit(log10(Qrange/Q), log10
  (unitcostrange /unitcost), 1)
log10ratiounitcostfit=polyval(p, log10
  (Qrange/Q));
ratiounitcostfit=10.
  ^log10ratiounitcostfit;
k=10^p(:,2)
R2 = 1 - sum((log10(unitcostrange
  /unitcost) - log10ratiounitcostfit).
  ^2)/sum((log10(unitcostrange
  /unitcost) - mean(log10(unitcostrange
  /unitcost))).^2)
xmax=max(Qrange/Q);
xmin=min(Qrange/Q);
xaxis=[xmin:1:xmax];
log10ratiounitcostfitplot=polyval(p,
  log10(xaxis));
ratiounitcostfitplot=10.
  ^log10ratiounitcostfitplot;
%plot%
figure(1)
plot(Qrange/Q,unitcostrange
  /unitcost,'ko','MarkerSize',3,
  'MarkerFaceColor',[0 0
  0],'MarkerEdgeColor',[0 0
  0],'linewidth',1);
hold on
plot(xaxis,ratiounitcostfitplot,'k-
  ','linewidth',1);
set(gca,'xtickmode',
  'auto','FontSize',10,
  'FontWeight','bold','FontName',
  'Times New Roman');

xlabel({'Q/Q_0'},'FontSize',14,
  'FontWeight','bold','FontName',
  'Times New Roman')
ylabel({'Unit Cost/Unit Cost_0'},
  'FontSize',14,'FontWeight',
  'bold','FontName', 'Times New Roman')
grid on
hold on
```

Answer: `MFOMCostnormal`

```
p = -0.4891 -0.1688; k = 0.6780; R2 = 0.9254
```

4) Equation for unit cost/unit cost$_0$ versus Q/Q_0 for O&M:

The regression of the trend line ($1 \le Q/Q_0 \le 10$):

$$\log\left(\frac{\text{unit cost}}{\text{unit cost}_0}\right) = -0.7635\log\left(\frac{Q}{Q_0}\right) - 0.022$$

$$r^2 = 0.9952$$

or the regression equation:

$$\frac{\text{Unit cost}}{\text{Unit cost}_0} = 0.9507\left(\frac{Q}{Q_0}\right)^{-0.7635} \quad r^2 = 0.9952$$

The regression of the trend line ($10 \le Q/Q_0 \le 100$):

$$\frac{\text{Unit cost}}{\text{Unit cost}_0} = -0.0008Q/Q_0 + 0.1540 \quad r^2 = 0.7018$$

Matlab code:

```
clear all
close all
%data%
Qrange=[1 2 4 6 8 10 20 40 60 80 100];
Q=Qrange(1);
totalcostrange=[290645.63 315291.2502
  361582.5003 410873.7505 457165.0007
  503456.2509 742912.50 1230825.00
  1706737.51 2194650.01 2670562.51];
totalcost=totalcostrange(1);
xmid=6;
%function%
unitcost=totalcost./Q./1000./365;
unitcostrange=totalcostrange./Qrange./
  1000./365;
x=Qrange/Q;
y=unitcostrange /unitcost;
x1=x(1:xmid);
y1=y(1:xmid);
```

```
x2=x(xmid:11);
y2=y(xmid:11);
%power
p=polyfit(log10(x1), log10(y1), 1)
log10ratiounitcostfit=polyval(p,
  log10(x1));
ratiounitcostfit=10.
  ^log10ratiounitcostfit;
k=10^p(:,2)
R2power = 1 - sum((log10(y1) -
  log10ratiounitcostfit).^2)/sum((log10
  (y1) - mean(log10(y1))).^2)
xmax=max(x1);
xmin=min(x1);
xaxis=[xmin:0.1:xmax];
log10ratiounitcostfitplot=polyval(p,
  log10(xaxis));
ratiounitcostfitplot=10.
  ^log10ratiounitcostfitplot;
%line
p2=polyfit((x2),(y2), 1)
z=polyval(p2, (x2));
R2linear = 1 - sum(((y2) - z).^2)/sum(((y2)
  - mean((y2))).^2)
xmax2=max(x2);
xmin2=min(x2);
xaxis2=[xmin2:0.1:xmax2];
z2=polyval(p2, xaxis2);
%plot%
figure(1)
plot(Qrange/Q,unitcostrange
  /unitcost,'ko',
  'MarkerSize',3,'MarkerFaceColor',
  [0 0 0],'MarkerEdgeColor',
  [0 0 0],'linewidth',1);
hold on
line1=plot(xaxis,
  ratiounitcostfitplot,'k-
  ','linewidth',1);
hold on
line2=plot(xaxis2,z2,'k--
  ','linewidth',1);
set(gca,'xtickmode','auto',
  'FontSize',10,'FontWeight',
  'bold','FontName',
  'Times New Roman');
legend([line1,line2],'Power
  trendline','Linear
  trendline','Location','Northeast');
xlabel({'Q/
  Q_0'},'FontSize',14,'FontWeight',
  'bold','FontName', 'Times New Roman')
ylabel({'Unit Cost/Unit
  Cost_0'},'FontSize',14,
  'FontWeight','bold','FontName',
  'Times New Roman')
grid on
hold on
```

Answer: `MFOMCostnormaltwoline`

```
p = -0.7635 -0.0220; k = 0.9507; R2power
  = 0.9952
p2 = -0.0008 0.1540; R2linear = 0.7018
```

Example 11.8

Given: Table 11.23 shows the capital cost and operation and maintenance cost of reverse osmosis and nanofiltration at different flow rates.

Table 11.23 Capital cost and operation and maintenance cost of reverse osmosis and nanofiltration at different flow rates.

Q (MGD)	2006 capital cost ($)	2006 operation and maintenance costs ($/year)
1	2 630 207.84	2 707 281.79
2	3 958 402.454	2 851 711.803
4	6 527 627.349	3 146 977.202
6	8 979 523.093	3 440 332.766
8	11 383 885.71	3 734 517.654
10	13 735 967.79	4 027 219.037
20	25 229 374.62	5 496 830.50
40	47 430 550.07	8 430 904.35
60	69 181 992.93	11 362 030.64
80	90 719 913.20	14 291 757.49
100	112 104 198.19	17 216 045.49

Find:

1) Calculate unit capital and O&M costs.
2) Plot unit cost/unit $cost_0$ versus Q/Q_0.
3) Determine correlation equations between unit cost/unit $cost_0$ for capital and O&M versus Q/Q_0, respectively.

Solution:

1) When Q = 1 MGD,

$$\text{Unit capital cost} = 2630207.84\$ \div 1\,\text{MGD} \times 1\,\text{million gal}/1000\,\text{kgal} \div 1\,\text{day} = 2630.21\$/\text{kgal}$$

$$\text{Unit O\&M costs} = 2707281.79\$/\text{year} \div 1\,\text{MGD} \times 1\,\text{million gal}/1000\,\text{kgal} \div 365\,\text{day/year}$$

$$\text{Unit O\&M costs} = 7.4172\,\$/\text{kgal}$$

When Q = 1 MGD,

$$Q_0 = 1\,\text{MGD}$$

$$\text{Unit capital cost}_0 = 2630.21\$/\text{kgal}$$

$$\text{Unit O\&M costs}_0 = 7.4172\$/\text{kgal}$$

$$\frac{Q}{Q_0} = 1$$

$$\frac{\text{Unit capital cost}}{\text{Unit capital cost}_0} = 1$$

$$\frac{\text{Unit O\&M costs}}{\text{Unit O\&M costs}_0} = 1$$

When Q = 20 MGD (Tables 11.24 and 11.25),

Table 11.24 Unit capital cost and unit operation and maintenance cost of reverse osmosis and nanofiltration at different flow rates.

Q (MGD)	Unit capital cost ($/kgal)	Unit O&M costs ($/kgal)
1	2630.21	7.4172
2	1979.20	3.9065
4	1631.91	2.1555
6	1496.59	1.5709
8	1422.99	1.2789
10	1373.60	1.103
20	1261.47	0.7530
40	1185.76	0.5775
60	1153.03	0.5188
80	1134.00	0.4894
100	1121.04	0.4717

Table 11.25 Unit cost ratio of reverse osmosis and nanofiltration at different flow ratios.

Q/Q_0	Unit capital cost/unit capital cost$_0$	Unit O&M costs/unit O&M costs$_0$
1	1.0000	1.0000
2	0.7525	0.5267
4	0.6204	0.2906
6	0.5690	0.2118
8	0.5410	0.1724
10	0.5222	0.1488
20	0.4796	0.1015
40	0.4508	0.0779
60	0.4384	0.0699
80	0.4311	0.0660
100	0.4262	0.0636

$$\frac{Q}{Q_0} = 20$$

$$\frac{\text{Unit capital cost}}{\text{Unit capital cost}_0} = 0.4796$$

$$\frac{\text{Unit O\&M costs}}{\text{Unit O\&M costs}_0} = 0.1015$$

2) Plot unit cost/unit cost$_0$ versus Q/Q_0 for capital and O&M, respectively (Figure 11.27).

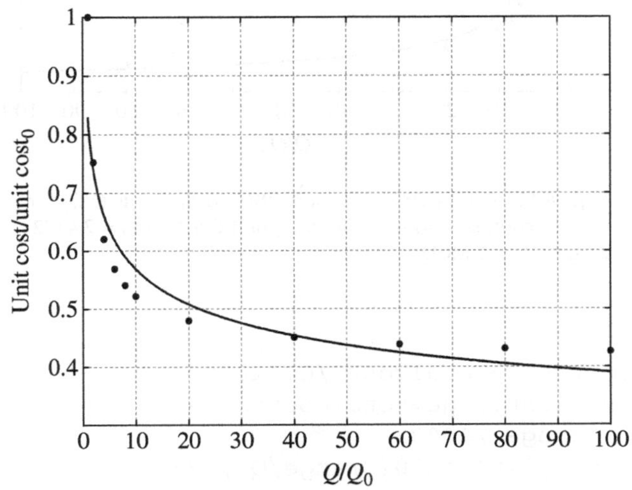

Figure 11.27 Unit capital cost/unit capital cost$_0$ versus Q/Q_0 for reverse osmosis and nanofiltration, unit capital cost$_0$ = 2630.21 $/kgal, Q_0 = 1 MGD.

3) 1) Equation for unit cost/unit cost$_0$ versus Q/Q_0 for capital:

The regression equation of the trend line:

$$\log\left(\frac{\text{unit cost}}{\text{unit cost}_0}\right) = -0.1634\log\left(\frac{Q}{Q_0}\right) - 0.0814$$

$$r^2 = 0.8909$$

or the equation of the regression (Figure 11.28):

$$\frac{\text{Unit cost}}{\text{Unit cost}_0} = 0.8291\left(\frac{Q}{Q_0}\right)^{-0.1634} \quad r^2 = 0.8909$$

Matlab code:

```
clear all
close all
%data%
Q=[1];
Qrange=[1 2 4 6 8 10 20 40 60 80 100];
totalcost=[2630207.84];
totalcostrange=[2630207.84 3958402.454
  6527627.349 8979523.093 11383885.71
  13735967.79 25229374.62 47430550.07
  69181992.93 90719913.20 112104198.19];
```

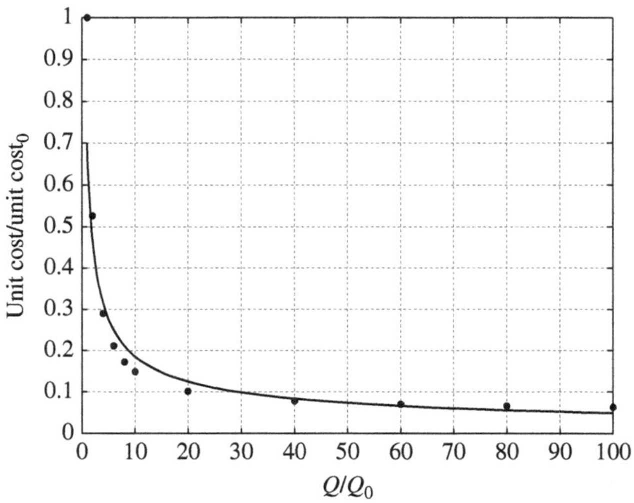

Figure 11.28 Unit O&M cost/unit O&M cost$_0$ versus Q/Q_0 for reverse osmosis and nanofiltration, unit O&M cost$_0$ = 7.4172 \$/kgal, Q_0 = 1 MGD.

```
%function%
unitcost=totalcost./Q./1000;
unitcostrange=totalcostrange./
  Qrange./1000;
p=polyfit(log10(Qrange/Q), log10
  (unitcostrange /unitcost), 1)
log10ratiounitcostfit=polyval(p, log10
  (Qrange/Q));
ratiounitcostfit=10.
  ^log10ratiounitcostfit;
k=10^p(:,2)
R2 = 1 - sum((log10(unitcostrange
  /unitcost) - log10ratiounitcostfit).
  ^2)/sum((log10(unitcostrange
  /unitcost) - mean(log10(unitcostrange
  /unitcost))).^2)
xmax=max(Qrange/Q);
xmin=min(Qrange/Q);
xaxis=[xmin:0.1:xmax];
log10ratiounitcostfitplot=polyval(p,
  log10(xaxis));
ratiounitcostfitplot=10.
  ^log10ratiounitcostfitplot;
%plot%
figure(1)
plot(Qrange/Q,unitcostrange
  /unitcost,'ko','MarkerSize',3,
  'MarkerFaceColor',
  [0 0 0],'MarkerEdgeColor',
  [0 0 0],'linewidth',1);
hold on
plot(xaxis,ratiounitcostfitplot,'k-
  ','linewidth',1);
set(gca,'xtickmode','auto',
  'FontSize',10,'FontWeight','bold',
  'FontName', 'Times New Roman');

xlabel({'Q/Q_0'},'FontSize',14,
  'FontWeight','bold','FontName',
  'Times New Roman')
ylabel({'Unit Cost/Unit Cost_0'},
  'FontSize',14,'FontWeight',
  'bold','FontName', 'Times New Roman')
grid on
hold on
```

Answer: `RONFConstructionCostnormal`

`p = -0.1634 -0.0814; k = 0.8291; R2 = 0.8909`

2) Equation for unit cost/unit cost$_0$ versus Q/Q_0 for O&M:

The regression of the trend line:

$$\log\left(\frac{\text{unit cost}}{\text{unit cost}_0}\right) = -0.5770\log\left(\frac{Q}{Q_0}\right) - 0.1540$$

$$r^2 = 0.9510$$

or the regression equation:

$$\frac{\text{Unit cost}}{\text{Unit cost}_0} = 0.7015\left(\frac{Q}{Q_0}\right)^{-0.5770} \qquad r^2 = 0.9510$$

Matlab code:

```
clear all
close all
%data%
Q=[1];
Qrange=[1  2  4  6  8  10 20 40 60 80 100];
totalcost=[2707281.79];
totalcostrange=[2707281.79 2851711.803
  3146977.202 3440332.766 3734517.654
  4027219.037 5496830.50 8430904.35
  11362030.64 14291757.49 17216045.49];
%function%
unitcost=totalcost./Q./1000./365;
unitcostrange=totalcostrange./Qrange./
  1000./365;
p=polyfit(log10(Qrange/Q), log10
  (unitcostrange /unitcost), 1)
log10ratiounitcostfit=polyval(p, log10
  (Qrange/Q));
ratiounitcostfit=10.
  ^log10ratiounitcostfit;
k=10^p(:,2)
```

```
R2 = 1 - sum((log10(unitcostrange
   /unitcost) - log10ratiounitcostfit).
   ^2)/sum((log10(unitcostrange
   /unitcost) - mean(log10(unitcostrange
   /unitcost))).^2)
xmax=max(Qrange/Q);
xmin=min(Qrange/Q);
xaxis=[xmin:0.1:xmax];
log10ratiounitcostfitplot=polyval(p,
   log10(xaxis));
ratiounitcostfitplot=10.
   ^log10ratiounitcostfitplot;
%plot%
figure(1)
plot(Qrange/Q,unitcostrange /unitcost,
   'ko','MarkerSize',3,
   'MarkerFaceColor',[0 0
   0],'MarkerEdgeColor',[0 0
   0],'linewidth',1);
hold on
plot(xaxis,ratiounitcostfitplot,'k-
   ','linewidth',1);
set(gca,'xtickmode','auto',
   'FontSize',10,'FontWeight','bold',
   'FontName', 'Times New Roman');

xlabel({'Q/Q_0'},'FontSize',14,
   'FontWeight','bold','FontName',
   'Times New Roman')
ylabel({'Unit Cost/Unit
   Cost_0'},'FontSize',14,
   'FontWeight','bold','FontName',
   'Times New Roman')
grid on
hold on
```

Answer: RONFOMCostnormal

```
p = -0.5770 -0.1540; k = 0.7015; R2 = 0.9510
```

11.11 Exercise

11.11.1 Questions

1) In designing a coagulation reactor, should additional jar tests be conducted to confirm the theoretical results?
2) Why is precipitation not an environmentally sustainable way to treat wastewater containing metals?
3) Why is a static mixer better than a coagulation tank?
4) List five reasons why activated sludge has fundamental faults in system design according to the SEE principles.
5) Rank new WW treatment technologies according to energy required while achieving treatment and discharge standards.
6) Rank three common sustainable separation technologies in wastewater and solid waste management.
7) How important is separation in energy recovery and reuse for biosolids in terms of nutrients, energy, and carbon (NEW) footprints?
8) Who are the stakeholders in your home county critical to leading the transition from traditional WWTP to energy-positive operation?
9) Formulate your strategies to convince WWTP managers to move toward an energy-neutral goal using separation as an example.

11.11.2 Calculation

1) If all the Cr of 75 mg/l and Zn of 35 mg/l in Example 11.1 are required to precipitate, the sludge production would be 2249.89 kg/day. What would be the total precipitation sludge of Cr for a treatment process of 0.5 MGD? What would be the volume of sludge produced daily? What are the treatment alternatives that could be used to remove Cr with less sludge?
2) In Example 11.6, the VOCs such as benzene, toluene, and trichloroethylene have concentration of 750, 1000, and 750 μg/l. If the required discharge standard concentrations of these pollutants are 10, 100, and 100 μg/l, then advanced oxidation processes such as UV/TiO_2, UV/O_3, UV/H_2O_2, or Fe^{2+}/H_2O_2 could be applied to destroy the VOCs rather than barely phase transfer such as air stripping. Answer the following:
   1) Rank the capital costs of commercial available products to achieve the requirements.
   2) Rank the AOPs in terms of OM costs.
   3) What would be the overall costs during the life cycle of 50 years for each AOP?
   4) In life cycle assessment, what would be the major environmental benefits of AOP over air stripping?
3) Develop a Matlab program to convert COD into energy when the end products are CH_4, H_2, and electricity, respectively. The COD changes from 30 to 3000 m/l.

11.11.3 Projects

11.11.3.1 Xiongan Project

1) Separation is critical to achieve zero water design. Use unit BOD, COD, N, and P data listed in this chapter to assess the separation strategies for Xiongan district in integrated wastewater management. The color-coded design of blue, green, gray, brown, and black should

be applied to design water resource recovery facilities in the district.

2) Estimate the cost saving of color-coded design versus traditional fit for all WWTP design. Separation strategies should include:
   a) Separation of gray water for irrigation.
   b) Separation of brown from black water with or without nutrient recovery.
   c) Separation of food wastes including FOG from DSW for anaerobic digestion in WRRF.
   d) Diversion of organic matter from wastewater in primary settler to anaerobic digester.

11.11.3.2 Community Projects

1) Choose a WWTP in your hometown city, contact its operations department, and get the WWTP process design. Calculate the following:
   a) Theoretical total maximum energy content in the average WW flow rate.
   b) The energy shortage of the current operating process for energy-neutral operation.
2) Use unit values compiled by Ishii and Boyer (2015) to estimate the following:
   1) How much struvite could be produced for the city that you have chosen?
   2) What would be the FP on C, N, P, water, and energy of your design?
3) Using knowledge and unit process data, propose two design alternatives to retrofit the current WWTP. Develop a process diagram and show the percentage of reduction of C, N, and P by your calculation of the two alternatives. Two design alternatives are with or without C, N, and P recovery. Use LCA methodology and assumptions made by them to calculate nine environmental impacts by using TRACI v2.1.
4) Carry out cost and benefit analysis to show how many years it will take for the retrofitted WWTP to recover the capital cost invested in your design alternatives.
5) Make an appointment with the local water authority, present the findings of your proposed design alternatives, and request the input of their chief engineers in charge of WWTP design.
6) Attend a local commissioner's meeting and work with your local water authority to obtain permission to introduce your design alternatives at the meeting. Work with the water authority engineering department to secure funding from the county budget to retrofit the WWTP.

References

Asano, T., Burton, F.L., Leverenz, H.L. et al. (2007). *Water reuse issues, technologies, and applications*. New York: McGraw-Hill.

Avallone, E.A. and Baumeister III, T. (1987). *Marks Standard Handbook for Mechanical Engineerse*, 9e. New York: McGraw-Hill.

Cornejo, P.K., Zhang, Q., and Mihelcic, J.R. (2016). How does scale of implementation impact the environmental sustainability of wastewater treatment integrated with resource recovery?. *Environmental Science and Technologies* 50 (13): 6680–6689.

Corominas, L., Flores-Alsina, X., Snip, L., and Vanrolleghem, P.A. (2012). Comparison of different modeling approaches to better evaluate greenhouse gas emissions from whole wastewater treatment plants. *Biotechnology and Bioengineering* 109: 2854–2863.

Crawford, G., Johnson, T.D., Johnson, B.R. et al. (2011). *CHEApet User's Manual*. Alexandria, VA: Water Environment Federation.

Eco-invent (2007). *Ecoinvent 2.0*. Dübendorf: Swiss Centre for Life Cycle Inventories.

EnviroSim Associates Ltd (2014). New development in BioWin 4.0. Retrieved 10 February 2014 from http://envirosim.com/files/downloads/WhatsNew40.pdf (accessed 21 December 2017).

Goel, R., Blackbourn, A., Khalil, Z. et al. (2012). Implementation of carbon footprint model in a dynamic process simulator. Proceedings of the Water Environment Federation's Annual Technical Exhibition and Conference, New Orleans, LA.

Guan, D.B., Hubacek, K., Tillotson, M. et al. (2014). Lifting China's Water Spell. *Environmental Science and Technologies* 48: 11048–11056.

Ishii, S.K.L. and Boyer, T.H. (2015). Life cycle comparison of centralized wastewater treatment and urine source separation with struvite precipitation: Focus on urine nutrient management. *Water Research* 79: 88–103.

ISO (2006). ISO 14040: 2006 Environmental Management: Life Cycle Assessment Principles and Framework, 2e. Geneva: International Organization for Standardization.

Johnston, A. (2011). Developing a greenhouse gas emission calculation tool for water utilities. MS thesis. Clemson University.

McCabe, W.L., Smith, J.C., and Harriott, P. (1993). *Unit Operations of Chemical Engineering*, pp. 501–736. New York: McGraw-Hill.

MWH (2012). MWH Launches MCO2™ Greenhouse Gas Calculation Tool. Retrieved 3 April 2012, from http://www.mwhglobal.com/blog/mwh-launches-mco2-

greenhousegascalculation-tool/ (accessed 21 December 2017).

NREL (2012). Simulating the value of concentrating solar power with thermal energy storage in a production cost model. Golden, CO: National Renewable Energy Laboratory.

Perry, J.H. (1984). *Chemical Engineers Handbook*, 6e. New York: McGraw-Hill.

Reffold, E.L., Choudhury, F., and Rayner, P.S. (2008). *Using Science to Create a Better Place: Greenhouse Gas Emissions of Water Supply and Demand Management Options*. Bristol: UK Environment Agency.

Shrestha, E., Ahmad, S., Johnson, W. et al. (2011). Carbon footprint of water conveyance versus desalination as alternatives to expand water supply. *Desalination* 280: 33–43.

Stokes, J.R. and Horvath, A. (2006). LCA methodology and case study life cycle energy assessment of alternative water supply systems. *International Journal of Life Cycle Assessment* 11: 335–343.

Stokes, J.R. and Horvath, A. (2011a). *Wastewater-Energy Sustainability Tool User's Manual*. Berkeley, CA: Department of Civil & Engineering, University of California Berkeley.

Stokes, J.R. and Horvath, A (2011b). *Water-Energy Sustainability Tool User's Manual*. Berkeley, CA: Department of Civil & Environmental Engineering, University of California.

Tampa Bay Water (2012). Energy water nexus – from greenhouse gases to utility energy optimization programs. Presentation for 2012 Energy Efficiency Global Forum, Water Research Foundation, Orlando, FL. Retrieved 3 October 2013, from http://collab.waterrf.org/Workshops/FLwater/Florida%20WaterEnergy%20Coalition%2Documents/Florida%20Water%20Energy%20Workshop%20Presentations/Dave%20Braciano%20Meribel%20Medina.pdf (accessed 21 December 2017).

Treybal, R. (1980). *Mass-Transfer Operations*, 3e, 275–313. New York: McGraw-Hill.

UKWIR (2008). Carbon Accounting. https://www.ukwir.org/eng/Search-UK-Water-Industry-Research-reports (accessed 23 January 2018).

US EPA (1980). Summary report: Control and treatment technology for the metal finishing industry; sulfide precipitation, technology transfer division. *EPA 625/8-80-003*. Washington, DC: US EPA.

US EPA (2014). US EPA Office of Research and Development. *EPA/600/R-12/554*. Washington, DC.

US EPA Design Manual (2004). United States Environmental Protection Agency design manual. Washington, DC: US EPA.

US EPA, Membrane Filtration Guidance Manual (2005). Office of Water (4601), *EPA 815-R-06-009*. November 2005.

Yoshida, H. (2014). Influence of data collection schemes on the Life Cycle Assessment of a municipal wastewater treatment plant. *Water Research* 56: 292–303.

Yu, X.M., Geng, Y., Heck, P., and Xue, B. (2015). A review of China's Rural Water Management. *Sustainability* 7: 5773–5792.

12

Treatment

Mineralization is a better strategy than phase transfer.

OBJECTIVES
To provide technical support of the UV disinfection to replace chlorination to comply with regulatory requirements To quantify UV dose for inactivating *Cryptosporidium* and *Giardia* To design high efficient UV reactor for disinfection To define and apply virus or bacteria sensitivity index of UV disinfection for engineering applications To provide simplified mathematical model quantifying bacteria shoulder effect during UV disinfection To present advanced oxidation processes such as UV/TiO_2 to replace phase transfer process To use the Hammett plot to differentiate the hole versus hydroxyl radical oxidation in UV/TiO_2 To assess and select different advanced oxidation processes (AOPs)

12.1 Principle 9

To meet the US EPA long-term regulation (LTR) and disinfection by-product (DBP) Stage 2 rules simultaneously, ultraviolet (UV) disinfection is the key enabling technology, because UV process could not only disinfect all the pathogenic microorganisms but also degrade micropollutants in water or wastewater.

Principle 9 Treatment should avoid adding chemicals or use minimal nontoxic chemicals and renewable energy to reduce footprints.

1) No toxic chemicals such as chlorine or aluminum should be added to drinking water.
2) Destructive technologies such as AOPs should be the first choice versus phase transfer technologies such as activated carbon.
3) Waste such as sludge produced per unit water produced should be minimal.

12.2 Challenges

The best disinfection strategy is to simultaneously meet the requirements of EPA LTR and DBP Stage 2 rules at the lowest cost. Although ozone is a highly effective disinfectant due to its high oxidation potential and most effective against viruses, it is less effective against *Cryptosporidium* oocysts. The required CT values at 10°C for inactivation of various microbes for each of the commonly used chemical disinfectants are shown in Table 12.1. The relative effectiveness of a disinfectant can be seen by its CT values. If temperature is higher, inactivation is expected to be greater.

The EPA (2005) proclaimed a Stage 2 Disinfectants and Disinfection Byproducts Rule (Stage 2 DBPR) with the Long Term 2 Enhanced Surface Water Treatment Rule (LT2ESWTR) to address the risk–risk trade-off between microbial disinfection and the DBPs formed by commonly used disinfectants. The Stage 2 DBPR aims to reduce peak DBP concentrations in the distribution system by modifying the Stage 1 DBPR monitoring requirements and procedures for compliance determination. When a public work system (PWS) assesses its disinfection strategy, it must consider both the disinfectant

Sustainable Environmental Engineering, First Edition. Walter Z. Tang and Mika Sillanpää.

Table 12.1 Required CT values (mg-min/l) for chemical disinfectants at 10°C (US EPA, 2003).

Disinfectant	4-log inactivation (99.99%) of viruses	3-log inactivation (99.9%) of *Giardia*	1-log inactivation (99.9%) of *Cryptosporidium*
Ozone	1	1.43	9.9
Chlorine	6	104	N/A
Chlorine dioxide	4.2	23	277

effectiveness against the target pathogen and the DBPs formed as a result of the disinfection. The term "log" means the order of magnitude reduction in concentration, e.g. 2-log, 3-log, and 4-log removal equals a 99, 99.9, and 99.99% reduction, respectively (Table 12.2).

12.3 Environmental Regulations

The LT2ESWTR and the Stage 2 DBPR were developed and promulgated together to address trade-offs risk between two different types of contaminants: microbial pathogens and DBPs. A PWS could encounter compliance issues with the Stage 2 DBPR when making changes to comply with the LT2ESWTR and vice versa. In addition to complying with the suite of microbial/disinfection by-product (M/DBP) rules simultaneously, a potable water system must also ensure that changes in treatment to comply with those rules do not adversely affect compliance with other drinking water regulations. Since there are many more regulations than LT2ESWTR and DBPR, simultaneous compliance refers to compliance with all existing Safe Drinking Water Act (SDWA) regulations. Here are the major environmental rules proclaimed by the US EPA (Table 12.3).

Since UV disinfection can simultaneously provide a barrier for *Cryptosporidium*, *Giardia*, bacteria, and virus, major PWS such as New York replaced chlorination in some WTPs with UV disinfection. In the next few decades, more and more PWS will install UV reactors to meet the EPA Stage2 DBPR and the LT2ESWTR. One of the main motivations for the industry operation and maintenance to switch to UV disinfection is because UV disinfection costs much less than ozonation, which was considered as a better technology than chlorination in terms of reducing DBP formation. The US EPA (2005) compared costs of three typical disinfection technologies and concluded that ozonation was the most expensive process (Figure 12.1).

Ozonation costs much more than chlorination, which is more expensive than UV disinfection. In terms of total annual cost, UV disinfection has its advantage over ozonation and chlorination (Table 12.4).

To comply with the EPA requirements, UV reactor design, validation, and operation involve many complicated calculations. To facilitate these calculations, Matlab codes are developed for engineering consulting communities and PWS engineers. Matlab can be used for complicated computations in stochastic engineering design, data analysis, modeling, simulation, quantification of uncertainty and sensitivity, system reliability, simulation, and process control. These Matlab codes are also useful for graduates, researchers, and scientists in designing and modeling the UV disinfection systems for water, wastewater, and water reuse industries.

The *Giardia* and *Cryptosporidium* outbreaks in the 1990s prompted regulatory action to develop disinfection

Table 12.2 Summary of microbial and DBP rules (US EPA, 2006).

Surface water treatment rules – minimum treatment requirements			
Regulation	***Giardia***	**Virus**	***Cryptosporidium***
SWTR	3-log removal and/or inactivation	4-log removal and/or inactivation	Not addressed
IESWTR and LT1ESWTR	No change from SWTR		2-log removal
LT2ESWTR	No change from SWTR		0- to 2.5-log additional treatment for filtered systems 2- or 3-log inactivation for unfiltered systems

DBP rules – MCLs based on running annual averages (RAAs) or locational RAAs (LRAAs)				
Regulation	**Total trihalomethanes (TTHM) (μg/l)**	**Five haloacetic acids (HAA5) (μg/l)**	**Bromate (μg/l)**	**Chlorite (μg/l)**
Stage 1 DBPR	80 as RAA	60 as RAA	10	1000
Stage 2 DBPR	80 as LRAA	60 as LRAA	No change from Stage 1	

Table 12.3 US EPA SDWA regulations (US EPA, 2006).

EPA rule	Contaminant of concern	Date of promulgation
Ground Water Rule (GWR)	Source water microbial pathogens	November 2006
Long Term 2 Enhanced Surface Water Treatment Rule (LT2ESWTR)	Source water microbial pathogens	January 2006
Stage 2 Disinfectants and Disinfection Byproducts Rule (Stage 2 DBPR)	Disinfection by-products	January 2006
Total Coliform Rule (TCR)	Distribution system microbial pathogens	June 1989
Stage 1 Disinfectants and Disinfection Byproducts Rule (Stage 1 D/DBPR)	Disinfectants and disinfection by-products	December 1998
Interim Enhanced Surface Water Treatment Rule (IESWTR)	Source water microbial pathogens	December 1998
Long Term 1 Enhanced Surface Water Treatment Rule (LT1ESWTR)	Source water microbial pathogens	January 2002
Filter Backwash Recycling Rule (FBRR)	Filter backwash (microbial pathogens)	June 2001
Surface Water Treatment Rule (SWTR)	Source water microbial pathogens	June 1989

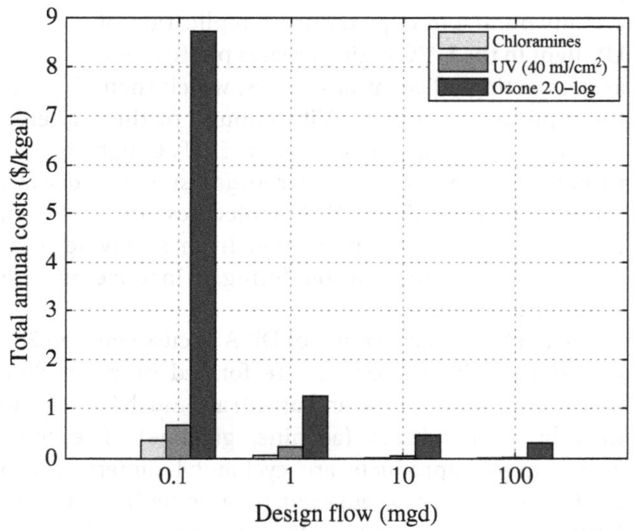

Figure 12.1 Cost comparison for alternative chemical disinfection strategies (US EPA, 2005).

processes that would specifically include inactivation of these protozoan pathogens in addition to conventional bacterial and virus pathogens. As a result, the US EPA recommends UV disinfection as a major strategy to simultaneously satisfy DBP reduction and the inactivation of *Cryptosporidium* and *Giardia lamblia* (Table 12.5).

The EPA LT2ESWTR aims to reduce disease incidents with *Cryptosporidium* and other pathogenic microorganisms in drinking water (US EPA, 2007). *Giardia* is an anaerobic flagellated protozoan parasite of the *phylum metamonada* in the supergroup "Excavata" that colonize and reproduce in the small intestines of several vertebrates, causing giardiasis. Their life cycle alternates between an actively swimming trophozoite and an infectious, resistant cyst. Cryptosporidiosis is a parasitic disease caused by *Cryptosporidium*, a protozoan parasite in the phylum Apicomplexa. It affects the intestines of mammals and is typically an acute short-term infection.

Table 12.4 Alternative disinfection total annual costs ($/kgal) for the Stage 2 D/DBP and LT2ESWT rules (US EPA, 2005).

	Design flow (MGD)			
Technology	**0.1**	**1**	**10**	**100**
Alternative chemical disinfection strategies				
Chloramines (NH_4 dose = 0.55 mg/l)	0.36	0.06	0.01	0.01
Chlorine dioxide	1.69	0.17	0.03	0.01
UV (40 mJ/cm^2)	0.66	0.23	0.05	0.02
UV (200 mJ/cm^2)	1.84	0.64		
Ozone (0.5-log *Cryptosporidium* inactivation)	8.3	1.05		
Ozone (1.0-log *Cryptosporidium* inactivation)	8.55	1.2	0.37	0.23
Ozone (2.0-log *Cryptosporidium* inactivation)	8.72	1.25	0.47	0.31

Table 12.5 Typical microbial removal/inactivation.

Regulatory requirements	Chemical inactivation credit
Viruses 4 log (99.99%)	Viruses 2 log (99%) plus 2.0 log with chemical disinfectant
Giardia 3 log (99.9%)	*Giardia* 2.5 log plus 0.5 log with chemical disinfectant
Crypto 2 log (99%)	Crypto 2 log (99%)
	Direct filtration only 2.0 log *Giardia* credit

It is spread through the fecal–oral route such as contaminated drinking water. The main symptom is self-limiting diarrhea in people with weak immune systems. In individuals with suppressed immune systems, the symptoms are particularly severe and often fatal. The most common waterborne diseases caused by the parasite are transmitted by microbial cysts (oocysts) and result in an infection of the intestinal epithelial tissue. Microbiological standards of *Giardia* and *Cryptosporidium* are listed in Table 12.6.

Table 12.6 Proposed maximum acceptable average concentration of protozoa in drinking water (US EPA, 2006).

Organism	Proposed counts	No. in m^3 drinking water
Cryptosporidium	2.6×10^{-5}/l	38
Giardia	5.5×10^{-6}/l	180

The European Union (EU) Council Directive 98/83/EC, dated 3 November 1998, in Article 5 requires that Member States set standards for water intended for human consumption. Every EU Member has adopted the 98/83/EC directive. Most of the EU Members have extended the directive to more rigorous regulations for surface water and groundwater under the direct influence of surface water. For comparison, Table 12.7 lists both the United States and EU standards.

The major difference between EU and the US standards is the UV dose. The EU selected a UV design dose of 40 mJ/cm^2 because it is known to achieve a 4-log (99.99%) inactivation of many harmful pathogens. UV disinfection systems must achieve a biodosimetrically verified UV dose of at least 40 mJ/cm^2 (at 254 nm) to obtain the DVGW, ÖVGW, and SVGW (Switzerland) certificates. These certificates were also recognized in Norway, Canada, and the United States. To ensure that a UV system is able to provide the required 40 mJ/cm^2, the DVGW standard defines a detailed microbiological testing method using *Bacillus subtilis* spores as a pathogenic surrogate (Figures 12.2, 12.3, and 12.4).

UV disinfection does not produce regulated DBPs in contrast to chlorination. In addition, it is effective against parasites such as *Cryptosporidium* and *G. lamblia* (Linden et al., 2002a) with a very short reaction time of fractions of a second. More importantly, neither pH nor temperature affects its efficiency significantly. As a result, UV disinfection finds wide applications in water, wastewater treatment, and water reclamation UVC light inactivates microorganisms by damaging their genetic material (deoxyribonucleic acid (DNA)) and rendering them unable to replicate. DNA is a double helix with two intertwined strands with base pairs attached by hydrogen bonds. The two strands are unwound and copied during replication of the cell. UV light in the UVC region creates photoproducts, such as connecting adjacent base pairs, which then obstruct the replication process. With enough of these altered base pairs, replication is prevented. UVC light is able to penetrate the cells of microorganisms and disrupt the structure of their DNA molecules. In doing so, the microorganism is prevented from surviving and/or reproducing, thereby rendering it inactive and no longer pathogenic.

The maximum absorption of DNA occurs between 260 and 265 nm. Photoproducts are formed by pyrimidine bases (thymine, cytosine, and uracil) at a much higher rate than by purine bases (adenine, guanine). The most common photoproducts are cyclobutyl dimers, which are formed when two adjacent pyrimidine bases join to form a four-carbon ring. Although pyrimidine dimers are often composed of two thymines, they may also be composed of two cytosines or a thymine and a cytosine. Other less common photoproducts include pyrimidine adducts, spore photoproducts, pyrimidine hydrates, and DNA–protein cross-links (Harm, 1980). The nucleotides in DNA are either purines (adenine and guanine) or pyrimidines (thymine and cytosine). The purines in RNA are the same as in DNA, but the pyrimidines are uracil and cytosine. The nucleotides absorb UV light at wavelengths from 200 to 300 nm. The UV absorption of DNA and RNA reflects their nucleotide composition and tends to have a peak near 260 nm and a local minimum near 230 nm. All purines and pyrimidines strongly absorb UV light, but the rate of UV-induced damage is greater with pyrimidines (Jagger, 1967). Absorbed UV light induces six types of damage in the pyrimidines of

Table 12.7 Disinfection standards of DVGW versus the US EPA.

UV disinfection	DVGW/ÖNORM	US EPA UVDGM
Test point	UVI set point method	Either UV-I set point method – or UV-I/UVT setpoint method – or UV dose calculation method from UVI and UVT
Lamp aging	Max. 70% (i.e. 30% aging)	Not specified (lamp specifics have to be proven)
Inlet conditions	Upstream with double DN600 bend ("worst case")	Not specified Hydraulic condition of installed UV reactor should be equal or better than validated UV reactor (typically validated with 90-degree elbow to simulate worst case)
UV dose	RED 40 mJ/cm^2	RED in relation to log credits
Operation points	Fixed	Variable
Interpolation–extrapolation	Not allowed	Interpolation allowed
Application	Compares performance of different reactors	Provides operation tools for different reactors
Disinfection focus	General disinfections. Suitable in all applications	Focused on *Giardia* and *Cryptosporidium* inactivation
Reactor validation	Experimental testing to determine the flow and UV transmission for a UV reactor at a RED of 40 mJ/cm^2	Experimental testing to determine the operating conditions under which a UV reactor delivers the dose required for inactivation credit of *Cryptosporidium, Giardia,* and viruses
Test microorganism	*Bacillus subtilis*	Typically *male-specific-2 (MS2)* or *T1 bacteriophage*

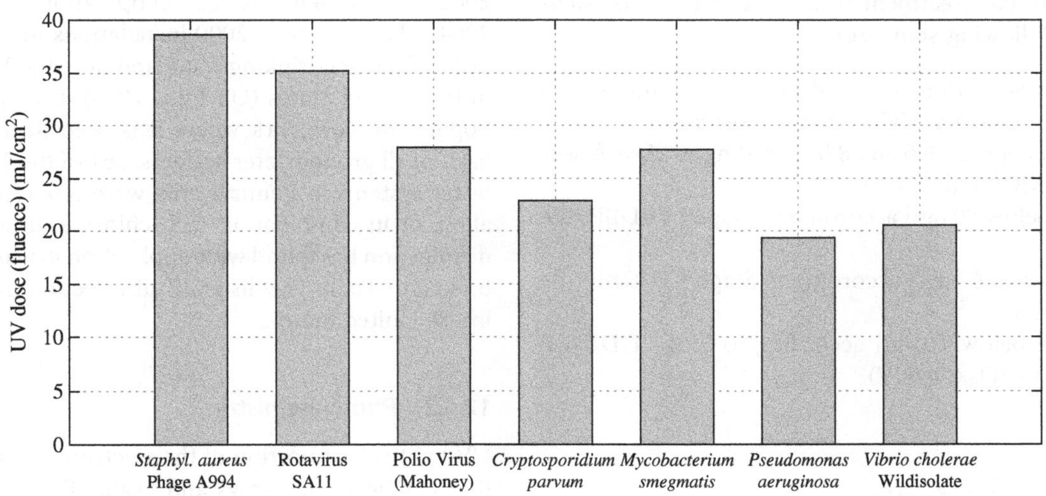

Figure 12.2 EU UV dose requirements for UV disinfection for drinking water (mJ/cm^2).

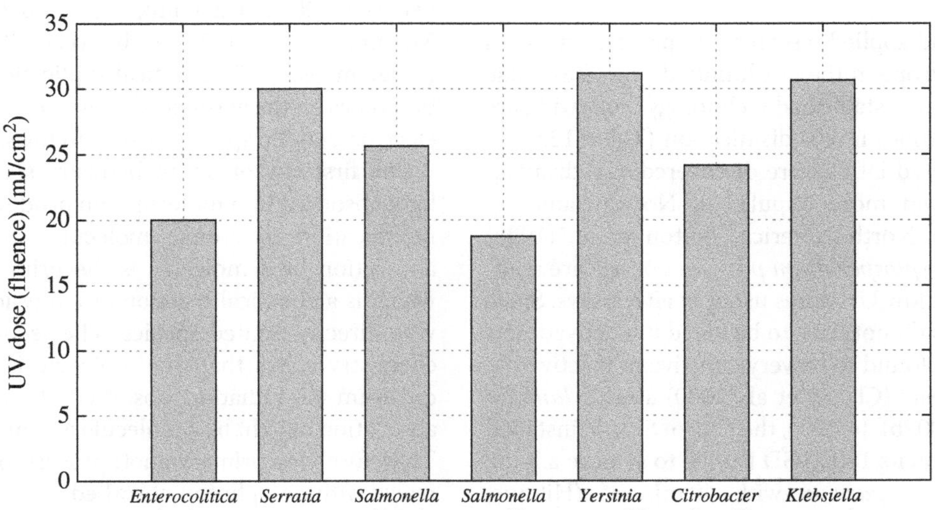

Figure 12.3 EU UV dose requirements for UV disinfection for drinking water (mJ/cm^2).

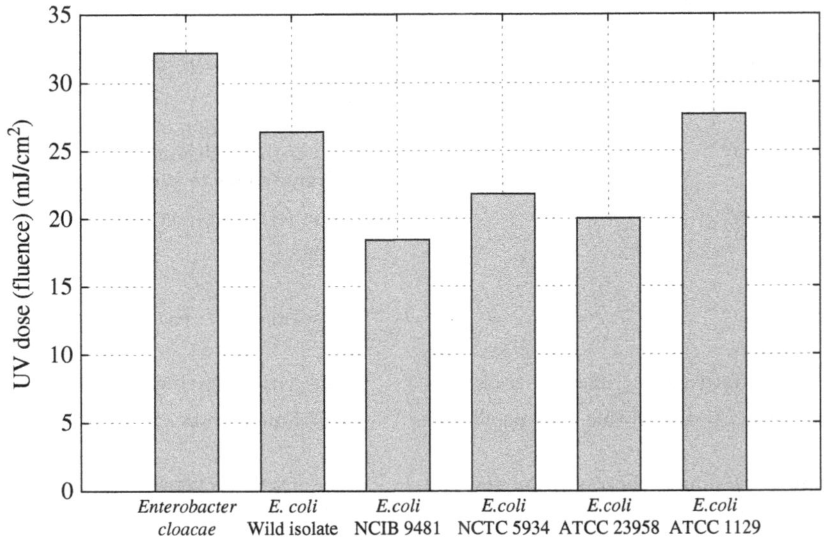

Figure 12.4 EU UV dose requirements for UV disinfection for drinking water (mJ/cm^2).

nucleic acid. For treatment technology, EPA also established the following standards:

1) Iron concentrations below 0.3 mg/l according to the SDWA secondary MCL (Fenton's reaction).
2) pH level between 6.6 and 8.5 according to SDWA secondary MCL (all AOPs).
3) THMs below 80 μg/l according to Stage 1 D/DBP rule (all AOPs).
4) HAA_5 below 60 μg/l according to Stage 1 D/DBP rule (all AOPs).
5) Bromate below 10 μg/l according to Stage 1 D/DBP rule (H_2O_2/O_3, O_3/UV).

12.4 UV Disinfection

12.4.1 History

Fundamental and applied research and practice in North America and Europe in the last hundred years has made UV disinfection an established technology. Following are the major milestones in UV disinfection (Table 12.8).

After chlorinated DBPs were discovered, UV disinfection became even more popular in Norway and the Netherlands. In North America, Bolton et al. (1998) reported that *Cryptosporidium parvum* oocysts are readily inactivated at low UV doses using *in vivo* assays. Since microorganisms do not have to be killed if inactivation is sufficient, UV is found to be very effective in inactivating *C. parvum* oocysts (Clancy et al., 2000) and *G. lamblia* (Linden et al., 2002b). In 2004, the City of Seattle installed UV disinfection in its 180 MGD PWTP to achieve a 3-log inactivation of *C. parvum* with 40 mJ/cm^2 (Hilmoe, 2002). This system has been in operation in Seattle since 2004. There are over 2000 installations in Europe using UV to disinfect drinking water and over 1000 installations in the United States (US EPA, 1996). UV disinfection is popular in New York where it is used to disinfect over 6.4% of all groundwater systems. 761 of the 10 700 public water systems in Pennsylvania were reported to use UV alone or in combination with chlorine. In addition, UV disinfection has found wide application in wastewater disinfection. Table 12.9 lists UV disinfection in wastewater in the United States.

12.4.2 Photochemistry

UV light is that portion of the electromagnetic spectrum that lies between X-rays and visible light (Figure 12.5). Four regions of the UV spectrum have been defined: (i) vacuum UV between 100 and 200 nm, (ii) UVC between 200 and 280 nm, (iii) UVB between 280 and 315 nm, and (iv) UVA between 315 and 400 nm (Meulemans, 1987). Practical application of UV disinfection relies on the germicidal ability of UVC and UVB. The electromagnetic spectrum is as follows.

The first law of photochemistry states that only the light absorbed by a molecule can produce photochemical modification in atoms, molecules, or radicals. Light absorption by a molecule is the primary photoprocess, which is a chemical reaction or a physical process involving directly excited species. The second law of photochemistry states that one molecule is excited for each quantum of radiation absorbed. In other words, the absorption of light by a molecule is a one-photon process. Therefore for a primary photoprocess, only one molecule reacts with one photon absorbed.

Table 12.8 History of UV disinfection.

Year	Progress in UV disinfection
1877	Discovered the germicidal properties of sunlight
1901	Mercury lamps as artificial UV light source
1906	The use of quartz as a UV-transmitting material
1910	The first drinking water UV disinfection application in Marseilles, France
1929	A link between UV disinfection and absorption of UV light by nucleic acid
1930s	The production of germicidal tubular fluorescent lamps
	UV disinfection in Germany, France, Norway, Austria, and Switzerland
1950s	Mechanisms of UV disinfection and the inactivation of microorganisms
1955	UV light for disinfecting municipal drinking water in Switzerland and Austria
1970s	Groundwater disinfection with a UV dose ranged from 20 to 40 mJ/cm^2 with medium-pressure (MP) UV lamps in the Netherlands
By 1985	Up to 600 such UV plants were installed in European countries
1986	US EPA design manual: UV disinfection for wastewater
1998	UV inactivating *Cryptosporidium parvum* oocysts
2002	UV inactivating *Giardia* lamblia
2003	NWRI/AWWARF: UV guidelines for wastewater reuse
2004	City of Seattle installed UV disinfection in its 180 MGD PWTP to achieve a 3-log inactivation of *Cryptosporidium parvum* with 40 mJ/cm^2
2004	UV disinfection with low-pressure (LP) lamps
2006	US EPA: UVDGM for water treatment
2007	US EPA: Long Term 2 Enhanced Surface Water Treatment Rule
2012	Over 25% of US wastewater treatment plants now disinfect with UV light

Table 12.9 UV disinfection of wastewater in the United States.

Classification	Definition (MGD)	Number of installations	Total flow rate treated (BGD)
Very large plants	>100	30	3.8
Large plants	>25	100	6
Medium plants	5–25	1600	10
Small plants	1–5	1500	2.6
Very small plants	<1	2500	0.58

12.4.3 UV Dose

UV intensity, expressed in milliwatts per square centimeter (mW/cm^2), multiplied by the exposure time, expressed in seconds (s), is a UV dose. If other units are used, their conversion can be achieved according to Table 12.10.

The UV transmittance (UVT) of the water directly influences UV dose delivery. If design UVT is overly conservative (i.e. low UVT), the capital costs may increase due to overdesign. On the other hand, inappropriately high design UVT values can result in frequent UV reactor off-specification operation, which could violate LT2ESWTR requirements. Therefore UVT range is critical because the UV reactor have to be validated in the range of UVT and flow combinations expected at the WTP to avoid off-specification operation. Specifying a matrix of flow and UVT conditions for the UV reactors to meet the required UV dose may be appropriate.

UV dose is calculated according to the absorbance, which is experimentally measured according to the UVT, which is the percentage of light passing through material (e.g. a water sample or quartz) over a specified distance. The UVT can be calculated using the Beer's law:

$$\%\,\mathrm{UVT} = 100\,\frac{I}{I_0} \tag{12.1}$$

where

UVT = UV transmittance at a specified wavelength (e.g. 254 nm) and path length (1 cm)
I = intensity of light transmitted through the sample (milliwatt per centimeter squared [mW/cm^2])
I_0 = intensity of light incident on the sample (mW/cm^2)

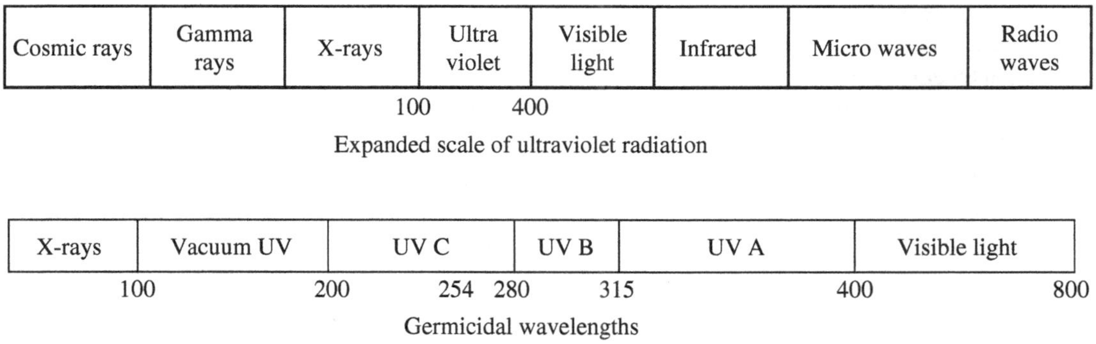

Figure 12.5 Electromagnetic spectrum.

Table 12.10 Conversion table for various units used in UV disinfection.

	Units	Equivalents
Energy	Joule (J)	
Power	Watt (W)	1 J/s
Intensity	W/cm^2	10^3 mW/cm^2
Fluence	mWs/cm^2	1 $mJ/cm^2 = 10^3$ $Ws/cm^2 = 10$ J/m^2

12.4.4 Absorption Coefficient

The absorption coefficient (a) describes how much light is lost as it travels through a medium. It can be determined experimentally and is reported in inverse centimeters. The absorption coefficient of pure distilled water is close to zero. UVT can also be calculated by relating it to UV absorbance using the following equation:

$$\%\,\text{UVT} = 10 \times 10^{-A} \tag{12.2}$$

where

UVT = UV transmittance at a specified wavelength (e.g. 254 nm) and path length (1 cm)
A = UV absorbance at a specified wavelength and path length (unitless)

Therefore, UV absorbance coefficient at 254 nm (A_{254}) can be related to UVT using Equation 12.3:

$$A_{254} = -\log\left[\frac{\text{UVT}\,(\%)}{100}\right] \tag{12.3}$$

where

UVT = the UV transmittance at a specified wavelength
A_{254} = the UV absorbance coefficient at a specified wavelength (at 254 nm) and path length (1 cm), (cm^{-1})

Natural organic matter, iron, nitrate, and manganese absorb UVC light and will increase the absorption coefficient of a water sample (Kolch, 1999). Absorption coefficients in drinking water would be expected in the range of 0.01–0.2 cm^{-1}. Maximum concentrations of iron and manganese have been recommended at 0.03 and 0.02 mg/l, respectively (Kolch, 1999), and turbidity should be less than 5 NTU (Table 12.11).

12.4.5 Fluence

Fluence is the total amount of radiant energy from all directions passing through a small sphere of the cross-sectional area dA, divided by dA. The fluence is the *fluence rate* times the *irradiation time* in seconds (SI unit: J/m^2). Fluence has commonly been called the UV dose. Dose is a term that is used to describe the total absorbed energy. In the case of microorganisms, only a small percentage of the UV light is absorbed, and the rest of it just passes through the organism. The term fluence is thus more appropriate, since it relates to the incident UV energy rather than the absorbed UV energy. *Fluence rate*, instead of *intensity* or *irradiance*, is the appropriate term for UV disinfection, since UV light can penetrate the microorganism from any direction. *Irradiance* is the appropriate term when a surface is irradiated by UV light coming from all directions above the surface. The radiometer that is used with UV apparatus measures the irradiance. In a well-designed bench setup, the fluence rate and the irradiance are identical.

To inactivate a given microorganism, the water must be exposed to certain UV dose in fluence, which is determined by variables associated with the design and operation of the UV disinfection reactor, as well as the characteristics of the water that is treated. Fluence is calculated as the product of light intensity at a given wavelength and exposure time. Intensity at any given point is determined by the lamp strength and the geometry of the reactor. The greater the thickness of water that

Table 12.11 UVT data collection and analysis[a] (US EPA, 2006).

Issue	Recommendation
Water quality events to capture in data collection	• Typical/average water quality conditions • Rainfall effects on source water • Reservoir turnover • Seasonal variations • Possible water quality blends if multiple source waters are used • Variation in upstream water treatment
Water quality sampling locations	Locations that are representative of potential UV facility location(s)
Sample type for various installation options	• Composite samples from operating filters or grab samples from the combined filtered water header should be collected for combined filter effluent installations • Grab samples from representative filter(s) for individual filter piping effluent installations • Grab samples from any locations downstream of clear well under consideration
Collection frequency and period	• Weekly for 1–2 months if water quality is stable • Weekly[b] for 6–12 months (or more) if water quality changes seasonally
Existing data for potential use	A_{254} is often collected in filtered waters to determine the specific UV absorbance (SUVA), and these measurements could be used in the data analysis. However, ultraviolet light absorbance at 254 nm (A_{254}) is typically filtered for the SUVA calculation, which would bias the A_{254} low (high UVT). Therefore, such data should only be used with this understanding
Recommended data analysis	• Cumulative frequency analysis • UVT occurrence with flows
Recommended data to provide to UV manufacturer	• Matrix of flows with corresponding UVTs • Target pathogen(s) and log inactivation • Design UVT[c] (corresponding to design flow) • Range of operating UVTs

[a] Existing A_{254} or UVT data may be available, which would reduce the sampling and analysis needed.
[b] More frequent samples may be needed to capture a water quality event (e.g. storm events).
[c] The design UVT is the UVT that will typically occur at the location of the facility.

the light travels through, the lower the intensity received. The exposure time is governed by the geometry and the hydrodynamics of the reactor. UV lamps are usually placed in such a way that the lowest fluence received by water in the reactor is sufficient to achieve the desired microorganisms reduction. Fluence is commonly measured in mW-s/cm^2, which is equivalent reduction to mJ/cm^2 or, in the SI units, J/m^2. Standard doses in a UV drinking water disinfection system are between 15 and 50 mW-s/cm^2.

UV intensity decreases due to attenuation and dissipation. The further from the source, the lower the strength of the light, because it is spread over a larger area (due to dissipation) and interacts with molecules in the water (attenuation). While dissipation is predicted simply by the geometry of the UV lamp, attenuation depends on the characteristics of the water. If concentration of materials that absorb UV light is high, then less UV will be transmitted. The amount of light absorbed per centimeter is expressed as the absorption coefficient (a). As this coefficient increases, transmissivity decreases exponentially; therefore the absorption coefficient of the water plays a very important role in the effectiveness of the UV disinfection reactor.

Example 12.1

Given: If the incoming light intensity is 100 mW/cm^2, the light intensity after the filtered water sample is 86 mW/cm^2.

Find:

Calculate the following by developing Matlab codes:

1) Plot UVT and A_{254} versus I, given $I = 1:99$ and $I_0 = 100$ mW/cm^2.
2) UVT and A_{254} at $I = 86$ mW/cm^2 and $I_0 = 100$ mW/cm^2.

Solution:

1) Following is the Matlab code:

```
clear
clc
% Parameters
% No parameters
% Get input from user
I = [1:99];
I_0 = 100;
I = 0:1:99;
% compute UVT
```

```
UVT = I/I_0
% compute the UV absorbance at 254 nm (A_254)
A_254 = -log (UVT)
hold on
plot(I,UVT,'-k','linewidth',2)
hold on
plot(I,A_254,'--k','linewidth',2)
hold on
set(gca,'xlim',[0 100],'FontWeight',
  'bold','FontName', 'Times New Roman');
xlabel('I, mJ/cm^2','FontSize',14,
  'FontWeight','bold','FontName',
  'Times New Roman')
ylabel('UVT or A_2_5_4','FontSize',14,
  'FontWeight','bold','FontName', 'Times
  New Roman')
hleg = legend('A_2_5_4','UVT')
box on
grid on
```

2) Figure 12.6 shows UVT versus *I*.
3) The answers to the problem (2) are:
   a) UVT = I/I_0 = 86/100 = 0.86 = 86%.
   b) A_{254} = −log (UVT) = −log 0.86 = 0.1508.

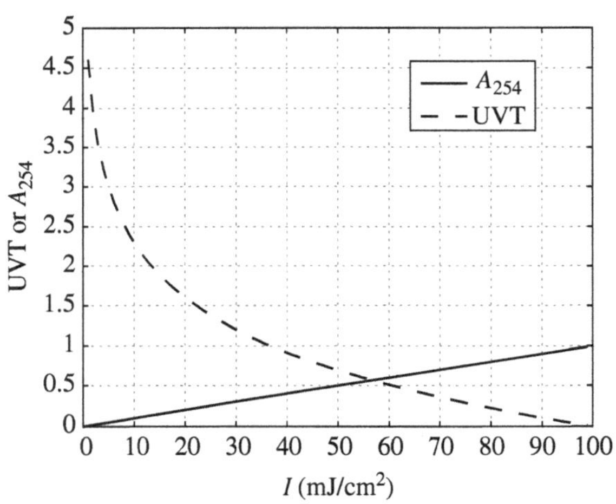

Figure 12.6 UVT or A_{254} versus *I*.

12.4.6 UV Dose-Response

Microbial response is a measure of the sensitivity of a microorganism to UV light and is unique to each microorganism. UV dose–response is determined by irradiating water samples containing the microorganism with various UV doses using a collimated beam apparatus and measuring the concentration of pathogenic microorganisms before and after exposure. The kinetics of microbial inactivation by UV is often referred as the Chick's law:

$$N = N_0 \, e^{-kt} \tag{12.4}$$

where N_0 is the initial concentration of microbes prior to applying UV, N is the number of microbes remaining after exposure time t to UV, I is the UV intensity, t is the exposure time, and k is a microbial inactivation rate constant. UV dose is defined as the product of UV intensity and time.

If a UV dose increase $2.30/k$, there is an order of magnitude reduction in the microbial population due to the first kinetics. UV dose–response relationships can be expressed as either the proportion of microorganisms inactivated or the proportion of microorganisms remaining as a function of UV dose. Microbial inactivation has a dose–response curve with a positive slope, while microbial survival has a dose–response curve with a negative slope. The microbial response is calculated using the following equation:

$$\text{Log inactivation} = \log_{10}\frac{N_0}{N} \tag{12.5}$$

where

N_0 = concentration of infectious microorganisms before exposure to UV light
N = concentration of infectious microorganisms after exposure to UV light

In the disinfection of water or wastewater, microbial response is expressed as log inactivation. Therefore, all dose–response curves presented (log inactivation vs. dose) have a positive slope with log inactivation on a logarithmic (base 10) scale and UV dose on a linear scale.

In the disinfection of bacteria or spores, both a shoulder at low doses or tailing at high doses is often observed. Shoulders at low doses can be explained using series-event inactivation kinetics (Severin et al., 1984). In series-event kinetics, inactivation of a microbe occurs only after sufficient DNA damage has occurred within the microbe. Accordingly, the onset of inactivation observed with a UV dose–response curve appears to require a threshold dose. Severin et al. (1984) and Harris et al. (1987) observed virus inactivation following Chick's law and bacterial inactivation followed series-event kinetics. Chang et al. (1985) observed viral inactivation and the inactivation of some bacterial pathogens following Chick's law, while the inactivation of cysts, spores, and other bacteria had shoulders.

Tailing occurs at high doses and can be attributed to the clumping of microbes and the clusters of microbes within particulates (Parker and Darby, 1995) and other microbes. Microbes trapped by particulate material and other microbes experience a lower UV dose compared with individual microbes in the bulk phase due to the UV absorbance of the particulate material and biomass. For example, the UV transmission through an *Escherichia coli* cell is 70% at 254 nm (Jagger, 1967). The UV transmission through particulates will depend on the composition of the particulate and the presence of UV absorbing compounds like iron.

Microbial response to UV light can vary significantly among microorganisms. The UV sensitivity of viruses and bacteriophage can vary by more than two orders of magnitude (Rauth, 1965). With bacteria, spore-forming and Gram-positive bacteria are more resistant to UV light than Gram-negative bacteria (Jagger, 1967). Among the pathogens of interest in drinking water, viruses are most resistant to UV disinfection followed by bacteria, *Cryptosporidium* oocysts, and *Giardia* cysts. UV dose–response is generally independent of the following factors:

1) **UV intensity:** UV dose–response follows the law of reflectivity over an intensity range from 1 to 200 mW/cm^2, where the same level of inactivation is achieved with a given UV dose regardless of whether that dose was obtained with a high UV intensity and low exposure time or vice versa. Nonreciprocity has been observed at low intensities where repair may compete with inactivation (Sommer et al., 1998).
2) **UV absorbance:** UV absorbance of the impurity in water should be considered when calculating UV dose. Increasing intensity or exposure time, however, may be necessary to achieve a constant UV dose as the absorbance of the water impurity changes.
3) **Temperature:** Temperature effects on dose–response are minimal and depend on the microorganism. For male-specific-2 (MS2) bacteriophage, inactivation is not temperature dependent. Severin et al. (1983) studied three microorganisms to determine the dose required to achieve 2-log inactivation as a function of temperature. For *E. coli* and *Candida parapsilosis*, the dose requires decreases by less than 10% as the temperature increases from 5 to 35°C, and for f2 bacteriophage, the dose requires decreases by less than 20% over the same temperature interval (Severin et al., 1983).
4) **pH:** Dose–response is independent of pH from pH 6 to pH 9.

Particle association and clumping of microorganisms affect UV dose–response. Small floc particles can enmesh and protect MS2 bacteriophage, and potentially other viruses, from exposure to UV light (Templeton et al., 2003). Similarly, the inactivation rate of particle-associated coliforms is slower than that of nonparticle-associated coliforms (Örmeci and Linden, 2003). Linden et al. (2002a) indicated that the UV dose–response of microorganisms added to filtered drinking waters is not altered by variation in turbidity that meets regulatory requirements for filtered effluents. For unfiltered waters, source water turbidity up to 10 nephelometric turbidity units (NTU) did not affect the UV dose–response of separately added (seeded) microorganisms (Oppenheimer et al., 2002; Passantino et al., 2004). The effect of particle enmeshment on the UV dose–response of seeded microorganisms in water has been studied by adding clays or natural particles. When coagulating suspensions containing kaolinite or montmorillonite clay using alum or ferric chloride, no difference was observed in the log inactivation of the seeded microorganisms (Mamane-Gravetz and Linden, 2004;Templeton et al., 2004). When humic acid particles and a coagulant were added to the suspensions, however, significantly less inactivation was reported (Templeton et al., 2004).

The validation protocol outlined in the UVDGM has two tiers. Tier 1 has a predetermined safety factor that is applied to the UV dose requirements listed in Table 12.12. The tier 1 safety factors are based on assumed uncertainties and corrections. The validation conditions should meet the criteria specified in the UVDGM in order for results to be practical. Filtered surface water, unfiltered surface water, and groundwater can be disinfected by UV if the UV reactor is validated according to the EPA guidelines. The purpose of UV disinfection is to provide an effective barrier against pathogens (such as *Cryptosporidium* and *Giardia*) and help minimize DBPs. To meet the EPA guidelines, first is to define the UV disinfection goals and select suitable comprehensive disinfection strategy for the entire treatment processes. Target pathogen(s), target log inactivation, and corresponding required UV dose should be identified. Once the design is complete, specific experimental procedures must be followed, and reports must be presented in terms of (i) UV dose, (ii) reactor design, (iii) reliability design, (iv) monitoring and alarm design, (v) field commissioning test, (vi) performance monitoring, and (vii) engineering report. To facilitate the plan, design, validation, and operation of a UV disinfection system, Matlab codes are developed to facilitate each step.

Table 12.12 UV dose (in mJ/cm^2) required by LT2ESWTR to inactivate target pathogens (US EPA, 2006).

Log inactivation	0.5	1	1.5	2	2.5	3	3.5	4
Cryptosporidium	1.6	2.5	3.9	5.8	8.5	12	—	—
Giardia	1.5	2.1	3.0	5.2	7.7	11	—	—
Virus	39	58	79	100	121	143	163	186

Example 12.2 The UV dose (in mJ/cm^2) required by EPA LT2ESWTR to inactivate a virus is clearly higher than both *Cryptosporidium* and *Giardia*. If a water treatment plant is to comply with the UV dose of 144 to achieve the required 3-log inactivation (99.9%), what would be the dose required for both *Cryptosporidium* and *Giardia* to achieve the same degree of log inactivation, respectively?

Solution:

1) The Matlab code is developed to plot the UV dose (in mJ/cm^2) required by LT2ESWTR against log inactivation of the target pathogens according to the above data (Figure 12.7):

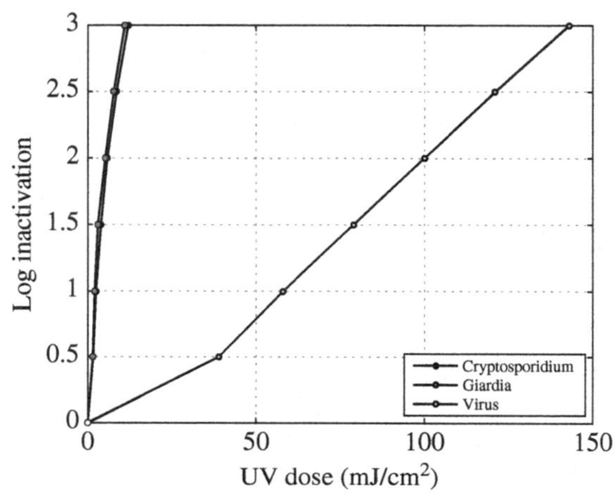

Figure 12.7 UV dose (in mJ/cm^2) required by EPA LT2ESWTR to inactivate target pathogens.

```
clc
clear
y=[0 0.5 1 1.5 2 2.5 3];
x = [0 1.6 2.5 3.9 5.8 8.5 12];
y2=y;
x2=[0 1.5 2.1 3 5.2 7.7 11];
y3=y;
x3=[0 39 58 79 100 121 143];
plot(x,y,'k-o','MarkerSize',3,'Marker
  FaceColor',[0 0 0],'MarkerEdgeColor',
  [0 0 0],'linewidth',1);
hold on
plot(x2,y2,'k-o','MarkerSize',3,'Marker
  FaceColor',[0.5 0.5 0.5],'MarkerEdge
  Color',[0 0 0],'linewidth',1);
hold on
plot(x3,y3,'k-o','MarkerSize',3,'Marker
  FaceColor',[1 1 1],'MarkerEdgeColor',
  [0 0 0],'linewidth',1);
hold on
set(gca,'xlim',[0 150],'FontWeight',
  'bold','FontName', 'Times New Roman');
set(gca,'ylim',[0 3],'FontWeight',
  'bold','FontName', 'Times New Roman');
xlabel('UV Dose,mJ/cm^2','FontSize',
  14,'FontWeight','bold','FontName',
  'Times New Roman')
ylabel({'Log Inactivation'},'Font
  Size',14,'FontWeight','bold',
  'FontName', 'Times New Roman')
legend('Cryptosporidium',
  'Giardia','Virus',
  'Location','SouthEast');
grid on
```

2) Polynomial regression is used to obtain three equations as follows (Table 12.13).
3) Using the above regression equations for both *Cryptosporidium* and *Giardia*, the UV doses required to achieve 3-log inactivation are as follows, respectively (Table 12.14):
   Cryptosporidium: $f(x) = (p1)x^2 + (p2)x + p3 = (-0.0154)4^2 + (0.437)4 - 0.0361 = 14.66$ mJ/cm^2.
   Giardia: $f(x) = (p1)x^2 + (p2)x + p3 = (-0.0210)4^2 + (0.500)4 - 0.0135 = 13.74$ mJ/cm^2.

The LT2ESWTR requires disinfection to achieve at least 3-log inactivation of *Giardia* and 4-log inactivation of viruses or 2- or 3-log inactivation of *Cryptosporidium*, depending on the source water concentration of *Cryptosporidium* as shown in Table 12.15.

12.5 Virus Sensitivity Index of UV Disinfection

12.5.1 Virus Sensitivity Index (VSI)

To predict the fluence required for different viruses and bacteria, Tang and Sillanpää (2015a) developed virus sensitivity index (VSI) through following mathematical analysis. Disinfection kinetic model of virus without shoulder or tailing effects can be described by the first-order kinetics as proposed by Chick (1908) and Watson (1908) as follows:

$$N = N_0 e^{-ktC} \tag{12.6}$$

where

N_0 = the initial concentration of microbes prior to applying UV
N = the number of microbes remaining after exposure time t to UV
k = a microbial inactivation rate constant (cm^2/mJ)
C = the concentration of disinfectant

Table 12.13 Polynomial regression equations for the EPA LT2ESWR UD dose requirements.

Target pathogens	Linear model Poly2	Goodness of fit
Cryptosporidium	Linear model Poly2: $f(x) = p1 \times x^2 + p2 \times x + p3$ Coefficients (with 95% confidence bounds): $p1 = -0.01546$ (−0.02179, −0.009125) $p2 = 0.4369$ (0.3575, 0.5164) $p3 = -0.03612$ (−0.2238, 0.1515)	SSE: 0.02723 *R*-square: 0.9961 Adjusted *R*-square: 0.9942 RMSE: 0.08251
Giardia	Linear model Poly2: $f(x) = p1 \times x^2 + p2 \times x + p3$ Coefficients (with 95% confidence bounds): $p1 = -0.02088$ (−0.03508, −0.00668) $p2 = 0.5$ (0.3369, 0.6631) $p3 = -0.0135$ (−0.3542, 0.3272)	SSE: 0.09154 *R*-square: 0.9869 Adjusted *R*-square: 0.9804 RMSE: 0.1513
Virus	Linear model Poly2: $f(x) = p1 \times x^2 + p2 \times x + p3$ Coefficients (with 95% confidence bounds): $p1 = 4.063e-05$ (−2.919e − 06, 8.417e − 05) $p2 = 0.01593$ (0.009338, 0.02252) $p3 = -0.05512$ (−0.2759, 0.1657)	SSE: 0.02837 *R*-square: 0.9959 Adjusted *R*-square: 0.9939 RMSE: 0.08421

Table 12.14 Bin requirements for filtered PWSs[a] (US EPA, 2006).

		And if the following filtration treatment is operating in full compliance with existing regulations, then the *additional* treatment requirements are			
***Cryptosporidium* concentration (oocysts/l)**	**Bin classification**	**Conventional filtration treatment (includes softening)**	**Direct filtration**	**Slow sand or diatomaceous earth filtration**	**Alternative filtration technologies**
<0.075	1	No additional treatment	No additional treatment	No additional treatment	No additional treatment
≥0.075 and <1.0	2	1-log treatment	1.5-log treatment	1-log treatment	As determined by the state
≥1.0 and <3.0	3	2-log treatment[a]	2.5-log treatment[a]	2-log treatment[a]	As determined by the state[a]
≥3.0	4	2.5-log treatment[a]	3-log treatment[a]	2.5-log treatment[a]	As determined by the state[a]

[a] PWSs must achieve at least 1 log of the required treatment using ozone, chlorine dioxide, UV light, membranes, bag/cartridge filters, or bank filtration.

Table 12.15 Requirements for unfiltered PWSs (US EPA, 2006).

Average *Cryptosporidium* concentration (oocysts/l)	Additional *Cryptosporidium* inactivation requirements
≤0.01	2 log
>0.01	3 log

For UV disinfection, the concentration is replaced by UV irradiance intensity (mW/cm^2). The product of UV intensity and the exposure time are defined as fluence (*H*), which has the unit of (mJ/cm^2). Therefore,

$$\log\frac{N_0}{N} = \log I = ktC = kH \tag{12.7}$$

Equation (12.7) can be applied to any virus species, i, and a reference species such as MS2-phage, r, as follows:

$$\log\left(\frac{N_0}{N}\right)_{\mathrm{i}} = \log I_{\mathrm{i}} = k_{\mathrm{i}}tC = k_{\mathrm{i}}H_{\mathrm{i}} \tag{12.8}$$

$$\log\left(\frac{N_0}{N}\right)_{\mathrm{r}} = \log I_{\mathrm{r}} = k_{\mathrm{r}}tC = k_{\mathrm{r}}H_{\mathrm{r}} \tag{12.9}$$

Dividing Equation (12.8) by Equation (12.9) on both sides, the following equation resulted:

$$\log\left(\frac{N_0}{N}\right)_{\mathrm{i}} = \frac{\log I_{\mathrm{i}}}{\log I_{\mathrm{r}}} = k_{\mathrm{i}}tC = \frac{k_{\mathrm{i}}H_{\mathrm{i}}}{k_{\mathrm{r}}H_{\mathrm{r}}} \tag{12.10}$$

To achieve the same level of $\log I$ for both virus and MS2 phage, the left side of the equation becomes unity 1. Inspection of Equation (12.10), a new concept of VSI, is defined as the ratio between the k_i of any virus and the k_r of a reference virus, MS2 phage, as follows:

$$\mathrm{VSI} = \frac{k_i}{k_r} \tag{12.11}$$

Therefore, the relative fluence required to achieve a given order of inactivation $\log I$, H_i/H_r, can be theoretically related to VSI according to the theoretical equation (12.10) as follows:

$$\frac{1}{\mathrm{VSI}} = \frac{H_i}{H_r} \tag{12.12}$$

Example 12.3 Relative UV fluence H_i/H_r required to inactivate viruses at different $\log I$ is given in Table 12.16.

Find:

1) General equation between H_i and $\log I$ as well as VSI.
2) Plot H_i versus $\log I$ and VSI using Matlab.

Solution:

1) To estimate the H_i to inactivate a given virus with a known VSI at any given $\log I$, the general equation would be as follows according to the regression analysis result at each Log I (Tang and Sillanpää, 2015a):

$$H_i = (21.635 \pm 2.566) \times (\log I)\left(\frac{1}{\mathrm{VSI}}\right) \tag{12.13}$$

Table 12.16 Relative UV fluence H_i/H_r required to inactivate viruses at different log *I*.

No	Viruses	Host	At 1 log *I*	At 2 log *I*	At 3 log *I*	VSI	1/VSI
1	Adenovirus type 40	PLC/PRF/5 cell line	2.88	2.56	2.52	0.404	2.47
2	Rotavirus	MA104 cells	1.05	1.95	2.28	0.489	2.04
3	Adenovirus type 2	PLC/ PRF/5 cell line	2.1	1.9	1.94	0.532	1.88
4	Adenovirus type 15	A549 cell line (ATCC CCL-185)	2.1	1.95	1.98	0.532	1.88
5	Adenovirus type 2	Human lung cell line	1.84	1.34	1.22	0.787	1.27
6	Adenovirus type 2	A549 cell line	1.05	1.1	1.3	0.851	1.18
7	MS2 phage	Different *Escherichia coli* strains	1	1	1	1	1
8	Poliovirus 1	CaCo2 cell line (ATCC HTB37)	0.37	0.42	0.46	2.362	0.42
9	Coxsackievirus B5	BGM cell line	0.5	0.44	0.44	2.362	0.42
10	Rotavirus SA-11	MA-104 cell line	0.48	0.46	0.42	2.383	0.42
11	B40-8 (phage)	*Bacteroides fragilis* HSP-40	0.63	0.44	0.37	2.489	0.4
12	Rotavirus SA-11	Monkey kidney cell line MA 104	0.42	0.37	0.44	2.489	0.4
13	*Staphylococcus aureus* phage A 994	*Staphylococcus aureus* 994	0.42	0.42	0.41	2.553	0.39
14	Virus: B40-8 phage	*B. fragilis*	0.58	0.42	0.37	2.574	0.39
15	Echovirus I	BGM cell line	0.42	0.4	0.41	2.574	0.39
16	Calicivirus feline	CRFK cell line	0.37	0.39	0.41	2.617	0.38
17	Coxsackievirus B3	BGM cell line	0.42	0.39	0.4	2.617	0.38
18	Poliovirus 1	BGM cell line	0.42	0.38	0.37	2.766	0.36
19	Calicivirus feline	CRFK cell line	0.26	0.37	0.37	2.851	0.35
20	Calicivirus canine	MDCK cell line	0.37	0.37	0.36	2.894	0.35
21	Hepatitis A HM175	FRhK-4 cell	0.27	0.33	0.36	3	0.33
22	Echovirus II	BGM cell line	0.37	0.34	0.33	3.085	0.32
23	Poliovirus type 1 LSc2ab	BGM cell	0.3	0.27	0.29	3.702	0.27
24	Poliovirus type Mahoney	Monkey kidney cell line Vero	0.16	0.17	0.23	4.936	0.2
25	PHI X 174 (phage)	*E. coli* WG 5	0.16	0.12	0.12	8.362	0.12
26	PHI X 174 (phage)	*E. coli* WG5	0.12	0.13	0.12	8.553	0.12
27	PHI X174 (phage)	*E. coli* ATCC 13706	0.11	0.09	0.08	12.383	0.08

Source: Tang and Sillanpää (2015a). Reproduced with permission of Taylor & Francis.

2) The Matlab code to plot H_i versus $\log I$ and VSI using Matlab is as follows:

```
clear all
close all
xi=0:0.1:5;yi=0:0.05:2.5;
[x,y]=meshgrid(xi,yi);
z=21.635*x.*y;
surf(x,y,z,'EdgeColor','none');
axis tight;
shading interp;
colorbar ('FontSize',14,'FontWeight',
  'bold','FontName', 'Times New Roman')
grid on;
set(gca,'xlim',[0 5],'FontWeight',
  'bold','FontName', 'Times New Roman');
set(gca,'ylim',[0 2.5],'FontWeight',
  'bold','FontName','Times New Roman');
xlabel('log I','FontSize',14,'Font
  Weight','bold','FontName', 'Times
  New Roman')
ylabel('1/VSI','FontSize',14,
  'FontWeight','bold','FontName', 'Times
  New Roman')
zlabel('H_i,mJ/cm^2 ','FontSize',14,
  'FontWeight','bold','FontName', 'Times
  New Roman')
grid on
```

Equation 12.13 can be visualized through the following 3D plot (Figure 12.8).

12.5.2 Applications of VSI

The concept VSI greatly facilitates the prediction of the required fluence during UV disinfection under different scenarios. For example, VSI does have the predictive power not only for inactivation rate constants at UV_{254} but also for the disinfection results reported from different light sources. Table 12.17 presents the inactivation rate constants reported by Love et al. (2010) when a solar simulator was used.

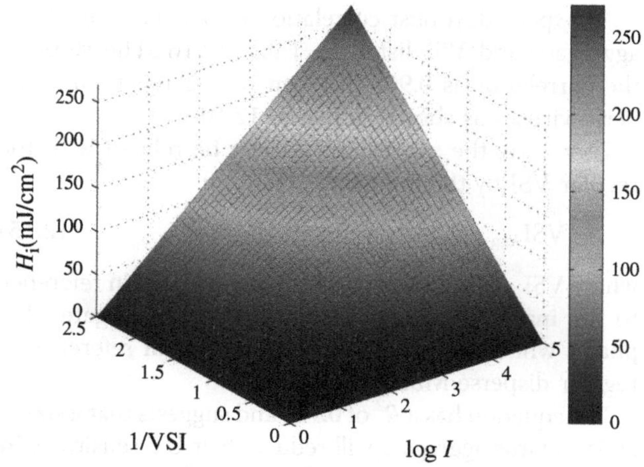

Figure 12.8 H_i (mJ/cm^2) predicted by log I and VSI. (*Source:* Tang and Sillanpää (2015a). Reproduced with permission of Taylor & Francis.)

Figure 12.9 shows a strong correlation between VSI under UV_{254} and VSI under sunlight with a slope of 1.398 and a R^2 of 0.990. Therefore, VSI under UV_{254} can also be used to predict the VSI under the sunlight using the following equation:

$$VSI_{Sunlight} = 1.398 \times VSI_{UV254} \tag{12.14}$$

where $VSI_{Sunlight}$ is the VSI of viruses inactivation rate constant under the sunlight in reference to that of MS2 phage under the sunlight, while VSI_{UV254} is the regular VSI defined by Equation (12.11).

The equation implies that most viruses will be about 40% more sensitive to sunlight than that to UV_{254}. Therefore, it can be expected that the required fluence of sunlight will be 40% less than UV_{254} to achieve the same level of $\log I$.

Naturally it can be expected that the VSI developed in this study should also have high predictive power for UV disinfection of virus aggregate by UV_{254}. For example, Mattle and Kohn (2012) investigated the virus size effect on the UV disinfection rate constants of 500 nm different aggregate viruses. Table 12.18 shows the inactivation rate constants due to viral aggregation as reported by them.

Table 12.17 VSI of viruses under simulated light in reference to MS2 phage by UV_{254}.

Viruses	Inactivation rate constant by UV_{254} (cm^2/mJ)	Inactivation rate constant by simulated sunlight (h^{-1})	VSI	Solar VSI in reference to UV_{254}	Solar VSI in reference to that MS2 under sunlight
MS2 phage	0.047	0.43	1	9.149	1
Phi X174 phage	0.582	7.310	12.383	155.532	17.00
Adenovirus type2	0.106	0.590	0.851	12.553	1.372
Poliovirus type 3	0.130	2.160	2.360	45.957	5.023

Source: Tang and Sillanpää (2015a). Reproduced with permission of Taylor & Francis.

As expected, robust correlation exists between VSI of aggregate and VSI listed in Table 12.16. The slope of the correlation is 0.901 between the aggregate and disperse viruses as shown in Figure 12.10.

Therefore, the aggregate VSI can be related with the regular VSI by the following equation:

$$VSI_{Agg} = 0.901 \times VSI \tag{12.15}$$

where VSI_{Agg} is the VSI of the aggregate virus in reference to the inactivation rate constant of the aggregate MS2 phage, while VSI is defined in this study in reference to regular disperse MS2 phage.

The equation has a R^2 of 0.972 and suggests that a size of 500 nm virus aggregate will reduce their UV sensitivity by 10%. In other words, it will take 10% higher fluence, H_i, to inactivate a given virus aggregate of 500 nm than itself in disperse form. Therefore, Equations (12.14) and (12.15) could be used to predict the ki under sunlight or virus aggregate of 500 nm of any other different viruses not studied, respectively. Figure 12.11 presents the predicted k_i values for both cases.

The major applications of VSI rooted in its dimensionless parameters such as H_i/H_r versus VSI. For example, the dataset of k_i values reported by Love et al. (2010) is in 1/h. Correlation analysis shows that the predicted k_i will not be affected by the units because they are canceled during data transformation. The dimensionless nature of variables cancels out all the other experimental parameters such as turbidity, transmittance, and reactor hydraulics. However, the slope of these equations reflects the nature of the fundamental disinfection process such as number of dimerization of pyrimidine, which forms pyrimidine *dimers* and disrupts DNA replication. Therefore, it should provide a powerful tool to quantify the effect of photoreactivation and dark repairing by dimensionless and statistical analysis.

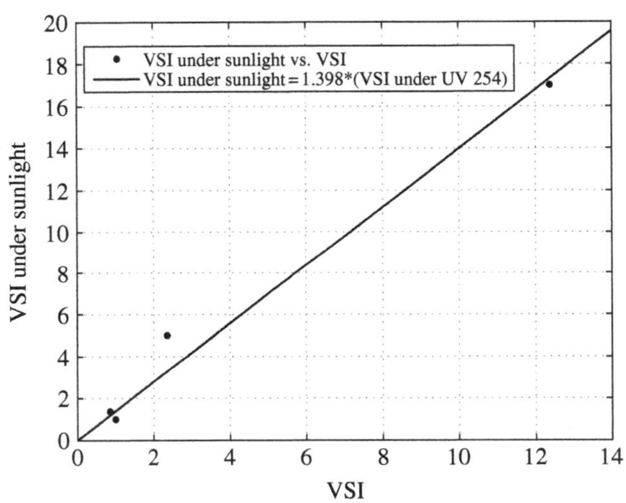

Figure 12.9 Correlation between VSI under sunlight and VSI. (*Source:* Tang and Sillanpää (2015a). Reproduced with permission of Taylor & Francis.)

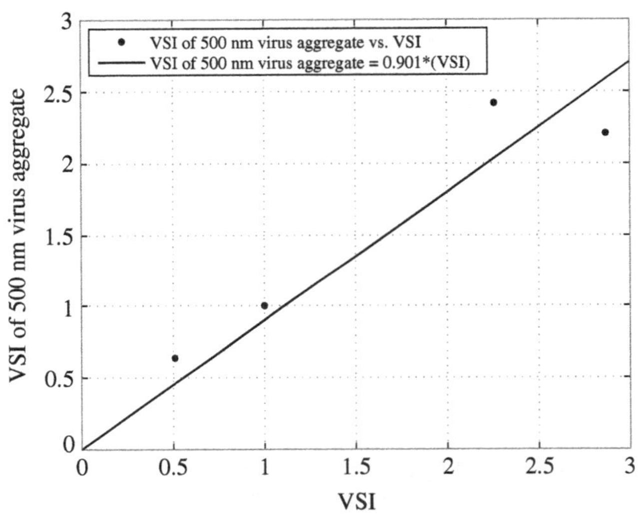

Figure 12.10 Correlation between aggregate VSI and VSI. (*Source:* Tang and Sillanpää (2015a). Reproduced with permission of Taylor & Francis.)

Table 12.18 VSI of aggregate viruses in reference disperse or aggregate MS2 phage

Viruses	Inactivation rate constant for disperse virus	Inactivation rate constant for aggregate virus	Regular VSI	VSI in reference to disperse MS2 phage	VSI in reference to aggregate MS2 phage
MS2 phage	0.0445	0.033	1	0.7	1
Adenovirus ST2, 15, 40, 41	0.024	0.021	0.51	0.45	0.64
Calicivirus feline, canine	0.106	0.08	2.26	1.7	2.42
Poliovirus type 1	0.135	0.073	2.87	1.55	2.21

Source: Tang and Sillanpää (2015a). Reproduced with permission of Taylor & Francis.

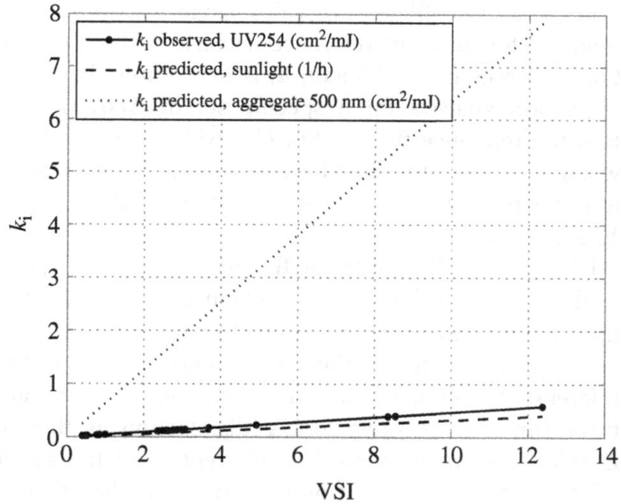

Figure 12.11 k_i predicted under sunlight or virus aggregate of 500 nm at UV_{254}. (*Source:* Tang and Sillanpää (2015a). Reproduced with permission of Taylor & Francis.)

12.6 Bacteria Sensitivity Index (BSI) with Shoulder Effect

12.6.1 Bacteria Sensitivity Index (BSI)

UV disinfection of bacteria usually has a shoulder effect. Shoulder effect during UV disinfection refers to the initial delay of inactivation of bacteria to achieve observable inactivation rate of log *I*. The phenomena were explained by a series hit event model (Severin et al., 1983) or the bacteria's ability to repair their DNA. Shoulder effect can be mathematically described as follows:

$$\frac{N_d}{N_0} = 1 - \left(1 - 10^{-k_H H}\right)^{10^d} \tag{12.16}$$

where N_0 is the initial bacterial concentration before UV disinfection and N_d is the bacterial concentration after it received UV fluence of *H*. When fluence equals zero, the shoulder effect *d* is the log (N_d/N_0), which will be referred to as log *I* for simplicity. k_H is the first-order initial rate constant (cm²/mJ) of the linear portion of log (N_d/N_0) versus *H*.

To accurately model the shoulder effect, two constants, *d* and k_H, are required. Although the two constants can be obtained through multivariate regression, the complicated mathematical model of double exponents prevents its wide application in the engineering design of UV disinfection reactors. So far, there has been no simple method that could quantify the shoulder effect, not to mention the predictive method of fluence required in designing UV disinfection reactors. Tang and Sillanpää (2016) presented a simple and effective model to systematically quantify the shoulder effect of bacteria during UV disinfection. Bacteria sensitivity index (BSI) is developed for the objectives: (i) to theoretically analyze the fluence required for any bacteria and a reference bacteria such as *E. coli*, (ii) to define a new concept of bacteria sensitivity index (BSI), (iii) to statistically derive a simple model for predicting the fluence of any bacteria with shoulder effect by using BSI, (iv) to assess the predictive power of BSI by comparing the predicted with experimental data of fluence required, and (v) to predict fluence required of bacteria with shoulder effects for other log *I*, at which no experimental fluence is reported by using BSI.

To simplify the mathematical model of Equation (12.16), a bacterial survival curve can be considered as two portions (Watson, 1908). First, the shoulder portion of the curve can be approximated as shoulder broadness (SB), H_0, which is the minimal fluence required to have observable log *I*. Mathematically, H_0 is the intercept of *H* when log (N_d/N_0) equals zero. The actual SB would typically have a parabolic shape. Since the actual SB, H_0, would be either less or greater than the H_0, which is obtained from the intercept from the linear portion of the survival curve, the predicted *H* could be either less or greater than the actual observed fluence *H*. Second, after the initial shoulder, the linear portion follows the first-order kinetics, which is characterized by the inactivation rate constant, k_H. As a result, a simple kinetic model is expressed as follows:

$$\log I = k_H \times H + b \tag{12.17}$$

where log *I* is log (N_d/N_0), k_H is the first-order initial constant (cm²/mJ) of the linear portion of log (N_d/N_0) versus *H*, and b is the intercept on the *y*-axis of log *I* of the bacteria survival dose curve. Equation (12.17) can be solved to get *H*:

$$H = k_i \times \log I + H_0 \tag{12.18}$$

where H_i is the fluence required at a given log *I*, k_i equals $1/k_H$, and H_0 is the shoulder broadness and equals b/k_H. Similarly, Equation (12.18) can be also applied to reference bacteria such as *E. coli* as follows:

$$H_r = k_{ir} \times \log I + H_{0r} \tag{12.19}$$

To obtain a simplified predictive model, Equations (12.18) and (12.19) can be rearranged as follows:

$$H - H_0 = k_i \times \log I \tag{12.20}$$

$$H_r - H_{0r} = k_{ir} \times \log I \tag{12.21}$$

When Equation (12.20) is divided by Equation (12.21) at both sides at the same inactivation log *I*, a simple linear equation is obtained:

$$\frac{H - H_0}{H_r - H_{0r}} = \frac{\Delta H_i}{\Delta H_r} = \frac{k_i}{k_{ir}} \tag{12.22}$$

Equation (12.22) suggests that the ratio between the fluence differences required for any bacteria and that of *E. coli* is proportional to the ratio of their inactivation rate constants at the linear portion if the same level inactivation rate of $\log I$ is to be achieved for any bacteria. Therefore, a new concept of BSI is defined as the ratio between the k_i of any bacteria for the linear portion of the dose–response curve and the k_{ir} of the reference bacteria such as *E. coli* as follows:

$$BSI = \frac{k_i}{k_{ir}} \quad (12.23)$$

With this definition, the ratio between the fluence differences required to achieve a specific inactivation log I, $\Delta H/\Delta H_r$, can be theoretically related to BSI as follows:

$$\frac{\Delta H_i}{\Delta H_r} = BSI \quad (12.24)$$

This equation suggests several important points between $\Delta H/\Delta H_r$ and BSI at a specific $\log I$. First, $\Delta H/\Delta H_r$ should be linearly proportional to BSI. Second, the slope of the equation should be one theoretically. In reality, however, uncertainty in measurement of H and the corresponding inactivation rate constants k_i, which depends upon the accuracy of fluence measured, will result in a slope different from one. Third, when the approximated shoulder H_0 from the intercept of the linear portion at the $\log I = 0$ is used to replace the actual SB, uncertainty of the coefficient, a, also increases. All these uncertain factors will affect the regression coefficient, a, which result its deviation from one. To determine the coefficient a, regression analysis was carried out according to the following linear model:

$$\frac{\Delta H_i}{\Delta H_r} = a \times (BSI) \quad (12.25)$$

In the US EPA UV disinfection guide (2003), *E. coli* is used as a bacterial indicator for long time due to its nonpathogenic nature. In addition, shoulder exists on the *E. coli* dose–response curve. Therefore, *E. coli* is used as the reference bacteria in defining the BSI. After regression analysis, the equation can be rewritten as follows:

$$H - H_0 = a \times BSI \times (H_r - H_{0r}) \quad (12.26)$$

Substitute Equation (12.21) into Equation (12.26):

$$H = a \times BSI \times (k_{ir} \times \log I + H_{0r} - H_{0r}) + H_0 \quad (12.27)$$

A predictive equation is obtained as follows:

$$H = a \times (k_{ir} \times \log I) \times BSI + H_0 \quad (12.28)$$

Regression coefficient, a, will change with fluence H required for different reference bacteria such as *E. coli*. Once correlation coefficient, a, is determined by linear regression analysis for a specific set of $\Delta H/\Delta H_r$ and BSI, the reference fluence SB, H_{0r}, will be canceled out when the fluence required for any bacteria at a given $\log I$ is predicted by using Equation (12.28) (Tables 12.19, 12.20, and 12.21).

Histogram of BSI using the fluence required for *E. coli* by the US EPA as the reference in Figure 12.12 shows that up to 6 bacteria have the same inactivation rate constants of the linear portion of the survival curve as that of the reference bacteria, *E. coli*. Three different bacteria are three times as sensitive to UV disinfection as that of *E. coli*. Only one bacteria, *Faecal streptococci*, has about 5.5 times higher inactivation rate constants than that of the reference bacteria, *E. coli*.

12.6.2 Shoulder Broadness Index (SBI)

As defined in theoretical analysis, SBI can be calculated by $\Delta H/H_{or}$, which is the ratio of fluence difference, ΔH, to the SB of *E. coli*, H_{or}.

When *E. coli* is used as reference bacteria, the fluence recommended by the US EPA is used in the correlation analysis to obtain the following equation:

$$\frac{SBI}{SBI_R} = (0.914 \pm 0.044) \times (BSI)_E \quad (12.29)$$

Equation (12.29) suggests that the ratio of SBI of any bacteria to the reference SBI_r of *E. coli* is a function of BSI only. Fluence required to achieve a specific $\log I$, however, is linearly proportional to $\log I$. Since $k_{ir} = 1$, therefore, Equation (12.29) can also be written as follows (Table 12.22):

$$H = (0.914 \pm 0.044) \times (\log I) \times (BSI)_E + H_0 \quad (12.30)$$

12.6.3 Transformation of H into $\Delta H/\Delta H_r$

$\Delta H/\Delta H_r$ at specific inactivation $\log I$ of *E. Coli* by using the US EPA requirement (US EPA, 2003) is presented in Table 12.23. According to Equation (12.25), $\Delta H/\Delta H_r$ should linearly correlate with BSI regardless of which set of reference fluence reported is used. Figure 12.13 plots a typical predicted output $\Delta H/\Delta H_r$ against the observed $\Delta H/\Delta H_r$ at 4 $\log I$. It shows that the correlation is very robust with $R = 0.9894$ between $\Delta H/\Delta H_r$ and BSI_E, which suggests that the theoretical equation is valid. In addition, it implies that approximations of SB H_0 and H_{or} do not result in major errors. Such robust linear relationship suggests that Equation (12.25) should have

Table 12.19 Inactivation kinetics of different bacteria with shoulder effect.

Fluence log I	Inactivation kinetics						References
	H_i						
	0	1	2	3	4	5	
Escherichia coli as reference bacteria							
E. coli	4	5	6	7	8	9	US EPA (2003)
	1.529	3	4.5	5.6	6.8	7.5	Sommer et al. (2000)
	0.923	2	3	4	5	6	Chang et al. (1985)
Bacteria for model development							
Faecal streptococci	4.000	9.5	14	20	26		Maya et al. (2003)
Faecal coliforms	3.000	6	9	13	18		Maya et al. (2003)
Klebsiella terrigena	1.5	4	6.2	9.2	12	15	Wilson et al. (1992)
Legionella pneumophila	1.6	3.2	5.3	7.2	9.5	11.6	Wilson et al. (1992)
S. typhimurium LT2	2.2	4	6	7.6	9.5	12	Chen et al. (2009)
Salmonella typhi ATCC 19430	1.500	1	3	4.8	6.7	8.6	Wilson et al. (1992)
S. typhi ATCC 6539	1.500	2.6	4.5	5.8	7	8	Chang et al. (1985)
Shigella sonnei ATCC 9290	2.000	2.6	5	7	8	9.1	Chang et al. (1985)
Staphylococcus aureus ATCC 25923	3.000	4.5	5.1	7.6	8		Chang et al. (1985)
Streptococcus faecalis ATCC 29212	7.000	8	9	10	11.5		Chang et al. (1985)
Yersinia enterocolitica ATCC 27729	0.500	1.6	2.6	3.5	4.2	5.6	Wilson et al. (1992)
Campylobacter jejuni ATCC 43429	0.700	2	3	4	5	6	Wilson et al. (1992)

Source: Tang and Sillanpää (2015b). Reproduced with permission of Elsevier.

Table 12.20 Fluence difference of different bacteria using *E. coli* as reference.

Fluence difference log I	$\Delta H = H - H_0$						References
	0	1	2	3	4	5	
ΔH_r, *Escherichia coli* as reference bacteria							
E. coli	4	1	2	3	4	5	US EPA (2003)
	1.529	1.471	2.971	4.071	5.271	5.971	Sommer et al. (2000)
	0.923	1.077	2.077	3.077	4.077	5.077	Chang et al. (1985)
$\Delta H/\Delta H_r$, bacteria for model development							
Faecal streptococci	4.000	5.5	5	5.333	5.500		Maya et al. (2003)
Faecal coliforms	3.000	3	3	3.333	3.75		Maya et al. (2003)
Klebsiella terrigena	1.5	2.5	2.35	2.567	2.625	2.700	Wilson et al. (1992)
Legionella pneumophila	1.6	1.6	1.85	1.867	1.975	2.000	Wilson et al. (1992)
S. typhimurium LT2	2.2	1.8	1.9	1.8	1.825	1.960	Chen et al. (2009)
Salmonella typhi ATCC 19430	1.500	0.2	0.75	3.3	1.300	1.420	Wilson et al. (1992)
S. typhi ATCC 6539	1.500	1.1	1.5	1.1	1.375	1.300	Chang et al. (1985)
Shigella sonnei ATCC 9290	2.000	0.6	1.5	1.467	1.5	1.420	Chang et al. (1985)
Staphylococcus aureus ATCC 25923	3.000	1.5	1.05	1.667	1.250		Chang et al. (1985)
Streptococcus faecalis ATCC 29212	7.000	1	1	1.533	1.125		Chang et al. (1985)
Yersinia enterocolitica ATCC 27729	0.500	1.1	1.05	1.0	0.925	1.02	Wilson et al. (1992)
Campylobacter jejuni ATCC 43429	0.700	1.3	1.15	1.1	1.080	1.06	Wilson et al. (1992)

Source: Tang and Sillanpää (2015b). Reproduced with permission of Elsevier.

Table 12.21 BSI using the fluence required for *E. coli* by the US EPA.

Number	Reference bacteria: *E. coli*	BSI_E EPA data $k_{Ir} = 1$	References
1	*Faecal streptococci*	5.495	Maya et al. (2003)
2	*Faecal coliforms*	3.745	Maya et al. (2003)
3	*Klebsiella terrigena*	2.703	Wilson et al. (1992)
4	*Legionella pneumophila*	2.000	Wilson et al. (1992)
5	*S. typhimurium LT2*	1.961	Chen et al. (2009)
6	*Salmonella typhi* ATCC 19430	1.934	Wilson et al. (1992)
7	*S. typhi* ATCC 6539	1.835	Chang et al. (1985)
8	*Shigella sonnei* ATCC 9290	1.499	Chang et al. (1985)
9	*Staphylococcus aureus* ATCC 25923	1.416	Chang et al. (1985)
10	*Streptococcus faecalis* ATCC 29212	1.250	Chang et al. (1985)
11	*Yersinia enterocolitica ATCC 27729*	1.100	Wilson et al. (1992)
12	*Campylobacter jejuni ATCC 43429*	1.100	Wilson et al. (1992)

Source: Tang and Sillanpää (2015b). Reproduced with permission of Elsevier.

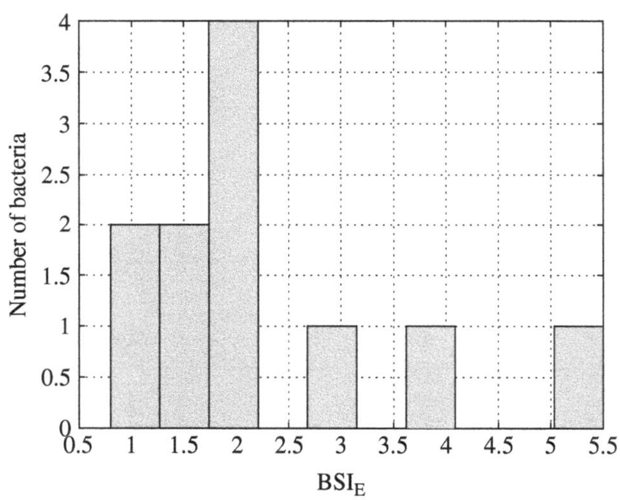

Figure 12.12 Histogram of BSI using the fluence required for *E. coli* by the US EPA. (*Source:* Tang and Sillanpää (2015b). Reproduced with permission of Elsevier.)

robust predictive power of the fluence requirement for a specific inactivation rate of log *I* as long as BSI is defined.

According to the results in Table 12.24, regression equation can be obtained to predict the required fluence of any bacteria with shoulder effect by using the fluence requirements of *E. coli* recommended by the US EPA, (2003) as follows:

$$H = (0.914 \pm 0.055) \times (\mathrm{BSI})_E \times (\log I) + H_0 \quad (12.31)$$

Not surprisingly, Equation (12.31) is the same as Equation (12.30) because it follows the same mathematical transformation despite the fact that the concept of SBI was used in deriving Equation (12.30). Figure 12.14 visually presents the predicted fluence by Equation (12.31) using a 3D plot. It shows that fluence difference between actual fluence required, *H*, and its shoulder H_0 can be predicted by using log *I* and SBI_E. The predicted fluence, therefore, can be used in reactor design of UV reactors to fulfill the US EPA validation requirements of UV disinfection systems. As a result, it would significantly reduce the trial and error experiments in deciding which fluence should be used to achieve a specific inactivation rate log *I* for a specific bacteria, providing the corresponding BSI is known.

12.6.4 Validation of the Models

Table 12.25 lists the statistics of the predicted fluence when the predicted fluence is compared with the observed fluence. It suggests that the US EPA recommended fluence of *E. coli* has robust predictive power to predict fluence required to achieve a specific log *I* of a specific bacterium.

12.6.5 Application of the Model

12.6.5.1 Experimental Data of UV Disinfection of ARBs

The developed model can be effectively applied to predict the fluence required for inactivation of antibiotic-resistant bacteria (ARBs), which are one of the most current concerns in protecting public health. One of the excellent fluence datasets at different log *I* levels was reported by McKinney and Pruden (2012). In their

Table 12.22 SBI = $\Delta H/H_{or}$, bacteria for model development.

Fluence difference	$\Delta H = H - H_0$						
log *I*	**0**	**1**	**2**	**3**	**4**	**5**	**References**
ΔH_r, *Escherichia coli* as reference bacteria							
E. coli	4	1	2	3	4	5	US EPA (2003)
$\Delta H/\Delta H_r$, bacteria for model development							
Faecal streptococci	4.000	5.5	5	5.333	5.500		Maya et al. (2003)
Faecal coliforms	3.000	3	3	3.333	3.75		Maya et al. (2003)
Klebsiella terrigena	1.5	2.5	2.35	2.567	2.625	2.700	Wilson et al. (1992)
Legionella pneumophila	1.6	1.6	1.85	1.867	1.975	2.000	Wilson et al. (1992)
S. typhimurium LT2	2.2	1.8	1.9	1.8	1.825	1.960	Chen et al. (2009)
Salmonella typhi ATCC 19430	1.500	0.2	0.75	3.3	1.300	1.420	Wilson et al. (1992)
S. typhi ATCC 6539	1.500	1.1	1.5	1.1	1.375	1.300	Chang et al. (1985)
Shigella sonnei ATCC 9290	2.000	0.6	1.5	1.467	1.5	1.420	Chang et al. (1985)
Staphylococcus aureus ATCC 25923	3.000	1.5	1.05	1.667	1.250		Chang et al. (1985)
Streptococcus faecalis ATCC 29212	7.000	1	1	1.533	1.125		Chang et al. (1985)
Yersinia enterocolitica ATCC 27729	0.500	1.1	1.05	1.0	0.925	1.02	Wilson et al. (1992)
Campylobacter jejuni ATCC 43429	0.700	1.3	1.15	1.1	1.080	1.06	Wilson et al. (1992)

Source: Tang and Sillanpää (2015b). Reproduced with permission of Elsevier.

Table 12.23 $\Delta H/\Delta H_r$ at specific inactivation log *I* of *E. coli* using the US EPA reference fluence.

Fluence difference	$\Delta H = H - H_0$						
log *I*	**0**	**1**	**2**	**3**	**4**	**5**	**References**
ΔH_r, *Escherichia coli* as reference bacteria							
E. coli	4	1	2	3	4	5	US EPA (2003)
	1.529	1.471	2.971	4.071	5.271	5.971	Sommer et al. (2000)
	0.923	1.077	2.077	3.077	4.077	5.077	Chang et al. (1985)
$\Delta H/\Delta H_r$, bacteria for model development							
Faecal streptococci	4.000	5.5	5	5.333	5.500		Maya et al. (2003)
Faecal coliforms	3.000	3	3	3.333	3.75		Maya et al. (2003)
Klebsiella terrigena	1.5	2.5	2.35	2.567	2.625	2.700	Wilson et al. (1992)
Legionella pneumophila	1.6	1.6	1.85	1.867	1.975	2.000	Wilson et al. (1992)
S. typhimurium LT2	2.2	1.8	1.9	1.8	1.825	1.960	Chen et al. (2009)
Salmonella typhi ATCC 19430	1.500	0.2	0.75	3.3	1.300	1.420	Wilson et al. (1992)
Salmonella typhi ATCC 6539	1.500	1.1	1.5	1.1	1.375	1.300	Chang et al. (1985)
Shigella sonnei ATCC9290	2.000	0.6	1.5	1.467	1.5	1.420	Chang et al. (1985)
Staphylococcus aureus ATCC 25923	3.000	1.5	1.05	1.667	1.250		Chang et al. (1985)
Streptococcus faecalis ATCC 29212	7.000	1	1	1.533	1.125		Chang et al. (1985)
Yersinia enterocolitica ATCC 27729	0.500	1.1	1.05	1.0	0.925	1.02	Wilson et al. (1992)
Campylobacter jejuni ATCC 43429	0.700	1.3	1.15	1.1	1.080	1.06	Wilson et al. (1992)

Source: Tang and Sillanpää (2015b). Reproduced with permission of Elsevier.

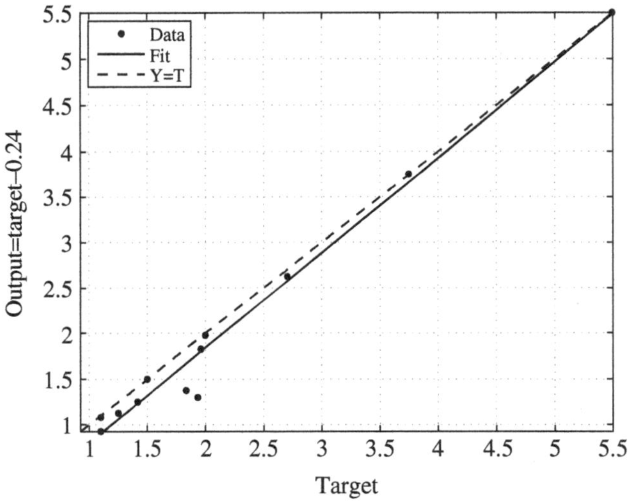

Figure 12.13 Regression plot between output and target for $\Delta H/\Delta H_r$ and BSI_E at 4 log *I*. (*Source:* Tang and Sillanpää (2015b). Reproduced with permission of Elsevier.)

study, four different ARBs include methicillin-resistant *Staphylococcus aureus* (MRSA), vancomycin-resistant *Enterococcus faecium* (VRE), *E. coli* SMS-3-5, and *Pseudomonas aeruginosa 01*. According to the authors, the fluence required to achieve a specific log *I* for each ARB is estimated in Table 12.26. The raw experimental data in Table 12.26 were used to obtain the inactivation constants k_i and its corresponding SB, H_0, for a given ARB. The obtained k_i was divided by the k_{ir} of *E. coli* to get their corresponding BSI, which were then used to predict the fluence required. The regression results of k_i as well as its corresponding SBI in reference to *E. coli* are listed in columns 7 and 8, respectively.

12.6.5.2 Error Analysis of Predicted *H* Compared with the Observed H

To assess the predictive power of Equation (12.31), the predicted fluence of different ARBs at 4 log *I* is compared with the observed experimental fluence reported by McKinney and Pruden (2012). For example, Figure 12.15 presents the comparison of the fluence predicted by Equation (12.31) with that observed at 1 log *I*.

Figure 12.15 shows that the predicted *H* closely correlate with the observed *H* with R^2 of 0.998. Naturally, it is very important to quantify the error percentage of the fit. For this purpose, the error percentage is defined as follows:

$$\text{Error\%} = \frac{H_{\text{predicted}} - H_{\text{observed}}}{H_{\text{observed}}} \times 100 \qquad (12.32)$$

Figure 12.16 presents the histogram of error % of the $H_{\text{predicted}}$ versus H_{observed}. It plots the distribution frequency at 1% error interval. It shows that 11 of 12 data points have error less than 4%. Only one data point has the error of 8%.

12.6.5.3 Prediction of Fluence Required at 5 log *I* for ARBs

One of the important applications of the developed model is to predict the fluence required at any level of log *I*, which was not experimentally determined. For example, McKinney and Pruden (2012) did not report the fluence required to achieve 5 log *I* inactivation rate of ARBs in their original experimental data. Therefore, Equation (12.31) is used to predict the fluence required at 5 log *I* as in Table 12.27.

According to the previous statistical analysis of the predicted *H* versus the observed *H* required for inactivation of different ARBs at different log *I* levels, it is reasonable to expect that the previously predicted *H* should have error no greater than 10%. The exact error percentage of the previously predicted *H* has to be experimentally determined. Table 12.27 also shows that for MRSA WW, the predicted *H* is 20.88 mJ/cm^2 due to high BSI value as well as the very broad shoulder of 10.28 mJ/cm^2. The easiest ARB is *P. aeruginosa* 01 PBS, which requires 5.8 mJ/cm^2 only due to low BSI value as well as no shoulder presented.

12.7 Emerging Treatment Technologies

In attempt to oxidize methyl tertiary butyl ether (MTBE) found in contaminated groundwater, Melin (2000) reported a comprehensive evaluation of advanced oxidation process (AOP) and other physicochemical processes as potential treatment technologies shown in Table 12.28, which lists all technologies if the technology and used energy have not yet attained commercially viable status.

Table 12.24 Correlation coefficients between $\Delta H/\Delta H_r$ and three sets of BSI.

log *I*	1	2	3	4	5	Average coefficient *a*	Standard deviation of *a*
$(BSI)_E$	0.860	0.858	0.979	0.958	0.917	0.914	0.055

Source: Tang and Sillanpää (2015b). Reproduced with permission of Elsevier.

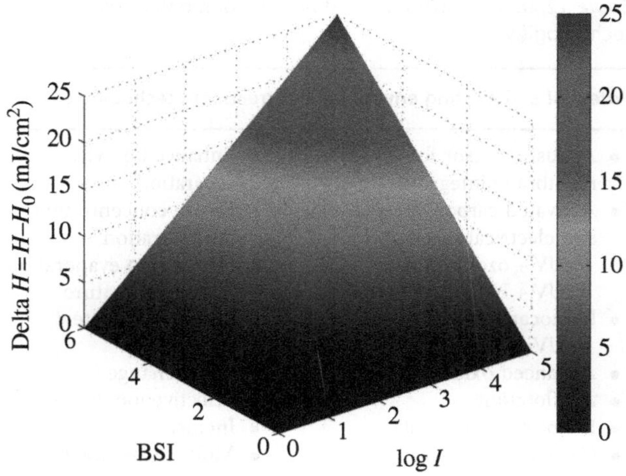

Figure 12.14 3D plot of $\Delta H = H - H_0$ as function of log I and SBI_E. (*Source:* Tang and Sillanpää (2015b). Reproduced with permission of Elsevier.)

UV Radiation. UV light exhibits excellent germicidal properties at certain wavelengths. In addition, UV radiation can transform molecular structures, making it useful in several industrial applications, such as pulp bleaching and in the curing of specialized coatings. This technology is well established with a number of reputable vendors of UV systems.

Ozonation. When molecular oxygen (O_3) is passed through an electric current, a portion of the oxygen is converted to ozone or O_3. Ozone is a highly reactive chemical that, under the proper circumstances, leads to AOPs. Worldwide, the ozone industry is mature, particularly in Europe, as well as in the United States.

Supercritical Water Oxidation. When the temperature and pressure of water are brought above its critical point (where pressure equals 22.1 MPa and temperature equals 374°C), water exhibits unique characteristics,

Table 12.25 Statistics of predicted fluence required.

log I	Number of observations	Error degrees of freedom	Root mean squared error	R-squared	Adjusted R-squared	F-statistic vs. constant model	p-Value
BSI_E							
1	12	10	0.989	0.946	0.941	176	1.13×10^{-7}
2	12	10	1.31	0.972	0.969	348	4.21×10^{-9}
3	12	10	1.51	0.989	0.988	928	3.40×10^{-11}
4	12	10	1.6	0.996	0.996	2450	2.75×10^{-13}
5	8	6	0.485	0.999	0.999	7090	1.89×10^{-11}
Average	11.2	9.2	1.179	0.980	0.979	2198	2.35×10^{-8}

Source: Tang and Sillanpää (2015b). Reproduced with permission of Elsevier.

Table 12.26 Fluence required for different ARBs.

Fluence log I	Observed fluence required at different log I levels						
	H_i						
	0	1	2	3	4	k_i	BSI = k_i/k_{ir}
E. coli as reference bacteria							
E. coli	4	5	6	7	8	1 (k_{ir})	1
Fluence required at different log I levels for each ARB							
MRSA WW	10.28	12	14.6	16.8	19	2.32	2.32
MRSA PBS	5.70	7.2	9.6	11	12.6	1.76	1.76
VRE WW	3.55	5	7.1	9	10	1.69	1.69
VRE PBS	3.35	4.8	7.1	9	10	1.75	1.75
E. coli SMS-3-5 WW	1.55	3.6	5.2	7.2	9.3	1.91	1.91
E. coli SMS-3-5 PBS	2.00	3.2	4.8	6.1	7.2	1.33	1.33
P. aeruginosa 01 WW	0	1.4	2.8	4.2	6.2	1.58	1.58
P. aeruginosa 01 PBS	0	1.2	2.5	3.8	5	1.27	1.27

Source: Tang and Sillanpää (2015b). Reproduced with permission of Elsevier.

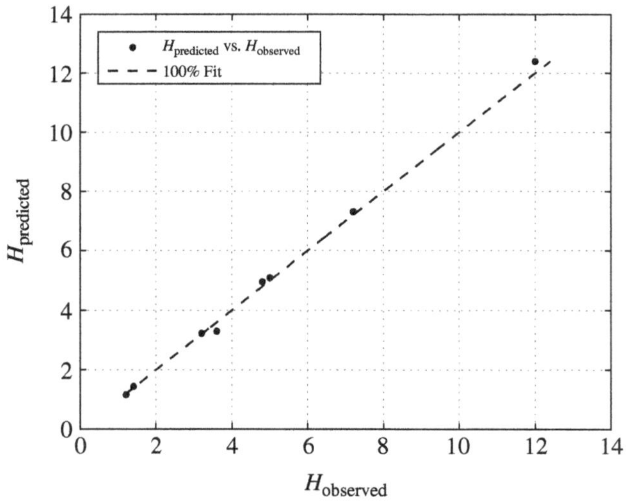

Figure 12.15 $H_{predicted}$ versus $H_{observed}$ at 1 log I. (*Source:* Tang and Sillanpää (2015b). Reproduced with permission of Elsevier.)

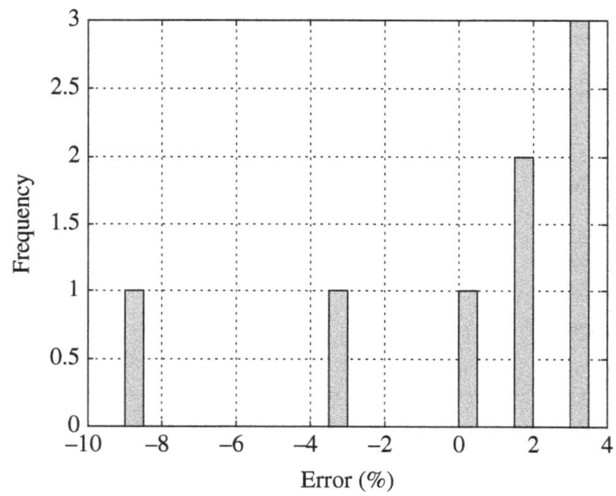

Figure 12.16 Histogram of error % at 1 log I. (*Source:* Tang and Sillanpää (2015b). Reproduced with permission of Elsevier.)

Table 12.27 Predicted fluence required at 5 log I for each ARB.

Name of ARB	BSI = k_i/k_{ir}	H_0	Predicted H (mJ/cm^2)
MRSA WW	2.32	10.28	20.88
MRSA PBS	1.76	5.70	13.74
VRE WW	1.69	3.55	11.27
VRE PBS	1.75	3.35	11.35
E. coli SMS-3-5 WW	1.91	1.55	10.28
E. coli SMS-3-5 PBS	1.33	2.00	8.08
P. aeruginosa 01 WW	1.58	0	7.22
P. aeruginosa 01 PBS	1.27	0	5.80

Table 12.28 Potential and existing environmental treatment technologies.

Potential and existing environmental treatment technologies

- Acoustic sensor for assessing membrane integrity
- Activated carbon absorption and electrical regeneration
  - UV + ozone
  - UV + H_2O_2
- Photocatalysis
  - UV + TiO_2
- Advanced oxidation processes
- Air flotation
- Biological treatment
- Capacitive deionization
- Dielectric heating system
- Microwave systems
- Radio-frequency (RF) systems
- Electrode watering
- Electrodialysis
- Electrohydraulic cavitation
- Spark-gap discharge (lithotripsy)
- Ultrasonic irradiation (sonolysis)
- Electrolysis
- Electrocoagulation
- Cerium-catalyzed oxidation process
- Electrokinetic remediation of groundwater
- Electron beam irradiation
- Electromagnetic leak detection
- Electro-osmosis dewatering
- Electrophoresis
- Electrostatic precipitation
- Enhanced gravity separation
- Freeze concentration/crystallization
- Heat pump evaporation
- High temperature plasma-arc heater/furnace
- Ion exchange
- Selective ion exchange
- Incineration
- Molten glass incinerator
- High temperature fluid wall reactor
- Infrared incinerator
- Induction incinerator
- Pyrolytic incinerator
- Infrared drying
- Low temperature (nonthermal) plasmas
- Mechanical vapor recompression
- Membrane separation
- Geotextile membranes
- "Smart" membranes
- Microfiltration
- Ultrafiltration
- Reverse osmosis
- Photocatalytic oxidation
- Phytoremediation
- Wetlands
- Ozonation
- Pyrolysis
- Supercritical water oxidation
- Ultraviolet radiation
- Vacuum pyrolysis
- Wet air oxidation

Source: From Melin (2000).

many of which are particularly advantageous for environmental treatment.

Sonolysis. Sonolysis is defined as the chemical effects produced by subjecting a chemical reaction to sound waves. Ultrasound, with frequencies roughly between 15 kHz and 10 MHz, has a drastic effect on chemical reactions. The mechanism that causes this effect is known as "acoustic cavitation." During the solution expansion phase, small vapor-filled bubbles are formed due to weak points in the solution, primarily at trapped gas pockets on particulate surfaces. Extreme environments are produced in and near the bubble as a result of the acoustic cavitation.

Electron Beam Irradiation. Beams of high energy electrons can be used to mineralize organic pollutants. In addition, when high energy electron beam is applied to a water or waste, the electrons generate various oxidative and reactive species, leading to AOPs.

Since all the emerging technologies have both advantages and disadvantages, the rankings were given in Table 12.29 in terms of applicability, interest, status, impact, and competitiveness.

Most of the AOP systems are available from the manufacturers for treating waters at small flow rate (e.g. 100 or 1000 gpm). Typical ranges of AOP reactor capacities are shown in Table 12.30. Smaller or larger AOP reactors can be custom-built. Most AOPs are modular processes, so more than one reactor can be employed in series (to obtain higher retention times) or parallel (to process larger volumes) mode to achieve the desired effluent goals for a given flow rate. Table 12.30 summarizes the reported MTBE removal efficiencies from field and pilot studies. Clearly, removal efficiencies will be a function of operating parameters (e.g. reactor residence time) and water quality parameters (e.g. alkalinity, NOM content). Table 12.30 presents some comparative removal efficiencies reported in literature for the various processes. In general, higher MTBE removal can be obtained under longer retention times and greater chemical dosages.

12.8 Design Considerations of UV Disinfection System

According to the EPA guidelines (US EPA, 2006), the following are important parameters:

1) Target pathogen and log inactivation.
2) UV dose required to achieve the log inactivation.
3) Minimal and maximal flow rate.
4) UV transmittance.
5) Fouling/aging factor.
6) Tie lamp aging to lamp life guarantee.
7) Tie sleeve fouling to cleaning method and frequency.
8) Power quality.
9) Preemptive response to flow increases.
10) Alarms and responses.
11) Level of redundancy.
12) Future considerations (expansion, new applications).

These design criteria are selected to minimize off-specification operation. UV disinfection reactor designs can be classified as either open-channel systems, closed-channel noncontact systems, or closed-channel contact systems. Open-channel systems are common in wastewater disinfection and consist of racks of UV lamps oriented horizontal or parallel within a wastewater flow. Flow is gravity fed and exposed to the atmosphere. Closed-channel noncontact systems involve water or wastewater flowing through UV-transmitting tubes, typically Teflon. Lamps are external to the tubes and the flow is either pressurized or gravity fed. Closed-channel contact/noncontact systems are usually for drinking water disinfection.

Since all the UV disinfection systems have to be validated to receive the EPA disinfection credit, it is very important to ensure that design parameters satisfy validation requirements. As a result, the following elements of UV design have to be coordinated with validation requirements:

1) Installation hydraulics.
2) The range of conditions for which the UV reactor performance has been validated such as flow rate, UVT,

Table 12.29 Suggested ranking of technologies.

Rank	Technology	Score					Total
		Applicability	Interest	Status	Impact	Compete	
1	Electrodewatering	3	5	5	4	4	21
2	Capacitive deionization	2	3	5	4	5	19
3	Dielectric heating	3	4	3	4	4	18
	Supercritical water oxidation	4	4	3	4	3	18
	Freeze–thaw conditioning	2	4	3	4	5	18
4	Electron beam irradiation	4	4	3	4	2	17
	Ion exchange selective recovery	2	3	3	4	5	17
5	Activated carbon-Electrosolv	4	2	3	2	4	15
6	Electrolysis electrokinetic remediation	1	4	3	2	4	14
7	Electrohydraulic cavitation by sonolysis	3	2	5	2	3	13
	Air flotation	3	3	0	2	4	12
	Incineration	3	3	0	3	3	12

Source: From Melin (2000).

Table 12.30 Range of AOP reactor capacities and MTBE removal efficiencies.

Technology	Range of reactor capacities	Reported MTRE removal efficiencies
H_2O_2/O_3	• 1–100 gpm, small portable site remediation HiPOx systems • 500–3000 gpm, potable and large remediation HiPOx systems	Up to 80% >99%
O_3/UV	• 5–1200 gpm, Ultrox system • Rayox-O process custom-built for desired capacity	No published remove efficiencies for drinking water systems
H_2O_2/ UV	• Single UV lamp reactor: 50–700 gpm • Up to 3000 gpm, Rayox tower system • 5–1200 gpm, Ultrox system	80–95% >99%
E-beam	40–100 gpm, mobile unit	>80% 60–99%
Hydrodynamic cavitation	100–2000 gpm, Hydrox system	>80%
TiO_2-catalyzed UV oxidation	Custom-built units	76–95%

Source: From Melin (2000).

target microorganism and log inactivation, and operating configurations (specific lamps on and off).
3) Headloss.
4) Calculated dose approach – dose equations developed.

Although the EPA guideline is very detailed, many issues are not addressed:

1) Type of reactor or reactor configuration.
2) Type of sensor.
3) Type of UV lamp.
4) UVT data collection.
5) Backup power/power quality mitigation.
6) Requirements related to mercury release.

12.8.1 UV Dose

The higher the flow rate, the shorter the hydraulic detention time, so the smaller the dose received by the water. An appropriate flow rate must be determined based on the other characteristics of the water and on the desired dose. Most important to achieve a given log inactivation of a target pathogen is the UV dose, which is the amount of UV received by the target pathogen as follows:

$$\text{UV dose} = I \times t \tag{12.33}$$

where

I = intensity (mW/cm^2)
t = hydraulic retention time
UV dose = (s) × (mW/cm^2)

There are four distributions to be considered: (i) UV dose distribution due to different hydraulic configurations of different UV reactors, (ii) hydraulic retention time, (iii) average intensity of the UV light, and (iv) distribution of UV sensitivity of pathogens. Following are different design methods in counting all these design parameters.

12.8.2 Hydraulic Retention Time

Hydraulic retention time depends upon the reactor volume and flow rate as follows:

$$t = \frac{V}{Q} \tag{12.34}$$

where

t = hydraulic retention time (s)
V = reactor volume (m^3)
Q = flow rate (m^3/s)

Since minimum and maximum flows differ greatly, minimum velocity and maximum velocity should also be considered. Major elements that should be considered in the hydraulic design of a UV closed vessel reactor are dispersion, turbulence, effective volume, residence time distribution, and flow rate (US EPA, 1996). With all reactors, the delivered dose will have the dose distribution. The narrower the dose distribution, the more efficient the reactor. For any stated dose, there is always some water that will receive a lower dose and some a higher dose. To achieve a specific log I, the amount of water receiving a lower dose than the target is critical. Just 1% of the water not receiving the target dose for inactivation will limit the performance to 1-log reduction. Average dose, as the name implies, is simply the average fluence throughout the reactor. It takes no account of dose distribution and so it could give a false view of reactor performance.

Average UV intensity is impacted by water quality such as UV transmittance, lamp output/power setting, lamp age, and quartz sleeve transmittance (coating). UVT is the determined factor in terms of UV intensity because the lower the UVT, the lower the UV dose that will reach the target pathogen. Generally, the lower the UVT, the closer the UV lamps must be placed to avoid dark space because only 50% of the light emitted by the lamp will penetrate the first 1 cm of the fluid being treated. With all reactors, the delivered dose will cover a range of doses (dose distribution). The average dose value will always be higher than an equivalent RED dose, often by as much as 70%.

12.8.3 UV Lamps

A typical mercury arc lamp consists of a hermetically sealed tube of UV-transmitting vitreous silica or quartz with electrodes at both ends (Phillips, 1983). The tube is filled with a small amount of mercury and an inert gas, usually argon. The electrodes are usually composed of tungsten with a mixture of alkaline earth metals to aid arc formation within the lamp. A gas discharge is struck by applying a high voltage across the electrodes. UV light is emitted from the lamp when mercury vapor, excited by the discharge, returns to a lower energy state. Argon is present to aid lamp starting, extend electrode life, and reduce thermal losses. Argon does not contribute to the spectral output of the lamp.

12.8.4 Turbidity

Turbidity is a measure of the quantity of particulates in a water or wastewater sample. It is determined by shining an infrared beam of light through a one centimeter-thick sample and measuring light detected by sensors placed at 90° to the beam. Turbidity is not necessarily correlated with the absorption coefficient. Turbidity is commonly reported in NTU. It is often thought to be a limiting factor in UV disinfection. However, work has shown that particles, as long as they are not UV absorbers, do not significantly reduce the overall irradiance by either shading or scattering but only when organisms are embedded within them (Linden and Darby, 1998; Emerick et al., 1999). Particle suspension can increase the apparent absorption coefficient – as measured by a spectrophotometer – by scattering rather than absorbing light (Linden and Darby, 1998). This effect can lead to underprediction of design capabilities.

UV LED modules have much more effective inactivation efficient than those obtained with a conventional mercury UV bench-scale apparatus. Both UV LEDs perform better than the conventional mercury lamp. According to Chen et al. (2009), the inactivation of *B. subtilis* spores should be comparable at the wavelengths of 254 and 279 nm (Table 12.31).

Microscreening/UV and pulsed UV are other two emerging UV reactor designs. All testing of these two systems to date has been under controlled laboratory or field conditions. Both these technologies require demonstration of efficacy and applicability in real-world treatment operations.

Table 12.31 Comparison of inactivation curves obtained with the apparatus designed for UV LEDs and a conventional bench-scale mercury lamp apparatus.

Fluence (J/m^2)	269 nm UV LED log RF	82 nm UV LED log RF ± 0.2	Mercury lamp log RF ± 0.1
100	1.6 ± 0.2	0.9	0.8
250	3.7 ± 1.4	2.7	1.9
400	5.8 ± 0.7	4.5	2.9

12.8.5 Typical Design Lives of Major UV Components

Table 12.32 provides typical design and guaranteed lives for major UV reactor components.

Table 12.32 Typical design and guaranteed lives of major UV components (US EPA, 2006).

Component	Design life	Guaranteed life
Low-pressure lamps (LP And LPHO)	12 000 h	8 000–12 000 h
MP lamps	8 000 h	4 000–8 000 h
Sleeve	8–10 years	1–3 years
Duty and reference UV sensors	3–10 years	1 year
UVT analyzer	3–5 years	1 year
Cleaning systems	3–5 years	1–3 years
Ballasts	10–15 years	1–5 years

12.9 Exercise

12.9.1 Questions

1) What is log *I*? What are the corresponding kill percentages at 1, 2, and 3 log *I*?
2) What is VSI and BSI? What design parameter of UV disinfection can you estimate by using these indices?
3) How does UV/TiO_2 degrade organic pollutants? What are the advantages?
4) How does pH change the dominant mechanism during degradation of organic pollutants in extreme acidic or alkaline conditions?

12.9.2 Calculations

1) If a virus has VSI of 3.5, what would be the fluence in mJ/cm^2 required to achieve 1, 2, 3, 4, and 5 log *I*? Can you use the developed equation to estimate the required fluence at 6 and 7 log *I*? Why?
2) The kinetics of UV disinfection of bacteria with shoulder effect is very complicated mathematically. The model of BSI provides much simpler mathematical treatment than double exponential model. Answer the following:
   a) In your opinion, what factors would contribute the error of the predicted fluence required using the simple model than the double exponential model?
   b) Which fluence would be more conservative if the predicted fluence is compared with the double exponential model? Why?

12.9.3 Projects

12.9.3.1 Xiongan Project

1) If WTP and WWTP employ UV disinfection, what are the major design differences?
2) If UV disinfection is going to be used to reclaim treated wastewater, what are the major design concerns?

12.9.3.2 Community Proposal Project

1) Contact local water utilities and inquire about disinfection processes.
2) Collect disinfection performance data, and compare them with both disinfection standards and DBP Stage 2 rules.
3) According to the compared data, make recommendations to the utility managers for using UV disinfection versus chlorination.
4) What would be the major barrier for the utility to switch from chlorination to UV disinfection?
5) Design a tubular plug-flow UV reactor to satisfy the discharge standards of your local utilities such as WTP, WWTP, or a water reclaiming plant.

References

Bolton, J.R., Dussert, B., Bukhari, Z. et al. (1998). Inactivation of *Cryptosporidium parvum* by medium-pressure Ultraviolet light in finished drinking water. Proceedings of the American Water Works Association Water Quality Technology Conference, San Diego, CA (1–5 November).

Chang, J.C.H., Ossoff, S.F., Lobe, D.C. et al. (1985). UV inactivation of pathogenic and indicator microorganisms. *Applied and Environmental Microbiology* 49 (6): 1361–1365.

Chen, Q., Tang, S., Jin, X. et al. (2009). Investigation of the genotoxicity of quinocetone, carbadox and olaquindox in vitro using Vero cells. *Food and Chemical Toxicology* 47: 328–334.

Chick, H. (1908). An investigation of the laws of disinfection. *The Journal of Hygiene* 8: 92–158.

Clancy, J.L., Bukhari, Z., Hargy, T.M. et al. (2000). Using UV to inactivate Cryptosporidium. *Journal AWWA* 92 (9): 97–104.

Emerick, R.W., Loge, F.J., Thompson, D., and Darby, J.L. (1999). Factors influencing ultraviolet disinfection performance Part II: association of Coliform Bacteria with wastewater particles. *Water Environment Research* 71 (6): 1178–1187.

Harm, W. (1980). *Biological Effects of Ultraviolet Radiation*. New York: Cambridge University Press.

Harris, D.G., Adams, V.D., Sorensen, D.L., and Curtis, M.S. (1987). Ultraviolet inactivation of selected bacteria and viruses with photoreactivation of bacteria. *Water Research* 21 (6): 687–692.

Hilmoe, D.J. (2002). Seattle's 180 MGD UV Decision, Workshop entitled "UV in drinking water treatment – Scientific foundation of successful application" (20–23 March 2002), Kiwa Water Research.

Jagger, J. (1967). *Ultraviolet Photobiology*. Englewood Cliffs: Prentice-Hall, Inc.

Kolch, A. (1999). Disinfecting drinking water with UV light. *Pollution Engineering* 50 (3): 19–21, 23.

Linden, K.G. and Darby, J.L. (1998). Ultraviolet disinfection of marginal effluents: determining ultraviolet absorbance and subsequent estimation of ultraviolet intensity. *Water Environment Research* 70 (2): 214–223.

Linden, K.G., Batch, L., and Schulz, C. (2002a). UV disinfection of filtered water supplies: water quality impacts on MS2 dose-response curves. Proceedings of the Annual Conference of the American Water Works Association, New Orleans, LA (16–20 June).

Linden, K.G., Shin, G.A., Faubert, G. et al. (2002b). UV disinfection of *Giardia lamblia* cysts in water. *Environmental Science and Technology* 36 (11): 2519–2522.

Love, D.C., Silverman, A., and Nelson, K.L. (2010). Human virus and bacteriophage inactivation in clear water by simulated sunlight compared to bacteriophage inactivation at a Southern California beach. *Environmental Science and Technology* 44: 6965–6970.

McKinney, C.W. and Pruden, A. (2012). Ultraviolet disinfection of antibiotic resistant bacteria and their antibiotic resistance genes in water and wastewater. *Environmental Science and Technology* 46 (24): 13393–13400.

Melin, G. (2000). Treatment technologies for removal of methyl tertiary butyl ether (MTBE) from drinking water: air stripping, advanced oxidation processes, granular activated carbon, and synthetic resin sorbents. *NWRI-99-06*. Center for Groundwater Restoration and Protection, National Water Research Institute.

Mamane-Gravetz, H. and Linden, K.G. (2004). Impact of particle aggregated microbes on UV disinfection. Proceedings of the American Water Works Association Water Quality Technology Conference, San Antonio, TX (14–17 November 2004).

Mattle, M.J., Kohn, T.(2012). Inactivation and tailing during UV254 disinfection of viruses: contributions of viral aggregation, light shielding within viral aggregates, and recombination. *Environmental Science and Technology* 46 (18): 10022–10030.

Maya, C., Beltrán, N., Jiménez, B., and Bonilla, P. (2003). Evaluation of the UV disinfection process in bacteria and amphizoic amoebae inactivation. *Water Science and Technology: Water Supply* 3 (4): 285–291.

Meulemans, C.C.E. (1987). The basic principles of UV-sterilization of water. In: Ozone + Ultraviolet Water Treatment, International Ozone Association, Paris. *Ozone: Science & Engineering, the Journal of the International Ozone Association* 9 (4): 299–313.

Oppenheimer, J., Gillogly, T., Stolarik, G., and Ward, R. (2002). Comparing the efficiency of low and medium pressure UV light for inactivating *Giardia muris* and *Cryptosporidium parvum* in waters with low and high levels of turbidity. Proceedings of AWWA annual conference, New Orleans, LA (16–20 June).

Örmeci, B. and Linden, K.G. (2003). Inactivation and detection of particle–associated microorganisms. Proceedings of the 2nd International Congress on Ultraviolet Technologies, Vienna, Austria (9–11 July).

Parker, J.A., Darby, J.L. (1995). Particle-associated coliform in secondary effluents: shielding from ultraviolet light disinfection. *Water Environment Research* 67: 1065–1075.

Passantino, L., Malley, J.P., Jr., Knudson, M. et al. (2004). Effect of low turbidity and algae on UV disinfection performance. *Journal – American Water Works Association* 96 (6): 128–137.

Phillips, R. (1983). *Sources and applications of ultraviolet radiation*. New York: Academic Press Inc.

Rauth, A.M. (1965). The physical state of viral nucleic acid and the sensitivity of viruses to ultraviolet light. *Biophysical Journal* 5: 257–273.

Severin, B.F., Suidan, M.T., and Engelbrecht, R.S. (1984). Mixing effects in UV disinfection. *Journal – Water Pollution Control Federation* 56 (7): 881–888.

Severin, B.F., Suldan, M.T., and Englebrecht, R.S. (1983). Effects of temperature on ultraviolet light disinfection. *Environmental Science and Technology* 17: 717–721.

Sommer, R., Haider, T., Cabaj, A. et al. (1998). Time dose reciprocity in UV disinfection of water. *Water Science and Technology* 38(12): 145–150.

Sommer, R., Lhotsky, M., Haider, T., and Cabaj, A. (2000). UV inactivation, liquid-holding recovery, and photoreactivation of Escherichia coli o157 and other pathogenic Escherichia coli strains in water. *Journal of Food Protection* 63 (8): 1015–1020.

Tang, W.Z. and Sillanpää, M. (2015a). Bacteria Sensitivity Index (BSI) of UV Disinfection of Bacteria with shoulder effect. *Journal of Environmental Chemical Engineering* 3 (4): 2588–2596.

Tang, W.Z. and Sillanpää, M.(2015b). Virus Sensitivity Index (VSI) of UV disinfection. *Environmental Technology* 36 (11): 1464–1475.

Templeton, M., Andrews, R.C., Hofmann, R., and Whitby, G.E. (2003). UV inactivation of floc-associated MS-2 coliphage. Proceeding of OMWA/OWWA conference, Hamilton.

Templeton, T.J., Lancto, C.A., Vigdorovich, V. et al. (2004). The Cryptosporidium oocyst wall protein is a member of a multigene family and has a homolog in *Toxoplasma*. *Infection and Immunity* 72: 980–987.

US EPA (1996). Ultraviolet Light Disinfection Technology in Drinking Water Application – An Overview. *EPA* 811-R-96-002. Washington, DC: US Environmental Protection Agency, Office of Ground Water and Drinking Water.

US EPA (2003). Ultraviolet Disinfection Guidance Manual. *EPA 815-D-03-007*. Washington, DC: US EPA.

US EPA (2005). Economic Analysis for Long Term 2 Enhanced Surface Water Treatment Rule. *Contract 68-C-02-026*. Arlington, VA: The Cadmus Group, Inc.

US EPA (2006). National Primary Drinking Water Regulations: Stage 2 Disinfectants and Disinfection Byproducts Rule. A Rule by the Environmental Protection Agency on 01/04/2006, Federal Register.

US EPA (2007). Simultaneously compliance guidance manual for long term 2 and stage 2 DBP rules. Washington, DC: Office of Water, US EPA.

Watson, H.E. (1908). A note on variation of the rate of disinfection with change in the concentration of the disinfectant. *The Journal of Hygiene* 8: 536–542.

Wilson, B.R., Roessler, P.F., van Dellen, E. et al. (1992). Coliphage MS2 as a UV water disinfection efficacy test surrogate for bacterial and viral pathogens. Proceedings of the American Water Works Association Water Quality Technology Conference, Toronto, Canada (15–19 November 1992).

13

Green Retrofitting and Remediation

Cost of inaction is far higher than the interruption cost by a disaster.

OBJECTIVES
To conduct energy auditing to improve energy efficiency in WWTPs To introduce innovative technologies of anaerobic digester to increase energy efficiency To identify innovative ways such as microsieving to divert organic matter in wastewater to anaerobic digester To determine the energy consumption of mechanical dewatering To quantify energy generation through anaerobic digestion of sludge To reduce the footprint of remediation projects through green remediation To present the US EPA green remediation procedure and tools

13.1 Principle 10

Wastewater treatment plant (WWTP) is the critical infrastructure in the water–food–energy nexus. Currently more than 15 000 and 5 000 WWTPs were designed and built according to the conventional design principle in the United States and China, respectively. However, 99% of these plants were currently energy negative. In the next decade, retrofitting these WWTPs to energy-positive water resource recovery facility (WRRF) could be the major task for SEE designers in both the developed world such as the United States and EU and the developing countries such as BRIICS. Due to different designs and combinations of different unit processes, retrofitting strategies will be different. In addition, solar energy could be added to the WRRF to reach the retrofitting targets. For developing countries, remediation of contaminated air, water, and soil is a paralleling challenge. In general, retrofitting and remediation should have the minimal footprints on energy and materials as follows:

Principle 10 Retrofitting and remediation should recover material, energy, water, or land with minimal footprints:

1) WWTPs should be systematically retrofitted to an energy-positive water resource recovery facility (WRRF).
2) WTP should gradually replace chlorination with ultraviolet disinfection and degradation.
3) Remediation of air, water, and soil should have minimal footprints on air, water, soil, and energy.

13.2 Challenges of WWTP Design

There are more than 60 000 water treatment plants (WTPs) and 15 000 WWTPs in the United States. These plants account for about 3–4% of electricity consumption of the United States. To power these plants, the cost is roughly \$10 billion. WWTPs purchase 60–80% of their total energy from electrical companies. The annual electricity bill to power three major WWTPs of 200 MGD for Miami-Dade County is \$24 million. Therefore, conventional WWTPs have lots of disadvantages, and retrofitting them should target to energy-positive WWTPs should be the major effort for water utilities to become sustainable.

WWTPs use aerobic bacteria with oxygen to degrade organic pollutants in wastewater and produce large amount of sludge. Sludge generated consumes one-third

Sustainable Environmental Engineering, First Edition. Walter Z. Tang and Mika Sillanpää.

of the total energy of a typical WWTP. The produced sludge is usually anaerobically digested to produce biogas of 70% methane. The efficiency of the tank process is very low because dead zones can occur throughout the tank, short circled, and energy used in activated sludge (AS) is intensive. Traditional WWTPs occupy unnecessary land. Nutrients such as nitrogen and phosphate are not recovered through this process. Chlorine is toxic and can produce disinfection by-products (DBP). Conventional WWTPs use tanks to treat wastewater (WW) and have the following disadvantages:

1) Large land footprint.
2) Large energy footprint.
3) Vulnerable to outdoor temperature.
4) Design philosophy is to treat WW, not to recover water, energy, nutrients, and materials.
5) Volatile organic compounds from aeration tanks of WWTPs cause odor and possibly smog.
6) Aerobic stabilization causes a higher financial burden (1/3 of WWTP cost comes from sludge disposal).

13.2.1 Energy Efficiency of Water and Wastewater Treatment

One of the urgent challenges faced by SEE designers is to retrofit the conventional WWTPs to energy positive. Sustainable designs should explore either containers or pipe reactors for communities with population equivalent (PE). Pipe reactors are better than tank units and might hold the key for sustainable WRRF. There are many innovative ways of designing WWTPs:

1) Modular pipe flow reactor.
2) Designed WWTP as an energy recovery facility, which can be 100% energy self-sufficient.
3) Employ the most advanced treatment technologies with minimal energy and carbon footprints possible.
4) Reclamation of treated WW (water reuse).
5) Nutrient recovery, such as nitrogen and phosphate.
6) The codigestion of sludge with food and agricultural waste to produce more methane than the energy needed to power the plant.

To improve energy efficiency, pipe reactors could be applied to many unit processes: has its advantages:

1) Static mixers may save a lot of energy compared to the traditional coagulation tanks
2) Ultraviolet (UV) disinfection kills bacteria/viruses.
3) Reverse osmosis
4) Membrane processes such as biological membrane reactors (MBR)

13.3 Anaerobic Digestion for Biogas Production

Modern design theories indicate that the best strategy to retrofit the WWTP is to optimize the coupling between carbon and nutrient removal process using anaerobic oxidation of organic matter while using Anammox to degrade ammonium. It skips the aerobic cycle and creates an anaerobic reaction and reduces nitrates to ammonia. As a result, it saves the energy for aeration in AS process. In aerobic process, organic compounds expressed as biological oxygen demand (BOD) are aerobically converted to the secondary waste such as AS that is expensive to be further treated to meet disposal standards. On the other hand, anaerobic digestion (AD) is the degradation of organic material by anaerobic microorganisms to produce biogas. Anaerobic digestion has four major stages:

1) Hydrolysis cleaves a chemical compound using water reaction and breaking down insoluble complex molecules to simple sugars, fatty acids, and amino acids.
2) Fermentation (acidogenesis) uses the products from hydrolysis and transform them into organic acids and alcohols.
3) Acetogenesis converts organic acids and alcohols into hydrogen, carbon dioxide, and acetic acid.
4) Methanogenesis makes the acetic acid react through methanogenic bacteria to produce, methane and carbon dioxide (Figure 13.1).

AD process produces biogas that is a mixture between methane and carbon dioxide, typically composed of 50–75% methane, 25–50% carbon dioxide, 5–10% hydrogen gas, and 1–2% nitrogen gas. Methane is the major component of natural gas (97% of the volume). It is a colorless, odorless gas and is flammable and has a calorific value of 10 kWh/Nm^3 (Nm^3 is normal cubic meter at standard temperature and pressure). AD undergoes reduction reactions (assimilation of electrons) and has relatively long sludge cycles (due to low growth rate of methanogens) As a result, long solid retention times are required (Table 13.1).

During the reduction processes, the relative size of organic matter is typically as shown in Table 13.2.

Anaerobic treatment is a biological process carried out in the absence of O_2 for the stabilization of organic materials by conversion to CH_4 and inorganic end products

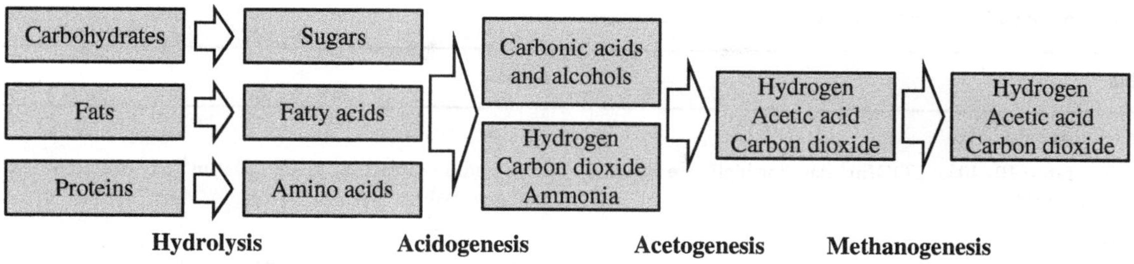

Figure 13.1 Degradation steps of anaerobic digestion process.

Table 13.1 Ultimate products of oxidation and reduction of common elements.

Chemical elements	Oxidation products	Reduction products
Carbon (C)	CO_2 (carbon dioxide)	CH_4 (methane)
Nitrogen (N)	NO_2 (nitrogen dioxide)	NH_3 (ammonia)
Phosphate (P)	P_2O_3 (diphosphorus trioxide)	H_4P (phosphonium)
Sulfur (S)	SO_2 (sulfur dioxide)	H_2S (hydrogen sulfide)

Table 13.2 Product size during anaerobic stages.

Anaerobic digestion stage	Products	Organic matter (in meters)
Hydrolysis	Super fatty acids, amino acids	$>10^{-4}$
Acidogenesis	Organic acids	$>10^{-7}$
Acetogenesis	Hydrogen gas, carbon dioxide	$>10^{-8}$
Methanogenesis	Hydrogen gas, carbon dioxide, methane gas	$>10^{-9}$

such as CO_2 and NH_3. Anaerobic fermentation breaks down organic matter without the need of an external electron to release energy. Anaerobic respiration requires external electron acceptors to release energy at a higher level than anaerobic fermentation. There are many advantages of anaerobic treatment:

1) Less energy requirement (no aeration needed).
2) Energy formation in the form of methane gas.
3) Less biomass (sludge) formation.
4) AD is the key to a low-carbon economy. It generates energy to heat the sludge, recycles digested sludge for agriculture, and protects environmental and human health.
5) It converts organic matter in the sludge as CH_4 to reduce energy or heat consumption and reduces greenhouse gas (GHG) emissions and the use of energy and material.
6) Hydrolysis at high temperature improves the gas production rate of AD.
7) CH_4 can be used for power generation and waste heat can be used to dry the sludge or to thermally hydrolyze food wastes for anaerobic digestion.
8) It reduces fertilizer application, energy consumption, and the need for land resources.

To reduce sludge production, organic matter should be diverted to anaerobic digester directly. As a result, the organic matter would be anaerobically digested to produce methane instead of fix free water into secondary waste such as sludge. The operation parameters of anaerobic and aerobic processes are compared in Table 13.3 which shows that the aerobic sludge production is from 4 to 10 times higher than that of anaerobic process. Furthermore, the anaerobic digested sludge is much easier to be be dewatered than the aerobic sludge (Tables 13.4 and 13.5).

13.3.1 Operation Guidelines for Wastewater Treatment Plants

1) In monitoring AD, the rule of thumb is that the higher the percentage of methane output, the better the performance of the plant,
2) On the other hand, higher the percentage of carbon dioxide, the worse the efficiency of the plant.

Thermophilic (~55°C) or hyperthermophilic (~73°C) could achieve a much higher rate of biogas production. At a given flow rate, the hydraulic retention time (HRT) is designed so that all the organic compounds are removed. Volume is calculated by the product of that flow rate and the HRT. Traditional AD needs up to 20 days of HRT. However, anaerobic membrane biological reactor separates HRT from solid retention time; as a result, AnMBR could produce significantly high amount of biogas than traditional AD.

Table 13.3 Comparison between anaerobic and aerobic processes.

Anaerobic	Aerobic
Organic loading rate	
High loading rates: 10–40 kg COD/m^3-day (for high rate reactors, e.g. AF, UASB, E/FBR)	Low loading rates: 0.5–1.5 kg COD/m^3-day (for activated sludge rate process)
Biomass yield	
Low biomass yield: 0.05–0.15 kg VSS/kg COD (biomass yield is not constant but depends on types of substrates metabolized)	High biomass yield: 0.35–0.45 kg VSS/kg COD (biomass yield is fairly constant irrespective of types of substrates metabolized)
Specific substrate utilization rate	
High rate: 0.75–1.5 kg COD/kg VSS-day	Low rate: 0.15–0.75 kg COD/kg VSS-day
Start-up time	
Long start-up: 1–2 months for mesophilic : 2–3 months for thermophilic	Short start-up: 1–2 weeks

Table 13.4 General conditions for sludge digestion.

Parameter		Range of values
Temperature	Optimum	35°C
	General range of operation	29°C
pH	Optimum	7.0–7.1
	General limits	6.7–7.4
Gas production	Per kg of volatile solids added	0.5–0.8 m^3
	Per kg of volatile solids destroyed	1–1.1 m^3
Gas composition	Methane	65–69%
	Carbon dioxide	31–35%
	Hydrogen sulfide	Trace
Volatile acid concentration as acetic acid	Normal operation	200–800 mg/l
	Maximum	Approx. 2000 mg/l
Alkalinity concentrations as $CaCO_3$	Normal operation	2000–3500 mg/l

Conventional WWTPs in the United States are implementing the biological membrane reactors (BMRs) technology to improve the efficiency of biogas production, because BMRs reduce HRT from 2 to 15 days to just 2 h and separate the organic compounds in water through biological membrane (Figure 13.2). Another important technology is the upflow anaerobic sludge blanket (UASB) reactors that use inflow rates in a vertical direction (upflow) with a high sludge concentration maintained in the reactor to result in long solid retention times (Figure 13.3).

To reduce the carbon, water, energy, nutrient, and ecological footprints of the WWTP, different innovative alternatives can be designed to save energy and recover nutrients. For example, to reduce organic load, a rotating cloth sieve can be an effective way to filter solid which is then diverted to AD to produce methane. AD does not require aeration, which significantly reduces electricity consumption. However, the mass transfer efficiency is low. For nonbiodegradable organic solid, the combustion of dry organic compounds can produce heat. Waste organic matter should be viewed as an energetic compound rather than the pollutant as it was viewed in the past. WWTPs can be retrofitted to energy positive

Table 13.5 Loadings and detention times for heated anaerobic digesters.

	Conventional single stage (unmixed)	First-stage high rate (completely mixed)
Loading (kg/m^3/day of VS)	0.3–0.8	1.5–3
Detection time (days)	30–90	10–15
Capacity of digester (m^3/population/equivalent)		
Primary only	0.03–0.06	0.011–0.015
Primary and secondary	0.12–0.18	0.02–0.04
Volatile solids reduction (%)	50–70	50

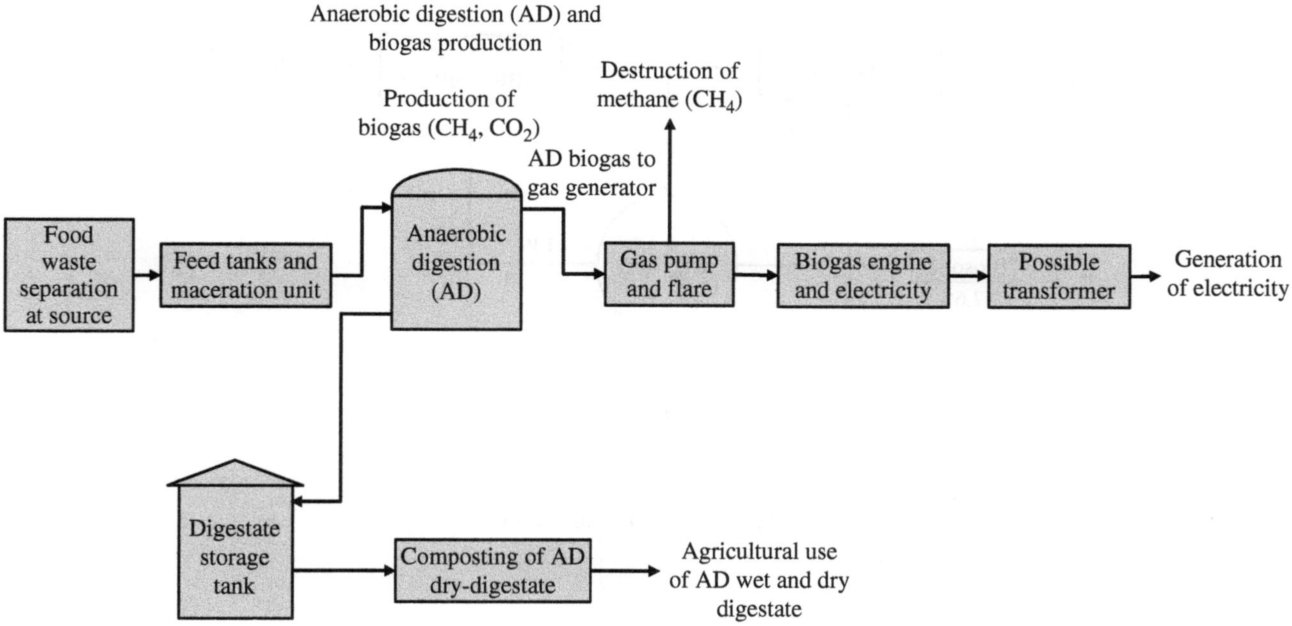

Figure 13.2 Anaerobic digestion process schematic.

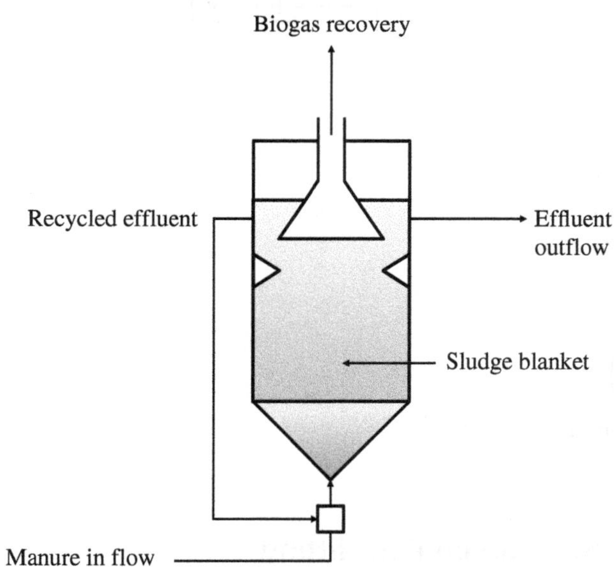

Figure 13.3 Upflow anaerobic sludge blanket.

if enough financial source were committed to renovate WWTPs. WW contains up to 99.9% water where the 0.01% reflects the biodegradable organics and nutrients. In addition to this, water contains pathogens and other toxic compounds that need to be removed. A typical WWTP uses 1200–2400 MJ per 1000 m^3, while a typical WW could contain up to 7000 MJ per 1000 m^3. Therefore, if the energy recovery efficiency is as low as 30%, a WWTP could reach energy neutral.

However, traditional activated sludge process employs aerobic bacteria to degrade organic matter in wastewater and produce four times larger amount of sludge than anaerobic digestion as follows

$$\text{Organic compound(s)} + O_2 + \text{bacteria} \rightarrow CO_2 + H_2O + \text{sludge (M.B.)}$$

Since sludge handling and disposal is very energy intensive and costly, the best retrofitting strategy is to divert organic matter from primary settler to anaerobic digester directly to produce more methane.

Figures 13.4 and 13.5 show that more than 60% BOD could be diverted directly to the anaerobic digester as the first stage. In the second stage, N is removed through anammox in a 10 MGD WWTP in Strass, Austria.

13.4 Best Practice Benchmark

In retrofitting the existing WWTP, it is very important to establish the unit consumption benchmark for each unit process of WWTP. Tarallo (2015) developed an excellent net zero energy guide for WWTP. The guide analyzed

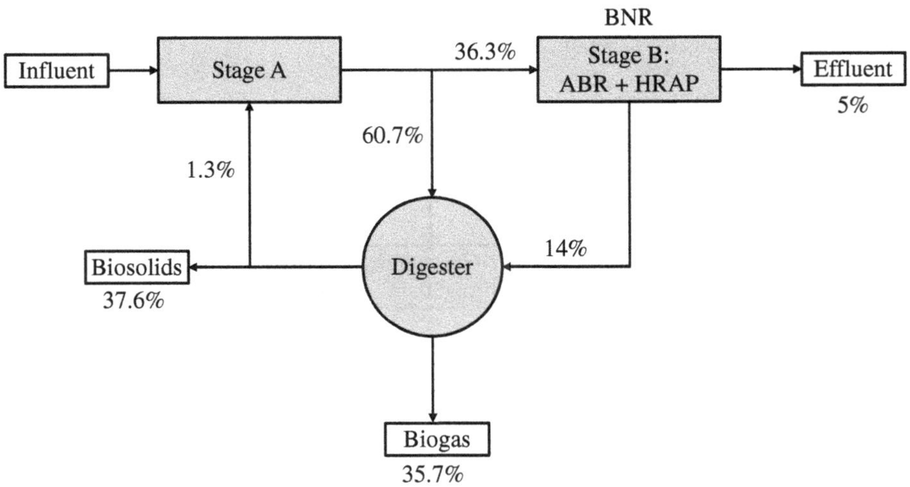

Figure 13.4 Treatment system (base case).

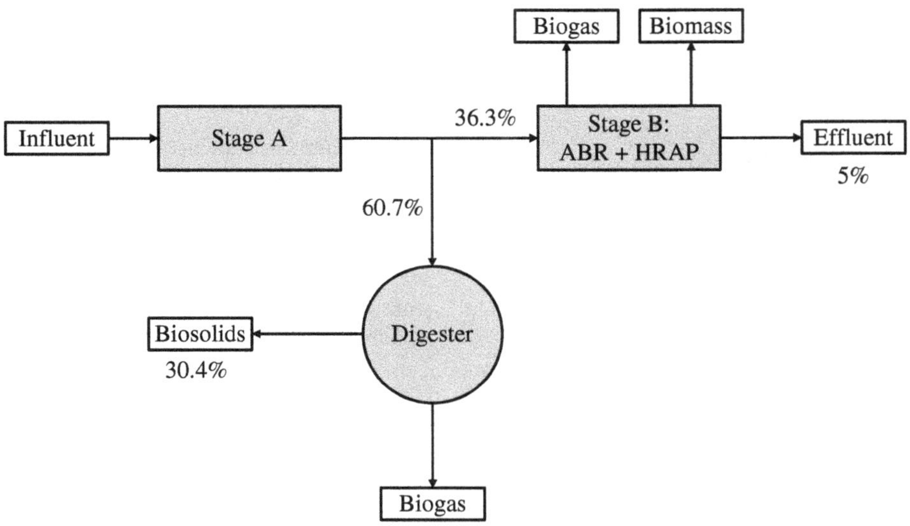

Figure 13.5 Treatment system (anaerobic stage B).

24 WWTPs in the state of New York and classified different WWTPs as typical versus the best practice according to the plant configuration. For an AS process as secondary treatment, configuration A1 in the guide (Tarallo, 2015) includes primary treatment, cothickening in gravity thickener, AD, and dewatering of sludge. According to each unit process, the typical and best performance in terms of energy efficiency is listed for 24 different treatment WWTPs. Table 13.6 shows a specific configuration referred as configuration A1.

Table 13.7 summarizes technical benchmark of 24 WWTPs in New York by statistically analyzing the energy consumption data reported by the Department of Environmental Protection of New York (Tarallo, 2015).

13.5 Green Retrofitting

13.5.1 Energy Auditing

The first step to improve plant energy efficacy is to conduct an energy auditing. Table 13.8 provides energy benchmark data for selected WW treatment processes. In general, biofilm reactor such as trickling filter consumes much less energy than AS WWTP and produces much less sludge due to higher solid retention time. However, the treatment capacity is limited for small WWTPs with flow rate less than 10 MGD. For AS system, nitrification could increase the energy consumption more than 30%.

Table 13.6 Energy consumption and production in configuration A1 (data from Tarallo, 2015).

Energy consumption and production	Units	Typical	Best practice	Percent
Major electricity using processes				
Biological reactor	kWh/MG	431	212	51
Odor control	kWh/MG	300	300	0
Influent pumping	kWh/MG	210	148	30
Anaerobic digestion	kWh/MG	129	11	92
Final clarifiers and RAS pumping	kWh/MG	103	75	27
Total facility electricity consumption and production				
Total electricity usage	kWh/MG	1416	937	34
Electricity generation	kWh/MG	0	0	—
Production as % of consumption (% neutrality)	%	0	0	—
Natural gas consumption				
Natural gas usage	MJ/MG	0	0	—

Table 13.7 Summary technical benchmark of WWTP unit processes (data from Tarallo, 2015).

MJ/MG Activity	Typical 10%	Best 10%	Typical 25%	Best 25%	Typical 50%	Best 50%	Typical 75%	Best 75%	Typical 90%	Best 90%	Typical Min	best Min	Typical Max	Best Max
Influent pump station									210	148	210	148	257	181
Screening and grit removal									188	188			188	188
Screening and grit removal (sidestream pump)	9	4	11	6	14	9.5	18.5	14	21	17	8.32	3.8	26.45	17.25
Primary clarifiers	84.5	95.5	85.5	96.3	88	98	89.5	100.5	91.5	102.5	84.4	95.8	93.2	102.9
Biological reactors and final clarifiers	46	36	49	39	54.5	45	80.5	51	125.1	106	45.9	37.8	126.9	119.9
Disinfection (discharge)	11.5	11.5	15.4	15	16.75	16.6	24.1	24	24.5	24.5	9.8	10.4	24.7	24.7
Gravity thickener	76	92	77	94	82	96	85.5	101	123	136	79.1	90.4	148.8	153.2
Gravity thickener (sidestream pump)	4	4.4	4.57	5.1	5	6.2	6.1	12	20.1	13.4	3.8	4.5	20.1	14.75
Mechanical thickener	43	33	46	36	50	47	79	100	108	125	44	37	132	132
Mechanical thickener (sidestream pump)	2.1	0.6	2.6	0.8	2.8	1.25	4	1.8	7	3.6	1.9	0.7	7.9	3.6
Lime stabilization (cake)	72	72	75	75	80	80	75	105	105	108	70.7	71.8	101.9	107.1
Anaerobic digestion	67	55	63	68	65	72	70	74	74	77.5	49.3	36.5	76.3	79
Anaerobic digestion (dewatering)	54	51	58	52	61	56	66.5	58	73	64	54.9	51.3	74.4	66.4
Flare	22.5	36	24	38	28	44	38	50	40	51	20.8	33.6	42.6	51.3
Boiler	17.4	15.2	17.8	15.4	19.7	16	24	19.5	24	20	17.5	14.2	25.4	20.7
Dewatering (cake)	49	48.5	52	50.5	56	54	62	60.5	68	65.5	49.6	49.4	69.8	69.5
Dewatering (sidestream pump)	4.53	2.1	4.3	2.2	5.1	2.4	5.8	2.7	11	4	17.5	2.14	11.1	5.9
Disinfection (discharge)	11.5	11.5	15.4	15	16.75	16.6	24.1	24	24.5	24.5	9.8	10.4	24.7	24.7
Gravity thickener	76	92	77	94	82	96	85.5	101	123	136	79.1	90.4	148.8	153.2
Gravity thickener (sidestream pump)	4	4.4	4.57	5.1	5	6.2	6.1	12	20.1	13.4	3.8	4.5	20.1	14.75
Mechanical thickener	43	33	46	36	50	47	79	100	108	125	44	37	132	132
Mechanical thickener (sidestream pump)	2.1	0.6	2.6	0.8	2.8	1.25	4	1.8	7	3.6	1.9	0.7	7.9	3.6
Lime stabilization (cake)	72	72	75	75	80	80	75	105	105	108	70.7	71.8	101.9	107.1

Table 13.8 Energy benchmarks for selected treatment processes.

Secondary treatment	Energy benchmark (kWh/MG)
Activated sludge with advanced treatment and nitrification	1900
Activated sludge with advanced treatment, no nitrification	1600
Activated sludge with no advanced treatment or nitrification	1400
No activated sludge, trickling filter	1000

Note: For WTPs the energy benchmark is approximately 1400 kWh/MG.

Figures 13.6 and 13.7 show the percentage of energy consumption in a typical WTP and a WWTP, respectively.

Figures 13.8 presents the energy balance in a typical WWTP.

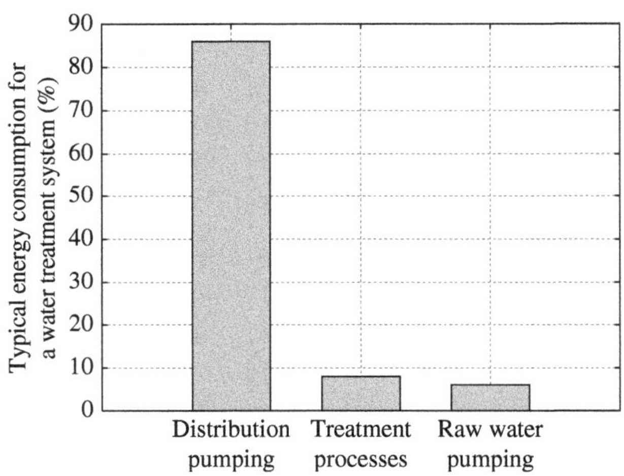

Figure 13.6 Typical energy consumption for a water treatment system.

Example 13.1

Given: Assume a facility charge of \$100.00/month. The demand charge is \$12/kW, and the metered demand is 6200 kW. During the month 1 000 000 kWh is consumed at a rate of \$0.10/kWh.

Find: The total charge for the month.

Solution:

Total Charge = \$100/month + \$12/kW*6200 +\$0.1/kWh*1 000 000 kWh/month = \$174,540/month

Table 13.9 Total charge for the month in Example 13.1.

Charge	Cost
Facility charge	\$100
Demand charge (\$12/kW × 6200 kW)	\$74 400
Electricity consumption charge (1 000 000 kWh×\$0.10/kWh)	\$100 000
Total	\$174 500

Example 13.2

Given: Assume that the minimum power factor (PF) set by the electric utility is 0.95. If the PF calculated based on the metered electricity usage is 0.86 and the metered demand is 1000 kW.

Find: Power demand.

Solution:

$$[1 + (\text{Set PF-metered PF})] \times \text{demand} = \text{power factor adjusted demand}$$

$$[1 + (0.95 - 0.86)] \times 1000 = 1090 \text{ kW}$$

Therefore, in this example the customer is paying a higher electricity bill for operating inefficiently. Rather than paying for 1000 kW, they must pay for 1090 kW.

Example 13.3

Given: A treatment plant has the following monthly data (Table 13.10).

Find: Average unit energy consumption for energy auditing.

Solution:

$$\text{Average energy benchmark (kWh/MG)} = 565520/571 = 990 \text{kWh/MG}$$

$$\text{Average energy benchmark (kWh/lb BOD)} = 565520/40360 = 14 \text{kWh/lb BOD}$$

$$\text{Average energy benchmark(kWh/lb TSS)} = 565520/40632 = 13.9 \text{kWh/lb TSS}$$

To estimate total energy content of domestic WW, it is very important to use the following daily per capita production data (Tables 13.11 and 13.12).

Table 13.13 can be used to convert different energy content equivalents: It shows that the highest energy recovery by anaerobic system is 7.3 kJ/g COD by anaerobic

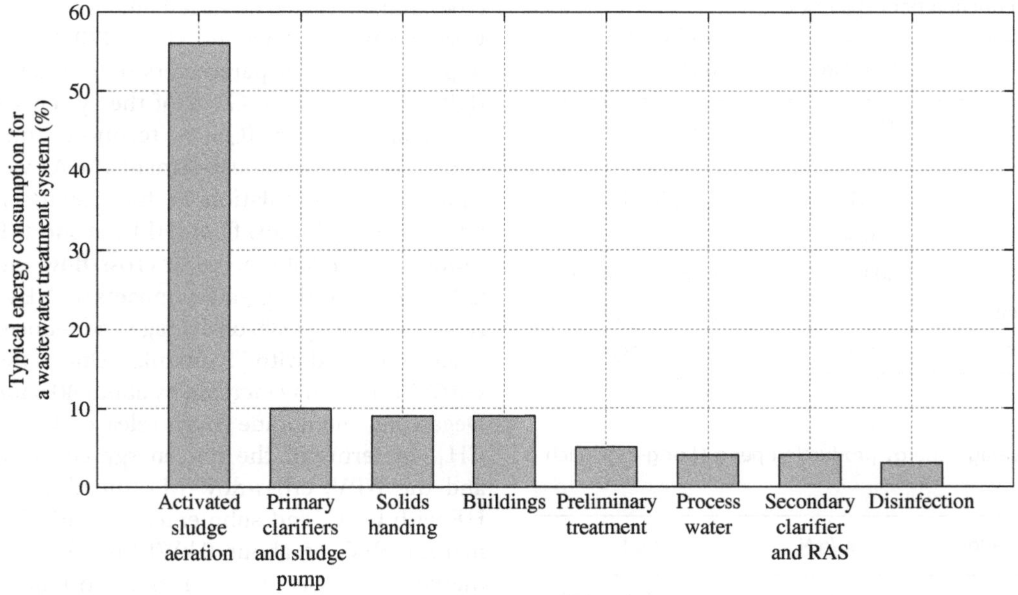

Figure 13.7 Typical energy consumption for a wastewater treatment plant.

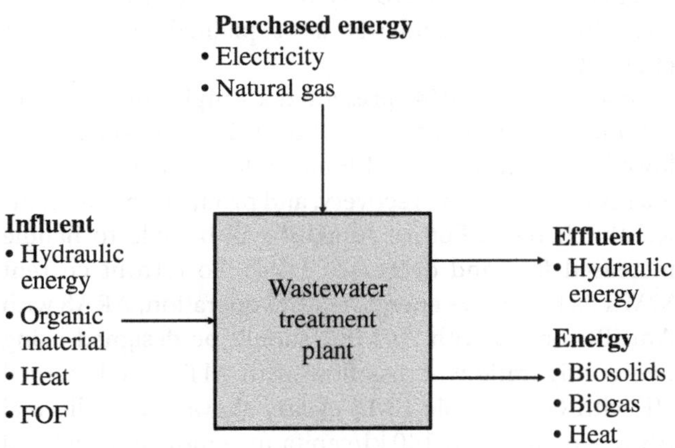

Figure 13.8 Wastewater treatment energy balance.

Table 13.10 Monthly energy consumption.

Month	Energy (kWh)	Flow (MG)	BOD (lb)	TSS (lb)
January	745 506	556	35 735	41 354
February	508 047	546	44 852	39 684
March	491 260	607	36 665	41 902
April	572 454	589	35 899	36 495
May	592 953	575	40 250	36 824
June	541 210	535	38 987	41 152
July	495 269	610	40 591	47 076
August	597 380	528	41 177	46 779
September	565 776	537	43 470	38 518
October	575 492	590	46 348	36 908
November	485 576	555	44 132	42 750
December	615 318	618	36 216	38 140
Average	565 520	571	40 360	40 632

baffled reactor (ABR) and the lowest is 0.48 kJ/g COD by microbial fuel cell (MFC).

Table 13.14 lists the percentages of COD removed by each technology and shows that Anaerobic Membrane Biological Reactor (AnMBR) is a much better process than the traditional AS process due to cost and environmental impacts. AnMBR not only improves discharge quality but also produces CH_4-rich biogas. AnMBR is one of the best alternatives to achieve energy-positive WWTP design because it integrates an AD with membrane filtration. It decouples solid retention time (SRT) from hydraulic retention time (HRT) that significantly reduces the reactor volume and environmental footprints. Under different temperatures and loading conditions, AnMBR is able to deliver reliable COD removal efficiency from 90 to 99% from WW. In CSTR or AF configurations of AnMBR systems, permeate pumping and membrane fouling control still consume significant energy. In terms of sensitivity, cost is the most sensitive to interest rate, flux, and membrane lifetime with median correlation of −0.50. Flux is the greatest contributor to global warming potential (GWP) due to tank volume and material increases with the flux. If specific gas demand increases, electricity consumption will increase.

After 150 combinations of different configurations such as reactor structure, fouling mitigation, membrane

Table 13.11 Production per capita per day.

Components	Standard	Unit
COD	180	g/capita/day
N	13	g/capita/day
P	2.1	g/capita/day
Flow rate	0.36	m^3/capita/day
COD	500	g/m^3/capita/day
N concentration	36	g N/m^3/capita/day
P concentration	5.8	g P/m^3/capita/day

Table 13.12 Average energy production per g-N or g-P by each technology.

Anaerobic technology	kJ/g N	kJ/g P
PBR	760 ± 250	640 ± 160
ATS	300 ± 160	1,600 ± 780
HRAP	210 ± 96	2,000 ± 1,300
CSTR		2,500 ± 1,600

Table 13.13 Standard energy content and conversion factors.

Component	Conversion factor	Unit
CH_4	803	kJ/mol CH_4
H_2	286	kJ/mol H_2
COD	13.9	kJ/g COD
kW/h	3600	kJ/kW/h

Table 13.14 Average percentage of energy recovery or COD removed.

	Average ± standard deviation	Average (min, max)
ABR	47.5 ± 4.5	90.3 (88.7, 92.5)
AnMBR	35.4 ± 26.8	86.7 (82, 90)
AFB	33.8 ± 12.9	82 (72, 89.7)
UASB	24.0 ± 11.4	67.6 (54, 85)
ASBR	17.7 ± 10.1	58.1 (33, 91)
MEC	14.3 ± 14.4	78 (33.7, 96.7)
MFC	1.6 ± 1.4	45.5 (19, 83)

Note: anaerobic baffled reactor (ABR) anaerobic membrane bioreactor (AnMBR) anaerobic fluidized bed reactor (AFB) upflow anaerobic sludge blanket (UASB) anaerobic sequencing batch reactor (ASBR) microbial electrolysis cell (MEC) microbial fuel cell (MFC)

type, and methane management for anaerobic systems were proposed, Shoener et al. (2016) quantified the impact of 22 input parameters on the net present value (NPV) and life cycle GWP of the systems. Quantitative sustainable design (QSD), techno-economic analysis (TEA), and life cycle assessment (LCA) were used with Monte Carlo simulation in their analysis. Membrane types included hollow fiber (HF), flat sheet (FS), and multitube (MT) in submerged or cross-flow with or without GAC. GAC has the greatest impacts on NPV and GWP. It reduces GWP by 58% on average and reduces cost by 25% when combined with FS modules. Due to the high cost of GAC, the cost may increase by about 40% for FS AnMBR. Degassing membrane may release 30–50% dissolved CH_4. In terms of the net energy consumption (NEC) and the NPV, cross-flow MT membranes, submerged HF with GAC, and submerged FS with GAC all have a neutral NEC of about 0 kWh/m^3. The NEC for submerged with gas sparging and cross-flow FS without GAC is about 0.8 and 1.6 kWh/m^3, respectively. Table 13.15 presents the desirable design standards for an energy-neutral AnMBR system together with the reasons for these standards as explained by Shoener et al. (2016).

Shoener et al. (2014) presented a simple conversion of aerobic biological nutrient removal (BNR) into ABR followed by High Rate Algal Pond (HRAP). The simulated energy consumption, recovery, and production are significantly changed. Future AnMBR will be able to handle increased flux and decreased HRT. To retrofit current WWTPs to achieve energy-neutral operation, ARAS with AnMBR or AB with AnMBR should be designed using two configurations, cross-flow with MT or submerged HF with GAC. Table 13.16 clearly shows the traditional AS waste energy of 140 kJ/capita in removing COD and NH_3–N. When an ABR is used to convert COD to CH_4 and HRAP is designed to produce biomass for fuel production, the total kJ/capita increased from barely 600 to 8900 kJ/capita.

13.5.1.1 Phototrophic System

The carbon limit appears to be important in an N-rich phototrophic system because cells require a C : N ratio of 6 : 1, while there is only 3 : 1 of C : N in WW.

There are four ways to achieve energy-neutral WWTP design:

1) Decrease the energy required for C, N, and P removal.
2) Increase biomass yield.
3) Reduce energy consumption during phototrophic process.
4) Increase fuel recovery efficacy.

Table 13.15 Desirable design standards for energy-neutral AnMBR.

AnMBR	Parameters	Standards	Reasons
CSTR	Cross-flow velocity	<0.5 m/s	Balance between fouling and hydrolysis
Submerged	Membrane cleaning	<7.5 m/h with HRT <1.6 h	Eliminate gas sparging
FS or MT	Membrane geometry	>0.02 m/s for cross-flow (m^3/h) and > 2000 m^2/m^3 material	Membrane material efficiency
Materials	Membrane lifetime	>10 years	Avoid fluorinated materials

Table 13.16 Desirable design standards for energy-neutral AnMBR.

	Stage B energy (kJ/capita)			
Condition	COD	NH_3–N	Digester biogas (kJ/capita)	Total (kJ/capita)
BNR	−140		740	600
ABR + HRAP	420	2200	600	3200
ABR + PBR	420	7900	600	8900

Comparing a phototrophic system with an anaerobic process based on maximum energy content of 13.9 kJ/g COD in WW, the potential to produce energy through a phototrophic system varies from 280 to 400% on the basis of cubic meters of WW treated. Indeed, energy recovery efficacy by an anaerobic process varies from 2 to 47% and 130 to 510% of energy production due to the use of nutrients for phototrophic cultivation. How biomass is converted to energy significantly affects energy yield as listed in Table 13.17 by Shoener et al. (2014). Therefore, following advanced technologies should be developed to promote synergy between effluent quality and energy production.

Enhanced primary sedimentation: To reduce the volume of AS, particulate organic compounds could be settled through enhanced coagulation. The settled sludge goes directly to the anaerobic sludge digester. As a result, maximum methane could be generated for energy recovery while significantly reducing sludge volume.

Codigestion with manure and biofuel: For anaerobic digestion, higher-influence COD leads to higher energy production and less energy consumption in terms of cubic meters of WW treated. For this reason, codigestion with manure and biofuel WW is the key to achieving energy positive. If AnMBR can add animal manure, CH_4 production can be tripled. When biodiesel is

Table 13.17 Biomass energy yield (kJ/g nutrient removed) by different energy conversion technologies.

Technology	Nutrient	HTL	Anaerobic digestion	Transesterification	Combustion
HRAP	N	75–160	32–160	34–100	90–130
	P	730–1600	32–1500	330–980	880–1300
PBR	N	270–590	120–580	120–370	330–500
	P	230–500	100–490	100–310	280–420
Stirred tank	P	900–1900	400–1900	400–1200	1100–1600
ATS	N	100–240	47–230	49–150	130–200
	P	580–1300	250–1200	260–790	700–1100

produced, the WW contains high energy waste known as glycerol after esterification. Glycerin can significantly increase CH_4 yield of anaerobic digestion by 300%.

Emerging technologies should be cost effective, predictable, controllable, and capable of achieving compliance with regulatory requirements. They could further contribute to energy-positive WWTP retrofitting.

Pyrolysis: Pyrolysis is a thermal process that uses high temperature and pressure in the absence of air to decompose organic material in the biosolids into gas, liquid, and solid (or char). Char is a carbon-based material formed by incomplete combustion of the organic material. The process yields a product that can be pelletized into solid fuel, which can be used with coal in power plants.

Gasification: Gasification is used to convert biosolids to either heat or electricity. It partially oxidizes organics (biomass) to carbon monoxide, hydrogen, and methane (syngas). New gasification technologies are emerging to use biosolids to generate electricity or hydrogen fuel.

Algae culture: Algae can recover nutrients from WW and use the biogenic carbon to generate biomass with high energy content. Algae-based WW solutions have the potential to manage carbon, phosphorus, and nitrogen life cycle issues and make significant net energy gains. Algal biofuels from WW could offset coal and imported oil, while nutrient recovery could reduce the use of fossil fuel in fertilizer manufacturing and improve environmental quality.

Microbial fuel cells and other microbial conversions: Anaerobic systems can be engineered to produce electrical energy by using microbial fuel cell technology to take advantage of the electrical potential in WW. Other microbial conversions of WW constituents to biofuel are possible, including butanol and methanol.

13.5.1.2 Renewable Energy for WWTPs

Renewable energy (RE) sources such as wind, solar, and hydropower for meeting the energy demands at WRRF can reduce facility requirements for external sources of energy. Large multi-megawatt wind turbines can be installed for utility-scale applications in the range of 100 kW to 1 MW. Photovoltaic systems are becoming more cost competitive with modular solar technologies. They can be used to power buildings as freestanding arrays of photovoltaic modules. Microhydropower technologies may also be installed at some WWTPs.

13.6 Sludge Processing and Disposal

The major by-product of all the biological process is the sludge produced. Due to no-regulations on sludge disposal legally, only 7% of sludge have been properly treated and disposed in China. In addition, to retrofit WWTPs, one of the key processes is anaerobic digestion of sludge to produce methane for heat and electricity. Waste sludge is classified based on the process from which it is produced. Raw or primary sludge is drawn from the primary sedimentation tanks and contains all the readily settleable matter from the WW plus 1% scum. It has high organic content, mainly fecal matter and food scraps, and is thus highly putrescible. Primary sludge accounts for 50–60% of the SS applied. Typical solid concentrations in raw primary sludge from settling municipal WW are 6–8%. The portion of volatile solids (VS) varies from 60 to 80% and therefore should be directly sent to anaerobic digester to generate methane. Waste-activated sludge is a mixed liquor solids that settle slowly, forming a rather bulky sludge of high water content. Therefore, AS contains only 0.4–1.5% SS with a volatile fraction of 0.7–0.8. Aerobically digested sludge contains 1.0–2.0% SS, and about 50% of the VS could be converted to gaseous end products. Trickling filter humus from secondary clarification from the biological filter media has a solids content of 46%. The solids content of anaerobically digested sludge ranges from 6 to 12%. The dry residue is 30–60% volatile depending on the mode of digester operation. Nonfilterable residue, expressed in milligrams per liter, is the solids content in sludge determined by filtration of the sample through a glass fiber filter. Since the filterable portion of sludge is very small, sludge solids are often determined by total residue on evaporation through drying in an oven at 103°C. Total solids (TS) is determined by evaporating to constant mass a measured amount of sludge, weighing the residue, and expressing this as a percentage of the original wet sludge mass. Sludge moisture content (P_M) is 100% minus percent of TS. VS content measured as the mass loss after incineration of the dried sludge solids from the TS test at 550°C reflects the organic content of the sludge. The fixed solid (FS) defines the mass of inorganic matter in the sludge that would require ultimate disposal after incineration. Since sludge commonly contains only between 1 and 10% solids by mass, its major component is water.

Sludge moisture–mass–volume relationships are very important calculation in sludge handling. For sludge containing 1% dry solids (moisture content, $P_M = 99\%$), 10 kg of dry solids is associated with 990 kg of water. The average density of water sludge solids is 1400 kg/m^3, and the density of water is 1000 kg/m^3. For 2% solids, the volume occupied with 20 kg of dry solids is 994 l. For any sludge, the volume, $\forall$, is given by

$$\forall = \frac{\text{Mass}}{\text{Density}}$$

If the sludge has a dry solids content less than 20% (that is, $P_M > 80\%$), then

$$\text{Density of wet sludge} \cong \text{density of water} = 1000\,\text{kg/m}^3$$

$$\text{Sludge volume, } \forall(\text{m}^3) = \frac{\text{Total mass of wet sludge}}{\text{Mass of dry sludge}} \times \frac{\text{mass of dry sludge}}{1000}$$

$$= \frac{100}{100 - P_M} \times \frac{\text{mass of dry sludge(kg)}}{1000} \quad (13.1)$$

For sludge of 2% solids content, 10 kg of dry solids would be contained in wet sludge with a volume given by

$$\forall = \frac{100}{100-98} \times \frac{10}{1000} = 0.5\,\text{m}^3$$

If sludge is concentrated so that the mass of dry solids, S_S, remains constant, but the moisture content is decreased from P_{M1} to P_{M2}, the ratio between the initial volume, $\forall_1$, and the final volume, $\forall_2$, is given by

$$\frac{\forall_1}{\forall_2} = \frac{100}{100 - P_{M1}} \times \frac{S_S}{1000} \times \frac{100 - P_{M2}}{100} \times \frac{1000}{S_S} = \frac{100 - P_{M2}}{100 - P_{M1}} \quad (13.2)$$

Thus, removing water from sludge of low solids content affords a dramatic reduction in volume. Doubling the solids content from 1 to 2% halves the volume of wet sludge, if the density of dry solids was assumed to be 1400 kg/m^3 for sludge of greater than 10% solids content and the liquid is assumed to be water (density 1000 kg/m^3) (Table 13.18).

Mechanically dewatered sludge varies in characteristics based on the type of sludge, chemical conditioning, and unit process employed. The density of dewatered cakes ranges from 15 to 40%. Waste solids production in primary and secondary processing can be estimated using the following calculation:

$$W_{ts} = W_{ps} + W_{ss} \quad (13.3)$$

Table 13.18 Density and volume of sludge.

Sludge solids content (%)	kg sludge/kg dry solids	m^3 sludge/ton dry solids
1	100	100
2	50	50
5	20	20
10	10	9.7
15	6.7	6.4
20	5.0	4.7
30	3.3	3.0
40	2.5	2.2

where

W_{ts} = total dry solids (kg/day)
W_{ps} = raw primary solids, kg/day = $f \times SS \times Q$
W_{ss} = secondary biological solids, kg/day = $(k \times BOD + 0.27SS)Q$
f = fraction of SS removed in primary settling
SS = suspended solids in unsettled WW (mg/l)
Q = daily WW flow (Ml/day)
k = fraction of applied BOD that appears as excess biological growth in waste-activated sludge or filter humus, assuming about 30 mg/l of BOD and SS remaining in the secondary effluent
BOD = concentration in applied WW (mg/l)

Example 13.4
Given: Waste-activated sludge with 0.4% solids content is gravity thickened to 2.0% with a removal efficiency of 95%.
Find: The quantity of underflow per 1.0 m^3 of slurry applied and the concentration of solids in the overflow. Assume a specific gravity of 1.05 for the dry solids.
Solution:

$$\text{Solids applied} = 1.0\,\text{m}^3 \times 1000\text{kg/m}^3 \times 0.004 = 4.0\text{kg}$$

$$\text{Underflow solids} = 0.95 \times 4 = 3.8\,\text{kg}$$

$$\text{Relative density}, \rho_u, \text{ of underflow} = \frac{98+2}{(98/1.0)+(2/1.05)} = 1.001$$

Volume of underflow (Equation 13.1)

$$= \frac{100 \times \text{mass of dry sludge (kg)}}{(100 - P_M) \times \text{density of water}\left(\frac{\text{kg}}{\text{m}^3}\right) \times \rho_u}$$

$$= \frac{3.8}{0.98 \times 100 \times 1.001} = 0.0039\ \text{m}^3$$

$$\text{Volume of overflow} = 1.0 - 0.0039 = 0.9961\ \text{m}^3$$

The concentration of solids in the overflow

$$= \frac{0.05 \times 4 \times 100}{0.9961 \times 1000 \times 1.0} = 0.02\% = 200\text{mg/l}$$

13.6.1 Design of Wastewater Sludge Thickeners

Example 13.5
Given: The daily quantity of primary sludge from a water treatment plant contains 500 kg of solids at a concentration of 4.5%. Size a gravity thickener based on a solids loading of 50 kg/m^2/day.

Find: Calculate the daily volumes of applied and thickened sludge, assuming an underflow of 8.0 and 95% solids capture. If the blanket of consolidation sludge in the tank has a depth of 1 m, estimate the solids retention time (Table 13.19).

Solution:

$$\text{Tank area required} = \frac{500\,\text{kg/day}}{50\,\text{kg/m}^2\text{/day}} = 10\,\text{m}^2$$

$$\text{Diameter} = \left(\frac{10 \times 4}{\pi}\right)^{0.5} = 3.5\,\text{m}$$

Use a depth of 3 m:

$$\text{Volume of applied sludge} = \frac{500}{0.045 \times 1000} = 11\,\text{m}^3\text{/day}$$

$$\text{Volume of thickened sludge} = \frac{500 \times 0.95}{(8.0/100) \times 1000} = 5.94\,\text{m}^3\text{/day}$$

$$\text{Solids retention time} = \frac{1 \times 10 \times 24}{5.94} = 40\,\text{h}$$

13.6.2 Suspended Solids Removal Efficiency

A relatively easy method of estimating solids capture by sludge thickening or dewatering unit is to measure solids concentrations in the process flows. The solids mass balance is given in Equations (13.4) and (13.5):

$$M_S = M_R + M_C \quad (13.4)$$

$$Q_S = Q_R + Q_C \quad (13.5)$$

where

M_S, Q_S = mass of solids and quantity of flow of feed sludge
M_R, Q_R = mass of solids and quantity of returns flow
M_C, Q_C = mass of solids and quantity of flow of thickened sludge or cake

The specific gravity of all flows can be assumed to be 1.0. Therefore, the mass of solids M in a process flow stream is the rate of flow Q times the solids concentration S:

$$M_S = Q \times S \quad (13.6)$$

Combining these equations results in the following relationships:

$$\frac{M_C}{M_S} = \frac{S_C(S_S - S_R)}{S_S(S_C - S_R)} \quad (13.7)$$

where

M_C/M_S = fraction of solids removal (solids capture)
S_C = solids concentration in thickened sludge or cake
S_S = solids concentration in feed sludge
S_R = solids concentration in return flow

$$\frac{Q_R}{Q_S} = \frac{S_C - S_S}{S_C - S_R} \quad (13.8)$$

where

Q_R/Q_S = fraction of feed sludge appearing as return flow

Example 13.6
Given: A scroll centrifuge dewaters an alum–lime sludge containing 8.0% solids at a feed rate of 5 m^3/m. The cake produced has a solids concentration of 55% and the centrate contains 9000 mg/l.
Find: Calculate the solids removal efficiency and the centrate flow.

Solution:

From Equation (13.9),

$$\text{Solids removal} = \frac{M_C}{M_S} = \frac{55(8.0 - 0.9)}{8.0(55 - 0.9)} = 90\% \quad (13.9)$$

Substituting into Equation (13.10),

$$Q_R = 5\,\text{m}^3 \times \frac{55 - 8.0}{55 - 0.9} = 4.34\,\text{m}^3\text{/h} \quad (13.10)$$

The three functions of a single-stage floating-cover digester are (i) anaerobic digestion of the volatile solids,

Table 13.19 Typical results from belt filter press dewatering of polymer flocculated wastewater sludge.

Type of sludge	Feed solids (%)	Hydraulic loading (gpm/m)	Solids loading (lb/m/h)	Cake solids (%)	Polymer dosage (lb/ton)
Anaerobically digested primary only	4–8	40–50	1000–1600	25–35	3–6
Anaerobically digested primary plus waste activated	2–5	40–50	500–1000	15–26	6–12
Aerobically digested without primary	1–3	30–45	200–500	11–22	8–14
Raw primary and waste activated	3–6	40–50	800–1200	16–25	4–10
Thickened waste activated	3–5	40–50	800–1000	14–20	6–8
Extended aeration waste activated	1–3	30–45	200–500	11–22	8–14
Heat treated primary plus waste activated	1–4	35–50	1000–1800	30–40	1–2

(ii) gravity thickening, and (iii) storage of the digested sludge (Table 13.20).

13.6.3 Anaerobic Digester Capacity

A maximum of 0.6 kg/m^3/day (1.3 kg/m^3/day) VS could be loaded for single-stage operation of AD. These loadings assume that the raw sludge is derived from domestic WW, the digestion temperature is in the range of 29–30°C, the VS reduction is 40–50%, and the digested sludge is removed frequently from the digester. The operation of a digester can be monitored by plotting the daily gas production per unit of raw sludge fed, the percentage of carbon dioxide in the digestion gases, or the concentration of volatile acids in the digesting sludge. A reduction in gas production, an increase in carbon dioxide percentage, and a rise in volatile acids concentration all indicate reduced activity of the acid-splitting methane-forming bacteria.

13.6.4 Aerobic Sludge Digestion

Aerobic sludge digestion (ASD) is to stabilize waste sludge solids by long-term aeration, thereby reducing the BOD and destroying VS. The most common application of ASD is in handling waste-activated sludge. Since dilute solids suspensions have a low rate of oxygen demand, the need for effective mixing rather than microbial metabolism usually governs the air supply required. The volume of air supplied for ASD is normally in the range of 0.9–1.8 m^3/m^3 of digester. Design criteria vary with the type of AS system, BOD loading, and the means provided for ultimate disposal of the digested sludge. Small AS plants without primary sedimentation are generally provided with 57–85 l of aerobic digester volume per design PE of the plant. For stabilizing waste-activated sludge with an SS concentration of 1.0% or less, the VS loading should be limited to 0.64 kg/m^3/day, and the aeration period 200–300 degree days computed by multiplying the digesting temperature in degrees Celsius by the sludge age. This equates to a minimum aeration period of 10 days at 20°C or 20 days at 10°C. VS and BOD reductions at these loadings are in the range of 30–50%, and the digested sludge can be disposed of without causing odors or other nuisance conditions.

Table 13.20 Loadings and detention times for heated anaerobic digesters.

	Conventional single stage (unmixed)	First-stage high rate (completely mixed)
Loading (kg/m^3/day of VS)	0.3–0.8	1.5–3
Detection time (days)	30–90	10–15
Capacity of digester (m^3/population/ equivalent)		
Primary only	0.03–0.06	0.011–0.015
Primary and secondary	0.12–0.18	0.02–0.04
Volatile solids reduction (%)	50–70	50

Example 13.7

Given: A town has 2000 people with an aerobic digester of 150 m^3 and BOD load on plant = 182 kg/day.

Find: Calculate the cubic meters of aerobic digester volume provided per design population equivalent, and estimate the volatile solids loading on the aerobic digester.

Solution:

$$\text{Volume provided} = \frac{150 \times 1000}{2000} = 75\ \text{l of digester volume/ design population equivalent}$$

Assume the estimated sludge produced per day is 0.42 g of SS/g of BOD load. For the digester volume of 150 m^3 and assuming 70% of the SS as volatile, the estimated volatile solids loading applied to the aerobic digester is

$$\frac{182 \times 0.70 \times 0.42}{150} = 0.36\,\text{kg/m}^3\text{/day}$$

Example 13.8

Given: Municipal wastewater contains 200 mg/l of BOD, 220 mg/l of suspended solids, 35 mg/l of nitrogen, and 0.16 mg/l of cadmium. Wastewater processing is primary sedimentation, and secondary activated sludge with the waste sludge is anaerobically digested prior to spreading on agricultural land. The plant effluent is discharged into a river. Based on operational data, 25% of the influent cadmium appears in the 100 mg of digested sludge solids produced per liter of wastewater processed. The fertilizer value of the dried sludge solids is 4.0% available nitrogen.

Find:

1) Calculate the concentration of cadmium in the plant effluent in milligrams per liter. If the concentration limit of cadmium in the river waste is 0.0013 mg/l to protect aquatic life, what dilution ratio of plant effluent to cadmium-free river water is needed?
2) Calculate the concentration of cadmium in the dried sludge solids in units of kilograms per ton. The digested sludge is being applied to land cultivated for growing soybeans. The recommended annual

nitrogen application is 200 kg/ha, and the maximum allowable cadmium loading is 0.5 kg/ha/year. Based on these criteria, compute the maximum sludge application in units of tons of dry solids per hectare.

3) Ferric chloride (waste pickle liquor), added to the wastewater for increasing phosphorus removal, also increases the cadmium removal to 70%. Calculate the data requested in question 1 through 2 for this degree of cadmium removal.

Solution:

1) The cadmium in effluent amounts to $0.75 \times 0.16 = 0.12$ mg/l. Assume the dilution ratio is X (i.e. when the effluent flow is 1, the river flow is X). Then

$$X \times 0 + 1 \times 0.12 = (1 + X) \times 0.0013$$

$$X = \text{Flow of river/flow of wastewater discharge} = 91.3$$

2) The concentration of cadmium in sludge solids is

$$= \frac{0.25 \times 0.16\,\text{mg/l} \times 1000\,\text{kg/ton}}{100\,\text{mg/l}} = 0.4\,\text{kg/ton}$$

The nitrogen content of the sludge solids equals

$$= \frac{100\text{mg/l} \times 0.04 \times 1000\text{kg/ton}}{100\,\text{mg/l}} = 40\,\text{kg/ton}$$

The maximum allowable sludge application rate is

$$\text{based on cadmium} = \frac{0.5}{0.4} = 1.25\,\text{ton/ha}$$

$$\text{based on nitrogen} = \frac{200}{40} = 5\,\text{ton/ha (use this loading)}$$

3) The cadmium in effluent is $0.30 \times 0.16 = 0.048$ mg/l. Dilution ratio required = 35.9.

Concentration of cadmium in sludge = 1.12 kg/ton. The maximum allowable sludge application rate is

$$\text{based on nitrogen} = 5\,\text{ton/ha(use this loading)}$$

$$\text{based on cadmium} = \frac{0.5}{1.12} = 0.45\,\text{ton/ha}$$

These computations illustrate that treatment for the removal of phosphate by chemical precipitation also increases the removal of cadmium. While the effluent quality then improves, the contamination of the sludge increases.

13.6.5 Retrofitting Strategies of WWTPs

It took more than ten years to retrofit the Strass WWTP in Austria from energy negative to energy positive in 2005. The major strategies were the planned retrofitting projects to reduce its energy footprint to 0.35 kWh/m^3 while innovative technologies are added. For Miami-Dade WASD to retrofit their three WWTPs to energy positive, Table 13.21 lists important technologies as well as the target time periods.

13.7 Green Remediation

To enhance the environmental, societal, and economic benefits of federal remediation programs, the US EPA provides sustainable strategies, approaches, and applications in two aspects:

1) Maximize sustainability, reduce energy and water use as well as toxic air emissions, promote carbon neutrality and industrial material reuse and recycling, and protect and preserve the land.
2) Efficient use of energy and water, use of EPA-designated recovered materials and bio-based materials, solid waste prevention, energy and water conservation, and use of alternative fuels.

Table 13.21 Retrofitting projects for three WWTPs of Miami-Dade Water and Sewer Department

Projects	Benefits	Targets
Diverted COD from primary settler to anaerobic digester	Decrease sludge production by 20% Increase biogas production by 10%	5 years
Anaerobic oxidation of ammonium (anammox)	Reduce sludge production by half Reduce aeration electricity by half	10 years
Fat–oil–grease and food wastes to increase biogas production (currently, food wastes are not diverted)	Increase biogas production 20–50% Decrease BOD to WWTPs	3 years
Membrane bioreactor to increase WWTP capacity	Increase WWTP capacity Increase solid retention time	5 years
Decentralized WWTPs	Increase irrigation rate	10 years
UV disinfection	Increase compliance of disinfection by-product rules Reduce operation costs	10 years
Leachate separated treated	Reduction of BOD and COD to WWTP Reduce toxicity of leachate Decrease	10 years

Green remediation considers all environmental effects of remedy implementation and incorporates alternatives to minimize the environmental footprint of cleanup actions. The environmental footprints of a cleanup project are to establish the parameters (metrics) to be quantified and to establish a straightforward process (methodology) for quantifying those metrics. Metric refers to a project parameter for which a quantitative value might be derived mathematically. EPA's five core elements of green remediation are energy, air and atmosphere, water, land and ecosystem, and materials and waste. Quantitative information on footprint reductions can be achieved by applying EPA's green remediation best management practices (BMPs). These BMPs should provide excellent technical reference data for China's huge markets of 500 billion USD in water and soil remediation in next decade. To ensure minimal footprints, green remediation metrics, quantification methods of these metrics, and the interpretation of the results are presented in the following sections.

13.7.1 Green Remediation Metrics and Methodology

EPA's green remediation metrics encourage environmentally friendly engineering design that applies to each of the core elements of greener cleanups and emphasizes parameters that a remedy project team likely has the ability to change. Materials and waste metrics include the following items:

1) Refined materials used on-site.
2) Percentage of refined materials from recycled or waste material.
3) Unrefined materials used on-site.
4) Percentage of unrefined materials from recycled or waste material.
5) On-site hazardous waste generated.
6) On-site nonhazardous waste generated.
7) Percentage of total potential on-site waste that is recycled or reused (Table 13.22).

Table 13.22 Green remediation metrics and their components.

Media of metrics	**Category**	**Metrics**
Water	Source	• Water from the public potable water supply • Extracted groundwater from each local aquifer • Surface water • Reclaimed water • Collected or diverted stormwater • Water from the public potable water supply • Extracted groundwater from each local aquifer • Surface water • Reclaimed water • Collected or diverted stormwater
	Use	• Equipment decontamination • Extraction and treatment • Injection for plume migration control • Chemical blending
	Fate	• Reuse in a public or domestic water supply • Use as industrial process water • Discharge to groundwater (specify the aquifer) • Discharge to fresh surface water • Irrigation • Discharges to brackish or saline water • Discharge to the atmosphere (i.e. as water vapor) • Discharge to the sanitary sewer
Energy		• Total energy use • Total energy voluntarily derived from renewable resources – On-site generation or use and biodiesel use – Voluntary purchase of renewable electricity Voluntary purchase of RECs
Air		• On-site NO_x, SO_x, and PM10 emissions • On-site HAP emissions • Total NO_x, SO_x, and PM10 emissions • Total HAP emissions • Total GHG emissions

The methodology consists of seven steps to quantify the metrics:

1) Set goals and scope of analysis.
2) Gather remedy information.
3) Quantify on-site materials and waste metrics.
4) Quantify on-site water metrics.
5) Quantify energy and air metrics.
6) Qualitatively describe affected ecosystem services.
7) Present results.

The results can be analyzed to identify large contributors to the metrics and evaluate opportunities for footprint reduction. In general, the methodology utilizes publicly available information and can be implemented using standard spreadsheet software developed by the US EPA (2012).

Step 1: Set Goals and Scope of Analysis

The goal of the analysis typically varies with the remedial stage and with other site-specific factors. The typical goals of footprint quantification, based on remedy stage, are:

1) Development of remedy alternatives is to identify components of various remedy alternatives that are large contributors to footprints.
2) Remedial design and remedy optimization is to:

a) Identify remedy components that are large contributors to the remedy footprint.
b) View remedy from different perspective to identify opportunities for improvement.
c) Identify design components that are large contributors to footprints during construction or during remedy operation.
d) Refine the conceptual site model that, if addressed, might help reduce footprints (e.g. improved characterization that refines the known contamination source area).
e) Identify potential opportunities for footprint reductions.

3) Quantify and document emissions.

The scope of the analysis significantly influences the data to be gathered in step 2 of the methodology. The boundary of study is also established at the outset of a footprint analysis. The study boundary is the conceptual boundary around a system or project that determines the processes that are included and the processes that are excluded.

Step 2: Gather Remedy Information

This step of the footprint methodology involves collecting information about remedy design, construction, and operation. The information includes design parameters, activities conducted, and types of materials used. The information will likely be relatively uncertain during the early design stage but will increase in certainty as the design stage proceeds and as the remedy is implemented. Information are similar to the design documents, remedial action progress reports, or conducting remedy optimization evaluations. Two types of screening limits of materials should be set and documented based on a specified percentage of the maximum contributor to a particular metric or based on a specified magnitude for a particular metric. Both types of limits should be applied to each of the green remediation metrics or grouping of metrics because an item that is a negligible contributor for one metric may be a significant contributor for another metric.

Step 3: Quantify On-site Materials and Waste Metrics

Quantification of materials on- or off-sites is an accounting of the significant materials used, the recycled content of those materials, various waste generated, and the portion of that waste that is recycled or reused. The primary challenges associated with this step include (i) converting the various quantities of materials and wastes into common units for each of the metrics, (ii) recognizing recycling and reuse when it occurs, and (iii) presenting information in a clear and concise manner. Table 13.23 provides materials typically involved in various remedy activities.

Table 13.23 Materials used in remediation projects.

Category	Materials	Examples
Wells	Grout	
	Well casing	PVC Steel
	Sand/gravel/bentonite	
Piping	Steel	
	Plastic	PVC HPDE Other
Building and foundations	Concrete	
	Steel	
Cutoff walls	Sheet pile	PVC Steel
	Slurry	Bentonite Grout Fill
Geomembranes/liners/caps	PVC, HDPE	
	Clay	
	Asphalt	
Treatment chemicals/materials	GAC	
	Oxidants	Hydrogen peroxide
	Acids	Sulfuric acid
	Bases	Sodium hydroxide
	Flocculants	Ferric chloride
	Polymer	
	Well rehabilitation chemicals	
Injection reagents	Oxidants	Permanganate
	Acids	Sulfuric acids
	Catalysts	Ferrous sulfate
	Zerovalent iron	
	Nutrients, electron donors	Vegetable oil
Other	Fertilizers	
	Mulch/compost	
	Process equipment	PVC HDPE Steel

Rules of Thumb and General Assistance for Quantifying Materials Use

Quantifying materials use in tons can be challenging during any phase of a remedy because many materials are ordered or described in different units, such as feet of pipe, square feet of plastic membrane, or gallons of chemical reagents. Some materials use can also be difficult to quantify if many small components are used. In addition, some commonly used construction materials include the actual recycled content, but the recycled content is not necessarily communicated to the purchaser. EPA (2012) documents provide (i) densities of common materials, (ii) approximate material content of commonly used aqueous chemical solutions, (iii) estimated materials use for piping and wiring runs, (iv) estimated materials use for well installation, (v) estimated materials use for process equipment and building construction, and (vi) reasonable assumptions for recycled content of steel, concrete, and asphalt, which are construction materials with a significant percentage of recycled content.

The following waste streams are commonly associated with remediation and may significantly contribute to on-site waste generation:

1) Drill cuttings and used drilling mud.
2) Excavated soil for off-site disposal.
3) Construction debris.
4) Treatment plant residue.
5) Spent GAC.

The following waste streams are also commonly associated with remediation but may be too small to merit inclusion in the waste footprint. These two items include package and disposal personal protection equipment.

Waste streams that exceed the screening limits would be included in the calculation of the footprint. Waste streams that fall below these limits would be excluded from the calculation of the footprint and should be described with an appropriate level of documentation and can be addressed through qualitative BMPs for green remediation.

Step 4: Quantify On-site Water Metrics

The methodology involves quantifying on-site water use with an emphasis on reducing on-site water use and returning water of high quality for reuse. Activities in this step include:

1) Distinguishing between various local and on-site water sources.
2) Identifying the quality of the water used from each source.
3) Quantifying the amount of water used from each source.
4) Identifying the use of the water from each source.
5) Distinguishing between various fates of the water from each source after use.

These items can be explained in more detail as follows:

Water resources – Common sources of water that may be used on-site in association with a remedy include potable water from a public water supply system, multiple groundwater resources, multiple surface water resources, captured stormwater, and reclaimed water.

Water quality – The quality of the water used is an important aspect of the water footprint because the potential beneficial uses of the water by the local community, economy, and environment are directly related to the water quality.

Volume used – The amount of water used from the indicated water resource refers to the volume that is used or diverted from that resource over the life cycle of a remedy.

Potential water uses – The uses of the water vary from remedy to remedy. Common uses of water resources associated with remediation include the following:

1) Extraction for treatment (e.g. in a groundwater P&T system).
2) Extraction for testing (e.g. for long-term pumping tests).
3) Injection for plume migration control.
4) Blending and injecting reagents for *in situ* bioremediation.
5) Blending and injecting reagents for *in situ* chemical oxidation (ISCO).
6) Blending of chemicals for treatment plant operation.
7) Makeup or backwash water from treatment plant operation.
8) Mixing grout or slurry.
9) Evapotranspiration from phytoremediation.
10) Purge water from sampling.
11) Equipment decontamination.
12) Dust control and general construction.

Potential fates of used water – Common fates of water after water use during remediation include the following:

1) Discharge to public water supply.
2) Discharge for use as industrial process water.
3) Discharge to groundwater.
4) Discharge to surface water.
5) Discharge for use as irrigation.
6) Discharge to a brackish or saline water body.
7) Discharge to the atmosphere through evapotranspiration or vaporization.

An on-site water metric is established for each combination of a water source, water use, and water fate that involves significant water use. Water uses from the various sources that fall below these limits would be excluded from the footprint but would be documented. These uses can be addressed through qualitative BMPs for green remediation.

Chemical solutions – Chemical solutions brought to the site from a chemical vendor can contain a significant amount of water. In the absence of other information, it is appropriate to consider the water content of the solution to be public water that is used on-site at the cleanup project.

Step 5: Quantify Energy and Air Metrics

The energy and air metrics as calculated by the methodology attempt to account for as much of the life cycle of the material or activity as is practicable. The methodology purposefully does not include energy use and air emissions resulting from the following aspects of the remedy:

1) The manufacturing of rental or temporary equipment for the site.
2) Office-related work and other off-site personnel activities (other than routine transportation between the site and the office).
3) Items or activities that are nonadditional.
4) Items or activities external to the remediation process associated with redeveloping a property.

Step 5 is divided into three parts to facilitate organizing, presenting, and calculating the energy and emission footprints.

Part 1: Inventory Remedy Transportation, Equipment Use, Materials and Waste, Public Water Use, and Off-site Services

Much of the information required for this inventory is based on the information previously gathered for the materials, waste, and water footprints. Additional information is gathered on transportation, equipment use, and off-site services required for the remedy.

In an effort to streamline the footprint analysis process, it is appropriate to screen the information to be included in the analysis. The material use, waste disposal, and public water use described in steps 2 and 3 can be included directly. The transportation, equipment use, and other off-site services are screened against the limits discussed. Items and activities that exceed the limits would be included, and items and activities that fall below the limits would be excluded. Items and activities that are excluded should be described with an appropriate level of documentation; these aspects can be addressed through qualitative BMPs for green remediation.

Part 2: Determine Remedy Energy Use

The second part of step 5 is to organize and refine the inventory information that was developed in step 2 and to use that information to determine the amount of energy involved. This part is divided into the following processes:

1) Determining fuel use:
   a) Converting personnel transportation into fuel use.
   b) Converting on-site equipment use into fuel use.
   c) Converting transportation of equipment, materials, and waste into fuel use.
   d) Gathering information about actual fuel use (when such information is available).
2) Determining electricity use:
   a) Converting electrical equipment and power requirements into electricity use.
   b) Gathering information about actual electricity use (when such information is available).

Part 3: Convert the Remedy Inventory into Energy and Air Metrics

The third part of step 5 is to convert the fuel, electricity, materials, and off-site services involved in the remedy into energy and air metrics. Footprint conversion factors are used for this purpose. One of the challenges of conducting a footprint analysis is establishing accurate conversion factors for the fuels, electricity, materials, and services used in a remedy.

Conversion factors for energy and air emissions from off-site electricity generation are dependent on the specific grid mix supplied to the site and are determined on a site-by-site basis. The type of information used to determine the conversion factors is generally available from the individual electricity service providers. Otherwise, related information at US state level is available online from the US Energy Information Administration (http://www.eia.gov/state) or can be accessed for US regions defined in the federally sponsored Emissions and Generation Resource Integrated Database (eGRID) (http://www.epa.gov/egrid).

Step 6: Qualitatively Describe Affected Ecosystem Services

The methodology suggests the use of qualitative descriptions of the effects of a remedy on land and ecosystem services such as nutrient uptake and erosion control. Concepts related to ecosystem services are available online from EPA's Ecosystem Services Research Program at https://www.epa.gov/eco-research/ecosystem-services.

Step 7: Present Results

In addition to the information gained through each step, documentation of a footprint analysis should be provided for the screening process, analytical assumptions, possible areas of uncertainty, and specific ideas of how to reduce the footprint.

Interpreting Footprint Results

Interpreting the results of a footprint analysis is influenced by a number of considerations, including the quality of the data used in the analysis, the trade-offs among footprint metrics, and the magnitude of the footprint. There are three main factors that influence data quality:

1) The quality of the remedy information input into the footprint analysis.
2) The accuracy or appropriateness of the formulas for quantifying the use of electricity, fuel, water, materials, and off-site services involved in a specific remedy.
3) The accuracy of the conversion factors used to convert the remedy inventory into green remediation parameters.

The methodology helps quantify metrics for more than 15 different parameters in 4 different green remediation core elements. Some of the parameters are local in nature (e.g. on-site HAP emissions or local water use), and others are more regional (e.g. NOx/SOx/PM10 emissions) or global in nature (e.g. total GHG emissions). A potential modification to an existing remedy may increase the values for some metrics and decrease the values for others.

Two main questions regarding the magnitude of a footprint that may arise when interpreting the results of a footprint analysis are (i) what is considered a large footprint or footprint reduction, and (ii) what is considered a significant difference between footprints of two or more potential modifications under consideration?

Data Quality: Transportation

Data quality of transportation and electricity generation are illustrated through example 13.9 and 13.10, respectively.

Example 13.9

Given: Consider the example of transporting 1.5 tons of emulsified vegetable oil (equivalent to ~400 gal) 500 miles via specialty freight and common freight. The fuel economies used are from Table 13.24, and assume that the weight of the freight negligibly affects fuel economy relative to other factors that affect fuel economy.

Table 13.24 Fuel efficiency.

Vehicle type	Fuel efficiency Units	Fuel efficiency Value
Truck	mpg	6
Truck common freight	gptm	0.029
Train	gptm	0.0025
Barge	gptm	0.0047
Aircraft	gptm	0.15

Find: Amount of diesel in gallons needed for specialty and common freights, respectively.

Solution:

1) Specialty freight (includes empty return trip).

Delivery trip	500 miles ÷ 6 miles /gal = 83 gal
Empty return trip	500 miles ÷ 6 miles/gal = 83 gal
Total	166 gal

Specialty freight (excludes empty return trip).

$$500\text{ miles} \div 6\text{ miles/gal} = 83\text{ gal}$$

2) Common freight "heavy load" (i.e. truck is fully loaded, included freight not related to the site):

$$1.5\text{ tons} \times 500\text{ miles} \times 0.029\text{ gal/ton-mile} = 22\text{ gal}$$

The fuel use differs by 750% depending on the mode of transportation assumed. This example also assumes that the distance transported is known and correct. In many cases, especially during the remedy design stages, the transport distance or mode of transportation may not be known, introducing additional data quality concerns.

Data Quality: Electricity Generation

Example 13.10

Given: Consider the following example of estimating the GHG emissions (measured in CO_2e) from a pumping and treatment system in Tacoma, WA, where electricity use is 100 000 kWh/year and represents the large majority of the GHG footprint for the remedy. Various sources of information on the fuel blend or generation mix for grid electricity supplied to the site may be found on the Internet or may be obtained from electricity providers. The sources include the 2007 generation mix for the eGRID (http://www.epa.gov/egrid) Northwest Power Pool (NWPP) subregion (where Washington is located), the 2009 generation mix for the NWPP from the Washington State Department of Commerce ("Commerce"), the 2007 Washington state generation mix from eGRID, the 2009 Washington state generation mix from Commerce, and the 2009 generation mix for Tacoma Power (local utility) provided by Commerce (Table 13.25).

Find: How electricity generation mixes affect its footprints.

Solution:

Electricity generation mixes vary considerably among the various sources of information. For example, the

Table 13.25 Energy mix in the United States.

Energy mix	NWPP (eGRID 2007)	NWPP (Commerce 2009)	WA (eGRID 2007)	WA (Commerce 2009)	Tacoma Power (Commerce 2009)
Generation mix					
Coal	31.96	44.29	8.00	17	3.83
Hydropower	48.37	34.76	73.72	64	87.64
Natural gas	12.78	17.46	6.81	13	1.51
Nuclear	3.0	1.41	7.58	4	6.68
Other	3.89	2.08	3.89	2	0.34
Total	100%	100%	100%	100%	100%
CO_2e emissions generation (lbs/kWh)	0.87	1.20	0.26	0.54	0.10
CO_2e footprint (lbs) for 100 000 kWh of electricity use	87 000	120 000	26 000	54 000	10 000

percentage of electricity generated from coal varies from 8 to 44%. The variation in the generation mix results in a variation in the footprint from 10 000 to 120 000 pounds of CO_2e per year, which is more than one order of magnitude. The conversion factor even varies by over 35% for the NWPP subregion depending on the year, the entity compiling the data, or both. The method for calculating the conversion factors for electricity generation therefore has a significant influence on the footprint results. The conversion factors for electricity are relatively straightforward to calculate compared with the conversion factors for other items associated with a remedy, and it is reasonable to expect that the conversion factors for manufactured goods also vary significantly based on factory location and several other factors. Table 13.26 lists all the equations for the quantification of remediation impacts on the environments.

13.7.2 Approaches to Reducing Footprints

Each approach should be tailored to a site's unique conditions and the expected operating parameters of a particular cleanup remedy. Some contributions to a metric may appear to be small relative to other contributions from the same project but are large when compared with comparable remedies implemented at other sites.

Experience and documentation gained through EPA's green remediation initiative, including illustrative profiles of green remediation strategies, are available online at http://www.cluin.org/greenremediation/tab_d.cfm)

The website suggests that the greatest opportunities for footprint reduction often relate to remedy optimization. Footprint reduction approaches can be linked to specific or general categories of EPA's green remediation BMPs. The agency compiled a series of quick-reference fact sheets describing BMPs forspecific remedial technologies, project phases, and issues common to many sites. Some of the BMPs described in these fact sheets include the following (Table 13.27).

13.7.2.1 Approaches to Reducing Materials and Waste Footprints

Pump and treat (P&T) remedies: The largest contributors to the materials and waste footprints for a P&T remedy are typically system construction, chemical use, and waste disposal. All these items are directly tied to the extraction rate and water quality, so the optimal extraction program would be a key focus for the footprint reduction of P&T systems.

In-situ Chemical Oxidation (ISCO) and *in situ* bioremediation: The largest contributors to the material footprint associated with these *in situ* remedies are the amount of reagents used/injected and the construction of permanent injection locations. Optimizing the number of injection locations and the reagent demand could lead to significant footprint reductions.

Excavation remedies: The largest contributions to material use and waste generation for excavation remedies are the disposal of excavated material in landfill and the use of clean fill for backfill. Consideration should be given to conducting the on-site treatment of impacted soils followed by the reuse of the treated soil on-site, using clean soil from the excavation to partially backfill the excavation, and locating crushed concrete or other reusable materials for fill.

13.7.2.2 Approaches to Reducing Water Footprints

Pumping and Treatment remedies: The most important factor in quantifying the water footprints for many

Table 13.26 Impact calculation.

Impacts	Equation	Note
Direct stationary emissions	$\underset{(\text{metric tons of } CO_2)}{\text{Total GHG emissions}} = \underset{(kgCO_2/MMBTU)}{\text{emission factor}} \times \underset{(MMBTU)}{\text{fuel consumed}} \times \underset{(\text{metric tons})}{0.001}$	Where % = percentage
Indirect emissions	$\underset{(\text{metric tons of } CO_2)}{\text{Total GHG emissions}} = \underset{(kWh)}{\text{electricity use}} \times \underset{(lbs\ CO_2/kWh)}{\text{electricity emission factor}} / \underset{(\text{lbs/metric ton})}{2204.62}$	Where kWh = kilowatts hour lbs = pounds
Mobile emissions	$\underset{(\text{metric tons of } CO_2)}{\text{Total } CO_2 \text{ emissions}} = \underset{(\text{gallons})}{\text{fuel consumed}} \times \underset{(kgCO_2/\text{gallon})}{\text{emission factor}} \times \underset{(\text{metric ton/kg})}{0.001}$	
	$\underset{(\text{gallons})}{\text{Total fuel use}} = \underset{(\text{miles})}{\text{mileage}} / \underset{(mpg)}{\text{fuel economy}}$	Where mpg = miles per gallon
	$\underset{(\text{metric tons of } CO_2)}{\text{Total } CO_2 \text{ emissions}} = \underset{(\text{miles})}{\text{distance traveled}} \times \underset{(kgCO_2/\text{miles})}{\text{emission factor}} \times \underset{(\text{metric ton/kg})}{0.001}$	
Total emissions	$\underset{\text{emissions}}{\text{Total GHG}} = \sum \underset{\text{combustion emissions}}{\text{direct stationary}} + \sum \underset{\text{emissions}}{\text{indirect}} + \sum \underset{\text{emissions}}{\text{mobile}} + \sum \underset{\text{processing emissions}}{\text{physical or chemical}}$	Where $\sum$ = sum of
Liquid waste production evaluation	$\underset{\text{production}}{\overset{\text{\% reduction}}{\text{in liquid waste}}} = \left(\sum \underset{\underset{(\text{gallons})}{\text{task (no BMPs)}}}{\overset{\text{liquid waste}}{\text{produced for each}}} - \sum \underset{\underset{(\text{gallons})}{\text{task(with BMPs)}}}{\overset{\text{liquid waste}}{\text{produced for each}}} \right) / \sum \underset{\underset{(\text{gallons})}{\text{task (no BMPs)}}}{\overset{\text{liquid waste}}{\text{produced for each}}} \times 100$	Where $\sum$ = sum of % = percentage
	$\underset{(\text{gallons})}{\text{Total impacted run-off production}} = \sum \underset{(\text{gallons})}{\text{impacted run-off generated for each task}}$	Where $\sum$ = sum of
	$\underset{\text{run-off production}}{\text{\% increase impacted}} = \underset{(\text{gallons})}{\overset{\text{impacted run-off}}{\text{generated during activities}}} / \underset{(\text{gallons})}{\overset{\text{impacted run-off in absence}}{\text{of remediation project}}} \times 100$	Where % = percentage
Solid waste recycling or salvaging/reuse evaluation	$\underset{\text{or salvaged}}{\overset{\text{\% solid waste}}{\text{that will be recycled}}} = \sum \underset{\underset{(\text{tons})}{\text{recycled or salvaged}}}{\overset{\text{solid waste generated}}{\text{for each task that will be}}} / \sum \underset{\underset{(\text{tons})}{\text{recycled or salvaged}}}{\overset{\text{liquid waste}}{\text{for each task that can be}}} \times 100$	Where % = percentage $\sum$ = sum of
Soil structure disruption evaluation	$\underset{(\text{cubic yard})}{\text{Total volume of disrupted soil}} = \sum \underset{(\text{cubic yard})}{\text{volume of disrupted soil for each task}}$	Where $\sum$ = sum of
Energy footprint	$\underset{\text{energy}}{\text{\% renewable}} = \sum \underset{(kWh)}{\overset{\text{renewable energy}}{\text{consumed for each task}}} / \sum \underset{(kWh)}{\overset{\text{total energy}}{\text{required for each task}}} \times 100$	Where % = percentage $\sum$ = sum of kWh = kilowatt hours
Material footprint	$\underset{\text{reused materials}}{\text{\% recycled or}} = \sum \underset{(\text{tons})}{\overset{\text{renewable energy}}{\text{utilized for each task}}} / \sum \underset{(\text{tons})}{\overset{\text{materials used materials}}{\text{required for each task}}} \times 100$	Where $\sum$ = sum of % = percentage
Footprint on cultural resources	Total number of disturbed cultural resources = $\sum$ disturbed cultural resources	Where $\sum$ = sum of
Impact on local economy	Total local economy expenditure = $\sum$ local economy expenditures	Where $\sum$ = sum of

P&T remedies is the change in water quality through the extraction, treatment, and discharge process.

In-situ Chemical Oxidation (ISCO) and *in situ* bioremediation: The largest contributors to the water footprint associated with these *in situ* remedies are reagent dilution, injection, and dispersal. Optimizing the number of injection locations, reagent demand, and delivery concentration could help reduce overall water use.

Excavation remedies: Large contributions to the water footprint for excavation and other earth-moving remedies could include water used for dust control or water from dewatering. Water used for dust control could be reduced by planning work for the appropriate season or weather condition.

13.7.2.3 Approaches to Reducing Energy and Air Footprints

Pumping and Treatment remedies: The largest contributors to the energy and air footprints are typically electricity, chemical use, process sampling, and waste disposal. Electricity, chemical use, and waste disposal are directly tied to the extraction rate, so establishing the optimal extraction rate would be a key focus for footprint reduction of P&T systems.

Table 13.27 EPA's green remediation best management practices (BMPs).

Category	BMPs
Excavation and surface restoration	• Incorporation of green requirements into product and service procurements • Install a modular renewable energy (RE) system to meet low energy demands of field equipment, other remedies, and construction or operational activities associated with site reuse • Reuse of wells and subsurface boreholes throughout investigations, remediation, and long-term monitoring • Select local providers for field operations • Cover soils with biodegradable tarps and mats, rather than spraying with water, to suppress dust while potentially enhancing soil fertility • Salvage uncontaminated and pest- or disease-free organic debris for use as infill, mulch, or compost • Installation of silt fences and basins to capture sediment runoff along sloped areas
Site investigation	• Identify local sources of energy efficient machinery and vehicles and alternative fuels • Limit the number of vehicles deployed on-site and renting electric, hybrid, or hydrogen fuel cell vehicles • Use rechargeable batteries for handheld data loggers and other field instruments • Explore options for reusing operational gray water and capturing rainwater for tasks such as irrigation or dust control • Use biodegradable hydraulic fluids on hydraulic equipment such as drill rigs • Reduce use of single-use plastic bags • Select test kits that generate less waste, such as soil samplers with reusable handles for coring syringes
Pump and treat technologies	• Establish an appropriate target capture zone and thoroughly evaluate the groundwater extraction needed to provide complete capture • Consider technologies that can operate in conjunction with P&T, such as *in situ* chemical oxidation, thermal remediation, or bioremediation in the source area • Plan for elimination of treatment train components that will become unnecessary as site conditions change • Conduct sufficient bench-scale tests to help optimize chemical dosage, which minimizes chemical use during treatment • Size pumps, fans, and motors appropriately and using energy efficient motors (such as National Electrical Manufacturers Association Premium®-labeled motors) • Insulate all pipes and equipment tied to treatment processes needing heat • Retain local laboratories or use an on-site laboratory program if possible to reduce the footprint associated with transportation of samples
Bioremediation	• Determine the on-site mass of contaminant parent and daughter products, other metabolic products, and existing microbial populations • Select the most suitable reagents or amendments and optimal concentrations or proportions • Identify unanticipated environmental impacts such as uncontrolled production of secondary by-products, suboptimal nutrient levels, or changes in nontargeted indigenous microbial populations • Maximize reuse of existing or new wells and boreholes for injections to avoid a range of wasted resources • Integrate a landfarm rain shield (such as a plastic tunnel) with rain barrels or a cistern to capture precipitation for potential on-site use • Evaluate feasibility of using pulsed rather than continuous injections when delivering air to increase energy efficiency • Investigate delivery of industrial by-products needed in high volumes by way of rail rather than trucks
Soil vapor extraction and air sparging	• Select vacuum pumps and blowers (including multiple low-flow blowers) that accommodate changes in operating requirements as treatment progresses • Consider barometric pumping, which can use barometric pressure differences to enhance air throughput if adequate response lag exists between the subsurface and atmosphere • Plan for cotreatment of SVE vapors with off-gases from other treatment systems, when concentrations allow, to gain efficiencies through economy of scale • Reclaim uncontaminated pumped water and treated groundwater for on-site use such as dust control, vegetation irrigation, or process input for other treatment systems • Treat condensate in on-site systems where contaminant types and concentrations permit • Regenerate adsorptive media such as GAC filters • Reduce monitoring frequency and intensity once the system is optimized
Clean fuel and emission technologies for site cleanup	• Replace gasoline engines with ones powered by diesel, which is more powerful and 30–35% more fuel efficient • Use solar or wind energy resources instead of diesel to generate electricity for operating small equipment such as groundwater circulation pumps

Table 13.27 (Continued)

Category	BMPs
	• Handle all materials used to absorb fuel spills in accordance with health and safety requirements and storing the material in noncombustible containers • Avoid rapid acceleration, braking, and excessive speeds, which can lower gas mileage as much as 30% on highways • Remove unneeded items in a vehicle; each 100 lb of extra weight can reduce gas mileage up to 2% • Replace (repower) aged engines or entire vehicles with cleaner burning ones • Improve lubricity of some engines, consequently reducing engine wear and tear
Integrating renewable energy into site cleanup	• Use microscale forms of renewable energy for small equipment and portable devices • Use hybrid systems that produce power from multiple renewable resources • Incorporate additional security measures to prevent damage or theft of system components • Investigate access to data that may be available from other organizations who routinely gather information at nearby met towers • Pursue opportunities to "scale up" and generate surplus electricity for credit or sale • Maintain in-house experts to assist with assessment and follow-up purchasing and maintenance of systems • Integrate geothermal applications in treatment system and building designs
Sites with leaking underground storage tank systems	• Use investigative techniques involving minimal land disturbance, field mobilization, and waste generation • Employ green purchasing techniques for products and services • Use surgical excavation techniques that minimize land disturbance • Prevent spillage and control odors and fugitive dust when emptying a tank and recycle all reusable fluids • Restore excavated areas quickly with native plants or pervious ground covers, depending on-site reuse • Optimize remedial operations through proper equipment sizing and frequent reassessment • Deploy long-term monitoring techniques that rely on remote sensing/control technology, with potential integration into site reuse
Landfill cover systems and energy production	• Design with the intent of maintaining natural settings and addressing potential effects of climate change • Maximize use of materials with recycled or bio-based content • Partner with other organizations to produce fuel • Consider a contact cover to serve as a biobarrier • Identify other activities that could maximize use of a covered area without jeopardizing the cover system • Schedule periodic inspection of cover system components and quickly complete needed repair • Use nondisruptive techniques and the site setting to monitor cover system performance

ISCO and *in situ* bioremediation: The largest contributors to energy and air footprints are typically the production and transport of the reagents to be injected. Therefore, optimizing the amount of reagent to be added would be a key focus for footprint reduction for *in situ* remedies that involve reagent injection.

Excavation remedies: The largest contributions to energy and air footprints for excavation remedies are the transport of the excavated material or backfill material from one location to another. Minimizing this horizontal transport distance and choosing the optimal mode of transportation would be key areas of focus in reducing the footprints of excavation remedies. Many footprint reduction approaches also result in cost reductions.

13.7.3 Evaluation Methods

13.7.3.1 Greenhouse Gas (GHG) Emissions Evaluation Fact Sheet

GHG emissions generated during project activities are measured by "metric tons/cubic yards COC-impacted soil" or "metric tons/gallons COC-impacted groundwater" for project activities. The six GHGs listed in the 1997 Kyoto Protocol of the United Nations Framework Convention on Climate Change include the following:

1) Carbon dioxide (CO_2)
2) Methane (CH_4)
3) Nitrous oxide
4) Sulfur hexafluoride
5) Hydrofluorocarbons (HFCs)
6) Perfluorocarbons (PFCs)

CO_2 is likely to be the primary contributor to GHG emissions at the majority of remediation sites; therefore the EPA assessment method only provides guidance to evaluate CO_2 emissions. The GHG emissions evaluation may include both direct and indirect GHG emissions. Direct GHG emissions are emissions from sources owned or controlled by a particular party (World Business Council for Sustainable Development/World Resources Institute, 2004; The Climate Registry, 2008). Indirect emissions are GHG emissions that are a consequence of a party but occur at sources owned or controlled by another party,such as electricity use (WBCSD/WRI, 2004; TCR, 2008).

Evaluation Approach

The remediation project team identifies sources of direct and indirect emissions and associated emission factors in order to perform GHG emissions evaluation. An emission factor is a factor allowing GHG emissions to be estimated from a unit of available activity data and absolute GHG emissions (WRI/WBCSD, 2004; TRC, 2008). Occupational health and safety associated with tasks performed during project activities should be emphasized throughout the project. The metric for occupational health and safety is accidents requiring treatment beyond first aid.

13.7.3.2 Future Land Use

The future land use of a remediation project should be evaluated. This stressor considers the anticipated future land use of the site following project completion, for example, unrestricted/residential or commercial land use. The stressor also considers whether institutional controls are required following project activities. The metric for future land use is land use. Throughout the project, the remediation project team seeks to obtain the least restrictive land use for the site following remediation. Actual land use for which the site is cleaned up, including related institutional controls and land use covenants, is recorded.

13.7.3.3 Green Building

A green building is a structure that is environmentally responsible and resource efficient. Examples of green buildings include buildings certified to the US Green Building Council LEED scheme and those with Energy Star certification. This stressor is only applicable to project planning and office-based task activities. The green building stressor does not have a metric. The remediation project team determines whether buildings occupied by team members are certified to the LEED, Energy Star, or other green building certifications.

13.7.3.4 Post-remediation Site Conditions

The benefit of post-remediation site conditions for a remediation project should be quantified. Post-remediation site conditions reflect those that benefit the local community. This stressor is only applicable to O&M/closure activities. The metric for post-remediation site conditions is beneficial use for the local community. The remediation project team identifies and implements opportunities to increase the local community's beneficial use of the site. Example 13.11 presents a hypothetical project illustrating how to quantify the sustainability rating.

Example 13.11

Given: A hypothetical project started incorporating sustainability for remedial investigation (RI) activities. The specific sustainability ratings of activities were *RI*(moderate), feasibility study (FS; low), remediation design and implementation (RDI; low), and *O&M/closure* (low).
Find: The rating of the project sustainability.

Solution:

Step 1: Calculate the activity-specific score.

Activity-specific score = activity-specific rating × weighting

where

"Low" activity-specific rating = 1
"Moderate" or "high" activity-specific rating = 0

Therefore, the following activity-specific scores were achieved for the remediation activities covered by the hypothetical project:

$$\text{RI}: 0 \times 1.25 = 0$$

$$\text{FS}: 1 \times 1 = 1$$

$$\text{RDI}: 1 \times 1.25 = 1.25$$

$$\text{O\&M/closure}: (\text{including post-remediation site conditions})$$

$$: 1 \times 1 = 1$$

Step 2: Calculate the project sustainability rating.
Project sustainability rating = sum of activity-specific ratings for remediation activities covered by the project.
In this example, the project sustainability rating equals

$$\text{Project sustainability rating} = 0 + 1 + 1.25 + 1 = 3.25$$

Step 3: Calculate the maximum sustainability rating.
Maximum project = sum of maximum activity-specific scores sustainability rating available for each activity covered by the project. In this example, the maximum sustainability rating equals

$$\text{Maximum sustainability rating} = 1.25 + 1 + 1.25 + 1 = 4.5$$

Step 4: Calculate the proportion of "low" scores.
Proportion of "low scores" = project sustainability rating maximum sustainability rating
In this example, the proportion of "low" scores is

$$\text{Proportion of "Low' scores"} = \frac{3.25}{4.5} = 0.72 \text{ or } 72\%$$

Answer: The project sustainability rating of this hypothetical example is "platinum."

13.8 Tools

There are two major tools in green remediation: One is sustainable remediation tool (SRT) and another is SiteWiseTM.

Sustainable remediation tool (SRT): SRT is an Excel-based tool designed to enable users to plan the future implementation of remediation technologies at a particular site, compare remediation approaches on the basis of sustainability metrics, and provide a means to evaluate the optimization of remediation technology systems already in place.

The SRT allows users to evaluate the sustainability impacts associated with various remediation technologies for soil and groundwater. The technologies include:

1) Excavation.
2) Soil vapor extraction.
3) *In situ* thermal.
4) Pump and treat.
5) Enhanced bioremediation.
6) Permeable reactive barriers.
7) In situ chemical oxidation.
8) Long-term monitoring/monitored natural attenuation.
9) Particulate matter (as PM_{10}).
10) Total energy consumed.
11) Change in resource service.
12) Technology cost.
13) Safety/accident risk.

The tool is structured into two tiers:

1) Tier 1, the simplest tier, comprises calculations based on rules of thumb that are widely used in the environmental remediation industry.
2) Tier 2 comprises calculations that are more detailed and incorporate site-specific factors.

The SRT is designed to be user-friendly with menu-driven work flows and on-screen help topics. The tool is available for download from AFCEE's sustainable remediation website at https://clu-in.org/products/tins/tinsone.cfm?num=66247610

SiteWiseTM includes a number of Excel spreadsheets designed to evaluate sustainability impact associated with different remediation projects. The remediation project is broken down into four phases in the tool: remedial investigation (RI), remedial action construction, remedial action operation, and long-term monitoring. Six different remedial alternatives may be considered at one time, enabling SiteWiseTM to be employed during the remedy selection process as well. Each phase is further broken down into following

1) Material production.
2) Transportation: personnel and equipment.
3) Equipment used: earthwork, pumps, and other electric.
4) Residual handling.

A number of stressors may be evaluated in SiteWiseTM, including GHG emissions; energy use; criteria pollutants emissions, including oxides of nitrogen, sulfur oxides, and particulate matter; water consumption; and worker safety. SiteWiseTM can be downloaded from the Navy's sustainable remediation website: https://www.navfac.navy.mil/navfac_worldwide/specialty_centers/exwc/products_and_services/ev/erb/gsr.html.

13.9 Exercise

13.9.1 Questions

1) Why is an activated sludge process so energy intensive yet is the most common wastewater treatment process?
2) What are the major ways to reduce sludge?
3) What are major strategies to retrofit WWTPs to energy positive?
4) Why is anaerobic sludge digestion a better sustainable design than mechanical dewatering of sludge?
5) According to energy benchmark data, which three processes in conventional WWTPs have the most potential to be retrofitted?
6) What are the major metrics for green remediation?
7) How is the footprint on energy calculated during remediation project?
8) How energy mix affect the energy footprint?

13.9.2 Calculation

1) Miami-Dade Water and Sewer Department (WASD) aims to increase its water reuse rate from 5 to 30 and 60% in 5–10 years to end ocean outfall of treated wastewater by 2025. The proposal is to reclaim 40 MGD of the treated wastewater effluent in the Miami-Dade South WWTP. The ultimate goal is to achieve zero energy, zero water, and zero nutrients.
   a) Since the discharge standards are extremely high, who are the stakeholders? How should two loops of the stakeholder involvement be designed during planning and implementation?
   b) Use the EPA five best WWTP examples reported by Tarallo (2015) to establish the best benchmark

of performance metrics for this new treatment plant to be designed and built in the Miami-Dade Country.

c) Using the 12 design principles, design a true WRRF that is energy neutral, 100% water reuse, and N and P recovery.

2) Miami-Dade County plans to retrofit three WWTPs into energy positive, in which eight projects are proposed as follows:
   a) COD is to be diverted from primary settler to anaerobic digester to decrease sludge production and increase biogas production.
   b) Anaerobic oxidation of ammonium (anammox) is proposed to reduce sludge production and aeration electricity.
   c) Fat–oil–grease and food wastes are diverted to anaerobic digester for more biogas production and decrease BOD to WWTP.
   d) Membrane bioreactor (MBR) will be used to increase WWTP capacity by increasing solid retention time.
   e) UV disinfection will be used to increase compliance of disinfection by-product rules and reduce operation costs.
   f) Leachate will be separated and treated to reduce BOD and COD to WWTP and toxicity of leachate.
   g) Food wastes will be diverted to anaerobic digesters to increase biogas production.

   According to different scenarios such as diverted 10–60% of COD from the primary settler to the anaerobic digester, please do the following:
   a) Theoretical values of energy balance in terms of kWh/m^3 or kWh/kg COD.
   b) Benchmark values of energy balance in terms of kWh/m^3 or kWh/kg COD.
   c) Best practices from the five best WWTPs.
   d) Compare WWTPs of Miami-Dade County with the previous three values, and develop the best design alternatives to achieve the energy-neutral design.

3) For a WWTP serving a population of one million, conventional activated sludge is used. Please compare the energy, water, and material footprints of the following different design alternatives using the US EPA TRACI 2.0 (https://cfpub.epa.gov/si/si_public_record_report.cfm?dirEntryId=227747):
   1) Sandbed drying.
   2) Mechanical dewatering.
   3) Anaerobic sludge digestion with AnMBR.
   4) Anaerobic digestion with 5% fat–oil–grease from kitchen and restaurants.

13.9.3 Projects

13.9.3.1 Xiongan Project

1) Please use daily per capita production data and energy content data to estimate the maximum energy that could be recovered by each of the different technologies discussed.
2) How would you design the WWTP to be energy positive using innovative technologies for WWTPs that serve 1, 2, and 5 million PE in 5, 10, and 20 years in Xiongan?
3) The first step in land development in Xiongan is to clean up the vast contaminated lakes and soil. Using green remediation metrics, please design the two different alternatives to clean up the site, and compare the water, energy, and material footprints of each alternative. Which alternative would you recommend and why?

13.9.3.2 Community Project Proposal

1) Energy auditing of WWTP serves two purposes: (i) baseline for developing energy efficiency projects and (ii) quantitative energy consumption of each unit process for identifying energy gaps between the WWTP and best practice benchmarks for proposing retrofitting projects. Please contact WWTPs of your own city and request permission to conduct energy auditing for their WWTPs. Once you get the permission, please request following information to conduct a comprehensive energy auditing of their plants (US EPA, 2013):

   Historical data from the facility supervisory control and data acquisition/control system (5–10 years):
   1) Average flow and load data
   2) Pump curves.
   3) Blower curves.
   4) Actual operational data (motor amperages).
   5) Electric/gas utility rates and bills.
   6) Diffuser manufacturer data.
   7) Miscellaneous plan drawings from previous construction projects (piping configurations, aeration diffuser layouts, etc.).

   In your energy auditing report, please answer the following:
   1) What is the current unit energy consumption in terms of kWh/1000 m^3 and kWh/kg COD removed?
   2) What are the unit processes that consume the most energy and the gap from the top 10% best practice WWTPs of its peer WWTPs?
   3) What are the energy retrofitting strategies would you recommend to improve the overall energy efficiency of the WWTP?
   4) What are the innovative technologies would you recommend to retrofit the WWTP to WRRF?

5) What are the life cycle cost and benefit of the retrofitting technologies in terms of capital and O&M costs if the WWTP is to be in operation in the next 20 years?

2) Within the Southern Florida Water Management District (SFWMD), water contaminated with nitrogen and phosphorus was discharged into the Everglades National Park (ENP), and, as a result, the ENP was almost ruined by exotic plants. In the federal court, the ENP sued the SFWMD for not performing its duty in protecting the ENP and won the case. As a result, every year, the restoration of the Southern Florida ecosystem started with one billion dollars per year from 1995 to 2015. In the plan, 13 elevated reservoirs were proposed, and the following questions were asked to stop the elevated reservoirs:
   a) How has the problem developed over time?
   b) What are the main direct and indirect causes of the problem's development, including link polarities?
   c) What are the consequences of the problem?
   d) What are main feedback processes?
   e) What kind of short-term policies do you think can be adopted to solve this problem?
   f) What kind of long-term policies can be adopted to solve this problem?
   g) What are the main hurdles in the success of these policies?

References

Shoener, B.D., Bradley, I.M., Cusick, R.D., and Guest, J.S. (2014). Energy positive domestic wastewater treatment: the roles of anaerobic and phototrophic technologies. *Environmental Science: Processes and Impacts* 16: 1204–1222.

Shoener, B.D., Zhong, C., Greiner, A.D. et al. (2016). Design of anaerobic membrane bioreactors for the valorization of dilute organic carbon waste streams. *Energy and Environmental Science* 9 (3): 1102–1112.

Tarallo, S. (2015). *A Guide to Net-Zero Energy Solutions for Water Resource Recovery Facilities*. Alexandria: Water Environment Research Foundation.

The Climate Registry (2008). General Reporting Protocol. Version 1.1, May.

US EPA (2012). Methodology for Understanding and Reducing a Project's Environmental Footprint. Washington, DC: US EPA.

US EPA (2013). Energy efficiency in water and wastewater facilities, Local Government Climate and Energy Strategy Series. Washington, DC: US EPA.

World Business Council for Sustainable Development/ World Resources Institute (2004). The Greenhouse Gas Protocol, a Corporate Accounting and Reporting Standard. Revised Edition, March.

14

Optimization through Modeling and Simulation

Optimization is to leverage key learnings in designing effective, efficient, and economic systems.

OBJECTIVES

To develop, calibrate, verify, and validate models of the Fenton process
To develop kinetic models using leachate treatment using the Fenton process as an example
To validate the Fenton kinetic model using computer simulation
To optimize conditions in the Fenton oxidation of leachate
To develop and apply dimensionless variables in mathematic models of the Fenton process
To improve the biodegradability and decrease the toxicity of leachate using the Fenton process

14.1 Principle

Mathematical model is the best way to describe a physical process so that the mass and energy efficiency of the process could be quantitatively predicted. Before a model could be used for this purpose, it has to be verified, calibrated, and validated. Models could be as simple as a black box at macro-scale or as detailed as some models of quantum chemistry or physics on micro-scale. In most SEE design, models at the elementary reaction level should be adequate. Fortunately, lots of commercial software are available. For example, ASM3biop is versatile-activated sludge simulation developed by the International Water Association (IWA) with an extension by the Swiss Federal Institute of Aquatic Science and Technology (EAWAG). It allows the holistic analysis of wastewater treatment plant (WWTP) through the modeling and dynamic simulation of the activated sludge process, sludge treatment, and biological phosphorous elimination. Alternatively, simulation model could be built by using Matlab and Simulink. The major benefit of mathematical model and simulation is that it could be used to optimize the process without expensive laboratory and pilot testing if model is the true presentation of a process or a system. Principle 11 generalizes the criteria for effective application of model and simulation.

Principle 11 Component, process, and system should be optimized through modeling and simulation under both global warming and cooling scenarios:

1) Models should be calibrated, verified, and validated for component, reactor, process, and system before its application in simulation and optimization.
2) Simulation should be conducted with computer tools such as Simulink so that time series data can be validated for different alternatives.
3) Design alternatives should be optimized with quantified uncertainty and sensitivity using validated models and computer software.

14.2 Introduction

To illustrate Principle 11 of optimization through modeling and simulation, the Fenton treatment of leachate is used as an example. Modern engineered sanitary landfill receives municipal solid wastes (MSWs) containing domestic, nonhazardous commercial and industrial wasted solids. There are four major different fates for precipitation that fell in

Sustainable Environmental Engineering, First Edition. Walter Z. Tang and Mika Sillanpää.

landfill: evaporation, transexpiration, runoff, and leaching through layers of MSW. When rainfall percolates through MSW, liquid reacts with the MSW through hydrolysis. Organic and inorganic dissolve into the liquid. Once MSW is hydrolyzed, biochemical reactions take place through acidogenesis, acetogenesis, and methanogenesis with the help of anaerobic bacteria. The intermediate products dissolve into the liquid in the form of excess liquid called leachate. When the landfills are properly designed and operated, leachate is collected to be treated onsite or discharged to a WWTP. Leachate quality can be best explained in terms of fundamental chemical and biochemical reactions during leachate formation. Two major challenges facing leachate treatment are nonbiodegradability and toxicity. These two characteristics have significant impacts on designing effective leachate treatment train to achieve discharge requirements by different environmental laws.

14.2.1 History of Landfill Leachate Quality

Table 14.1 summarizes the current understanding of leachate quality which is critical in the selection of treatment processes during design process.

14.2.2 Leachate Characteristics

An anaerobic degradation scheme for organic material in a sanitary landfill is shown in Figure 14.1. In young landfills containing large amounts of biodegradable organic matter (OM), rapid anaerobic fermentation takes place, resulting in volatile fatty acids (VFA) as the main fermentation products. Acid fermentation is enhanced by high moisture or water content in the solid waste. This early phase of a landfill's lifetime is referred to as the acidogenic phase and leads to the release of large quantities of free VFA, as much as 95% of the organic content. As a landfill matures, the methanogenic phase occurs. Methanogenic

Table 14.1 History of landfill leachate quality.

Year	Authors	Leachate quality topics
1973	Farquhar and Rover	Eight phases of MSW decomposition and their products
1977	Chian and Dewall	Fulvic and humic end products in leachate
1980	Cameron and Koch	Leachate toxicity
1982	Zehnder	Identification of hydrolic, acidogenic, acetogenic, and methanogenic bacteria
1982	Bookfer and Ham	Cellulosic material decomposition
1983	Harmsen	Acids during leachate decomposition
1983	Ehrig	Water balance for estimation of leachate quantity and toxicity of ammonia
1986	Pohland and Harper	Quantification of hydrolic, acidogenic, acetogenic, and methanogenic phases in MSW decomposition
1989	Weis	Quantification of C, H, and S content, Cu complexation capacity and phenolic end products
1989	Belevi and Baccini	Long-term leachate quality
1990	Barlaz, Ham, and Schaefer	Methane production from leachate
1991	Bolton and Evans	Cd speciation
1993	Gounavis et al.	Colloidal particles in Zn, Pb, and Cr migration
1993	Assmuth and Strandberg	Xenobiotic pollutants
1994	Christensen	Attenuation of leachate pollutants
1995	Robinson	Adsorbable organic halogen (AOX)
1995	Reardon	Anaerobic corrosion of iron
1995	Flyhammer et al.	Heavy metals in leachate
1996	Clement et al.	Different leachate toxicities
1997	Gibbson	Halogenated aliphatic compounds in leachate
1997	Krug and Hum	BTEX in the US leachate
1999	KruBozhurt et al.	Reoxygeneration of post closure and heavy metal release
2001	Christophersen	BTEX in Danish leachate

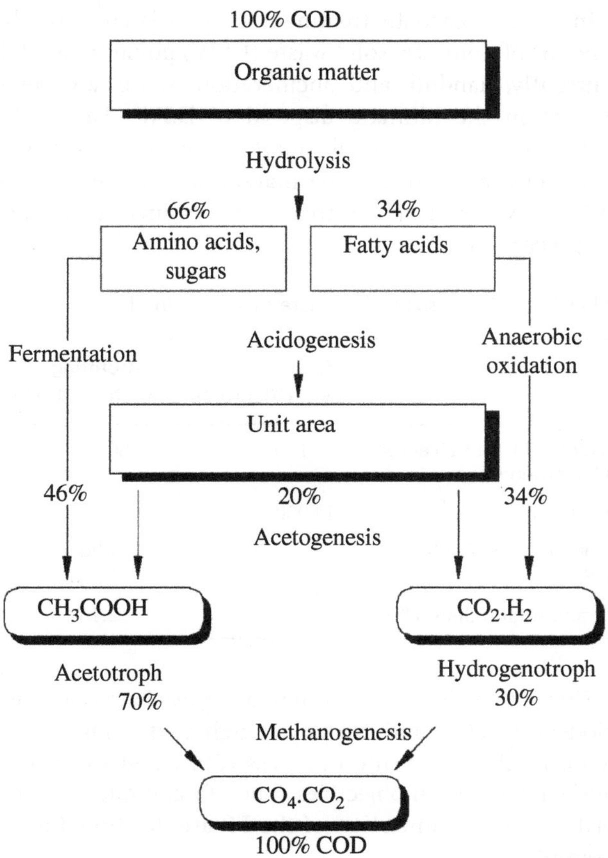

Figure 14.1 Anaerobic degradation of organic compounds of leachate.

microorganisms develop in the waste, and the VFA are converted to biogas (CH_4, CO_2). The organic fraction in the leachate becomes dominated by refractory (nonbiodegradable) compounds such as humic substances (Figure 14.1).

Landfills undergo at least five stages of waste degradation, accompanied by the formation of various compounds and emissions: (i) for aerobic, water and carbon dioxide are the main products, with carbon dioxide released as gas or absorbed into water to form carbonic acid; (ii) for acidogenic, carbon dioxide, hydrogen, ammonia, and organic acids; (iii) for acetogenic, acetic acid and derivatives, carbon dioxide, and hydrogen; (iv) for methanogenic, typical landfill gas has approximately 60% methane and 40% carbon dioxide; and (v) for aerobic, carbon dioxide and water. The acidogenic phase occurs when the landfill containing large amounts of biodegradable OM is a few years old, resulting in the so-called young leachate with high values of chemical oxygen demand (COD) and biologic oxygen demand (BOD) levels, while the ratio BOD/COD is higher than 0.7 and pH is at low value due to the high concentrations of VFA. Acidic fermentation is enhanced by a high moisture or water content in the solid waste. The methanogenic phase is specific to landfills older than 10 years, and the leachate generated is referred to as old. Methanogenic microorganisms developed in the waste degrade VFA, which are converted to biogas (CH_4, CO_2), while pH >7 and the ratio BOD/COD is stabilized on levels. Refractory (nonbiodegradable) compounds such as humic substances become the dominant organic fraction in the leachate.

There are many factors affecting the quality of leachates, such as landfill age, precipitation, seasonal weather variation, waste type, and composition. In particular, the composition of landfill leachates varies greatly depending on the age of the landfill (Silva et al., 2004). The characteristics of the landfill leachate can usually be represented by the basic parameters COD, BOD, the ratio of BOD/COD, pH, suspended solids (SS), ammonium nitrogen (NH–N), total Kjeldahl nitrogen (TKN), and heavy metals. Table 14.2 presents typical physicochemical characteristics of leachate. The leachate is usually slightly alkaline pH (>7), dark in color, low in BOD_5 (151 mg/l), and high in COD (2183 mg/l) with a BOD_5/COD ratio of 0.07, consistent with typical characteristics of stabilized leachate.

Using the EDS analysis of a transmission electronic microscope, the elements of raw leachate are presented in Table 14.3. The energy-dispersive spectroscopy (EDS) analysis for the elemental composition of leachate sample (Table 14.3) showed highest concentration for the elemental carbon (35 ± 11% (w/w)) that included inorganic as well as organic carbon. It shows that carbon, oxygen, sodium, and chlorine occupy 35.3, 21.5, 27.4, and 12.4% of major weight of leachate sample, respectively.

Table 14.2 Physicochemical characteristics of a typical leachate sample.

Parameter	Average	Standard deviation
pH (S.U.)	7.51	0.09
DO (mg/l)	2.6	
Conductivity (mS/cm)	28.6	1.4
TDS (mg/l)	19 066	757
BOD_5 (mg/l)	151	50
COD (mg/l)	2 183	46.8
BOD_5/COD	0.07	—
TOC (mg/l)	778	73
Alkalinity (mg/l as $CaCO_3$)	2 345	35

Table 14.3 Elemental composition by weight % in raw leachate as measured by the EDS analysis.

Element	Raw leachate
C	35.3 ± 11.9
O	21.5 ± 8.7
Na	27.4 ± 12.1
Si	0.3 ± 0.3
S	0.7 ± 1.3
Cl	12.4 ± 10.9
K	2.0 ± 2.1
Ca	0.3 ± 0.4

14.3 Challenges and Opportunities

In addition to toxicity and recalcitrant nature, leachate is also a significant source of new and emerging pollutants reported from WWTP discharge. Emerging compounds identified in untreated leachates samples were detected at nanogram or microgram per liter levels in both surface and groundwater:

1) Chlorinated alkylphosphates.
2) Carcinogenic flame retardant (TCPP).
3) Neurotoxin plasticizer (NBBS).
4) Insect repellent (DEET).
5) PFCs.
6) Pharmaceuticals.
7) Personal care products such as NSAIDs (ibuprofen and naproxen).
8) Polycyclic musk compounds.

Since leachate has gone through an extended period of anaerobic degradation, the organic residuals as listed earlier could be very difficult to be degraded further. However, the current practice all over the United States is transporting leachate to local WWTPs if the leachate flow rate is less than 5% of that of a WWTP. Due to the non-biodegradable and toxic nature, even 5% of the total wastewater flow rate could consume a significant amount of energy. More importantly, the emerging toxic pollutants may go through WWTP without degradation. For example, a pilot study of advanced treatment of wastewater effluent was conducted to reclaim the treated effluent to replenish the Biscayne aquifer at the South Dade WWTP. However, the nonbiodegradable toxic organic compounds from leachate could not be removed to meet the stringent discharge standards for the Everglades National Park. The unit cost of the reclaimed wastewater reached as high as $6/gal. As a result, the $10 million dollar pilot project plus $70 million engineering design fee was suspended indefinitely.

In China, leachate treatment is associated with the amount of domestic solid waste (DSW) going to landfill. Currently, landfill and incineration share an equal market in the ultimate disposal of DSW. Table 14.4 estimates the potential market of new leachate treatment facilities and the market size for retrofitting 70% of existing leachate treatment systems in the 13th Five-Year Plan.

Table 14.4 Market sizes of leachate treatment in China.

	Total new leachate facility	Retrofitting leachate facility
Volume in 13th Five-Year Plan (1000 tons)	170	180
Unit cost ($/tons)	14 000	7 000
Total market size (billion USD)	2.38	1.26
Grand total (billion USD)		3.46

Since the major difficulty in treating leachate is its non-biodegradability and toxicity, Singh and Tang (2013) reported that the Fenton process (FP) could simultaneously achieve both objectives. Therefore, modeling, simulation, and optimization of the FP are illustrated as an example.

14.4 Modeling of the Fenton Process

Fenton oxidation (FO) uses hydrogen peroxide (H_2O_2) as an oxidizing agent and a reduced form of iron (Fe^{2+}) as a catalyst to generate hydroxyl radicals ($^{\bullet}OH$) in acidic condition as shown in Equations (14.1)–(14.4), and the $^{\bullet}OH$ effectively mineralizes or transforms refractory OM into more biodegradable OM that can be subsequently treated using biological processes (Bae et al., 1997):

$$Fe^{2+} + H_2O_2 \xrightarrow{k_i} Fe^{3+} + OH^- + \bullet OH \quad k_i = 51\,M^{-1}s^{-1} \tag{14.1}$$

$$Fe^{2+} + \bullet OH \xrightarrow{k_{t1}} Fe^{3+} + OH^- \quad k_{t1} = 3 \times 10^8\,M^{-1}s^{-1} \tag{14.2}$$

$$\bullet OH + H_2O_2 \xrightarrow{k_{t2}} \bullet OOH + H_2O \quad k_{t2} = 2.7 \times 10^7\,M^{-1}s^{-1} \tag{14.3}$$

$$RH + OH\bullet \rightarrow HO{-}R\bullet \quad k_3 = 10^9 - 10^{10}\,M^{-1}s^{-1} \tag{14.4}$$

where k_i is the initiation rate constant of hydroxyl radical generation and $k_{t1,t2}$ is the termination rate constants of hydroxyl radicals quenching reaction by Fe^{2+} and H_2O_2, respectively. RH is an unsaturated organic compound.

One of the effective ways to oxidize leachate is using the FP to reduce COD and increase BOD, which represents the biodegradability of leachate. To effectively apply the FP, accurately describing the process by a mathematical model can help experiments to be designed and carried out more efficiently. Tang (2004) developed a mathematical model for the FO of unsaturated organic pollutants such as chlorophenols and demonstrated the theoretical optimal ratio of H_2O_2/Fe^{2+} that can be expressed as follows:

$$\left(\frac{H_2O_2}{Fe^{2+}}\right)_{Optimal} = \frac{k_{(Fe^{2+})(OH\bullet)}}{k_{(H_2O_2)(OH\bullet)}} = \frac{k_{t1}}{k_{t2}} = 11 \qquad (14.5)$$

Such an optimal ratio provides a theoretical framework for FO to produce a maximum amount of hydroxyl radical. Since hydroxyl radical is a powerful oxidative agent (2.80 V/SHE) and a highly reactive chemical species able to attack organic pollutants with second-order rate constants ranging from 10^7 to 10^{10} l/mol/s (Tang, 2004), its high reactivity together with its low selectivity makes the hydroxyl radical a very powerful oxidation species for the mineralization of persistent organic pollutants. In addition to Fenton process, other processes such as chemical, photochemical, sonochemical, and electrochemical processes can be used to produce hydroxyl radicals ($^{\bullet}OH$). No matter which advanced oxidation process (AOP), the generation of hydroxyl radicals is the key parameter to control and optimize to improve the efficiency of the process and to elucidate the degradation mechanisms.

To accurately describe the FP, quantification of hydroxyl radical concentration is crucial. However, the very short lifetime of $^{\bullet}OH$ (about 10^{-9} s) makes its detection and quantification difficult, so indirect detection methods such as UV-Vis spectrophotometry (Zhao et al., 2015), luminescence (Tsai et al., 2001), fluorescence (Xiang et al., 2011), and other electrochemical and HPLC methods (Si et al., 2014) have been reported after hydroxyl radical has been scavenged. The hydroxyl radical concentration was then inferred from the product concentration instead of measuring radical concentration directly. Although electron paramagnetic resonance (EPR) could not directly measure short-lived species such as hydroxyl radical, this technique could detect longer-lived radical adducts generated by using spin trap agents at concentration as low as 10^{-14} M (Haywood, 2013). For this purpose, 5,5-dimethyl-1-pyrroline *N*-oxide (DMPO) is the most commonly used compound, especially because its reaction with the hydroxyl radical leads to the formation of relatively stable oxygen-centered aminoxyl radicals detectable by EPR (Buettner, 1993).

The Fenton reaction includes chemical Fenton, photo-Fenton, electro-Fenton, and photoelectron-Fenton. Under conditions used in AOPs, steady-state hydroxyl radical concentration could be as high as 10^{-6} M. As a result, the stability of DMPO and its hydroxyl radical spin adduct (DMPO–OH) in the presence of high concentration of iron and hydrogen peroxide becomes a critical issue in the accurate quantification of hydroxyl radical concentration because of the formation of by-products (paramagnetic or not) affecting the reliability of DMPO–OH (and thus $^{\bullet}OH$) detection (Makino et al., 1990; Buettner, 1993; Li et al., 2007).

14.4.1 Kinetic Model of DMPO–OH EPR Signal

Hydroxyl radicals generated from the Fenton reaction (Reaction 14.1) could also react with Fe^{2+} and H_2O_2 as described in termination Reactions (14.2) and (14.3). However, when DMPO is in excess, all the hydroxyl radicals generated by H_2O_2 and Fe^{2+} would quickly attack the electron-rich carbon of the DMPO in Reaction (14.4) to form the classic paramagnetic compound 2-hydroxy-5,5-dimethyl-1-pyrrolidinyloxy (DMPO–OH; see Scheme 1, compound **2**), thus limiting side Reactions (14.2) and (14.3):

$$\text{DMPO} + \bullet\text{OH} \xrightarrow{k_D} \text{DMPO}-\text{OH} \quad k_D = 3.4 \times 10^9\,\text{M}^{-1}\text{s}^{-1} \qquad (14.6)$$

where k_D is the formation rate constant of paramagnetic DMPO–OH compound.

The stability of DMPO under attack from a hydroxyl radical depends upon the relative amount of DMPO and Fenton reagent concentrations. In the presence of high concentration of Fe^{3+} under acidic conditions, it was previously reported that DMPO–OH could lead to the formation of two diamagnetic intermediates (Makino et al., 1992): 1-hydroxy-5,5-dimethyl-1-pyrrolid-2-one (HDMPN, Scheme 1, compound **3**) and its tautomer 2-hydroxy-5,5-dimethyl-1-pyrroline-*N*-oxide (HDMPO, Scheme 1, compound **4**), as described in Reaction (14.7):

$$\text{DMPO}-\text{OH} + \text{Fe}^{3+} \xrightarrow{k_I} \text{HDMPN} \leftrightarrows \text{HDMPO} \qquad (14.7)$$

where k_I is the formation rate constants for diamagnetic tautomers HDMPN and HDMPO.

Moreover, if there is an excess of hydroxyl radicals in the system, these EPR-silent compounds quickly react with •OH to form the stable paramagnetic radicals 5,5-dimethyl-2-oxopyrroline-*N*-oxyl (DMPOX, Scheme 1, compound **5**) and 2-dihydroxy-5,5-dimethyl-1-pyrrolidinyloxy (HDMPO–OH, Scheme 1, compound **6**) (Reactions 14.8 and 14.9):

$$\text{HDMPN} + \bullet\text{OH} \xrightarrow{k_H} \text{HDMPO}-\text{OH} \qquad (14.8)$$

$$\text{HDMPO} + \bullet\text{OH} \xrightarrow{k_X} \text{DMPOX} \qquad (14.9)$$

where k_H and k_X are the formation rate constants for the paramagnetic products HDMPO–OH and DMPOX, respectively.

Therefore, if DMPO is not present in extra-excess, additional EPR signals might appear, thus preventing a reliable detection/quantification of hydroxyl radicals. The critical question to elucidate is therefore at which concentration DMPO could be considered to be in excess. For this purpose, a kinetic model was developed based on the previous elementary reactions. With the aid of this model, experiments were designed to answer this critical question by monitoring DMPO–OH and additional signals using an EPR and UV-Vis spectrophotometer. When the ratio of H_2O_2 and Fe^{2+} was kept at an optimal 10 (Tang, 2004), the conditions under which DMPO is in excess, i.e. trapping all the hydroxyl radicals in the system, without further Reactions (14.8) and (14.9), can be found starting from Reaction (14.10):

$$\frac{d[\bullet OH]}{dt} = k_i[H_2O_2][Fe^{2+}] - k_D[DMPO][\bullet OH] - k_{t1}[\bullet OH][Fe^{2+}] - k_{t2}[\bullet OH][H_2O_2] - k_H[HDMPN][\bullet OH] - k_X[HDMPO][\bullet OH] \quad (14.10)$$

At a steady-state concentration of hydroxyl radical, Equation 14.10 should equal zero to derive the steady-state hydroxyl radical concentration. The final form of hydroxyl radical steady-state concentration can then be expressed as follows:

$$[\bullet OH]_{SS} = \frac{k_i[H_2O_2][Fe^{2+}]}{k_D[DMPO] + k_{t1}[Fe^{2+}] + k_{t2}[H_2O_2] + k_H[HDMPN] + k_X[HDMPO]} \quad (14.11)$$

To accurately quantify DMPO–OH concentration, the analytical conditions must be found and defined so that a stable quartet EPR signal can be monitored over the analytical time without further degradation to HDMPN. In terms of elementary reactions, the rate of Reactions (14.8) and (14.9) should approach zero as follows:

$$k_H[HDMPN] = 0 \quad (14.12)$$

$$k_X[HDMPO] = 0 \quad (14.13)$$

Therefore, the steady-state hydroxyl radical concentration can be simplified as follows:

$$[\bullet OH]_{SS} = \frac{k_i[H_2O_2][Fe^{2+}]}{k_D[DMPO] + k_{t1}[Fe^{2+}] + k_{t2}[H_2O_2]} \quad (14.14)$$

Moreover, the reaction rate of DMPO–OH can be expressed as follows:

$$\frac{d[DMPO-OH]}{dt} = k_D[DMPO][\bullet OH]_{SS} - k_I[DMPO-OH][Fe^{3+}] = k_D[DMPO]\frac{k_i[H_2O_2][Fe^{2+}]}{k_D[DMPO] + k_{t1}[Fe^{2+}] + k_{t2}[H_2O_2]} - k_I[DMPO-OH][Fe^{3+}] \quad (14.15)$$

The generation of a constant quartet EPR signal relies upon two conditions: the first is that the rate of reaction (14.7) approaches zero. Therefore,

$$k_I[DMPO-OH] = 0 \quad (14.16)$$

The second condition is to make sure the change rate of $d[DMPO–OH]/dt$ is not a function of [DMPO]. The only possible way to satisfy this condition is for the term k_D[DMPO] to be significantly greater than $k_{t1}[Fe^{2+}] + k_{t2}[H_2O_2]$ so that the latter can be negligible. Mathematically, the following conditions must be satisfied:

$$k_D[DMPO] \gg k_{t1}[Fe^{2+}] + k_{t2}[H_2O_2] \quad (14.17)$$

This equation can be divided by $k_D[H_2O_2]$ at both sides, and the suitable analytical conditions become

$$\frac{[DMPO]}{[H_2O_2]} \gg \frac{k_{t1}[Fe^{2+}]}{k_D[H_2O_2]} + \frac{k_{t2}}{k_D} \quad (14.18)$$

The sign of significantly greater means that the right-hand ratio of reaction (14.18) should be negligible compared with the left-hand ratio. In other words, the left-hand ratio should be at minimum two to three orders smaller than the right-hand ratio for Equations (14.17) and (14.18) to truly hold. Therefore, at the initial H_2O_2 and Fe^{2+} concentrations of 1 and 0.1 mM (ratio of 10), respectively, the required ratio of initial DMPO to H_2O_2 concentrations, [DMPO] : $[H_2O_2]$, should be at minimum 1.68–16.8. As a result, the system should generate a stable quartet EPR signal without any additional noise. Fontmorin et al. (2016) designed their experiments based upon these theoretical analyses, and results were used to determine the optimal [DMPO] : [H_2O_2] ratio more accurately. In this study, $H_2O_2 : Fe^{2+}$ ratio was kept constant; however, the aforementioned model would help to predict an optimal concentration of DMPO for any other concentration of the Fenton reagent since initial conditions are included in Equation (14.18).

Fontmorin et al. (2016) showed the impact of the Fenton reagent concentration with H_2O_2 and Fe^{2+} concentrations varying from 1 to 100 mM and from 0.1 to 10 mM, respectively, with [DMPO] : [H_2O_2] ratio varying from 1 to 100. A typical spectrum of DMPO–OH is presented in Figure 14.2a: it is composed of a characteristic 1 : 2 : 2 : 1 quartet with hyperfine couplings $a_N = a_{\beta H} =$ 1.50 mT (computer simulation, Figure 14.2b).

The amplitude of the DMPO–OH signal was followed over time at different concentrations of Fenton's reagent, and results are presented in Figure 14.3.

As depicted, the concentration of Fenton's reagents had a significant impact on both the intensity and stability of DMPO–OH signal. Indeed, when H_2O_2 and Fe^{2+} concentrations were 1 and 0.1 mM, respectively (i.e. [DMPO] : [H_2O_2] = 100 > 16.8), the peak amplitude of DMPO–OH signal was statistically stable because no clear downward trend appears over the time of experiment (60 min). The variations from one point to another could be attributed to the instability of the reaction (no control of pH after the initial adjustment). Nevertheless, the general trend of the system together with the nature of the EPR background (DMPO–OH quartet, data not shown) suggested an appropriate concentration in DMPO compared with the concentration of Fenton reagent for the detection of radicals, as predicted by the theoretical model. The increase in Fenton reagent concentration to 10 mM H_2O_2 and 1 mM Fe^{2+} (i.e. [DMPO] : [H_2O_2] = 10 > 1.68) led to a different behavior. The higher intensity measured during the very early minutes suggested the generation of higher concentration of hydroxyl radicals, as expected after the increase of the Fenton reagent. However, an important decrease in the DMPO–OH peak amplitude was then observed during the first 20 min, followed by a slower decrease until the end of the experiment. The fast decrease measured during the early stage of the reaction was unlikely to have been related to the lifetime of DMPO–OH, generally reported around

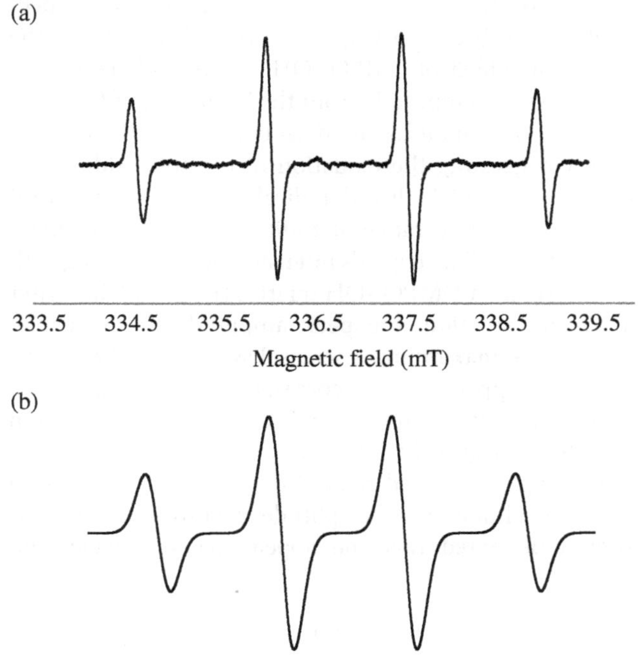

Figure 14.2 (a) Typical DMPO–OH signal obtained with 1 mM H_2O_2, 0.1 mM Fe^{2+}, and 0.1 M DMPO and (b) simulation of DMPO–OH ($a_N = a_{\beta H} = 1.50$ mT). (*Source:* Fontmorin et al. (2016). Reproduced with permission of Elsevier.)

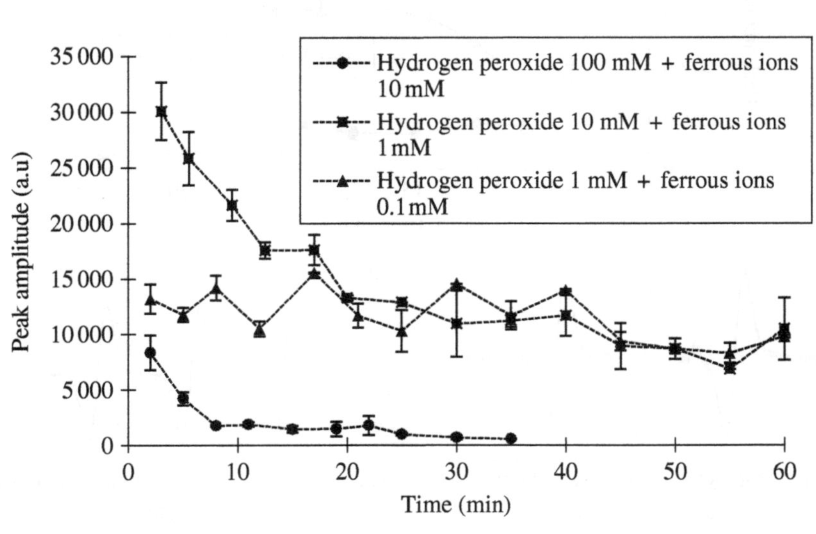

Figure 14.3 Time courses of DMPO–OH peak amplitude for various concentrations in the Fenton reagent and constant 0.1 M DMPO concentration. (*Source:* Fontmorin et al. (2016). Reproduced with permission of Elsevier.)

30 min. Even though the typical DMPO–OH quartet was the only signal detected in EPR spectra without the apparition of any additional line, the fast decrease of the signal intensity suggested the possible quenching or degradation of the spin adduct. According to the kinetic model developed, this observation would also suggest that the left-hand ratio of Equation (14.18) should be more than two orders greater than the right-hand ratio to provide a stable quartet signal. To confirm the possible quenching of DMPO–OH, H_2O_2 and Fe^{2+} concentrations were increased again to 100 and 10 mM, respectively ([DMPO] : [H_2O_2] = 1). In this case, the quenching effect of DMPO–OH was very obvious, with a lower signal amplitude from the beginning of the reaction and becoming hardly detectable after 10 min, thus clearly suggesting the oxidation of the spin adduct due to the excess of hydroxyl radicals. Moreover, the EPR revealed the appearance of additional lines consisting a 1 : 1 : 1 triplet. The triplet's intensity increased along with the decrease of DMPO–OH quartet (Figure 14.4a). After a 35-min reaction, the peak amplitude of the triplet reached its maximum, whereas DMPO–OH had completely disappeared. The corresponding EPR spectrum is presented in Figure 14.4b (a_N = 1.51 mT, simulation depicted in Figure 14.4c).

The triplet observed remained stable until the 60-min reaction, although its amplitude slightly decreased. It should be noted that the appearance of background EPR signals in case of an excessively high initial concentration of Fe^{2+} as described (Bosnjakovic and Schlick, 2006). However, the signals reported and already attributed to HDMPO–OH and DMPOX are different than the triplet observed by Fontmorin et al. (2016), thus implying the generation of a different by-product. To understand the mechanisms leading to the decrease of DMPO–OH in the aforementioned two systems and to the formation of the triplet in the last system, similar analyses were performed by UV-Vis spectrophotometry. The contribution of each reagent (H_2O_2, Fe^{2+} and DMPO) was first analyzed, and, as shown in Figure 14.5a, the peak measured in visible part (λ_{max} = 520 nm) only appeared in the presence of both the Fenton reagent and DMPO. Time series of UV-Vis absorbance were then followed for the last two Fenton systems described previously, and the corresponding results are presented in Figure 14.5b and c. With 10 mM H_2O_2 and 1 mM Fe^{2+}, EPR and UV-Vis signals are following opposite trends. Indeed, the DMPO–OH EPR signal intensity decreased during the first 20 min, whereas the UV-Vis signal increased during the same period of time (Figure 14.5b). This result suggests the generation of EPR-silent by-products (diamagnetic) related to the quenching of DMPO–OH, which can most likely be explained by the reaction of DMPO–OH with the generated Fe^{3+} to form the tautomers HDMPN and HDMPO (Bosnjakovic and Schlick, 2006). This result also confirms that a [DMPO] : [H_2O_2] ratio of 1.68 is not high

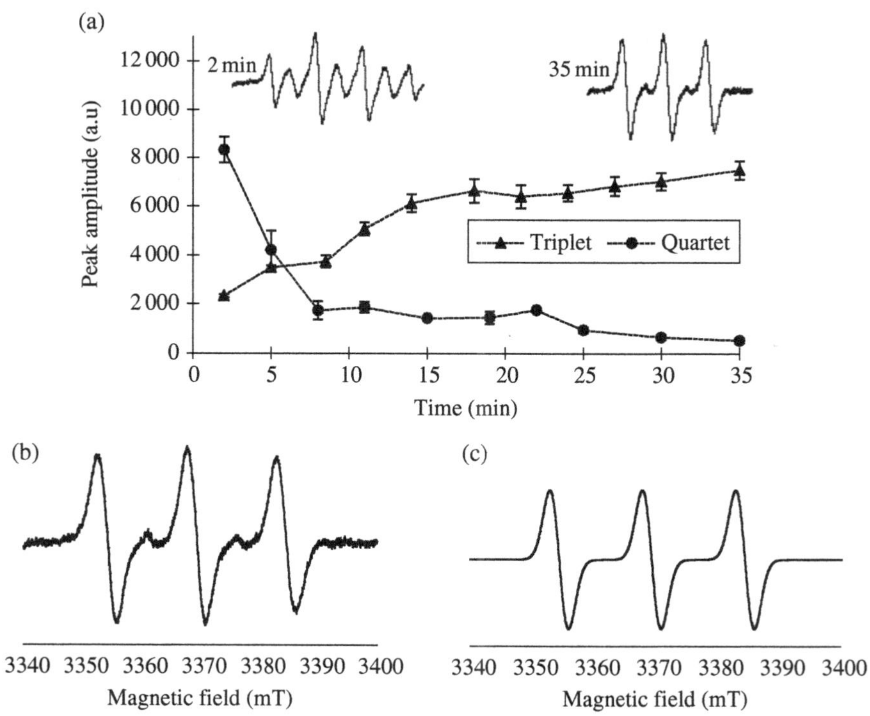

Figure 14.4 (a) Time courses of quartet (DMPO–OH) and additional triplet signals amplitudes during the Fenton reaction with 0.1 M H_2O_2, 10 mM Fe^{2+}, and 0.1 M DMPO, (b) triplet observed after 35-min reaction in these conditions, and (c) computer simulation of triplet in (b) (a_N = 1.51 mT). (*Source:* Fontmorin et al. (2016). Reproduced with permission of Elsevier.)

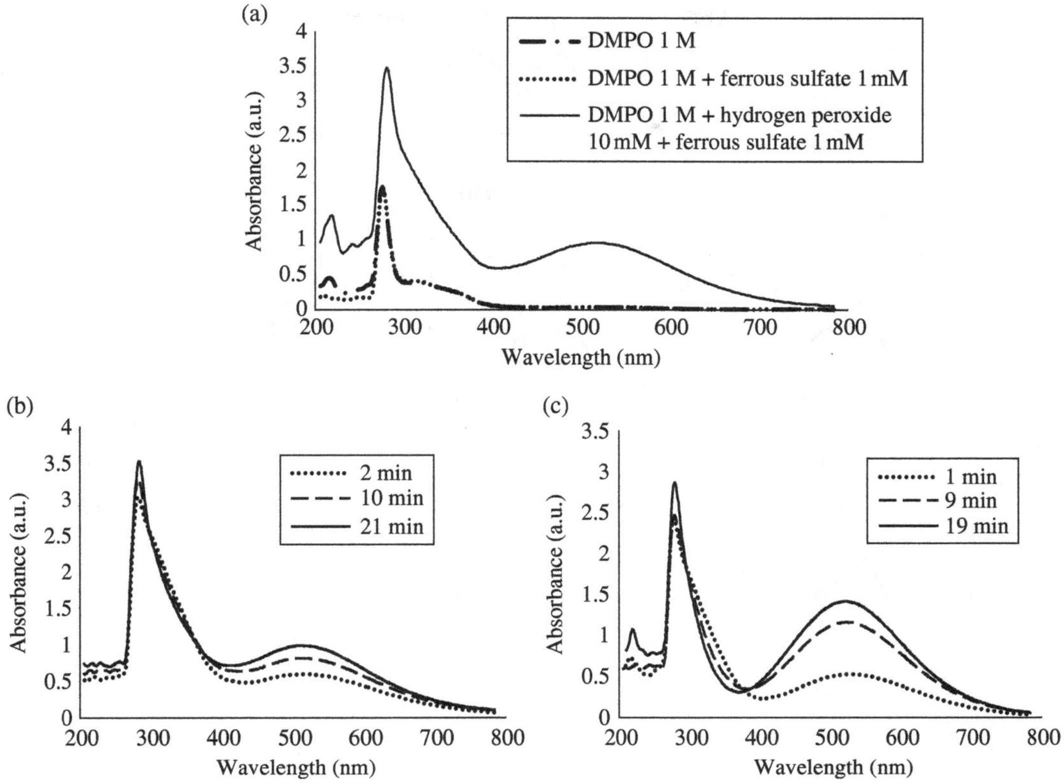

Figure 14.5 (a) UV-Vis spectra of DMPO, DMPO and ferrous sulfate, Fenton reagent, and DMPO, (b) UV-Vis spectra during Fenton reaction with 10 mM H_2O_2, 1 mM Fe^{2+}, and 0.1 M DMPO, and (c) UV-Vis spectra during Fenton reaction with 100 mM H_2O_2, 10 mM Fe^{2+}, and 0.1 M DMPO (diluted 10 times). (*Source:* Fontmorin et al. (2016). Reproduced with permission of Elsevier.)

enough to provide a stable EPR signal. It implies that under these conditions, Reactions (14.8) and (14.9) barely took place in the system, while reaction (14.7) is not negligible. Increasing the concentration of Fenton reagent to 100 mM H_2O_2 and 10 mM Fe^{2+} led to the increase of the absorbance in the visible range over the first 20 min (Figure 14.5c). It can reasonably be assumed that the higher concentration in Fe^{2+}, and therefore in the generated Fe^{3+}, enhanced the production of the two tautomers.

In the presence of excessive hydroxyl radicals, the oxidation of HDMPN and HDMPO can lead to the formation of HDMPO–OH and DMPOX (Scheme 1). Since DMPOX generates a 1 : 1 : 1 triplet ($a_N = 0.72$ mT) of 1 : 2 : 1 triplet ($a_{\gamma H} = 0.41$ mT, 2H) and that HDMPO–OH generates a 1 : 1 : 1 triplet ($a_N = 1.53$ mT) of 1 : 3 : 3 : 1 quartet ($a_{\gamma H} = 0.12$ mT, 3H) (Bosnjakovic and Schlick, 2006), the presence of HDMPO–OH and DMPOX should be excluded. The computer simulation of the triplet ($a_N = 1.51$ mT, Figure 14.4b) suggests the absence of the contribution of hydrogen nuclei in β and γ positions. The possibility of degradation of DMPO–OH spin adduct via C–N bond cleavage and the opening of the pyrroline ring have also been suggested and assigned to a similar triplet signal. However, neither a structure nor a mechanism was proposed. Such a degradation pathway was also discussed in a thermodynamic study stating that this mechanism may play a crucial role in the stability of spin adducts. The detailed mechanism related to the DMPO–OH ring opening remains unclear; moreover, the degradation product that could be assigned to the triplet observed (Scheme 1, compound **7**) because of the absence of hydrogen nuclei in α and β positions is thermodynamically highly unstable and is thus unlikely to generate a stable signal over time. In the case of 2,5,5-trimethyl-1-pyrroline-*N*-oxide (M_3PO) tautomers, the appearance of the EPR triplet signal could be explained by the dimerization reaction occurring between HDMPN and HDMPO, thus leading to the formation of a stable diamagnetic dimer, as described in the proposed mechanism presented in Scheme 2. In the presence of excessive hydroxyl radicals compared with the initial concentration of DMPO, the dimer could be further oxidized to generate the corresponding paramagnetic compound **8** associated with the additional triplet. Indeed, as shown in Scheme 2, the paramagnetic dimerization product does not present hydrogen nuclei in β position, and the contribution of hydrogen in γ position is highly unlikely due to steric effect after dimerization (Figure 14.6).

According to these experimental results, the theoretical model proposed can be adjusted as follows: to ensure a reliable detection of hydroxyl radicals, the left-hand ratio of Equation (14.18) should be three orders higher than

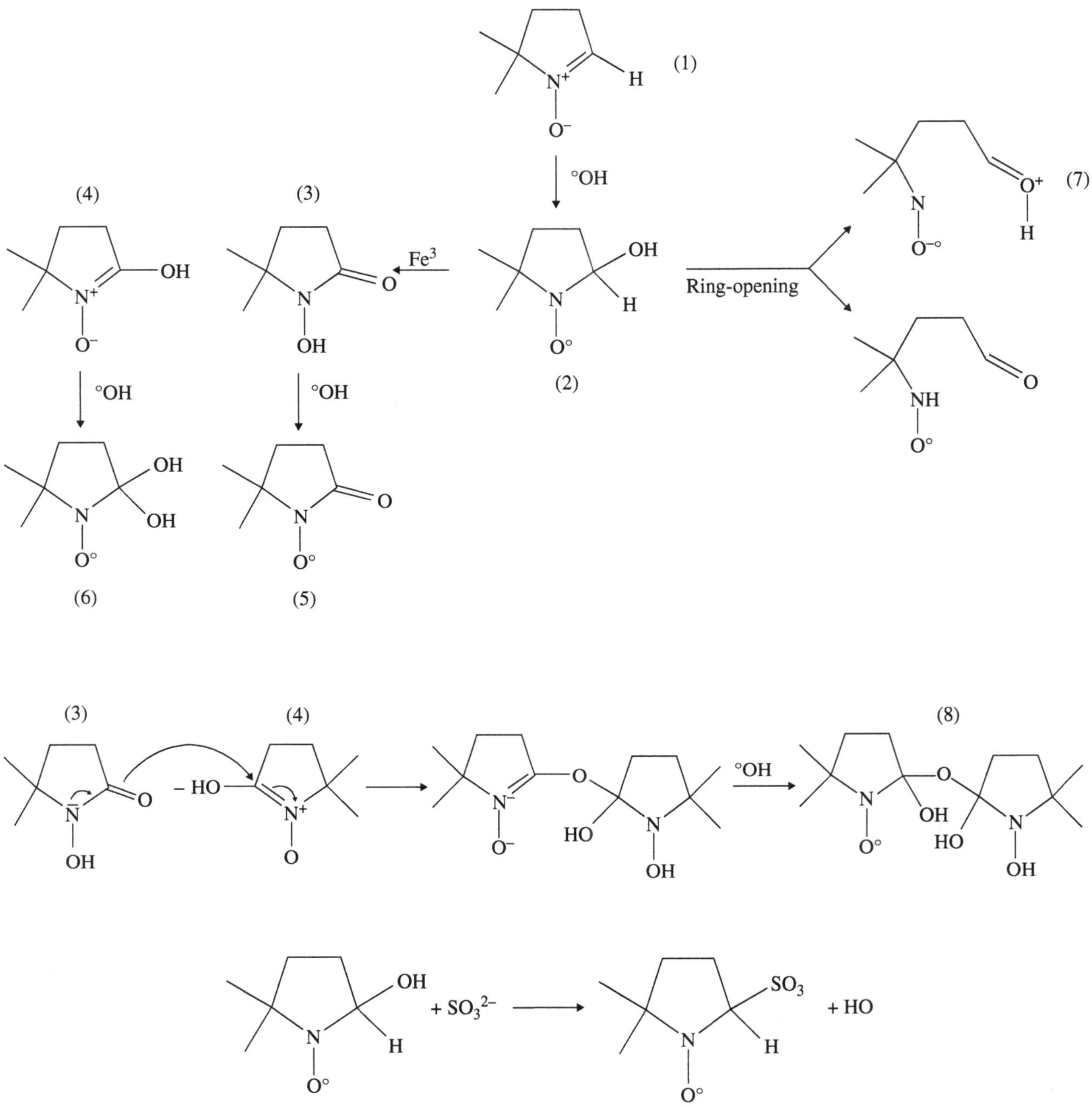

Figure 14.6 Reaction schemes of the products formed during Fenton oxidation of DMPO. (*Source:* Fontmorin et al. (2016). Reproduced with permission of Elsevier.)

the right-hand ratio. In other words, the [DMPO] : $[H_2O_2]$ ratio should be at least 16.8. The observed reactions described earlier suggest that the DMPO concentration should therefore be carefully selected according to Fenton's reagent concentration and expected magnitude of hydroxyl radical concentration as demonstrated in the previous theoretical kinetic analysis. The condition illustrated by Equation (14.18) could be applied to any concentration of the Fenton reagent during EPR quantification hydroxyl radical concentration as standard calibration procedure. Therefore, the impact of DMPO concentration on EPR signals was validated at a constant ratio of H_2O_2/Fe^{2+} but different the Fenton reagent concentrations. Initial concentration of H_2O_2 and Fe^{2+} was chosen at 5 and 0.5 mM, respectively. DMPO concentration varied from 5 to 100 mM, corresponding to the ratio of [DMPO] to [H_2O_2] varying from 1 to 20. The time series signals of DMPO–OH are presented in Figure 14.7a. As observed, DMPO concentration had a significant impact on DMPO–OH signal intensity. Using DMPO 100 mM, which is 20 times the initial H_2O_2 concentration, the signal intensity remained statistically stable until 25 min (apart from the instability of the system itself). The peak-to-peak amplitude then decreased slowly until the end of the experiment, which is consistent with the lifetime of DMPO–OH. As shown in Figure 14.7b, the corresponding EPR background after 10 min only composted of the DMPO–OH quartet, showing that no other paramagnetic by-product was formed and that the adduct was not further oxidized. Decreasing DMPO concentration from 100 to 50 mM ([DMPO] : [H_2O_2] = 10) also led to a decrease in DMPO–OH signal intensity, even though the 1 : 2 : 2 : 1 quartet was the only signal detected (Figure 14.7c).

Therefore, under these conditions, it can be assumed that DMPO–OH is partially degraded to HDMPN and HDMPO and that hydroxyl radicals are not in excess to generate the paramagnetic dimer. Finally, when DMPO concentration was lowered to 5 mM (DMPO] : [H_2O_2] = 1), the EPR background evolved into a combination of quartet and

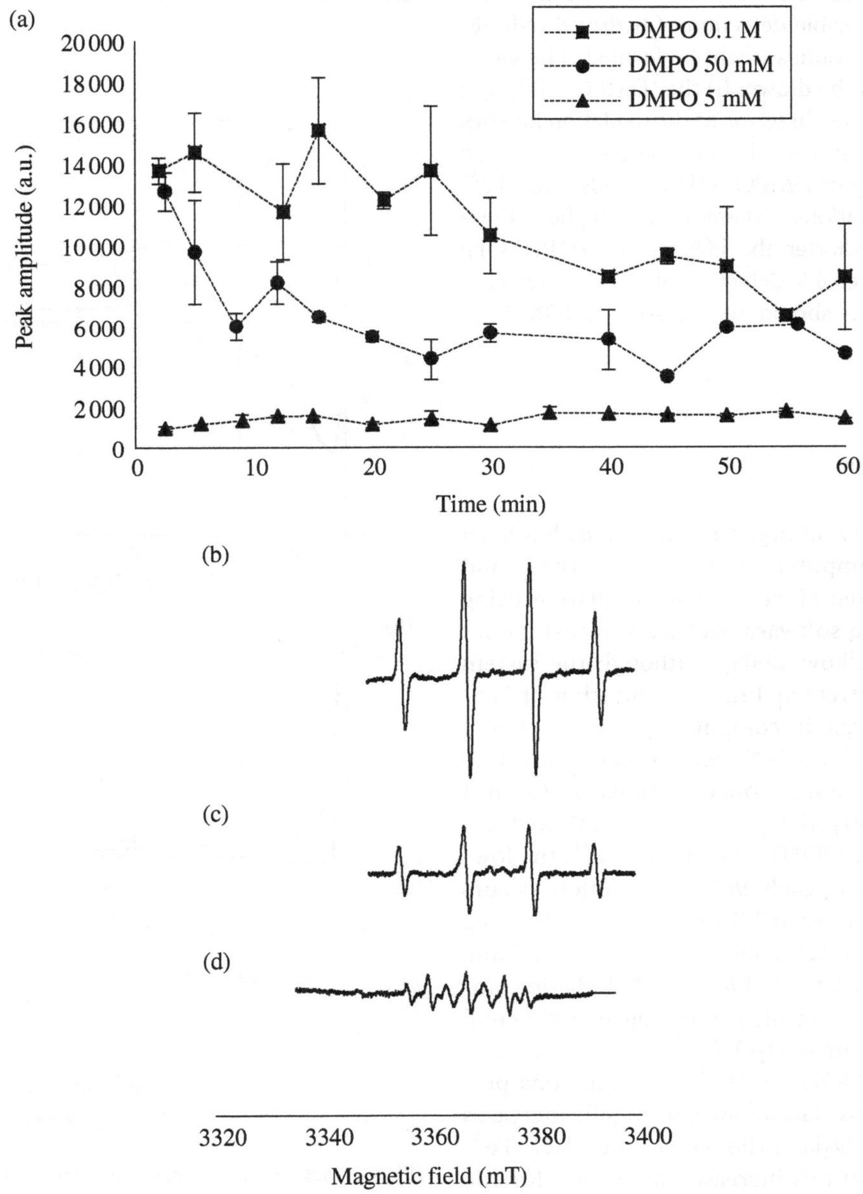

Figure 14.7 (a) Impact of DMPO concentration on DMPO–OH peak area during the time for Fenton reaction with 5 mM H_2O_2 and 0.5 mM Fe^{2+}, (b) EPR signal obtained with 0.1 M DMPO after 9-min reaction, (c) EPR signal obtained with 50 mM DMPO after 9-min reaction, and (d) EPR signal obtained with 5 mM DMPO after 9-min reaction. (*Source:* Fontmorin et al. (2016). Reproduced with permission of Elsevier.)

triplet (Figure 14.7d). These results are consistent with both the theoretical model and the results presented in Figure 14.7. It confirms that in order to achieve a reliable detection of hydroxyl radicals, the concentration ratio between DMPO and Fenton reagent should be carefully chosen, which is true for any concentration of Fenton reagent. Indeed, it should be noted that with the most concentrated system, increasing DMPO concentration from 0.1 to 0.5 M led to a significant decrease in the triplet and an increase in the quartet. The influence of DMPO concentration on the stability of EPR signal shows that the "excess concentration" in DMPO should be carefully adopted to each system, even in the absence of additional lines. The absence of additional lines does not necessarily imply a reliable detection of hydroxyl radicals, since EPR-silent compounds might be formed. Therefore, three conclusions can be drawn: (i) the [DMPO] : $[H_2O_2]$: $[Fe^{2+}]$ ratio holds the key in terms of the oxidation kinetics of DMPO–OH, and at a fixed ratio, the kinetics are the same; (ii) the stability of DMPO–OH depends upon Fe^{3+} and $^{\bullet}OH$ concentrations wherein the higher these concentrations, the shorter the lifetime of DMPO–OH; and (iii) to ensure a reliable detection of hydroxyl radicals, DMPO concentration should be set so that [DMPO] : $[H_2O_2]$ > 16.8.

14.5 Simulation

Fenton oxidation (FO) of organic compounds has been simulated using computer models. Duesterberg and Waite (2006) reported kinetic model simulation using the kinetic modeling software package Presto-Kinetics (Telgmann and Wulkow, 2005). Although the system simulated has an extremely low concentration of Fenton reagents and organic compounds, the simulation results of optimal H_2O_2/Fe^{2+} ratio range from 4 to 40. Their system contains concentrations of 400 nM Fe^{2+}, 400 nM to 400 μM H_2O_2, and 100 nM to 4 μM HCOOH. When $[HCOOH]_0$ is less than Fe^{2+}, the lowest time required to reach 90% degradation occurs when H_2O_2/Fe^{2+} is between 2.5 and 5. Tang and Huang (1997) came to a similar conclusion when examining the oxidation of a variety of chlorinated aliphatic compounds by Fenton's reagent, presenting experimental evidence for an optimal H_2O_2/Fe^{2+} ratio between 5 and 11. When $[HCOOH]_0 > [Fe^{2+}]_0$, simulations predict the formic acid oxidation rate continually increases with increasing $[H_2O_2]_0$, although as the H_2O_2/Fe^{2+} ratio rises, the rate of this increase diminishes. Kinetic simulations thus offer potential operating information for both ratios of starting concentrations, indicating either an optimum H_2O_2 concentration or an upper limit above which additional H_2O_2 provides little further benefit to improve the oxidation rate. At low initial organic concentrations (relative to the Fenton reagents), the availability of hydroxyl radicals for attack on the substrate is subject to competing reactions with Fe^{2+} and H_2O_2, both of which eventually lead to the chain-terminating oxidation of Fe^{2+} to Fe^{3+}. In the presence of oxygen, time required to achieve 90% oxidation of HCOOH at pH 3 is plotted against $(H_2O_2)_0/(HCOOH)_0$ and $(H_2O_2)_0/(Fe^{2+})_0$ in Figure 14.8a and b, respectively. Figure 14.8a shows that the shortest time to achieve 90% oxidation of HCOOH at pH 3 occurs at the $(H_2O_2)_0/(Fe^{2+})_0$ ratio between 5 and 11, which further confirms that theoretical $(H_2O_2)_0/(Fe^{2+})_0$ ratio should be at 11 (Tang, 2004).

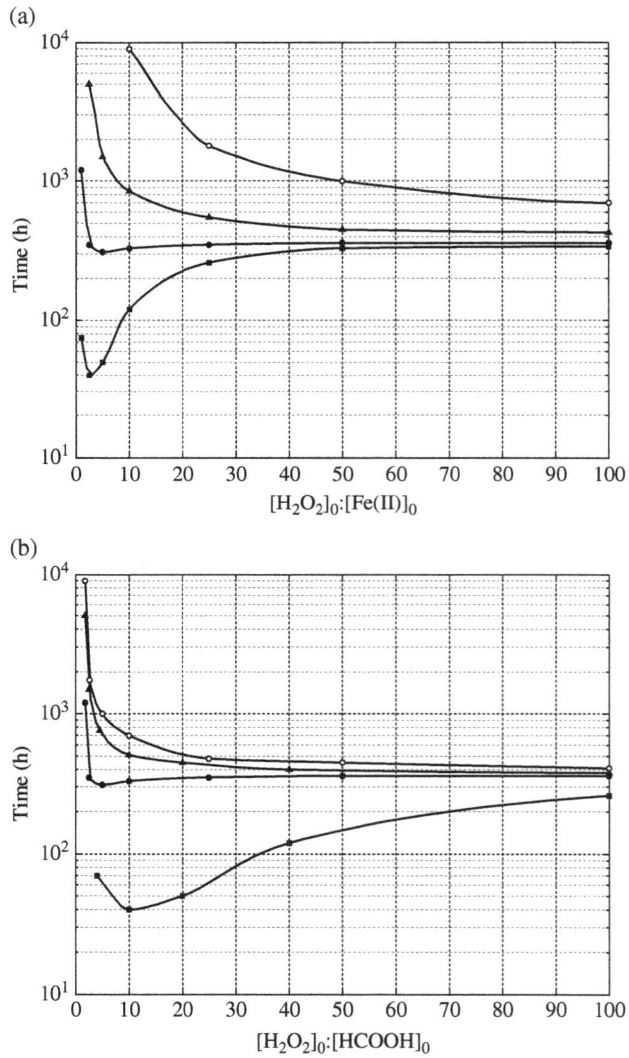

Figure 14.8 Results of kinetic simulations showing the time required to reach 90% oxidation of HCOOH at pH 3 and in the presence of oxygen. Initial conditions $(Fe(II))_0$ = 400 nM and $(HCOOH)_0$ = 100 nM, 400 nM, 1 μM, and 4 μM. (*Source:* Duesterberg and Waite (2006). Reproduced with permission of American Chemical Society.)

At pH 4, the shortest time required to reach 90% oxidation of HCOOH in the presence of oxygen is 1.51 h at theoretical $(H_2O_2)_0/(Fe^{2+})_0$ ratio of 11 as shown in Figure 14.9.

The optimal $(H_2O_2)_0/(Fe^{2+})_0$ ratio of 11 is based upon the assumption that organic substrates are directly hydroxylated without chain-promoting or chain-terminating reactions that require different kinetic models to be simulated. Duesterberg and Waite (2006) presented the elegant simulation results in Figure 14.10. In FO of formic acid as a chain-propagating organic and acetone as a chain-terminating organic, the time required to oxidize 90% against the $(H_2O_2)_0/(Fe^{2+})_0$ ratio is completely different. In oxidation of formic acid, the chain-propagating organic would facilitate the second-stage generation of hydroxyl radical by H_2O_2/Fe^{3+}; as a result, the time required to achieve 90% oxidation of formic acid decreases with increasing the $(H_2O_2)_0/(Fe^{2+})_0$ ratio. For a chain-terminating organic, there is an optimal $(H_2O_2)_0/(Fe^{2+})_0$ ratio of 4. When the $(H_2O_2)_0/(Fe^{2+})_0$ ratio increases from 4 to 40, the time required increased from 2 to 13.38 h before the time leveled off with the $(H_2O_2)_0/(Fe^{2+})_0$ ratio.

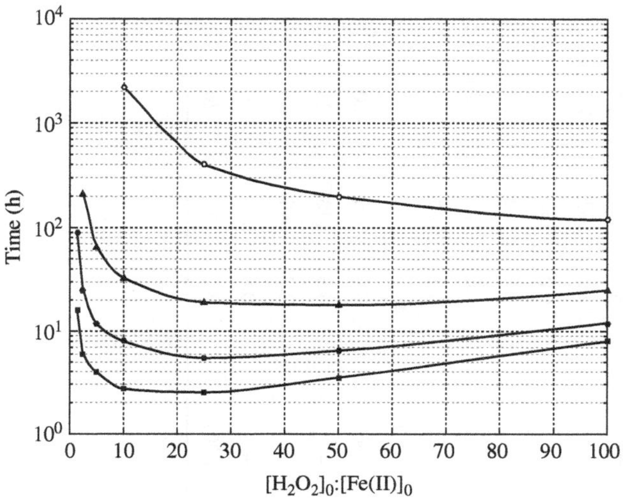

Figure 14.9 Results of kinetic simulations showing the time required to reach 90% oxidation of HCOOH at pH 4 and in the presence of oxygen. (*Source:* Duesterberg and Waite (2006). Reproduced with permission of American Chemical Society.)

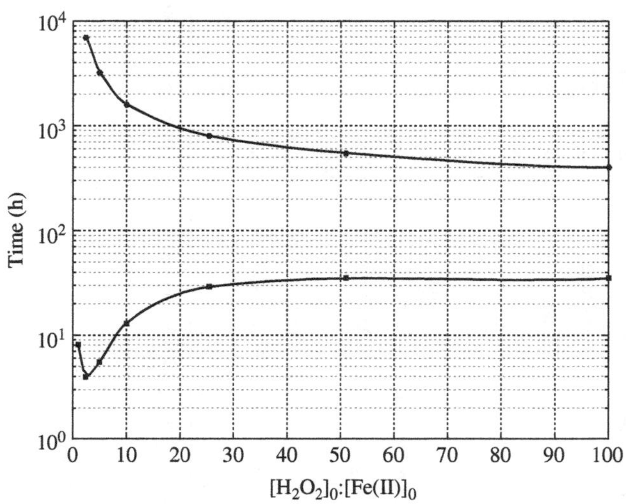

Figure 14.10 Time required to oxidize 90% of a chain-propagating organic, formic acid, and a chain-terminating organic acetone. (*Source:* Duesterberg and Waite (2006). Reproduced with permission of American Chemical Society.)

14.6 Optimization

14.6.1 Fenton Oxidation of Landfill Leachate

FO of leachate is a multistep process. At first, leachate pH is lowered to the range of 2.5–4.0 because in this pH range $^{\bullet}OH$ is more efficiently generated in addition to keeping Fe^{2+} and Fe^{3+} ions soluble (Kochany and Lipczynska-Kochany, 2009). Sulfuric acid is generally used to lower the pH and added at the start of reaction or continuously controlled the pH at the desired level (Zhang et al., 2006). Afterward, Fe^{2+} salt (ferrous sulfate) is added to the leachate and mixed at 80–400 rpm followed by gradual addition of H_2O_2 while continuing the rapid mixing for 30 s to 60 min. H_2O_2 is commonly added in single step (Lopez et al., 2004; Mohajeri et al., 2010). The effect of multistep H_2O_2 addition on leachate oxidation has also been reported (Rivas et al., 2003; Goi et al., 2010). At the end of this stage, pH is increased to the neutral pH range using sodium hydroxide or lime solution. Under neutral pH conditions, the Fe^{3+} forms ferric-hydroxo complexes, and a slow mixing (20–100 rpm) of solution for 10–30 min flocculates ferric-hydroxo complexes that subsequently precipitate for approximately 30 min to several days (Kim and Huh, 1997; Cotman and Gotvajn, 2010). The analysis of supernatant provides the treatment efficiency of the process.

FO has great potential in removing OM that contributes to the COD and color of landfill leachate. A statistical analysis of the data of leachate treatment efficiency and the optimum operating conditions as obtained from peer-reviewed publications on FO landfill leachate is presented in Table 14.5 by Singh and Tang (2013). Raw and pretreated leachates have been subjected to FO. Biological treatment and coagulation are the most common pretreatment processes used with FO. Biological treatment processes effectively remove low MW OM, whereas FO removes high MW refractory OM (Gulsen and Turan, 2004). FO has

Table 14.5 Statistical analysis of leachate treatment efficiencies and optimum operating conditions observed in peer-reviewed literature.

		Treatment efficiencies				Optimum operating conditions					
Type of leachate used for Fenton oxidation	Parameter	Initial COD (mg/l)	COD removal efficiency (%)	Color removal efficiency (%)	Increase in BOD/COD (%)	pH	Reaction time (min)	Temperature (°C)	H_2O_2/Fe^{2+} (mg/l)/(mg/l)	H_2O_2/COD_0 (mg/l)/(mg/l)	Fe^{2+}/COD_0 (mg/l)/(mg/l)
Each leachate	#[a]	37	37	3	5	35	33	24	37	37	37
	Average	4 085	67	89	940	3.4	241	27	5.4	9.1	2.2
	Max	34 920	95	95	2275	6.0	2880	50	60.7	73.1	14.9
	Min	93	31	79	58	2.5	10	20	0.5	0.0	0.01
	Std Dev[b]	6 943	14	9	915	0.9	582	8.4	10.8	17.0	3.3
	Median	1 500	68	95	920	3.0	60	25	1.8	1.2	0.9
Raw leachate	#[a]	25	25	1	4	25	24	18	25	25	25
	Average	4 180	65	79	845	3.2	234	28	5.0	11.6	2.4
	Max	34 920	86	79	2275	4.5	2880	50	30.4	73.1	12.0
	Min	93	31	79	58	2.5	10	20	0.8	0.1	0.0
	Std Dev[b]	7 181	14	—	1030	0.6	631	9.1	6.2	19.5	2.9
	Median	1 800	66	79	525	3.0	60	25	2.4	2.4	1.3
Biologically treated leachate	#[a]	8	8	2	—	6	6	4	8	8	8
	Average	922	73	95	—	4.2	132	24	1.1	4.3	2.5
	Max	1 500	95	95	—	6.0	331	25	2.0	30.1	14.9
	Min	220	56	95	—	2.5	30	20	0.5	0.0	0.1
	Std Dev[b]	492	12	0	—	1.6	121	2.5	0.5	10.4	5.0
	Median	1 050	71	95	—	4.2	110	25	0.9	0.8	0.9
Coagulation-treated leachate	#[a]	4	4	—	1	40	3	2	4	4	4
	Average	9 823	68	—	1307	3.4	515	20	16.8	2.9	0.5
	Max	22 400	80	—	—	4.5	1440	20	60.7	9.6	1.4
	Min	900	47	—	—	2.7	45	20	0.8	0.1	0.0
	Std Dev[b]	9 736	15	—	—	0.8	801	0	29.3	4.5	0.6
	Median	7 995	73	—	1,307	3.2	60	20	2.8	1.1	0.2

Source: Singh and Tang (2013). Reproduced with permission of Elsevier.
Note:
—, no data.
[a] Number of studies.
[b] Standard deviation.

been generally used as a polishing step with coagulation treatment due to its effectiveness in removing similar types of OM (Goi et al., 2010).

Raw and biological- and coagulation-treated leachate showed a median COD removal of 66, 71, and 73%, respectively. Reduced COD_0 values of pretreated leachates lead to an increased percentage of COD removal as compared with raw leachate (Kang and Hwang, 2000); however, the percentage of COD removal does not necessarily translate to the amount of COD removed. Zhang et al. (2006) observed increased amounts of COD removal at higher COD_0, but the percentage of COD removal efficiency decreased under the same Fenton-operating conditions. An average of 89% color removal was observed. FO greatly improved biodegradability of leachates, which can be represented by the ratio of BOD_5 and COD (BOD/COD). An increase from less than 0.01 to a range of 0.15–0.70 with a median increase of 0.50 was observed for BOD/COD values, which shows FO suitability as a pretreatment process for the biological treatment of landfill leachate.

FO may also help reduce the toxicity of leachate. Goi et al. (2010) observed a twofold toxicity reduction for Daphnia magna, and Cotman and Gotvajn (2010) observed 24 and 32% reduction for heterotrophic and nitrifying microorganisms, respectively; however, the removal was not significant enough, and leachate remained extremely toxic. Although $^{\bullet}OH$ has a strong oxidizing capacity, the generated $^{\bullet}OH$ during FO cannot oxidize ammonia-N that largely contributes to the toxicity of leachate (Gotvajn et al., 2011). Therefore, FO cannot be used as a primary method to remove the toxicity of leachate. Addition of Fe^{2+} salt and chemicals used for pH adjustment also increases the total dissolved solids (TDS) of treated leachate (Deng and Englehardt, 2006).

14.6.2 Optimization Fenton Oxidation of Leachate

Landfills are designed to dispose high quantities of waste at economical cost with potentially fewer environmental effects; however, improper landfill management may pose serious environmental threats through the discharge of high-strength leachate. Although modern landfill engineering and operating practices minimize leachate generation, proper leachate discharge remains one of the great challenges for landfill operators due to its complicated and changing characteristics over time. Multiple factors influencing leachate characteristics include waste age, climatic conditions, waste composition, landfill design, and operating practice; however, selecting leachate treatment method, which is often a combination of multiple processes, is mostly associated with the state of landfill stabilization.

Leachate treatability is evaluated as a function of landfill age or by the ratio of biochemical oxygen demand (BOD_5) to COD with values close to 1, representing young leachate and values of 0.1 or less described as stabilized leachate (Diamadopoulos et al., 1997). Young leachate generally contains low molecular weight and more biodegradable OM, so it is subjected to standard biological treatment processes. Once waste decomposition reaches the stable methanogenic phase, landfill leachate OM is dominated by biologically refractory OM (such as humic and fulvic-like OM), reducing the effectiveness of biological treatment processes and necessitating other physicochemical processes.

The most common physicochemical processes used for landfill leachate treatment are coagulation and flocculation (Amokrane et al., 1997; Tatsi et al., 2003), activated carbon (AC) adsorption (Maranon et al., 2009; Singh et al., 2012a), chemical oxidation (Rivas et al., 2004; Tizaoui et al., 2007), and membrane-based technologies (Trebouet et al., 2001; Tabet et al., 2002). These physicochemical processes are often used as pretreatment or posttreatment steps prior to biological processes. Although FO reactions were discovered more than 100 years ago, their application for oxidizing toxic OM in leachate was first reported only in the early 1990s (Huang et al., 1993). Landfill leachate treatment using FO should be carried out under optimal operating conditions such as pH, reaction time, reaction temperature, and reagent dose for a specific leachate to achieve the most cost-effective treatment. Optimal conditions were generally determined based on OM or COD removal efficiencies from leachate (Deng and Englehardt, 2006); however, a wide range of optimum conditions have been reported in literature primarily due to variation in leachate characteristics and experimental conditions.

Singh and Tang (2013) developed a database for analysis of the optimum FO parameters such as pH, reaction time, reaction temperature, and reagent doses as determined by previous studies along with the experimental conditions such as initial BOD, COD (COD_0), and color and COD removal efficiencies. The reagent weight ratios (H_2O_2/Fe^{2+}), reagent doses (H_2O_2/COD_0 and Fe^{2+}/COD_0), COD loading factor ($L_{COD} = COD_0/H_2O_2$), and oxidation efficiencies ($\eta = COD_{ox}/H_2O_2$) were calculated for each study. The mean, maximum, minimum, standard deviation, and median for each parameter were calculated. A relationship between dimensionless parameters η and L_{COD} was developed using the optimum operating conditions obtained by previous researchers and can be used to predict η under optimum operating conditions for landfill leachate treatment. The relationship between η and L_{COD} was cross-validated using the leave-one-out cross-validation technique, in which one observation is left out at a time, while regression is carried out. The uncertainty in the relationship was evaluated by applying Monte Carlo simulation using Crystal Ball (v11.1) software. During the uncertainty analysis, the mean (η_{mean}) and standard deviation (σ) of η at a 95% confidence interval for a range of L_{COD} were determined. The L_{COD} range of 0.01–34 was used as observed in previous peer-reviewed studies. The validation and uncertainty analysis procedures were adopted from Tang and Wang (2010).

14.6.3 Optimum Operating Conditions

14.6.3.1 pH

Obtaining an optimum pH is the first most important task in FO process, and typically it varies in a narrow range of 2.5–4.0. The pH below the optimum range affects the oxidation process by producing fewer $^{\bullet}OH$, scavenging $^{\bullet}OH$ by H^+, and reducing Fe^{2+} regeneration (Tang and Huang, 1997). The pH above the optimum range affects the oxidation reaction by reducing the production of $^{\bullet}OH$ radicals, self-decomposition of H_2O_2, formation of ferric-hydroxo complexes, scavenging of $^{\bullet}OH$ radicals by alkalinity (CO_3^{2-} and HCO_3^-), and reduction in oxidation potential of $^{\bullet}OH$ radicals (Buxton et al., 1988).

A statistical analysis of optimum pH conditions determined from the previous studies showed a range of 2.5–6.0 with a median of 3.0 (Table 14.5). An optimum pH range of 2.5–4.5 was observed for raw and coagulation-treated leachate with a median of 3.0, whereas biologically treated leachate ranged from 2.5 to 6.0 with a median of 3.5. The probable reason for a higher optimum pH of biologically treated leachate is the variability in OM characteristics as compared with raw and coagulation-treated leachate. Lau et al. (2001) and Li et al. (2010) observed that coagulation dominates the FP as compared with oxidation while using Fenton reagents for biologically treated leachate possibly due to the presence of a higher relative concentration of negatively charged OM as compared with raw and coagulation-treated leachate.

Figure 14.11 shows the effect of optimum pH conditions on COD removal efficiencies for FO of raw and pretreated leachates. At each optimum pH condition in the range of 2.5–6.0, a wide variability in COD removal efficiency was observed, so one conclusive pH condition is difficult to determine; however, a pH range of 2.5–4.0 showed maximum COD removal efficiencies and varied in the range of 31–95%. The variability in optimum pH conditions among different studies is most likely caused by the differences in experimental conditions and the type of OM present in leachate.

14.6.3.2 Reaction Time

Oxidation of OM using Fenton reagents is a two-stage process. In the first stage, OM rapidly (~10 min) is quickly oxidized due to the rapid generation of $^{\bullet}OH$ by H_2O_2 and Fe^{2+}. In the second stage, Fe^{3+} slowly regenerates Fe^{2+}, which subsequently generates $^{\bullet}OH$ by reacting with hydrogen peroxide. Therefore, the FP can be operated for minutes to days. A reaction time of 10 min to 2 days with a median of 60 min has been previously studied for leachate treatment as shown in Table 14.5. The large variation in reaction time is generally caused by the second-stage reactions. Zhang et al. (2006) concluded that most of the OM in the leachate rapidly degrades within 30 min while observing the reaction for 300 min, whereas Goi et al. (2010) observed the reactions for 24 h to achieve maximum OM removal.

14.6.3.3 Effect of Reaction Time on Fenton Oxidation

The kinetics of OM removal from landfill leachate using FO was evaluated by conducting a Fenton experiment on leachate COD_0 679 mg/l and L_{COD} 0.75 at initially adjusted pH of 2.0, 3.5, and 6.0 by Singh et al. (2013). The percentage of COD removal obtained during the FO is shown in Figure 14.12 with a fit to the exponential rise to maximum. Fenton reaction kinetics showed a two-stage reaction at each pH conditions, and a COD removal of 26–46% was observed within 1 min of reaction that later stabilized. For an initial 1 min (first stage) of FO reaction, the catalyst Fe^{2+} helped in rapidly generating $^{\bullet}OH$ from H_2O_2, and $^{\bullet}OH$ rapidly oxidized OM present in leachate. However, after 1 min of reaction (second stage), the COD removal was not as fast, and only 5–9% of additional COD removal was observed. At this stage, the concentration of $^{\bullet}OH$ reached a steady state due to limited availability of Fe^{2+}. Although Fe^{2+} are regenerated during the Fenton reaction by Fe^{3+}, the regeneration reaction rate of Fe^{2+} (10^{-3} to 10^{-2}/Ms) is approximately four times slower than Fe^{2+} consumption (76/Ms) (Umar et al., 2010). Similar OM removal kinetics have been reported by Zhang et al. (2006). In addition, they observed the first stage of FO reaction for approximately 60 min when the pH was consistently maintained at 3.5. Singh et al. (2013) did not maintain the initially adjusted pH during the oxidation reaction, and a change in pH was reported for an initial 1 min of reaction that later stabilized in the range of 2.3–2.7. The change in pH during the initial 1 min of reaction might have led to shorter FO duration as compared with 60 min as observed by Zhang et al. (2006). The OM removal results also show that the most of the OM removal by oxidation occurred when pH was changing. Very few previously published studies have reported a change in pH during FO of leachate even when pH was not maintained at an initially adjusted level (Wang et al., 2009; Cortez et al., 2011).

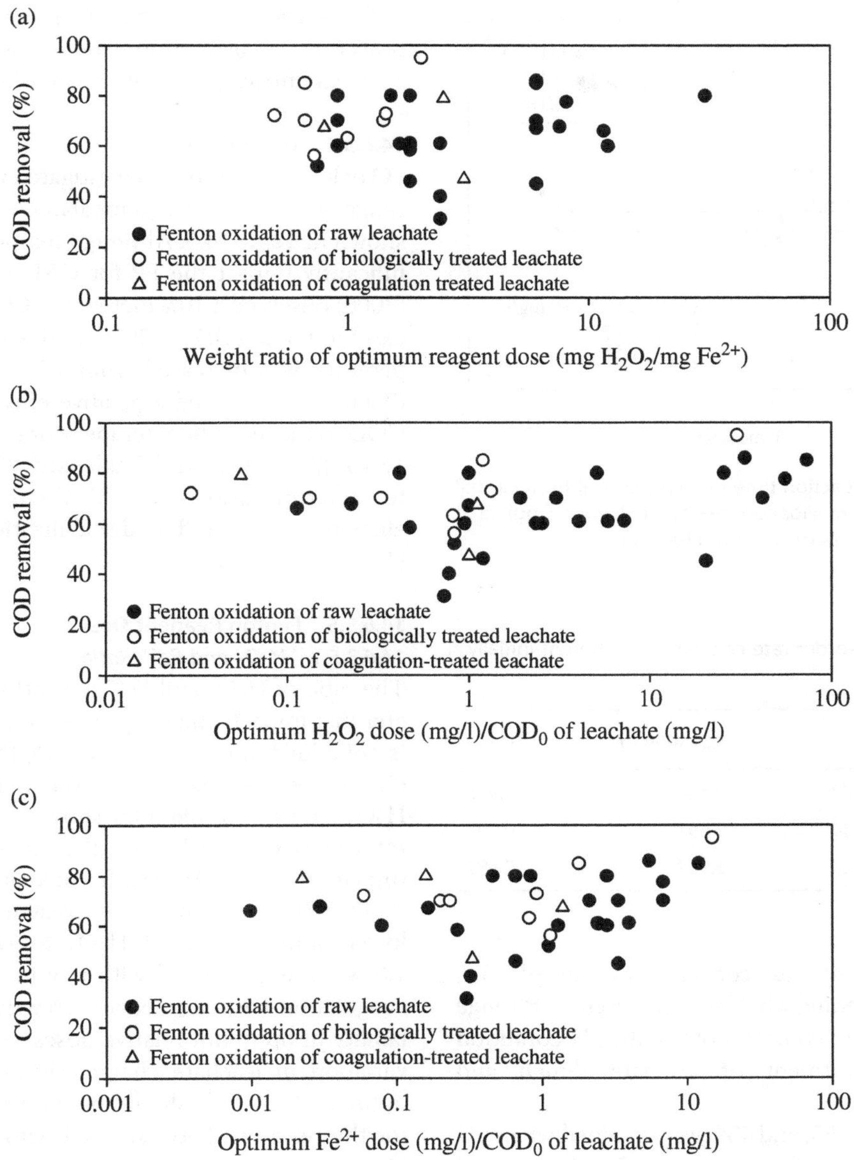

Figure 14.11 Effect of ratio of optimum (a) H_2O_2 and Fe^{2+} dose, (b) H_2O_2 and initial COD, and (c) Fe^{2+} and initial COD on COD removal efficiency during landfill leachate treatment using Fenton oxidation as determined by previous studies. (*Source:* Singh and Tang (2013). Reproduced with permission of Elsevier.)

The kinetics of OM removal by $OH^{\bullet}$ can be expressed by pseudo-first-order kinetics as shown in Equations (14.19) and (14.20):

$$-\frac{dC}{dt} = k_{obs}C \tag{14.19}$$

$$\frac{C_0 - C_t}{C_0} = K[1 - \exp(k_{obs}t)] \tag{14.20}$$

where

C is the concentration of OM represented by COD
k_{obs} is the pseudo-first-order rate constant
C_0 and C_t are the concentration of OM at time 0 and t, respectively
K is a constant dependent on k_{obs} and amount of OM present in samples

The COD removal kinetic parameters at initially adjusted pH 2.0, 3.5, and 6.0 are shown in Table 14.6. The experimental results show a strong correlation with pseudo-first-order kinetics with a coefficient of determination (r^2) of 0.99. Zhang et al. (2006) also described a similar observation while studying FO in a continuously stirred tank reactor. Among the three pH conditions

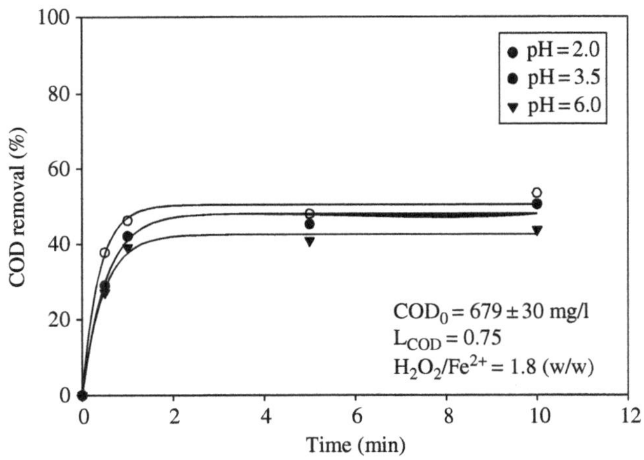

Figure 14.12 Effect of reaction time on OM removal from landfill leachate using the Fenton oxidation treatment. (*Source:* Singh et al. (2013). Reproduced with permission of Elsevier.)

Table 14.6 Pseudo-first-order rate constants at different initially adjusted pH levels.

Initial pH	k	k_{obs} (min^{-1})	r^2
2.0	48.02	2.132	0.995
3.5	50.44	2.691	0.996
6.0	42.49	1.967	0.997

evaluated, the greatest rate constant was for pH 3.5, which also complemented what have observed a pH range of 3.0–4.0 as the most common optimum pH condition for the Fenton treatment of leachate (Singh and Tang, 2013).

COD removal of 50, 53, and 43% was obtained when pH was adjusted at the start of experiment to 2.0, 3.5, and 6.0, respectively. Although COD removal efficiency was observed to be dependent on the initially adjusted pH in the range of 2–6, the reaction is rather independent of pH when pH falls below 3. In the range of pH 3–4, the reaction rate increases until the reaction reaches a steady state due to the formation of ferrous hydroxides ($Fe(OH)_2$), which react 10 times faster than Fe^{2+} (Pignatello et al., 2006). Based on the results of kinetic experiments and total COD removal, further experiments were conducted at an initially adjusted pH 3.5.

The increase in reaction time did not always translate into higher COD removal. Gotvajn et al. (2011) obtained 86% COD removal for COD_0 of 3064 mg/l within 10 min of reaction time, whereas Kim and Huh (1997) observed 67% COD removal efficiency for COD_0 of 2000 mg/l in 2 days. The characteristics of raw and pretreated leachates also did not show any conclusive impact on change in the reaction time of the FP for maximum OM removal; however, a reaction time of 10–30 min showed COD removal efficiencies in the range of 60–95%.

14.6.3.4 Temperature

FO of leachate has been investigated within a temperature range of 20–40 °C with a median temperature of 25 °C as shown in Table 14.5. Although temperature does not significantly impact the FP for OM removal, Zhang et al. (2006) observed a 10% increase in COD removal efficiencies for initial COD_0 of 3000 mg/l when the reaction temperature was increased from 13 to 37°C. Zhang et al. (2006) also observed a positive effect on the amount of COD removed when temperature was increased from 17 to 37°C. Deng and Englehardt (2006) observed that temperatures greater than 50°C reduce the OM removal efficiency due to floc destabilization and insufficient H_2O_2 decomposition.

14.6.3.5 Fenton Reagent Dose

14.6.3.5.1 H_2O_2 and Fe^{2+} Ratio

The ratio of H_2O_2 and Fe^{2+} is another important parameter that impacts the FO process. If either of the reagents is not added in their optimal ratio, the OM removal efficiency decreases due to scavenging of •OH by either H_2O_2 or Fe^{2+}. Table 14.5 shows the optimum relative ratio of H_2O_2 and Fe^{2+} in the range of 0.5–60.7 (w/w) (molar ratio 0.8–101.2) with a median of 1.2 (w/w) (molar ratio: 2.7). The biologically treated leachate showed a lower optimal ratio of H_2O_2 to Fe^{2+} doses (median: 1.6 w/w) as compared with raw (median: 3.0 w/w) and coagulation-treated (median: 4.6 w/w) leachate. The variability in optimum relative doses might be the result of variation in leachate characteristics. Theoretically, the optimal H_2O_2/Fe^{2+} should be the ratio of rate constants for the reactions between •OH with Fe^{2+} and •OH with H_2O_2 for unsaturated organic matters. The rate constant (k_{t1}) for the consumption of Fe^{2+} by •OH radicals is approximately an order higher than the rate constant (k_{t2}) for consumption of H_2O_2 by •OH; therefore, it is not surprising that most reported H_2O_2/Fe^{2+} ratios in previous studies are greater than 1 (M/M) (0.6 w/w).

For a range of 0.5–60.7 (w/w) of optimum relative concentrations of H_2O_2 and Fe^{2+}, the COD removal efficiencies varied from 31 to 95%. An optimal H_2O_2/Fe^{2+} in the range of 0.6–10.0 (w/w) was observed in 36 out of 41 (88%) studies, and 28 out of these 36 (78%) studies showed a COD removal efficiency in the range of 60–80%. Maximum variability (31–95%) in COD removal percentage was observed at relative concentration of 2–2.5 (w/w). The biologically treated leachate used fewer relative doses of H_2O_2/Fe^{2+} while showing comparable COD removal efficiency of 65–95% for raw and coagulation-treated leachate. The variability in COD removal efficiencies between studies might be caused by the variation in initial concentration and type of OM present in leachate and the differences in methods of determining optimum conditions. Cotman and Gotvajn (2010) observed a declining COD removal percentage from 85 to 45% when initial COD concentration was increased from 1400 to 5000 mg/l while maintaining the relative dose of 6.1 (w/w).

14.6.3.5.2 H_2O_2 and COD Ratio

The effectiveness of the FP also depends on the absolute concentrations of both reagents. Although H_2O_2 is the main responsible reagent in oxidation of OM, overdosed H_2O_2 scavenges $^{\bullet}OH$ as shown in Equation (14.20) and reduces the oxidation efficiency. The excess H_2O_2 may also self-decompose and cause sludge floatation due to oxygen off-gassing. Optimum H_2O_2 concentrations (Table 14.5) showed a median H_2O_2 dose of 1.2 mg H_2O_2/mg COD_0. Raw and biological and coagulation-treated leachate showed a median optimum dose of H_2O_2/COD_0 as 2.4, 0.8, and 1.1, respectively. Raw leachate required a higher optimum dose of H_2O_2/COD_0 as compared with the pretreated leachates possibly due to the nonproductive decomposition of H_2O_2 or higher COD_0 in raw leachate. Optimum H_2O_2/COD_0 doses varied to an order of three with a COD removal efficiency ranging from 31 to 95%. A ratio of H_2O_2/COD_0 of less than 10 was used in 34 of 42 (81%) studies for the COD_0 ranging from 93 to 177 000 mg/l. High H_2O_2/COD_0 doses did not necessarily show greater COD removal efficiency. Trujillo et al. used optimum H_2O_2/COD_0 dose of 1.0 for COD_0 of 177 000 mg/l and observed 75% COD removal, whereas Cotman and Gotvajn (2010) and Gotvajn et al. (2011) used H_2O_2/COD_0 in the range of 20–73 for COD_0 in the range of 1400–5000 mg/l and observed COD removal percentage in the range of 45–95%. Maximum variability in COD removal efficiency of 31–85% was observed at H_2O_2/COD_0 in the range of 0.7–1.2; however, the variability in optimal H_2O_2/COD_0 doses among these studies is possibly due to changes in leachate characteristics, variation in methods for determining optimal dosages, or failure to maintain the optimum conditions during leachate treatment.

14.6.3.5.3 Fe^{2+} and COD Ratio

Initial OM oxidation rate is a function of the Fe^{2+} concentration, so increasing the concentration of Fe^{2+} increases the rate of OM oxidation. If Fe^{2+} is overdosed, the excess Fe^{2+} starts competing with OM for $^{\bullet}OH$ and decreases the OM oxidation efficiency. The coagulation process also starts dominating at higher Fe^{2+} doses (Neyens and Baeyens, 2003), in addition to producing extra sludge and increasing TDS of effluent. Optimum Fe^{2+} dose in the range of 0.01–14.9 mg Fe^{2+}/mg COD_0 with a median of 0.8 Fe^{2+}/COD_0 as shown in Table 14.5. Raw leachate (median 1.3) required higher optimum Fe^{2+}/COD_0 (w/w) as compared with biologically (median 0.5) and coagulation (median 0.2)-pretreated leachates. Optimum Fe^{2+} doses varied to an order of three with a COD removal efficiency ranging from 31 to 95%. Fe^{2+}/COD_0 doses in the range of 0.15–5.0 (w/w) were used in 31 out of 42 (74%) studies showing COD removal efficiency in the range of 31–85% for COD_0 in the range of 93–177 000 mg/l. Similar to doses of H_2O_2/COD_0, high doses of Fe^{2+}/COD_0 did not necessarily show significant improvement in COD removal efficiencies. Cotman and Gotvajn (2010) and Gotvajn et al. (2011) used Fe^{2+}/COD_0 in the range of 3.3–14.9 for COD_0 in the range of 1400–5000 mg/l and observed COD removal in the range of 45–95%, whereas Trujillo et al. used 0.18 as optimum Fe^{2+}/COD_0 dose for COD_0 of 177 000 mg/l and observed 75% COD removal efficiency.

14.6.3.6 Generalized Fenton Dosing for Landfill Leachate Treatment

Although studies on FO of landfill leachate in the past were generally focused on determining the optimum Fenton operating conditions, these studies report a large variability in terms of COD removal efficiencies under optimum conditions, so a generalized relationship between the reagent dose and COD removal efficiency was not obtained. Theoretically, 1 mg/l H_2O_2 can oxidize 0.47 mg/l COD, and theoretical η can be defined as Equation (14.21) (Kang and Hwang, 2000):

$$\eta = COD_{ox}/0.47{*}H_2O_2 \qquad (14.21)$$

where η is the oxidation efficiency.

However, this relationship does not always hold true for landfill leachate. When COD_0 is considerably higher than H_2O_2, the observed η may be greater than 100%, but when concentration of H_2O_2 and COD_0 is comparable, η gradually decreases due to the remaining refractory OM that cannot be oxidized or to the scavenging of generated $OH\bullet$ by leachate alkalinity (Kang and Hwang, 2000). Therefore, the selection of an optimum H_2O_2 dose is very much dependent upon the COD_0, and, to determine the η in landfill leachate, it is important to consider the H_2O_2 dose with respect to COD_0.

The impact of L_{COD} (COD_0/H_2O_2) on η was evaluated from the data obtained from previous studies published in peer-reviewed literature. L_{COD} and the oxidation efficiency, η, were linearly correlated, and as L_{COD} increased from 0.01 to 34, η increased from 0.01 to 24. The linear relationship obtained between L_{COD} and η as shown in Equation (14.21) had a coefficient of determination (r^2) of 0.99:

$$\eta = 0.732L_{COD} - 0.081 \qquad (14.22)$$

More importantly, the relationship (Equation 14.22) between L_{COD} and η holds for FO of raw and biological and coagulation-treated leachates. Ideally, the slope of Equation (14.22) should be 1; however, due to the presence of •OH scavengers in landfill leachate, the slope of Equation (14.22) was less than 1.

14.6.3.7 Total COD Removal Under Different L_{COD}

The effectiveness of FO and FP treatment for total COD removal from landfill leachate can be seen for leachate of three different COD_0 1092, 546, and 273 mg/l under Fenton operating conditions: L_{COD} in the range of 0.25–1.0, reagent dose ratio (H_2O_2/Fe^{2+}) of 1.8 (w/w), and initially adjusted pH 3.5. Figure 14.13 shows that COD removal increased with a decrease in L_{COD} for all three leachates. Lower L_{COD} corresponds to higher amount of H_2O_2 used in the reaction that provided more O_2 to oxidize OM; therefore, increased COD removal was observed with the decrease in L_{COD}. FO in all three leachates showed COD removal in the range of 38–52% at L_{COD} 0.25, which was approximately twice as much as at L_{COD} 1.0. COD removal due to FO also showed dependence on COD_0, and leachate with COD_0 1092 and 273 mg/l showed a COD removal of 52 and 38%, respectively, at L_{COD} 0.25, i.e. higher OM oxidation was obtained with increased availability of OM. Zhang et al. (2006) also observed a similar trend of increased COD oxidation with the increase in H_2O_2 dose and at higher COD_0.

FP includes oxidation as well as coagulation–flocculation processes. During FO, OM was mineralized or transformed from high molecular weight OM (such as humic and fulvic like) to smaller OM. The smaller OM remains in the solution removed by coagulation and flocculation processes under neutral pH conditions. FP treatment of all three leachate showed more than 70 and 43% of COD removal at L_{COD} of 0.25 and 1.0, respectively. The highest OM removal was observed for leachate with the highest COD_0. Additionally, the COD/COD_0 after FO and FP treatment of all three leachates was observed to be linearly dependent upon L_{COD} with a slope of 0.29 and 0.37, respectively, for L_{COD} in the range of 0.25–1.0. The slope <1 for FO and FP treatments shows that the decrease in L_{COD} does not proportionally remove OM from leachate because of the presence of OM that are harder to be oxidized by OH radicals, and the amount of H_2O_2 applied at lower L_{COD} is less efficiently utilized than those at higher L_{COD}. Although none of the previous studies selected the Fenton reagent doses based on L_{COD}, mathematical conversion of doses used by Zhang et al. (2006) into L_{COD} also showed a linear dependence of COD removal on L_{COD}.

The effect of COD removal by FO, FP, and coagulation–flocculation during the Fenton treatment of three leachates with COD_0 1092, 575, and 373 mg/l, respectively, was also evaluated for L_{COD} in the range of 0.25–1.0. As shown in Figure 14.14 by the average COD removal for all three leachates by each process, FO was a more dominant process than coagulation–flocculation at lower L_{COD} of 0.25, and an average of 46% COD was removed by FO, whereas 25% COD was removed by the coagulation–flocculation process. The possible reason for greater oxidation efficiency at low L_{COD} was the increased availability of •OH within the system. To maintain the constant H_2O_2/Fe^{2+} ratio, a lower amount of Fe^{2+} was used at higher L_{COD} as compared with Fe^{2+} used at lower L_{COD}, which caused less OM removal by coagulation processes at higher L_{COD}. The

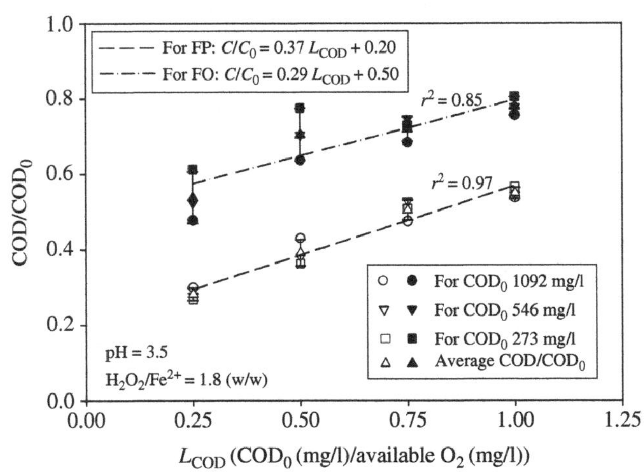

Figure 14.13 Effect of L_{COD} on COD removal by FP and FO. The filled and unfilled symbols represent COD/COD_0 after FP and FO treatment, respectively. (*Source:* Singh et al. (2013). Reproduced with permission of Elsevier.)

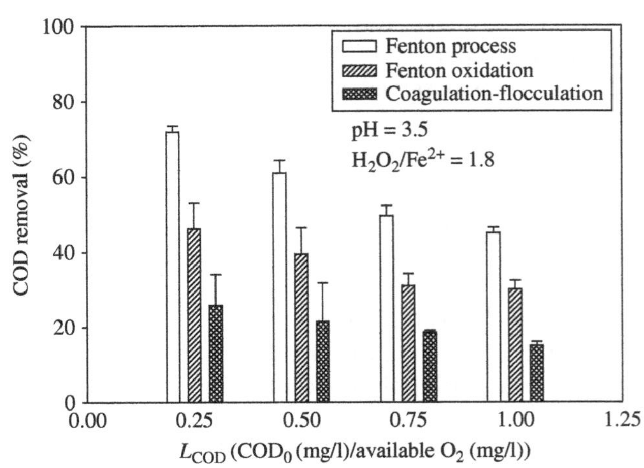

Figure 14.14 Effect of L_{COD} on COD removal by FO, coagulation, and FP. The bar and error bar represents average and standard deviation of COD removal of three leachates with COD_0 1092, 575, and 373 mg/l at each L_{COD}. (*Source:* Singh et al. (2013). Reproduced with permission of Elsevier.)

L_{COD} 1.0 showed a COD reduction of 15%, which was 10% less COD removal than at L_{COD} 0.25. Kang and Hwang (2000) also observed increased OM removal at higher Fe^{2+} doses in the coagulation process.

14.6.3.8 Effect of L_{COD} on COD Removal Efficiency

Leachate with the highest COD_0 (1092 mg/l) showed η_{FO} in the range of 0.15–0.26 for L_{COD} variation in the range of 0.25–1.0, in addition to showing relatively greater η_{FO} among the three leachates at each L_{COD} (Figure 14.15). For all three initial COD_0 concentrations of leachate, η_{FO} increases from 0.11 ± 0.03, 0.15 ± 0.05, 0.20 ± 0.02, to 0.22 ± 0.02 at L_{COD} 0.25, 0.50, 0.75, and 1.0, respectively. The higher COD_0, which corresponds to a relatively greater amount of OM present to be oxidized and higher L_{COD}, which corresponds to a lower amount of available O_2 or $^{\bullet}OH$ for oxidation of OM, leads to the increased possibility of $^{\bullet}OH$ radical reaction with OM. As a result, η_{FO} increases with L_{COD}. The average η_{FO} for all three initial COD_0 concentrations of leachate also showed a linear relationship with the increase in L_{COD} in the range of 0.25–1.0, which implies that the parameter L_{COD} derived from the COD_0 and applied H_2O_2 dose can help in predicting the OM oxidation efficiency for any leachate.

The COD removal efficiency due to FP (η_{FP}) also showed greatest utilization of available O_2 at higher COD_0 and at COD_0 1092 mg/l; the η_{FP} varied in the range of 0.20–0.49 for L_{COD} ranging from 0.25 to 1.0. Similar to the FO, the FP also showed an increase in η_{FP} with the increase in L_{COD}. At L_{COD} 0.25, 0.50, 0.75, and 1.0, the average η_{FP} increases from 0.17 ± 0.02, 0.30 ± 0.03, 0.36 ± 0.03, to 0.47 ± 0.01, respectively. As discussed earlier,

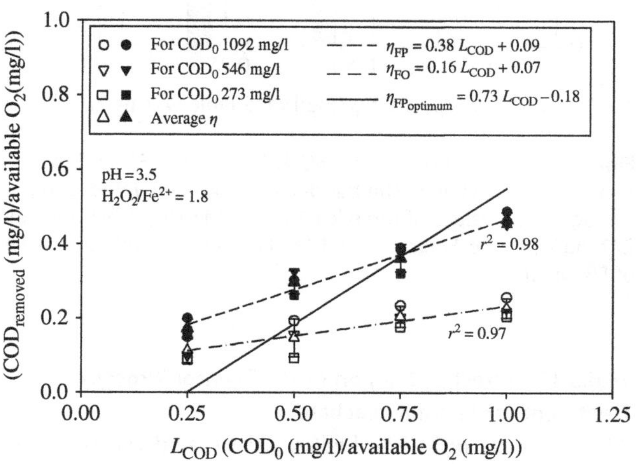

Figure 14.15 Effect of L_{COD} on η_{FO}, η_{FP}, and $\eta_{FP_{optimum}}$ as obtained by model equation (14.25) for L_{COD} in the range of 0.25–1.0 for three landfill leachates with COD_0 1092, 546, and 273 mg/l, respectively. The filled and unfilled symbols represent η_{FP} and η_{FO}, respectively. (*Source:* Singh et al. (2013). Reproduced with permission of Elsevier.)

the higher the L_{COD}, the greater the possibility of OM reacting with $^{\bullet}OH$ radicals, leading to the completely mineralization of OM or its transformation into smaller OM. These smaller OM may coagulate during the coagulation–flocculation process, thereby showing increased η_{FP} with the increase in L_{COD}. Additionally, η_{FP} was observed to be linearly increasing with L_{COD} ranging from 0.25 to 1.0 with a slope of 0.38, so the application of L_{COD} for selecting Fenton oxidant H_2O_2 dose can be useful in predicting the treatment efficiency of landfill leachate.

L_{COD} and η_{FP} were plotted against the relationship between L_{COD} and $\eta_{FP_{optimum}}$ (Equation 14.25) as obtained by Singh and Tang (2013) for L_{COD} in the range of 0.25–1.0. For L_{COD} in the range of 0.25–0.75, $\eta_{FP_{optimum}}$ was lower as compared with η_{FP} obtained by experimental data; however, for L_{COD} in the range of 0.75–1.0, the $\eta_{FP_{optimum}}$ was higher than the η_{FP}. The experiments were conducted at pH 3.5 and the H_2O_2/Fe^{2+} dose ratio of 1.8, which is the median of previous studies as reported by Singh and Tang (2013) without determining the optimum Fenton conditions of each leachate. Figure 14.15 shows that the line of η_{FP} versus L_{COD} intercepts the line of $\eta_{FP_{optimumal}}$ versus L_{COD} at L_{COD} 0.75. This suggests that η_{FP} is the same as the $\eta_{FP_{optimum}}$ obtained at the optimum conditions by previous studies at L_{COD} 0.75. Therefore, L_{COD} 0.75 could be used as an optimum dose for all three different initial COD_0 of leachate. As a result, it could be concluded that the universal relationship between $\eta_{FP_{optimum}}$ and L_{COD} under optimum conditions could be applied in determining the optimum Fenton oxidant dose without performing multiple experiments according to its initial COD_0 and the desired $\eta_{FP_{optimum}}$.

14.6.3.9 Effect of L_{COD} on Biodegradability

Lopez et al. (2004) found the effectiveness of the Fenton reagent process for pretreating landfill leachate in order to improve its biodegradability (defined as BOD_5/COD) to be ≥0.5, a value compatible with subsequent aerobic biological treatment. In their study, the leachate analyzed had a low concentration of heavy metals, a pH value of 8.2, a BOD_5/COD ratio of 0.2, and high contents of NH_4–N (5210 mg/l) and alkalinity (2174 mg/l) and therefore was classified as "old" and nonbiodegradable. To maximize the effectiveness of the FP (i.e. minimize scavenging effects), it was necessary to determine the optimal operational H_2O_2/Fe^{2+} mass ratio using a constant concentration of H_2O_2 of 6300 mg/l. This ratio was determined to be 12 (6300/500). Maintaining this mass ratio, the leachate was treated with increasing amounts of both reagents in order to determine the maximum amount of COD removed, which was determined to be 60% removal for dosages of 10 000 mg/l of H_2O_2 and

830 mg/l of Fe^{2+}. Also maintaining the mass ratio of 12, experiments were conducted in order to determine the achievement of a BOD_5/COD ratio of ≥0.5, which was achieved with dosages of 3300 mg/l of H_2O_2 and 275 mg/l of Fe^{2+} as indicated in Figure 14.16. Lopez et al. (2004) concluded that their results were encouraging and should be tested on a larger scale as the results indicated a technologically simple and cost-effective method for improving the biodegradability of leachate.

A common approach to landfill leachate treatment is to transfer leachate to a wastewater treatment facility where it is subjected to low-cost biological treatment after mixing with municipal sewage (Renou et al., 2008). However, the low biodegradability of stabilized leachate as compared with municipal sewage affects the treatment performance of WWTPs that are often built for the treatment of municipal wastewater (Diamadopoulos et al., 1997). Therefore, a single or combination of physicochemical processes that help improve leachate biodegradability is often used as a pretreatment step in biological processes (Lin and Chang, 2000). The most common physicochemical processes that help reduce refractory OM from landfill leachate treatment are coagulation (Comstock et al., 2010), anion exchange (Singh et al., 2012a), AC adsorption (Maranon et al., 2009; Singh et al., 2012b), and chemical oxidation (Rivas et al., 2004; Tizaoui et al., 2007).

In the past decade, the applicability of AOPs for removing OM and improving the biodegradability of leachate has attracted significant interest from various researchers (Umar et al., 2010). Singh et al., 2013 evaluated the biodegradability of leachate by analyzing the BOD_5 of raw and treated leachate samples. Figure 14.17 showed that a twofold increase in BOD_5 at each L_{COD} in the range of 0.25–1.0 during Fenton oxidation (FO), with the maximum increase of 2.4 times at L_{COD} of 0.25. When L_{COD} range varied from 0.5 to 1.0 of 0.50–1.0 $BOD_5/BOD_{5,0}$ was increased from 1.9 to 2.4. The smaller L_{COD} corresponds to a higher amount of available O_2 or $^{\bullet}OH$ for oxidation, so more OM may be transformed into biodegradable form as compared with the higher L_{COD}. OM that were resistant to reaction with $^{\bullet}OH$ did not transform into biodegradable form, so the application of a higher amount of H_2O_2 at lower L_{COD} would not provide additional benefit in terms of improving the biodegradability of leachate. FP showed smaller $BOD_5/BOD_{5,0}$ as compared with FO at each L_{COD} and varied in the range of 1.4–2.1. The increase in BOD_5 after FO was greater than after FP because after FO, the larger OM that are transformed into smaller OM coagulate with each other at an increased pH. Although most of the coagulated OM was precipitated during the precipitation process, the coagulated OM that are still present in the sample are harder to biologically degrade, thus showing less BOD_5 by FP than FO.

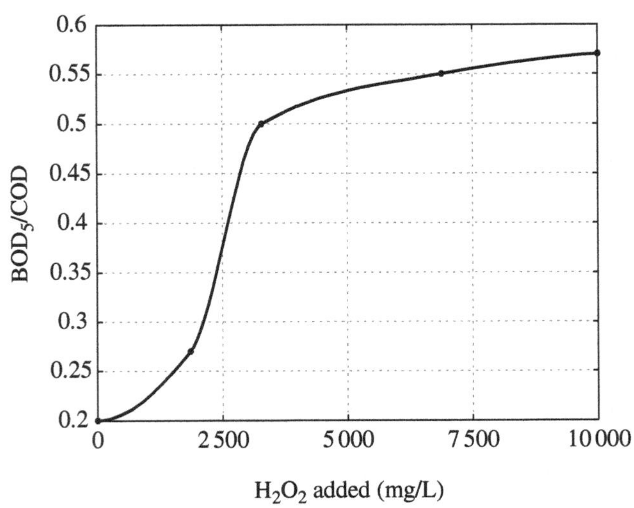

Figure 14.16 Optimal dosage of H_2O_2 for BOD_5/COD ratio of >0.5. (*Source:* Data from Lopez et al. (2004).)

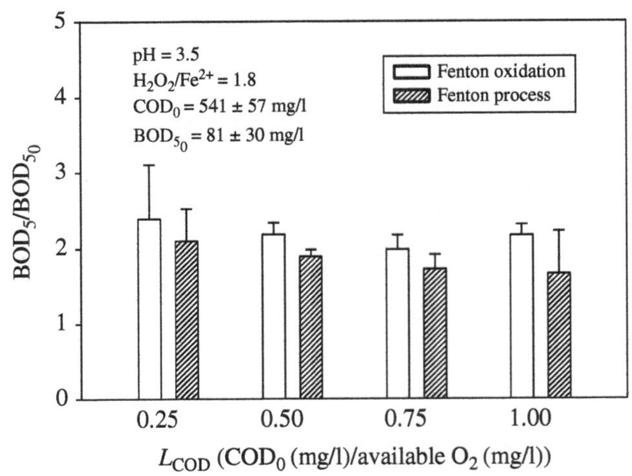

Figure 14.17 Effect of L_{COD} on BOD_5/BOD_{5_0} after FO and FP treatment of leachates. The bar diagrams are generated using average BOD_5 values of three leachates with COD_0 1092, 546, and 273 mg/l. (*Source:* Singh et al. (2013). Reproduced with permission of Elsevier.)

14.6.3.10 Effect of L_{COD} on Cost of Fenton Process Treatment for Landfill Leachate

The amounts of H_2O_2, $FeSO_4{\cdot}7H_2O$, acid (H_2SO_4), and base (NaOH) used for pH adjustment for oxidation of leachate can be used to estimate the required amount of each chemical in leachate treatment. The unit price for each chemical was collected from the commercial grade chemical suppliers, and the cost associated with each chemical and total cost was calculated using Equations (14.23) and (14.24):

$$\text{Chemical cost} = \text{unit chemical cost} \times \text{chemical consumption} \tag{14.23}$$

$$\text{Total cost} = \$_{H_2O_2} + \$_{Fe^{2+}} + \$_{H_2SO_4} + \$_{NaOH} \tag{14.24}$$

where total and chemical cost was in $/m^3 of leachate to be treated, unit chemical cost in $/m^3 of chemical to be used for leachate treatment, and chemical consumption in m^3 of chemical to be used /m^3 of leachate to be treated.

The average commercial grade cost of H_2O_2 (35% w/w), $FeSO_4 \cdot 7H_2O$, H_2SO_4 (98%), and NaOH (97%) was found to be $510, 75, 200, and 325/ton, respectively. The cost of H_2O_2 (35% w/w) was found to be the major cost-contributing chemical in FP. The chemical consumption cost at each L_{COD} for the three leachates was calculated as shown in Figure 14.18. Leachate with the highest COD_0 and lowest L_{COD} (COD_0 1092 mg/l at L_{COD} of 0.25) was observed to be the most expensive treatment case with a total chemical cost of $17.11/m^3, whereas leachate with low COD_0 (273 mg/l) and the highest L_{COD} (1.0) required a chemical cost of $1.30/m^3. Additionally, the total chemical cost for all three leachates exponentially decreased with an increase in L_{COD}. The L_{COD} 0.25 required an average of $10.3/m^3 of leachate, whereas at L_{COD} 1.0, the estimated chemical cost was $2.9/m^3. The high cost at low L_{COD} was primarily due to the amount of H_2O_2 used for the Fenton reaction and relatively high usage of H_2SO_4 and NaOH for pH adjustment as compared with the Fenton treatment of L_{COD} in the range of 0.50–1.0.

The average cost associated with treatment of three leachates using FP was also evaluated with respect to η_{FP}. For L_{COD} smaller than approximately 0.47, the η_{FP} decreased, while the treatment cost exponentially increased.

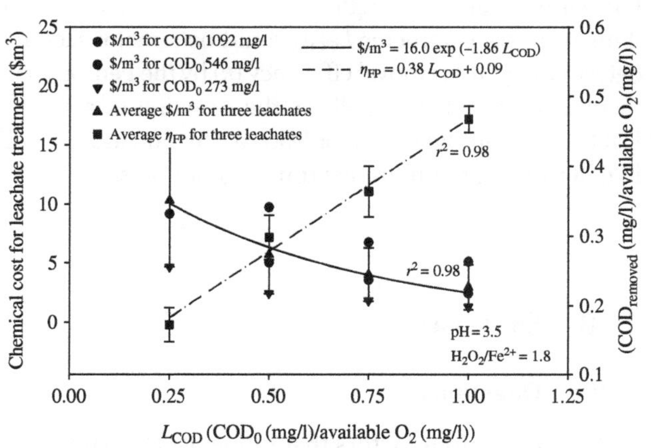

Figure 14.18 Chemical cost associated with the treatment of three leachates using FP and average η_{FP} for L_{COD} in the range of 0.25–1.0. (*Source:* Singh et al. (2013). Reproduced with permission of Elsevier.)

However, for L_{COD} greater than 0.47, the treatment cost associated with chemical usage decreased and η_{FP} increased. Therefore, L_{COD} less than 0.47 would not be cost effective due to decrease in η_{FP} and increase in treatment cost. The total chemical cost was also calculated in terms of per m^3 of leachate treated per mg/l COD removed. The average cost for all three initial leachate COD_o varied in the range of 8.2×10^{-3}–2.0×10^{-2} ($/m^3 of leachate/mg/l COD removed) for L_{COD} in the range of 0.25–1.0. Although not enough information has been published on the cost analysis of leachate treatment using FP, Rivas et al. (2004) observed a comparable treatment cost of $0.008/m^3 of leachate/mg/l of COD removed for H_2O_2 at a cost of $300/ton.

Dimensionless parameter L_{COD} could be used to predict the optimal Fenton oxidant doses without performing multiple experiments. To establish a universal dimensionless equation between the optimal efficiency of Fenton and the L_{COD}, Singh et al. (2013) conducted Fenton treatment of three leachates for L_{COD} in the range of 0.25–1.0. The kinetics of FO showed that the oxidation process reached stable stage within 1 min at the start of reaction and showed a maximum COD removal of 46% at pH 3.5, depicting FO as a very quick efficient process for OM removal from landfill leachate. The lower the L_{COD}, the greater the COD removal observed, and a maximum of 52 and 70% COD removal was observed at L_{COD} 0.25, by FO and FP, respectively. However, the usability of Fenton oxidation was greater at L_{COD} 1.0 than at L_{COD} 0.25 as observed by higher η_{FO} and η_{FP}. The comparison of experimental results with the model equation $\eta_{FP\,optimum} = 0.733L_{COD} - 0.182$ as developed by Singh and Tang (2013) showed L_{COD} 0.75 as an optimum L_{COD} for any leachate. Therefore, the selection of Fenton oxidant dose in terms of L_{COD} and comparison of experimental results with the equation $\eta_{FP\,optimum} = 0.733L_{COD} - 0.182$ may eliminate the practice of performing multiple experiments to determine optimum Fenton operating conditions. Leachate biodegradability with respect to L_{COD} showed an increase in BOD_5 in the range of 1.9–2.4 after FO for L_{COD} in the range of 0.25–1.0. Additionally, FO showed greater biodegradability improvement than FP. The evaluation of leachate treatment cost (chemical) showed an exponential increase with the decrease in L_{COD}, and L_{COD} greater than 0.47 showed cost-effective treatments by decreasing cost and increasing COD removal efficiencies.

14.7 Validation and Uncertainty

The validation and uncertainty analysis of the relationship between η and L_{COD} determined the mean (η_{mean}) and the standard deviation (σ) for a confidence interval of 95%. For a level of confidence of 95%, η lies between

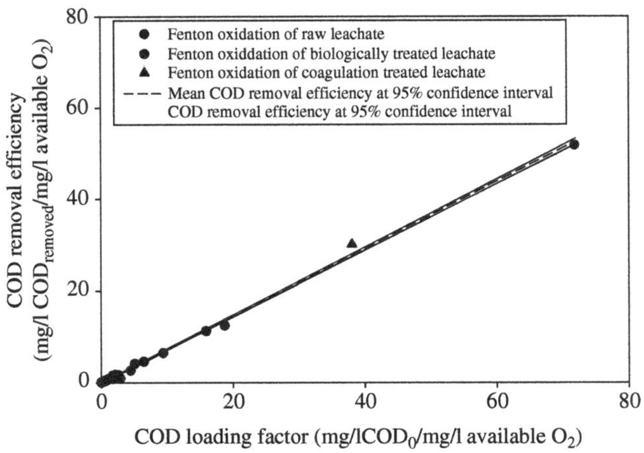

Figure 14.19 Effect of COD loading factor on COD removal efficiency observed by previous studies and COD removal efficiency at 95% confidence interval. (*Source:* Singh and Tang (2013). Reproduced with permission of Elsevier.)

η_{mean} and is two times the σ that is $\eta_{mean} \pm 2\sigma$ (Tang and Wang, 2010). The impact of L_{COD} on η_{mean} and $\eta_{mean} \pm 2\sigma$ is shown in Figure 14.19. The uncertainty analysis showed a smaller window of error between $\eta_{mean} + 2\sigma$ and $\eta_{mean} - 2\sigma$, and the predicted η was approximately the same as observed η from previous studies. Therefore, Equation (14.22) may be used to determine η for any type of landfill leachate at the L_{COD} in the range of 0.01–34, while optimal operating conditions for other parameters are maintained. Maximum OM removal by FO is obtained under optimum conditions of pH, H_2O_2/Fe^{2+}, reaction time, temperature, and reagent doses (H_2O_2/COD_0 and Fe^{2+}/COD_0). Conventionally these optimum conditions are determined by multiple laboratory or pilot-scale experiments.

FO of raw and biological and coagulation-treated leachate showed an average COD removal of 67% for an average COD_0 of 4024 mg/l, but the medians of COD_0 and COD removal efficiency were 1500 mg/l and 70%, respectively. The pretreated leachates showed a higher average of COD removal efficiencies (>70%) compared with that of raw leachates (65%). The optimal operating conditions generally varied in a range such that pH varied from 2.5 to 4.0 and H_2O_2/Fe^{2+} ratio from 0.6 to 10.0 (w/w) with COD removal efficiencies ranging from 31 to 95%. Biologically treated leachate showed a relatively lower optimum ratio of H_2O_2/Fe^{2+} doses (median: 1.6 w/w) as compared with raw (median: 3.0 w/w) and coagulation-treated (median: 4.6 w/w) leachate. Median H_2O_2 and Fe^{2+} doses were 1.2 H_2O_2/COD_0 and 0.8 Fe^{2+}/COD_0, respectively. Additionally, raw leachate required higher optimum doses of H_2O_2 and Fe^{2+} as compared with the pretreated leachates. L_{COD} and η showed a strong linear relationship such that when L_{COD} increased from 0.01 to 34, η increased as $\eta = 0.732L_{COD} - 0.081$. This linear relationship between L_{COD} and η holds for FO of raw and biological- and coagulation-treated leachates. As a result, the relationship can be used as a universal equation to predict Fenton η for L_{COD} in the given range.

The effectiveness of the FP depends on the selection of optimum Fenton operating conditions such as pH, reaction time, reaction temperature, H_2O_2 and Fe^{2+} doses, and their ratio, so previous studies on landfill leachate treatment were primarily focused on determining optimum Fenton operating conditions for specific leachates that required the conducting of multiple experiments. More often than not, previously published literature does not provide a good rationale in selecting Fenton doses used to determine optimum dose, so a wide range of Fenton optimum conditions have been reported by previous studies that do not provide confirmed universal operating conditions for landfill leachate treatment (Deng and Englehardt, 2006). Singh and Tang (2013) statistically analyzed Fenton optimum operating conditions for landfill leachate treatment reported by previous researchers and developed a linear relationship (Equation 14.25) between dimensionless parameter COD removal efficiencies ($\eta_{FP_{optimum}}$) and COD loading factor (L_{COD}):

$$\eta_{FP_{optimum}} = 0.733L_{COD} - 0.182 \qquad (14.25)$$

where $\eta_{FP_{optimum}}$ and L_{COD} *are* defined as $\eta_{FP_{optimum}} = COD_{removed_{optimum}}/\text{available } O_2$ and $L_{COD} = COD_0/\text{available } O_2$, respectively. The "available O_2" is the amount of reactive O_2 equivalent to the added H_2O_2. Theoretically, 1 mg/l of H_2O_2 has 0.47 mg/l of available O_2 for oxidation.

During Fenton treatment of landfill leachate, H_2O_2 is the main responsible reagent that generates •OH to oxidize OM, and the concentration of catalyst Fe^{2+} can always be determined according to the optimal H_2O_2/Fe^{2+}. Therefore, selecting the dose of H_2O_2 is the most critical parameter in the Fenton treatment. Singh et al., (2013) concluded that the dimensionless parameter L_{COD} could be a major factor in determining OM removal efficiency (η) by the Fenton treatment. As a result, L_{COD} allows the selection of a dose of H_2O_2 based on the COD_0 of leachate by Equation (14.25) in determining optimum Fenton reagent doses.

14.8 Exercise

14.8.1 Questions

1) List at least four factors that influence leachate generation rate.
2) Why is leachate control necessary?

3) List at least four conditions that will influence leachate chemical characteristics.
4) Please explain why leachate is toxic.
5) What does it imply when the theoretical molar ratio of H_2O_2/Fe^{2+} should be 11?
6) Should leachate be treated with wastewater in WWTP?
7) Why is equation $\eta_{FP_{optimum}} = 0.733L_{COD} - 0.182$ considered as an universal equation? What are the limits of this equation?
8) What type of organic compound would be oxidized the most at this ratio at pH 3.5?

14.8.2 Calculations

1) If H_2O_2 and Fe^{2+} were dosed at 1 and 0.1 mM, what would be the steady-state concentration of hydroxyl radical at pH 3.5? If a leachate has the same water quality as listed in Table 14.2 and has COD of 1500 mg/l, what treatment efficiency would you expect at pH 3.5?
2) In Miami-Dade Southern WWWTP, there are two treatment systems of leachate. One treats 1.2 MGD that is discharged to WWTP of 75 MGD WWTP, while 0.32 MGD leachate was treated and sent to the 35 MGD WWTP. If the leachate quality is the same as those listed in Table 14.2, please design a leachate treatment process of 1.52 MGD to meet the following standards:
   a) BOD_5 less than 20 mg/l, TSS less than 20 mg/l, total N less than 5, and total P less than 0.4 mg/l for deep well injection.
   b) BOD_5 less than 10 mg/l, TSS less than 10 mg/l, total N less than 2, and total P less than 0.02 mg/l for water reuse, what are the additional unit processes should be added to the above treatment train?
   c) For water reuse, UV disinfection is usually used to reduce the production of disinfection by-products (DBPs). What are the major barriers if the treated leachate is going to be disinfected with UV? What are the technologies you would recommend to improve the system so that UV disinfection would be effective?

14.8.3 Projects

14.8.3.1 Xiongan Project

1) One of the major challenges is to remediate contaminated water bodies within Xiongan district. 85% of Baiyangding that has 366 km^2 and 143 lakes locate within Xiongan district. However, major effluent of 250 000 m^3/day from Baoding WWTP was discharged to the Baiyangding. Since the Baoding WWTP has only secondary treatment, the discharge concentrations of N and P are 20 and 2 mg/l, respectively. Please propose a WRRF to reach discharge standards of N less than 2 and P less than 0.02 mg/l.
2) It is proposed to build another WWTP of 160 000 m^3/day. What would be the best treatment train in WRRF so that discharge concentration of BOD_5, TSS, and total N and P could be less than 10, 10, 2, and 0.02 mg/l and energy neutral? Please design this WRRF.
3) In Xiongan district, leachate from open dumping contaminates surface water and groundwater. Use the ASM3biop models to predict the performance of WWTPs using activated sludge process. Please assess the performance improvement of the following technologies:
   a) Without leachate.
   b) With leachate of 5% flow rates.
   c) If UV disinfection is used to eliminate virus and bacteria of treated wastewater, what effect of leachate would most likely to have on the UV disinfection process?
   d) How would you eliminate or minimize the leachate effect?
4) Develop two remediation alternatives of leachate for Xiongan for 5, 10, 20, and 50 years. Please conduct a life cycle cost and benefit analysis of each leachate treatment alternative and recommend one as the best SEE design.

14.8.3.2 Community Project

1) Miami-Dade Water and Sewer Department is assessing different alternatives of leachate treatment. If the leachate has typical water quality as follows and has an average flow rate of 2 MGD, estimate the following (Table 14.7):
   a) Unit energy consumption reduced due to leachate is diverted away from the WWTP.
   b) Electricity consumption reduced due to decreasing loading of COD to WWTP.

Table 14.7 Physicochemical characteristics of leachate.

Parameter	Average	Standard deviation
pH (S.U.)	7.51	0.09
DO (mg/l)	2.6	
Conductivity (mS/cm)	28.6	1.4
TDS (mg/l)	19 066	757
BOD_5 (mg/l)	151	50
COD (mg/l)	2 183	46.8
BOD_5/COD	0.07	—
TOC (mg/l)	778	73
Alkalinity (mg/l as $CaCO_3$)	2 345	35

c) Amount of sludge that could have been produced in WWTP.

2) Please contact your local solid waste management authority agency and obtain the following information:
   a) Amount of leachate generated each day in local landfills.
   b) How is it treated?
   c) Design SEE alternatives to improve the leachate treatment system.
3) Use computer models such as ASM3biop to evaluate retrofitting strategies of existing WWTP to energy positive.

References

Amokrane, A., Comel, C., and Veron, J. (1997). Landfill leachates pre-treatment by coagulation-flocculation. *Water Research* 31 (11): 2775–2782.

Bae, W., Jones, P.G., and Inouye, M. (1997). CspA, the major cold-shock protein of Escherichia coli, negatively regulates its own gene expression. *Journal of Bacteriology* 179: 7081–7088.

Bosnjakovic, A., and Schlick, S. (2006). Spin trapping by 5,5-dimethylpyrroline- N–oxide in Fenton media in the presence of nafion perfluorinated membranes: limitations and potential. *The Journal of Physical Chemistry B* 110: 10720–10728.

Buettner, G.R. (1993). The spin trapping of superoxide and hydroxyl free radicals with DMPO (5,5-dimethylpyrroline-N-oxide): more about iron. *Free Radical Research* 19: s79–s87.

Buxton, G.V., Greenstock, C.L., Helman, W.P., and Ross, A.B.(1988). Critical review of rate constants for reactions of hydrated electrons, hydrogen atoms and hydroxyl radicals in aqueous solution. *Journal of Physical and Chemical Reference Data* 17: 513–886.

Comstock, S.E.H., Boyer, T.H., Graf, K.C., and Townsend, T. G. (2010). Effect of landfill characteristics on leachate organic matter properties and coagulation treatability. *Chemosphere* 81: 976–983.

Cortez, S., Teixeira, P., Oliveira, R., and Mota, M. (2011). Evaluation of Fenton and ozone-based advanced oxidation processes as mature landfill leachate pre-treatments. *Journal of Environmental Management* 92: 749–755.

Cotman, M. and Gotvajn, A.Z. (2010). Comparison of different physico-chemical methods for the removal of toxicants from landfill leachate. *Journal of Hazardous Materials* 178: 298–305.

Deng, Y. and Englehardt, J.D. (2006). Treatment of landfill leachate by the Fenton process. *Water Research* 40: 3683–3694.

Diamadopoulos, E., Samaras, P., Dabou, X., and Sakellaropoulos, G.P. (1997). Combined treatment of landfill leachate and domestic sewage in a sequencing batch reactor. *Water Science and Technology* 36: 61–68.

Duesterberg, C.K. and Waite, T.D. (2006). Process optimization of Fenton oxidation using kinetic modeling. *Environmental Science and Technology* 40: 4189–4195.

Fontmorin, J.M., Burgos-Castillo, R.C., Tang, W.Z., and Sillanpää, M. (2016). Stability of 5,5-dimethyl-1-pyrroline-N-oxide as a radical spin trap agent for quantification of hydroxyl radical concentration in processes based on Fenton reaction. *Water Research* 99: 24–32.

Goi, N., Morishita, K., Taniguchi, A. et al. (2010). Evaluation of percutaneous permeation of flurbiprofen and ketoprofen after application of transdermal patches using a lateral sectioning approach in hairless rats. *Pharmaceutical Development and Technology* 15: 658–665.

Gotvajn, A.Z., Končan, J.Z., and Cotman, M. (2011). Fenton's oxidative treatment of municipal landfill leachate as an alternative to biological process. *Desalination* 275: 269–275.

Gulsen, H., Turan, M. (2004). Treatment of Sanitary Landfill Leachate Using a Combined Anaerobic Fluidized Bed Reactor and Fenton's Oxidation. *Environmental Engineering Science* 21 (5): 627–636.

Haywood, R. (2013). Spin-trapping: theory and applications. In: *Encyclopedia of Biophysics*, pp. 2447–2453. Berlin/Heidelberg: Springer.

Huang, C.P., Dong, C., and Tang, W.Z. (1993). Advanced chemical oxidation: its present role and potential future in hazardous waste treatment. *Waste Management* 13: 361–377.

Lau, I.W.C., Wang, P., and Fang, H.H.P. (2001). Organic removal of anaerobically treated leachate by Fenton coagulation. *Journal of Environmental Engineering* 127 (7): 666–669.

Kang, Y.W. and Hwang, K.Y. (2000). Effects of reaction conditions on the oxidation efficiency in the Fenton process. *Water Research* 34: 2786–2790.

Kim, Y. and Huh, I.(1997). Enhancing biological treatability of landfill leachate by chemical oxidation. *Environmental Engineering Science* 14 (1): 73–79.

Kochany, J. and Lipczynska-Kochany, E. (2009). Utilization of landfill leachate parameters for pretreatment by Fenton reaction and struvite precipitation a comparative study. *Journal of Hazardous Materials* 166: 248–254.

Li, J.P., Wu, Z.W., Jiang, Z.H., and He, J.H. (2010). Can global warming strengthen the East Asian summer monsoon?. *Journal of Climate* 23: 6696–6705.

Li, L., Abe, Y., Kanagawa, K. et al. (2007). Iron-chelating agents never suppress Fenton reaction but participate in quenching spin-trapped radicals. *Analytica Chimica Acta* 599: 315–319.

Lin, S.H. and Chang, C.C. (2000). Treatment of landfill leachate by combined electro-fenton oxidation and sequencing batch reactor method. *Water Research* 34: 4243–4249.

Lopez, A., Pagano, M., Volpe A., and Di Pinto, A.C. (2004). Fenton's pre-treatment of mature landfill leachate. *Chemosphere* 54 (7): 1005–1010.

Makino, K., Hagi, A., Ide, H. et al. (1992). Mechanistic studies on the formation of aminoxyl radicals from 5,5-dimethyl-1-pyrroline-N-oxide in Fenton systems. Characterization of key precursors giving rise to background ESR signals. *Canadian Journal of Chemistry* 70: 2818–2827.

Makino, K., Hagiwara, T., Imaishi, H. et al. (1990). DMPO spin trapping in the presence of Fe ion. *Free Radical Research Communications* 9: 233–240.

Maranon, E., Castrillon, L., Fernandez-Nava, Y. et al. (2009). Tertiary treatment of landfill leachates by adsorption. *Waste Management and Research* 27: 527–533.

Mohajeri, S., Aziz, H.A., Isa, M.H. et al. (2010). Statistical optimization of process parameters for landfill leachate treatment using electro-Fenton technique. *Journal of Hazardous Materials* 176: 749–758.

Neyens, E. and Baeyens, J. (2003). A review of thermal sludge pre-treatment processes to improve dewaterability. *Journal of Hazardous Materials* 98 (1–3): 51–67.

Pignatello, J.J., Oliveros, E., and MacKay, A. (2006). Advanced oxidation processes for organic contaminant destruction based on the Fenton reaction and related chemistry. *Critical Reviews in Environment Science and Technology* 36: 1–84.

Renou, S., Givaudan, J.G., Poulain, S. et al. (2008). Landfill leachate treatment: review and opportunity. *Journal of Hazardous Materials* 150: 468–493.

Rivas, F.J., Beltran, F., Carvalho, F. et al. (2004). Stabilized leachates: sequential coagulation-flocculation plus chemical oxidation process. *Journal of Hazardous Materials* 116: 95–102.

Rivas, F.J., Beltran, F., Gimeno, O., and Carvalho, F. (2003). Fenton-like oxidation of landfill leachate. *Journal of Environmental Science and Health, Part A* 38: 371–379.

Si, J.C., Lu, L., Gao, Z.F. et al. (2014). A sensitive electrochemical method based on Fenton-induced DNA oxidation for detection of hydroxyl radical. *Analytical Methods* 6: 6536–6540.

Silva, A.C., Dezotti, M., and Jr Sant'Anna, G.L. (2004). Treatment and detoxification of a sanitary landfill leachate. *Chemosphere* 55: 207–214.

Singh, S.K. and Tang, W.Z. (2013). Statistical analysis of optimum Fenton oxidation conditions for landfill leachate treatment. *Waste Management* 33 (1): 81–88.

Singh, S.K., Tang, W.Z., and Tachiev, G. (2013). Fenton treatment of landfill leachate under different COD loading factors. *Waste management* 33 (10): 2116–2122.

Singh, S.K., Townsend, T.G., and Boyer, T.H. (2012a). Evaluation of coagulation (FeCl3) and anion exchange (MIEX) for stabilized landfill leachate treatment and high-pressure membrane pretreatment. *Separation and Purification Technology* 96: 98–106.

Singh, S.K., Townsend, T.G., Mazyck, D., and Boyer, T.H. (2012b). Equilibrium and intraparticle diffusion of stabilized landfill leachate onto micro- and meso-porous activated carbon. *Water Research* 46: 491–499.

Tabet, K., Moulin, P., Vilomet, J.D. et al. (2002). Purification of landfill leachate with membrane processes: preliminary studies for an industrial plant. *Separation Science and Technology* 37: 1041–1063.

Tang, W.Z. (2004). *Physicochemical treatment of hazardous wastes*. Boca Raton, FL: Lewis Publishers.

Tang, W.Z. and Huang, C.P. (1997). Stoichiometry of Fenton's reagent in the oxidation of chlorinated aliphatic organic pollutants. *Environmental Technology* 18: 13–23.

Tang, W.Z. and Wang, F. (2010). Quantitative structure activity relationship (QSAR) of chlorine effects on E(LUMO) of disinfection by-product: chlorinated alkanes. *Chemosphere* 78: 914–921.

Tatsi, A.A.A., Zouboulis, A.I., Matis, K.A., and Samaras, P (2003). Coagulation-flocculation pretreatment of sanitary landfill leachates. *Chemosphere* 53: 737–744.

Telgmann, R. and Wulkow, M. (2005). *Presto-Kinetics–Simulation of Kinetic Models*. Rastede: Computing in Technology GmbH.

Tizaoui, C., Bouselmi, L., Mansouri, L., and Ghrabi, A. (2007). Landfill leachate treatment with ozone and ozone/hydrogen peroxide systems. *Journal of Hazardous Materials* 140: 316–324.

Trebouet, D., Schlumpf, J.P., Jaouen, P., and Quemeneur, F. (2001). Stabilized landfill leachate treatment by combined physicochemical–nanofiltration processes. *Water Research* 35 (12): 2935–2942.

Tsai, C.H., Stern, A., Chiou, J.F. et al. (2001). Rapid and specific detection of hydroxyl radical using an ultra weak chemiluminescence analyzer and a low level chemiluminescence emitter: application to hydroxyl radical-scavenging ability of aqueous extracts of food constituents. *Journal of Agricultural and Food Chemistry* 49: 2137–2141.

Umar, M., Aziz, H.A., Yusoff, M.S. (2010). Trends in the use of Fenton, electro-Fenton and photo-Fenton for the treatment of landfill leachate. *Waste Management* 30: 2113–2121.

Wang, X.J., Chen, S.L., Go, X.Y., and Wang, K.Y. (2009). Pilot study on the advanced treatment of landfill leachate using a combined coagulation, Fenton oxidation and biological aerated filter process. *Waste Management* 29: 1354–1358.

Xiang, Q., Yu, J., and Wong, P.K. (2011). Quantitative characterization of hydroxyl radicals produced by various photocatalysts. *Journal of Colloid and Interface Science* 357: 163–167.

Zhang, H., Choi, H.J., and Huang, C.P. (2006). Treatment of landfill leachate by Fenton's reagent in a continuous stirred tank reactor. *Journal of Hazardous Materials* 136: 618–623.

Zhao, H., Gao, J., Zhou, W. et al. (2015). Quantitative detection of hydroxyl radicals in Fenton system by UV-vis spectrophotometry. *Analytical Methods* 7: 5447–5453.

15

Life Cycle Cost and Benefit Analysis

The essence of sustainable design is to optimally balance capital, and operation and maintenance costs.

OBJECTIVES

To conduct life cycle cost and benefit analysis of an EEIS design
To optimize pipe size with consideration of capital and O&M costs using innovative technologies
To quantify monetary benefit of EEIS project in terms of environmental tax
To select different phosphorus recovery technologies according to capital and O&M costs
To bring a technical concept to market quickly and efficiently in a rapidly changing technical environment
To introduce basic concepts of business development, strategy, value and vision for SEE industry
To identify business opportunities and create a new business of SEE industry
To explain public and private partnership (PPP) to finance a business venture related to SEE

15.1 Principle

To reduce life cycle cost and maximize benefit over the life expectancy of an EEIS, standardized and modularized technologies play a key role in all the environmental engineering infrastructure (EEI) systems. Up-front capital costs, operating costs, and total expected present value costs will be significantly reduced using modular components. For decentralized WRRF or even centralized WRRFs, emerging innovative technologies in anaerobic digestion, biomass conversion, and nutrient recovery technologies might significantly increase system reliability and resiliency. For example, molten salt fuel cell (MSFC) technologies could generate hydrogen, heat, and electricity without any emission while reducing the physical and environmental footprint of a WRRF. A standardized modular system could also be more resilient to influent variation of flow and wastewater quality by employing different modules. For remote areas and external acute stressors such as hurricanes and earthquake, modular system could be readily deployed. Furthermore, the economies of scale of large centralized wastewater treatment plant (WWTP) could diminish if it were compared with an integrated standardized modularized system. For SEE, standards on modular engineering design are extremely important. The following is a table of Principle 12.

Principle 12 Standardized and modularized components or systems should be designed with real-time monitoring and controlling during operation to optimize capital and operation costs.

1) Components and units should be standardized and modulated for massive production.
2) Final design alternatives should be determined by balancing capital vs. operation/maintenance costs.
3) Constructability should meet LEED standards in terms of material, water, and energy consumption.

15.2 Challenges and Opportunities

Traditional environmental engineering (EE) evoluted from civil engineering and inherits tank design as huge inefficient treatment processes. SEE attempts to change the historical mindset of traditional EE to embrace new challenges and opportunities under climate change. Since unit processes in the past are not modular, no uniform industrial standards were established. On other hand, international standards and protocol of telecommunication device using the protocol of Internet of things (IoT) made it possible for all the electronic devices to be connected. In developing countries such as China, WWTPs mostly use activated

Sustainable Environmental Engineering, First Edition. Walter Z. Tang and Mika Sillanpää.

sludge and are becoming major pollution sources if sludge was not properly disposed due to high operations and maintenance (O&M) costs. After energy-intensive activated sludge treatment, huge quantity of sludge is generated. Its treatment became a major financial burden for local water utilities. In addition, WTPs and WWTPs in China were designed without consideration of life cycle assessment on environments as well as footprints. Although 4800 WWTPs in China are relative new, these plants are not efficient because all of them were designed by using activated sludge with huge aeration tanks and produce large amount of sludge. Most of these WWTPs do not have anaerobic digesters to generate biogas to power the plants, and retrofitting these WWTPs are the major challenges in China and other developing countries. At the same time, the developed countries also need upgrade or retrofit current WTPs and WWTPs to extend their serve life. When the plants approach to its design life, effluent quality deteriorated so much that it is either beyond repair or out of capacity. For example, the City of Opa-locka WTP reached its design life of 50 years. The water quality deteriorated so much that Florida Department of Environmental Protection denied its water withdrawal permit. The WTP was shut down in 1980. In Miami-Dade County, the WWTP could not satisfy peak flow capacity, and the US Environmental Protection Agency (EPA) fined two million dollars for violation of overflow rules and imposed three concert decrees to the Miami-Dade county. The county has to sell 13.6 billion dollar bond to retrofit its aging WWTP and sewage and pump system. Yet, if local governmental agencies are not up to modern design innovation, sustainability issue will come back again and cause financial difficulty of the local governments. For these reasons, life cycle cost and benefit analysis becomes crucial.

Every four years, the US EPA prepares a report to Congress, based on a survey of states, on the unfunded capital costs for projects to address water quality or water quality-related health problems due to combined sewer overflows. As of 2008, states had identified the need for \$63.6 billion to address combined sewer system overflows and \$42.3 billion for stormwater management. One approach to address combined sewer overflows is to replace a single combined sewer system with separate storm and sanitary sewer systems. Directing storm sewer flows to waterways and sanitary sewer flows to treatment plants can ensure that flow rate to WWTPs does not exceed capacity during periods of heavy rain or snowmelt. However, system separation is expensive and leaves untreated stormwater flowing directly to rivers, lakes, and estuaries. Another approach is to construct underground facilities that can hold excess wastewater from combined sewer systems until WWTPs has the capacity to handle it.

With different SEE design alternatives, life cycle cost and benefit analysis is one of the most important ways in selecting final design alternative. While monetary damage by sewer overflow and stormwater pollution of rivers may be difficult to quantify, environmental tax to be imposed by the Chinese Ministry of Environmental Protection (CMEP) provides easy way to quantify the monetary benefits of SEE design alternatives. Following are list of environmental tax of major air, water, solid wastes, and noise pollutants to be effective on 1 January 2018 (Table 15.1).

Water pollutants are further divided as class I and II as follows (Tables 15.2, 15.3, 15.4, 15.5, and 15.6).

15.3 Optimum Pipe Size

Optimal pipe size should be determined by the capital cost and its O&M costs during its life cycle. From a short-sighted point of view, the best pipe size is clearly the

Table 15.1 Environmental tax items and rates by CMEP.

Taxable items			Tax unit	Amount taxed (\$)
Air pollutants			Per pollution equivalent	0.17–1.71
Water pollutants			Per pollution equivalent	0.20–2.00
Solid waste	Coal gangue		Per ton	0.71
	Tailings		Per ton	2.14
	Hazardous waste		Per ton	142.86
	Other solid wastes (including semisolid and liquid waste)		Per ton	3.57
Noise pollution	Industrial noise	1–3 dB above standard	Per month	50.00
		4–6 dB above the standard	Per month	100.00
		7–9 dB above the standard	Per month	200.00
		10–12 dB above the standard	Per month	400.00
		13–15 dB above the standard	Per month	800.00
		16 dB above the standard	Per month	1600.00

Table 15.2 Class I water pollutants pollution equivalent values.

Pollutants	Pollution equivalent values (kg)
1. Total mercury	0.000 5
2. Total cadmium	0.005
3. Total chromium	0.04
4. Chromium(VI)	0.02
5. Total arsenic	0.02
6. Total lead	0.025
7. Total nickel	0.025
8. Benzo(*a*)pyrene	0.000 000 3
9. Total beryllium	0.01
10. Total silver	0.02

Table 15.3 Class II water pollutants pollution equivalent values.

Pollutants	Pollution equivalent values (kg)
11. Suspended solids (SS)	4
12. Biochemical oxygen demand (BOD_5)	0.5
13. Chemical oxygen demand (COD)	1
14. Total organic carbon (TOC)	0.49
15. Petroleum and derivatives	0.1
16. Animal and vegetable oils	0.16
17. Volatile phenols	0.08
18. Total cyanides	0.05
19. Sulfides	0.125
20. Ammonia nitrogen	0.8
21. Fluorides	0.5
22. Formaldehyde	0.125
23. Aniline	0.2
24. Nitrobenzene	0.2
25. Linear alkylbenzene sulfonate (LAS)	0.2
26. Total copper	0.1
27. Total zinc	0.2
28. Total manganese	0.2
29. Color developing reagent (CD-2)	0.2
30. Total phosphorus	0.25
31. Elemental phosphorus (measured in P)	0.05
32. Organophosphorus pesticides (measured in P)	0.05
33. Dimethoate	0.05
34. Parathion methyl	0.05
35. Malathion	0.05
36. Parathion	0.05
37. Pentachlorophenol sodium pentachlorophenate (in terms of PCP)	0.25
38. Chloroform	0.04
39. Adsorbable organic halide (AOX) (measured in Cl)	0.25
40. Carbon tetrachloride	0.04
41. Trichloroethylene	0.04
42. Tetrachloroethylene	0.04
43. Benzene	0.02
44. Methylbenzene	0.02
45. Ethylbenzene	0.02
46. *ortho*-Xylene	0.02
47. *para*-Xylene	0.02
48. *meta*-Xylene	0.02
49. Chlorobenzene	0.02
50. *ortho*-Dichlorobenzene	0.02
51. *p*-Dichlorobenzene	0.02
52. *p*-Nitrochlorobenzene	0.02
53. 2,4-Dinitrochlorobenzene	0.02
54. Phenol	0.02
55. *m*-Cresol	0.02
56. 2,4-Dichlorophenol	0.02
57. 2,4,6-Trichlorophenol	0.02
58. Fat-dibutyl phthalate	0.02
59. Dioctyl phthalate	0.02
60. Acrylonitrile	0.125
61. Total selenium	0.02

Note:
1) The basis for the classification of class I and II pollutants is the "Integrated Wastewater Discharge Standard" (GB8978-1996).
2) Only one item is levied environmental protection taxes among the chemical oxygen demand (COD), the biochemical oxygen demand (BOD_5), and the total organic carbon (TOC) from a single discharge outlet.

littlest size that will suit the current application. From a sensible stance, ideal channel size can have numerous implications with legitimate thought of the application. Ideal can mean monetarily productive over the life of a framework. Therefore, optimal pipe size can minimize cost of capital and operation over the lifetime of a pipe.

Table 15.4 Pollution equivalent values for pH values, color, coliform group numbers, and the amount of residual chlorine.

Pollutants		Pollution equivalent values	Unit
1. pH values	0–1, 13–14	0.06	Tons of sewage
	1–2, 12–13	0.125	Tons of sewage
	2–3, 11–12	0.25	Tons of sewage
	3–4, 10–11	0.5	Tons of sewage
	4–5, 9–10	1	Tons of sewage
	5–6	5	Tons of sewage
2. Color		5	Tons of water times
3. Coliform group numbers (above standards)		3.3	Tons of sewage
4. Amount of residual chlorine (hospital wastewater disinfected with chlorine)		3.3	Tons of sewage

Note:
1) Taxes are levied either on coliform group numbers or on the amount of residual chlorine.
2) pH 5–6 refers to [a pH value] greater than or equal to 5 and less than 6, pH 9–10 refers to [a pH value] greater than 9 and less than or equal to 10, and so forth.

Table 15.5 Pollution equivalent values for livestock husbandry, small business, and the tertiary industry.

Type		Pollution equivalent values	Unit
Livestock farms	1. Cattle	0.1	Head
	2. Pigs	1	Head
	3. Chicken, ducks, and other poultry	30	Birds
4. Small businesses		1.8	Tons of sewage
5. Catering, entertainment, and service industries		0.5	Tons of sewage
6. Hospitals	Disinfected	0.14	Beds
		2.8	Tons of sewage
	Not disinfected	0.07	Beds
		1.4	Tons of sewage

Note:
1) This chart only applies to the calculation of pollution equivalent numbers of small polluters such as livestock husbandry industry, small businesses, and the tertiary industry for which actual monitoring or materials accounting cannot be conducted.
2) Tax is only levied against livestock breeding farms with stock of more than 50 heads of cattle, 500 pigs, or 5000 chicken or ducks.
3) Where there are more than 20 beds in a hospital, follow this chart to calculate the pollution equivalent values.

Table 15.6 Atmospheric pollutant equivalent values.

Pollutants	Pollution equivalent values (kg)
1. Sulfur dioxide	0.95
2. Nitrogen oxides	0.95
3. Carbon monoxide	16.7
4. Chlorine gas	0.34
5. Hydrogen chloride	10.75
6. Fluoride	0.87
7. Hydrogen cyanide	0.005
8. Sulfuric acid mist	0.6
9. Chromic acid mist	0.0007
10. Mercury and its compounds	0.0001
11. The general dust	4
12. Asbestos dust	0.53
13. Glass cotton dust	2.13
14. Carbon black dust	0.59
15. Lead and its compounds	0.02
16. Cadmium and its compounds	0.03
17. Beryllium and its compounds	0.0004
18. Nickel and its compounds	0.13
19. Tin and its compounds	0.27
20. Dust	2.18
21. Benzene	0.05
22. Toluene	0.18
23. *p*-Xylene	0.27
24. Benzo(*a*)pyrene	0.000002
25. Formaldehyde	0.09
26. Acetaldehyde	0.45
27. Acrolein	0.06
28. Methanol	0.67
29. Phenol	0.35
30. Asphalt smoke	0.19
31. Aniline	0.21
32. Chlorobenzene	0.72
33. Nitrobenzene	0.17
34. Acrylonitrile	0.22
35. Chlorethylene	0.55
36. Phosgene	0.04
37. hydrogen sulfide	0.29
38. Ammonium	9.09
39. Trimethylamine	0.32
40. Methyl mercaptan	0.04
41. Methyl sulfide	0.28
42. Dimethyl disulfide	0.28
43. Styrene	25
44. carbon disulfide	20

Example 15.1

Given: Water is to be continuously pumped at a rate of 200 gal/min through 960 ft of schedule 40 alloy steel pipe. The fluid has a specific gravity of 0.8 and a viscosity of 1.1 cP at the pumping temperature. There is no flow independent head, i.e. static head, in the process. If annual cost escalation is taken as 5%, annual maintenance cost is 4% of capital equipment cost, and energy cost is 7¢/kWh, what is the most economic pipe size for this operation? The system is expected to have a useful life of 15 years.

To transport a flow rate of 200 gpm, steel pipe of 960 ft is to be used ($\varepsilon = 0.00\,015$ ft). If $G = 0.8$; $\mu = 1.1$ cP; $i = 0.05$; maintenance allowance = 0.04; power cost is \$0.07 per kWh; $n = 15$ years.

Find: The most economic pipe size d based on lowest life cycle cost presents value of the system.

Solution:

1) For $Q = 200$ gpm, from the nominal pipe size equation in Table 15.7,

$$d = \left(\frac{Q}{1.2}\right)^{1/3} - 2$$

$$d = \left(\frac{200}{1.2}\right)^{1/3} - 2$$

$$d = 3.503 \text{ inches}$$

From Table 15.8 select a 3-inch NPS schedule 40 pipe with $d = 3.068$ inches as a trial size. Later, for comparison, examine 2.5-inch NPS and 4-inch NPS as economic alternatives.

2) With the given fluid properties, required flow rate, and pipe material and length, calculate

$$V = 8.7 \text{ ft/s}$$

$$N_{Re} = 149\,694$$

$$\frac{\varepsilon}{D} = 0.00059$$

$$h_f = 87 \text{ ft of fluid}$$

3) Input power to a pump is given by the formula presented below. Using the assumed pump and motor operating efficiencies, determine the required horsepower to overcome the system friction in the 3-inch NPS schedule 40 pipe to be

$$\text{HP} = \frac{Q h_f G}{3960 \eta_p \eta_m} = \frac{(200)(87)(0.8)}{(3960)(0.60)(0.90)} = 6.51 \text{ hp}$$

4) The annual operating cost for this system is

$$C_o = \frac{(\$0.07)}{\text{kWh}} \times \frac{\text{kWh}}{1.341 \text{ hph}} \times \frac{6.51 \text{ hp}}{1} \times \frac{8760 \text{ h}}{1}$$
$$= \$2977 (\text{round} - \text{up to } \$3000)$$

5) Obtain a manufacturer's quotation for the pump/motor set. For this iteration (probably 7.5 hp and 90 ft of head), assume this value to be \$13 000. From this equipment cost, we can estimate an annual maintenance cost to be (0.04) (\$13 000) = \$520. The value of the recurring annual cost C_o in step 4 becomes C_o = \$3000 + \$520 = \$3520.

6) From a detailed piping arrangement of the proposed system, estimate the material and labor cost to install the trial 3-inch NPS pipe system. For this example, assume this cost is \$21 000. The initial capital cost is the sum of the piping and equipment cost, or C_i = \$13 000 + \$21 000 = \$34 000.

7) Total life cycle cost for the 3-inch NPS size system consists of the sum of the first cost and the present net value of the recurring annual costs:

$$C = C_i + C_o \frac{(1+i)^n - 1}{i(1+i)^n} = \$34\,000 + \$3\,520$$
$$\times \frac{(1+0.05)^{15} - 1}{0.05(1+0.05)^{15}} = \$70540$$

8) When steps 2 through 7 are repeated for 2.5- and 4-inch NPS schedule 40 pipe, Table 15.9 summarized the results.

Comment: The alternative with the lowest present net value is the 4-inch NPS. Therefore it is the most economic pipe size. Final pipe size could be traded off against speed and weight drop.

The same calculations are presented in Matlab solution as follows.

Example 15.2

Given: $Q = 200$ gpm; $L = 960$ ft of steel pipe ($\varepsilon = 0.00\,015$ ft); $G = 0.8$; $\mu = 1.1$ cP; $i = 0.05$; maintenance allowance = 0.04; power cost is \$0.07 per kWh; $n = 15$ years.

Find: The most economic pipe size based on lowest life cycle cost presents value of the system. Methodology: The stepwise cost analysis approach just described will be used. In the interest of brevity, for this example, many of the specific calculation details will not be presented.

Solution:

For $Q = 200$ gpm, from the nominal pipe size equation in Table 15.7,

$$d = \left(\frac{Q}{1.2}\right)^{1/3} - 2$$

$$d = \left(\frac{200}{1.2}\right)^{1/3} - 2$$

$$d = 3.503 \text{ inches}$$

Table 15.7 Quick reference summary chart and collection of pipe size formulae.

Explanation	Equation	Range and limitations
Pressurized flow of liquids		
Darcy–Weisbach frictional head loss	$d = 12\left(\frac{0.0311fLQ^2}{h_f}\right)^{0.2}$	Q in gal/min
Liquid general flow equation	$d = 0.64\sqrt{\frac{Q}{V}}$	Q in gal/min
Nominal pipe size for nonviscous flow	$d = \left(\frac{Q}{1.2}\right)^{1/3} - 2$	$Q > 100$ gal/min
Nominal pipe size for nonviscous flow	$d = 0.25\sqrt{Q}$	$Q \le 100$ gal/min
Pump suction size to limit frictional head loss	$d = \sqrt{0.0744Q}$	Q in gal/min
Pressurized flow of gases		
Gas general flow equation	$d = 0.29\sqrt{\frac{QT}{PV}}$	Q in standard ft^3/min; T in °R; P in lb/in^2 absolute
Gas general flow equation	$d = \left[\frac{Q}{8306\sqrt{\frac{P_1^2 - P_2^2}{GTLZf}}}\right]^{0.4}$	Isothermal fully turbulent flow; L in miles; Q in standard ft^3/day
Minimum diameter to limit erosional gas flow	$d = 0.001\sqrt{Q}\left(\frac{ZTG}{P}\right)^{0.25}$	Q in standard ft^3/day; T in °R; P in lb/in^2 absolute
Weymouth gas flow equation	$d = \left[\frac{W^2GL}{786.6(P_1^2 - P_2^2)}\right]^{0.1876}$	Isothermal fully turbulent flow; L in miles
Two-phase flow		
Relief valve discharge flashing flow	$d = \left[\frac{W\sqrt{ZT/M_w}}{408.245MP}\right]^{0.5}$	Isothermal flow of an ideal gas for selected Mach number M
Suitable pipe size for gas/liquid flow	$d = 0.2256\frac{\frac{W_g}{\rho_g} + \frac{W_l}{\rho_l}}{V}$	With known two-phase mixture velocity V
Steam and vapor		
Steam and vapor general flow equation	$d = 1.75\sqrt{\frac{Wv_gx}{V}}$	
Gravity flow		
Manning formula for maximum flow	$d = \left[1525\frac{Q_n}{\sqrt{S}}\right]^{0.375}$	Based on $y/d = 0.983$; Q in ft^3/s
Manning formula for maximum flow	$d = \left[1639\frac{Q_n}{\sqrt{S}}\right]^{0.375}$	Based on $y = d$; Q in ft^3/s
Minimum opening size of self-venting side entry overflow	$d = 0.92Q^{0.4}$	Q in gal/min; $2 \le d \le 18$; $y/d = 0.75$
Minimum pipe size for self-venting vertical pipe flow	$d = \left(\frac{Q}{27.8r^{5/3}}\right)^{3/8}$	Where r is ratio of annual flow area to pipe cross sectional area; Q in gal/min

Table 15.8 Inside pipe diameter of NPS pipe.

Nominal size	Schedule	d	Nominal size	Schedule	d
1	10	1.097	14	10	13.624
	40	1.049		30	13.250
	80	0.957		60	12.814
1.5	10	1.682	16	10	15.500
	40	1.610		30	15.250
	80	1.500		60	14.688
2	10	2.157	18	10	17.500
	40	2.067		30	17.124
	80	1.939		60	16.500
2.5	10	2.635	20	10	19.500
	40	2.469		30	19.000
	80	2.323		60	18.376
3	10	3.260	22	10	21.500
	40	3.068		ST	21.250
	80	2.900		XS	21.000
4	10	4.260	24	10	23.500
	40	4.026		20	23.250
	80	3.826		40	22.625
6	10	6.357	26	ST	25.250
	40	6.065		XS	25.000
	80	5.761		—	24.500
8	10	8.329	28	ST	27.250
	40	7.981		XS	27.000
	80	7.625		—	26.500
10	10	10.420	30	ST	29.250
	40	10.020		20	29.000
	80	9.564		30	28.750
12	10	12.390	32	ST	31.250
	40	12.000		XS	31.000
	80	11.750		-—	30.500

Table 15.9 Summary of economic variable for three alternative pipe systems.

Parameter (variable, units)	Pipe size, nominal pipe size		
Nominal pipe size	2.5 inch	3 inch	4 inch
Diameter (d, inches)	2.469 inch	3.068 inch	4.026 inch
Mean fluid velocity (V, ft/s)	13.4	8.7	5.0
Reynolds number (Re)	186012	149694	114074
Relative roughness (ε/D)	0.00073	0.00059	0.00045
Colebrook friction factor (f)	0.02001	0.01976	0.01975
Head loss due to friction (h_f, ft)	260	87	22
Power input (HP, hp)	19.45	6.51	1.65
Annual recurring costs (C_o, \$)	\$9 120	\$3 520	\$1 150
Initial capital cost (C_i, \$)	\$30 000	\$34 000	\$38 000
Life cycle cost present value (C, \$)	\$124 660	\$70 540	\$49 940

Matlab code:

Function:

```
function[d,drange]
  =myfundQexamplechaptertable531
  (Q,Qrange)
d=(Q./1.2).^(1./3)-2
drange=(Qrange./1.2).^(1./3)-2;
figure(1)
semilogx(Qrange/Q, drange/d,'k-',
  'linewidth',2)
set(gca,'xtickmode','auto',
  'FontSize',10,'FontWeight',
  'bold','FontName', 'Times New Roman');
xlabel({'Q/Q_0'},'FontSize',14,
  'FontWeight','bold','FontName','Times
  New Roman')
ylabel({'d/d_0'},'FontSize',14,
  'FontWeight','bold','FontName',
  'Times New Roman')
grid on
hold on
end
```

Input:

```
clear
close
Q=200;
logrange=[1:10000];
Qrange=Q.*logrange./100;
[d,drange]
  =myfundQexamplechaptertable531
  (Q,Qrange);
```

Answer: d = 3.5032 (Figure 15.1)

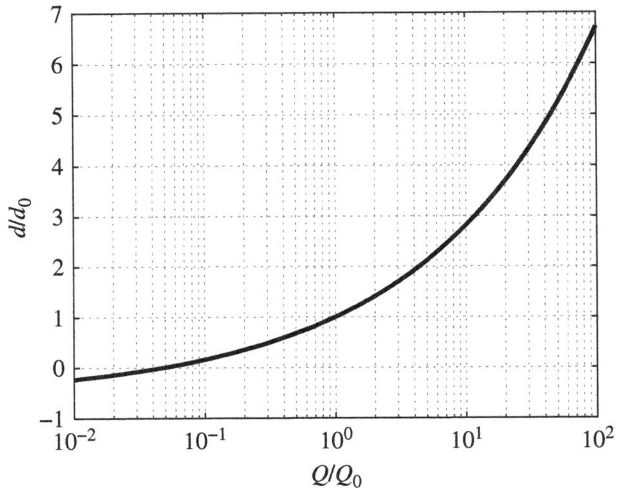

Figure 15.1 d/d_0 vs. Q/Q_0, $Q_0 = 200$ gpm.

Example 15.3
Given: Q = 90 gpm, nominal pipe size for nonviscous flow:

$$d = 0.25\sqrt{Q}$$

Find: d, d/d_0 vs. Q/Q_0.

Solution:

Matlab code:
Function:

```
function[d,drange]
  =myfundQexamplechaptertable532
  (Q,Qrange)
d=0.25*(Q).^(0.5)
drange=0.25*(Qrange).^(0.5);
figure(1)
semilogx(Qrange/Q, drange/d,'k-',
  'linewidth',2)
set(gca,'xtickmode','auto',
  'FontSize',10,'FontWeight',
  'bold','FontName', 'Times New Roman');
xlabel({'Q/Q_0'},'FontSize',14,
  'FontWeight','bold','FontName',
  'Times New Roman')
ylabel({'d/d_0'},'FontSize',14,
  'FontWeight','bold','FontName',
  'Times New Roman')
grid on
hold on
end
```

Input:

```
clear
close
Q=90;
logrange=[1:10000];
Qrange=Q.*logrange./100;
[d,drange]
  =myfundQexamplechaptertable532(Q,
  Qrange);
```

Answer: d = 2.3717 (Figure 15.2)

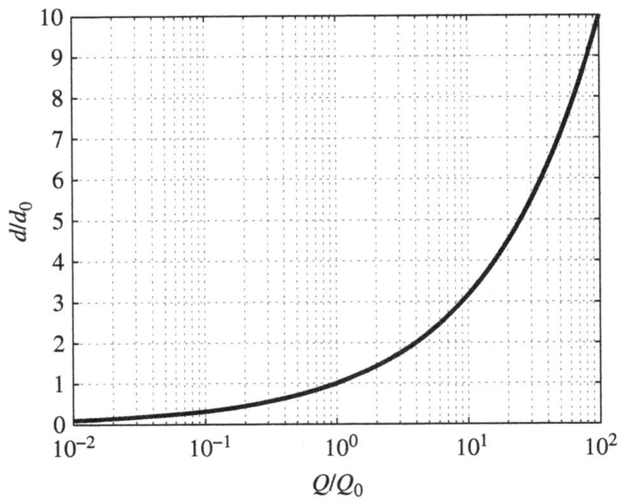

Figure 15.2 d/d_0 vs. Q/Q_0, $Q_0 = 90$ gpm.

Example 15.4
Given: Q = 100 gpm, pump suction size to limit frictional head loss:

$$d = \sqrt{0.0744Q}$$

Find: d, d/d_0 vs. Q/Q_0.

Solution:

Matlab code:
Function:

```
function[d,drange]
  =myfundQexamplechaptertable533(Q,
  Qrange)
d=(0.0744*Q).^(0.5)
drange=(0.0744*Qrange).^(0.5);
figure(1)
semilogx(Qrange/Q, drange/d,'k-',
  'linewidth',2)
set(gca,'xtickmode','auto',
  'FontSize',10,'FontWeight',
  'bold','FontName', 'Times New Roman');
xlabel({'Q/Q_0'},'FontSize',14,
  'FontWeight','bold','FontName',
  'Times New Roman')
```

```
ylabel({'d/d_0'},'FontSize',14,
   'FontWeight','bold','FontName',
   'Times New Roman')
grid on
hold on
end
```

Input:

```
clear
close
Q=100;
logrange=[1:10000];
Qrange=Q.*logrange./100;
[d,drange]
   =myfundQexamplechaptertable533
   (Q,Qrange);
```

Answer: d = 2.7276 (Figure 15.3)

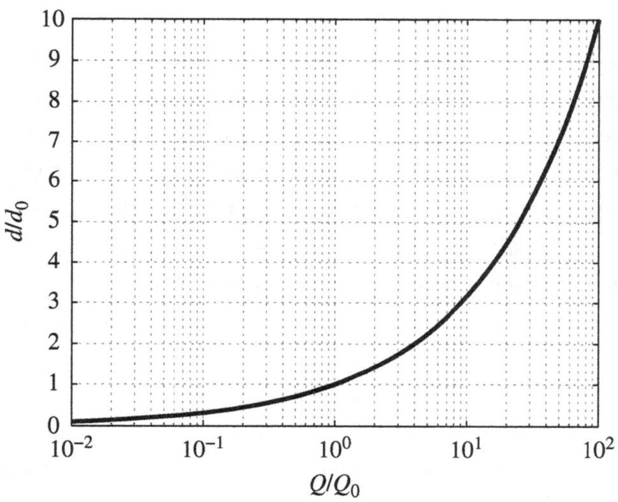

Figure 15.3 d/d_0 vs. Q/Q_0, $Q_0 = 100$ gpm.

15.4 Advanced Oxidation Process Costs

15.4.1 UV Disinfection

The US EPA (2005) provides the following cost analysis of ultraviolet (UV) disinfection systems: For flow rate less than 1 MGD, low-pressure (LP) UV lamp is usually used. For all systems, when flow rate is greater than 1 MGD, cost estimates reflect either LP or medium-pressure (MP) lamp systems. Equipment, replacement parts, labor, and power requirements are usually based upon manufacturers' information. Other associated costs by UV equipment costs represent only a portion of the total process costs. Additional process costs include instrumentation and controls, interstage pumping, piping and valves, and housing. For small systems (flows <1.0 MGD), additional process costs are estimated by using the capital cost multiplier as shown in Table 15.10. Indirect capital costs (for both large and small systems) include pilot testing, training, and spare parts. Pilot testing cost assumptions for UV are presented in Table 15.10.

Table 15.10 Capital cost multiplier.

Design flows (MGD)	Capital cost multiplier
<0.1–1.2	1 000
2–17	5 000
17–76	10 000
>76	200 000

Capital cost multipliers used for UV disinfection differ from those recommended by National Drinking Water Advisory Council (NDWAC). For flows less than 1 MGD, the capital cost multiplier is 1.2. For flows greater than or equal to 1 MGD, the capital cost multiplier is 1.36 for systems using a dose of 40 mJ/cm^2 and 1.76 for a dose of 200 mJ/cm^2. Systems less than 1 MGD require a smaller capital cost multiplier than other treatment technologies because small UV systems do not need significant area (i.e. new building not needed), equipment installation is not complex, and plant modifications are minor compared with other technologies. The capital cost multiplier of 1.36 used for 40 mJ/cm^2 systems is a revised multiplier based on actual data from facilities. The lower-cost multiplier was used because lower installation costs and less site work are necessary compared with other treatment technologies.

Indirect capital costs for systems using a dose of 40 mJ/cm^2, pilot testing, operator training, housing, and a spare parts inventory are included as indirect capital costs. Pilot testing was assumed to be $1 000 for systems with a design flow of less than 2 MGD, $5 000 from 2 to 10 MGD, $10 000 for 10–25 MGD, and $200 000 for systems with a design flow greater than 25 MGD. Operator training was assumed to be $1 000 for small systems and ranges from $3 000 to $25 000 for larger systems. Housing costs were based on the estimated UV system footprint size multiplied by a median housing unit costs of $150/ft^2 based on actual UV costs. Footprint sizes ranged from 335 to 22 000 ft^2. Also, based on data reported from 18 actual UV facilities, it was assumed that 39% of facilities would not require an additional building; therefore the housing costs were reduced by this percentage to reflect a national average cost. The spare parts inventory costs were based on a 10% backup of system equipment including lamps, sleeves, and sensors, with the exception of ballasts and ultraviolet transmittance (UVT) monitors that were based on a 5% and one unit backup of system equipment, respectively.

The O&M costs reflect labor hours, replacement parts, and lamp operating information provided by the manufacturer. The number of lamps, sensors, and ballasts is different, depending on the different manufacturer. Costs for replacement parts for each manufacturer were based on the following replacement intervals:

1) LP lamps replaced annually.
2) MP lamps replaced every 6 months.
3) Sleeves replaced every 8 years.
4) Intensity sensors and reference sensors replaced every 5 years.
5) Ballasts and UVT monitors replaced every 10 years.

For systems treating less than 2 MGD, 1 h of labor per month plus an additional 2 h per lamp replacement was assumed. For systems treating more than 2 MGD, labor hours were estimated by manufacturers for the following tasks: daily operation, lamp replacement (annually for LP lamps and every 6 months for MP lamps), quarterly sensor calibration, and cleaning once per month for UV systems that do not use automatic cleaning. Labor costs were derived from the labor hours estimate and assumed labor rate. Table 15.11 represents the US EPA's estimate of UV disinfection cost.

Table 15.11 UV disinfection cost summary (40 mJ/cm^2 without UPS) (EPA 2005).

Design flow (MGD)	**0.007**	**0.022**	**0.037**	**0.091**	**0.18**	**0.27**	**0.36**	**0.68**	**1**
Average flow (MGD)	0.0015	0.0054	0.0095	0.025	0.054	0.084	0.11	0.23	0.35
Capital cost summary									
Total capital cost	10 195	13 034	15 834	25 596	40 597	54 386	66 790	99 661	310 154
Indirect capital costs	3 686	3 704	3 722	3 794	3 934	4 102	4 296	5 200	6 206
Training	1 000	1 000	1 000	1 000	1 000	1 000	1 000	1 000	1 000
Treatability testing	1 000	1 000	1 000	1 000	1 000	1 000	1 000	1 000	1 000
Spare parts	1 686	1 704	1 722	1 794	1 934	2 102	2 296	3 200	4 206
Direct capital cost[1]	6 509	9 330	12 112	21 802	36 662	50 284	62 493	94 461	303 947
Subtotal process cost	5 424	7 775	10 094	18 168	30 552	41 903	52 078	78 717	223 491
I&C (incl. HVAC)	—	—	—	—	—	—	—	—	40 000
Pipes and valves	—	—	—	—	—	—	—	—	17 717
Adjusted pumping	—	—	—	—	—	—	—	—	1 564
Adjusted housing	—	—	—	—	—	—	—	—	20 210
UV reactors	5 424	7 775	10 094	18 168	30 552	41 903	52 078	78 717	128 000
Electrical	—	—	—	—	—	—	—	—	16 000
Annual O&M cost summary									
Total O&M cost	3 350	3 380	3 769	4 549	4 736	6 115	6 493	8 152	9 016
Replacement parts	3 000	3 000	3 377	4 000	4 000	5 200	5 400	6 400	6 800
Power/electricity	50	80	91	180	320	420	524	960	1 400
Labor $	300	300	300	369	416	495	569	792	816
UV (dose = 40 mJ/cm^2, UV254 = 0.05, turbidity = 0.1 NTU, Alk = 60 mg/l, hardness = 100 mg/l)	737.1	215.9	139.4	68.7	37.9	31.9	27.4	17.7	23.4
UV (dose = 200 mJ/cm^2, UV254 = 0.05, turbidity = 0.1 NTU, Alk = 60 mg/l, hardness = 100 mg/l)	1 870.7	562.2	368.8	190.9	117.8	97.7	84.9	58.6	64.3
UV with UPS (dose = 40 mJ/cm^2, UV254 = 0.05, turbidity = 0.1 NTU, Alk = 60 mg/l, hardness = 100 mg/l)	129.7	45.9	31.6	19.3	14.2	12.3	11.5	8.4	16.7
UV with UPS (dose = 200 mJ/cm^2, UV254 = 0.05, turbidity = 0.1 NTU, Alk = 60 mg/l, hardness = 100 mg/l)	2 140.2	640.5	413.8	208.9	126.7	103.8	89.9	61.5	67.6

The US EPA reported a detail cost of treating methyl tertiary-butyl ether (MTBE) using granular activated carbon (GAC) and advanced oxidation process (AOP). Relevant assumptions and costs for these polishing systems using Calgon Filtrasorb 600 carbon are listed in Tables 15.12 and 15.13. Costs generally are between \$0.03 and \$3.00 per 1000 l (\$0.10–\$10.00 per 1000 gal). Factors that influence the cost to implementing UV/oxidation include:

1) Types and concentration of contaminants (as they affect oxidizer selection, oxidizer dosage, UV light intensity, and treatment time).
2) Degree of contaminant destruction required.
3) Desired water flow rates.
4) Requirements for pretreatment and/or posttreatment.

Tables 15.14 and 15.15 provide a sample calculation of total capital costs, total annual costs, and unit treatment costs of H_2O_2/O_3 system for MTBE removal. As these tables indicate, the capital costs provided by the vendors were the bases for estimating the complete installed system costs. Piping, valves, and electrical work were estimated at 30% of the system equipment costs. Site work was estimated at 10% of equipment costs, engineering was estimated at 15% of equipment costs, and contractor O&M was estimated at 15% of equipment costs. A contingency of 20% of the total costs was then added.

Table 15.12 Cost of hydrogen peroxide removal.

H_2O_2 treatment systems (centaur carbon)			
Flow rate (gal/min)	**Capital costs (\$)**	**Annual O&M costs (\$)**	**Total amortized operating costs (\$/1000 gal)**
60	18 000	39 000	1.27
600	89 000	44 000	0.60
6000	667 000	194 000	0.10

Table 15.13 Cost of oxidation by-product removal.

Treatment for TBA, TBF, acetone, and other by-products (Bio-GAC)			
Flow rate (gal/min)	**Capital costs (\$)**	**Annual O&M costs (\$)**	**Total amortized operating costs (\$/1000 gal)**
60	16 700	42 200	1.38
600	67 800	42 700	0.15
6000	667 000	204 800	0.08

Note:
60 gpm systems consist of one 500-lb GAC vessel.
600 gpm systems consist of one 5000-lb GAC vessel.
6000 gpm systems consist of ten 5000-lb GAC vessels.
O&M costs include carbon replacement, equipment, analytical sampling, oversight during change-outs, and general system O&M.
Capital costs include pining, valves, electrical (30%), site work (10%), contractor O&P (15%), engineering (15%), and contingency (20%).

Table 15.14 Costs of H_2O_2/O_3 system for MTBE removal (Applied Process Technology, Inc).

Items	Cost (\$)
Advanced oxidation unit[a]	750 000
Piping, valves, electrical (30%)	225 000
Site work (10%)	75 000
Contractor O&P (15%)	157 500
Engineering (15%)	181 125
Subtotal	**1 388 625**
Contingency (20%)	277 725
Total capital	**1 666 400**
Amortized capital	**134 290**
Annual O&M	**129 692**
Total annual cost	**263 981**
Total cost/1000 gal treated	**0.84**

[a] Cost of oxidation unit from vendor.

Table 15.15 Summary of annual O&M costs of H_2O_2/O_3 system for MTBE removal.

Summary of annual O&M costs[a]				
Item	**Unit**	**Quantity**	**Unit cost (\$)**	**Cost (\$)**
Replacement parts[b]	Lump sum	1	11 250	11 250
Labor[c]	Hour	724	80	57 920
Analytical costs[d]	Sample	208	200	41 600
Chemical costs[e]	\$/1000 gal	315 360	0.03	9 461
Power (\$0.08/kWh)[f]	kWh	118 263	0.08	9 461
Annual O&M				129 692

Note:
System parameters: 600 gpm.
200 μg/l influent MTBE concentration.
20 μg/l effluent MTBE concentration.
[a] Amortization based on 30 year period at 7% discount rate.
[b] Replacement is based on vendor's estimate of 1.5% capital cost.
[c] Breakdowns of labor costs are based on a rate of \$80/h.
[d] Sampling conducted weekly at four locations.
[e] Chemical costs based on dosages and prices estimated by vendor.
[f] Power is based on consumption estimates provided by vendor, priced at \$0.08/kWh.

Table 15.16 presents a summary of the capital costs for three AOP technologies evaluated under the scenarios

Table 15.16 Capital costs of AOPs.

Flow (gpm)	Influent (μg/l)	Effluent (μg/l)	Removal efficiency (%)	Calgon Carbon Corporation ($)	Applied Process Technology, Inc. ($)	Oxidation Systems, Inc. ($)	Hydroxyl Systems, Inc. ($)
60	20	5	75.00	177 700	444 400	134 400	277 700
60	20	0.5	97.50	188 900	444 400	242 800	426 600
60	200	20	90.00	177 700	444 400	156 600	426 600
60	200	5	97.50	188 900	444 400	242 800	439 900
60	200	0.5	99.75	188 900	533 200	260 000	586 600
60	2000	20	99.00	188 900	533 200	260 000	586 600
60	2000	5	99.75	188 900	533 200	260 000	517 700
60	2000	0.5	99.98	266 600	622 100	260 000	691 000
600	20	5	75.00	266 600	1 666 400	356 600	1 142 000
600	20	0.5	97.50	488 800	1 777 400	461 200	1 344 200
600	200	20	90.00	337 700	1 666 400	356 600	1 344 200
600	200	5	97.50	488 800	1 777 400	461 200	1 730 800
600	200	0.5	99.75	695 400	1 777 400	792 100	2 035 200
600	2000	20	99.00	695 400	1 777 400	482 100	2 035 200
600	2000	5	99.75	811 000	1 888 500	482 100	2 628 400
600	2000	0.5	99.98	1 299 800	1 888 500	482 100	3 092 800
6000	20	5	75.00	999 800	7 998 500	1 446 400	9 711 500
6000	20	0.5	97.50	2 666 200	8 887 200	4 151 000	9 711 500
6000	200	20	90.00	1 666 400	7 998 500	3 209 400	11 426 700
6000	200	5	97.50	3 332 700	8 887 200	4 151 000	14 703 900
6000	200	0.5	99.75	4 221 400	8 887 200	4 339 200	17 298 900
6000	2000	20	99.00	4 999 000	8 887 200	4 339 200	17 298 900
6000	2000	5	99.75	6 665 400	8 887 200	4 339 200	22 344 600
6000	2000	0.5	99.98	9 998 100	9 775 900	4 339 200	26 288 300

Note:
Capital costs include equipment costs (provided by vendor), piping, valves, electrical (30%), site work (10%), contractor O&P (15%), engineering (15%), and contingency (20%).
Costs exclude polishing treatment. Refer to Tables 15.12 and 15.13 for additional treatment costs.

identified. The capital costs include the complete treatment system and installation.

Table 15.17 presents a summary of the annual O&M costs for the three AOP technologies evaluated under the same scenarios described above. The O&M costs consist of replacement parts, labor, analytical, chemical, and electrical costs. The replacement part costs are based on vendor estimates and include replacement parts, such as UV lamps, and spare parts. The labor costs include labor for sampling of water, system operation, and general maintenance for the specific type of AOP. The maintenance and sampling labor rate used was \$80/h. Analytical costs are based on weekly sampling of the influent and the effluent from each reactor and are estimated at \$200 per sample. Chemical costs include H_2O_2, O_3, and TiO_2 as they apply to the technology and were provided by the vendor (both dosage and costs). The electrical cost was based on power consumption and was estimated by the vendors based on \$0.08/kWh. Amortized annual capital costs and annual O&M costs were combined to determine the total amortized operating costs for each system per 1000 gal of treated water as presented in Table 15.18. The equipment was amortized at a discount rate of 7% over a 30-year period.

The cost estimates provided by Calgon Carbon Corporation were for their H_2O_2/MP-UV system. The cost per 1000 gal of treated water ranged from \$0.32 (6000 gpm, 20 μg/l) to \$4.11 (60 gpm, 2000 μg/l). Calgon had among the lowest capital costs with relative

Table 15.17 Annual O&M costs of AOPs.

Flow (gpm)	Influent (μg/l)	Effluent (μg/l)	Removal efficiency (%)	Calgon Carbon Corporation ($)	Applied Process Technology, Inc. ($)	Oxidation Systems, Inc. ($)	Hydroxyl Systems, Inc. ($)
60	20	5	75.00	54 400	47 162	60 200	74 831
60	20	0.5	97.50	63 700	48 800	62 900	78 531
60	200	20	90.00	58 900	47 700	60 400	79 131
60	200	5	97.50	63 700	48 800	62 900	80 131
60	200	0.5	99.75	70 600	50 900	66 300	90 231
60	2000	20	99.00	81 600	60 800	71 700	100 131
60	2000	5	99.75	94 200	61 500	74 600	95 731
60	2000	0.5	99.98	108 000	63 900	75 200	107 431
600	20	5	75.00	157 800	123 400	167 700	265 300
600	20	0.5	97.50	248 300	139 900	174 500	277 900
600	200	20	90.00	198 200	129 800	167 700	280 100
600	200	5	97.50	264 000	139 900	174 500	330 500
600	200	0.5	99.75	343 500	155 700	208 100	349 500
600	2000	20	99.00	422 000	193 300	202 700	373 400
600	2000	5	99.75	487 700	203 500	233 100	452 200
600	2000	0.5	99.98	551 500	222 500	239 400	483 700
6000	20	5	75.00	930 600	464 500	1 101 900	2 389 400
6000	20	0.5	97.50	1 429 600	628 100	1 177 200	2 515 500
6000	200	20	90.00	1 190 400	527 500	1 109 800	2 535 500
6000	200	5	97.50	1 631 800	628 100	1 177 200	3 103 200
6000	200	0.5	99.75	1 989 000	785 900	1 512 900	3 197 800
6000	2000	20	99.00	1 657 400	1 061 700	1 359 300	3 431 800
6000	2000	5	99.75	3 386 100	1 156 400	1 662 700	4 346 400
6000	2000	0.5	99.98	4 210 900	1 351 600	1 725 800	4 504 000

Note:
O&M costs include replacement parts (based on vendor's estimate), labor costs at $80/h, analytical costs for sampling conducted weekly at $200 per sample, chemical costs based on dose and price estimated by vendor, and power based on consumption estimates provided by vendor, priced at $0.08/kWh.
Costs exclude polishing treatment. Refer to Tables 15.12 and 15.13 for additional treatment costs.

high O&M costs. The costs prepared by Calgon were based on meeting the specified effluent concentration of MTBE. However, by-products produced as a result of the oxidation process would require further treatment to meet drinking water standards. Calgon provided the most complete analyses on by-product formations and quantified by-product formation based on the data extrapolated from an actual study and provided estimates for the 600 gpm scenario. In addition, Calgon calculated the hydrogen peroxide residual remaining in the treated water. Because these concentrations are high (>10 mg/l), an additional treatment step will be required for H_2O_2 removal. Calgon recommended using Centaur carbon for removal of the excess H_2O_2 and a biologically activated carbon system for removal of acetone, formaldehyde, and other acids prior to distribution of the treated drinking water.

15.5 Recovery of N and P

Phosphorus (P) as one of the essential elements of life could be depleted in the next hundred years if it is not sustainably managed because P can neither be produced synthetically nor substituted by any other elements (Kabbe, 2015). In addition, it is unevenly distributed on earth. For example, Morocco has about 75% of the world P rock deposit, while the United States and China have

Table 15.18 Total amortized operating costs (per 1000 gal treated) for AOPs.

Flow (gpm)	Influent (μg/l)	Effluent (μg/l)	Removal efficiency (%)	Calgon Carbon Corporation ($)	Applied Process Technology, Inc. ($)	Oxidation Systems, Inc. ($)	Hydroxyl Systems, Inc. ($)
60	20	5	75.00	2.18	2.63	2.25	3.08
60	20	0.5	97.50	2.50	2.68	2.61	3.58
60	200	20	90.00	2.32	2.65	2.32	3.60
60	200	5	97.50	2.50	2.68	2.61	3.67
60	200	0.5	99.75	2.72	2.98	2.77	4.36
60	2000	20	99.00	3.07	3.29	2.94	4.67
60	2000	5	99.75	3.47	3.31	3.03	4.36
60	2000	0.5	99.98	4.11	3.62	3.05	5.17
600	20	5	75.00	0.57	0.82	0.62	1.13
600	20	0.5	97.50	0.91	0.90	0.67	1.22
600	200	20	90.00	0.71	0.84	0.62	1.23
600	200	5	97.50	0.96	0.90	0.67	1.49
600	200	0.5	99.75	1.27	0.95	0.86	1.63
600	2000	20	99.00	1.52	1.07	0.77	1.70
600	2000	5	99.75	1.75	1.13	0.86	2.11
600	2000	0.5	99.98	2.08	1.19	0.88	2.32
6000	20	5	75.00	0.32	0.35	0.39	1.01
6000	20	0.5	97.50	0.52	0.43	0.48	1.05
6000	200	20	90.00	0.42	0.37	0.43	1.10
6000	200	5	97.50	0.60	0.43	0.48	1.36
6000	200	0.5	99.75	0.74	0.48	0.59	1.46
6000	2000	20	99.00	0.65	0.56	0.54	1.53
6000	2000	5	99.75	1.24	0.59	0.64	1.95
6000	2000	0.5	99.98	1.59	0.68	0.66	2.10

Note: Amortized at 7% over 30 years.

another 5% of the world reserve each. The rest of the world has the rest of 15%. As a result, the recovery of phosphorus from sludge or wastewater is the third important priority after water and energy recoveries. Up to 90% of the phosphorus entering the WWTP is transferred into the sewage sludge (Pinnekamp et al., 2013). Therefore, the recovery of phosphorus from the sewage sludge line offers a higher potential of phosphorus recovery in comparison with a recovery process from the effluent of the WWTP. To effectively quantify the yield efficiency, following definitions are used.

15.5.1 Yield Coefficients

The overall phosphorus yield is the phosphorus mass flow recovered in the final recycling product divided by the total phosphorus mass flow that enters the WWTP. Therefore this yield shows which proportion of the total phosphorus content available in the wastewater can be recovered with the corresponding process:

$$\eta_{\text{Overall}} = \frac{\dot{m}_{\text{total P, recovered}}}{\dot{m}_{\text{total P, raw wastewater}}} \tag{15.1}$$

The sludge specific phosphorus release yield is the phosphorus mass flow in the sludge liquor phase (as dissolved PO_4–P) divided by the mass flow of phosphorus in the digested sludge. Therefore this yield shows which proportion of the phosphorus available in the digested sludge can be dissolved:

$$\eta_{\text{Release from sludge}} = \frac{\dot{m}_{\text{PO}_4-\text{P, sludge liquor}}}{\dot{m}_{\text{total P, digested sludge}}} \tag{15.2}$$

The sludge specific phosphorus yield is the phosphorus mass flow recovered in the final recycling product divided by the total mass flow of phosphorus in the digested sludge. Therefore this yield shows which proportion of the phosphorus available in the digested sludge can be recovered with the corresponding process:

$$\eta_{\text{Sludge}} = \frac{\dot{m}_{\text{total P, recovered}}}{\dot{m}_{\text{total P, digested sludge}}} \tag{15.3}$$

The liquor specific phosphorus yield is the phosphorus mass flow recovered in the final recycling product divided by the total mass flow of phosphorus in the sludge liquor phase. Therefore this yield shows which proportion of the phosphorus available in the sludge liquor phase can be recovered with the corresponding process:

$$\eta_{\text{Liquor}} = \frac{\dot{m}_{\text{total P, recovered}}}{\dot{m}_{\text{total P, sludge liquor}}} \tag{15.4}$$

Almost all processes developed for phosphorus recovery focus on this fraction of the phosphorus originally contained in the raw wastewater. Elimination of dissolved phosphate from wastewater and the transfer into sewage sludge are at least partly taking place by bacterial growth (38–45%) (Pinnekamp et al., 2007). The phosphorus is used by the microorganisms in the activated sludge as nutrient and included into the biomass. The phosphorus needs to be extracted in dissolved form out of the sludge matrix to produce mineral fertilizers. During digestion of the sludge, its biomass is degraded, and therefore biologically bound phosphorus released into the liquid phase of the sludge is water-soluble orthophosphate. This is especially the case for phosphorus taken up by phosphorus-accumulating bacteria due to anaerobic conditions during digestion, which lead to an enhanced release of phosphorus from the bacteria. The amount of phosphorus dissolved after the digestion is estimated to account for up to 23% of the phosphorus contained originally in the wastewater (Cullen et al., 2013).

Four processes have been developed to recover phosphorus in a European Union project. One process applies a reactor in which the crystallization of a mineral phosphorus product occurs directly in the sludge (AirPrex®). The other three processes are applied on the process water after sludge dewatering by mechanical solid–liquid separation like centrifugation (Struvia process, Crystalactor, Pearl® process). These three processes are also applicable to industrial wastewater containing significant concentration of dissolved orthophosphate. The AirPrex process can be designed as one-reactor or two-reactor layout. The process mechanisms of pH increase by aeration and stripping, crystallization due to dosing of magnesium chloride, and sedimentation and harvesting of the mineral phosphorus product remain the same. Dewatering is done after treating the sludge in the AirPrex reactor.

The most relevant process mechanism is the controlled crystal growth in the separated sludge water, leading to high quality products of specific size, of easy dewaterability, and of high purity. The reactor design of these processes can be summarized as follows:

- Crystalactor: Cylindrical, upflow fluidized bed reactor with large crystallization surface, mainly on surface of sand seedings (in principle also other seedings possible), classification according to particle size in the fluidized bed, operational conditions leading to low supersaturation (metastable region), and recycle flow for dilution.
- Pearl process: Upflow fluidized bed reactor with zones of increasing diameter, self-seeding systems (struvite crystals are the seedings themselves), classification according to changes in diameter, operational conditions leading to low supersaturation (metastable region), and recycle flow for dilution.
- Struvia process: Continuous stirred tank reactor with integrated solid–liquid separation by calming zone and lamellar packing (Turboflo™) or with additional lamella settler (Turbomix®). In the latter case, part of the struvite harvested from the bottom of the settler is recycled into the mixing reactor in order to reduce the amount of nucleation and to obtain crystals with larger particle sizes.

The implementation of struvite recovery from the sludge water (no matter if done prior or after dewatering) does not reduce the efficiency of technologies aiming to recover the phosphorus from the solid phase of the sludge, as done with acid leaching or thermochemical treatment of sludge ash. Phosphorus extraction is started by addition of sulfuric acid (2.8 l/m^3 H_2SO_4 (96%)) to adjust a pH around 4. On the one hand, lowering the pH leads to a transformation of not readily soluble phosphate species into better soluble phosphate species ($H_2PO_4^-$ /HPO_4^{2-}) (Falbe and Regitz, 1995). On the other hand, pH values below 5 can induce cell destruction and therefore result in phosphorus release (Kunst, 1991). After a reaction time of 0.5 h, a PO_4–P concentration of around 500 mg/l was achieved, which means that roughly 66% of the total phosphorus content of the sludge was brought into solution. In a phosphorus precipitation reactor, after the first separation step, the remaining liquor that contains most of nutrients (phosphorus, ammonium, calcium, and magnesium) is pumped in the phosphorus recovery unit, consisting of two redundant stainless steel reactors with 16 m^3 each operated in batch mode. The precipitation process is initiated by dosage of magnesium hydroxide (0.15

l $Mg(OH)_2$ (53%) per m^3). Due to the alkaline properties of the hydroxide, the pH is slightly raised to pH 6–6.5. After a reaction time of 0.5 h, sodium hydroxide (1.74 l NaOH (50%) per m^3) is dosed to the solution to raise the pH to 9.3–9.4. After 0.5 h reaction time, the precipitated nutrients are separated from the process solution by a centrifuge and the addition of polymer. The dosage of magnesium is under stoichiometric with reference to phosphate in solution to force precipitation of calcium and prevent scaling in the subsequent ammonium stripping unit.

About 40% of the phosphate in the product is bound as struvite ($MgNH_4PO_4{\cdot}6H_2O$), 5% as vivianite ($Fe_3(PO_4){\cdot}8H_2O$), and 55% as calcium phosphate (supposedly hydroxyapatite with low calcium content: $Ca_9(PO_4)_5(HPO_4){\cdot}OH$). The product obtained directly after dewatering has a solid content of 35–40%, which increases quickly by air-drying to greater than 80%. Table 15.19 shows site-specific data for the installation of the process in Gifhorn.

Table 15.20 shows significant amount of chemicals required in the processes. As a result, both the capital cost and the O&M cost are usually very high. All technologies contain an initial sludge treatment step – the dissolution of P from the solid into the aqueous phase – to increase the concentration of dissolved phosphorus in the aqueous phase.

According to the different process components, several chemicals are consumed. All processes include a precipitation step and hence require the dosing of cation ions such calcium or magnesium. The three processes with an extraction process apply an acidic pH. For the pH increase before precipitation or crystallization, dosing of caustic soda is required. Processes without acidic treatment need a slight adjustment of pH but not necessarily depending on the properties of the sludge. A summary of types of chemicals applied in the processes is given in Table 15.21.

Table 15.19 Site-specific data for the Gifhorn process.

Site-specific data for the Gifhorn process	Data
Size of the WWTP (p.e.)	50 000
Throughput of digested sludge (m^3/day)	108 (with 2% TS)
P extraction rate from sludge	66.1 % (for pH 4)
Liquor specific phosphorus yield (%)	93.3%
Sludge specific phosphorus yield (%)	48.7%
Energy consumption including dewatering (kWh/m^3 sludge)	7.6
Energy consumption without dewatering (kWh/m^3 sludge)	4.5
Energy consumption including dewatering (kWh/kg P in product)	21.6
Energy consumption without dewatering (kWh/kg P in product)	12.9
H_2SO_4 (96%) (kg/kg P recovered)	14.6
Na_2S (15%) (kg/kg P recovered)	8.9
NaOH (50%) (kg/kg P recovered)	2.3
Polymer (100%) (kg/t_{DM})	14
Molar ratio of magnesium to phosphate (—)	0.19
Percentage of P in the product (weight%)	11.0–12.0

By adding magnesium components and increasing the pH, most of the processes precipitate or crystallize MAP (struvite). The product of AirPrex, Crystalactor, Pearl, and Struvia processes is a nearly pure struvite. The processes Crystalactor and Struvia are designed to produce calcium phosphate as an alternative product. The product of the Gifhorn and Stuttgarter process contains mainly struvite but also calcium and iron phosphates. The operational experiences differ significantly for the different technologies in terms of plant

Table 15.20 All the P recovery process components.

	Extraction process	Metal elimination process	Precipitation process
AirPrex	×	×	Airlift reactor
Crystalactor®	×	×	Upflow reactor
Pearl	×	×	Upflow reactor
Struvia (Turbomix/Turboflo)	×	×	Continuous stirred tank reactor
Gifhorn	Dosing of H_2SO_4	NAS precipitation	Precipitation and centrifugation
Stuttgarter	Dosing of H_2SO_4	Complexation with citric acid	Precipitation and sedimentation
Budenheim	CO_2 extraction	×	Precipitation

Table 15.21 Summary of chemicals used for the different concepts.

	Extraction process	Metal elimination process	Precipitation process
AirPrex	×	×	$MgCl_2$ (33%), NaOH (30%), or $CaCl_2$
Crystalactor®	×	×	$Mg(OH)_2$, NaOH, or $MgCl_2$ upflow reactor
Pearl	×	×	$MgCl_2$ (33%), NaOH
Struvia (Turbomix/Turboflo)	×	×	$MgCl_2$ (33%)
Gifhorn	H_2SO_4 (96%)	NAS precipitation (15%)	$Mg(OH)_2$ (53%), NaOH (50%)
Stuttgarter	H_2SO_4 (78%)	Citric acid (50%)	MgO (92%), NaOH (22%)
Budenheim	CO_2	×	Lime

size and duration of operation. Table 15.22 gives an overview of the duration and sizes of installations of the processes. N and P recovery has positive impact on the operation of the WWTP, which are a reduced phosphorus load for the biological wastewater treatment and a reduction of the amount of sludge to be disposed. All processes that produce struvite additionally reduce the load of ammonium for the biological wastewater treatment. Furthermore all of them reduce the risk of encrustations in the sludge treatment part of the WWTP. However, this positive effect is mainly relevant for WWTPs using enhanced biological phosphorus removal and anaerobic digestion since otherwise the redissolution of phosphorus during the digestion is relatively low.

Table 15.22 Overview of size and duration of operation.

	Duration	Scale
AirPrex	>2 years	Full scale, several sites
Crystalactor®	>2 years	Full scale, several sites
Pearl	>2 years	Full scale, several sites
Struvia (Turbomix/ Turboflo)	>1 year	Pilot scale, one site
Gifhorn	Months/>2 years	Full scale, one site
Stuttgarter	>2 years	One demonstration site
Budenheim	>1 year	Lab scale

15.5.2 Capital Cost of P Recovery Systems

To estimate the capital and O&M costs of different process for P recovery, the US EPA provides the following examples. Table 15.23 lists the capital cost of eight recovery systems (Tables 15.24 and 15.25; Figures 15.4 and 15.5).

15.5.3 Activated Sludge

Example 15.5

Given: Table 15.26 shows the capital cost and operations and maintenance (O&M) cost of one-stage activated sludge process (AS) at different flow rates.

Find:

1) Calculate unit capital and O&M costs.
2) Plot unit cost/unit $cost_0$ vs. Q/Q_0.
3) Determine correlation equations between unit cost/unit $cost_0$ for capital and O&M vs. Q/Q_0, respectively (Tables 15.27 and 15.28).

Solution:

1) When $Q = 1$ MGD,

$$\text{Unit Capital Cost} = 7.36\times10^6\$ \div 1\,\text{MGD}\times1\,\text{million gal}/10^3\,\text{kgal} \div 1\,\text{day} = 7360.00\$/\text{kgal}$$

$$\text{Unit O\&M Costs} = 0.93\times10^6\$/\text{year} \div 1\,\text{MGD}\times1\,\text{million gal}/10^3\,\text{kgal} \div 365\,\text{day/year}$$

$$\text{Unit O\&M Costs} = 2.5479\$/\text{kgal}$$

When $Q = 1$ MGD,

$$Q_0 = 1\ \text{MGD}$$

$$\text{Unit Capital Cost}_0 = 7360.00\$/\text{kgal}$$

$$\text{Unit O\&M costs}_0 = 2.5479\$/\text{kgal}$$

$$\frac{Q}{Q_0} = 1$$

$$\text{Unit Capital Cost/Unit Capital Cost}_0 = 1$$

$$\text{Unit O\&M costs/unit O\&M costs}_0 = 1$$

When $Q = 10$ MGD,

$$\frac{Q}{Q_0} = 10$$

$$\text{Unit Capital Cost/Unit Capital Cost}_0 = 0.5371$$

$$\text{Unit O\&M Costs/Unit O\&M Costs}_0 = 0.4441$$

Table 15.23 Capital cost of the eight designs.

Process	Flow (MGD)	Total construction cost (1978 \$×10⁶)	Total indirect cost (1978 \$×10⁶)		Total capital cost (1978 \$×10⁶)	Total capital cost (2004 \$×10⁶)
			Engineering	Contingency		
AS	1.0	2.24	0.34	0.34	2.91	7.36
	10.0	12.03	1.80	1.80	15.64	39.53
	20.0	19.95	2.99	2.99	25.94	65.56
	50.0	38.95	5.84	5.84	50.64	127.99
	100.0	64.60	9.69	9.69	83.98	212.28
AO	1.0	2.72	0.41	0.41	3.54	8.94
	10.0	14.27	2.14	2.14	18.55	46.89
	20.0	23.51	3.53	3.53	30.56	77.25
	50.0	45.48	6.82	6.82	59.12	149.45
	100.0	74.92	11.24	11.24	97.40	248.29
AAO	1.0	2.95	0.44	0.44	3.84	9.69
	10.0	16.59	2.49	2.49	21.57	54.52
	20.0	27.90	4.19	4.19	36.27	91.68
	50.0	55.47	8.32	8.32	72.11	182.28
	100.0	93.29	13.99	13.99	121.28	306.55
AAO + Al	1.0	2.97	0.45	0.45	3.86	9.76
	10.0	16.63	2.49	2.49	21.62	54.65
	20.0	27.95	4.19	4.19	36.34	91.84
	50.0	55.55	8.33	8.33	72.22	182.54
	100.0	93.42	14.01	14.01	121.45	306.98
AAO + Al + S	1.0	3.08	0.46	0.46	4.00	10.12
	10.0	17.05	2.56	2.56	22.17	56.03
	20.0	28.68	4.30	4.30	37.28	94.24
	50.0	57.39	8.61	8.61	74.61	188.58
	100.0	97.06	14.56	14.56	126.18	318.94
AAO + Al + S + F	1.0	3.30	0.50	0.50	4.29	10.84
	10.0	17.87	2.68	2.68	23.23	58.72
	20.0	30.45	4.57	4.57	39.58	100.06
	50.0	62.81	9.42	9.42	81.65	206.39
	100.0	104.81	15.72	15.72	136.25	344.41
AAO + Al + S + C	1.0	3.38	0.51	0.51	4.40	11.11
	10.0	20.05	3.01	3.01	26.07	65.88
	20.0	34.38	5.16	5.16	44.69	112.97
	50.0	72.39	10.86	10.86	94.11	237.87
	100.0	125.06	18.76	18.76	162.58	410.95
AAO + Al + F + UF	1.0	/	/	/	/	12.27
	10.0	/	/	/	/	72.79
	20.0	/	/	/	/	113.18
	50.0	/	/	/	/	277.31
	100.0	/	/	/	/	486.18

Table 15.24 The O&M cost estimates for the eight processes.

Process	Flow (MGD)	Maintenance (2004 \$×10[6])	Taxes and insurance (2004 \$×10[6])	Labor (2004 \$×10[6])	Electricity (2004 \$×10[6])	Chemicals (2004 \$×10[6])	Residuals management (2004 \$×10[6])	Total O&M cost (2004 \$×10[6])
AS	1.0	0.29	0.15	0.41	0.05	0	0.03	0.93
	10.0	1.58	0.79	1.11	0.32	0	0.33	4.13
	20.0	2.62	1.31	1.7	0.55	0	0.66	6.84
	50.0	5.12	2.56	3.11	1.13	0	1.65	13.57
	100.0	8.49	4.25	5.16	1.95	0	3.31	23.16
AO	1.0	0.36	0.18	0.45	0.07	0	0.03	1.09
	10.0	1.88	0.94	1.24	0.45	0	0.34	4.85
	20.0	3.09	1.55	1.88	0.78	0	0.68	7.98
	50.0	5.98	2.99	3.45	1.60	0	1.71	15.73
	100.0	9.93	4.97	5.76	2.77	0	3.42	26.85
AAO	1.0	0.39	0.19	0.48	0.07	0	0.04	1.17
	10.0	2.18	1.09	1.33	0.45	0	0.38	5.43
	20.0	3.67	1.83	2.1	0.78	0	0.75	9.13
	50.0	7.29	3.65	3.85	1.62	0	1.88	18.29
	100.0	12.26	6.13	6.61	2.8	0	3.77	31.57
AAO + Al	1.0	0.39	0.2	0.49	0.07	0.005	0.2	1.35
	10.0	2.19	1.09	1.33	0.45	0.05	1.99	7.1
	20.0	3.67	1.84	2.1	0.78	0.1	3.97	12.46
	50.0	7.3	3.65	3.85	1.62	0.24	9.93	26.59
	100.0	12.28	6.14	6.61	2.81	0.48	19.85	48.17
AAO + Al + S	1.0	0.4	0.2	0.5	0.07	0.007	0.23	1.41
	10.0	2.24	1.12	1.35	0.46	0.07	2.26	7.5
	20.0	3.77	1.88	2.12	0.79	0.15	4.51	13.22
	50.0	7.54	3.77	3.89	1.62	0.37	11.29	28.48
	100.0	12.76	6.38	6.69	2.81	0.75	22.57	51.96
AAO + Al + S + F	1.0	0.43	0.22	0.52	0.08	0.007	0.23	1.49
	10.0	2.35	1.17	1.42	0.48	0.07	2.33	7.82
	20.0	4	2	2.24	0.84	0.15	4.65	13.88
	50.0	8.26	4.13	4.14	1.76	0.37	11.64	30.3
	100.0	13.78	6.89	7.26	3.08	0.75	23.27	55.03
AAO + Al + S + C	1.0	0.44	0.22	0.61	0.08	0.008	0.24	1.60
	10.0	2.64	1.32	1.53	0.51	0.08	2.37	8.44
	20.0	4.52	2.26	2.34	0.89	0.16	4.74	14.91
	50.0	9.51	4.76	4.21	1.87	0.41	11.85	32.61
	100.0	16.44	8.22	7.1	3.31	0.81	23.7	59.58
AAO + Al + F + UF	1.0	0.49	0.25	0.66	0.09	0.02	0.25	1.76
	10.0	2.91	1.46	1.50	0.62	0.19	2.5	9.18
	20.0	4.53	2.26	2.33	1.12	0.38	5	15.62
	50.0	11.09	5.55	4.25	2.46	0.95	12.5	36.80
	100.0	19.45	9.72	7.39	4.49	1.90	25	67.95

Table 15.25 Estimated effluent quality of the eight reference designs.

Process	TBOD (mg/l)	TSS (mg/l)	TP (mg/l)	P removal (%)
Influent	174	172	7.5	/
Effluent of AS	22	20	5.86	21.8
Effluent of AO	11–20	20	4.12	45.1
Effluent of AAO	11	20	2.95	60.7
Effluent of AAO + M	10	20	1.00	86.7
Effluent of AAO + M+ S	5–10	5	0.325	95.7
Effluent of AAO + M+ S + F	5	1	0.145	98.1
Effluent of AAO + M+ S + C	<1	<1	0.10	98.7
Effluent of AAO + M + F + UF	<1	<1	0.05	99.3

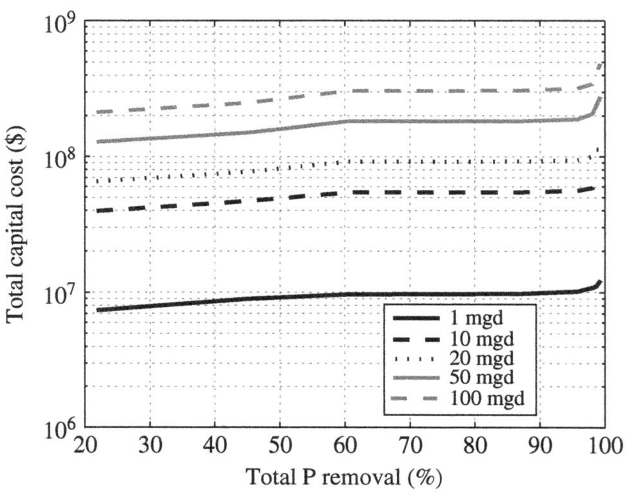

Figure 15.4 Capital cost curves of the eight designs.

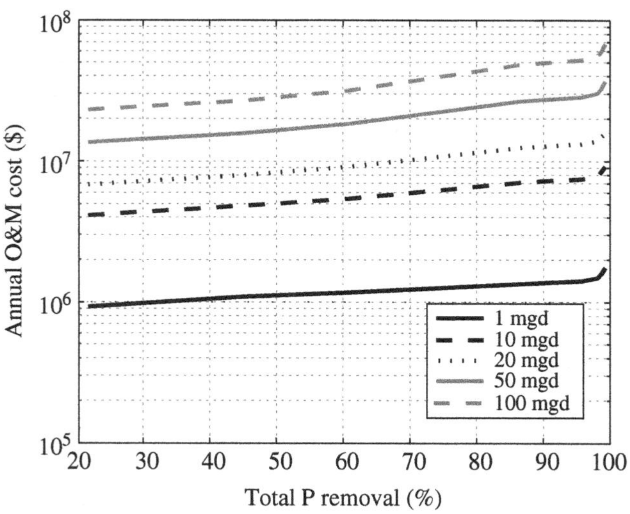

Figure 15.5 Annual O&M costs for the eight designs.

Table 15.26 Capital cost and O&M cost of AS.

Flow (MGD)	Total capital cost (2004 $\$\times10^6$)	Total O&M cost (2004 $\$\times10^6$)
1	7.36	0.93
10	39.53	4.13
20	65.56	6.84
50	127.99	13.57
100	212.28	23.16

Table 15.27 Unit capital cost and unit operations and maintenance cost of AS at different flow rates.

Q (MGD)	Unit capital cost ($/kgal)	Unit O&M costs ($/kgal)
1	7360.00	2.5479
10	3953.00	1.1315
20	3278.00	0.9370
50	2559.80	0.7436
100	2122.80	0.6345

Table 15.28 Unit cost ratio of AS at different flow ratios.

Q/Q_0	Unit capital cost/ unit capital $cost_0$	Unit O&M costs/ unit O&M $costs_0$
1	1.0000	1.0000
10	0.5371	0.4441
20	0.4454	0.3677
50	0.3478	0.2918
100	0.2884	0.2490

2) Plot unit cost/unit $cost_0$ vs. Q/Q_0 for capital and O&M, respectively (Figure 15.6).

3) a) Equation for unit cost/unit $cost_0$ vs. Q/Q_0 for capital: The regression equation of the trend line:

$$\log(\text{unit cost/unit cost}_0) = -0.2700\log(Q/Q_0) + 0.0000 \quad r^2 = 1.0000$$

or the equation of the regression:

$$\text{Unit cost/unit cost}_0 = (Q/Q_0)^{-0.27} \quad r^2 = 1.0000$$

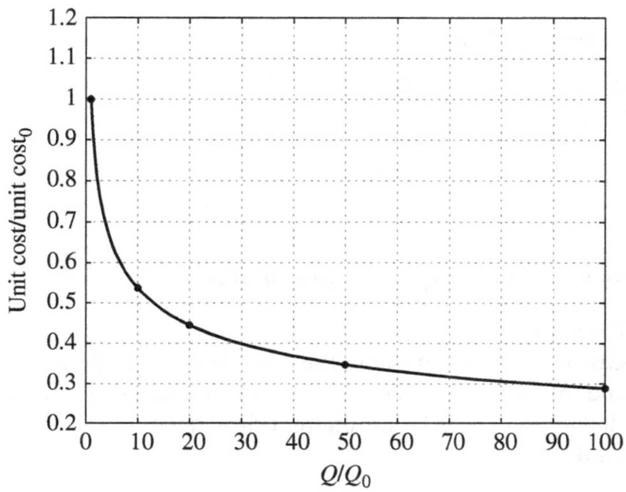

Figure 15.6 Unit capital cost/unit capital $cost_0$ vs. Q/Q_0 for AS, unit capital $cost_0$ = 7360.00\$/kgal, Q_0 = 1 MGD.

Matlab code:

```
clear all
close all
%data%
Qrange=[1 10 20 50 100];
Q=Qrange(1);
totalcostrange=[7.36 39.53 65.56
  127.99 212.28];
totalcostrange=totalcostrange.*10^6;
totalcost=totalcostrange(1);
unitcost=totalcost./Q./1000;
unitcostrange=totalcostrange./
  Qrange./1000;
p=polyfit(log10(Qrange/Q), log10
  (unitcostrange /unitcost), 1)
log10ratiounitcostfit=polyval
  (p, log10(Qrange/Q));
ratiounitcostfit=10.
  ^log10ratiounitcostfit;
k=10^p(:,2)
R2 = 1 - sum((log10(unitcostrange
  /unitcost) - log10ratiounitcostfit).
  ^2)/sum((log10(unitcostrange
  /unitcost) - mean(log10(unitcostrange
  /unitcost))).^2)
xmax=max(Qrange/Q);
xmin=min(Qrange/Q);
xaxis=[xmin:0.1:xmax];
log10ratiounitcostfitplot=polyval
  (p, log10(xaxis));
ratiounitcostfitplot=10.
  ^log10ratiounitcostfitplot;
%plot%
figure(1)
plot(Qrange/Q,unitcostrange
  /unitcost,'ko','MarkerSize',3,
  'MarkerFaceColor',[0 0 0],
  'MarkerEdgeColor',[0 0 0],
  'linewidth',1);
hold on
plot(xaxis,ratiounitcostfitplot,
  'k-','linewidth',1);
set(gca,'xtickmode','auto',
  'FontSize',10,'FontWeight','bold',
  'FontName', 'Times New Roman');
xlabel({'Q/Q_0'},'FontSize',14,
  'FontWeight','bold','FontName',
  'Times New Roman')
ylabel({'Unit Cost/Unit Cost_0'},
  'FontSize',14,'FontWeight','bold',
  'FontName', 'Times New Roman')
grid on
hold on
```

Answer: `ASCapitalCost` (Figure 15.7)

p = -0.2700 0.0000; k = 1.0000; R^2 = 1.0000

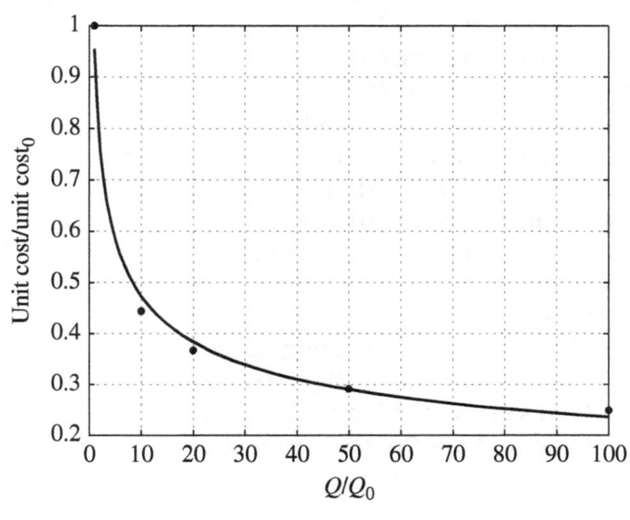

Figure 15.7 Unit O&M cost/unit O&M $cost_0$ vs. Q/Q_0 for AS, unit O&M $cost_0$ = 2.5479\$/kgal, Q_0 = 1 MGD.

b) Equation for unit cost/unit $cost_0$ vs. Q/Q_0 for O&M:

The regression equation of the trend line:

$$\log(\text{unit cost/unit cost}_0) = -0.3040\log(Q/Q_0) - 0.0203 \quad r^2 = 0.9903$$

or the equation of the regression:

$$\text{Unit cost/unit cost}_0 = 0.9543(Q/Q_0)^{-0.3040} \quad r^2 = 0.9903$$

Matlab code:

```
clear all
close all
%data%
Qrange=[1 10 20 50 100];
Q=Qrange(1);
totalcostrange=[0.93 4.13 6.84
  13.57 23.16];
totalcostrange=totalcostrange.*10^6;
totalcost=totalcostrange(1);
unitcost=totalcost./Q./1000./365;
unitcostrange=totalcostrange./Qrange./
  1000./365;
p=polyfit(log10(Qrange/Q), log10
  (unitcostrange /unitcost), 1)
log10ratiounitcostfit=polyval
  (p, log10(Qrange/Q));
ratiounitcostfit=10.
  ^log10ratiounitcostfit;
k=10^p(:,2)
R2 = 1 - sum((log10(unitcostrange
  /unitcost) - log10ratiounitcostfit).
  ^2)/sum((log10(unitcostrange
  /unitcost) - mean(log10(unitcostrange
  /unitcost))).^2)
xmax=max(Qrange/Q);
xmin=min(Qrange/Q);
xaxis=[xmin:0.1:xmax];
log10ratiounitcostfitplot=polyval
  (p, log10(xaxis));
ratiounitcostfitplot=10.
  ^log10ratiounitcostfitplot;
%plot%
figure(1)
plot(Qrange/Q,unitcostrange
  /unitcost,'ko','MarkerSize',3,
  'MarkerFaceColor',[0 0 0],
  'MarkerEdgeColor',[0 0 0],
  'linewidth',1);
hold on
plot(xaxis,ratiounitcostfitplot,
  'k-','linewidth',1);
set(gca,'xtickmode','auto',
  'FontSize',10,'FontWeight','bold',
  'FontName', 'Times New Roman');
xlabel({'Q/Q_0'},'FontSize',14,
  'FontWeight','bold','FontName',
  'Times New Roman')
ylabel({'Unit Cost/Unit Cost_0'},
  'FontSize',14,'FontWeight','bold',
  'FontName', 'Times New Roman')
grid on
hold on
```

Answer: `ASOMCost`

```
p = -0.3040 -0.0203; k = 0.9543;
R^2 = 0.9903
```

15.5.4 Two-Stage Activated Sludge

Example 15.6

Given: Table 15.29 shows the capital cost and operations and maintenance (O&M) cost of two-stage activated sludge process (AO) at different flow rates.

Find:

1) Calculate unit capital and O&M costs.
2) Plot unit cost/unit $cost_0$ vs. Q/Q_0.
3) Determine correlation equations between unit cost/unit $cost_0$ for capital and O&M vs. Q/Q_0, respectively (Tables 15.30 and 15.31).

Solution:

1) When $Q = 1$ MGD,

$$\text{Unit capital cost} = 8.94 \times 10^6\$ \div 1\,\text{MGD} \times 1\,\text{million gal}/10^3\,\text{kgal} \div 1\,\text{day} = 8940.00\$/\text{kgal}$$

Table 15.29 Capital cost and O&M cost of AO.

Flow (MGD)	Total capital cost (2004 $\$\times10^6$)	Total O&M cost (2004 $\$\times10^6$)
1	8.94	1.09
10	46.89	4.85
20	77.25	7.98
50	149.45	15.73
100	248.29	26.85

Table 15.30 Unit capital cost and unit operations and maintenance cost of AO at different flow rates.

Q (MGD)	Unit capital cost ($/kgal)	Unit O&M costs ($/kgal)
1	8940.00	2.9863
10	4689.00	1.3288
20	3862.50	1.0932
50	2989.00	0.8619
100	2482.90	0.7356

Table 15.31 Unit cost ratio of AO at different flow ratios.

Q/Q_0	Unit capital cost/unit capital $cost_0$	Unit O&M costs/unit O&M $costs_0$
1	1.0000	1.0000
10	0.5245	0.4450
20	0.4320	0.3661
50	0.3343	0.2886
100	0.2777	0.2463

$$\text{Unit O\&M costs} = 1.09 \times 10^6 \$/\text{year} \div 1\,\text{MGD} \times 1\,\text{million gal}/10^3\,\text{kgal} \div 365\,\text{day/year}$$

$$\text{Unit O\&M costs} = 2.9863\$/\text{kgal}$$

When $Q = 1$ MGD,

$Q_0 = 1$ MGD

Unit capital $cost_0 = 8940.00\$/\text{kgal}$

Unit O&M $costs_0 = 2.9863\$/\text{kgal}$

$$\frac{Q}{Q_0} = 1$$

Unit capital cost/unit capital $cost_0 = 1$

Unit O&M costs/unit O&M $costs_0 = 1$

When $Q = 10$ MGD,

$$\frac{Q}{Q_0} = 10$$

Unit capital cost/unit capital $cost_0 = 0.5245$

Unit O&M costs/unit O&M $costs_0 = 0.4450$

2) Plot unit cost/unit $cost_0$ vs. Q/Q_0 for capital and O&M, respectively (Figure 15.8).

3) a) Equation for unit cost/unit $cost_0$ vs. Q/Q_0 for capital: The regression equation of the trend line:

$$\log(\text{unit cost/unit cost}_0) = -0.2788\log(Q/Q_0) - 0.0008 \quad r^2 = 1.0000$$

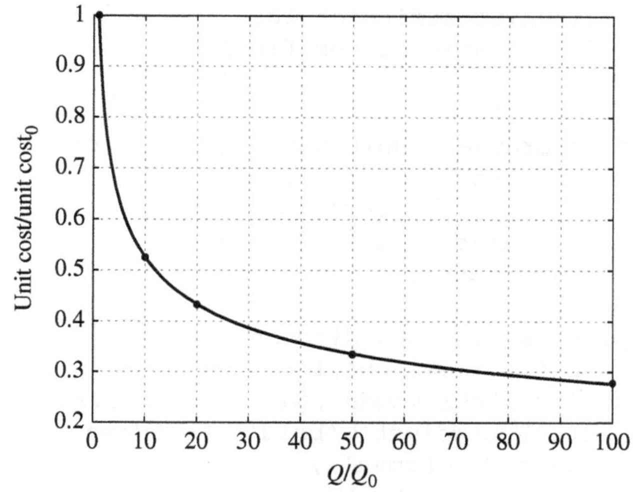

Figure 15.8 Unit capital cost/unit capital $cost_0$ vs. Q/Q_0 for AO, unit capital $cost_0 = 8940.00\$/\text{kgal}$, $Q_0 = 1$ MGD.

or the equation of the regression:

$$\text{Unit cost/unit cost}_0 = 0.9981(Q/Q_0)^{-0.2788}$$
$$r^2 = 1.0000$$

Matlab code:

```
clear all
close all
%data%
Qrange= [1 10 20 50 100];
Q=Qrange(1);
totalcostrange= [8.94 46.89 77.25
  149.45 248.29];
totalcostrange=totalcostrange.*10^6;
totalcost=totalcostrange(1);
unitcost=totalcost./Q./1000;
unitcostrange=totalcostrange./
  Qrange./1000;
p=polyfit(log10(Qrange/Q), log10
  (unitcostrange /unitcost), 1)
log10ratiounitcostfit=polyval
  (p, log10(Qrange/Q));
ratiounitcostfit=10.
  ^log10ratiounitcostfit;
k=10^p(:,2)
R2 = 1 - sum((log10(unitcostrange
  /unitcost) - log10ratiounitcostfit).
  ^2)/sum((log10(unitcostrange
  /unitcost) - mean(log10(unitcostrange
  /unitcost))).^2)
xmax=max(Qrange/Q);
xmin=min(Qrange/Q);
xaxis= [xmin:0.1:xmax];
log10ratiounitcostfitplot=polyval
  (p, log10(xaxis));
```

```
ratiounitcostfitplot=10.
  ^log10ratiounitcostfitplot;
%plot%
figure(1)
plot(Qrange/Q,unitcostrange /unitcost,
  'ko','MarkerSize',3,
  'MarkerFaceColor',[0 0 0],
  'MarkerEdgeColor',[0 0 0],
  'linewidth',1);
hold on
plot(xaxis,ratiounitcostfitplot,
  'k-','linewidth',1);
set(gca,'xtickmode','auto','FontSize',
  10,'FontWeight','bold','FontName',
  'Times New Roman');
xlabel({'Q/Q_0'},'FontSize',14,
  'FontWeight','bold','FontName',
  'Times New Roman')
ylabel({'Unit Cost/Unit Cost_0'},
  'FontSize',14,'FontWeight',
  'bold','FontName', 'Times New Roman')
grid on
hold on
```

Answer: AOCapitalCost (Figure 15.9)
p = -0.2788 -0.0008; k = 0.9981; R^2 = 1.0000

b) Equation for unit cost/unit $cost_0$ vs. Q/Q_0 for O&M:

The regression equation of the trend line:

$$\log(\text{unit cost/unit cost}_0) = -0.3068\log(Q/Q_0) - 0.0192 \quad r^2 = 0.9914$$

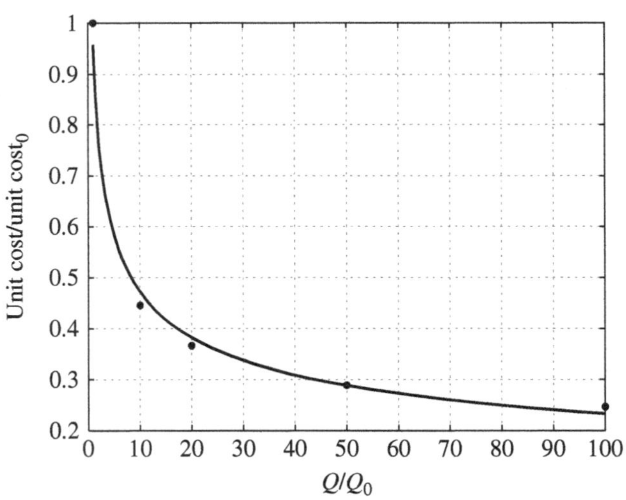

Figure 15.9 Unit O&M cost/unit O&M $cost_0$ vs. Q/Q_0 for AO, unit O&M $cost_0$ = 2.9863$/kgal, Q_0 = 1 MGD.

or the equation of the regression:

$$\text{Unit cost/unit cost}_0 = 0.9569(Q/Q_0)^{-0.3068}$$
$$r^2 = 0.9914$$

Matlab code:

```
clear all
close all
%data%
Qrange=[1 10 20 50 100];
Q=Qrange(1);
totalcostrange=[1.09 4.85 7.98
  15.73 26.85];
totalcostrange=totalcostrange.*10^6;
totalcost=totalcostrange(1);
unitcost=totalcost./Q./1000./365;
unitcostrange=totalcostrange./Qrange./
  1000./365;
p=polyfit(log10(Qrange/Q), log10
  (unitcostrange /unitcost), 1)
log10ratiounitcostfit=polyval
  (p, log10(Qrange/Q));
ratiounitcostfit=10.
  ^log10ratiounitcostfit;
k=10^p(:,2)
R2 = 1 - sum((log10(unitcostrange
  /unitcost) - log10ratiounitcostfit).
  ^2)/sum((log10(unitcostrange
  /unitcost) - mean(log10
  (unitcostrange /unitcost))).^2)
xmax=max(Qrange/Q);
xmin=min(Qrange/Q);
xaxis=[xmin:0.1:xmax];
log10ratiounitcostfitplot=polyval
  (p, log10(xaxis));
ratiounitcostfitplot=10.
  ^log10ratiounitcostfitplot;
%plot%
figure(1)
plot(Qrange/Q,unitcostrange /unitcost,
  'ko','MarkerSize',3,
  'MarkerFaceColor',[0 0 0],
  'MarkerEdgeColor',[0 0 0],
  'linewidth',1);
hold on
plot(xaxis,ratiounitcostfitplot,
  'k-','linewidth',1);
set(gca,'xtickmode','auto',
  'FontSize',10,'FontWeight',
  'bold','FontName', 'Times New Roman');
xlabel({'Q/Q_0'},'FontSize',14,
  'FontWeight','bold','FontName',
  'Times New Roman')
```

```
ylabel({'Unit Cost/Unit Cost_0'},
   'FontSize',14,'FontWeight','bold',
   'FontName', 'Times New Roman')
grid on
hold on
```

Answer: AOOMCost

p = -0.3068 -0.0192; k = 0.9569;
R^2 = 0.9914

15.5.5 Three-Stage Activated Sludge

Example 15.7
Given: Table 15.32 shows the capital cost and operations and maintenance (O&M) cost of three-stage activated sludge process (AAO) at different flow rates.
Find:

1) Calculate unit capital and O&M costs.
2) Plot unit cost/unit $cost_0$ vs. Q/Q_0.
3) Determine correlation equations between unit cost/unit $cost_0$ for capital and O&M vs. Q/Q_0, respectively (Tables 15.33 and 15.34).

Table 15.32 Capital cost and O&M cost of AAO.

Flow (MGD)	Total capital cost (2004 $\$\times10^6$)	Total O&M cost (2004 $\$\times10^6$)
1	9.69	1.17
10	54.52	5.43
20	91.68	9.13
50	182.28	18.29
100	306.55	31.57

Table 15.33 Unit capital cost and unit operations and maintenance cost of AAO at different flow rates.

Q (MGD)	Unit capital cost ($/kgal)	Unit O&M costs ($/kgal)
1	9690.00	3.2055
10	5452.00	1.4877
20	4584.00	1.2507
50	3645.60	1.0022
100	3065.50	0.8649

Table 15.34 Unit cost ratio of AAO at different flow ratios.

Q/Q_0	Unit capital cost/unit capital $cost_0$	Unit O&M costs/unit O&M $costs_0$
1	1.0000	1.0000
10	0.5626	0.4641
20	0.4731	0.3902
50	0.3762	0.3126
100	0.3164	0.2698

Solution:

1) When $Q = 1$ MGD,

$$\text{Unit capital cost} = 9.69\times10^6\$ \div 1\,\text{MGD}\times1\,\text{million gal}/10^3\,\text{kgal} \div 1\,\text{day} = 9690.00\$/\text{kgal}$$

$$\text{Unit O\&M costs} = 1.17\times10^6\$/\text{year} \div 1\,\text{MGD}\times1\,\text{million gal}/10^3\,\text{kgal}\div365\,\text{day/year}$$

$$\text{Unit O\&M costs} = 3.2055\$/\text{kgal}$$

When $Q = 1$ MGD,

$Q_0 = 1$ MGD

Unit capital $cost_0 = 9690.00\$/\text{kgal}$

Unit O&M $costs_0 = 3.2055\$/\text{kgal}$

$$\frac{Q}{Q_0} = 1$$

Unit capital cost/unit capital $cost_0 = 1$

Unit O&M costs/unit O&M $costs_0 = 1$

When $Q = 10$ MGD,

$$\frac{Q}{Q_0} = 10$$

Unit capital cost/unit capital $cost_0 = 0.5626$

Unit O&M costs/unit O&M $costs_0 = 0.4641$

2) Plot unit cost/unit $cost_0$ vs. Q/Q_0 for capital and O&M, respectively (Figure 15.10).

3) a) Equation for unit cost/unit $cost_0$ vs. Q/Q_0 for capital:
The regression equation of the trend line:

$$\log(\text{unit cost/unit cost}_0) = -0.2499\log(Q/Q_0) + 0.0001 \quad r^2 = 1.0000$$

or the equation of the regression:

$$\text{Unit cost/unit cost}_0 = 1.0001(Q/Q_0)^{-0.2499} \quad r^2 = 1.0000$$

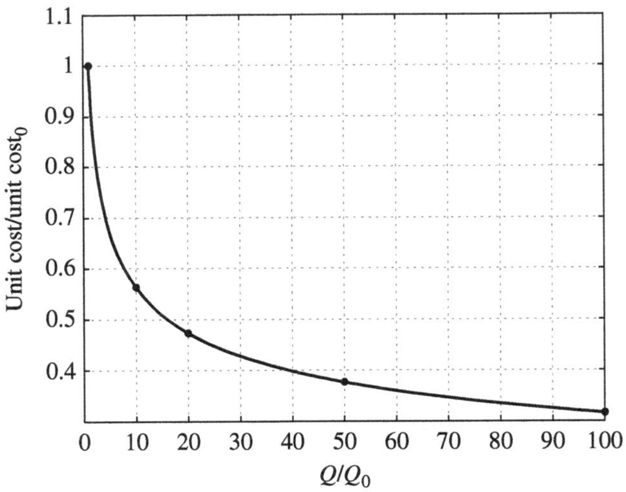

Figure 15.10 Unit capital cost/unit capital cost$_0$ vs. Q/Q_0 for AAO, unit capital cost$_0$ = 9690.00$/kgal, Q_0 = 1 MGD.

Matlab code:

```
clear all
close all
%data%
Qrange=[1 10 20 50 100];
Q=Qrange(1);
totalcostrange=[9.69 54.52 91.68
  182.28 306.55];
totalcostrange=totalcostrange.*10^6;
totalcost=totalcostrange(1);
unitcost=totalcost./Q./1000;
unitcostrange=totalcostrange./
  Qrange./1000;
p=polyfit(log10(Qrange/Q), log10
  (unitcostrange /unitcost), 1)
log10ratiounitcostfit=polyval
  (p, log10(Qrange/Q));
ratiounitcostfit=10.
  ^log10ratiounitcostfit;
k=10^p(:,2)
R2 = 1 - sum((log10(unitcostrange
  /unitcost) - log10ratiounitcostfit).
  ^2)/sum((log10(unitcostrange
  /unitcost) - mean(log10(unitcostrange
  /unitcost))).^2)
xmax=max(Qrange/Q);
xmin=min(Qrange/Q);
xaxis=[xmin:0.1:xmax];
log10ratiounitcostfitplot=polyval
  (p, log10(xaxis));
ratiounitcostfitplot=10.
  ^log10ratiounitcostfitplot;
%plot%
figure(1)
plot(Qrange/Q,unitcostrange
  /unitcost,'ko',
  'MarkerSize',3,'MarkerFaceColor',
  [0 0 0],'MarkerEdgeColor',
  [0 0 0],'linewidth',1);
hold on
plot(xaxis,ratiounitcostfitplot,
  'k-','linewidth',1);
set(gca,'xtickmode','auto',
  'FontSize',10,'FontWeight',
  'bold','FontName', 'Times New Roman');
xlabel({'Q/Q_0'},'FontSize',14,
  'FontWeight','bold','FontName',
  'Times New Roman')
ylabel({'Unit Cost/Unit Cost_0'},
  'FontSize',14,'FontWeight','bold',
  'FontName', 'Times New Roman')
grid on
hold on
```

Answer: AAOCapitalCost (Figure 15.11)

p = -0.2499 0.0001; k = 1.0001; R^2 = 1.0000

b) Equation for unit cost/unit cost$_0$ vs. Q/Q_0 for O&M:

The regression equation of the trend line:

$$\log(\text{unit cost/unit cost}_0) = -0.2865\log(Q/Q_0) - 0.0194 \quad r^2 = 0.9901$$

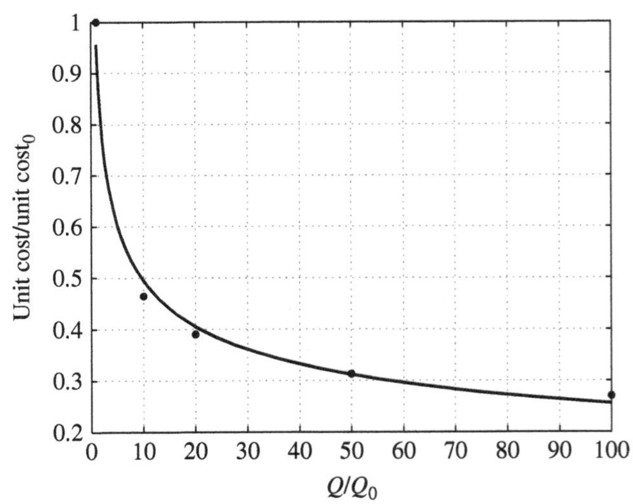

Figure 15.11 Unit O&M cost/unit O&M cost$_0$ vs. Q/Q_0 for AAO, unit O&M cost$_0$ = 3.2055$/kgal, Q_0 = 1 MGD.

or the equation of the regression:

$$\text{Unit cost/ unit cost}_0 = 0.9564(Q/Q_0)^{-0.2865}$$
$$r^2 = 0.9901$$

Matlab code:

```
clear all
close all
%data%
Qrange= [1 10 20 50 100];
Q=Qrange(1);
totalcostrange= [1.17 5.43 9.13
  18.29 31.57];
totalcostrange=totalcostrange.*10^6;
totalcost=totalcostrange(1);
unitcost=totalcost./Q./1000./365;
unitcostrange=totalcostrange./Qrange./
  1000./365;
p=polyfit(log10(Qrange/Q), log10
  (unitcostrange /unitcost), 1)
log10ratiounitcostfit=polyval
  (p, log10(Qrange/Q));
ratiounitcostfit=10.
  ^log10ratiounitcostfit;
k=10^p(:,2)
R2 = 1 - sum((log10(unitcostrange
  /unitcost) - log10ratiounitcostfit).
  ^2)/sum((log10(unitcostrange
  /unitcost) - mean(log10(unitcostrange
  /unitcost))).^2)
xmax=max(Qrange/Q);
xmin=min(Qrange/Q);
xaxis= [xmin:0.1:xmax];
log10ratiounitcostfitplot=polyval
  (p, log10(xaxis));
ratiounitcostfitplot=10.
  ^log10ratiounitcostfitplot;
%plot%
figure(1)
plot(Qrange/Q,unitcostrange
  /unitcost,'ko',
  'MarkerSize',3,'MarkerFaceColor',
  [0 0 0],'MarkerEdgeColor',
  [0 0 0],'linewidth',1);
hold on
plot(xaxis,ratiounitcostfitplot,
  'k-','linewidth',1);
set(gca,'xtickmode','auto',
  'FontSize',10,'FontWeight',
  'bold','FontName',
  'Times New Roman');
xlabel({'Q/Q_0'},'FontSize',14,
  'FontWeight','bold','FontName',
  'Times New Roman')
ylabel({'Unit Cost/Unit Cost_0'},
  'FontSize',14,'FontWeight',
  'bold','FontName', 'Times New Roman')
grid on
hold on
```

Answer: AAOOMCost

p = -0.2865 -0.0194; k = 0.9564; R^2 = 0.9901

15.5.6 Three-Stage Activated Sludge with Alum Addition

Example 15.8

Given: Table 15.35 shows capital cost and operations and maintenance (O&M) cost of three-stage activated sludge with aluminum addition (AAO + Al) at different flow rates.

Find:

1) Calculate unit capital and O&M costs.
2) Plot unit cost/unit cost_0 vs. Q/Q_0.
3) Determine correlation equations between unit cost/ unit cost_0 for capital and O&M vs. Q/Q_0, respectively (Tables 15.36 and 15.37).

Table 15.35 Capital cost and O&M cost of AAO + Al.

Flow (MGD)	Total capital cost (2004 \$×10^6)	Total O&M cost (2004 \$×10^6)
1	9.76	1.35
10	54.65	7.10
20	91.84	12.46
50	182.54	26.59
100	306.98	48.17

Table 15.36 Unit capital cost and unit operations and maintenance cost of AAO + Al at different flow rates.

Q (MGD)	Unit capital cost (\$/kgal)	Unit O&M costs (\$/kgal)
1	9760.00	3.6986
10	5465.00	1.9452
20	4592.00	1.7068
50	3650.80	1.4570
100	3069.80	1.3197

Table 15.37 Unit cost ratio of AAO + AI at different flow ratios.

Q/Q_0	Unit capital cost/unit capital $cost_0$	Unit O&M costs/unit O&M $costs_0$
1	1.0000	1.0000
10	0.5599	0.5259
20	0.4705	0.4615
50	0.3741	0.3939
100	0.3145	0.3568

Solution:

1) When $Q = 1$ MGD,

$$\text{Unit capital cost} = 9.76 \times 10^6\$ \div 1\,\text{MGD} \times 1\,\text{million gal}/10^3\,\text{kgal} \div 1\,\text{day} = 9760.00\$/\text{kgal}$$

$$\text{Unit O\&M costs} = 1.35 \times 10^6\$/\text{year} \div 1\,\text{MGD} \times 1\,\text{million gal}/10^3\,\text{kgal} \div 365\,\text{day/year}$$

$$\text{Unit O\&M costs} = 3.6986\$/\text{kgal}$$

When $Q = 1$ MGD,

$$Q_0 = 1\,\text{MGD}$$

$$\text{Unit capital cost}_0 = 9760.00\$/\text{kgal}$$

$$\text{Unit O\&M costs}_0 = 3.6986\$/\text{kgal}$$

$$\frac{Q}{Q_0} = 1$$

$$\text{Unit capital cost/unit capital cost}_0 = 1$$

$$\text{Unit O\&M costs/unit O\&M costs}_0 = 1$$

When $Q = 10$ MGD,

$$\frac{Q}{Q_0} = 10$$

$$\text{Unit capital cost/unit capital cost}_0 = 0.5599$$

$$\text{Unit O\&M costs/unit O\&M costs}_0 = 0.5259$$

2) Plot unit cost/unit $cost_0$ vs. Q/Q_0 for capital and O&M, respectively (Figure 15.12).

3) a) Equation for unit cost/unit $cost_0$ vs. Q/Q_0 for capital:

The regression equation of the trend line:

$$\log(\text{unit cost/unit cost}_0) = -0.2512\log(Q/Q_0) - 0.0003 \quad r^2 = 1.0000$$

or the equation of the regression:

$$\text{Unit cost/unit cost}_0 = 0.9993(Q/Q_0)^{-0.2512} \quad r^2 = 1.0000$$

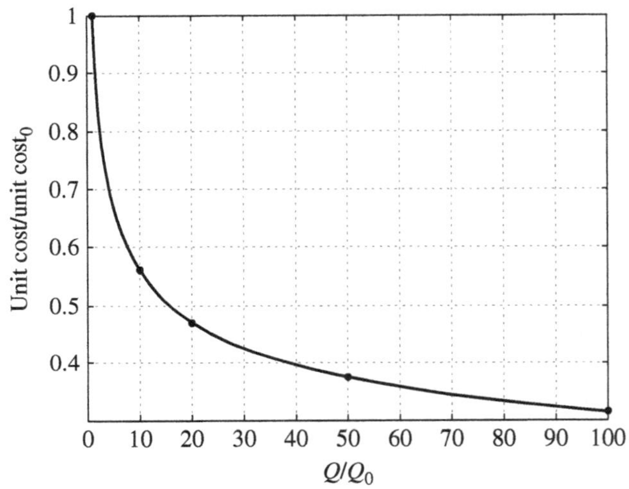

Figure 15.12 Unit capital cost/unit capital $cost_0$ vs. Q/Q_0 for AAO + AI, unit capital $cost_0 = 9760.00\$/kgal$, $Q_0 = 1$ MGD.

Matlab code:

```
Qrange=[1 10 20 50 100];
Q=Qrange(1);
totalcostrange=[9.76 54.65 91.84
  182.54 306.98];
totalcostrange=totalcostrange.*10^6;
totalcost=totalcostrange(1);
unitcost=totalcost./Q./1000;
unitcostrange=totalcostrange./
  Qrange./1000;
p=polyfit(log10(Qrange/Q), log10
  (unitcostrange /unitcost), 1)
log10ratiounitcostfit=polyval(p, log10
  (Qrange/Q));
ratiounitcostfit=10.
  ^log10ratiounitcostfit;
k=10^p(:,2)
R2 = 1 - sum((log10(unitcostrange
  /unitcost) - log10ratiounitcostfit).
  ^2)/sum((log10(unitcostrange
  /unitcost) - mean(log10(unitcostrange
  /unitcost))).^2)
xmax=max(Qrange/Q);
xmin=min(Qrange/Q);
xaxis=[xmin:0.1:xmax];
log10ratiounitcostfitplot=polyval
  (p, log10(xaxis));
ratiounitcostfitplot=10.
  ^log10ratiounitcostfitplot;
%plot%
figure(1)
plot(Qrange/Q,unitcostrange
  /unitcost,'ko','MarkerSize',3,
  'MarkerFaceColor',[0 0 0],
```

```
    'MarkerEdgeColor', [0 0 0],
    'linewidth',1);
hold on
plot(xaxis,ratiounitcostfitplot,
    'k-','linewidth',1);
set(gca,'xtickmode','auto',
    'FontSize',10,'FontWeight','bold',
    'FontName', 'Times New Roman');
xlabel({'Q/Q_0'},'FontSize',14,
    'FontWeight','bold','FontName',
    'Times New Roman')
ylabel({'Unit Cost/Unit Cost_0'},
    'FontSize',14,'FontWeight',
    'bold','FontName', 'Times New Roman')
grid on
hold on
```

Answer: AAOAlCapitalCost (Figure 15.13)

p = -0.2512 -0.0003; k = 0.9993;
R^2 = 1.0000

b) Equation for unit cost/unit $cost_0$ vs. Q/Q_0 for O&M:

The regression equation of the trend line:

$$\log(\text{unit cost/unit cost}_0) = -0.2261\log(Q/Q_0) - 0.0220 \quad r^2 = 0.9795$$

or the equation of the regression:

$$\text{Unit cost/unit cost}_0 = 0.9505(Q/Q_0)^{-0.2261} \quad r^2 = 0.9795$$

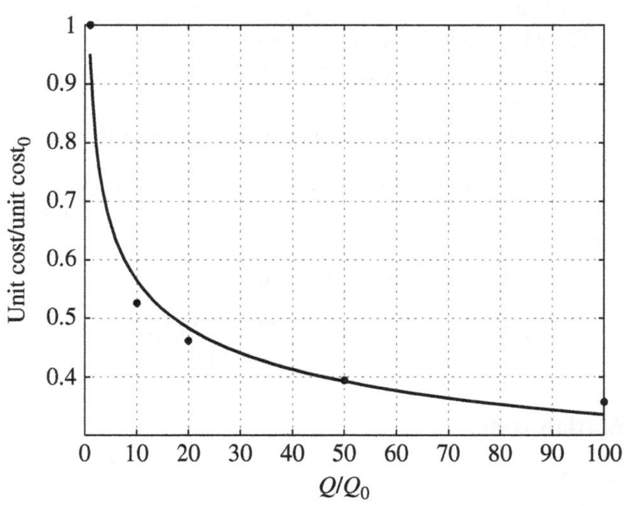

Figure 15.13 Unit O&M cost/unit O&M $cost_0$ vs. Q/Q_0 for AAO + Al, unit O&M $cost_0$ = 3.6986\$/kgal, Q_0 = 1 MGD.

Matlab code:

```
clear all
close all
%data%
Qrange= [1 10 20 50 100];
Q=Qrange(1);
totalcostrange= [1.35 7.1 12.46
    26.59 48.17];
totalcostrange=totalcostrange.*10^6;
totalcost=totalcostrange(1);
unitcost=totalcost./Q./1000./365;
unitcostrange=totalcostrange./Qrange./
    1000./365;
p=polyfit(log10(Qrange/Q), log10
    (unitcostrange /unitcost), 1)
log10ratiounitcostfit=polyval
    (p, log10(Qrange/Q));
ratiounitcostfit=10.
    ^log10ratiounitcostfit;
k=10^p(:,2)
R2 = 1 - sum((log10(unitcostrange
    /unitcost) - log10ratiounitcostfit).
    ^2)/sum((log10(unitcostrange
    /unitcost) - mean(log10(unitcostrange
    /unitcost))).^2)
xmax=max(Qrange/Q);
xmin=min(Qrange/Q);
xaxis= [xmin:0.1:xmax];
log10ratiounitcostfitplot=polyval
    (p, log10(xaxis));
ratiounitcostfitplot=10.
    ^log10ratiounitcostfitplot;
%plot%
figure(1)
plot(Qrange/Q,unitcostrange
    /unitcost,'ko','MarkerSize',3,
    'MarkerFaceColor', [0 0 0],
    'MarkerEdgeColor', [0 0 0],
    'linewidth',1);
hold on
plot(xaxis,ratiounitcostfitplot,
    'k-','linewidth',1);
set(gca,'xtickmode','auto',
    'FontSize',10,'FontWeight',
    'bold','FontName', 'Times New Roman');
xlabel({'Q/Q_0'},
    'FontSize',14,'FontWeight','bold',
    'FontName', 'Times New Roman')
ylabel({'Unit Cost/Unit Cost_0'},
    'FontSize',14,'FontWeight','bold',
    'FontName', 'Times New Roman')
grid on
hold on
```

Answer: `AAOAlOMCost`

```
p = -0.2261 -0.0220; k = 0.9505;
R^2 = 0.9795
```

15.5.7 Three-Stage Activated Sludge with Alum and Tertiary Clarifier

Example 15.9

Given: Table 15.38 shows the capital cost and operations and maintenance (O&M) cost of three-stage activated sludge with aluminum addition and tertiary clarifier (AAO + Al + S) at different flow rates.

Find:

1) Calculate unit capital and O&M costs.
2) Plot unit cost/unit $cost_0$ vs. Q/Q_0.
3) Determine correlation equations between unit cost/unit $cost_0$ for capital and O&M vs. Q/Q_0, respectively (Tables 15.39 and 15.40).

Solution:

1) When $Q = 1$ MGD,

$$\text{Unit capital cost} = 10.12 \times 10^6\$ \div 1\,\text{MGD} \times 1\,\text{million gal}/10^3\,\text{kgal} \div 1\,\text{day} = 10120.00\$/\text{kgal}$$

Table 15.38 Capital cost and O&M cost of AAO + Al + S.

Flow (MGD)	Total capital cost (2004 $\$\times10^6$)	Total O&M cost (2004 $\$\times10^6$)
1	10.12	1.41
10	56.03	7.50
20	94.24	13.22
50	188.58	28.48
100	318.94	51.96

Table 15.39 Unit capital cost and unit operations and maintenance cost of AAO + Al + S at different flow rates.

Q (MGD)	Unit capital cost ($/kgal)	Unit O&M costs ($/kgal)
1	10 120.00	3.8630
10	5 603.00	2.0548
20	4 712.00	1.8110
50	3 771.60	1.5605
100	3 189.40	1.4236

Table 15.40 Unit cost ratio of AAO + Al + S at different flow ratios.

Q/Q_0	Unit capital cost/unit capital $cost_0$	Unit O&M costs/unit O&M $costs_0$
1	1.0000	1.0000
10	0.5537	0.5319
20	0.4656	0.4688
50	0.3727	0.4040
100	0.3152	0.3685

$$\text{Unit O\&M costs} = 1.41 \times 10^6\$/\text{year} \div 1\,\text{MGD} \times 1\,\text{million gal}/10^3\text{kgal} \div 365\,\text{day/year}$$

$$\text{Unit O\&M costs} = 3.8630\$/\text{kgal}$$

When $Q = 1$ MGD,

$Q_0 = 1$ MGD

Unit capital $cost_0 = 10120.00\$/\text{kgal}$

Unit O&M $costs_0 = 3.8630\$/\text{kgal}$

$$\frac{Q}{Q_0} = 1$$

Unit capital cost/unit capital $cost_0 = 1$

Unit O&M costs/unit O&M $costs_0 = 1$

When $Q = 10$ MGD,

$$\frac{Q}{Q_0} = 10$$

Unit capital cost/unit capital $cost_0 = 0.5537$

Unit O&M Costs/Unit O&M $Costs_0 = 0.5319$

2) Plot unit cost/unit $cost_0$ vs. Q/Q_0 for capital and O&M, respectively (Figure 15.14).
   a) Equation for unit cost/unit $cost_0$ vs. Q/Q_0 for capital:
      The regression equation of the trend line:

$$\log(\text{unit cost/unit cost}_0) = -0.251\log(Q/Q_0) - 0.0025 \quad r^2 = 0.9998$$

or the equation of the regression:

$$\text{Unit cost/unit cost}_0 = 0.9942\,(Q/Q_0)^{-0.2510} \quad r^2 = 0.9998$$

Matlab code:

```
clear all
close all
%data%
Qrange= [1 10 20 50 100];
```

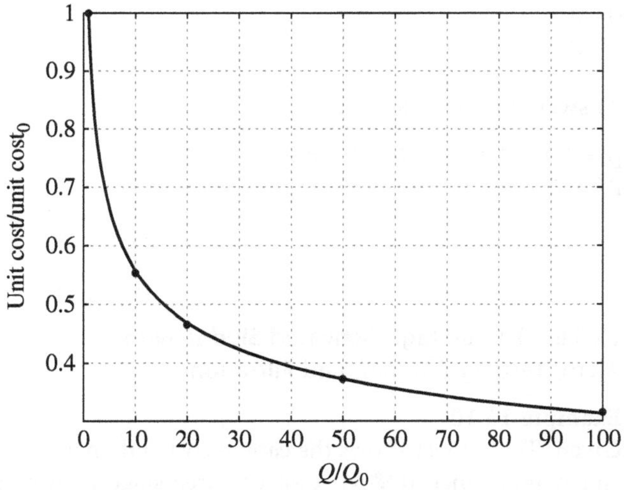

Figure 15.14 Unit capital cost/unit capital cost$_0$ vs. Q/Q_0 for AAO + Al + S, unit capital cost$_0$ = 10120.00\$/kgal, Q_0 = 1 MGD.

```
Q=Qrange(1);
totalcostrange=[10.12 56.03 94.24
  188.58 318.94];
totalcostrange=totalcostrange.*10^6;
totalcost=totalcostrange(1);
unitcost=totalcost./Q./1000;
unitcostrange=totalcostrange./
  Qrange./1000;
p=polyfit(log10(Qrange/Q), log10
  (unitcostrange /unitcost), 1)
log10ratiounitcostfit=polyval
  (p, log10(Qrange/Q));
ratiounitcostfit=10.
  ^log10ratiounitcostfit;
k=10^p(:,2)
R2 = 1 - sum((log10(unitcostrange
  /unitcost) - log10ratiounitcostfit).
  ^2)/sum((log10(unitcostrange
  /unitcost) - mean(log10(unitcostrange
  /unitcost))).^2)
xmax=max(Qrange/Q);
xmin=min(Qrange/Q);
xaxis=[xmin:0.1:xmax];
log10ratiounitcostfitplot=polyval
  (p, log10(xaxis));
ratiounitcostfitplot=10.
  ^log10ratiounitcostfitplot;
%plot%
figure(1)
plot(Qrange/Q,unitcostrange /unitcost,
  'ko','MarkerSize',3,
  'MarkerFaceColor',[0 0 0],
  'MarkerEdgeColor',[0 0 0],
  'linewidth',1);
hold on
plot(xaxis,ratiounitcostfitplot,
  'k-','linewidth',1);
set(gca,'xtickmode','auto',
  'FontSize',10,'FontWeight',
  'bold','FontName', 'Times New Roman');
xlabel({'Q/Q_0'},'FontSize',14,
  'FontWeight','bold','FontName',
  'Times New Roman')
ylabel({'Unit Cost/Unit Cost_0'},
  'FontSize',14,
  'FontWeight','bold','FontName',
  'Times New Roman')
grid on
hold on
```

Answer: `AAOAlSCapitalCost` (Figure 15.15)

`p = -0.2510 -0.0025; k = 0.9942;` R^2 `= 0.9998`

b) Equation for unit cost/unit cost$_0$ vs. Q/Q_0 for O&M:

The regression equation of the trend line:

$$\log(\text{unit cost/unit cost}_0) = -0.2193\log(Q/Q_0) - 0.0230 \quad r^2 = 0.9764$$

or the equation of the regression:

$$\text{Unit cost/unit cost}_0 = 0.9485\ (Q/Q_0)^{-0.2193} \quad r^2 = 0.9764$$

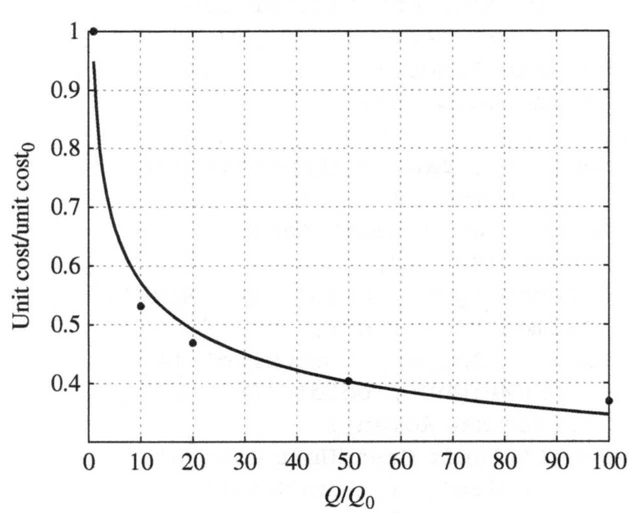

Figure 15.15 Unit O&M cost/unit O&M cost$_0$ vs. Q/Q_0 for AAO + Al + S, unit O&M cost$_0$ = 3.8630\$/kgal, Q_0 = 1 MGD.

Matlab code:

```
clear all
close all
%data%
Qrange=[1 10 20 50 100];
Q=Qrange(1);
totalcostrange=[1.41 7.5 13.22
  28.48 51.96];
totalcostrange=totalcostrange.*10^6;
totalcost=totalcostrange(1);
unitcost=totalcost./Q./1000./365;
unitcostrange=totalcostrange./Qrange./
  1000./365;
p=polyfit(log10(Qrange/Q), log10
  (unitcostrange /unitcost), 1)
log10ratiounitcostfit=polyval
  (p, log10(Qrange/Q));
ratiounitcostfit=10.
  ^log10ratiounitcostfit;
k=10^p(:,2)
R2 = 1 - sum((log10(unitcostrange
  /unitcost) - log10ratiounitcostfit).
  ^2)/sum((log10(unitcostrange
  /unitcost) - mean(log10(unitcostrange
  /unitcost))).^2)
xmax=max(Qrange/Q);
xmin=min(Qrange/Q);
xaxis=[xmin:0.1:xmax];
log10ratiounitcostfitplot=polyval
  (p, log10(xaxis));
ratiounitcostfitplot=10.
  ^log10ratiounitcostfitplot;
%plot%
figure(1)
plot(Qrange/Q,unitcostrange
  /unitcost,'ko','MarkerSize',3,
  'MarkerFaceColor',[0 0 0],
  'MarkerEdgeColor',[0 0 0],
  'linewidth',1);
hold on
plot(xaxis,ratiounitcostfitplot,
  'k-','linewidth',1);
set(gca,'xtickmode','auto',
  'FontSize',10,
  'FontWeight','bold','FontName',
  'Times New Roman');
xlabel({'Q/Q_0'},'FontSize',14,
  'FontWeight','bold','FontName',
  'Times New Roman')
ylabel({'Unit Cost/Unit Cost_0'},
  'FontSize',14,'FontWeight',
  'bold','FontName',
  'Times New Roman')
grid on
hold on
```

Answer: AAOAlSOMCost

p = -0.2193 -0.0230; k = 0.9485;
R^2 = 0.9764

15.5.8 Three-Stage Activated Sludge with Alum, Tertiary Clarifier, and Filtration

Example 15.10
Given: Table 15.41 shows the capital cost and operations and maintenance (O&M) cost of three-stage activated sludge with aluminum addition, tertiary clarifier, and filtration (AAO + Al + S + F) at different flow rates.
Find:

1) Calculate unit capital and O&M costs.
2) Plot unit cost/unit cost_0 vs. Q/Q_0.
3) Determine correlation equations between unit cost/unit cost_0 for capital and O&M vs. Q/Q_0, respectively (Tables 15.42 and 15.43).

Table 15.41 Capital cost and O&M cost of AAO + Al + S + F.

Flow (MGD)	Total capital cost (2004 $\$\times10^6$)	Total O&M cost (2004 $\$\times10^6$)
1	10.84	1.49
10	58.72	7.82
20	100.06	13.88
50	206.39	30.30
100	344.41	55.03

Table 15.42 Unit capital cost and unit operations and maintenance cost of AAO + Al + S + F at different flow rates.

Q (MGD)	Unit capital cost ($/kgal)	Unit O&M costs ($/kgal)
1	10 840.00	4.0822
10	5 872.00	2.1425
20	5 003.00	1.9014
50	4 127.80	1.6603
100	3 444.10	1.5077

Table 15.43 Unit cost ratio of AAO + Al + S + F at different flow ratios.

Q/Q_0	Unit capital cost/unit capital $cost_0$	Unit O&M costs/unit O&M $costs_0$
1	1.0000	1.0000
10	0.5417	0.5248
20	0.4615	0.4658
50	0.3808	0.4067
100	0.3177	0.3693

Solution:

1) When $Q = 1$ MGD,

$$\text{Unit capital cost} = 10.84 \times 10^6\$ \div 1\,\text{MGD} \times 1\,\text{million gal}/10^3\,\text{kgal} \div 1\,\text{day} = 10\,840.00\$/\text{kgal}$$

$$\text{Unit O\&M costs} = 1.49 \times 10^6\$/\text{year} \div 1\,\text{MGD} \times 1\,\text{million gal}/10^3\,\text{kgal} \div 365\,\text{day/year}$$

$$\text{Unit O\&M costs} = 4.0822\$/\text{kgal}$$

When $Q = 1$ MGD,

$Q_0 = 1$MGD

Unit capital $cost_0 = 10840.00\$/\text{kgal}$

Unit O&M $costs_0 = 4.0822\$/\text{kgal}$

$$\frac{Q}{Q_0} = 1$$

Unit capital cost/unit capital $cost_0 = 1$

Unit O&M costs/unit O&M $costs_0 = 1$

When $Q = 10$ MGD,

$$\frac{Q}{Q_0} = 10$$

Unit capital cost/unit capital $cost_0 = 0.5417$

Unit O&M costs/unit O&M $costs_0 = 0.5248$

2) Plot unit cost/unit $cost_0$ vs. Q/Q_0 for capital and O&M, respectively (Figure 15.16).

a) Equation for unit cost/unit $cost_0$ vs. Q/Q_0 for capital:

The regression equation of the trend line:

$$\log(\text{unit cost/unit cost}_0) = -0.2473\log(Q/Q_0) - 0.2473 - 0.0072 \quad r^2 = 0.9978$$

or the equation of the regression:

$$\text{Unit cost/unit cost}_0 = 0.9837\,(Q/Q_0)^{-0.2473} \quad r^2 = 0.9978$$

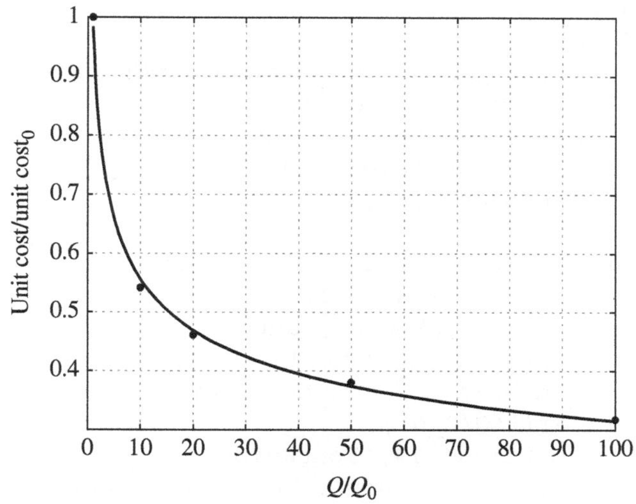

Figure 15.16 Unit capital cost/unit capital $cost_0$ vs. Q/Q_0 for AAO + Al + S + F, unit capital $cost_0 = 10840.00\$/\text{kgal}$, $Q_0 = 1$ MGD.

Matlab code:

```
clear all
close all
%data%
Qrange=[1 10 20 50 100];
Q=Qrange(1);
totalcostrange=[10.84 58.72 100.06
  206.39 344.41];
totalcostrange=totalcostrange.*10^6;
totalcost=totalcostrange(1);
unitcost=totalcost./Q./1000;
unitcostrange=totalcostrange./
  Qrange./1000;
p=polyfit(log10(Qrange/Q), log10
  (unitcostrange /unitcost), 1)
log10ratiounitcostfit=polyval
  (p, log10(Qrange/Q));
ratiounitcostfit=10.
  ^log10ratiounitcostfit;
k=10^p(:,2)
R2 = 1 - sum((log10(unitcostrange
  /unitcost) - log10ratiounitcostfit).
  ^2)/sum((log10(unitcostrange
  /unitcost) - mean(log10(unitcostrange
  /unitcost))).^2)
xmax=max(Qrange/Q);
xmin=min(Qrange/Q);
xaxis=[xmin:0.1:xmax];
log10ratiounitcostfitplot=polyval
  (p, log10(xaxis));
ratiounitcostfitplot=10.
  ^log10ratiounitcostfitplot;
```

```
%plot%
figure(1)
plot(Qrange/Q,unitcostrange /unitcost,
  'ko','MarkerSize',3,
  'MarkerFaceColor',[0 0 0],
  'MarkerEdgeColor',[0 0 0],
  'linewidth',1);
hold on
plot(xaxis,ratiounitcostfitplot,
  'k-','linewidth',1);
set(gca,'xtickmode','auto',
  'FontSize',10,'FontWeight','bold',
  'FontName', 'Times New Roman');
xlabel({'Q/Q_0'},
  'FontSize',14,'FontWeight',
  'bold','FontName', 'Times New Roman')
ylabel({'Unit Cost/Unit Cost_0'},
  'FontSize',14,'FontWeight',
  'bold','FontName', 'Times New Roman')
grid on
hold on
```

Answer: `AAOAlSFCapitalCost` (Figure 15.17)

`p = -0.2473 -0.0072; k = 0.9837;` R^2 `= 0.9978`

b) Equation for unit cost/unit $cost_0$ vs. Q/Q_0 for O&M:

The regression equation of the trend line:

$$\log(\text{unit cost/unit cost}_0) = -0.2180\log(Q/Q_0) - 0.0255 \quad r^2 = 0.9711$$

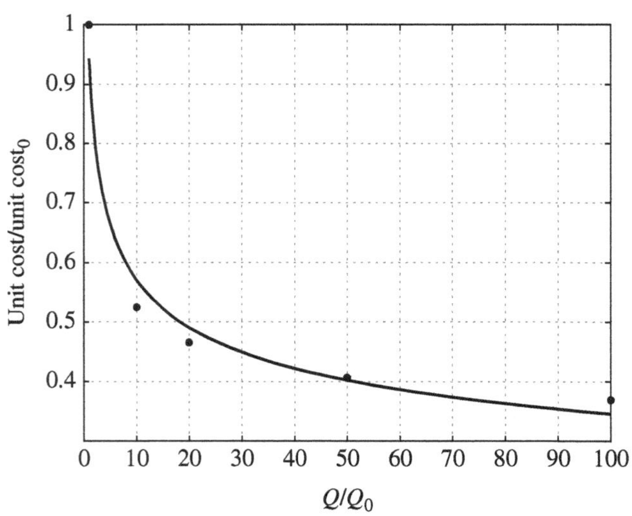

Figure 15.17 Unit O&M cost/unit O&M $cost_0$ vs. Q/Q_0 for AAO + Al + S + F, unit O&M $cost_0$ = 4.0822$/kgal, Q_0 = 1 MGD.

or the equation of the regression:

$$\text{Unit cost/unit cost}_0 = 0.9430(Q/Q_0)^{-0.2180}$$
$$r^2 = 0.9711$$

Matlab code:

```
clear all
close all
%data%
Qrange=[1 10 20 50 100];
Q=Qrange(1);
totalcostrange=[1.49 7.82 13.88
  30.3 55.03];
totalcostrange=totalcostrange.*10^6;
totalcost=totalcostrange(1);
unitcost=totalcost./Q./1000./365;
unitcostrange=totalcostrange./Qrange./
  1000./365;
p=polyfit(log10(Qrange/Q), log10
  (unitcostrange /unitcost), 1)
log10ratiounitcostfit=polyval
  (p, log10(Qrange/Q));
ratiounitcostfit=10.
  ^log10ratiounitcostfit;
k=10^p(:,2)
R2 = 1 - sum((log10(unitcostrange
  /unitcost) - log10ratiounitcostfit).
  ^2)/sum((log10(unitcostrange
  /unitcost) - mean(log10(unitcostrange
  /unitcost))).^2)
xmax=max(Qrange/Q);
xmin=min(Qrange/Q);
xaxis=[xmin:0.1:xmax];
log10ratiounitcostfitplot=polyval
  (p, log10(xaxis));
ratiounitcostfitplot=10.
  ^log10ratiounitcostfitplot;
%plot%
figure(1)
plot(Qrange/Q,unitcostrange /unitcost,
  'ko','MarkerSize',3,
  'MarkerFaceColor',[0 0 0],
  'MarkerEdgeColor',[0 0 0],
  'linewidth',1);
hold on
plot(xaxis,ratiounitcostfitplot,
  'k-','linewidth',1);
set(gca,'xtickmode','auto',
  'FontSize',10,'FontWeight',
  'bold','FontName', 'Times New Roman');
xlabel({'Q/Q_0'},
  'FontSize',14,'FontWeight',
  'bold','FontName', 'Times New Roman')
```

```
ylabel({'Unit Cost/Unit Cost_0'},
   'FontSize',14,'FontWeight',
   'bold','FontName', 'Times New Roman')
grid on
hold on
```

Answer: AAOAlSFOMCost

p = -0.2180 -0.0255; k = 0.9430;
R^2 = 0.9711

15.5.9 Three-Stage Activated Sludge with Tertiary Clarifier and Activated Aluminum Absorption

Example 15.11
Given: Table 15.44 shows the capital cost and operations and maintenance (O&M) cost of three-stage activated sludge with tertiary clarifier and activated aluminum absorption (AAO + Al + S + C) at different flow rates.
Find:

1) Calculate unit capital and O&M costs.
2) Plot unit cost/unit cost$_0$ vs. Q/Q_0.
3) Determine correlation equations between unit cost/unit cost$_0$ for capital and O&M vs. Q/Q_0, respectively (Tables 15.45 and 15.46).

Solution:

1) When $Q = 1$ MGD,

$$\text{Unit capital cost} = 11.11 \times 10^6\$ \div 1\,\text{MGD} \times 1\,\text{million gal}/10^3\,\text{kgal} \div 1\,\text{day} = 11110.00\$/\text{kgal}$$

$$\text{Unit O\&M costs} = 1.60 \times 10^6\$/\text{year} \div 1\,\text{MGD} \times 1\,\text{million gal}/10^3\,\text{kgal} \div 365\,\text{day/year}$$

$$\text{Unit O\&M costs} = 4.3836\$/\text{kgal}$$

Table 15.44 Capital cost and O&M cost of AAO + Al + S + C.

Flow (MGD)	Total capital cost (2004 $\$\times10^6$)	Total O&M cost (2004 $\$\times10^6$)
1	11.11	1.60
10	65.88	8.44
20	112.97	14.91
50	237.87	32.61
100	410.95	59.58

Table 15.45 Unit capital cost and unit operations and maintenance cost of AAO + Al + S + C at different flow rates.

Q (MGD)	Unit capital cost (\$/kgal)	Unit O&M costs (\$/kgal)
1	11 110.00	4.3836
10	6 588.00	2.3123
20	5 648.50	2.0425
50	4 757.40	1.7868
100	4 109.50	1.6323

Table 15.46 Unit cost ratio of AAO + Al + S + C at different flow ratios.

Q/Q_0	Unit capital cost/unit capital cost$_0$	Unit O&M costs/unit O&M costs$_0$
1	1.0000	1.0000
10	0.5930	0.5275
20	0.5084	0.4659
50	0.4282	0.4076
100	0.3699	0.3724

When $Q = 1$ MGD,

$Q_0 = 1$ MGD
Unit capital cost$_0$ = 11 110.00\$/kgal
Unit O&M costs$_0$ = 4.3836\$/kgal

$$\frac{Q}{Q_0} = 1$$

Unit capital cost/unit capital cost$_0$ = 1
Unit O&M costs/unit O&M costs$_0$ = 1

When $Q = 10$ MGD,

$$\frac{Q}{Q_0} = 10$$

Unit capital cost/unit capital cost$_0$ = 0.5930
Unit O&M costs/unit O&M costs$_0$ = 0.5275

2) Plot unit cost/unit cost$_0$ vs. Q/Q_0 for capital and O&M, respectively (Figure 15.18).
   a) Equation for unit cost/unit cost$_0$ vs. Q/Q_0 for capital:
      The regression equation of the trend line:

$$\log(\text{unit cost/unit cost}_0) = -0.2159\log(Q/Q_0) - 0.0051 \quad r^2 = 0.9986$$

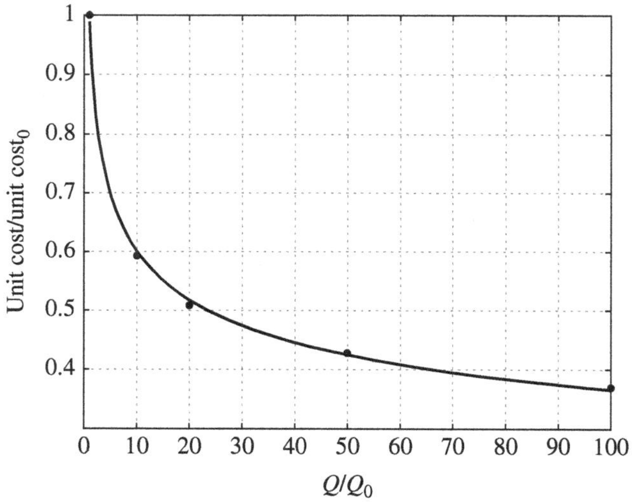

Figure 15.18 Unit capital cost/unit capital cost$_0$ vs. Q/Q_0 for AAO + Al + S + C, unit capital cost$_0$ = 11110.00$/kgal, Q_0 = 1 MGD.

or the equation of the regression:
Unit cost/unit cost$_0$ = $0.9882\ (Q/Q_0)^{-0.2159}$
$r^2 = 0.9986$

Matlab code:

```
clear all
close all
%data%
Qrange=[1 10 20 50 100];
Q=Qrange(1);
totalcostrange=[11.11 65.88 112.97
  237.87 410.95];
totalcostrange=totalcostrange.*10^6;
totalcost=totalcostrange(1);
unitcost=totalcost./Q./1000;
unitcostrange=totalcostrange./
  Qrange./1000;
p=polyfit(log10(Qrange/Q), log10
  (unitcostrange /unitcost), 1)
log10ratiounitcostfit=polyval(p, log10
  (Qrange/Q));
ratiounitcostfit=10.
  ^log10ratiounitcostfit;
k=10^p(:,2)
R2 = 1 - sum((log10(unitcostrange
  /unitcost) - log10ratiounitcostfit).
  ^2)/sum((log10(unitcostrange
  /unitcost) - mean(log10(unitcostrange
  /unitcost))).^2)
xmax=max(Qrange/Q);
xmin=min(Qrange/Q);
xaxis=[xmin:0.1:xmax];
log10ratiounitcostfitplot=polyval(p,
  log10(xaxis));
ratiounitcostfitplot=10.
  ^log10ratiounitcostfitplot;
%plot%
figure(1)
plot(Qrange/Q,unitcostrange
  /unitcost,'ko','MarkerSize',3,
  'MarkerFaceColor',[0 0 0],
  'MarkerEdgeColor',[0 0 0],
  'linewidth',1);
hold on
plot(xaxis,ratiounitcostfitplot,'k-',
  'linewidth',1);
set(gca,'xtickmode','auto',
  'FontSize',10,
  'FontWeight','bold','FontName',
  'Times New Roman');
xlabel({'Q/Q_0'},
  'FontSize',14,'FontWeight',
  'bold','FontName', 'Times New Roman')
ylabel({'Unit Cost/Unit Cost_0'},
  'FontSize',14,'FontWeight',
  'bold','FontName', 'Times New Roman')
grid on
hold on
```

Answer: `AAOAlSCCapitalCost` (Figure 15.19)

```
p = -0.2159 -0.0051; k = 0.9882;
R^2 = 0.9986
```

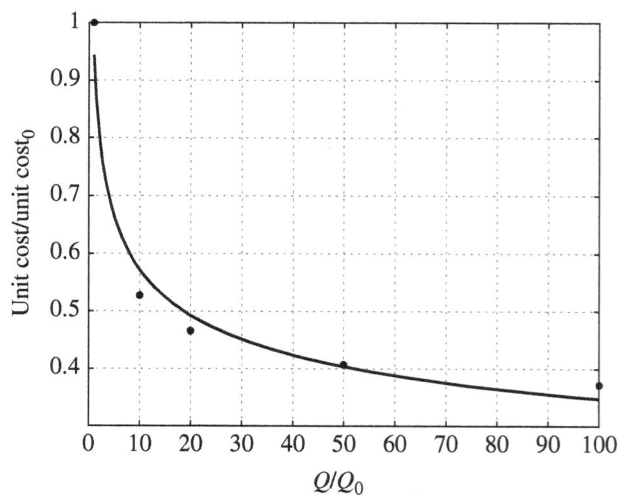

Figure 15.19 Unit O&M cost/unit O&M cost$_0$ vs. Q/Q_0 for AAO + Al + S + C, unit O&M cost$_0$ = 4.3836$/kgal, Q_0 = 1 MGD.

b) Equation for unit cost/unit $cost_0$ vs. Q/Q_0 for O&M:

The regression equation of the trend line:

$$\log(\text{unit cost/unit cost}_0) = -0.2167\log(Q/Q_0) - 0.0256 \quad r^2 = 0.9704$$

or the equation of the regression:

$$\text{Unit cost/unit cost}_0 = 0.9428(Q/Q_0)^{-0.2167} \quad r^2 = 0.9704$$

Matlab code:

```
clear all
close all
%data%
Qrange= [1 10 20 50 100];
Q=Qrange(1);
totalcostrange= [1.6 8.44 14.91
  32.61 59.58];
totalcostrange=totalcostrange.*10^6;
totalcost=totalcostrange(1);
unitcost=totalcost./Q./1000./365;
unitcostrange=totalcostrange./Qrange./
  1000./365;
p=polyfit(log10(Qrange/Q), log10
  (unitcostrange /unitcost), 1)
log10ratiounitcostfit=polyval
  (p, log10(Qrange/Q));
ratiounitcostfit=10.
  ^log10ratiounitcostfit;
k=10^p(:,2)
R2 = 1 - sum((log10(unitcostrange
  /unitcost) - log10ratiounitcostfit).
  ^2)/sum((log10(unitcostrange
  /unitcost) - mean(log10
  (unitcostrange /unitcost))).^2)
xmax=max(Qrange/Q);
xmin=min(Qrange/Q);
xaxis= [xmin:0.1:xmax];
log10ratiounitcostfitplot=polyval
  (p, log10(xaxis));
ratiounitcostfitplot=10.
  ^log10ratiounitcostfitplot;
%plot%
figure(1)
plot(Qrange/Q,unitcostrange
  /unitcost,'ko','MarkerSize',3,
  'MarkerFaceColor',[0 0 0],
  'MarkerEdgeColor',[0 0 0],
  'linewidth',1);
hold on
plot(xaxis,ratiounitcostfitplot,
  'k-','linewidth',1);
set(gca,'xtickmode','auto',
  'FontSize',10,'FontWeight',
  'bold','FontName',
  'Times New Roman');
xlabel({'Q/Q_0'},
  'FontSize',14,'FontWeight',
  'bold','FontName', 'Times New Roman')
ylabel({'Unit Cost/Unit Cost_0'},
  'FontSize',14,'FontWeight',
  'bold','FontName',
  'Times New Roman')
grid on
hold on
```

Answer: AAOAlSCOMCost

```
p = -0.3040 -0.0203; k = 0.9543;
R^2 = 0.9903
```

15.5.10 Three-Stage Activated Sludge with Tertiary Clarifier and Activated Absorption

Example 15.12

Given: Table 15.47 shows the capital cost and operations and maintenance (O&M) cost of three-stage activated sludge with aluminum addition, filter, and membrane methods (AAO + Al + F + UF) at different flow rates.

Find:

1) Calculate unit capital and O&M costs.
2) Plot unit cost/unit $cost_0$ vs. Q/Q_0.
3) Determine correlation equations between unit cost/unit $cost_0$ for capital and O&M vs. Q/Q_0, respectively (Tables 15.48 and 15.49).

Table 15.47 Capital cost and O&M cost of AAO + Al + F + UF.

Flow (MGD)	Total capital cost (2004 $\$\times10^6$)	Total O&M cost (2004 $\$\times10^6$)
1	12.27	1.76
10	72.79	9.18
20	113.18	15.62
50	277.31	36.80
100	486.18	67.95

Table 15.48 Unit capital cost and unit operations and maintenance cost of AAO + AI + F + UF at different flow rates.

Q (MGD)	Unit capital cost ($/kgal)	Unit O&M costs ($/kgal)
1	12 270.00	4.8219
10	7 279.00	2.5151
20	5 659.00	2.1397
50	5 546.20	2.0164
100	4 861.80	1.8616

Table 15.49 Unit cost ratio of AAO + AI + F + UF at different flow ratios.

Q/Q_0	Unit capital cost/unit capital $cost_0$	Unit O&M costs/unit O&M $costs_0$
1	1.0000	1.0000
10	0.5932	0.5216
20	0.4612	0.4438
50	0.4520	0.4182
100	0.3962	0.3861

Solution:

1) When $Q = 1$ MGD,

$$\text{Unit capital cost} = 12.27\times10^6\$ \div 1\text{ MGD}\times1\text{ million gal}/10^3\text{ kgal} \div 1\text{ day} = 12\,270.00\$/\text{kgal}$$

$$\text{Unit O\&M costs} = 1.76\times10^6\$/\text{year} \div 1\text{ MGD}\times1\text{ million gal}/10^3\text{ kgal}\div365\text{ day/year}$$

$$\text{Unit O\&M costs} = 4.8219\$/\text{kgal}$$

When $Q = 1$ MGD,

$Q_0 = 1$ MGD

Unit capital $cost_0 = 12\,270.00\$/kgal$

Unit O&M $costs_0 = 4.8219\$/kgal$

$$\frac{Q}{Q_0} = 1$$

Unit capital cost/unit capital $cost_0 = 1$

Unit O&M costs/unit O&M $costs_0 = 1$

When $Q = 10$ MGD,

$$\frac{Q}{Q_0} = 10$$

Unit capital cost/unit capital $cost_0 = 0.5932$

Unit O&M costs/unit O&M $costs_0 = 0.5216$

2) Plot unit cost/unit $cost_0$ vs. Q/Q_0 for capital and O&M, respectively (Figure 15.20).

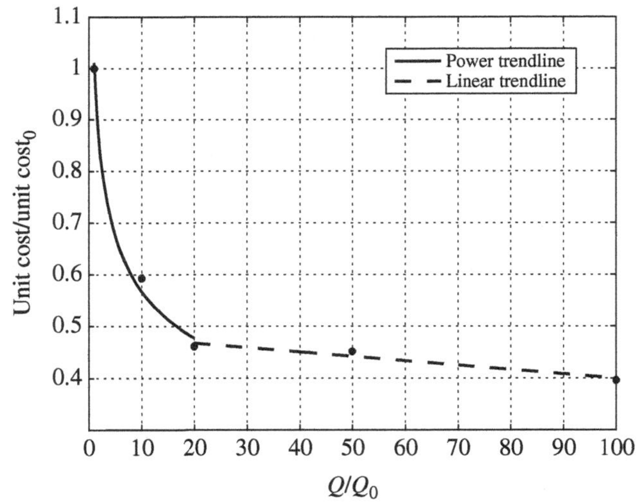

Figure 15.20 Unit capital cost/unit capital $cost_0$ vs. Q/Q_0 for AAO + AI + F + UF, unit capital $cost_0 = 12\,270.00\$/kgal$, $Q_0 = 1$ MGD.

a) Equation for unit cost/unit $cost_0$ vs. Q/Q_0 for capital:

The regression equation of the trend line ($1 \le Q/Q_0 \le 20$):

$$\log(\text{unit cost/unit cost}_0) = -0.2504\log(Q/Q_0) + 0.0044 \quad r^2 = 0.9897$$

or the equation of the regression:

$$\text{Unit cost/unit cost}_0 = 1.0103\,(Q/Q_0)^{-0.2504} \quad r^2 = 0.9897$$

The regression equation of the trend line ($20 \le Q/Q_0 \le 100$):

$$\text{Unit cost/unit cost}_0 = -0.0008Q/Q_0 + 0.4843 \quad r^2 = 0.9392$$

Matlab code:

```
clear all
close all
%data%
Qrange=[1 10 20 50 100];
Q=Qrange(1);
totalcostrange=[12.27 72.79 113.18
  277.31 486.18];
totalcostrange=totalcostrange.*10^6;
totalcost=totalcostrange(1);
xmid=3;
%function%
unitcost=totalcost./Q./10^3;
unitcostrange=totalcostrange./
  Qrange./10^3;
x=Qrange/Q;
```

```
y=unitcostrange /unitcost;
x1=x(1:xmid);
y1=y(1:xmid);
x2=x(xmid:end);
y2=y(xmid:end);
%power
p=polyfit(log10(x1), log10(y1), 1)
log10ratiounitcostfit=polyval(p,
  log10(x1));
ratiounitcostfit=10.
  ^log10ratiounitcostfit;
k=10^p(:,2)
R2power = 1 - sum((log10(y1) -
  log10ratiounitcostfit).^2)/sum((log10
  (y1) - mean(log10(y1))).^2)
xmax=max(x1);
xmin=min(x1);
xaxis=[xmin:0.1:xmax];
log10ratiounitcostfitplot=polyval(p,
  log10(xaxis));
ratiounitcostfitplot=10.
  ^log10ratiounitcostfitplot;
%line
p2=polyfit((x2),(y2), 1)
z=polyval(p2, (x2));
R2linear = 1 - sum(((y2) - z).^2)/sum(((y2)
  - mean((y2))).^2)
xmax2=max(x2);
xmin2=min(x2);
xaxis2=[xmin2:0.1:xmax2];
z2=polyval(p2, xaxis2);
%plot%
figure(1)
plot(Qrange/Q,unitcostrange
  /unitcost,'ko','MarkerSize',3,
  'MarkerFaceColor',[0 0 0],
  'MarkerEdgeColor',[0 0 0],
  'linewidth',1);
hold on
line1=plot(xaxis,
  ratiounitcostfitplot,'k-',
  'linewidth',1);
hold on
line2=plot(xaxis2,z2,'k--',
  'linewidth',1);
set(gca,'xtickmode','auto',
  'FontSize',10,'FontWeight',
  'bold','FontName', 'Times New Roman');
legend([line1,line2],'Power
  trendline','Linear
  trendline','Location','Northeast');
xlabel({'Q/Q_0'},
  'FontSize',14,'FontWeight',
  'bold','FontName', 'Times New Roman')
ylabel({'Unit Cost/Unit Cost_0'},
  'FontSize',14,'FontWeight','bold',
  'FontName', 'Times New Roman')
grid on
hold on
```

Answer: `AAOAlFUFCapitalCosttwoline` (Figure 15.21)

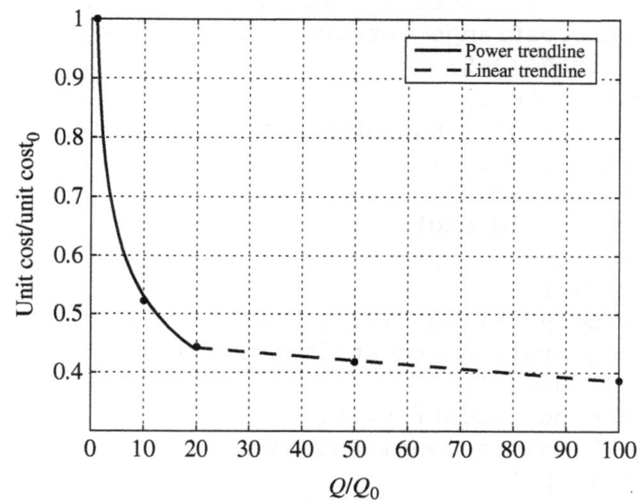

Figure 15.21 Unit O&M cost/unit O&M cost$_0$ vs. Q/Q_0 for AAO + Al + F + UF, unit O&M cost$_0$ = 4.8219$/kgal, Q_0 = 1 MGD.

```
p = -0.2504 0.0044; k = 1.0103;
R²power = 0.9897
p2 = -0.0008 0.4843; R²linear = 0.9392
```

b) Equation for unit cost/unit cost$_0$ vs. Q/Q_0 for O&M:

The regression equation of the trend line ($1 \le Q/Q_0 \le 20$):

$$\log(\text{unit cost/unit cost}_0) = -0.2741\log(Q/Q_0) - 0.0016 \quad r^2 = 0.9989$$

or the equation of the regression:

$$\text{Unit cost/ unit cost}_0 = 0.9963\,(Q/Q_0)^{-0.2741} \quad r^2 = 0.9989$$

The regression equation of the trend line ($20 \le Q/Q_0 \le 100$):

$$\text{Unit cost/unit cost}_0 = -0.0007Q/Q_0 + 0.4564 \quad r^2 = 0.9939$$

Matlab code:

```
clear all
close all
%data%
```

```
Qrange=[1 10 20 50 100];
Q=Qrange(1);
totalcostrange=[1.76 9.18 15.62
  36.8 67.95];
totalcostrange=totalcostrange.*10^6;
totalcost=totalcostrange(1);
xmid=3;
%function%
unitcost=totalcost./Q./1000./365;
unitcostrange=totalcostrange./Qrange./
  1000./365;
x=Qrange/Q;
y=unitcostrange /unitcost;
x1=x(1:xmid);
y1=y(1:xmid);
x2=x(xmid:end);
y2=y(xmid:end);
%power
p=polyfit(log10(x1), log10(y1), 1)
log10ratiounitcostfit=polyval(p,
  log10(x1));
ratiounitcostfit=10.
  ^log10ratiounitcostfit;
k=10^p(:,2)
R2power = 1 - sum((log10(y1) -
  log10ratiounitcostfit).^2)/sum((log10
  (y1) - mean(log10(y1))).^2)
xmax=max(x1);
xmin=min(x1);
xaxis=[xmin:0.1:xmax];
log10ratiounitcostfitplot=polyval(p,
  log10(xaxis));
ratiounitcostfitplot=10.
  ^log10ratiounitcostfitplot;
%line
p2=polyfit((x2),(y2), 1)
z=polyval(p2, (x2));
R2linear = 1 - sum(((y2) - z).^2)/sum(((y2)
  - mean((y2))).^2)
xmax2=max(x2);
xmin2=min(x2);
xaxis2=[xmin2:0.1:xmax2];
z2=polyval(p2, xaxis2);
%plot%
figure(1)
plot(Qrange/Q,unitcostrange
   /unitcost,'ko','MarkerSize',3,
   'MarkerFaceColor',[0 0 0],
   'MarkerEdgeColor',[0 0 0],
   'linewidth',1);
hold on
line1=plot(xaxis,
  ratiounitcostfitplot,'k-',
  'linewidth',1);
hold on
line2=plot(xaxis2,z2,'k--',
  'linewidth',1);
set(gca,'xtickmode','auto',
  'FontSize',10,'FontWeight',
  'bold','FontName', 'Times New Roman');
legend([line1,line2],'Power
  trendline','Linear
  trendline','Location','Northeast');
xlabel({'Q/Q_0'},
  'FontSize',14,'FontWeight',
  'bold','FontName', 'Times New Roman')
ylabel({'Unit Cost/Unit Cost_0'},
  'FontSize',14,'FontWeight',
  'bold','FontName', 'Times New Roman')
grid on
hold on
```

Answer: AAOAlFUFOMCosttwoline

```
p = -0.2741 -0.0016; k = 0.9963;
R^2power = 0.9989
p2 = -0.0007 0.4564; R^2linear = 0.9939
```

15.6 Entrepreneur in SEE

To truly sustainably manage urban water and wastewater infrastructure, SEE attempts to change the paradigm of traditional design of GI, WTP, and WWTPs. More importantly, innovation and entrepreneur also play a critical role in creative design. Entrepreneur refers to individuals who create their own small business or individuals working for someone else that develop new ventures within their current institution. Before you create your new business, you need to answer: (i) Do I have what it takes to be successful? (ii) Am I starting a business where real and lasting opportunities exist? Managing a small business includes accounting, personnel management, and marketing. Successful entrepreneurs should be (i) opportunity seekers, (ii) future oriented, (iii) committed to being the best, (iv) market driven and customer oriented, (v) team players, (vi) realistic, (vii) tolerant of the tedium, and (viii) resilient. In addition, they should have experience in EE business including design, construction, and O&M of EEIS. Also they should clearly understand the nature of the products, services, customer needs, employees, suppliers, and seasonality in SEE industry.

Designing for environments involves (i) ease of disassembly, ease of cleaning, inspection, part replacement, and reassembly, (ii) reusable modular components, and (iii) few fasteners and interfaces. Since the 1970s, products were designed to minimize waste, resources, and energy to achieve a sustainable society. Later, circular economy attempts to minimize material, energy, and water flow without restricting economic growth or social and technical progress. Currently, zero water, wastes, and energy (3Z) is popular in building sustainable communities across the United States. One of the approaches is from cradle to cradle in product design by minimizing wastes through following strategies:

1) Product design should be optimized for durability, remanufacturing, and recycling.
2) Remanufacturing and preserving the stable frame of a product after use and replacing only worn-out parts.
3) Leasing instead of selling wherein the manufacturer's interest lies in durability.
4) Manufacturers should guarantee low pollution use and easy reuse.

Traditional box thinking sets controllable and measurable limits or useful restraints. On the other hand, thinking out of the box refers to knowledge and skill to do creative work, techniques for taking new perspectives on problems and for incubating and persevering on difficult problems, and taking risks with solutions to problems and the desire to solve the problem. Breaking out of box includes the following: (i) relook at the box you think you are in; (ii) look within to solutions you have never considered or can reconsider from the past; (iii) visit other boxes, within or without your organization; (iv) experiment at least part of the time with having no boxes; (v) encourage the use of virtual or transparent wall material for your box; and (vi) teach others the benefits of out-of-the-box thinking. In general, out-of-the-box thinking skills include the following:

1) Common assumption that creativity is something you have or haven't got.
2) Appropriate techniques can bring out anyone's creative potential.
3) Creativity to the extent that it is now a practical business discipline.
4) Traditional management practices predate the current frenetic pace of change.
5) Creativity is one of a new set of skills necessary to manage in today's business environment.
6) Different ways to think outside the box.

15.6.1 Business Plan

The most important step in entrepreneurial world is a good business plan. Typical business plan should have the following components as shown in Table 15.50.

15.6.2 Finance of Environmental Infrastructure

In the United States, government loan programs, such as the Clean Water and Drinking Water State Revolving Loan Funds, will often carry lower interest rates than private bond issues. Capital acquisition costs will be different for the State Revolving Loan Fund (SRLF) financing and private capital. Government loan programs will have loan application and ongoing reporting-related administrative costs. Private capital acquisition costs typically include financial advisory services, bond counsel, underwriting fees, rating agency fees, closing costs and fees, and bond insurance and will have a mix of recurring costs including those for reporting, accounting, and general administration. Further major project capitalization costs include contributions to specified reserves (e.g. reserve account needs related to annual principal and interest payments, for emergency repairs, and for replacements) or meeting coverage covenants imposed by the financing agency. There are no specific federal SRF requirements for reserves or coverage covenants, although many state SRF programs require one or the other. Through coverage covenants, state SRFs can require that O&M expenses are met and net annual revenues must equal some increment above 100% (e.g. 120%) of the annual debt service payments for principal and interest. Finance cost includes construction, e-engineering and technical services, pilot studies, environmental review and permitting, bidding and contracts, administration and legal services, commissioning costs, and construction management (Table 15.51).

15.6.3 EEI Financing

Raising awareness of the true cost of water may be an effective way to involve the stakeholders in financing an EEIS project. Current water rates do not reflect the true cost of supplying drinking water. Replacing the nation's aging pipes will require additional local investment through higher water rates. Also, a large portion of public supply water is used for watering lawns, flushing toilets, and washing clothes, which are referred as gray water and do not require potable water. Therefore, municipalities should promote the construction of separate lines for potable and nonpotable uses. In the United States, the SRF under the Clean Water Act authorizes minimum federal funding of $20 billion over 5 years. States are allowed to issue bonds for water infrastructure projects at an estimated $6–7 billion annually. The ASCE recommended that a federal Water Infrastructure Trust Fund should be established to finance the national shortfall in funding of infrastructure systems under the Clean Water Act and the Safe Drinking Water Act.

Table 15.50 Business plan for SEE designers.

Executive summary	Topics	Questions
Table of contents		
Overview of the concept	1 Description of market demand and size according to the TDPs of SEE	a. What are the market sizes at city, state, country, and global scales?
	2 Description of EEIS design	b. Why innovative design of EEIS is better than traditional design?
	3 Identify options of EEIS	c. Is to quantify your design or products significantly better than traditional design in terms of cost or environmental impacts?
	4 Market analysis of each EEIS design option	a. Is it for new design or retrofit?
		b. How big is the market?
		c. How many WWTPs will be impacted?
		d. What is the competition?
		e. What are the drivers of your design, energy efficiency, or material recovery?
		f. Will the government accept?
		g. How to satisfy the regulatory requirements?
Business models	1 Build, operation, and transfer (BOT)	a. What is the time frame?
	2 Public–private partnership (PPP)	b. What are the percentage from government and your private side?
	3 Private finance	c. What are the return rates?
Management team	1 Team leader	a. How many people and in which area?
	2 Technical leader and specialists	b. What is the specific expertise required?
	3 Financial analyst	c. What are the experiences?
	4 Contract specialist	d. What are the bidding specifics?
	5 Legal specialist	e. What are the relevant laws?
Execution plan	1 Milestones	a. What are the key achievements in a specific time frame?
	2 Timeline	b. What are the major deliverables in a specific time frame?
	3 Financial plan	a. Cash flow
		b. Capital outlay
		c. Funding source
		d. Payback or breakeven analysis
		e. Facilities/resources
Marketing strategy	1 How to obtain approval?	a. Is the product matured?
	2 To whom to present?	b. Who are the consumers?
	3 Format for presentation	a. Who are the stakeholders?
		b. How do the stakeholders get involved?
		c. Short synopsis and briefing
		d. Detailed written plan
		e. Who are the stakeholders?
		f. Why are they important to the decision process?
		g. What do they want to hear?
		h. What to deliver?
Technological risk	1 Risk identification	a. How reliable is the technology?
	2 Risk management strategy	b. What is plan B if plan A did not go as planned?
	3 Risk communication	c. Are communications between technical and financial management people efficient and effective?

Table 15.51 Procedure in cost and benefit analysis.

Step	Action
1	Account fully for all project capital costs
2	Account fully for operations and maintenance costs
3	Account for the impacts new projects may have on overall utility system costs and revenues
4	Develop a capital financing strategy
5	Determine current revenue adequacy and develop future revenue strategy

15.6.4 Financial Planning

Financial forecast contains income statement, balance sheet, and cash flow statement. The debt to equity ratio is also important. The income statement should include sales, cost of goods sold, administrative expenses, expenses, accounts receivable, and inventory. In addition, the most likely, most pessimistic, most optimistic, and expected value should be estimated. Cash flow statement should cover combination of the income statement and the balance sheet forecasts, the changes in the cash balance with time, the cash flow statement, the income statement (the monthly values), and the balance sheet (the period-to-period changes). A cash flow statement should include total and net operating cash inflows, priority outflows, discretionary outflows, and total financial flows.

15.7 Innovation in SEE

15.7.1 Innovative Technologies

The trillion-dollar environmental markets in both China and the US require innovative products, technology, process, and systems from the SEE sector. Integrated approach and new concepts emerging from sustainable developments will present huge opportunities for an entrepreneur. For example, element management is practiced in Denmark, waste refinery is booming in European countries, and circular economy is implemented as eco-friendly industrial park. Waste-to-energy becomes more and more dominant in China form in replacing facility landfills. Recognizing the damage by air, water, and soil pollution in China, the ambitious goals to build new smart community and urban city will provide new generation of EEIS designers' golden opportunity to be the leaders of innovation, creation, and entrepreneur. Indeed, China's investment in EEI is at its reflection point due to consistent low 2% of GDP having been spent in environmental protection annually because GDP was the only performance evaluation criteria of their leaders. Goldman Sachs predicted that about \$1.2 trillion will be spent from 2016 to 2020 on EEI in areas such as solid waste, sludge treatment and disposal, air pollution, and soil remediation. This presents about 60% increase from \$0.7 trillion spent in China from 2011 to 2015. In China, 0.77 billion people live in urban cities and this number is expected to be 1.00 billion by 2020. Currently, ecological civilization is on the core agenda for the current government in the next 5 years. New regulations such as 10 rules on air, water, and soil pollution control impose significant financial responsibilities for national and local governmental agencies to finance EEI at unpreceded pace in the following environmental technologies:

1) Biological denitrification and phosphorus removal technologies.
2) Membrane separation and manufacturing technologies and equipment.
3) Manufacturing technology of anaerobic biological reactors.
4) High-concentration organic wastewater treatment technology and equipment.
5) Standardized and modularized water and wastewater treatment equipment with high efficiency.
6) Water-saving technologies and equipment.
7) Water treatment agents.
8) Monitoring instruments.
9) Natural water-body rehabilitation technology.
10) Artificial intelligence, cloud computing, big data, and wireless technologies.

Among all these technologies, consumer products related to water, air, and solid waste are also important markets for SEE entrepreneur to establish their business in China for one billion people without safe drinking water and decent sanitary infrastructure.

15.7.2 Innovative Consumer Products

15.7.2.1 SteriPEN

SteriPEN is a small pen-like electronic device that emits an UV light and purifies either half a liter or one full liter of water at a time. The device operates off batteries and works with clear water by killing the DNA of harmful microbes and bacteria. A light wand sticks out of one end of the device – which gets submerged into the bottle of water. By pushing the button either once or twice, you can choose to purify a full liter of water or a half liter. The light turns on and stays on while you agitate the water with the light wand until the dose is complete. Once the water is safe to drink (about 1 min later), the light turns off and flashes green – if it flashes red, you have to repeat the dose because something went wrong (if you accidentally lift the wand out of the water during a treatment, a sensor will flag it that the water was not properly treated and flash red). The product website is http://www.steripen.com/.

Most models are reusable for up to 8000 l for over 7 years of safe drinking water. SteriPEN avoids plastic pollution by keeping thousands of plastic water bottles out of the waste cycle. UV light destroys over 99.9% of bacteria, viruses, and protozoa like *Giardia* and *Cryptosporidium*. SteriPEN is certified effective by the Water Quality Association and has won multiple industry awards. The major advantage is that travelers do not have to purchase bottled water by carrying Nalgene bottles with the SteriPEN.

15.7.2.2 Drinkable Book™

Dr. Theresa Dankovich from Carnegie Mellon teaming with WATERisLIFE introduced The Drinkable Book, the first ever instructional manual that provides safe water, sanitation, and hygiene education and serves as a tool to kill deadly waterborne diseases by providing the reader with an opportunity to create clean, drinkable water from each page and teaching methods and ways to share the message through training, storytelling, and discussions in communities and schools worldwide where there is a desperate need. Each book can provide a user with clean water for up to 4 years. WATERisLIFE has introduced a campaign to move into full production of the Drinkable Book. WATERisLIFE is working in partnership with the Drinkable Book on research and production. It plans to begin global distribution of the Drinkable Book in 2016.

SunSpring™ is a solar-powered microbiological water treatment system. It is a lifesaver for entire communities, medical facilities, or schools. The SunSpring is Gold Seal Certified to the US EPA standard for microbiological water purifiers. The SunSpring physically removes particulate matter, turbidity, bacteria, viruses, and cysts from virtually any water source, including rivers, lakes, streams, ponds, shallow wells, rainwater, unsafe municipal supplies, and springs. This ultra-purification technology is self-contained, is designed for community-level water output, and has an automated self-cleaning maintenance system. Each SunSpring filtration unit can filter up to 5000 gal of clean water per day. WATERisLIFE has installed the first unit in Northern Haiti, providing clean water to 2000 residents per day. The company's website is http://waterislife.com/clean-water/new-technology.

15.7.3 Future of SEE

By 2050, human population would reach 10 billion on this planet. 70% of the people will live in urban areas. If the design criterion is 100 gal/capita/day, the total WRRF capacity would be 0.7 trillion gallon/day. Since the average flow rate of WWTPs in the United States is 3 MGD, there would have 233 000 WWTPs by 2050. Therefore, the market for SEE designers to either retrofit the old WWTPs or to build new WRRFs is in trillion US dollars. In the United States and China alone over the next several decades, SEE designers are fortunate to have a trillion-dollar market to apply the TDPs of the SEE to design sustainable EEIS. If all the 15 000 WWTPs in the United States and 5 000 WWTPs in China were to be retrofitted to WRRFs, these two markets could be $0.5 trillion dollars, respectively. The SEE TDPs could also transform the design of air, and solid waste management infrastructure systems is another half trillion market. Therefore, a trillion-dollar market is waiting for the future SEE designers to take the lead. The roadmap could be simple: after a decade, 100% water reuse rate by 2030, energy neutral or positive by 2040. By 2050, the standardized and modulated WRRFs could be significantly more economic than current WWTP. As a result, nutrients and materials were recovered at the WRRFs, which could be revenue generating facilities instead of a liability of a water utility by 2050.

To achieve these ambitious targets, fast adoption of innovative technologies would increase both the reliability and resiliency of EEIS. For example, 3D printing could produce a model of a unit process or a system of minimal footprints in just a few days. Mobile wireless technologies could monitor environmental parameters at real time. Supervisory control and data acquisition (SCADA) could monitor and optimize the integrated EEIS in real time. One important application of SCADA is to forecast the infiltration/inflow through sanitary sewer system according to anticipated precipitation. As computer and internet technologies advances, deep learning, machine learning, and artificial intelligence could help SEE designers to achieve performance metrics and energy benchmarks of EEIS. Internet of Things (IoT) could connect millions of water meters through the Internet and eliminate manual reading of water meters. Sensor technologies and automatic control system could adjust the precise ratio of NH_3:NO_2^- to achieve the maximal efficiency of Anammox. Virtual reality changes the way we educate and train plant personnel. Mining data pattern from big data of water distribution and use could identify major leak to reduce water loss. Cloud computing would increase computing capacity and allow SEE designer's access sophisticated software, which might not be accessible to current engineers. Blockchain technology significantly reduces the cost of the governmental bidding and procurement. For these reasons, SEE designers should keep an open mind and track all the new technology breakthrough and apply them to their design. When SEE designers systematically apply the TDPs and continuously update their knowledge on innovative technologies, human society and economic prosperity could be decoupled from environmental degradation. Only in this way human would coexist with the nature in harmony!

15.8 Exercise

15.8.1 Questions

1) Optimal pipe diameter is determined by what costs?
2) AOP as emerging technology is to replace phase transfer processes such as carbon adsorption. Why AOP should be a better choice than carbon adsorption?
3) What are the major causes preventing WWTPs to recover N and P?
4) What are the key elements in developing a new technology?
5) Why WTP and WWTP design have not changed for last 100 years, yet iPhone has its new product every year?
6) What are the major areas in SEE markets that you would like to devote to develop, design, manufacture, and market new products?
7) What are the market sizes of your new products in developed and developing countries, respectively?

15.8.2 Calculations

1) Moving bed biofilm reactors (MBBR) separates solid retention time (SRT) from hydraulic retention time (HRT) by growing nitrification and denitrification bacteria on the biofilm fillers. It could significantly increase total nitrogen removal and reduce sludge production. If the aeration tank has volume of 30 000 m^3 for a WWTP of 100 MGD, 60% will be filled with the filter media of the biofilm reactors, and the unit cost of the filter media is \$500/m^3, answer the following questions:
   a) What would be the total cost of the filter media of the biofilm reactor if the aeration tanks be filled with 60% of the biofilm reactors?
   b) If the N and P concentration of wastewater effluent is significantly decreased from 17 and 1 mg/l to 1 and 0.02 mg/l, respectively, 40 MGD of the discharge effluent could be reused for irrigation and recharge the aquifer. How many years of the retrofitting capital cost could be recovered due to the water reuse if the reclaimed water is valued at \$3/1000 gal?
2) If Miami-Dade Water and Sewer Department (WASD) is to design and build the state-of-the-art 40 MGD West Dade WRRF with nutrient recovery, estimate the capital and O&M costs in recovering nutrients of eight different processes by using the examples presented in this chapter. If the West Dade WRRF is to use the reclaimed water for irrigation or aquifer recharge, conduct the life cycle cost and benefit analysis and estimate the years of return of capital cost of this WRRF.

15.8.3 Projects

To effectively design SEEI system, six criteria are listed as follows:

1) All the GI, WTP, and WWTP have to be modulated and standardized so that further expansion and retrofitting are easy to achieve. When the industry shifts to pipe reactors, critical technologies such as FO, MBR, MF, NF, RO, and SPVR will be designed according to the same industrial standards.
2) Decentralized WWT should be encouraged to break utility monopoly. As a result, competition will brew innovation and race with time. If there was no competition, Tesla will not invent his alternative current electrical generator, Wright brothers would delay in building their first airplane, and Bill Gates would not launch his first Window operating system so quickly. Under competition, each player will be challenged and pressurized to deliver her/his maximal potential.
3) Students should be introduced to at least three critical industrial trend setting technologies in designing WRRFs. More importantly, the students should be motivated to stay on the technology as early and as long as possible. Throughout the students' curriculum, students should be motivated so much that by the end of graduation, meaningful creative design or innovation should be expected under the mentoring of professors.
4) According to SEE students' academic strength and personal passion, each student should select at least one critical technology. Product development as well as innovative design or solving one community environmental unsustainable problems by using modern design tools should start from this course. Students should stay with the problem by following peer-reviewed journal papers and current industrial products.
5) Complicated environmental problem should be broken into several pieces. Each group will then be in charge to solve smaller problems under the leadership of team leader. Different students have different passion and skills. More importantly, the team needs to incorporate electrical engineers for electrical control by SCADA, mechanical engineers for innovative design of critical technologies, and computer scientists to optimize the process and automatic control and monitoring of WTP or WWTP.
6) SEE students need to learn how to effectively pitch their innovative products to the financial market. Chinese industrial parks served as powerful bridges between academy and industry. Since most industrial parks are located so close to major universities, many would be garage inventors that are nurtured in the

industrial parks. SEE students should select one product to collect information on its design, improvement, and manufacture. All the key components need to be updated as technology advances.

According to the aforementioned six criteria, develop a comprehensive business plan for Xiongan and your hometown that you have chosen.

15.8.3.1 Xiongan Project

1) Develop a business plan for Xiongan to achieve zero water, waste, and energy in 5, 10, 20, and 50 years.
   a) What are the major design variables?
   b) What are the major challenges?
   c) What are the major strategies to achieve 3Z?
2) Design integrated air, water, solid waste, and soil management system that can be used for Xiongan in the next 5, 10, and 50 years using the TDPs of SEE. Your design has to be assessed according to life cycle cost and benefit analysis. Write one milestone for your venture to establish an innovative company and make it specific to your product or service.
3) Propose milestone for your venture and make it specific to your product or service.
4) Develop a list of events for your venture and develop a flow diagram showing the relationship between the events for your venture.
5) Discuss how the events selected relate to the assumptions you have developed. Do you have at least one event for every assumption?
6) Which events are time sensitive? Why?
7) What skills, abilities, and knowledge should an SEE entrepreneur have?
8) What kind of education, work experience, and motivation should an SEE entrepreneur have?

Project 1:
The major environmental challenges in Xiongan district are the following:

1) Current heavy and vast pollution of air, water, and soil.
2) Energy structure is coal-based power.
3) Old industrial and agricultural plants and farms.

   Using TDPs that you learned and computer skills you mastered, develop design alternatives for retrofitting current WWTPs in the district and remediation of contaminated air, water, and soil.

Project 2:

1) Develop alternatives that include the type of technologies and how it will be installed and operated on building, community, city, and the special district. Develop renewable energy alternatives to replace electricity from coal due to its high pollution rates.
2) Develop specifications, drawings, and calculations of each alternative in achieving zero energy with minimal environmental impacts using LCA and life cycle cost and benefit analysis.

Project 3:

1) Design zero water on house, community, city, and the special district scales.
2) Design color-coded system for blue, green, gray, brown, and black water to achieve zero water design.
3) Each alternative should be assessed using the LCA.
4) Provide any applicable specifications or drawings to demonstrate the system design and location.

Project 4:
Develop a comprehensive resource recovery strategy plan:

1) Type of resources such as water, energy, and nutrients to be recovered.
2) Descriptions and specification on system used.
3) Estimated annual quantity recovered.
4) Estimated impact on total building resource production (i.e. approximately what percent of the identified resource is recovered).
5) What are the strategies that support both sustainable communities and green infrastructure?
6) How to organize stakeholders.
7) Use three color-coded strategies to reduce use of potable water for irrigation purposes.
8) What are the efficiency standards for plumbing fixtures?
9) How much deep-well geothermal energy is available?
10) What are the recommendations according to life cycle assessment and cost–benefit analysis, respectively? Why is the recommendation of different alternatives?

15.8.3.2 Community Project Proposal

1) Students should compile the inventory data from their corresponding hometown on following:
   a) Annual industrial production.
   b) Annual agricultural production.
   c) City population and predicted population in the next 5, 10, 20, and 50 years.
2) Use the US EPA emission factor handbook to estimate the following:
   a) Total annual air, water, and soil pollutants.
   b) What are the net pollutants that may subject to environmental tax and estimate environmental tax on each category of pollutants?
   c) Estimate the market size for air, water, and soil pollutants.

3) Develop a business plan on retrofitting GI, WTP, and WWTPs in your local community according to the business plan outline in this chapter. Do the same as above for your community.

15.8.3.3 Course Project and Beyond

The Xiongan project has a much bigger purpose. It attempts to develop true sustainable modern urban smart city to achieve zero water, waste, and energy design. It is a flagship of new urban design and construction for China's One Belt and Road initiative. Therefore, in the EEIS design, theoretical potential, benchmark data, design standards, and actual technical parameters should be carefully evaluated so that design standards, codes, and criteria can be developed for zero water, waste, and energy according to the TDPs. The key to make major contribution to SEE design is to apply the TWPs of SEE in different EEIS design on different spatial and temporal scales. More importantly, all the developed technical standards, codes, and criteria should be rapidly adopted across the country to replace the traditional unsustainable EE design to decouple environmental deterioration from economic development. As a result, we can coexist with the nature in harmony for generations to come.

References

Cullen, N., Baur, R., and Schauer, P. (2013). Three years of operation of North America's first nutrient recovery facility. Water Science & Technology 68 (4): 763–768.

Falbe, J. and Regitz, M. (1995). Römpp Chemie Lexikon, 9. Auflage, Band 4, S. 3386 ff. Stuttgart: Thieme Verlag.

Kabbe, C. (2015) Sustainable sewage sludge management fostering phosphorus recovery and energy efficiency. P-REX project report. Berlin: Kompetenzzentrum Wasser Berlin gGmbH.

Kunst, S. (1991). Untersuchungen zur biologischen Phosphorelimination im Hinblick auf ihre abwassertechnische Nutzung. Hannover: Veröffentlichungen des Institutes für Siedlungswasserwirtschaft und Abfallwirtschaft, Heft 77.

Pinnekamp J., Baumann, P., Cornel, P. et al. (2013). Stand und Perspektiven der Phosphorrückgewinnung aus Abwasser und Klärschlamm – Teil 1, KA Korrespondenz Abwasser. *Abfall, 60/10,* 837–844.

Pinnekamp, J., Montag, D., Gethke, K. et al. (2007). Rückgewinnung eines schadstofffreien, mineralischen Kombinationsdüngers Magnesiumammoniumphosphat – MAP aus Abwasser und Klärschlamm. *UBA-Texte 25/07,* Dessau-Rosslau.

US Environmental Protection Agency (2005). UV disinfection guidance manual. EPA. Office of Water.

Index

q

r

W